# Preface

Knowledge about genes (genetics) and genomes (genomics) of different organisms continues to advance at a brisk pace. All manifestations of life are determined by genes and their interactions with the environment. A genetic component contributes to the cause of nearly every human disease. More than a thousand diseases result from alterations in single known genes.

Classical genetics, developed during the first half of the last century, and molecular genetics, developed during the second half, have merged into a fascinating scientific endeavor. This has provided both a theoretical foundation and a broad repertoire of methods to explore cellular mechanisms and to understand normal processes and diseases at the molecular level.

Deciphering the genomes of many different organisms, including bacteria and plants, by determining the sequence of the individual building blocks—the nucleotide bases of deoxyribonucleic acid (DNA)—will augment our understanding of normal and abnormal functions. The new knowledge holds promise for the design of pharmaceutical compounds aimed at individual requirements. This will pave the way to new approaches to therapy and prevention. Insights are gained into how organisms are related by evolution.

Students in biology and medicine face an enormous task when attempting to acquire the new knowledge and to interpret it within a conceptual framework. Many good textbooks are available (see General References, p. 421). This Color Atlas differs from standard textbooks by using a visual approach to convey important concepts and facts in genetics. It is based on carefully designed color plates, each accompanied by a corresponding explanatory text on the opposite page.

In 1594 Mercator first used the term "atlas" for a collection of maps. Although maps of genes are highly important in genetics, the term atlas in the context of this book refers to illustrations in general. Here they provide the basis for an introduction, hopefully stimulating interest in an exciting field of study.

This second edition has been extensively revised, rewritten, updated, and expanded. A new section on genomics (Part II) has been added. Twenty new plates deal with a variety of topics such as the molecular bases of genetics, regulation and expression of genes, genomic imprinting, mutations, chromosomes, genes predisposing to cancer, ion channel diseases, hearing and deafness, a brief guide to genetic diagnosis, human evolution, and many others. The Chronology of Important Advances in Genetics and the Definitions of Genetic Terms have been updated. As in the first edition, references are included for further reading. Here and in the list of general references, the reader will find access to more detailed information than can be presented in the limited space available. Websites for further information are included.

A single-author book cannot provide all the details on which scientific knowledge is based. However, it can present an individual perspective suitable as an introduction. In making the difficult decisions about which material to include and which to leave out, I have relied on 25 years' experience of teaching medical students at preclinical and clinical levels. I have attempted to emphasize the intersection of theoretical fundaments and the medical aspects of genetics, taking a broad viewpoint based on the evolution of living organisms.

All the color plates were produced as computer graphics by Jürgen Wirth, Professor of Visual Communication at the Faculty of Design, University of Applied Sciences, Darmstadt. He created the plates from hand drawings, sketches, photographs, and photocopies assembled by the author. I am deeply indebted to Professor Jürgen Wirth for his most skilful work, the pleasant cooperation, and his patience with all of the author's requests. Without him this book would not have been possible.

*Essen, November 2000* *E. Passarge*

# Acknowledgements

In updating, revising, and rewriting this second edition, I received invaluable help from many colleagues who generously provided information and advice, photographic material, and other useful suggestions in their areas of expertise: Hans Esche, Essen; Ulrich Langenbeck, Frankfurt; Clemens Müller-Reible, Würzburg; Maximilian Muenke, Bethesda, Maryland; Stefan Mundlos, Berlin; Alfred Pühler, Bielefeld; Gudrun Rappold, Heidelberg; Helga Rehder, Marburg; Hans Hilger Ropers, Berlin; Gerd Scherer, Freiburg; Evelyn Schröck, Bethesda, Maryland; Eric Schulze-Bahr, Münster; Michael Speicher, München; Manfred Stuhrmann-Spangenberg, Hannover; Gerd Utermann, Innsbruck; and Douglas C Wallace and Marie Lott, Atlanta.

In addition, the following colleagues at our Department of Human Genetics, University of Essen Medical School, made helpful suggestions: Beate Albrecht, Karin Buiting, Gabriele Gillessen-Kaesbach, Cornelia Hardt, Bernhard Horsthemke, Frank Kaiser, Dietmar Lohmann, Hermann-Josef Lüdecke, Eva-Christina Prott, Maren Runte, Frank Tschentscher, Dagmar Wieczorek, and Michael Zeschnigk.

I thank my wife, Dr. Mary Fetter Passarge, for her careful reading and numerous helpful suggestions. Liselotte Freimann-Gansert and Astrid Maria Noll transcribed the many versions of the text. I am indebted to Dr. Clifford Bergman, Ms Gabriele Kuhn, Mr Gert Krüger, and their co-workers at Thieme Medical Publishers Stuttgart for their excellent work and cooperative spirit.

# About the Author

The author is a medical scientist in human genetics at the University of Essen, Medical Faculty, Germany. He graduated in 1960 from the University of Freiburg with an M.D. degree. He received training in different fields of medicine in Hamburg, Germany, and Worcester, Massachusetts/USA between 1961 and 1963. During a residency in pediatrics at the University of Cincinnati, Children's Medical Center, he worked in human genetics as a student of Josef Warkany (1963-66), followed by a research fellowship in human genetics at the Cornell Medical Center New York with James German (1966-68). Thereafter he established cytogenetics and clinical genetics at the Department of Human Genetics, University of Hamburg (1968 – 1976). In 1976 he became founding chairman of the Department of Human Genetics, University of Essen, from which he will retire in 2001. The author's special research interests are the genetics and the clinical delineation of hereditary disorders, including chromosomal and molecular studies, documented in more than 200 peer-reviewed research articles. He is a former president of the German Society of Human Genetics, secretary-general of the European Society of Human Genetics, and a member of various scientific societies in Europe and the USA. He is a corresponding member of the American College of Medical Genetics. The practice of medical genetics and teaching of human genetics are areas of the author's particular interests.

# Table of Contents (Overview)

# Table of Contents in Detail

# Introduction

Each of the approximately $10^{14}$ cells of an adult human contains a program with life-sustaining information in its nucleus. This allows an individual to interact with the environment not only through the sensory organs by being able to see, to hear, to taste, to feel heat, cold, and pain, and to communicate, but also to remember and to integrate the input into cognate behavior. It allows the conversion of atmospheric oxygen and ingested food into energy production and regulates the synthesis and transport of biologically important molecules. The immune defense against unwarranted invaders (e.g., viruses, bacteria, fungi) is part of the program. The shape and mobility of bones, muscles, and skin could not be maintained without it. The fate of each cell is determined by the control of cell division and differentiation into different types of cells and tissues, including cell-to-cell interactions and intracellular and extracellular signal transduction. Finally, such different areas as reproduction or the detoxification and excretion of molecules that are not needed depend on this program as well as many other functions of life.

This cellular program is genetically determined. It is transferred from one cell to both daughter cells at each cell division and from one generation to the next through specialized cells, the germ cells (oocytes and spermatozoa). The integrity of the genetic program must be maintained without compromise, yet it should also be adaptable to long-term changes in the environment. This is an enormous task. It is no wonder, therefore, that errors in maintaining and propagating the genetic program occur frequently in all living systems despite the existence of complex systems for damage recognition and repair.

All these biological processes are the result of biochemical reactions performed by biomolecules called proteins. Proteins are involved in the production of almost all molecules (including other proteins) in living cells. Proteins are made up of dozens to several hundreds of amino acids linearly connected to each other as a polypeptide, subsequently to be arranged in a specific three-dimensional structure, often in combination with other polypeptides. Only this latter feature allows biological function. Genetic information is the cell's blueprint to make the proteins that a specific cell typically makes. Most cells do not produce all possible proteins, but a selection depending on the type of cell.

Each of the 20 amino acids used by living organisms has a code of three specific chemical structures, the nucleotide bases, that are part of a large molecule, DNA (deoxyribonucleic acid). DNA is a read-only memory of the genetic information system. In contrast to the binary system of strings of ones and zeros used in computers ("bits", which are then combined into "bytes" that are eight binary digits long), the genetic code in the living world uses a quaternary system of four nucleotide bases with chemical names having the initial letters A, C, G, and T (see Part I, Fundamentals). With a quaternary code used in living cells the bytes (called "quytes" by The Economist in a Survey of the Human Genome, July 1, 2000) are shorter: three only, each called a triplet codon. Each linear sequence of amino acids in a protein is encoded by a corresponding sequence of codons in DNA (genetic code). The genetic code is universal and is used by all living cells, including plants and also by viruses. Each unit of genetic information is called a gene. This is the equivalent of a single sentence in a text. In fact, genetic information is highly analogous to a text and is amenable to being stored in computers.

Depending on the organizational complexity of the organism, the number of genes may be small as in viruses and bacteria (10 genes in a small bacteriophage or 4289 genes in *Escherichia coli*), medium (6241 genes in yeast; 13601 in *Drosophila*, 18 424 in a nematode), or large (about 80000 in humans and other mammals). Since many proteins are involved in related functions of the same pathway, they and their corresponding genes can be grouped into families of related function. It is estimated that the human genes form about 1000 gene families. Each gene family arose by evolution from one ancestral gene or from a few. The entirety of genes and DNA in each cell of an organism is called the genome. By analogy, the entirety of proteins of an organism is called the proteome. The corresponding fields of study are termed genomics and proteomics, respectively.

Genes are located in chromosomes. These are individual, paired bodies consisting of DNA and special proteins in the cell nucleus. One chromosome of each homologous pair is derived from the mother and the other from the father. Man has 23 pairs. While the number and size of chromosomes in different organisms vary, the total amount of DNA and the total number of genes are the same for a particular class of organism. Genes are arranged linearly along each chromosome. Each gene has a defined position (gene locus) and an individual structure and function. As a rule, genes in higher organisms are structured into contiguous sections of coding and noncoding sequences called exons (coding) and introns (noncoding), respectively. Genes in multicellular organisms vary with respect to overall size (a few thousand to over a million base pairs), number and size of exons, and regulatory DNA sequences that determine their state of activity, called the expression (most genes in differentiated, specialized cells are permanently turned off). It is remarkable that more than 90% of the total of 3 billion ($3 \times 10^9$) base pairs of DNA in higher organisms do not carry any coding information (see Part II, Genomics).

The linear text of information contained in the coding sequences of DNA in a gene cannot be read directly. Rather, its total sequence is first transcribed into a structurally related molecule with a corresponding sequence of codons. This molecule is called RNA (ribonucleic acid) because it contains ribose instead of the deoxyribose of DNA. From this molecule the introns (from the noncoding parts) are then removed by special enzymes, and the exons (the coding parts) are spliced together into the final template, called messenger RNA (mRNA). From this molecule the corresponding encoded sequence of amino acids (polypeptide) is read off in a complex cellular machinery (ribosomes) in a process called translation.

Genes with the same, a similar, or a related function in different organisms are the same, similar, or related in certain ways. This is expressed as the degree of structural or functional similarity. The reason for this is evolution. All living organisms are related to each other because their genes are related. In the living world, specialized functions have evolved but once, encoded by the corresponding genes. Therefore, the structures of genes required for fundamental functions are preserved across a wide variety of organisms, for example functions in cell cycle control or in embryonic development and differentiation. Such genes are similar or identical even in organisms quite distantly related, such as yeast, insects, worms, vertebrates, mammals, and even plants. Such genes of fundamental importance do not tolerate changes (mutations), because this would compromise function. As a result, deleterious mutations do not accumulate in any substantial number. Similar or identical genes present in different organisms are referred to as conserved in evolution. All living organisms have elaborate cellular systems that can recognize and eliminate faults in the integrity of DNA and genes (DNA repair). Mechanisms exist to sacrifice a cell by programmed cell death (apoptosis) if the defect cannot be successfully repaired.

Unlike the important structures that time has evolutionarily conserved, DNA sequences of no or of limited direct individual importance differ even among individuals of the same species. These individual differences (genetic polymorphism) constitute the genetic basis for the uniqueness of each individual. At least one in 1000 base pairs of human DNA differs among individuals (single nucleotide polymorphism, SNP). In addition, many other forms of DNA polymorphism exist that reflect a high degree of individual genetic diversity.

Individual genetic differences in the efficiency of metabolic pathways are thought to predispose to diseases that result from the interaction of many genes, often in combination with particular environmental influences. They may also protect one individual from an illness to which another is prone. Such individual genetic differences are targets of individual therapies by specifically designed pharmaceutical substances aimed at high efficacy and low risk of side effects (pharmacogenomics). The Human Genome Project should greatly contribute to the development of an individual approach to diagnostics and therapy (genetic medicine).

Human populations of different geographic origins also are related by evolution (see section on human evolution in Part II). They are often mistakenly referred to as races. Modern mankind originated in Africa about 200 000 years ago and had migrated to all parts of the world by about 100 000 years ago. Owing to regional adaptation to climatic and other conditions, and favored by geographic isolation, different ethnic groups evolved. They are recognizable by literally superficial features, such as color of the skin, eyes, and hair, that betray the low degree of human genetic variation between different populations. Genetically speaking, *Homo sapiens* is one rather homogeneous species of recent origin. Of the total genetic variation, about 85% is interindividual within a given group, only 15% is among different groups (populations). In contrast, chimpanzees from one group in West Africa are genetically more diverse than all humans ever studied. As a result of evolutionary history, humans are well adapted to live peacefully in relatively small groups with a similar cultural and linguistic background. Unfortunately, humans are not yet adapted to global conditions. They tend to react with hostility to groups with a different cultural background in spite of negligible genetic differences. Genetics does not provide any scientific basis for claims that favor discrimination, but it does provide direct evidence for the evolution of life on earth. Genetics is the science concerned with the structure and function of all genes in different organisms (analysis of biological variation). New investigative methods and observations, especially during the last 10 to 20 years, have helped to integrate this field into the mainstream of biology and medicine. Today, it plays a central, unifying role comparable to that of cellular pathology at the beginning of the last century. Genetics is relevant to virtually all medical specialties. Knowledge of basic genetic principles and their application in diagnosis are becoming an essential part of medical education today.

Johann Gregor Mendel

## Classical Genetics Between 1900 and 1953

*(see chronological table on p. 13)*

In 1906, the English biologist William Bateson (1861 – 1926) proposed the term *genetics* for the new biological field devoted to investigating the rules governing heredity and variation. Bateson referred to heredity and variation when comparing the similarities and differences, respectively, of genealogically related organisms, two aspects of the same phenomenon. Bateson clearly recognized the significance of the Mendelian rules, which had been rediscovered in 1900 by Correns, Tschermak, and DeVries.

The Mendelian rules are named for the Augustinian monk Gregor Mendel (1822–1884), who conducted crossbreeding experiments on garden peas in his monastery garden in Brünn (Brno, Czech Republic) well over a century ago. In 1866, Mendel wrote that heredity is based on individual factors that are independent of each other (see Brink and Styles, 1965; Mayr, 1982). Transmission of these factors to the next plant generation, i.e., the distribution of different traits among the offspring, occurred in predictable proportions. Each factor was responsible for a certain trait. The term *gene* for such a heritable factor was introduced in 1909 by the Danish biologist Wilhelm Johannsen (1857 – 1927).

Starting in 1902, Mendelian inheritance was systematically analyzed in animals, plants, and also in man. Some human diseases were recognized as having a hereditary cause. A form of brachydactyly (type A1, McKusick 112500) observed in a large Pennsylvania sibship by W. C. Farabee (PhD thesis, Harvard University, 1902) was the first condition in man to be described as being transmitted by autosomal dominant inheritance (Haws and McKusick, 1963).

In 1909, Archibald Garrod (1857 – 1936), later Regius Professor of Medicine at Oxford University, demonstrated that four congenital metabolic diseases (albinism, alkaptonuria, cystinuria, and pentosuria) are transmitted by autosomal recessive inheritance (Garrod, 1909). Garrod was the first to recognize that there are biochemical differences among individuals that do not lead to illness but that have a genetic basis. However, the relationship of genetic and biochemical findings revealed by this concept was ahead of its time: the far-reaching significance for the genetic individuality of man was not recognized (Bearn, 1993). Certainly part of the reason was that the nature of genes and how they function was completely unclear. Early genetics was not based on chemistry or on cytology (Dunn, 1965; Sturtevant, 1965). Chromosomes in mitosis (Flemming, 1879) and meiosis (Strasburger, 1888) were observed; the term chromosome was coined by Waldeyer in 1888, but a functional relationship between genes and chromosomes was not considered. An exception was the prescient work of Theodor Boveri (1862 – 1915) about the genetic individuality of chromosomes (in 1902).

Genetics became an independent scientific field in 1910 with the study of the fruit fly *(Drosophila melanogaster)* by Thomas H. Morgan at Columbia University in New York. Subsequent systematic genetic studies on *Drosophila* over many years (Dunn, 1965; Sturtevant, 1965; Whitehouse, 1973) showed that genes are arranged linearly on chromosomes. This led to the chromosome theory of inheritance (Morgan, 1915).

Thomas H. Morgan

The English mathematician Hardy and the German physician Weinberg recognized that Mendelian inheritance accounts for certain regularities in the genetic structure of populations (1908). Their work contributed to the successful introduction of genetic concepts into plant and animal breeding. Although genetics was well established as a biological field by the end of the third decade of last century, knowledge of the physical and chemical nature of genes was sorely lacking. Structure and function remained unknown.

That genes can change and become altered was recognized by DeVries in 1901. He introduced the term mutation. In 1927, H. J. Muller determined the spontaneous mutation rate in *Drosophila* and demonstrated that mutations can be induced by roentgen rays. C. Auerbach and J. M. Robson (1941) and, independently, F. Oehlkers (1943) observed that certain chemical substances also could induce mutations. However, it remained unclear what a mutation actually was, since the physical basis for the transfer of genetic information was not known.

The complete lack of knowledge of the structure and function of genes contributed to misconceptions in the 1920s and 30s about the possibility of eliminating "bad genes" from human populations (eugenics). However, modern genetics has shown that the ill conceived eugenic approach to eliminating human genetic disease is also ineffective.

Thus, incomplete genetic knowledge was applied to human individuals at a time when nothing was known about the structure of genes. Indeed, up to 1949 no essential genetic findings had been gained from studies in man. Quite the opposite holds true today.

Today, it is evident that genetically determined diseases generally cannot be eradicated. No one is free from a genetic burden. Every individual carries about five or six severe genetic defects that are inapparent, but that may show up in offspring.

With the demonstration in the fungus *Neurospora* that one gene is responsible for the formation of one enzyme ("one gene, one enzyme", Beadle and Tatum, 1941), the close relationship of genetics and biochemistry became apparent, quite in agreement with Garrod's concept of inborn errors of metabolism. Systematic studies in microorganisms led to other important advances in the 1940s: genetic recombination was demonstrated in bacteria (Lederberg and Tatum, 1946) and viruses (Delbrück and Bailey, 1947). Spontaneous mutations were observed in bacterial viruses (bacteriophages; Hershey, 1947). The study of genetic phenomena in microorganisms turned out to be as significant for the further development of genetics as the analysis of *Drosophila* had been 35 years earlier (for review, see Cairns et al., 1978). A very influential, small book entitled "What Is Life?" by the physicist E. Schrödinger (1944) defined genes in molecular terms. At that time, elucidation of the molecular biology of the gene became a central theme in genetics.

Oswald T. Avery

## Genetics and DNA

A major advance occurred in 1944 when Avery, MacLeod, and McCarty at the Rockefeller Institute in New York demonstrated that a chemically relatively simple long-chained nucleic acid (deoxyribonucleic acid, DNA) carried genetic information in bacteria (for historical review, see Dubos, 1976; McCarty, 1985). Many years earlier, F. Griffith (in 1928) had observed that permanent (genetic) changes can be induced in pneumococcal bacteria by a cell-free extract derived from other strains of pneumococci ("transforming principle"). Avery and his co-workers showed that DNA was this transforming principle. In 1952, Hershey and Chase proved that genetic information is transferred by DNA alone. With this knowledge, the question of its structure became paramount.

This was resolved most elegantly by James D. Watson, a 24-year-old American on a scholarship in Europe, and Francis H. Crick, a 36-year-old English physicist, at the Cavendish Laboratory of the University of Cambridge. Their findings appeared in a three-quarter-page article on April 25, 1953 in *Nature* (Watson and Crick, 1953). In this famous article, Watson and Crick proposed that the structure of DNA is a double helix. The double helix is formed by two complementary chains with oppositely oriented alternating sugar (deoxyribose) and monophosphate molecules. Inside this helical molecule lie paired nucleotide bases, each pair consisting of a purine and a pyrimidine. The crucial feature is that the base pairs lie inside the molecule, not outside. This insight came from construction of a model of DNA that took into account stereochemical considerations and the results of previous X-ray diffraction studies by M. Wilkins and R. Franklin. That the authors fully recognized the significance for genetics of the novel structure is apparent from the closing statement of their article, in which they state, "It has not escaped our notice that the specific pairing we have postulated immediately suggests a possible copying mechanism for the genetic material." Vivid, albeit different, accounts of their discovery have been given by the authors (Watson, 1968; Crick, 1988).

The elucidation of the structure of DNA is regarded as the beginning of a new era of molecular biology and genetics. The description of DNA as a double-helix structure led directly to an understanding of the possible structure of genetic information.

When F. Sanger determined the sequence of amino acids of insulin in 1955, he provided the first proof of the primary structure of a protein. This supported the notion that the sequence of amino acids in proteins could correspond with the sequential character of DNA. However, since DNA is located in the cell nucleus and protein synthesis occurs in the cytoplasm, DNA could not act directly. It turned out that DNA is first transcribed into a chemically similar messenger molecule (messenger ribonucleic acid, mRNA) (Crick, Barnett, Brenner, Watts-Tobin 1961) with a corresponding nucleotide sequence, which is transported into the cytoplasm. In the cytoplasm, the mRNA then serves as a template for the amino acid sequence to be formed. The genetic code for the synthesis of proteins from DNA and messenger RNA was determined in the years 1963–1966 (Nirenberg, Mathaei, Ochoa, Benzer, Khorana, and others). Detailed accounts of these developments have been presented by Chargaff (1978), Judson (1996), Stent (1981), Watson and Tooze (1981), Crick (1988), and others.

## Important Methodological Advances in the Development of Genetics after About 1950

From the beginning, genetics has been a field strongly influenced by the development of new experimental methods. In the 1950s and 1960s, the groundwork was laid for biochemical genetics and immunogenetics. Relatively simple but reliable procedures for separating complex molecules by different forms of electrophoresis, methods for synthesizing DNA in vitro (Kornberg, 1956), and other approaches were applied to questions in genetics. The development of cell culture methods was of particular importance for the genetic analysis of humans. Pontecorvo introduced the genetic analysis of cultured eukaryotic cells (somatic cell genetics) in 1958. The study of mammalian genetics, with increasing significance for studying human genes, was facilitated by methods for fusing cells in culture (cell hybridization, T. Puck, G. Barski, B. Ephrussi, 1961) and the development of a cell culture medium for selecting certain mutants in cultured cells (HAT medium, J. Littlefield, 1964). The genetic approach that had been so successful in bacteria and viruses could now be applied in higher organisms, thus avoiding the obstacles of a long generation time and breeding experiments. A hereditary metabolic defect of man (galactosemia) was demonstrated for the first time in cultured human cells in 1961 (Krooth). The correct number of chromosomes in man was determined in 1956 (Tjio and Levan; Ford and Hamerton). Lymphocyte cultures were introduced for chromosomal analysis (Hungerford et al., 1960). The replication pattern of human chromosomes was described (J. German, 1962). These developments further paved the way for expansion of the new field of human genetics.

## Human Genetics

The medical aspects of human genetics (medical genetics) came to attention when it was recognized that sickle cell anemia is hereditary (Neel, 1949) and caused by a defined alteration of normal hemoglobin (Pauling, Itano, Singer, and Wells 1949), and again when it was shown that an enzyme defect (glucose-6-phosphatase

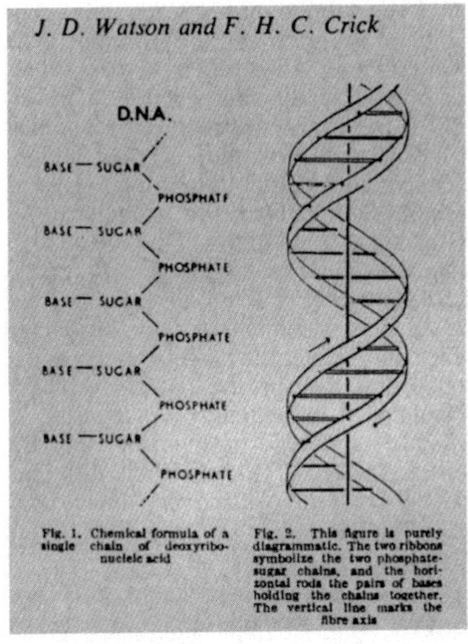


DNA structure 1953

Watson (left) and Crick (right) in 1953

deficiency, demonstrated in liver tissue by Cori and Cori in 1952) was the cause of a hereditary metabolic disease in man (glycogen storage disease type I, or von Gierke disease). The American Society of Human Genetics and the first journal of human genetics *(American Journal of Human Genetics)* were established in 1949. In addition, the first textbook of human genetics appeared (Curt Stern, *Principles of Human Genetics*, 1949).

In 1959, chromosomal aberrations were discovered in some well-known human disorders (trisomy 21 in Down syndrome by J. Lejeune, M. Gautier, R. Turpin; 45,X0 in Turner syndrome by Ford et al.; 47,XXY in Klinefelter syndrome by Jacobs and Strong). Subsequently, other numerical chromosome aberrations were shown to cause recognizable diseases in man (trisomy 13 and trisomy 18, by Patau et al. and Edwards et al. in 1960, respectively), and loss of small parts of chromosomes were shown to be associated with recognizable patterns of severe developmental defects (Lejeune et al., 1963; Wolf, 1964; Hirschhorn, 1964). The Philadelphia chromosome, a characteristic structural alteration of a chromosome in bone marrow cells of patients with adult type chronic myelogenous leukemia, was described by Nowell and Hungerford in 1962. The central role of the Y chromosome in establishing gender in mammals became apparent when it was realized that individuals without a Y chromosome are female and individuals with a Y chromosome are male, irrespective of the number of X chromosomes present. These observations further promoted interest in a new subspecialty, human cytogenetics.

Since early 1960, important knowledge about genetics in general has been obtained, often for the first time, by studies in man. Analysis of genetically determined diseases in man has yielded important insights into the normal function of genes in other organisms as well. Today, more is known about the general genetics of man than about that of any other species. Numerous subspecialties of human genetics have arisen, such as biochemical genetics, immunogenetics, somatic cell genetics, cytogenetics, clinical genetics, population genetics, teratology, mutational studies, and others. The development of the field has been well summarized by Vogel and Motulsky (1997) and McKusick (1992).

## Genetics and Medicine

Most disease processes can be viewed as resulting from environmental influences interacting with the individual genetic makeup of the affected individual. A disease is genetically determined if it is mainly or exclusively caused by disorders in the genetic program of cells and tissues. More than 3000 defined human genetic diseases are known to be due to a mutation at a single gene locus (monogenic disease) and to follow a Mendelian mode of inheritance (McKusick 1998). They differ as much as the genetic information in the genes involved and may be manifest in essentially all age groups and organ systems. An important category of disease results from genetic predisposition interacting with precipitating environmental factors (multigenic or multifactorial diseases). This includes many relatively common chronic diseases (e.g., high blood pressure, hyperlipidemia, diabetes mellitus, gout, psychiatric disorders, certain congenital malformations). Further categories of genetically determined diseases are nonhereditary disorders in somatic cells (different forms of cancer) and chromosomal aberrations.

Due to new mutations and small family size in developed countries, genetic disorders usually do not affect more than one member of a family. About 90% occur as isolated cases within a family. Thus, their genetic origin cannot be recognized by familial aggregation. Instead, they must be recognized by their clinical features. This may be difficult in view of the many different functions of genes in normal tissues and in disease. Since genetic disorders affect all organ systems and age groups and are frequently not recognized, their contribution to the causes of human diseases appears smaller than it actually is. Genetically determined diseases are not a marginal group, but make up a substantial proportion of diseases. More than one-third of all pediatric hospital admissions are for diseases and developmental disorders that, at least in part, are caused by genetic factors (Weatherall 1991). The total estimated frequency of genetically determined diseases of different categories in the general population is about 3.5 – 5.0% (see Table 1).

The large number of individually rare genetically determined diseases and the overlap of diseases with similar clinical manifestations

Table **1** Frequency of genetically determined diseases

| Type of genetic disease | Frequency per 1000 individuals |
|---|---|
| 1. Monogenic diseases, total | 4.5 – 15.0 |
| Autosomal dominant | 2 – 9.5 |
| Autosomal recessive | 2 – 3.5 |
| X-chromosomal | 0.5 – 2 |
| 2. Chromosomal aberrations | 5 – 7 |
| 3. Multifactorial disorders* | 70 – 90 |
| 4. Congenital malformations | 19 – 22 |
| Total | ca. 80 – 115 |

* Contribution of genetic factors variable. (Data based on Weatherall, 1991.)

but different etiology (principle of genetic or etiological heterogeneity) cause additional diagnostic difficulties. This must be considered during diagnosis to avoid false conclusions about a genetic risk.

In 1966 Victor A. McKusick introduced a catalog of human phenotypes transmitted according to Mendelian inheritance (McKusick catalog, currently in its 12th edition; McKusick 1998). This catalog and the 1968 – 1973 Baltimore Conferences organized by McKusick (Clinical Delineation of Birth Defects) have contributed substantially to the systematization and subsequent development of medical genetics. The extent of medical genetics is reflected by the initiation of several new scientific journals since 1965 (*Clinical Genetics, Journal of Medical Genetics, Human Genetics, Annales de Génétique, American Journal of Medical Genetics, Cytogenetics and Cell Genetics, European Journal of Human Genetics, Prenatal Diagnosis, Clinical Dysmorphology,* and others).

In recent years, considerable, previously unexpected progress in clarifying the genetic etiology of human diseases, and thereby in furnishing insights into the structure and function of normal genes, has been achieved by molecular methods.

## Molecular Genetics

The discovery in 1970 (independently by H. Temin and D. Baltimore) of reverse transcriptase, an unusual enzyme complex in RNA viruses (retroviruses), upset the dogma—valid up to that time—that the flow of genetic information went in one direction only, i.e., from DNA to RNA and from there to the gene product (a peptide). Not only is the existence of reverse transcriptase an important biological finding, but the enzyme provides a means of obtaining complementary DNA (cDNA) that corresponds to the coding regions of an active gene. Therefore, it is possible to analyze a gene directly without knowledge of its gene product, provided it is expressed in the tissue examined.

In addition, enzymes that cleave DNA at specific sites (restriction endonucleases or, simply, restriction enzymes) were discovered in bacteria (W. Arber, 1969; D. Nathans and H. O. Smith, 1971). With appropriate restriction enzymes, DNA can be cut into pieces of reproducible and defined size, thus allowing easy recognition of an area to be studied. DNA fragments of different origin can be joined and their properties analyzed. Methods for producing multiple copies of DNA fragments and sequencing them (determining the sequence of their nucleotide bases) were developed between 1977 and 1985. These methods are collectively referred to as recombinant DNA technology (see Chronology at the end of this introduction).

In 1977, recombinant DNA analysis led to a completely new and unexpected finding about the structure of genes in higher organisms, but also in yeast and *Drosophila:* Genes are not continuous segments of coding DNA, but are usually interrupted by noncoding segments (see Watson and Tooze 1981; Watson et al., 1992). The size and sequence of coding DNA segments, or exons (a term introduced by Gilbert in 1978), and noncoding segments, or introns, are specific for each individual gene (exon/intron structure of eukaryotic genes).

With the advent of molecular genetic DNA analysis, many different types of polymorphic DNA markers, i.e., individual heritable differences in the nucleotide sequence, have been mapped to specific sites on chromosomes (physical map). As a result, the chromosomal position of a gene of interest can now be determined (mapped) by analyzing the segregation of a disease locus in relation to the polymorphic DNA markers (linkage analysis). Once the chromosomal location of a gene is known, the latter can be isolated and its structure can be characterized (positional cloning, a term introduced by F. Collins). The advantage of such a direct analysis is that nothing needs to be known about the gene of interest aside from its approximate location. Prior knowledge of the gene product is not required.

Another, complementary, approach is to identify a gene with possible functional relevance to a disorder (a candidate gene), determine its chromosomal position, and then demonstrate mutations in the candidate gene in patients with the disorder. Positional cloning and identification of candidate genes have helped identify genes for many important diseases such as achondroplasia, degenerative retinal diseases, cystic fibrosis, Huntington chorea and other neurodegenerative diseases, Duchenne muscular dystrophy and other muscular diseases, mesenchymal diseases with collagen defects (osteogenesis imperfecta), Marfan syndrome (due to a defect of a previously unknown protein, fibrillin), immune defects, and numerous tumors.

The extensive homologies of genes that regulate embryological development in different organisms and the similarities of genome structures have contributed to leveling the boundaries in genetic analysis that formerly existed for different organisms (e.g., *Drosophila* genetics, mammalian genetics, yeast genetics, bacterial genetics, etc.). Genetics has become a broad, unifying discipline in biology, medicine, and evolutionary research.

## The Dynamic Genome

Between 1950 und 1953, remarkable papers appeared entitled "The origin and behavior of mutable loci in maize" (*Proc Natl Acad Sci.* 36: 344–355, 1950), "Chromosome organization and genic expression" (*Cold Spring Harbor Symp Quant Biol.* 16: 13–45, 1952), and "Introduction of instability at selected loci in maize" (*Genetics* 38: 579–599, 1953). Here the author, Barbara McClintock of Cold Spring Harbor Laboratory, described mutable loci in Indian corn plants (maize) and their effect on the phenotype of corn due to a gene that is not located at the site of the mutation. Surprisingly, this gene can exert a type of remote control. In addition, other genes can change their location and cause mutations at distant sites.

In subsequent work, McClintock described the special properties of this group of genes, which she called controlling genetic elements (*Brookhaven Symp Biol.* 8: 58–74, 1955). Different controlling elements could be distinguished according to their effects on other genes and the mutations caused. However, her work received little interest (for review see Fox Keller 1983; Fedoroff and Botstein 1992).

Thirty years later, at her 1983 Nobel Prize lecture ("The significance of responses of the genome to challenge," Science 226: 792–801, 1984), things had changed. Today we know that the genome is not rigid and static. Rather, it is flexible and dynamic because it contains parts that can move from one location to another (mobile genetic elements, the current designation). The precision of the genetic information depends on its stability, but complete stability would also mean static persistence. This would be detrimental to the development of new forms of life in response to environmental changes. Thus, the genome is subject to alterations, as life requires a balance between the old and the new.

## The Human Genome Project

A new dimension has been introduced into biomedical research by the Human Genome Project (HGP) and related programs in many other organisms (see Part II, Genomics). The main goal of the HGP is to determine the entire sequence of the 3 billion nucleotide pairs in the DNA of the human genome and to find all the genes within it. This is a daunting task. It is comparable to deciphering each individual 1-mm-wide letter along a 3000-km-long text strip. As more than 90% of DNA is not part of genes, other approaches aimed at expressed (active) genes are taken. The completion of a draft covering about 90% of the genome was announced in

June 2000 (*Nature* June 29, 2000, pp. 983–985; *Science* June 30, pp. 2304–2307). The complete sequence of human chromosomes 22 and 21 was published in late 1999 and early 2000, respectively. Conceived in 1986 and officially begun in 1990, the HGP has progressed at a brisk pace. It is expected to be completed in 2003, several years ahead of the original plan (for a review see Lander and Weinberg, 2000, and Part II, Genomics).

### Ethical and Societal Aspects

From its start the Human Genome Project devoted attention and resources to ethical, legal, and social issues (the ELSI program). This is an important part of the HGP in view of the far-reaching consequences of the current and expected knowledge about human genes and the genome. Here only a few areas can be mentioned. Among these are questions of validity and confidentiality of genetic data, of how to decide about a genetic test prior to the first manifestation of a disease (presymptomatic genetic testing), or whether to test for the presence or absence of a disease-causing mutation in an individual before any signs of the disease can be expected (predictive genetic testing). How does one determine whether a genetic test is in the best interest of the individual? Does she or he benefit from the information, could it result in discrimination? How are the consequences defined? How is (genetic) counseling done and informed consent obtained? The use of embryonic stem cells is another area that concerns the public. Careful consideration of benefits and risks in the public domain will aid in reaching rational and balanced decisions.

### Education

Although genetic principles are rather straightforward, genetics is opposed by some and misunderstood by many. Scientists should seize any opportunity to inform the public about the goals of genetics and genomics and the principal methods employed. Genetics should be highly visible at the elementary and high school levels. Human genetics should be emphasized in teaching in medical schools.

## Selected Introductory Reading


Bearn, A.G.: Archibald Garrod and the Individuality of Man. Oxford University Press, Oxford, 1993

Brink, R.A., Styles, E.D., eds.: Heritage from Mendel. University of Wisconsin Press, Madison, 1967.

Cairns, J.: Matters of Life and Death. Perspectives on Public Health, Molecular Biology, Cancer, and the Prospects for the Human Race. Princeton Univ. Press, Princeton, 1997.

Cairns, J., Stent, G.S., Watson, J.D., eds.: Phage and the Origins of Molecular Biology. Cold Spring Harbor Laboratory Press, New York, 1978.

Chargaff, E.: Heraclitean Fire: Sketches from a Life before Nature. Rockefeller University Press, New York, 1978.

Clarke, A.J., ed.: The Genetic Testing of Children. Bios Scientific Publishers, Oxford, 1998.

Coen, E.: The Art of Genes: How Organisms Make Themselves. Oxford Univ. Press, Oxford, 1999.

Crick, F.: What Mad Pursuit: A Personal View of Scientific Discovery, Basic Books, New York, 1988.

Dawkins, R.: The Selfish Gene. 2nd ed., Oxford Univ. Press, Oxford, 1989.

Dobzhansky, T.: Genetics of the Evolutionary Process. Columbia Univ. Press, New York, 1970.

Dubos, R.J.: The Professor, the Institute, and DNA: O.T. Avery, his life and scientific achievements. Rockefeller Univ. Press, New York, 1976.

Dunn, L.C.: A Short History of Genetics. McGraw-Hill, New York, 1965.

Fedoroff, N., Botstein, D., eds.: The Dynamic Genome: Barbara McClintock's Ideas in the Century of Genetics. Cold Spring Harbor Laboratory Press, New York, 1992.

Fox Keller, E.A.: A Feeling for the Organism: the Life and Work of Barbara McClintock. W.H. Freeman, New York, 1983.

Haws, D.V., McKusick, V.A.: Farabee's brachydactyly kindred revisited. Bull. Johns Hopkins Hosp. 113: 20 –30, 1963.

Harper, P.S., Clarke, A.J.: Genetics, Society, and Clinical Practice. Bios Scientific Publishers, Oxford, 1997.

Holtzman, N.A., Watson, M.S., ed.: Promoting Safe and Effective Genetic Testing in the

United States. Final Report of the Task Force on Genetic Testing. National Institute of Health, Bethesda, September 1997.

Judson, H.F.: The Eighth Day of Creation. Makers of the Revolution in Biology. Expanded Edition. Cold Spring Harbor Laboratory Press, New York, 1996.

Lander, E.S., Weinberg, R.A.: Genomics: Journey to the center of biology. Pathways of discovery. Science **287**:1777 – 1782, 2000.

Mayr, E.: The Growth of Biological Thought: Diversity, Evolution, and Inheritance. Harvard University Press, Cambridge, Massachusetts, 1982.

McCarty, M.: The Transforming Principle, W.W. Norton, New York, 1985.

McKusick, V.A.: Presidential Address. Eighth International Congress of Human Genetics: The last 35 years, the present and the future. Am. J. Hum. Genet. **50**:663 – 670, 1992.

McKusick, V.A.: Mendelian Inheritance in Man: A Catalog of Human Genes and Genetic Disorders, 12th ed. Johns Hopkins University Press, Baltimore, 1998.

Online Version OMIM: (http://www.ncbi.nlm.nih.gov/Omim/).

Miller, O.J., Therman, E.: Human Chromosomes. 4th ed. Springer Verlag, New York, 2001.

Neel, J.V.: Physician to the Gene Pool. Genetic Lessons and Other Stories. John Wiley & Sons, New York, 1994.

Schmidtke, J.: Vererbung und Vererbtes – Ein humangenetischer Ratgeber. Rowohlt Taschenbuch Verlag, Reinbek bei Hamburg, 1997.

Schrödinger, E.: What Is Life? The Physical Aspect of the Living Cell. Penguin Books, New York, 1944.

Stebbins, G.L.: Darwin to DNA: Molecules to Humanity. W.H. Freeman, San Francisco, 1982.

Stent, G.S., ed.: James D. Watson. The Double Helix: A Personal Account of the Discovery of the Structure of DNA. A New Critical Edition Including Text, Commentary, Reviews, Original Papers. Weidenfeld & Nicolson, London, 1981.

Sturtevant, A.H.: A History of Genetics. Harper & Row, New York, 1965.

Vogel, F., Motulsky, A.G.: Human Genetics: Problems and Approaches, 3rd ed. Springer Verlag, Heidelberg, 1997.

Watson, J.D.: The Double Helix. A Personal Account of the Discovery of the Structure of DNA. Atheneum, New York–London, 1968.

Watson, J.D.: A Passion fot DNA. Genes, Genomes, and Society. Cold Spring Harbor Laboratory Press, 2000

Watson J.D. and Crick F.H.C.: A structure for deoxyribonucleic acid. Nature 171: 737, 1953.

Watson, J.D., Tooze, J.: The DNA Story: A Documentary History of Gene Cloning. W.H. Freeman, San Francisco, 1981.

Weatherall, D.J.: The New Genetics and Clinical Practice, 3rd ed. Oxford Univ. Press, Oxford, 1991.

Whitehouse, H.L.K.: Towards the Understanding of the Mechanisms of Heredity. 3rd ed. Edward Arnold, London, 1973.

# Chronology

## Advances that Contributed to the Development of Genetics

(This list contains selected events and should not be considered complete, especially for the many important developments during the past several years.)

1839 Cells recognized as the basis of living organisms *(Schleiden, Schwann)*

1859 Concepts of evolution *(Charles Darwin)*

1865 Rules of inheritance by distinct "factors" acting dominantly or recessively *(Gregor Mendel)*

1869 "Nuclein," a new acidic, phosphorus-containing, long molecule *(F. Miescher)*

1879 Chromosomes in mitosis *(W. Flemming)*

1883 Quantitative aspects of heredity *(F. Galton)*

1889 Term "nucleic acid" introduced *(R. Altmann)*

1892 Term "virus" introduced *(R. Ivanowski)*

1897 Enzymes discovered *(E. Büchner)*

1900 Mendel's discovery recognized *(H. de Vries, E.Tschermak, K. Correns, independently)*
ABO blood group system *(Landsteiner)*

1902 Some diseases in man inherited according to Mendelian rules *(W. Bateson, A. Garrod)*
Individuality of chromosomes *(T. Boveri)*
Chromosomes and Mendel's factors are related *(W. Sutton)*
Sex chromosomes *(McClung)*

1906 Term "genetics" proposed *(W. Bateson)*

1908 Population genetics *(Hardy, Weinberg)*

1909 Inborn errors of metabolism *(Garrod)*
Terms "gene," "genotype," "phenotype" proposed *(W. Johannsen)*
Chiasma formation during meiosis *(Janssens)*
First inbred mouse strain DBA *(C. Little)*

1910 Beginning of *Drosophila* genetics *(T. H. Morgan)*
First *Drosophila* mutation *(white-eyed)*

1911 Sarcoma virus *(Peyton Rous)*

1912 Crossing-over *(Morgan and Cattell)*
Genetic linkage *(Morgan and Lynch)*
First genetic map *(A. H. Sturtevant)*

1913 First cell culture *(A. Carrel)*

1914 Nondisjunction *(C. B. Bridges)*

1915 Genes located on chromosomes (chromosomal theory of inheritance) *(Morgan, Sturtevant, Muller, Bridges)*

1922 Characteristic phenotypes of different trisomies in the plant *Datura stramonium (F. Blakeslee)*

1924 Blood group genetics *(Bernstein)*
Statistical analysis of genetic traits *(Fisher)*

1926 Enzymes are proteins *(J. Sumner)*

1927 Mutations induced by X-rays *(H. J. Muller)*
Genetic drift *(S. Wright)*

1928 Euchromatin/heterochromatin *(E. Heitz)*
Genetic transformation in pneumococci *(F. Griffith)*

1933 Pedigree analysis *(Haldane, Hogben, Fisher, Lenz, Bernstein)*
Polytene chromosomes *(Heitz and Bauer, Painter)*

1935 First cytogenetic map in *Drosophila (C. B. Bridges)*

1937 Mouse H2 gene locus *(P. Gorer)*

1940 Polymorphism *(E. B. Ford)*
Rhesus blood groups *(Landsteiner and Wiener)*

1941 Evolution through gene duplication *(E. B. Lewis)*
Genetic control of enzymatic biochemical reactions *(Beadle and Tatum)*
Mutations induced by mustard gas *(Auerbach)*

1944 DNA as the material basis of genetic information *(Avery, MacLeod, McCarty)*
"What is life? The Physical Aspect of the Living Cell." An influential book *(E. Schrödinger)*

1946 Genetic recombination in bacteria *(Lederberg and Tatum)*

1947 Genetic recombination in viruses *(Delbrück and Bailey, Hershey)*

1949 Sickle cell anemia, a genetically determined molecular disease *(Neel, Pauling)*
Hemoglobin disorders prevalent in areas of malaria *(J. B. S. Haldane)*
X chromatin *(Barr and Bertram)*

1950 A defined relation of the four nucleotide bases *(Chargaff)*

1951 Mobile genetic elements in Indian corn *(Zea mays) (B. McClintock)*

1952 Genes consist of DNA *(Hershey and Chase)*
Plasmids *(Lederberg)*
Transduction by phages *(Zinder and Lederberg)*
First enzyme defect in man *(Cori and Cori)*
First linkage group in man *(Mohr)*
Colchicine and hypotonic treatment in chromosomal analysis *(Hsu and Pomerat)*
Exogenous factors as a cause of congenital malformations *(J. Warkany)*

1953 DNA structure *(Watson and Crick, Franklin, Wilkins)*
Nonmendelian inheritance *(Ephrussi)*
Cell cycle *(Howard and Pelc)*
Dietary treatment of phenylketonuria *(Bickel)*

1954 DNA repair *(Muller)*
Leukocyte drumsticks *(Davidson and Smith)*
Cells in Turner syndrome are X-chromatin negative *(Polani)*

1955 Amino acid sequence of insulin *(F. Sanger)*
Lysosomes *(C. de Duve)*
Buccal smear *(Moore, Barr, Marberger)*
5-Bromouracil, an analogue of thymine, induces mutations in phages *(A. Pardee and R. Litman)*

1956 46 Chromosomes in man *(Tijo and Levan, Ford and Hamerton)*
DNA synthesis in vitro *(Kornberg)*
Genetic heterogeneity *(Harris, Fraser)*

1957 Amino acid sequence of hemoglobin molecule *(Ingram)*
Cistron, the smallest nonrecombinant unit of a gene *(Benzer)*
Genetic complementation *(Fincham)*
DNA replication is semiconservative *(Meselson and Stahl, Taylor, Delbrück, Stent)*
Genetic analysis of radiation effects in man *(Neel and Schull)*

1958 Somatic cell genetics *(Pontocorvo)*
Ribosomes *(Roberts, Dintzis)*
Human HLA antigens *(Dausset)*
Cloning of single cells *(Sanford, Puck)*
Synaptonemal complex, the area of synapse in meiosis *(Moses)*

1959 First chromosomal aberrations described in man: trisomy 21 *(Lejeune, Gautier, Turpin)*, Turner syndrome: 45,XO *(Jacobs)*, Klinefelter syndrome: 47 XXY *(Ford)*
Isoenzymes *(Vesell, Markert)*
Pharmacogenetics *(Motulsky, Vogel)*

1960 Phytohemagglutinin-stimulated lymphocyte cultures *(Nowell, Moorehead, Hungerford)*

1961 The genetic code is read in triplets *(Crick, Brenner, Barnett, Watts-Tobin)*
The genetic code determined *(Nirenberg, Mathaei, Ochoa)*
X-chromosome inactivation *(M. F. Lyon,* confirmed by *Beutler, Russell, Ohno)*
Gene regulation, concept of operon *(Jacob and Monod)*
Galactosemia in cell culture *(Krooth)*
Cell hybridization *(Barski, Ephrussi)*
Thalidomide embryopathy *(Lenz, McBride)*

1962 Philadelphia chromosome *(Nowell and Hungerford)*
Xg, the first X-linked human blood group *(Mann, Race, Sanger)*
Screening for phenylketonuria *(Guthrie, Bickel)*
Molecular characterization of immunoglobulins *(Edelman, Franklin)*
Identification of individual human chromosomes by $^3$H-autoradiography *(German, Miller)*
Replicon *(Jacob and Brenner)*
Term "codon" for a triplet of (sequential) bases *(S. Brenner)*

1963 Lysosomal storage diseases *(C. de Duve)*
First autosomal deletion syndrome (cri-du-chat syndrome) *(J. Lejeune)*

1964 Excision repair *(Setlow)*
MLC test *(Bach and Hirschhorn, Bain and Lowenstein)*
Microlymphotoxicity test *(Terasaki and McClelland)*
Selective cell culture medium HAT *(Littlefield)*
Spontaneous chromosomal instability *(German, Schroeder)*
Cell fusion with Sendai virus *(Harris and Watkins)*
Cell culture from amniotic fluid cells *(H. P. Klinger)*
Hereditary diseases studied in cell cultures *(Danes, Bearn, Krooth, Mellman)*
Population cytogenetics *(Court Brown)*
Fetal chromosomal aberrations in spontaneous abortions *(Carr, Benirschke)*

1965 Limited life span of cultured fibroblasts *(Hayflick, Moorehead)*

1966 Catalog of Mendelian phenotypes in man *(McKusick)*

1968 HLA-D the strongest histocompatibility system *(Ceppellini, Amos)*
Repetitive DNA *(Britten and Kohne)*
Biochemical basis of the ABO blood group substances *(Watkins)*
DNA excision repair defect in xeroderma pigmentosum *(Cleaver)*
Restriction endonucleases *(H. O. Smith, Linn and Arber, Meselson and Yuan)*
First assignment of an autosomal gene locus in man *(Donahue, McKusick)*
Synthesis of a gene in vitro *(Khorana)*

1970 Reverse transcriptase (*D. Baltimore, H. Temin*, independently)
Synteny, a new term to refer to all gene loci on the same chromosome *(Renwick)*
Enzyme defects in lysosomal storage diseases *(Neufeld, Dorfman)*
Individual chromosomal identification by specific banding stains *(Zech, Casperson, Lubs, Drets and Shaw, Schnedl, Evans)*
Y-chromatin *(Pearson, Bobrow, Vosa)*
Thymus transplantation for immune deficiency *(van Bekkum)*

1971 Two-hit theory in retinoblastoma *(A. G. Knudson)*

1972 High average heterozygosity *(Harris and Hopkinson, Lewontin)*
Association of HLA antigens and diseases

1973 Receptor defects in the etiology of genetic defects, genetic hyperlipidemia *(Brown, Goldstein, Motulsky)*
Demonstration of sister chromatid exchanges with BrdU *(S. A. Latt)*
Philadelphia chromosome as translocation *(J. D. Rowley)*

1974 Chromatin structure, nucleosome *(Kornberg, Olins and Olins)*
Dual recognition of foreign antigen and HLA antigen by T lymphocytes *(P. C. Doherty and R. M. Zinkernagel)*
Clone of a eukaryotic DNA segment mapped to a specific chromosomal location *(D. S. Hogness)*

1975 Asilomar conference
First protein-signal sequence identified *(G. Blobel)*
Southern blot hybridization *(E. Southern)*
Monoclonal antibodies *(Köhler and Milstein)*

1976 Overlapping genes in phage $\phi$ X174 *(Barell, Air, Hutchinson)*
First transgenic mouse *(R. Jaenisch)*
Loci for structural genes on each human chromosome known *(Baltimore Conference on Human Gene Mapping)*

1977 Genes contain coding and noncoding DNA segments ("split genes") (exon/intron structure) (*R. J. Roberts, P. A. Sharp*, independently)
First recombinant DNA molecule that contains mammalian DNA
Methods to sequence DNA *(F. Sanger; Maxam and Gilbert)*
X-ray diffraction analysis of nucleosomes *(Finch et al.)*

1978 β-Globulin gene structure *(Leder, Weissmann, Tilghman and others)*
Mechanisms of transposition in bacteria
Production of somatostatin with recombinant DNA
Introduction of "chromosome walking" to find genes

1979 First genetic diagnosis using DNA technology *(Y. H. Kan)*

1980 Genes for embryonic development in *Drosophila* studied by mutational screen *(C. Nüsslein-Volhard and others)*

1981 Sequencing of a mitochondrial genome *(S. Anderson, S. G. Barrell, A. T. Bankier)*

1982 Tumor suppressor genes *(H. P. Klinger)*
Prions (proteinaceous infectious particles) proposed as cause of some chronic progressive central nervous system diseases (kuru, scrapie, Creutzfeldt-Jakob disease) *(S. B. Prusiner)*

1983 Cellular oncogenes *(H. E. Varmus and others)*
HIV virus *(L. Montagnier, R. Gallo)*

1984 Localization of the gene for Huntington disease *(Gusella)*
Identification of the T-cell receptor *(Tonegawa)*
Variable DNA sequences as "genetic fingerprints" *(A. Jeffreys)*
*Helicobacter pylori* (B. Marshall)

1985 Polymerase chain reaction *(Mullis, Saiki)*
Characterization of the gene for clotting factor VIII *(Gitschier)*
Sequencing of the AIDS virus
Localization of the gene for cystic fibrosis
Hypervariable DNA segments
Genomic imprinting in the mouse *(B. Cattanach)*

1986 First cloning of human genes. First identification of a human gene based on its chromosomal location (positional cloning) *(Royer-Pokora et al.)*
RNA as catalytic enzyme *(T. Cech)*

1987 Fine structure of an HLA molecule *(Björkman, Strominger et al.)*
Cloning of the gene for Duchenne muscular dystrophy *(Kunkel)*
Knockout mouse *(M. Capecchi)*
A genetic map of the human genome *(H. Donis-Keller et al.)*
Mitochondrial DNA and human evolution *(R. L. Cann, M. Stoneking, A. C. Wilson)*

1988 Start of the Human Genome Project
Successful gene therapy in vitro
Molecular structure of telomeres at the ends of chromosomes *(E. Blackburn and others)*

1989 Cloning of a defined region of a human chromosome obtained by microdissection *(Lüdecke, Senger, Claussen, Horsthemke)*

1990 Evidence for a defective gene causing inherited breast cancer *(Mary-Claire King)*

1991 Cloning of the gene for cystic fibrosis and for Duchenne muscular dystrophy
Odorant receptor multigene family *(Buck and Axel)*
Complete sequence of a yeast chromosome
Increasing use of microsatellites as polymorphic DNA markers

1992 Trinucleotide repeat expansion as a new class of human pathogenic mutations
High density map of DNA markers on human chromosomes
X chromosome inactivation center identified
*p53* knockout mouse *(O. Smithies)*

1993 Gene for Huntington disease cloned

1994 Physical map of the human genome in high resolution
Mutations in fibroblast growth factor receptor genes as cause of achondroplasia and other human diseases
Identification of genes for breast cancer

1995 Master gene of the vertebrate eye, *sey* (small-eye) *(W. J. Gehring)*
STS-band map of the human genome *(T. J. Hudson et al.)*

1996 Yeast genome sequenced
Mouse genome map with more than 7000 markers *(E. S. Lander)*

1997 Sequence of *E. coli* *(F. R. Blattner et al.)*, Helicobacter pylori *(J. F. Tomb)*
Mammal cloned by transfer of an adult cell nucleus into an enucleated oocyte *(Wilmut)*
Embryonic stem cells

1998 Nematode *C. elegans* genome sequenced

1999 First human chromosome (22) sequenced
Ribosome crystal structure

2000 *Drosophila* genome sequenced
First draft of the complete sequence of the human genome
First complete plant pathogen *(Xylella fastidiosa)* genome sequence
*Arabidopsis thaliana*, the first plant genome sequenced

(For more complete information, see references and *http://www.britannica.com* and click on the science channel, Lander and Weinberg, 2000.)

# Fundamentals

# Molecular Basis of Genetics

## The Cell and Its Components

Cells are the smallest organized structural units able to maintain an individual, albeit limited, life span while carrying out a wide variety of functions. Cells have evolved on earth during the past 3.5 billion years, presumably orginating from suitable early molecular aggregations. Each cell originates from another living cell as postulated by R. Virchow in 1855 (*"omnis cellula e cellula"*). The living world consists of two basic types of cells: *prokaryotic cells,* which carry their functional information in a circular genome without a nucleus, and *eukaryotic cells,* which contain their genome in individual chromosomes in a nucleus and have a well-organized internal structure. Cells communicate with each other by means of a broad repertoire of molecular signals. Great progress has been made since 1839, when cells were first recognized as the "elementary particles of organisms" by M. Schleiden and T. Schwann. Today we understand most of the biological processes of cells at the molecular level.

### A. Eukaryotic cells

A eukaryotic cell consists of cytoplasm and a nucleus. It is enclosed by a plasma membrane. The cytoplasm contains a complex system of inner membranes that form cellular structures (organelles). The main organelles are the mitochondria (in which important energy-delivering chemical reactions take place), the endoplasmic reticulum (consisting of a series of membranes in which glycoproteins and lipids are formed), the Golgi apparatus (for certain transport functions), and peroxisomes (for the formation or degradation of certain substances). Eukaryotic cells contain lysosomes, in which numerous proteins, nucleic acids, and lipids are broken down. Centrioles, small cylindrical particles made up of microtubules, play an essential role in cell division. Ribosomes are the sites of protein synthesis.

### B. Nucleus of the Cell

The eukaryotic cell nucleus contains the genetic information. It is enclosed by an inner and an outer membrane, which contain pores for the transport of substances between the nucleus and the cytoplasm. The nucleus contains a nucleolus and a fibrous matrix with different DNA–protein complexes.

### C. Plasma membrane of the cell

The environment of cells, whether blood or other body fluids, is water-based, and the chemical processes inside a cell involve water-soluble molecules. In order to maintain their integrity, cells must prevent water and other molecules from flowing in or out uncontrolled. This is accomplished by a water-resistant membrane composed of bipartite molecules of fatty acids, the *plasma membrane.* These molecules are phospholipids arranged in a double layer (bilayer) with a fatty interior. The plasma membrane itself contains numerous molecules that traverse the lipid bilayer once or many times to perform special functions. *Different types of membrane proteins* can be distinguished: (i) transmembrane proteins used as channels for transport of molecules into or out of the cell, (ii) proteins connected with each other to provide stability, (iii) receptor molecules involved in signal transduction, and (iv) molecules with enzyme function to catalyze internal chemical reactions in response to an external signal. (Figure redrawn from Alberts et al., 1998.)

### D. Comparison of animal and plant cells

Plant and animal cells have many similar characteristics. One fundamental difference is that plant cells contain chloroplasts for photosynthesis. In addition, plant cells are surrounded by a rigid wall of cellulose and other polymeric molecules and contain vacuoles for water, ions, sugar, nitrogen-containing compounds, or waste products. Vacuoles are permeable to water but not to the other substances enclosed in the vacuoles. (Figures in A, B and D adapted from de Duve, 1984.)

### References

Alberts, B. et al.: Essential Cell Biology. An Introduction to the Molecular Biology of the Cell. Garland Publishing Co., New York, 1998.

de Duve, C.: A Guided Tour of the Living Cell. Vol. I and II. Scientific American Books, Inc., New York, 1984.

Lodish, H. et al.: Molecular Cell Biology (with an animated CD-ROM). 4th ed. W.H. Freeman & Co., New York, 2000.

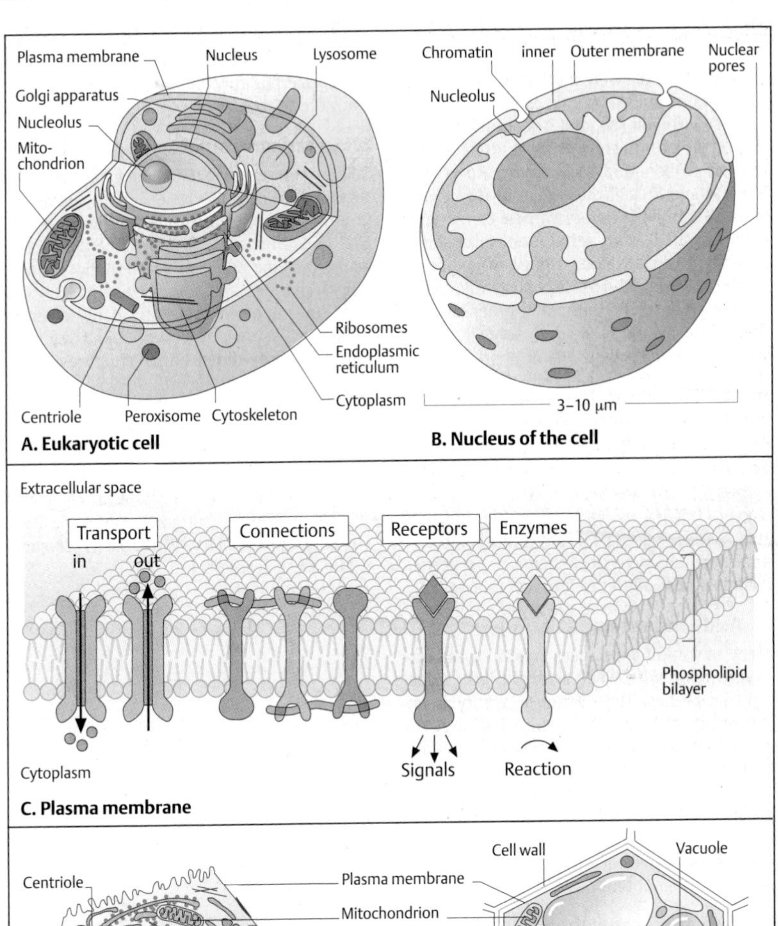


A. Eukaryotic cell

B. Nucleus of the cell

C. Plasma membrane

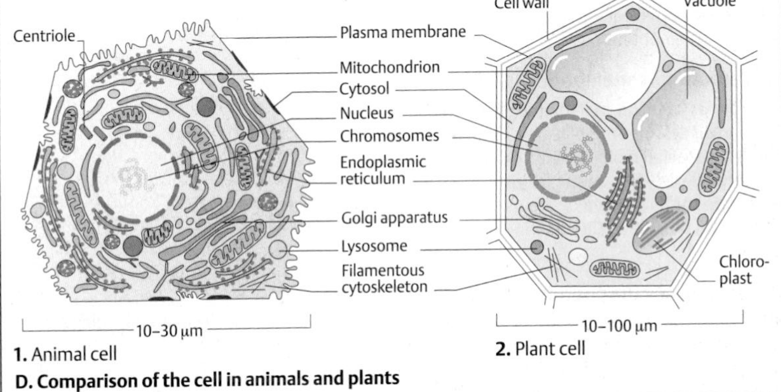


1. Animal cell

2. Plant cell

D. Comparison of the cell in animals and plants

## Some Types of Chemical Bonds

Close to 99% of the weight of a living cell is composed of just four elements: carbon (C), hydrogen (H), nitrogen (N), and oxygen (O). Almost 50% of the atoms are hydrogen atoms; about 25% are carbon, and 25% oxygen. Apart from water (about 70% of the weight of the cell) almost all components are carbon compounds. Carbon, a small atom with four electrons in its outer shell, can form four strong covalent bonds with other atoms. But most importantly, carbon atoms can combine with each other to build chains and rings, and thus large complex molecules with specific biological properties.

### A. Compounds of hydrogen (H), oxygen (O), and carbon (C)

Four simple combinations of these atoms occur frequently in biologically important molecules: hydroxyl (–OH; alcohols), methyl ($-CH_3$), carboxyl (–COOH), and carbonyl (C=O; aldehydes and ketones) groups. They impart to the molecules characteristic chemical properties, including possibilities to form compounds.

### B. Acids and esters

Many biological substances contain a carbon-oxygen bond with weak acidic or basic (alkaline) properties. The degree of acidity is expressed by the pH value, which indicates the concentration of $H^+$ ions in a solution, ranging from $10^{-1}$ mol/L (pH 1, strongly acidic) to $10^{-14}$ mol/L (pH 14, strongly alkaline). Pure water contains $10^{-7}$ moles $H^+$ per liter (pH 7.0). An ester is formed when an acid reacts with an alcohol. Esters are frequently found in lipids and phosphate compounds.

### C. Carbon–nitrogen bonds (C—N)

C–N bonds occur in many biologically important molecules: in amino groups, amines, and amides, especially in proteins. Of paramount significance are the amino acids (cf. p. 30), which are the subunits of proteins. All proteins have a specific role in the functioning of an organism.

### D. Phosphate compounds

Ionized phosphate compounds play an essential biological role. $HPO_4^{2-}$ is a stable inorganic phosphate ion from ionized phosphoric acid. A phosphate ion and a free hydroxyl group can form a phosphate ester. Phosphate compounds play an important role in energy-rich molecules and numerous macromolecules because they can store energy.

### E. Sulfur compounds

Sulfur often serves to bind biological molecules together, especially when two sulfhydryl groups (–SH) react to form a disulfide bridge (–S–S–). Sulfur is a component of two amino acids (cysteine and methionine) and of some polysaccharides and sugars. Disulfide bridges play an important role in many complex molecules, serving to stabilize and maintain particular three-dimensional structures.

### References

Alberts, B. et al.: Molecular Biology of the Cell. 3rd ed. Garland Publishing Co., New York, 1994.

Koolman, J., Röhm K.H.: Color Atlas of Biochemistry. Thieme, Stuttgart – New York, 1996.

Stryer, L.: Biochemistry, 4th ed. W.H. Freeman & Co., New York, 1995.

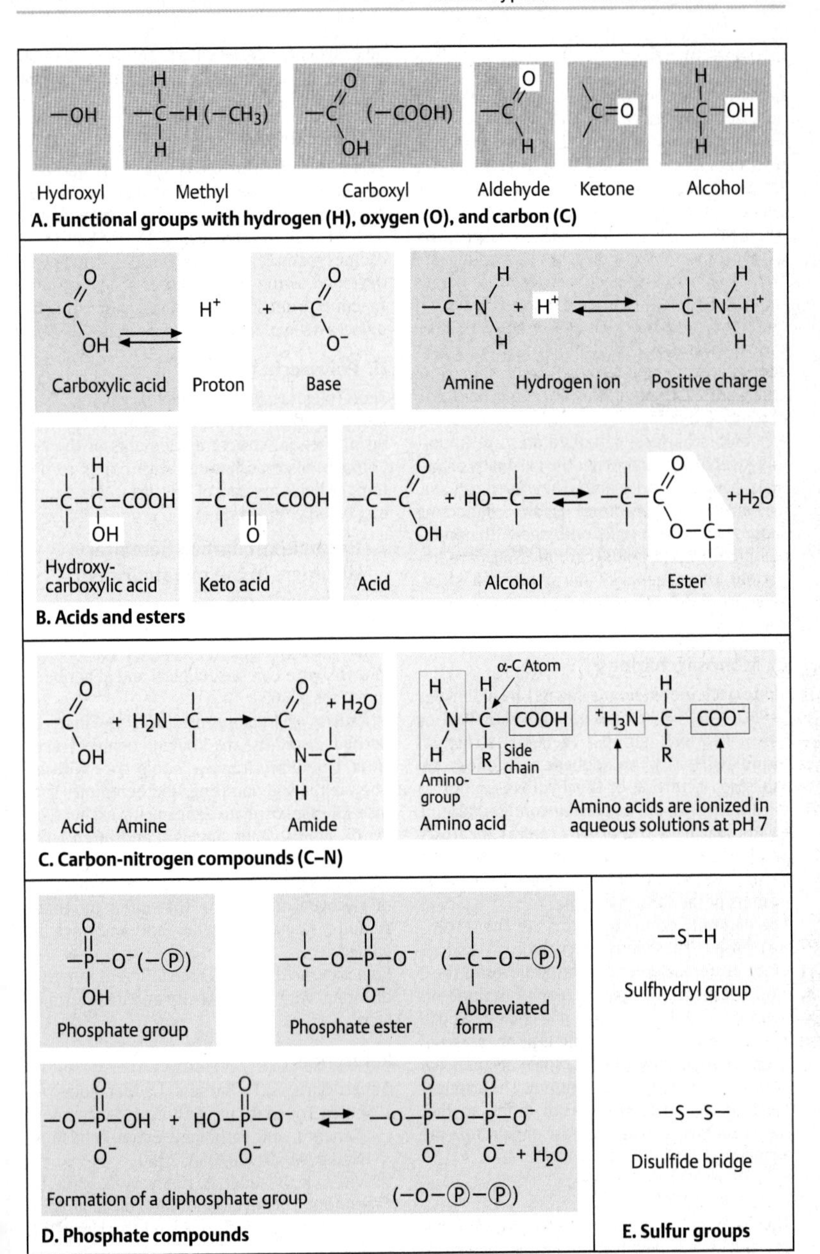

—OH
Hydroxyl
—C—H (—CH3)
Methyl
(—COOH)
Carboxyl
Aldehyde
C=O
Ketone
—C—OH
Alcohol
A. Functional groups with hydrogen (H), oxygen (O), and carbon (C)
Carboxylic acid
Proton
Base
Amine
Hydrogen ion
Positive charge
Hydroxy-carboxylic acid
Keto acid
Acid
Alcohol
Ester
+H2O
B. Acids and esters
+ H2N—C—
+ H2O
Acid
Amine
Amide
α-C Atom
COOH
Side chain
Amino-group
Amino acid
+H3N
COO−
Amino acids are ionized in aqueous solutions at pH 7
C. Carbon-nitrogen compounds (C–N)
—P—O−(—Ⓟ)
Phosphate group
Phosphate ester
(—C—O—Ⓟ)
Abbreviated form
Formation of a diphosphate group
(—O—Ⓟ—Ⓟ)
D. Phosphate compounds
—S—H
Sulfhydryl group
—S—S—
Disulfide bridge
E. Sulfur groups

# Carbohydrates

Carbohydrates in their various chemical forms and their derivatives are an important group of biomolecules for genetics. They provide the basic framework of DNA and RNA. Their flexibility makes them especially suitable for transferring genetic information from cell to cell.

Along with nucleic acids, lipids, and proteins, carbohydrates are one of the most important classes of biomolecules. Their main functions can be classified into three groups: (i) to deliver and store energy, (ii) to help form DNA and RNA, the information-carrying molecules (see pp. 34 and 38), (iii) to help form cell walls of bacteria and plants. Carbohydrates are often bound to proteins and lipids.

As polysaccharides, carbohydrates are important structural elements of the cell walls of animals, bacteria, and plants. They form cell surface structures (receptors) used in conducting signals from cell to cell. Combined with numerous proteins and lipids, carbohydrates are important components of numerous cell structures. Finally, they function to transfer and store energy in intermediary metabolism.

## A. Monosaccharides

Monosaccharides (simple sugars) are aldehydes (–C=O, –H) or ketones (>C=O) with two or more hydroxy groups (general structural formula $(CH_2O)_n$). The aldehyde or ketone group can react with one of the hydroxy groups to form a ring. This is the usual configuration of sugars that have five or six C atoms (pentoses and hexoses). The C atoms are numbered. The D- and the L-forms of sugars are mirror-image isomers of the same molecule.

The naturally occurring forms are the D-(dextro) forms. These further include β- and α-forms as stereoisomers. In the cyclic forms the C atoms of sugars are not on a plane, but three-dimensionally take the shape of a chair or a boat. The β-D-glucopyranose configuration (glucose) is the energetically favored, since all the axial positions are occupied by H atoms. The arrangement of the –OH groups can differ, so that stereoisomers such as mannose or galactose are formed.

## B. Disaccharides

These are compounds of two monosaccharides. The aldehyde or ketone group of one can bind to an α-hydroxy or a β-hydroxy group of the other. Sucrose and lactose are frequently occurring disaccharides.

## C. Derivatives of sugars

When certain hydroxy groups are replaced by other groups, sugar derivatives are formed. These occur especially in polysaccharides. In a large group of genetically determined syndromes, complex polysaccharides can not be degraded owing to reduced or absent enzyme function (mucopolysaccharidoses, mucolipidoses) (see p. 356).

## D. Polysaccharides

Short (oligosaccharides) and long chains of sugars and sugar derivatives (polysaccharides) form essential structural elements of the cell. Complex oligosaccharides with bonds to proteins or lipids are part of cell surface structures, e.g., blood group antigens.

### Examples of human hereditary disorders in the metabolism of carbohydrates

*Diabetes mellitus:* a heterogeneous group of disorders characterized by elevated levels of blood glucose, with complex clinical and genetic features (see p. 362).

*Disorders of fructose metabolism:* Three inherited disorders are known: benign fructosuria, hereditary fructose intolerance with hypoglycemia and vomiting, and hereditary fructose 1,6-bisphosphate deficiency with hypoglycemia, apnea, lactic acidosis, and often lethal outcome in newborn infants.

*Glycogen storage diseases:* a group of disorders of glycogen metabolism that differ in clinical symptoms and the genes and enzymes involved.

*Galactose metabolism:* Three different inherited disorders with acute toxicity and long-term effects.

## References

Gilbert-Barness, E., Barness, L.: Metabolic Diseases. Foundations of Clinical Management, Genetics, and Pathology. Eaton Publishing, Natick, MA 01760, USA, 2000.

Scriver, C. R., Beaudet, A. L., Sly, W. S., Valle, D., editors: The Metabolic and Molecular Bases of Inherited Disease. 8th ed., McGraw-Hill, New York, 2001.

β-D-Glucose α-D-Glucose L-Glucose

Stereoisomers

Glucose Mannose Galactose

Isomers of glucose

**A. Monosaccharides**

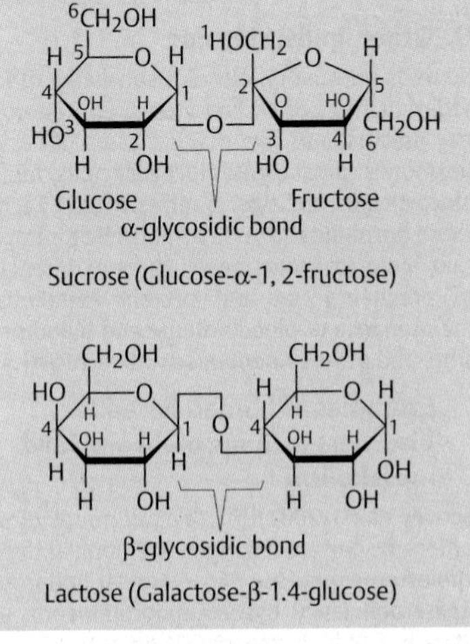


Sucrose (Glucose-α-1, 2-fructose)

Lactose (Galactose-β-1.4-glucose)

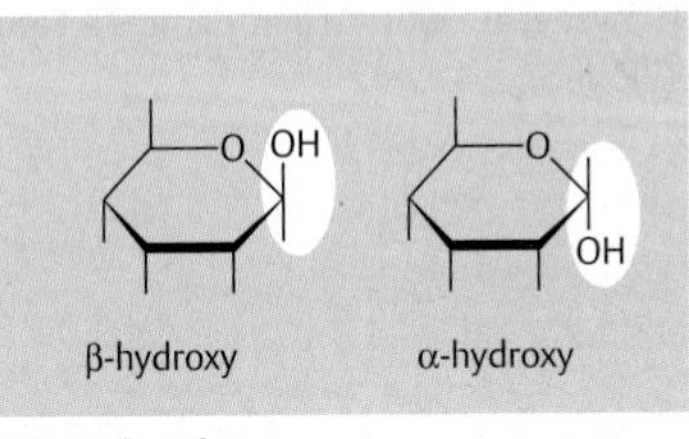


**B. Disaccharides**

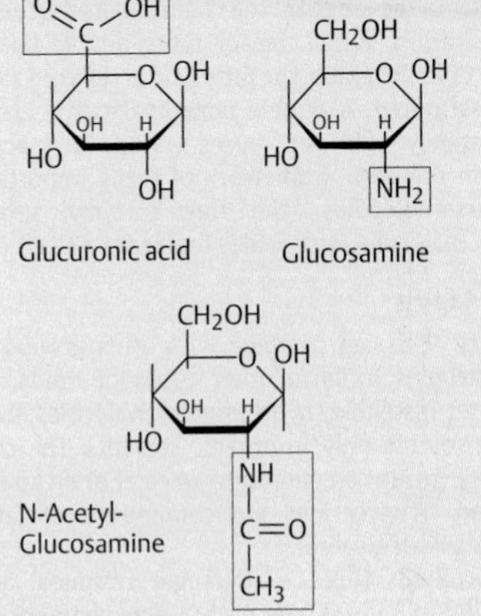


**C. Sugar derivatives**

α-1,6 bonds at branching points

All others are α-1,4 bonds

**D. Polysaccharides**

## Lipids (Fats)

Lipids usually occur as large molecules (macromolecules). They are essential components of membranes and precursors of other important biomolecules, such as steroids for the formation of hormones and other molecules for transmitting intercellular signals. In addition to fatty acids, compounds with carbohydrates (glycolipids), phosphate groups (phospholipids), and other molecules are especially important. A special characteristic is their pronounced polarity, with a hydrophilic (water-attracting) and a hydrophobic (water-repelling) region. This makes lipids especially suited for forming the outer limits of the cell (cell membrane).

### A. Fatty acids

Fatty acids are composed of a hydrocarbon chain with a terminal carboxylic acid group. Thus, they are polar, with a hydrophilic (–COOH) and a hydrophobic end (–$CH_3$), and differ in the length of the chain and its degree of saturation. When one or more double bonds occur in the chain, the fatty acid is referred to as unsaturated. A double bond makes the chain relatively rigid and causes a kink. Fatty acids form the basic framework of many important macromolecules. The free carboxyl group (–COOH) of a fatty acid is ionized (–$COO^-$).

### B. Lipids

Fatty acids can combine with other groups of molecules to form other types of lipids. As water-insoluble (hydrophobic) molecules, they are soluble only in organic solvents. The carboxyl group can enter into an ester or an amide bond. Triglycerides are compounds of fatty acids with glycerol.

Glycolipids (lipids with sugar residues) and phospholipids (lipids with a phosphate group attached to an alcohol derivative) are the structural bases of important macromolecules. Their intracellular degradation requires the presence of numerous enzymes, disorders of which have a genetic basis and lead to numerous genetically determined diseases.

Sphingolipids are an important group of molecules in biological membranes. Here, sphingosine, instead of glycerol, is the fatty acid-binding molecule. Sphingomyelin and gangliosides contain sphingosine. Gangliosides make up 6% of the central nervous system lipids. They are degraded by a series of enzymes. Genetically determined disorders of their catabolism lead to severe diseases, e.g., Tay–Sachs disease due to defective degradation of ganglioside GM2 (deficiency of β-*N*-acetylhexosaminidase).

### C. Lipid aggregates

Owing to their bipolar properties, fatty acids can form lipid aggregates in water. The hydrophilic ends are attracted to their aqueous surroundings; the hydrophobic ends protrude from the surface of the water and form a surface film. If completely under the surface, they may form a micelle, compact and dry within. Phospholipids and glycolipids can form two-layered membranes (lipid membrane bilayer). These are the basic structural elements of cell membranes, which prevent molecules in the surrounding aqueous solution from invading the cell.

### D. Other lipids: steroids

Steroids are small molecules consisting of four different rings of carbon atoms. Cholesterol is the precursor of five major classes of steroid hormones: prostagens, glucocorticoids, mineralocorticoids, androgens, and estrogens. Each of these hormone classes is responsible for important biological functions such as maintenance of pregnancy, fat and protein metabolism, maintenance of blood volume and blood pressure, and development of sex characteristics.

#### Examples of human hereditary disorders in lipoprotein and lipid metabolism

Scriver et al. (2001) list several groups of disorders. Important examples are familial hypercholesterolemia (see 358), familial lipoprotein lipase deficiency, dysbetalipoproteinemia, and disorders of high-density lipoprotein.

(Scriver et al., 2001; Gilbert-Barness & Barness, 2000).

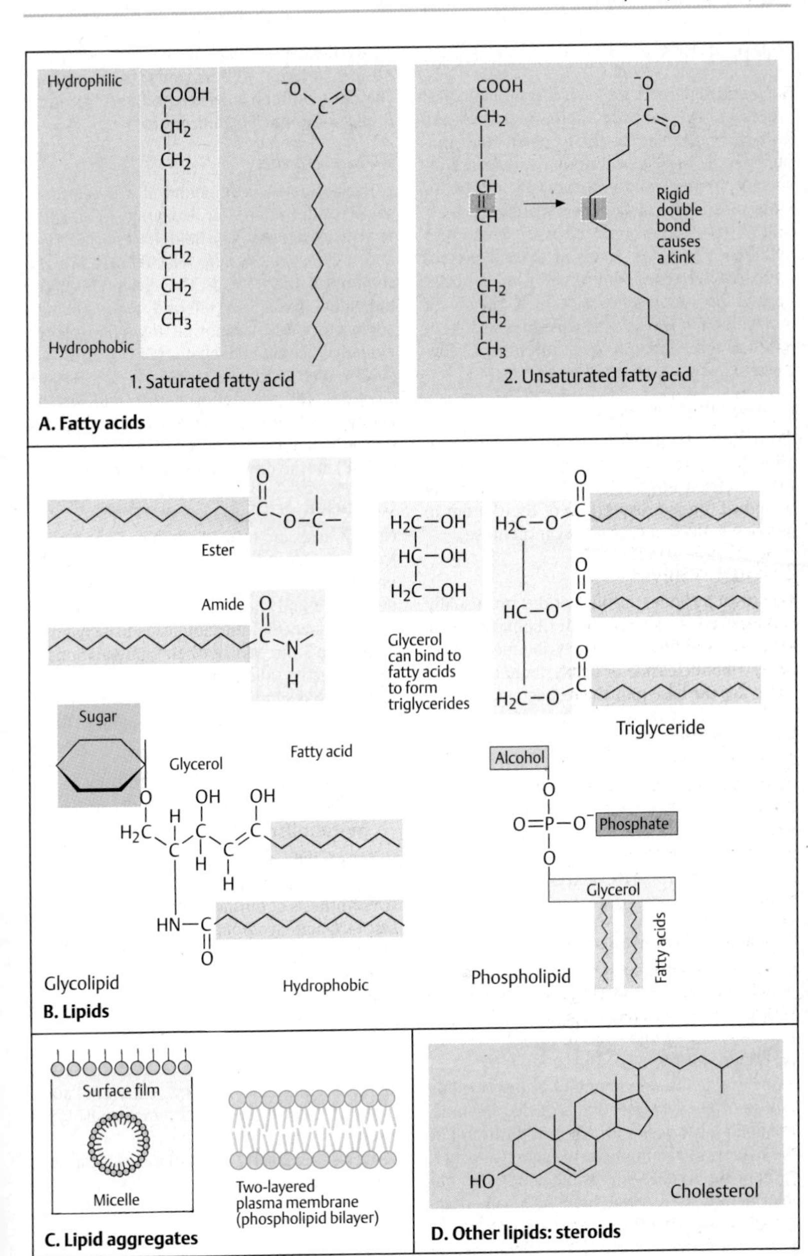
Hydrophilic
COOH
CH2
CH2
CH2
CH2
CH3
Hydrophobic
1. Saturated fatty acid
COOH
CH2
CH
CH
CH2
CH2
CH3
Rigid double bond causes a kink
2. Unsaturated fatty acid
A. Fatty acids
Ester
Amide
H2C—OH
HC—OH
H2C—OH
Glycerol can bind to fatty acids to form triglycerides
H2C—O
HC—O
H2C—O
Triglyceride
Sugar
Glycerol
Fatty acid
OH
OH
H2C
HN—C
Glycolipid
Hydrophobic
Alcohol
O=P—O⁻
Phosphate
Glycerol
Fatty acids
Phospholipid
B. Lipids
Surface film
Micelle
Two-layered plasma membrane (phospholipid bilayer)
C. Lipid aggregates
HO
Cholesterol
D. Other lipids: steroids

# Nucleotides and Nucleic Acids

Nucleotides participate in almost all biological processes. They are the subunits of DNA and RNA, the molecules that carry genetic information (see p. 34). Nucleotide derivatives are involved in the biosynthesis of numerous molecules; they convey energy, are part of essential coenzymes, and regulate numerous metabolic functions. Since all these functions are based on genetic information of the cells, nucleotides represent a central class of molecules for genetics. Nucleotides are composed of three integral parts: phosphates, sugars, and purine or pyramidine bases.

## A. Phosphate groups

Phosphate groups may occur alone (monophosphates), in twos (diphosphates) or in threes (triphosphates). They are normally bound to the hydroxy group of the C atom in position 5 of a five-C-atom sugar (pentose).

## B. Sugar residues

The sugar residues in nucleotides are usually derived from either ribose (in ribonucleic acid, RNA) or deoxyribose (in deoxyribonucleic acid, DNA) (ribonucleoside or deoxyribonucleoside). These are the base plus the respective sugar.

## C. Nucleotide bases of pyrimidine

Cytosine (C), thymine (T), and uracil (U) are the three pyrimidine nucleotide bases. They differ from each other in their side chains ($-NH_2$ on C4 in cytosine, $-CH_3$ on C5 in thymine, O on C4 in uracil) and in the presence or absence of a double bond between N3 and C4 (present in cytosine).

## D. Nucleotide bases of purine

Adenine (A) and guanine (G) are the two nucleotide bases of purine. They differ in their side chains and a double bond (between N1 and C6).

## E. Nucleosides

A nucleoside is a compound of a sugar residue (ribose or deoxyribose) and a nucleotide base. The bond is between the C atom in position 1 of the sugar (as in compounds of sugars) and an N atom of the base (N-glycosidic bond). The nucelotides of the various bases are named according to whether they are a ribonucleoside or a deoxyribonucleoside, e.g., adenosine or deoxyadenosine, guanosine or deoxyguanosine, uridine (occurs only as a ribonucleoside), cytidine or deoxycytidine. Thymidine occurs only as a deoxynucleoside.

## F. Nucleotides

A nucleotide is a compound of a five-C-atom sugar residue (ribose or deoxyribose) attached to a nucleotide base (pyrimidine or purine base) and a phosphate group. Nucleotides are the subunits of nucleic acids. The nucleotides of the individual bases are referred to as follows: adenylate (AMP, adenosine monophosphate), guanolyte monophosphate (GMP), uridylate (UMP), and cytidylate (CMP) for the ribonucleotides (5' monophosphates) and deoxyadenylate (dAMP), deoxyguanylate (dGMP), deoxythymidylate (dTMP), and deoxycytidylate (dCMP) for the deoxyribonucleotides.

## G. Nucleic acids

Nucleic acids are formed when nucleotides are joined to each other by means of phosphodiester bridges between the 3' C atom of one nucleotide and the 5' C atom of the next. The linear sequence of nucleotides is usually given in the 5' to 3' direction with the abbreviations of the respective nucleotide bases. For instance, ATCG would signify the sequence adenine (A), thymine (T), cytosine (C), and guanine (G) in the 5' to 3' direction.

### Examples of human hereditary disorders in purine and pyrimidine metabolism

*Hyperuricemia and gout:* A group of disorders resulting from genetically determined excessive synthesis of purine precursors.

*Lesch–Nyhan syndrome:* A variable, usually severe infantile X-chromosomal disease with marked neurological manifestations resulting from hypoxanthine–guanine phosphoribosyltransferase deficiency.

*Adenosine deaminase deficiency:* A heterogeneous group of disorders resulting in severe infantile immunodeficiency. Different autosomal recessive and X-chromosomal types exist.

Scriver et al., 2001; Gilbert-Barness & Barness, 2000).

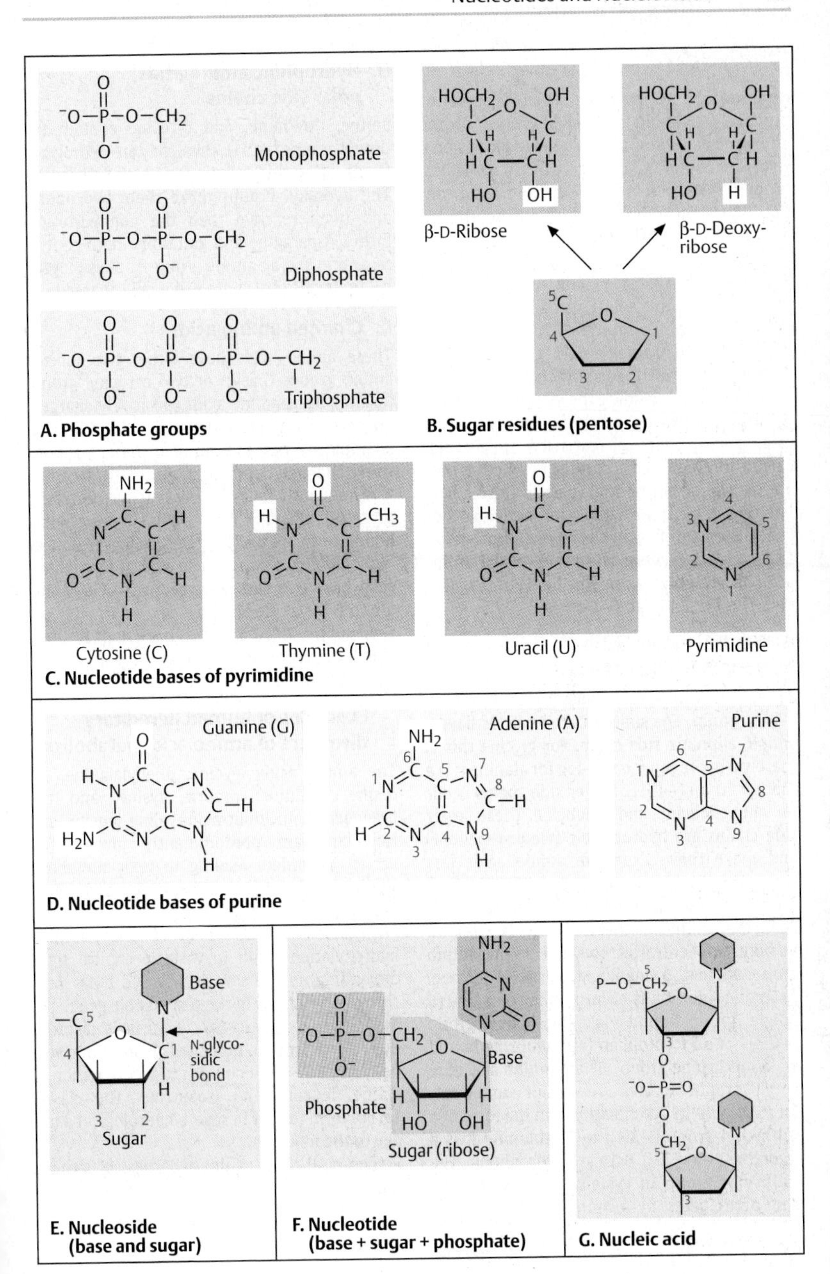
Monophosphate
Diphosphate
Triphosphate
A. Phosphate groups
β-D-Ribose
β-D-Deoxyribose
B. Sugar residues (pentose)
Cytosine (C)
Thymine (T)
Uracil (U)
Pyrimidine
C. Nucleotide bases of pyrimidine
Guanine (G)
Adenine (A)
Purine
D. Nucleotide bases of purine
Base
N-glycosidic bond
Sugar
E. Nucleoside (base and sugar)
Phosphate
Base
Sugar (ribose)
F. Nucleotide (base + sugar + phosphate)
G. Nucleic acid

## Amino Acids

Amino acids are the basic structural units of proteins. A defined linear sequence of the amino acids and a specific three-dimensional structure confer quite specific physicochemical properties to each protein. An amino acid consists of a "central" carbon with one bond to an amino group ($-NH_2$) one to a carboxyl group ($-COOH$) one to a hydrogen atom, and the fourth to a variable side chain. Amino acids are ionized in neutral solutions, since the amino group takes on a proton ($-NH_3^+$) and the carboxyl group dissociates ($-COO^-$). The side chain determines the distinguishing characteristics of an amino acid, including the size, form, electrical charge or hydrogen-bonding ability, and the total specific chemical reactivity. Amino acids can be differentiated according to whether they are neutral or not neutral (basic or acidic) and whether they have a polar or nonpolar side chain. Each amino acid has its own three-letter and one-letter abbreviations. Essential amino acids in vertebrates are His, Ile, Leu, Lys, Met, Phe, Thr, Tyr, and Val.

### A. Neutral amino acids, nonpolar side chains

All neutral amino acids have a $-COO^-$ and an $-NH_3^+$ group. The simplest amino acids have a simple aliphatic side chain. For glycine this is merely a hydrogen atom ($-H$); for alanine it is a methyl group ($-CH_3$). Larger side chains occur on valine, leucine, and isoleucine. These larger side chains are hydrophobic (water-repellent) and make their respective amino acids less water-soluble than do hydrophilic (water-attracting) chains. Proline has an aliphatic side chain that, unlike in other amino acids, is bound to both the central carbon and to the amino group, so that a ringlike structure is formed. Aromatic side chains occur in phenylalanine (a phenyl group bound via a methylene group ($-CH_2-$) and tryptophan (an indol ring bound via a methylene group). These amino acids are very hydrophobic. Two amino acids contain sulfur (S) atoms. In cysteine this is in the form of a sulfhydryl group ($-SH$); in methionine it is a thioether ($-S-CH_3$). Both are hydrophobic. The sulfhydryl group in cysteine is very reactive and participates in forming disulfide bonds ($-S-S-$). These play an important role in stabilizing the three-dimensional forms of proteins.

### B. Hydrophilic amino acids, polar side chains

Serine, threonine, and tyrosine contain hydroxyl groups ($-OH$). Thus, they are hydrolyzed forms of glycine, alanine, and phenylalanine. The hydroxyl groups make them hydrophilic and more reactive than the nonhydrolyzed forms. Asparagine and glutamine both contain an amino and an amide group. At physiological pH their side chains are negatively charged.

### C. Charged amino acids

These amino acids have either two ionized amino groups (basic) or two carboxyl groups (acidic). Basic amino acids (positively charged) are arginine, lysine, and histidine. Histidine has an imidazole ring and can be uncharged or positively charged, depending on its surroundings. It is frequently found in the reactive centers of proteins, where it takes part in alternating bonds (e.g., in the oxygen-binding region of hemoglobin). Aspartic acid and glutamic acid each have two carboxyl groups ($-COOH$) and are thus (as a rule) acidic.
Seven of the 20 amino acids have slightly ionizable side chains, making them highly reactive (Asn, Glu, His, Cys, Tyr, Lys, Arg).

### Examples of human hereditary disorders of amino acid metabolism

The amino acids glycine, phenylalanine, tyrosine, histidine, proline, lysine, and the branched chain amino acids valine, leucine, and isoleucine are predominantly involved in various disorders leading to toxic metabolic symptoms due to an increase or decrease of their plasma concentration.
*Phenylketonuria:* Disorders of phenylalanine hydroxylation result in variable clinical signs depending on the severity, caused by a spectrum of mutations in the responsible gene.
*Maple syrup urine disease:* A variable disorder due to deficiency of branched chain α-keto acid dehydrogenase, which leads to accumulation of valine, leucine, and isoleucine. The classic severe form results in severe neurological damage to the infant.
Scriver et al., 2001; Gilbert-Barness & Barness, 2000).

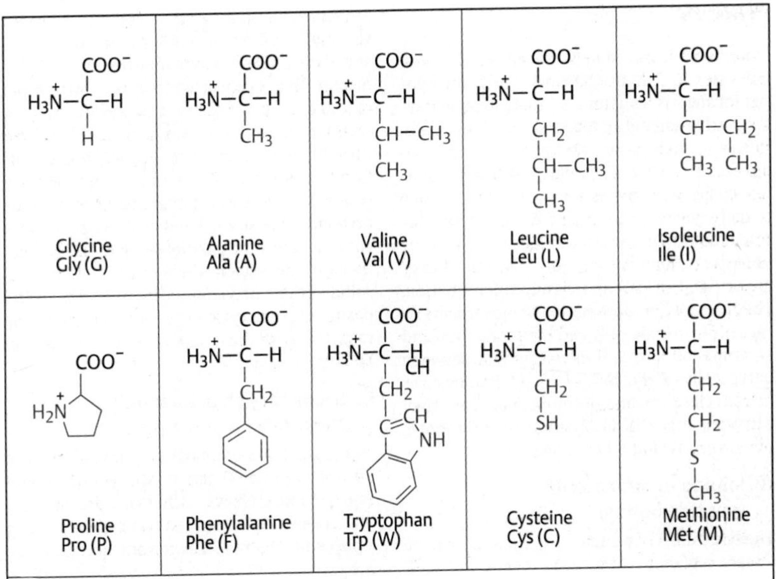

Glycine
Gly (G)

Alanine
Ala (A)

Valine
Val (V)

Leucine
Leu (L)

Isoleucine
Ile (I)

Proline
Pro (P)

Phenylalanine
Phe (F)

Tryptophan
Trp (W)

Cysteine
Cys (C)

Methionine
Met (M)

**A. Neutral amino acids, nonpolar side chains**

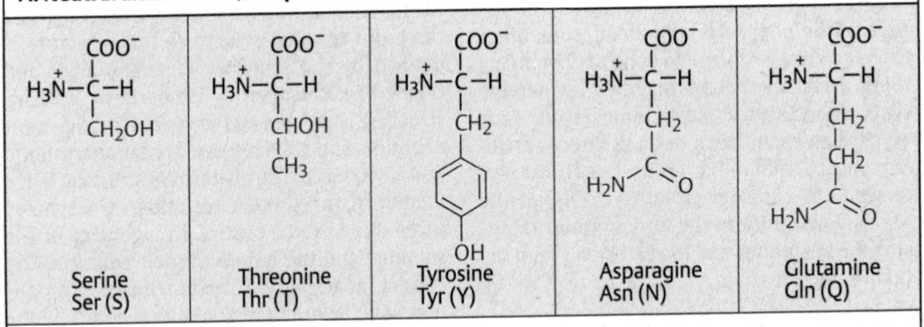

Serine
Ser (S)

Threonine
Thr (T)

Tyrosine
Tyr (Y)

Asparagine
Asn (N)

Glutamine
Gln (Q)

**B. Neutral amino acids, polar side chains**

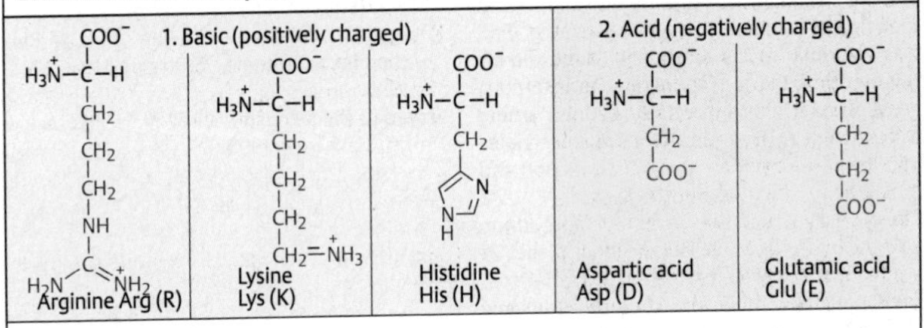

1. Basic (positively charged)

Arginine Arg (R)

Lysine
Lys (K)

Histidine
His (H)

2. Acid (negatively charged)

Aspartic acid
Asp (D)

Glutamic acid
Glu (E)

**C. Charged amino acids**

# Proteins

Proteins are involved in practically all chemical processes in living organisms. Their universal significance is apparent in that, as enzymes, they drive chemical reactions in living cells. Without enzymatic catalysis, the macromolecules involved would not react spontaneously. All enzymes are the products of one or more genes. Proteins also serve to transport small molecules, ions, or metals. Proteins have important functions in cell division during growth and in cell and tissue differentiation. They control the coordination of movements by regulating muscle cells and the production and transmission of impulses within and between nerve cells. They control blood homeostasis (blood clotting) and immune defense. They carry out mechanical functions in skin, bone, blood vessels, and other areas.

## A. Joining of amino acids (peptide bonds)

The basic units of proteins, amino acids, can be joined together very easily owing to their dipolar ionization (zwitterions). The carboxyl group of one amino acid binds to the amino group of the next (a peptide bond, sometimes also referred to as an amide bond). When many amino acids are bound together by peptide bonds, they form a polypeptide chain. Each polypeptide chain has a defined direction, determined by the amino group ($-NH_2$) at one end and the carboxyl group ($-COOH$) at the other. By convention, the amino group represents the beginning, and the carboxyl group the end of a peptide chain.

## B. Primary structure of a protein

The determination of the complete amino acid sequence of insulin by Frederick Sanger in 1955 was a landmark accomplishment. It showed for the first time that a protein, in genetic terms a gene product, has a precisely defined amino acid sequence. The amino acid sequence yields important information about the function and evolutionary origin of a protein.

The primary structure of a protein is its amino acid sequence in a one-dimensional plane. As are many other proteins, insulin is synthesized from precursor molecules: preproinsulin and proinsulin. Preproinsulin consists of 110 amino acids including 24 amino acids of a leader sequence at the amino end. The leader sequence directs the molecule to the correct site in the cell and is then removed to yield proinsulin. This is converted to insulin by removal of the connecting peptide (C peptide) consisting of amino acids 31–65. Amino acids 1–30 form the B chain; the remaining (66–86) amino acids form the A chain. The A and the B chains are connected by two disulfide bridges joining the cysteines in position 7 and position 20 of the A chain to those of positions 7 and 19, respectively, of the B chain. The A chain contains a disulfide bridge between positions 6 and 11. The positions of the cysteines reflect the spatial arrangements of the amino acids, called the secondary structure.

## C. Secondary structural units, the α helix and the β sheet

Two basic units of global proteins are α helix formation (α helix) and a flat sheet (β pleated sheet). Panel C shows a schematic drawing of a unit of one α helix between two β-sheets, called a βαβ unit (Figure redrawn from Stryer, 1995).

## D. Tertiary structure of insulin

All functional proteins assume a well-defined three-dimensional structure. This structure is defined by the sequence of amino acids and their physicochemical properties. Tertiary structure is defined by the spatial arrangement of amino acid residues that are far apart in the linear sequence. The quaternary structure is the folding of the protein resulting in a specific three-dimensional spatial arrangement of the subunits and the nature of their contacts. The correct quaternary structure ensures proper function. (Figure from Koolman & Röhm, 1996).

## References

Koolman, J., Röhm, K.-H.: Color Atlas of Biochemistry. Thieme, Stuttgart–New York, 1996.

Stryer, L.: Biochemistry, 4th ed. W.H. Freeman & Co., New York, 1995.

Peptide bond

Amino acid 1

Amino acid 2

form a peptide

$(+H_2O)$

Amino end ($NH_2$)

Carboxy end (COOH)

Peptide composed of five amino acids (pentapeptide)

**A. Joining of amino acids (peptide bond)**

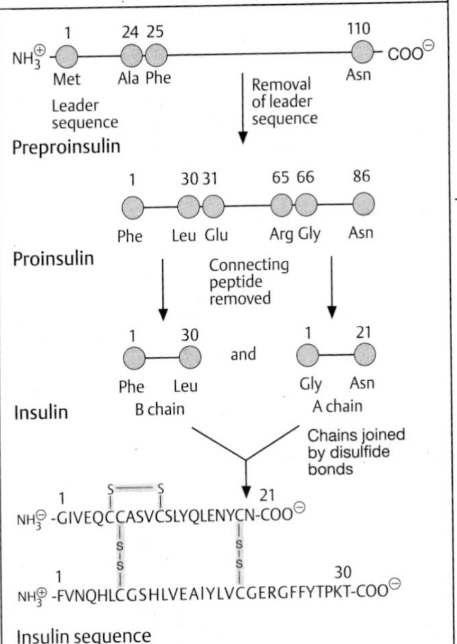


**B. Primary structure of a protein**

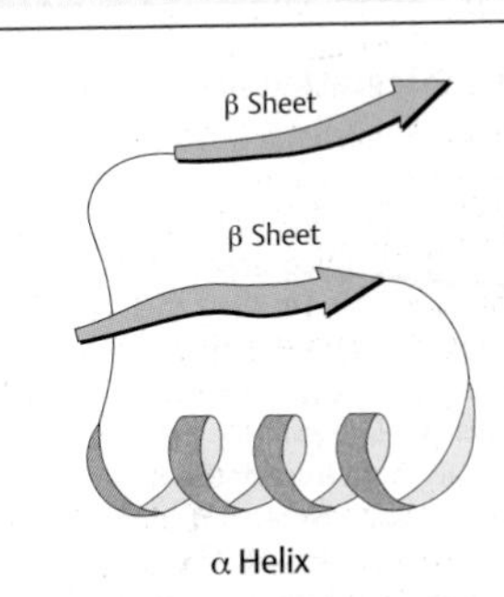


**C. Secondary structural units, the α helix and the β sheet**

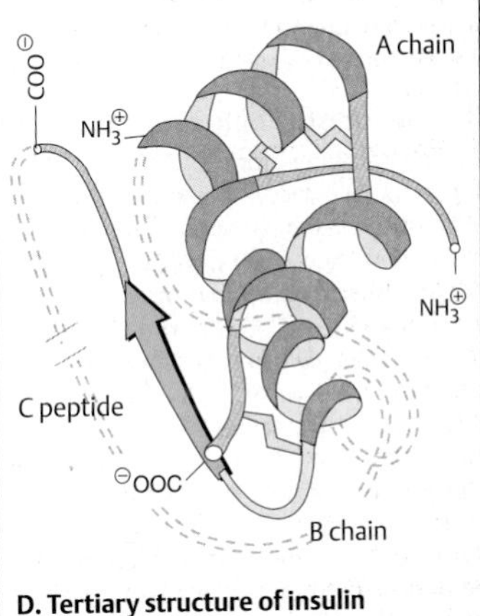


**D. Tertiary structure of insulin**

## DNA as Carrier of Genetic Information

Although DNA was discovered in 1869 by Friedrich Miescher as a new, acidic, phosphorus-containing substance made up of very large molecules that he named "nuclein", its biological role was not recognized. In 1889 Richard Altmann introduced the term "nucleic acid". By 1900 the purine and pyrimidine bases were known. Twenty years later, the two kinds of nucleic acids, RNA and DNA, were distinguished. An incidental but precise observation (1928) and relevant investigations (1944) indicated that DNA could be the carrier of genetic information.

### A. The observation of Griffith

In 1928 the English microbiologist Fred Griffith made a remarkable observation. While investigating various strains of *Pneumococcus*, he determined that mice injected with strain S (smooth) died (1). On the other hand, animals injected with strain R (rough) lived (2). When he inactivated the lethal S strain by heat, there were no sequelae, and the animal survived (3). Surprisingly, a mixture of the nonlethal R strain and the heat-inactivated S strain had a lethal effect like the S strain (4). And he found normal living pneumococci of the S strain in the animal's blood. Apparently, cells of the R strain were changed into cells of the S strain (transformed). For a time, this surprising result could not be explained and was met with skepticism. Its relevance for genetics was not apparent.

### B. The transforming principle is DNA

Griffith's findings formed the basis for investigations by Avery, MacLeod, and McCarty (1944). Avery and co-workers at the Rockefeller Institute in New York elucidated the chemical basis of the transforming principle. From cultures of an S strain (1) they produced an extract of lysed cells (cell-free extract) (2). After all its proteins, lipids, and polysaccharides had been removed, the extract still retained the ability to transform pneumococci of the R strain to pneumococci of the S strain (transforming principle) (3).

With further studies, Avery and co-workers determined that this was attributed to the DNA alone. Thus, the DNA must contain the corresponding genetic information. This explained Griffith's observation. Heat inactivation had left the DNA of the bacterial chromosomes intact. The section of the chromosome with the gene responsible for capsule formation (S gene) could be released from the destroyed S cells and be taken up by some R cells in subsequent cultures. After the S gene was incorporated into its DNA, an R cell was transformed into an S cell (4). Page 90 shows how bacteria can take up foreign DNA so that some of their genetic attributes will be altered correspondingly.

### C. Genetic information is transmitted by DNA alone

The final evidence that DNA, and no other molecule, transmits genetic information was provided by Hershey and Chase in 1952. They labeled the capsular protein of bacteriophages (see p. 88) with radioactive sulfur ($^{35}S$) and the DNA with radioactive phosphorus ($^{32}P$). When bacteria were infected with the labeled bacteriophage, only $^{32}P$ (DNA) entered the cells, and not the $^{35}S$ (capsular protein). The subsequent formation of new, complete phage particles in the cell proved that DNA was the exclusive carrier of the genetic information needed to form new phage particles, including their capsular protein. Next, the structure and function of DNA needed to be clarified. The genes of all cells and some viruses consist of DNA, a long-chained threadlike molecule.

### References

Avery, O.T., MacLeod, C.M., McCarty, M.: Studies on the chemical nature of the substance inducing transformation of pneumococcal types. J. Exp. Med. **79**:137–158, 1944.

Griffith, F., The significance of pneumoccocal types. J. Hyg. **27**:113–159, 1928.

Hershey, A.D., Chase, M.: Independent functions of viral protein and nucleic acid in growth of bacteriophage. J. Gen. Physiol. **36**:39–56, 1952.

Judson, M.F.: The Eighth Day of Creation. Makers of the Revolution in Biology. Expanded Edition. Cold Spring Harbor Laboratory Press, New York, 1996.

McCarty, M.: The Transforming Principle. Discovering that Genes are made of DNA. W.W. Norton & Co., New York–London, 1985.

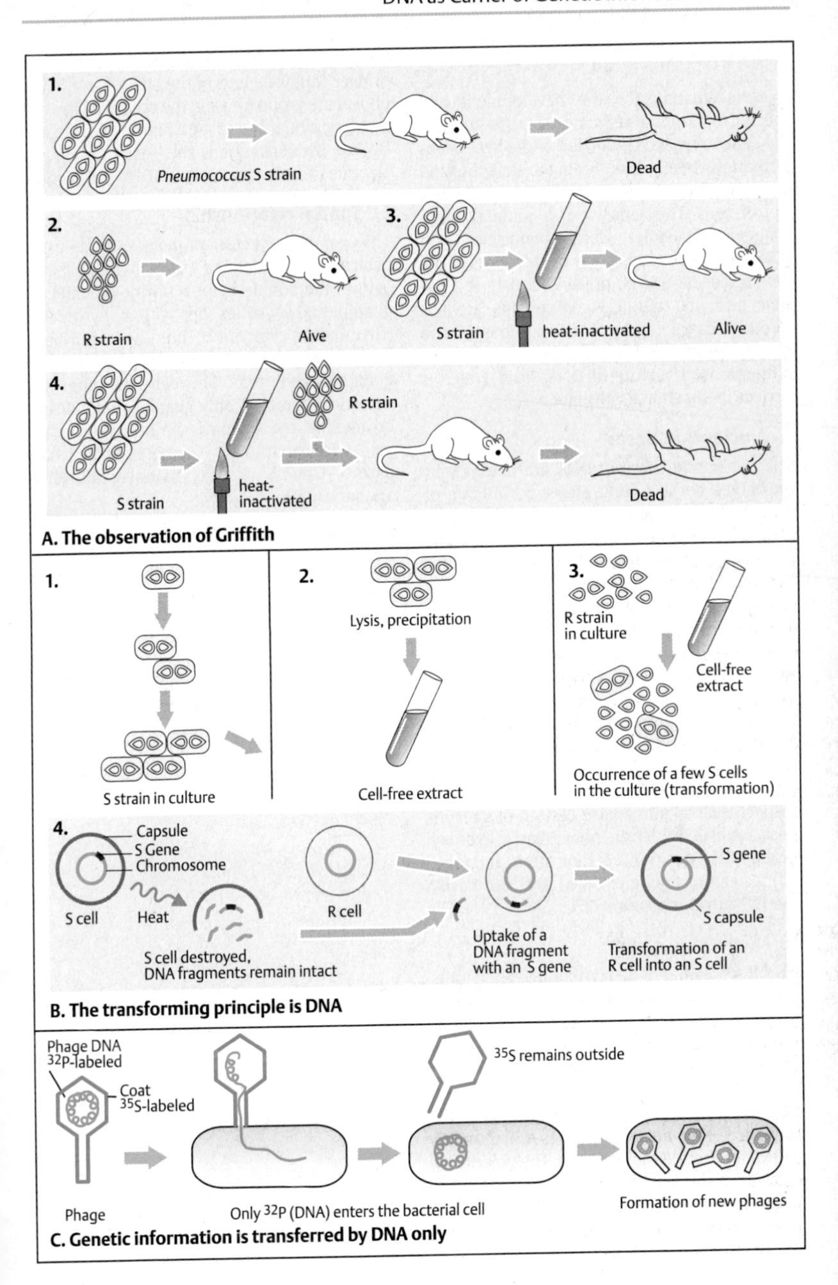


A. The observation of Griffith

B. The transforming principle is DNA

C. Genetic information is transferred by DNA only

# DNA and Its Components

The information for the development and specific functions of cells and tissues is stored in the genes. A gene is a portion of the genetic information, definable according to structure and function. Genes lie on chromosomes in the nuclei of cells. They consist of a complex long-chained molecule, *d*eoxyribo*n*ucleic *a*cid (DNA). In the following, the constituents of the DNA molecule will be presented. DNA is a nucleic acid. Its chemical components are nucleotide bases, a sugar (deoxyribose), and phosphate groups. They determine the three-dimensional structure of DNA, from which it derives its functional consequence.

## A. Nucleotide bases

The nucleotide bases in DNA are heterocyclic molecules derived from either pyrimidine or purine. Five bases occur in the two types of nucleic acids, DNA and RNA. The purine bases are adenine (A) and guanine (G). The pyrimidine bases are thymine (T) and cytosine (C) in DNA. In RNA, uracil (U) is present instead of thymine. The nucleotide bases are part of a subunit of DNA, the nucleotide. This consists of one of the four nucleotide bases, a sugar (deoxyribose), and a phosphate group. The nitrogen atom in position 9 of a purine or in position 1 of a pyrimidine is bound to the carbon in position 1 of the sugar (*N*-glycosidic bond).
Ribonucleic acid (RNA) differs from DNA in two respects: it contains ribose instead of deoxyribose (unlike the latter, ribose has a hydroxyl group on the position 2 carbon atom) and uracil (U) instead of thymine. Uracil does not have a methyl group at position C5.

## B. Nucleotide chain

DNA is a polymer of deoxyribonucleotide units. The nucleotide chain is formed by joining a hydroxyl group on the sugar of one nucleotide to the phosphate group attached to the sugar of the next nucleotide. The sugars linked together by the phosphate groups form the invariant part of the DNA. The variable part is in the sequence of the nucleotide bases A, T, C, and G. A DNA nucleotide chain is polar. The polarity results from the way the sugars are attached to each other. The phosphate group at position C5 (the 5′ carbon) of one sugar joins to the hydroxyl group at position C3 (the 3′ carbon) of the next sugar by means of a phosphate diester bridge. Thus, one end of the chain has a 5′ triphosphate group free and the other end has a 3′ hydroxy group free (5′ end and 3′ end, respectively). By convention, the sequence of nucleotide bases is written in the 5′ to 3′ direction.

## C. Spatial relationship

The chemical structure of the nucleotide bases determines a defined spatial relationship. Within the double helix, a purine (adenine or guanine) always lies opposite a pyrimidine (thymine or cytostine). Three hydrogen-bond bridges are formed between cytosine and guanine, and two between thymine and adenine. Therefore, only guanine and cytosine or adenine and thymine can lie opposite and pair with each other (complementary base pairs G–C and A–T). Other spatial relationships are not usually possible.

## D. DNA double strand

DNA forms a double strand. As a result of the spatial relationships of the nucleotide bases, a cytosine will always lie opposite to a guanine and a thymine to an adenine. The sequence of the nucleotide bases on one strand of DNA (in the 5′ to 3′ direction) is complementary to the nucleotide base sequence (or simply the base sequence) of the other strand in the 3′ to 5′ direction. The specificity of base pairing is the most important structural characteristic of DNA.

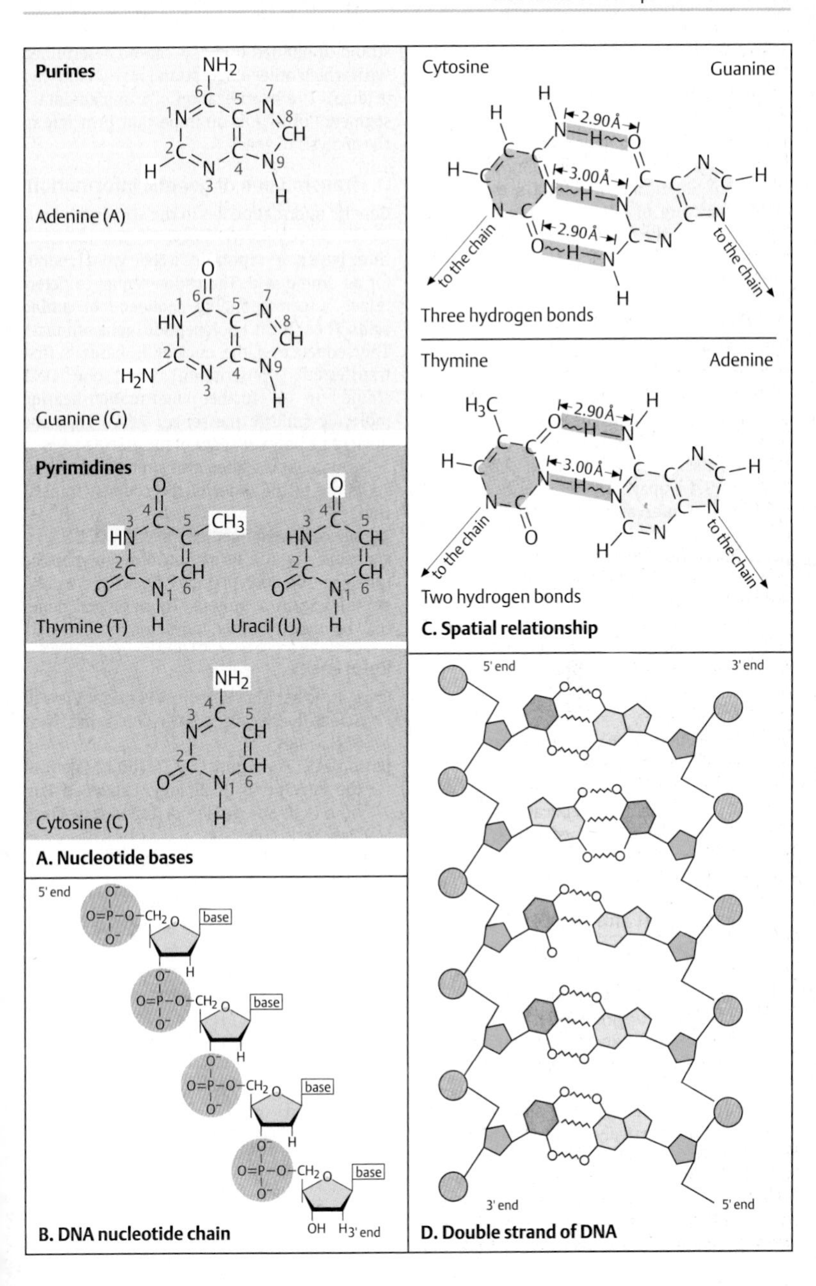
Purines
Adenine (A)
Guanine (G)
Pyrimidines
Thymine (T)
Uracil (U)
Cytosine (C)
A. Nucleotide bases
5' end
base
OH
3' end
B. DNA nucleotide chain
Cytosine
Guanine
2.90 Å
3.00 Å
2.90 Å
to the chain
to the chain
Three hydrogen bonds
Thymine
Adenine
2.90 Å
3.00 Å
to the chain
to the chain
Two hydrogen bonds
C. Spatial relationship
5' end
3' end
3' end
5' end
D. Double strand of DNA

# DNA Structure

In 1953, James Watson and Francis Crick recognized that DNA must exist as a double helix. This structure explains both important functional aspects: replication and the transmission of genetic information. The elucidation of the structure of DNA is considered as the beginning of the development of modern genetics. With it, gene structure and function can be understood at the molecular level.

## A. DNA as a double helix

The double helix is the characteristic structural feature of DNA. The two helical polynucleotide chains are wound around each other along a common axis. The nucleotide base pairs (bp), either A–T or G–C, lie within. The diameter of the helix is 20 Å ($2 \times 10^{-7}$ mm). Neighboring bases lie 3.4 Å apart. The helical structure repeats itself at intervals of 34 Å, or every ten base pairs. Because of the fixed spatial relationship of the nucleotide bases within the double helix and opposite each other, the two chains of the double helix are exactly complementary. The form illustrated here is the so-called B form (B-DNA). Under certain conditions, DNA can also assume other forms (Z-DNA, A-DNA, see p. 41).

## B. Replication

Since the nucleotide chains lying opposite each other within the double helix are strictly complementary, each can serve as a pattern (template) for the formation (replication) of a new chain when the helix is opened. DNA replication is semiconservative, i.e., one completely new strand will be formed and one strand retained.

## C. Denaturation and renaturation

The noncovalent hydrogen bonds between the nucleotide base pairs are weak. Nevertheless, DNA is stable at physiological temperatures because it is a very long molecule. The two complementary strands can be separated (denaturation) by means of relatively weak chemical reagents (e.g., alkali, formamide, or urea) or by careful heating. The resulting single-stranded molecules are relatively stable. With cooling, complementary single strands can reunite to form double-stranded molecules (renaturation). Noncomplementary single strands do not unite. This is the basis of an important method of identifying nucleic acids: With a single strand of defined origin, it can be determined with which other single strand it will bind (hydridize). The hybridization of complementary segments of DNA is an important principle in the analysis of genes.

## D. Transmission of genetic information

Genetic information lies in the sequence of nucleotide base pairs (A–T or G–C). A sequence of three base pairs represents a code word (codon) for an amino acid. The codon sequence determines a corresponding sequence of amino acids. These form a polypeptide (gene product). The sequence of the nucleotide bases is first transferred (transcription) from one DNA strand to a further information-bearing molecule (mRNA, messenger RNA). Then the nucleotide base sequence of the mRNA serves as a template for a sequence of amino acids corresponding to the order of the codons (translation).

A gene can be defined as a section of DNA responsible for the formation of a polypeptide (one gene, one polypeptide). One or more polypeptides form a protein. Thus, several genes may be involved in the formation of a protein.

## References

Crick, F.: What Mad Pursuit. A Personal View of Scientific Discovery. Basic Books, Inc., New York, 1988.

Judson, H.F.: The Eighth Day Creation. Makers of the Revolution in Biology. Expanded Edition. Cold Spring Harbor Laboratory Press, New York, 1996.

Stent, G.S., ed.: The Double Helix. Weidenfeld & Nicolson, London, 1981.

Watson, J.D.: The Double Helix. A Personal Account of the Structure of DNA. Atheneum, New York, 1968.

Watson, J.D., Crick, F.H.C.: Molecular structure of nucleic acid. Nature **171**:737–738, 1953.

Watson, J.D., Crick, F.H.C.: Genetic implications of the structure of DNA. Nature **171**:964–967, 1953.

Wilkins, M.F.H., Stokes, A.R., Wilson, H.R.: Molecular structure of DNA. Nature **171**:738–740, 1953.

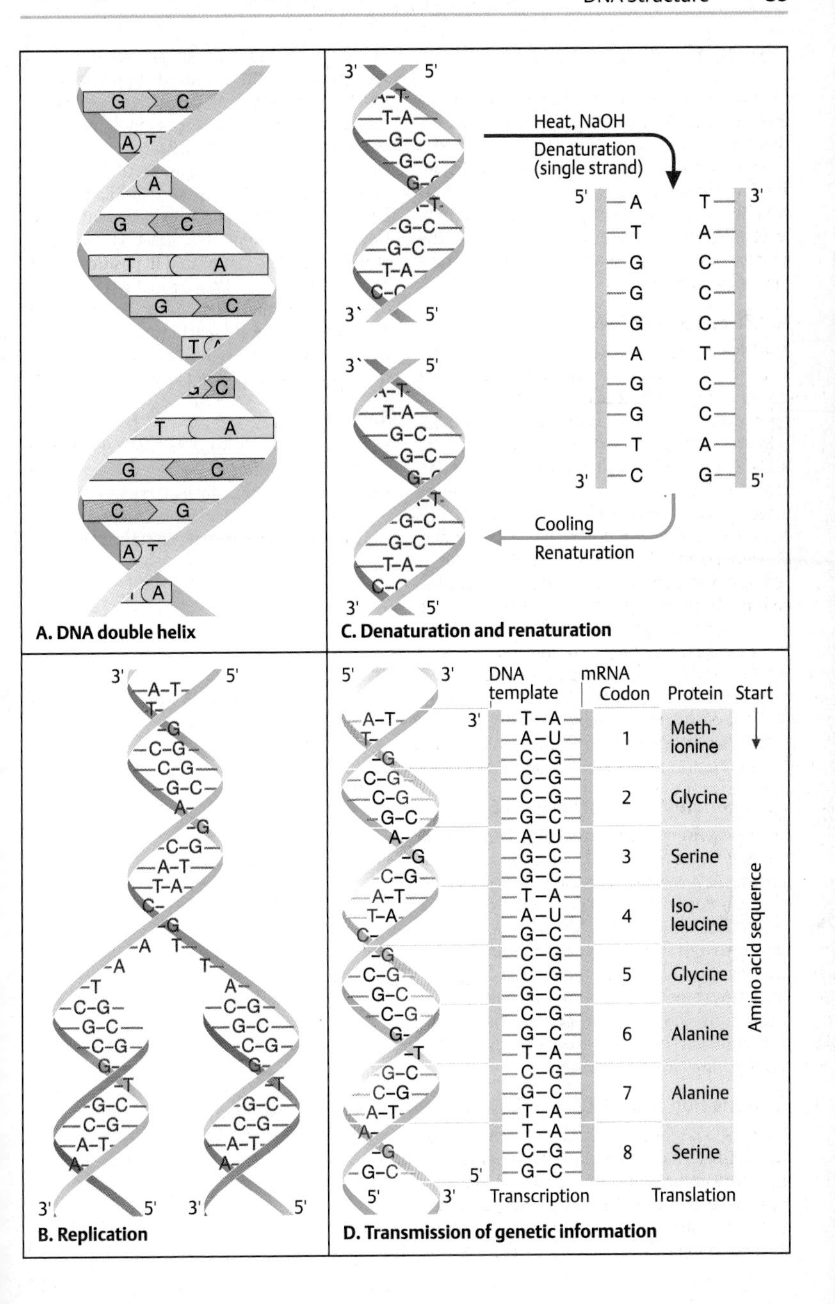


A. DNA double helix

B. Replication

C. Denaturation and renaturation

D. Transmission of genetic information

## Alternative DNA Structures

Gene expression and transcription can be influenced by changes of DNA topology. However, this type of control of gene expression is relatively universal and nonspecific. Thus, it is more suitable for permanent suppression of transcription, e.g., in genes that are expressed only in certain tissues or are active only during the embroyonic period and later become permanently inactive.

### A. Three forms of DNA

The DNA double helix does not occur as a single structure, but rather represents a structural family of different types. The original classic form, determined by Watson and Crick in 1953, is B-DNA. The essential structural characteristic of B-DNA is the formation of two grooves, one large (major groove) and one small (minor groove). There are at least two further, alternative forms of the DNA double helix, Z-DNA and the rare form A-DNA. While B-DNA forms a right-handed helix, Z-DNA shows a left-handed conformation. This leads to a greater distance (0.77 nm) between the base pairs than in B-DNA and a zigzag form (thus the designation Z-DNA). A-DNA is rare. It exists only in the dehydrated state and differs from the B form by a 20-degree rotation of the perpendicular axis of the helix. A-DNA has a deep major groove and a flat minor groove (Figures from Watson et al, 1987).

### B. Major and minor grooves in B-DNA

The base pairing in DNA (adenine–thymine and guanine–cytosine) leads to the formation of a large and a small groove because the glycosidic bonds to deoxyribose (dRib) are not diametrically opposed. In B-DNA, the purine and pyrimidine rings lie 0.34 nm apart. DNA has ten base pairs per turn of the double helix. The distance from one complete turn to the next is 3.4 nm. In this way, localized curves arise in the double helix. The result is a somewhat larger and a somewhat smaller groove.

### C. Transition from B-DNA to Z-DNA

B-DNA is a perfect regular double helix except that the base pairs opposite each other do not lie exactly at the same level. They are twisted in a propeller-like manner. In this way, DNA can easily be bent without causing essential changes in the local structures.

In Z-DNA the sugar–phosphate skeleton has a zigzag pattern; the single Z-DNA groove has a greater density of negatively charged molecules. Z-DNA may occur in limited segments in vivo. A segment of B-DNA consisting of GC pairs can be converted into Z-DNA when the bases are rotated 180 degrees. Normally, Z-DNA is thermodynamically relatively unstable. However, transition to Z-DNA is facilitated when cytosine is methylated in position 5 (C5). The modification of DNA by methylation of cytosine is frequent in certain regions of DNA of eukaryotes. There are specific proteins that bind to Z-DNA, but their significance for the regulation of transcription is not clear.

### References

Stryer, L.: Biochemistry, 4th ed. W.H. Freeman & Co., New York, 1995.

Watson, J.D. et al.: Molecular Biology of the Gene. 3 rd ed. Benjamin/Cummings Publishing Co., Menlo Park, California, 1987.

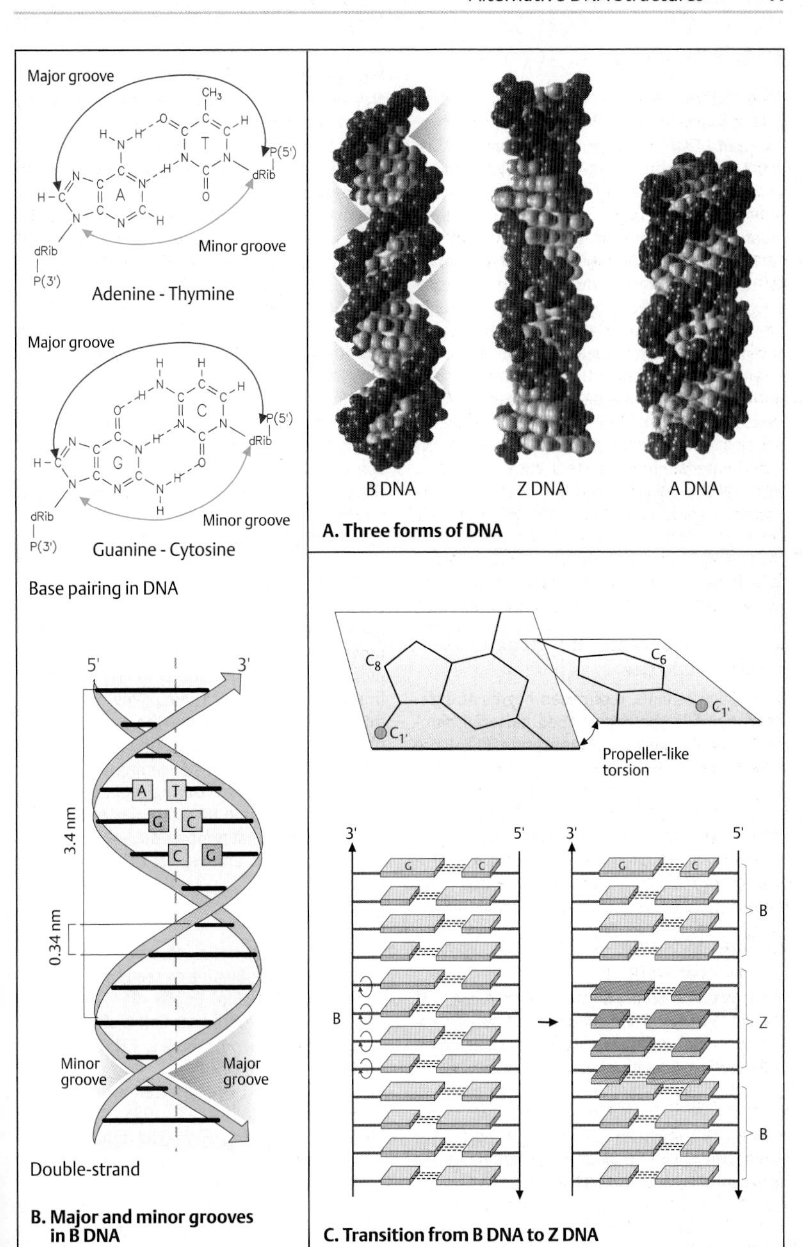


A. Three forms of DNA

B. Major and minor grooves in B DNA

C. Transition from B DNA to Z DNA

## DNA Replication

DNA synthesis involves a highly coordinated action of many proteins. Precision and speed are required. The two new DNA chains are assembled at a rate of about 1000 nucleotides per second in *E. coli*. The principal enzymatic proteins are polymerases, which carry out template-directed synthesis; helicases, which separate the two strands to generate the replication fork (see D); primases, which initiate chain synthesis at preferred sites; initiation proteins, which recognize the origin of replication point; and proteins that remodel the double helix. The entire complex is called the replisome.

In their paper elucidating the structure of DNA, Watson and Crick (1953) noted in closing, "It has not escaped our attention that this structure immediately suggests a copying mechanism for the genetic material," at that time an unsolved problem. Although biochemically complex, DNA replication is genetically relatively simple. During replication, each strand of DNA serves as a template for the formation of a new strand (semiconservative replication).

### A. Prokaryote replication begins at a single site

In prokaryote cells, replication begins at a defined point in the ring-shaped bacterial chromosome, the origin of replication (1). From here, new DNA is formed at the same speed in both directions until the DNA has been completely duplicated and two chromosomes are formed. Replication can be visualized by autoradiography after the newly replicated DNA has incorporated tritium ($^3$H)-labeled thymidine (2).

### B. Eukaryote replication begins at several sites

DNA synthesis occurs during a defined phase of the cell cycle (S phase). This would take a very long time if there were only one starting point. However, replication of eukaryotic DNA begins at numerous sites (replicons) (1). It proceeds in both directions from each replicon until neighboring replicons fuse (2) and all of the DNA is duplicated (3). The electron micrograph (4) shows replicons at three sites.

### C. Scheme of replication

New DNA is synthesized in the 5′ to 3′ direction, but not in the 3′ to 5′ direction. A new nucleotide cannot be attached to the 5′-OH end of the new nucleotide chain. Only at the 3′ end can nucleotides be attached continuously. New DNA at the 5′ end is replicated in small segments. This represents an obstacle at the end of a chromosome (telomere, see p. 180).

### D. Replication fork

At the replication fork, each of the two DNA strands serves as a template for the synthesis of new DNA. First, the double helix at the replication fork region is unwound by an enzyme system (topoisomerases). Since the parent strands are antiparallel, DNA replication can proceed continuously in only one DNA strand (5′ to 3′ direction) (leading strand). Along the 3′ to 5′ strand (lagging strand), the new DNA is formed in small segments of 1000–2000 bases (Okazaki fragments). In this strand a short piece of RNA is required as a primer to start replication. This is formed by an RNA polymerase (primase). The RNA primer is subsequently removed; DNA is inserted into the gap by polymerase I and, finally, the DNA fragments are linked by DNA ligase. The enzyme responsible for DNA synthesis (DNA polymerase III) is complex and comprises several subunits. There are different enzymes for the leading and lagging strands in eukaryotes. During replication, mistakes are eliminated by a complex proof-reading mechanism that removes any incorrectly incorporated bases and replaces them with the correct ones.

### References


Cairns, J.: The bacterial chromosome and its manner of replication as seen by autoradiography. J. Mol. Biol. **6**:208–213, 1963.

Lodish, H. et al.: Molecular Cell Biology. 4th ed. Scientific American Books, F.H. Freeman & Co., New York, 2000.

Marx, J.: How DNA replication originates. Science **270**:1585–1587, 1995.

Meselson, M., Stahl, F.W.: The replication of DNA in Escherichia coli. Proc. Natl. Acad. Sci. **44**:671–682, 1958.

Watson, J.D. et al.: Molecular Biology of the Gene, 3rd ed. Benjamin/Cummings Publishing Co., Menlo Park, California, 1987.

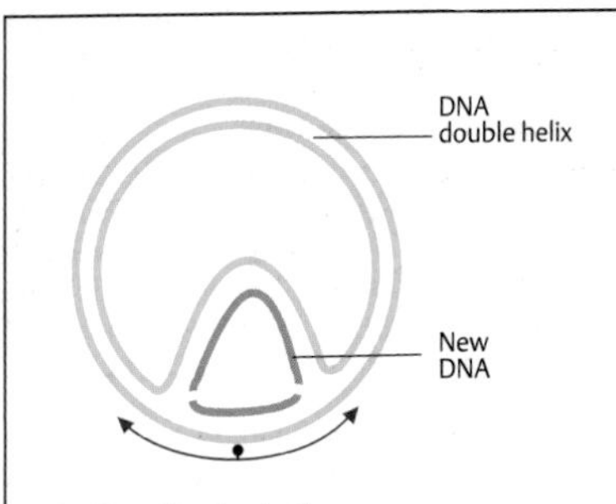


1. DNA replication in the bacterial chromosome

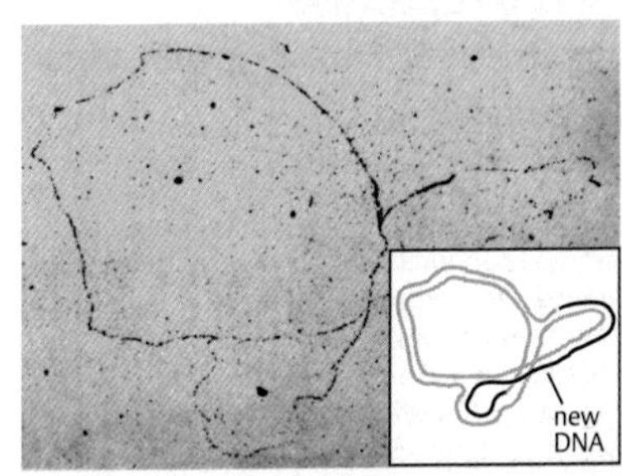


2. Prokaryotic replication in an autoradiogram in *E. coli* (J. Cairns)

**A. Prokaryotic replication begins at one site**

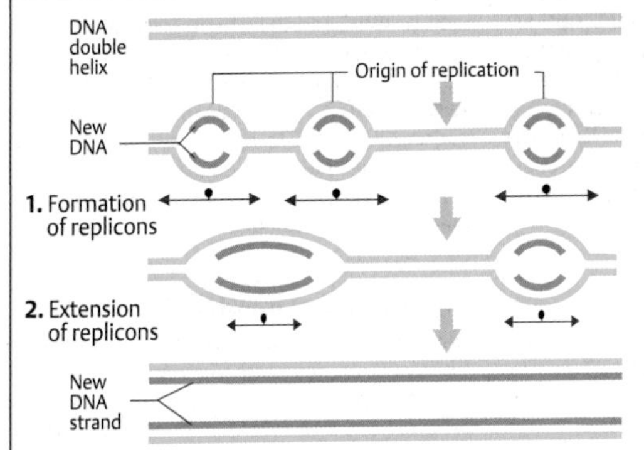


3. Replication completed

**B. Eukaryotic replication begins at several sites**

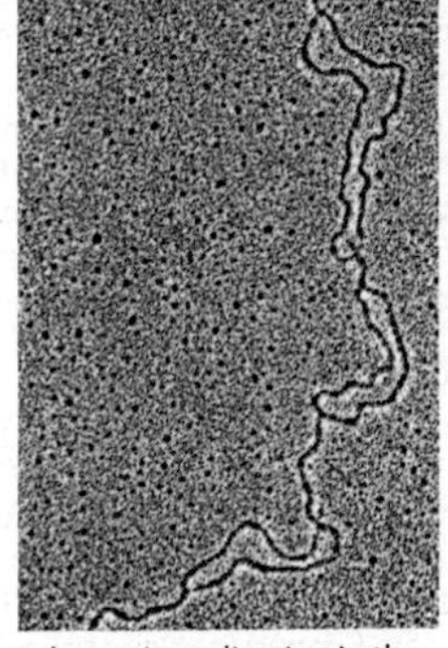

4. Eukaryotic replication in the EM (D. S. Hogness)

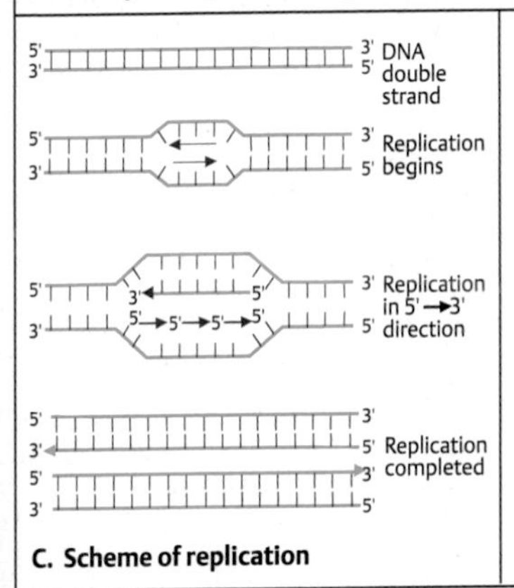


**C. Scheme of replication**

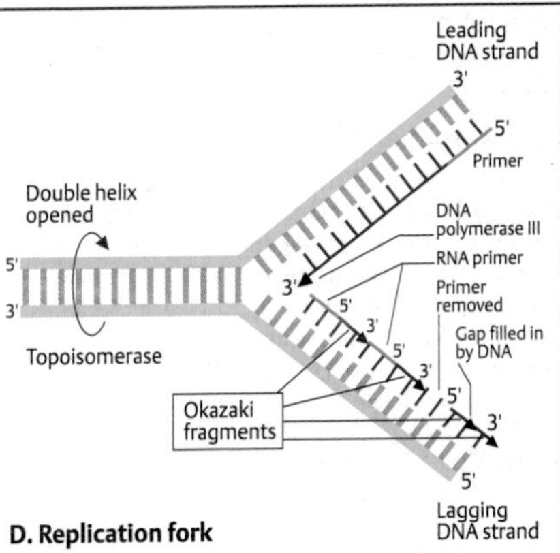


**D. Replication fork**

# The Flow of Genetic Information: Transcription and Translation

The information contained in the nucleotide sequence of a gene must be converted into useful biological function. This is accomplished by proteins, either directly, by being involved in a biochemical pathway, or indirectly, by regulating the activity of a gene. The flow of genetic information is unidirectional and requires two major steps: *transcription* and *translation*. First, the information of the coding sequences of a gene is transcribed into an intermediary RNA molecule, which is synthesized in sequences that are precisely complementary to those of the coding strand of DNA (transcription). During the second major step the sequence information in the messenger RNA molecule (mRNA) is translated into a corresponding sequence of amino acids (translation). The length and sequence of the amino acid chain specified by a gene results in a polypeptide with a biological function (gene product).

## A. Transcription

First, the nucleotide sequence of one strand of DNA is transcribed into a complementary molecule of RNA (messenger RNA, mRNA). The DNA helix is opened by a complex set of proteins. The DNA strand in the 3′ to 5′ direction (coding strand) serves as the template for the transcription into RNA, which is synthesized in the 5′ to 3′ direction. It is called the RNA sense strand. RNA transcribed under experimental conditions from the opposing DNA strand is called antisense RNA.

## B. Translation

During translation the sequence of codons made up of the nucleotide bases in mRNA is converted into a corresponding sequence of amino acids. Translation occurs in a reading frame which is defined at the start of translation (start codon). Amino acids are joined in the sequence determined by the mRNA nucleotide bases by a further class of RNA, transfer RNA (tRNA). Each amino acid has its own tRNA, which has a region that is complementary to its codon of the mRNA (anticodon). The codons 1, 2, 3, and 4 of the mRNA are translated into the amino acid sequence methionine (Met), glycine (Gly), serine (Ser), and isoleucine (Ile), etc. Codon 1 is always AUG (start codon).

## C. Stages of translation

Translation (protein synthesis) in eukaryotes occurs outside of the cell nucleus in ribosomes in the cytoplasm. Ribosomes consist of subunits of numerous associated proteins and RNA molecules (ribosomal RNA, rRNA; p. 204). Translation begins with initiation (1): an initiation complex comprising mRNA, a ribosome, and tRNA is formed. This requires a number of initiation factors (IF1, IF2, IF3, etc.). Then elongation (2) follows: a further amino acid, determined by the next codon, is attached. A three-phase elongation cycle develops, with codon recognition, peptide binding to the next amino acid residue, and movement (translocation) of the ribosome three nucleotides further in the 3′ direction of the mRNA. Translation ends with termination (3), when one of three mRNA stop codons (UAA, UGA, or UAG) is reached. The polypeptide chain formed leaves the ribosome, which dissociates into its subunits. The biochemical processes of the stages shown here have been greatly simplified.

## D. Structure of transfer RNA (tRNA)

Transfer RNA has a characteristic, cloverleaf-like structure, illustrated here by yeast phenylalanine tRNA (1). It has three single-stranded loop regions and four double-stranded "stem" regions. The three-dimensional structure (2) is complex, but various functional areas can be differentiated, such as the recognition site (anticodon) for the mRNA codon and the binding site for the respective amino acid (acceptor stem) on the 3′ end (acceptor end).

## References

Brenner, S., Jacob. F., Meselson, M.: An unstable intermediate carrying information from genes to ribosomes for protein synthesis. Nature **190**:576–581, 1961.

Ibba, M., Söll, D.: Quality control mechanisms during translation. Science **286**:1893–1897, 1999.

Watson J.D. et al.: Molecular Biology of the Gene. 3rd ed. Benjamin/Cummings Publishing Co., Menlo Park, California, 1987.

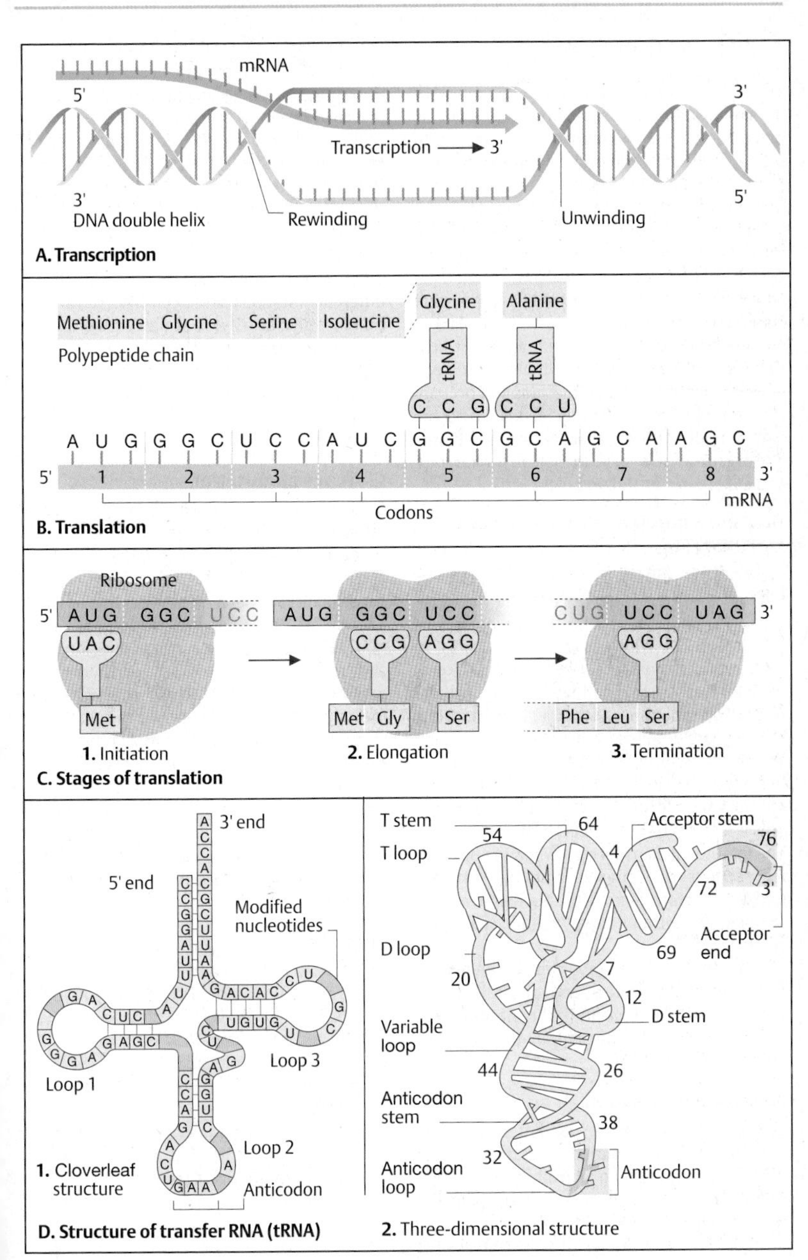
mRNA
5'
3'
Transcription → 3'
3'
DNA double helix
Rewinding
Unwinding
5'
A. Transcription
Glycine
Alanine
Methionine
Glycine
Serine
Isoleucine
Polypeptide chain
tRNA
tRNA
C C G
C C U
A U G G G C U C C A U C G G C G C A G C A A G C
5'
1 2 3 4 5 6 7 8
3'
mRNA
Codons
B. Translation
Ribosome
5' AUG GGC UCC
UAC
Met
AUG GGC UCC
CCG AGG
Met Gly
Ser
CUG UCC UAG 3'
AGG
Phe Leu Ser
1. Initiation
2. Elongation
3. Termination
C. Stages of translation
3' end
5' end
Modified nucleotides
Loop 1
Loop 2
Loop 3
Anticodon
1. Cloverleaf structure
T stem
T loop
Acceptor stem
Acceptor end
D loop
D stem
Variable loop
Anticodon stem
Anticodon loop
Anticodon
54
64
4
76
72
3'
69
20
7
12
44
26
38
32
2. Three-dimensional structure
D. Structure of transfer RNA (tRNA)

## Genes and Mutation

The double helix structure of DNA is the basis of both replication and transcription as seen in the preceding pages. The information transmitted during replication and transcription is arranged in units called *genes*. The term gene was introduced in 1909 by the Danish biologist Wilhelm Johannsen (along with the terms genotype and phenotype). Until it was realized that a gene consists of DNA, it was defined in somewhat abstract terms as a factor (Mendel's term) that confers certain heritable properties to a plant or an animal. However, it was not apparent how mutations could be related to the structure of a gene. The discovery that mutations also occur in bacteria and other microorganisms paved the way to understanding their nature (see p. 84). The organization of genes differs in prokaryotes and eukaryotes as shown below.

### A. Transcription in prokaryotes and eukaryotes

Transcription differs in unicellular organisms without a nucleus, such as bacteria (prokaryotes, 1), and in multicellular organisms (eukaryotes, 2), which have a cell nucleus. In prokaryotes, the mRNA serves directly as a template for translation. The sequences of DNA and mRNA correspond in a strict 1 : 1 relationship, i.e., they are colinear. Translation begins even before transcription has completely ended.

In contrast, a primary transcript of RNA precursor mRNA) is formed first in eukaryotic cells. This is a preliminary form of the mature mRNA. The mature mRNA is formed when the noncoding sections are removed from the primary transcript, before it leaves the nucleus to act as a template for forming a polypeptide (RNA processing).

The reason for these important differences is that functionally related genes generally lie together in prokaryotes and that noncoding segments (introns) are present in the genes of eukaryotes (see p. 50).

### B. DNA and mutation

Coding DNA and its corresponding polypeptide are colinear. An alteration (mutation) of the DNA base sequence may lead to a different codon. The position of the resulting change in the sequence of amino acids corresponds to the position of the mutation (1). Panel B shows the gene for the protein tryptophan synthetase A of *E. coli* bacteria and mutations at four positions. At position 22, phenylalanine (Phe) has been replaced by leucine (Leu); at position 49, glutamic acid (Glu) by glutamine (Gln); at position 177, Leu by arginine (Arg). Every mutation has a defined position. Whether it leads to incorporation of another amino acid depends on how the corresponding codon has been altered. Different mutations at one position (one codon) in different DNA molecules are possible (2). Two different mutations have been observed at position 211: glycine (Gly) to arginine (Arg) and Gly to glutamic acid (Glu). Normally (in the wildtype), codon 211 is GGA and codes for glycine (3). A mutation of GGA to AGA leads to a codon for arginine; a mutation to GAA leads to a codon for glutamic acid (4).

### C. Types of mutation

Basically, there are three different types of mutation involving single nucleotides (point mutation): substitution (exchange), deletion (loss), and insertion (addition). With substitution, the consequences depend on how a codon has been altered. Two types of substitution are distinguished: transition (exchange of one purine for another purine or of one pyrimidine for another) and transversion (exchange of a purine for a pyrimidine, or vice versa). A substitution may alter a codon so that a wrong amino acid is present at this site but has no effect on the reading frame (missense mutation), whereas a deletion or insertion causes a shift of the reading frame (frameshift mutation). Thus the sequences that follow no longer code for a functional gene product (nonsense mutation).

### References


Alberts, B. et al.: Molecular Biology of the Cell. 3rd ed. Garland Publishing, New York, 1994.

Alberts, B. et al.: Essential Cell Biology. An Introduction to the Molecular Biology of the Cell. Garland Publishing, New York, 1998.

Lodish, H. et al.: Molecular Cell Biology. 4th ed. Scientific American Books, F.H. Freeman & Co., New York, 2000.

Watson, J.D. et al.: Molecular Biology of the Gene, 3rd ed. Benjamin/Cummings Publishing Co., Menlo Park, California, 1987.

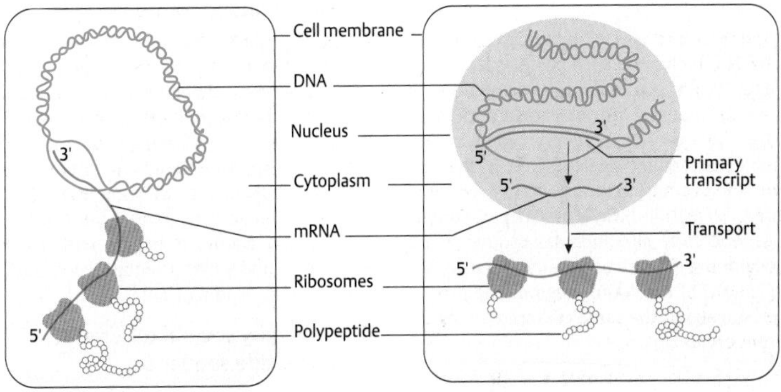


1. Prokaryote

2. Eukaryote

**A. Transcription and translation in prokaryotes and eukaryotes**

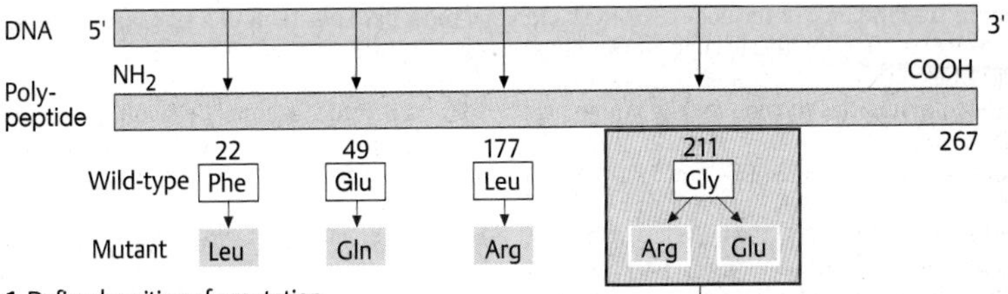


1. Defined position of a mutation

**B. DNA and mutation**

2. Different mutations of one codon

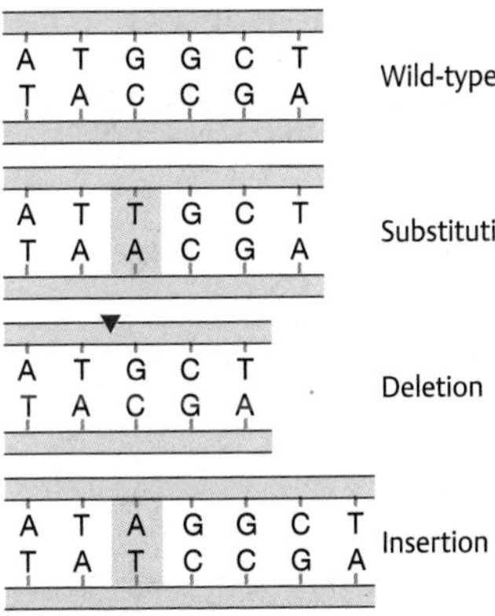


**C. Types of mutation**

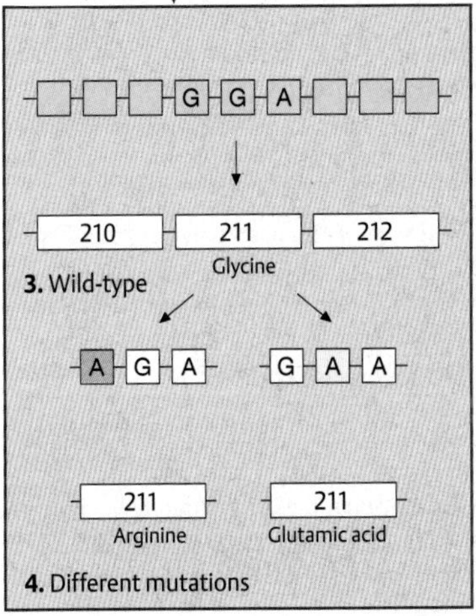

## Genetic Code

The genetic code is the set of biological rules by which DNA nucleotide base pair sequences are translated into corresponding sequences of amino acids. Genes do not code for proteins directly, but do so through a messenger molecule (messenger RNA, mRNA). A code word (codon) for an amino acid consists of a sequence of three nucleotide base pairs (triplet codon). The genetic code also includes sequences for the beginning (start codon) and for the end (stop codon) of the coding region. The genetic code is universal; the same codons are used by different organisms.

### A. Genetic code in mRNA for all amino acids

Each codon corresponds to one amino acid, but one amino acid may be coded for by different codons (redundancy of the code). For example, there are two possibilities to code for the amino acid phenylalanine: UUU and UUC, and there are six possibilities to code for the amino acid serine: UCU, UCC, UCA, UCG, AGU, and AGC. Many amino acids are determined by more than one codon. The greatest variation is in the third position (at the 3′ end of the triplet). The genetic code was elucidated in 1966 by analyzing how triplets transmit information from the genes to proteins. mRNA added to bacteria could be directly converted into a corresponding protein.Synthetic RNA polymers such as polyuridylate (poly (U)), polyadenylate (poly(A)), and polycytidylate (poly(C)) could be directly translated into polyphenylalanine, polylysine, and polyproline in extracts of *E. coli* bacteria. This showed that UUU must code for phenylalanine, AAA for lysine, and CCC for proline. By further experiments with mixed polymers of different proportions of two or three nucleotides, the genetic code was determined for all amino acids and all nucleotide compositions.

### B. Abbreviated code

Sequences of amino acids are designated with the single-letter abbreviations ("alphabetic code").

The start codon is AUG (methionine). Stop codons are UAA, UAG, and UGA. The only amino acids that are encoded by a single codon are methionine (AUG) and tryptophan (UGG).

### C. Open reading frame (ORF)

A segment of a nucleotide sequence can correspond to one of three reading frames (e.g., A, B, or C); however, only one is correct (open reading frame). In the example shown, the reading frames B and C are interrupted by a stop codon after three and five codons, respectively. Thus they cannot serve as reading frames for a coding sequence. On the other hand, A must be the correct reading frame: It begins with the start codon AUG and yields a sequence without stop codons (open reading frame).

### D. Coding by several different nucleotide sequences

Since the genetic code has redundancy, it is possible that different nucleotide sequences code for the same amino acid sequence. However, the differences are limited to one (or at most two) positions of a given triplet codon.

### References

Alberts, B. et al.: Essential Cell Biology. An Introduction to the Molecular Biology of the Cell. Garland Publishing, New York, 1998.

Crick, F.H.C. et al: General nature of the genetic code for proteins. Nature 192: 1227–1232, 1961.

Lodish, H. et al.: Molecular Cell Biology. 4th ed. Scientific American Books, F.H. Freeman & Co., New York, 2000.

Rosenthal, N.: DNA and the genetic code. New Eng. J. Med. 331: 39–41, 1995.

Singer, M., Berg, P.: Genes and Genomes: a changing perspective. Blackwell Scientific Publications, Oxford–London, 1991.

| Nucleotide base | | | | | |
|---|---|---|---|---|---|
| First | Second | | | | Third |
| | Uracil (U) | Cytosine (C) | Adenine (A) | Guanine (G) | |
| Uracil (U) | F Phenylalanine (Phe) | S Serine (Ser) | Y Tyrosine (Tyr) | C Cysteine (Cys) | U |
| | F Phenylalanine (Phe) | S Serine (Ser) | Y Tyrosine (Tyr) | C Cysteine (Cys) | C |
| | L Leucine (Leu) | S Serine (Ser) | Stop Codon | Stop Codon | A |
| | L Leucine (Leu) | S Serine (Ser) | Stop Codon | W Tryptophan (Trp) | G |
| Cytosine (C) | L Leucine (Leu) | P Proline (Pro) | H Histidine (His) | R Arginine (Arg) | U |
| | L Leucine (Leu) | P Proline (Pro) | H Histidine (His) | R Arginine (Arg) | C |
| | L Leucine (Leu) | P Proline (Pro) | Q Glutamine (Gln) | R Arginine (Arg) | A |
| | L Leucine (Leu) | P Proline (Pro) | Q Glutamine (Gln) | R Arginine (Arg) | G |
| Adenine (A) | I Isoleucine (Ile) | T Threonine (Thr) | N Asparagine (Asn) | S Serine (Ser) | U |
| | I Isoleucine (Ile) | T Threonine (Thr) | N Asparagine (Asn) | S Serine (Ser) | C |
| | I Isoleucine (Ile) | T Threonine (Thr) | K Lysine (Lys) | R Arginine (Arg) | A |
| | Start (Methionine) | T Threonine (Thr) | K Lysine (Lys) | R Arginine (Arg) | G |
| Guanine (G) | V Valine (Val) | A Alanine (Ala) | D Aspartic acid (Asp) | G Glycine (Gly) | U |
| | V Valine (Val) | A Alanine (Ala) | D Aspartic acid (Asp) | G Glycine (Gly) | C |
| | V Valine (Val) | A Alanine (Ala) | E Glutamic acid (Glu) | G Glycine (Gly) | A |
| | V Valine (Val) | A Alanine (Ala) | E Glutamic acid (Glu) | G Glycine (Gly) | G |

**A. Genetic code for all amino acids in mRNA**

| | Codons |
|---|---|
| Start | AUG |
| Stop | UAA, UAG, UGA |
| A (Ala) | GCU, GCC, GCG, GCA |
| C (Cys) | UGU, UGC |
| D (Asp) | GAU, GAC |
| E (Glu) | GAG, GAA |
| F (Phe) | UUU, UUC |
| G (Gly) | GGU, GGC, GGG, GGA |
| H (His) | CAU, CAC |
| I (Ile) | AUU, AUC, AUA |
| K (Lys) | AAG, AAA |
| L (Leu) | CUU, CUC, CUG, CUA, UUG, UUA |
| M (Met) | AUG |
| N (Asn) | AAU, AAC |
| P (Pro) | CCU, CCC, CCG, CCA |
| Q (Gln) | CAG, CAA |
| R (Arg) | CGU, CGC, CGG, CAA, AGG, AGA |
| S (Ser) | UCU, UCC, UCG, UCA, AGU, AGC |
| T (Thr) | ACU, ACC, ACG, ACA |
| V | GUU, GUC, GUG, GUA |
| W (Trp) | UGG |
| Y (Tyr) | UAU, UAC |
| B (Asx) | Asn or Asp |
| Z (Glx) | Gln or Glu |

**B. Abbreviated code**

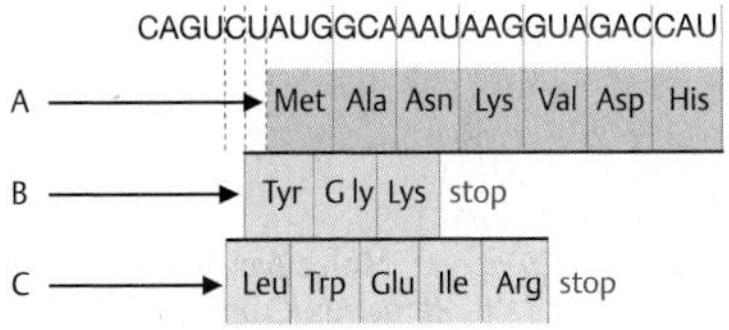

CAGUCUAUGGCAAAUAAGGUAGACCAU

A → Met Ala Asn Lys Val Asp His

B → Tyr Gly Lys stop

C → Leu Trp Glu Ile Arg stop

**C. Open reading frame (ORF)**

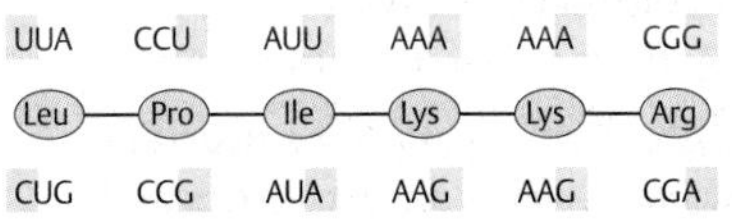

| UUA | CCU | AUU | AAA | AAA | CGG |
|---|---|---|---|---|---|
| Leu | Pro | Ile | Lys | Lys | Arg |
| CUG | CCG | AUA | AAG | AAG | CGA |

**D. Coding by several different nucleotide sequences**

# The Structure of Eukaryotic Genes

Eukaryotic genes consist of coding and noncoding segments of DNA, called exons and introns, respectively. At first glance it seems to be an unnecessary burden to carry DNA without obvious functions within a gene. However, it has been recognized that this has great evolutionary advantages. When parts of different genes are rearranged on new chromosomal sites during evolution, new genes may be constructed from parts of previously existing genes.

## A. Exons and introns

In 1977, it was unexpectedly found that the DNA of a eukaryotic gene is longer than its corresponding mRNA. The reason is that certain sections of the initially formed primary RNA transcript are removed before translation occurs. Electron micrographs show that DNA and its corresponding transcript (RNA) are of different lengths (1). When mRNA and its complementary single-stranded DNA are hybridized, loops of single-stranded DNA arise because mRNA hybridizes only with certain sections of the single-stranded DNA. In (2), seven loops (A to G) and eight hybridizing sections are shown (1 to 7 and the leading section L). Of the total 7700 DNA base pairs of this gene (3), only 1825 hybridize with mRNA. A hybridizing segment is called an exon. An initially transcribed DNA section that is subsequently removed from the primary transcript is an intron. The size and arrangement of exons and introns are characteristic for every eukaryotic gene (exon/intron structure). (Electron micrograph from Watson et al., 1987).

## B. Intervening DNA sequences (introns)

In prokaryotes, DNA is colinear with mRNA and contains no introns (1). In eukaryotes, mature mRNA is complementary to only certain sections of DNA because the latter contains introns (2). (Figure adapted from Stryer, 1995).

## C. Basic eukaryotic gene structure

Exons and introns are numbered in the 5′ to 3′ direction of the coding strand. Both exons and introns are transcribed into a precursor RNA (primary transcript). The first and the last exons usually contain sequences that are not translated. These are called the 5′ untranslated region (5′ UTR) of exon 1 and the 3′ UTR at the 3′ end of the last exon. The noncoding segments (introns) are removed from the primary transcript and the exons on either side are connected by a process called splicing. Splicing must be very precise to avoid an undesirable change of the correct reading frame. Introns almost always start with the nucleotides GT in the 5′ to 3′ strand (GU in RNA) and end with AG. The sequences at the 5′ end of the intron beginning with GT are called splice donor site and at the 3′ end, ending with AG, are called the splice acceptor site. Mature mRNA is modified at the 5′ end by adding a stabilizing structure called a "cap" and by adding many adenines at the 3′ end (polyadenylation) (see p. 50).

## D. Splicing pathway in GU–AG introns

RNA splicing is a complex process mediated by a large RNA-containing protein called a spliceosome. This consists of five types of small nuclear RNA molecules (snRNA) and more than 50 proteins (small nuclear riboprotein particles). The basic mechanism of splicing schematically involves autocatalytic cleavage at the 5′ end of the intron resulting in lariat formation. This is an intermediate circular structure formed by connecting the 5′ terminus (UG) to a base (A) within the intron. This site is called the branch site. In the next stage, cleavage at the 3′ site releases the intron in lariat form. At the same time the right exon is ligated (spliced) to the left exon. The lariat is debranched to yield a linear intron and this is rapidly degraded. The branch site identifies the 3′ end for precise cleavage at the splice acceptor site. It lies 18–40 nucleotides upstream (in 5′ direction) of the 3′ splice site. (Figure adapted from Strachan and Read, 1999).

### References

Lewin, B.: Genes VII. Oxford Univ. Press, Oxford, 2000.

Strachan, T., Read A.P.: Human Molecular Genetics. 2nd ed. Bios Scientific Publishers, Oxford, 1999.

Stryer, L.: Biochemistry, 4th ed. W.H. Freeman & Co., New York, 1995.

Watson, J.D. et al.: Molecular Biology of the Gene, 3rd ed. Benjamin/Cummings Publishing Co., Menlo Park, California, 1987.

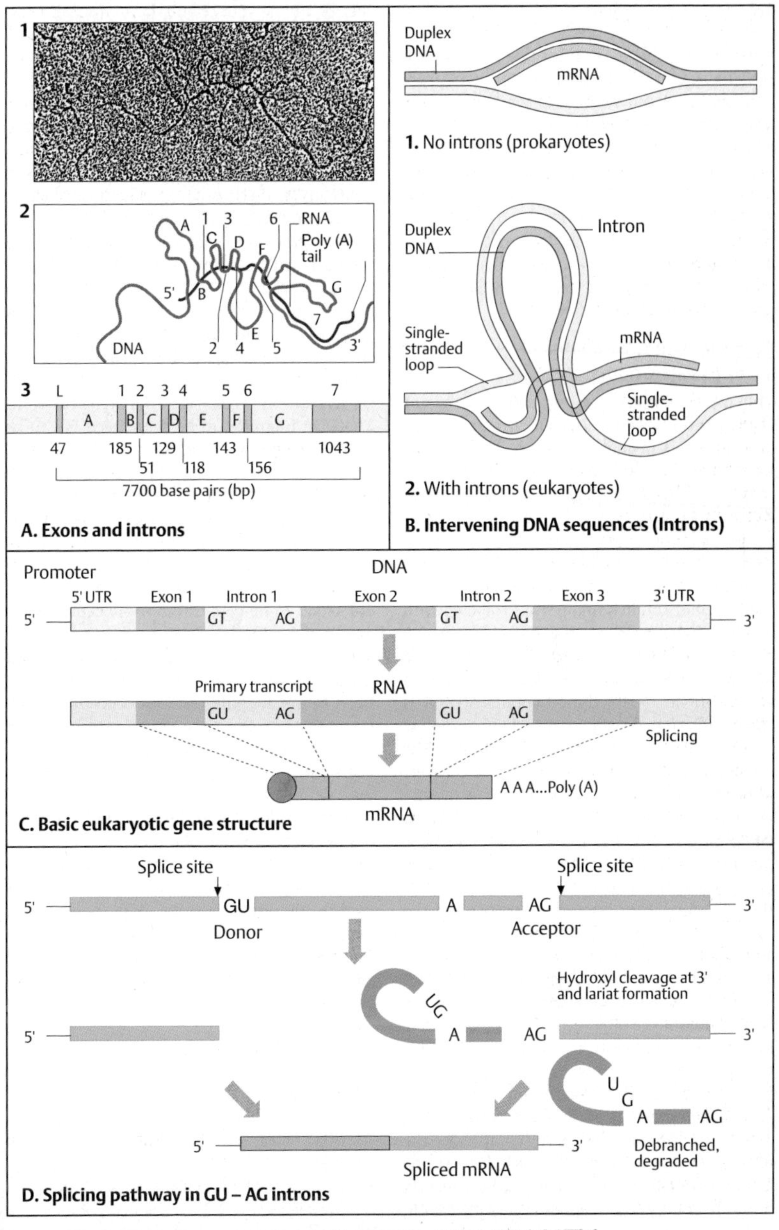


A. Exons and introns

B. Intervening DNA sequences (Introns)

C. Basic eukaryotic gene structure

D. Splicing pathway in GU – AG introns

## *DNA Sequencing*

Knowledge of the nucleotide sequence of a gene provides important information about its structure, function, and evolutionary relationship to other similar genes in the same or different organisms. Thus, the development in the 1970s of relatively simple methods for sequencing DNA has had a great impact on genetics. Two basic methods for DNA sequencing have been developed: a chemical cleavage method (A. M. Maxam and W. Gilbert, 1977) and an enzymatic method (F. Sanger, 1981). A brief outline of the underlying principles follows.

### A. Sequencing by chemical degradation

This method utilizes base-specific cleavage of DNA by certain chemicals. Four different chemicals are used in four reactions, one for each base. Each reaction produces a set of DNA fragments of different sizes. The sizes of the fragments in a reaction mixture are determined by positions in the DNA of the nucleotide that has been cleaved. A double-stranded or single-stranded fragment of DNA to be sequenced is processed to obtain a single strand labeled with a radioactive isotope at the 5′ end (1). This DNA strand is treated with one of the four chemicals for one of the four reactions. Here the reaction at guanine sites (G) by dimethyl sulfate (DMS) is shown. Dimethyl sulfate attaches a methyl group to the purine ring of G nucleotides. The amount of DMS used is limited so that on average just one G nucleotide per strand is methylated, not the others (shown here in four different positions of G). When a second chemical, piperidine, is added, the nucleotide purine ring is removed and the DNA molecule is cleaved at the phosphodiester bond just upstream of the site without the base. The overall procedure results in a set of labeled fragments of defined sizes according to the positions of G in the DNA sample being sequenced. Similar reactions are carried out for the other three bases (A, T, and C, not shown). The four reaction mixtures, one for each of the bases, are run in separate lanes of a polyacrylamide gel electrophoresis. Each of the four lanes represents one of the four bases G, A, T, or C. The smallest fragment will migrate the farthest downward, the next a little less far, etc. One can then read the sequence in the direction opposite to migration to obtain the sequence in the 5′ to 3′ direction (here TAGTCGCAGTACCGTA).

### B. Sequencing by chain termination

This method, now much more widely used than the chemical cleavage method, rests on the principle that DNA synthesis is terminated when instead of a normal deoxynucleotide (dATP, dTTP, dGTP, dCTP), a dideoxynucleotide (ddATP, ddTTP, ddGTP, ddCTP) is used. A dideoxynucleotide (ddNTP) is an analogue of the normal dNTP. It differs by lack of a hydroxyl group at the 3′ carbon position. When a dideoxynucleotide is incorporated during DNA synthesis, no bond between its 3′ position and the next nucleotide is possible because the ddNTP lacks the 3′ hydroxyl group. Thus, synthesis of the new chain is terminated at this site. The DNA fragment to be sequenced has to be single-stranded (1). DNA synthesis is initiated using a primer and one of the four ddNTPs labeled with $^{32}P$ in the phosphate groups or, for automated sequencing, with a fluorophore (see next plate). Here an example of chain termination using ddATP is shown (3). Wherever an adenine (A) occurs in the sequence, the dideoxyadenine triphosphate will cause termination of the new DNA chain being synthesized. This will produce a set of different DNA fragments whose sizes are determined by the positions of the adenine residues occurring in the fragment to be sequenced. Similar reactions are done for the other three nucleotides. The four parallel reactions will yield a set of fragments with defined sizes according to the positions of the nucleotides where the new DNA synthesis has been terminated. The fragments are separated according to size by gel electrophoresis as in the chemical method. The sequence gel is read in the direction from small fragments to large fragments to derive the nucleotide sequence in the 5′ to 3′ direction. An example of an actual sequencing gel is shown between panel A and B.

### References

Brown, T.A.: Genomes. Bios Scientific Publ., Oxford, 1999.

Rosenthal, N.: Fine structure of a gene—DNA sequencing. New Eng. J. Med. **332**:589–591, 1995.

Strachan, T., Read, A.P.: Human Molecular Genetics. 2nd ed. Bios scientific Publishers,

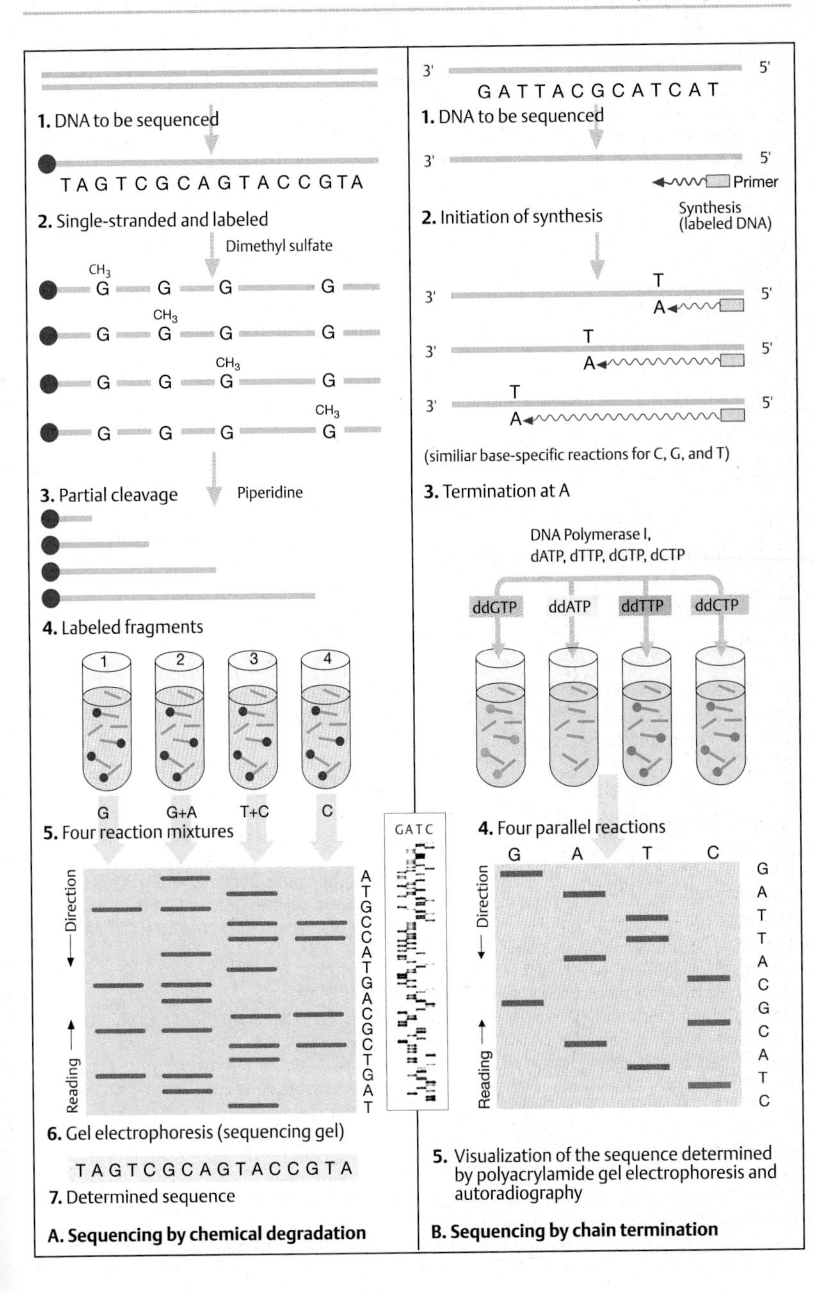
1. DNA to be sequenced
TAGTCGCAGTACCGTA
2. Single-stranded and labeled
Dimethyl sulfate
CH3
G G G G
3. Partial cleavage
Piperidine
4. Labeled fragments
1 2 3 4
G G+A T+C C
5. Four reaction mixtures
Direction
Reading
ATGCCATGACGCTGAT
GATC
6. Gel electrophoresis (sequencing gel)
TAGTCGCAGTACCGTA
7. Determined sequence
A. Sequencing by chemical degradation
3' 5'
GATTACGCATCAT
1. DNA to be sequenced
Primer
2. Initiation of synthesis
Synthesis (labeled DNA)
T
A
(similiar base-specific reactions for C, G, and T)
3. Termination at A
DNA Polymerase I, dATP, dTTP, dGTP, dCTP
ddGTP ddATP ddTTP ddCTP
4. Four parallel reactions
G A T C
GATTACGCATC
5. Visualization of the sequence determined by polyacrylamide gel electrophoresis and autoradiography
B. Sequencing by chain termination

# Automated DNA Sequencing

Large-scale DNA sequencing requires automated procedures based on fluorescence labeling of DNA and suitable detection systems. In general, a fluorescent label can be used either directly or indirectly. Direct fluorescent labels, as used in automated sequencing, are fluorophores. These are molecules that emit a distinct fluorescent color when exposed to UV light of a specific wavelength. Examples of fluorophores used in sequencing are fluorescein, which fluoresces pale green when exposed to a wavelength of 494 nm; rhodamine, which fluoresces red at 555 nm; and aminomethylcumarin acetic acid, which fluoresces blue at 399 nm. In addition, a combination of different fluorophores can be used to produce a fourth color. Thus, each of the four bases can be distinctly labeled.

Another approach is to use PCR-amplified products (thermal cycle sequencing, see A). This has the advantage that double-stranded rather than single-stranded DNA can be used as the starting material. And since small amounts of template DNA are sufficient, the DNA to be sequenced does not have to be cloned beforehand.

## A. Thermal cycle sequencing

The DNA to be sequenced is contained in vector DNA (1). The primer, a short oligonucleotide with a sequence complementary to the site of attachment on the single-stranded DNA, is used as a starting point. For sequencing short stretches of DNA, a universal primer is sufficient. This is an oligonucleotide that will bind to vector DNA adjacent to the DNA to be sequenced. However, if the latter is longer than about 750 bp, only part of it will be sequenced. Therefore, additional internal primers are required. These anneal to different sites and amplify the DNA in a series of contiguous, overlapping chain termination experiments (2). Here, each primer determines which region of the template DNA is being sequenced.

In thermal cycle sequencing (3), only one primer is used to carry out PCR reactions, each with one dideoxynucleotide (ddA, ddT, ddG, or ddC) in the reaction mixture. This generates a series of different chain-terminated strands, each dependent on the position of the particular nucleotide base where the chain is being terminated (4). After many cycles and with electrophoresis, the sequence can be read as shown in the previous plate. One advantage of thermal cycle sequencing is that double-stranded DNA can be used as starting material. (Illustration based on Figures 4.5 and 4.6 in Brown, 1999).

## B. Automated DNA sequencing (principle)

Automated DNA sequencing involves four fluorophores, one for each of the four nucleotide bases. The resulting fluorescent signal is recorded at a fixed point when DNA passes through a capillary containing an electrophoretic gel. The base-specific fluorescent labels are attached to appropriate dideoxynucleotide triphosphates (ddNTP). Each ddNTP is labeled with a different color, e.g., ddATP green, ddCTP blue, ddGTP yellow, and ddTTP red (1). (The actual colors for each nucleotide may be different.) All chains terminated at an adenine (A) will yield a green signal; all chains terminated at a cytosine (C) will yield a blue signal, and so on. The sequencing reactions based on this kind of chain termination at labeled nucleotides (2) are carried out automatically in sequencing capillaries (3). The electrophoretic migration of the ddNTP-labeled chains in the gel in the capillary pass in front of a laser beam focused on a fixed position. The laser induces a fluorescent signal that is dependent on the specific label representing one of the four nucleotides. The sequence is electronically read and recorded and is visualized as alternating peaks in one of the four colors, representing the alternating nucleotides in their sequence positions. In practice the peaks do not necessarily show the same maximal intensity as in the schematic diagram shown here. (Illustration based on Brown, 1999, and Strachan and Read, 1999).

### References

Brown, T.A.: Genomes. Bios Scientific Publ., Oxford, 1999.

Rosenthal, N.: Fine structure of a gene—DNA sequencing. New Eng. J. Med. **332**:589–591, 1995.

Strachan, T., Read, A.P.: Human Molecular Genetics. 2nd ed. Bios Scientific Publishers, Oxford, 1999.

Wilson, R.K., et al.: Development of an automated procedure for fluorescent DNA sequencing. Genomics **6**:626–636, 1990.

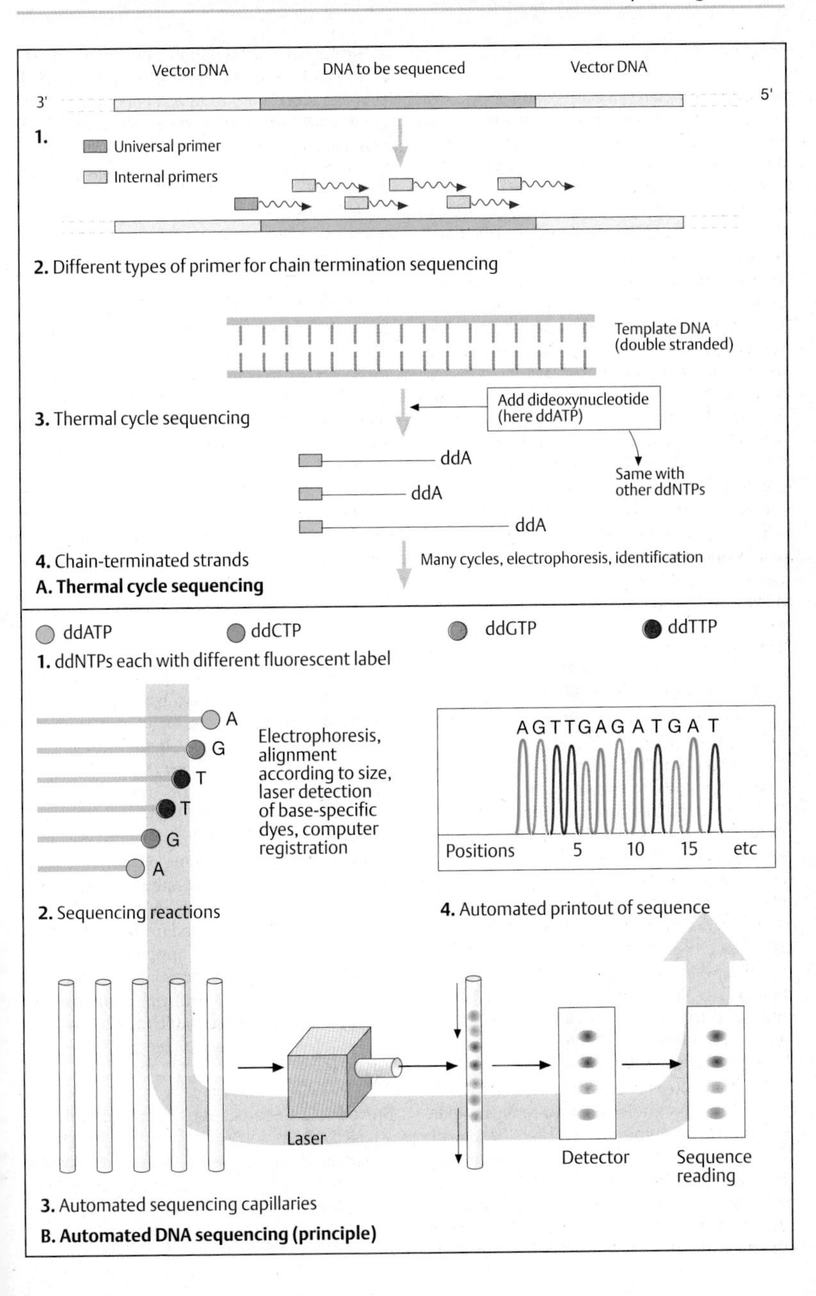
Vector DNA
DNA to be sequenced
Vector DNA
3'
5'
1.
Universal primer
Internal primers
2. Different types of primer for chain termination sequencing
Template DNA
(double stranded)
Add dideoxynucleotide
(here ddATP)
3. Thermal cycle sequencing
ddA
ddA
ddA
Same with
other ddNTPs
4. Chain-terminated strands
Many cycles, electrophoresis, identification
ddATP
ddCTP
ddGTP
ddTTP
1. ddNTPs each with different fluorescent label
A
G
T
T
G
A
Electrophoresis,
alignment
according to size,
laser detection
of base-specific
dyes, computer
registration
AGTTGAGATGAT
Positions 5 10 15 etc
2. Sequencing reactions
4. Automated printout of sequence
Laser
Detector
Sequence
reading
3. Automated sequencing capillaries

**A. Thermal cycle sequencing**

**B. Automated DNA sequencing (principle)**

## DNA Cloning

To obtain sufficient amounts of a specific DNA sequence (e.g., a gene of interest) for study, it must be selectively amplified. This is accomplished by DNA cloning, which produces a homogeneous population of DNA fragments from a mixture of very different DNA molecules or from all the DNA of the genome. Here procedures are required to identify DNA from the correct region in the genome, to separate it from other DNA, and to multiply (clone) it selectively. Identification of the correct DNA fragment utilizes the specific hybridization of complementary single-stranded DNA (molecular hybridization). A short segment of single-stranded DNA, a probe, originating from the sequence to be studied, will hybridize to its complementary sequences after these have been denatured (made single-stranded, see Southern blot analysis, p. 62). After the hybridized sequence has been separated from other DNA, it can be cloned. The selected DNA sequences can be amplified in two basic ways: in cells (cell-based cloning) or by cell-free cloning (see polymerase chain reaction, PCR, p. 66).

### A. Cell-based DNA cloning

Cell-based DNA cloning requires four initial steps. First, a collection of different DNA fragments (here labeled 1, 2, and 3) are obtained from the desired DNA (target DNA) by cleaving it with a restriction enzyme (see p. 64). Since fragments resulting from restriction enzyme cleavage have a short single-stranded end of a specific sequence at both ends, they can be ligated to other DNA fragments that have been cleaved with the same enzyme. The fragments produced in step 1 are joined to DNA fragments containing the origin of replication (OR) of a replicon, which enables them to replicate (2). In addition, a fragment may be joined to a selectable marker, e.g., a DNA sequence containing an antibiotic resistance gene. The recombinant DNA molecules are transferred into host cells (bacterial or yeast cells). Here the recombinant DNA molecules can replicate independently of the host cell genome (3). Usually the host cell takes up only one (although occasionally more than one) foreign DNA molecule. The host cells transformed by recombinant (foreign) DNA are grown in culture and multiplied (propagation, 4). Selective growth of one of the cell clones allows isolation of one type of recombinant DNA molecule (5). After further propagation, a homogeneous population of recombinant DNA molecules is obtained (6). A collection of different fragments of cloned DNA is called a clone library (7, see DNA libraries). In cell-based cloning, the replicon-containing DNA molecules are referred to as vector molecules. (Figure adapted from Strachan and Read, 1999).

### B. A plasmid vector for cloning

Many different vector systems exist for cloning DNA fragments of different sizes. Plasmid vectors are used to clone small fragments. The experiment is designed in such a way that incorporation of the fragment to be cloned changes the plasmid's antibiotic resistance to allow selection for these recombinant plasmids. A formerly frequently used plasma vector (pBR322) is presented. This plasmid contains recognition sites for the restriction enzymes *Pst*I, *Eco*RI, and *Sal*I in addition to genes for ampicillin and tetracycline resistance (1). If a foreign DNA fragment is incorporated into the plasmid at the site of the *Eco*RI recognition sequence, then tetracycline and ampicillin resistance will be retained (2). If the enzyme *Pst*I is used to incorporate the fragment to be used, ampicillin resistance is lost (the bacterium becomes ampicillin sensitive), but tetracycline resistance is retained. If the enzyme *Sal*I is used to incorporate the fragment, tetracycline resistance disappears (the bacterium becomes tetracycline sensitive), but ampicillin resistance is retained. Thus, depending on how the fragment has been incorporated, recombinant plasmids containing the DNA fragment to be cloned can be distinguished from nonrecombinant plasmids by altered antibiotic resistance. Cloning in plasmids (bacteria) has become less important since yeast artificial chromosomes (YACs) have become available for cloning relatively large DNA fragments (see p. 110).

### References

Brown, T.A.: Genomes. Bios Scientific Publ., Oxford, 1999.

Strachan, T., Read, A.P.: Human Molecular Genetics. 2nd ed. Bios Scientific Publishers, Oxford, 1999.

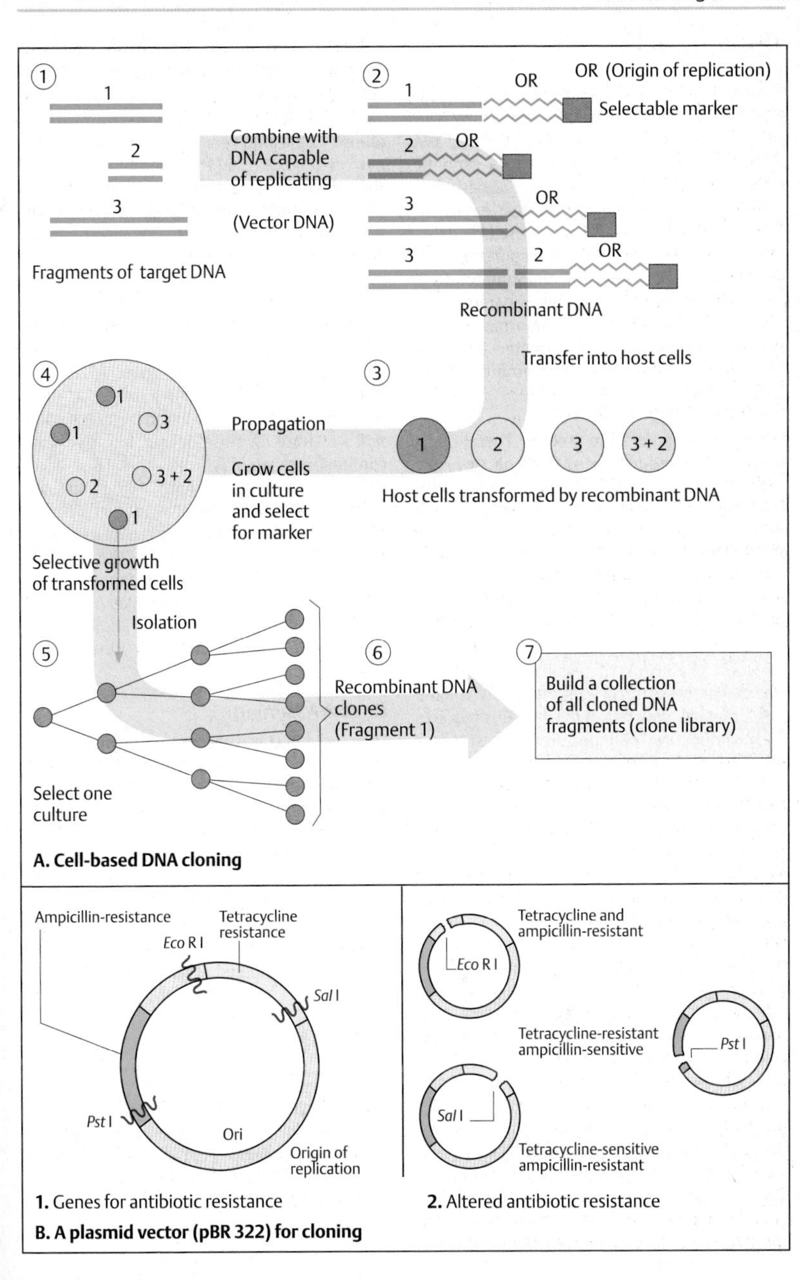


**A. Cell-based DNA cloning**

**1.** Genes for antibiotic resistance

**2.** Altered antibiotic resistance

**B. A plasmid vector (pBR 322) for cloning**

# cDNA Cloning

cDNA is a single-stranded segment of DNA that is complementary to the mRNA of a coding DNA segment or of a whole gene. It can be used as a probe (cDNA probe as opposed to a genomic probe) for the corresponding gene because it is complementary to coding sections (exons) of the gene. If the gene has been altered by structural rearrangement at a corresponding site, e.g., by deletion, the normal and mutated DNA can be differentiated. Thus, the preparation and cloning of cDNA is of great importance. From the cDNA sequence, essential inferences can be made about a gene and its gene product.

## A. Preparation of cDNA

cDNA is prepared from mRNA. Therefore, a tissue is required in which the respective gene is transcribed and mRNA is produced in sufficient quantities. First, mRNA is isolated. Then a primer is attached so that the enzyme reverse transcriptase can form complementary DNA (cDNA) from the mRNA. Since mRNA contains poly(A) at its 3′ end, a primer of poly(T) can be attached. From here, the enzyme reverse transcriptase can start forming cDNA in the 5′ to 3′ direction. The RNA is then removed by ribonuclease. The cDNA serves as a template for the formation of a new strand of DNA. This requires the enzyme DNA polymerase. The result is a double strand of DNA, one strand of which is complementary to the original mRNA. To this DNA, single sequences (linkers) are attached that are complementary to the single-stranded ends produced by the restriction enzyme to be used. The same enzyme is used to cut the vector, e.g., a plasmid, so that the cDNA can be incorporated for cloning.

## B. Cloning vectors

The cell-based cloning of DNA fragments of different sizes is facilitated by a wide variety of vector systems. Plasmid vectors are used to clone small DNA fragments in bacteria. Their main disadvantage is that only 5–10 kb of foreign DNA can be cloned. A plasmid cloning vector that has taken up a DNA fragment (recombinant vector), e.g., pUC8 with 2.7 kb of DNA, must be distinguished from one that has not. In addition, an ampicillin resistance gene ($Amp^+$) serves to distinguish bacteria that have taken up plasmids from those that have not. Several unique restriction sites in the plasmid DNA segment where a DNA fragment might be inserted serve as markers along with a marker gene, such as the *lacZ* gene. The uptake of a DNA fragment by a plasmid vector disrupts the plasmid's marker gene. Thus, in the recombinant plasmid the enzyme β-galactosidase will not be produced by the disrupted *lagZ* gene, whereas in the plasmid without a DNA insert (nonrecombinant) the enzyme is produced by the still intact *lacZ* gene. The activity of the gene and the presence or absence of the enzyme are determined by observing a difference in color of the colonies in the presence of an artificial substrate sugar. β-Glactosidase splits an artificial sugar (5-bromo-4-chloro-3-indolyl-β-D-galactopyranoside) that is similar to lactose, the natural substrate for this enzyme, into two sugar components, one of which is blue. Thus, bacterial colonies containing nonrecombinant plasmids with an intact *lacZ* gene are blue. In contrast, colonies that have taken up a recombinant vector remain pale white. The latter are grown in a medium containing ampicillin (the selectable marker for the uptake of plasmid vectors). Subsequently, a clone library can be constructed. (Figure adapted from Brown, 1999).

## C. cDNA cloning

Only those bacteria become ampicillin resistant that have incorporated a recombinant plasmid. Recombinant plasmids, which contain the gene for ampicillin resistance, transform ampicillin-sensitive bacteria into ampicillin-resistant bacteria. In an ampicillin-containing medium, only those bacteria grow that contain the recombinant plasmid with the desired DNA fragment. By further replication in these bacteria, the fragment can be cloned until there is enough material to be studied. (Figures after Watson et al., 1987).

### References

Brown, T.A.: Genomes. Bios Scientific Publ., Oxford, 1999.

Watson, J.D., et al.: Molecular Biology of the Gene, 3 rd ed. Benjamin/Cummings Publishing Co., Menlo Park, California, 1987.

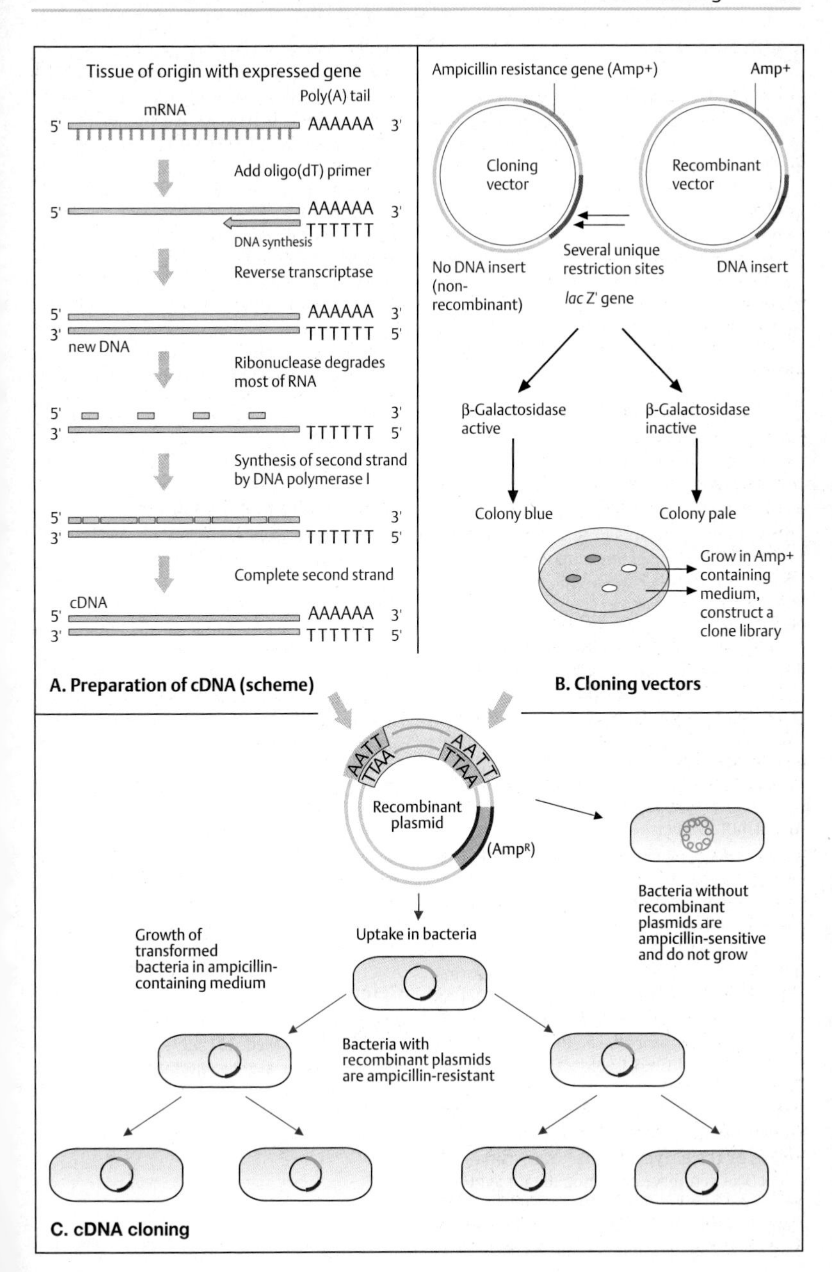
Tissue of origin with expressed gene
Poly(A) tail
mRNA
5' AAAAAA 3'
Add oligo(dT) primer
5' AAAAAA 3'
TTTTTT
DNA synthesis
Reverse transcriptase
5' AAAAAA 3'
3' TTTTTT 5'
new DNA
Ribonuclease degrades most of RNA
5' 3'
3' TTTTTT 5'
Synthesis of second strand by DNA polymerase I
5' 3'
3' TTTTTT 5'
Complete second strand
cDNA
5' AAAAAA 3'
3' TTTTTT 5'
Ampicillin resistance gene (Amp+)
Amp+
Cloning vector
Recombinant vector
No DNA insert (non-recombinant)
Several unique restriction sites
lac Z' gene
DNA insert
β-Galactosidase active
β-Galactosidase inactive
Colony blue
Colony pale
Grow in Amp+ containing medium, construct a clone library
AATT
TTAA
AATT
TTAA
Recombinant plasmid
(AmpR)
Bacteria without recombinant plasmids are ampicillin-sensitive and do not grow
Uptake in bacteria
Growth of transformed bacteria in ampicillin-containing medium
Bacteria with recombinant plasmids are ampicillin-resistant

**A. Preparation of cDNA (scheme)**

**B. Cloning vectors**

**C. cDNA cloning**

# DNA Libraries

A DNA library is a collection of DNA fragments that in their entirety represent the genome, that is, a particular gene being sought and all remaining DNA. It is the starting point for cloning a gene of unknown chromosomal location. To produce a library, the total DNA is digested with a restriction enzyme, and the resulting fragments are incorporated into vectors and replicated in bacteria. A sufficient number of clones must be present so that every segment is represented at least once. This is a question of the size of the genome being investigated and the size of the fragments. Plasmids and phages are used as vectors. For larger DNA fragments, yeast cells may be employed. There are two different types of libraries: genomic DNA and cDNA.

## A. Genomic DNA library

Clones of genomic DNA are copies of DNA fragments from all of the chromosomes (1). They contain coding and noncoding sequences. Restriction enzymes are used to cleave the genomic DNA into many fragments. Here four fragments are schematically shown, containing two genes, A and B (2). These are incorporated into vectors, e.g., into phage DNA, and are replicated in bacteria. The complete collection of recombinant DNA molecules, containing all DNA sequences of a species or individual, is called a genomic library. To find a particular gene, a screening procedure is required (see B).

## B. cDNA library

Unlike a genomic library, which is complete and contains coding and noncoding DNA, a cDNA library consists only of coding DNA sequences. This specificity offers considerable advantages over genomic DNA. However, it requires that mRNA be available and does not yield information about the structure of the gene. mRNA can be obtained only from cells in which the respective gene is transcribed, i.e., in which mRNA is produced (1). In eukaryotes, the RNA formed during transcription (primary transcript) undergoes splicing to form mRNA (2, see p. 50). Complementary DNA (cDNA) is formed from mRNA by the enzyme reverse transcriptase (3). The cDNA can serve as a template for synthesis of a complementary DNA strand, so that complete double-stranded DNA can be formed (cDNA clone). Its sequence corresponds to the coding sequences of the gene exons. Thus it is well suited for use as a probe (cDNA probe). The subsequent steps, incorporation into a vector and replication in bacteria, correspond to those of the procedure to produce a genomic library. cDNA clones can only be won from coding regions of an active (mRNA-producing) gene; thus, the cDNA clones of different tissues differ according to genetic activity. Since cDNA clones correspond to the coding sequences of a gene (exons) and contain no noncoding sections (introns), cloned cDNA is the preferred starting material when further information about a gene product is sought by analyzing the gene. The sequence of amino acids in a protein can be determined from cloned and sequenced cDNA. Also, large amounts of a protein can be produced by having the cloned gene expressed in bacteria or yeast cells.

## C. Screening of a DNA library

Bacteria that have taken up the vectors can grow on an agar-coated Petri dish, where they form colonies (1). A replica imprint of the culture is taken on a membrane (2), and the DNA that sticks to the membrane is denatured with an alkaline solution (3). DNA of the gene segment being sought can then be identified by hybridization with a radioactively (or otherwise) labeled probe (4). After hybridization, a signal appears on the membrane at the site of the gene segment (5). DNA complementary to the labeled probe is located here; its exact position in the culture corresponds to that of the signal on the membrane (6). A probe is taken from the corresponding area of the culture (5). It will contain the desired DNA segment, which can now be further replicated (cloned) in bacteria. By this means, the desired segment can be enriched and is available for subsequent studies.

### References

Rosenthal, N.: Stalking the gene—DNA libraries. New Eng. J. Med. **331**:599–600, 1994.

Watson, J.D. et al.: Recombinant DNA. 2nd ed. Scientific American Books, New York, 1992.

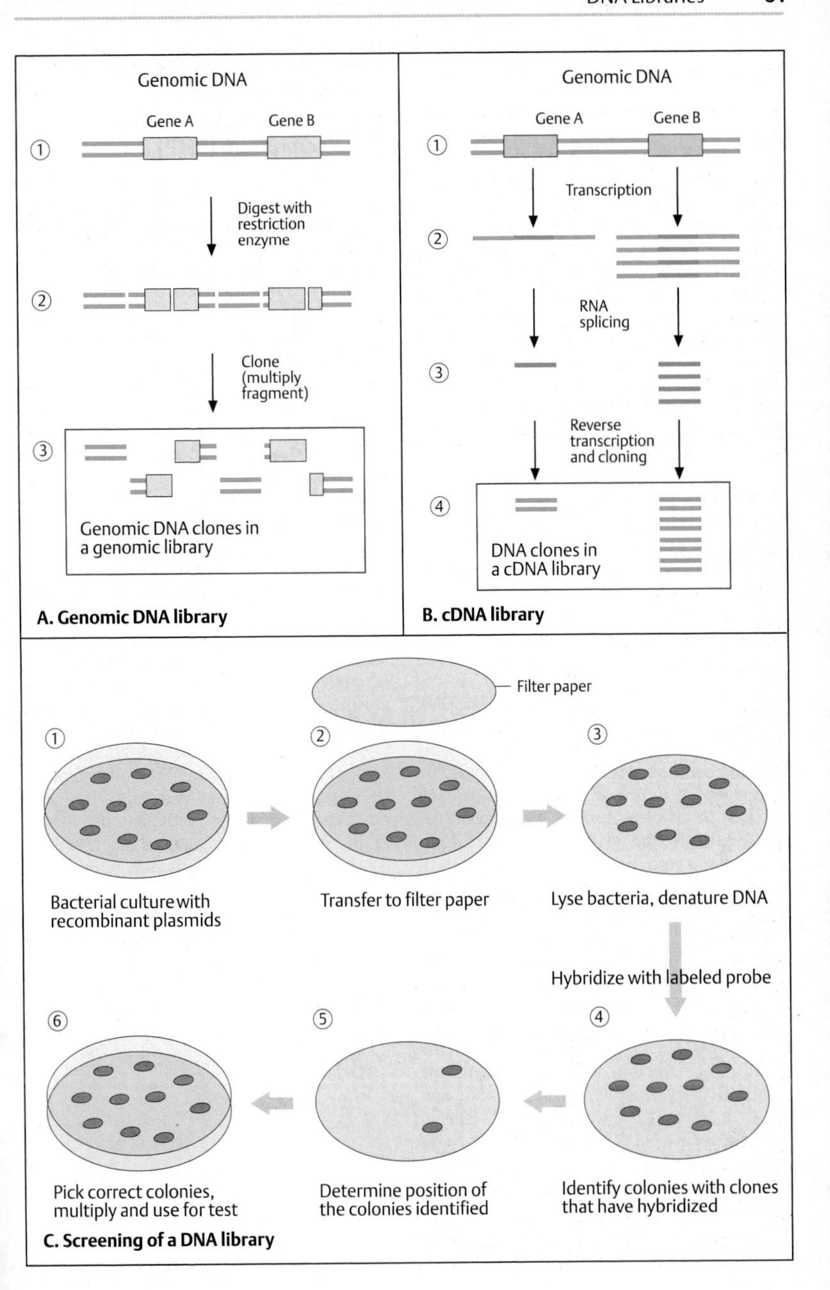


A. Genomic DNA library

B. cDNA library

C. Screening of a DNA library

## *Restriction Analysis by Southern Blot Analysis*

Restriction endonucleases are DNA-cleaving enzymes with defined sequences as targets (see next plate). They are often simply called restriction enzymes. Since each enzyme cleaves DNA only at its specific recognition sequence, the total DNA of an individual present in nucleated cells can be cut into pieces of manageable and defined size in a reproducible way. Individual DNA fragments can then be selected, ligated into suitable vectors, multiplied, and examined. Owing to the uneven distribution of recognition sites, the DNA fragments differ in size. A starting mixture of DNA fragments is sorted according to size. Two procedures detect target DNA or RNA fragments after they have been arranged by size in gel electrophoresis—the Southern blot hybridization for DNA (named after E. Southern who developed this method 1975) and the Northern blot hybridization for RNA (a word play on Southern, not named after a Dr. Northern). Immunoblotting (Western blot) detects proteins by an antibody-based procedure.

### A. Southern blot hybridization

The analysis starts with total DNA (1). The DNA is isolated and cut with restriction enzymes (2). One of the not yet identified fragments contains the gene being sought or part of the gene. The fragments are sorted by size in a gel (usually agarose) in an electric field (electrophoresis) (3). The smaller the fragment, the faster it migrates; the larger, the slower it migrates. Next, the blot is carried out: The fragments contained in the gel are transferred to a nitrocellulose or nylon membrane (4). There the DNA is denatured (made single-stranded) with alkali and fixed to the membrane by moderate heating (~ 80 °C) or UV cross-linkage. The sample is incubated with a probe of complementary single-stranded DNA (genomic DNA or cDNA) from the gene (5). The probe hybridizes solely with the complementary fragment being sought, and not with others (6). Since the probe is labeled with radioactive $^{32}P$, the fragment being sought can be identified by placing an X-ray film on the membrane, where it appears as a black band on the film after development (autoradiogram) (6). The size, corresponding to position, is determined by running DNA fragments of known size in the electrophoresis.

### B. Restriction fragment length polymorphism (RFLP)

In about every 100 base pairs of a DNA segment, the nucleotide sequence differs in some individuals (DNA polymorphism). As a result, the recognition sequence of a restriction enzyme may be present on one chromosome but not the other. In this case the restriction fragment sizes differ at this site (restriction fragment length polymorphism, RFLP). An example is shown for two 5 kb (5000 base pair) DNA segments. In one, a restriction site in the middle is present (allele 1); in the other (allele 2) it is absent. With a Southern blot, it can be determined whether in this location an individual is homozygous 1–1 (two alleles 1, no 5 kb fragment), heterozygous 1–2 (one allele each, 1 and 2), or homozygous 2–2 (two alleles 2). If the mutation being sought lies on the chromosome carrying the 5 kb fragment, the presence of this fragment indicates presence of the mutation. The absence of this fragment would indicate that the mutation is absent.

It is important to understand that the RFLP itself is unrelated to the mutation. It simply distinguishes DNA fragments of different sizes from the same region. These can be used as markers to distinguish alleles in a segregation analysis (see p. 134). In addition to RFLPs, other types of DNA polymorphism can be detected by Southern blot hybridization, although polymerase chain reaction-based analysis of microsatellites is now used more frequently (see p. 244).

### References

Brown, T.A.: Genomes. Bios Scientific Publ., Oxford, 1999.

Housman, D.: Human DNA polymorphism. New Engl. J. Med. **332**:318–320, 1995.

Strachan, T., Read, A.P.: Human Molecular Genetics. 2nd ed. Bios Scientific Publishers, Oxford, 1999.

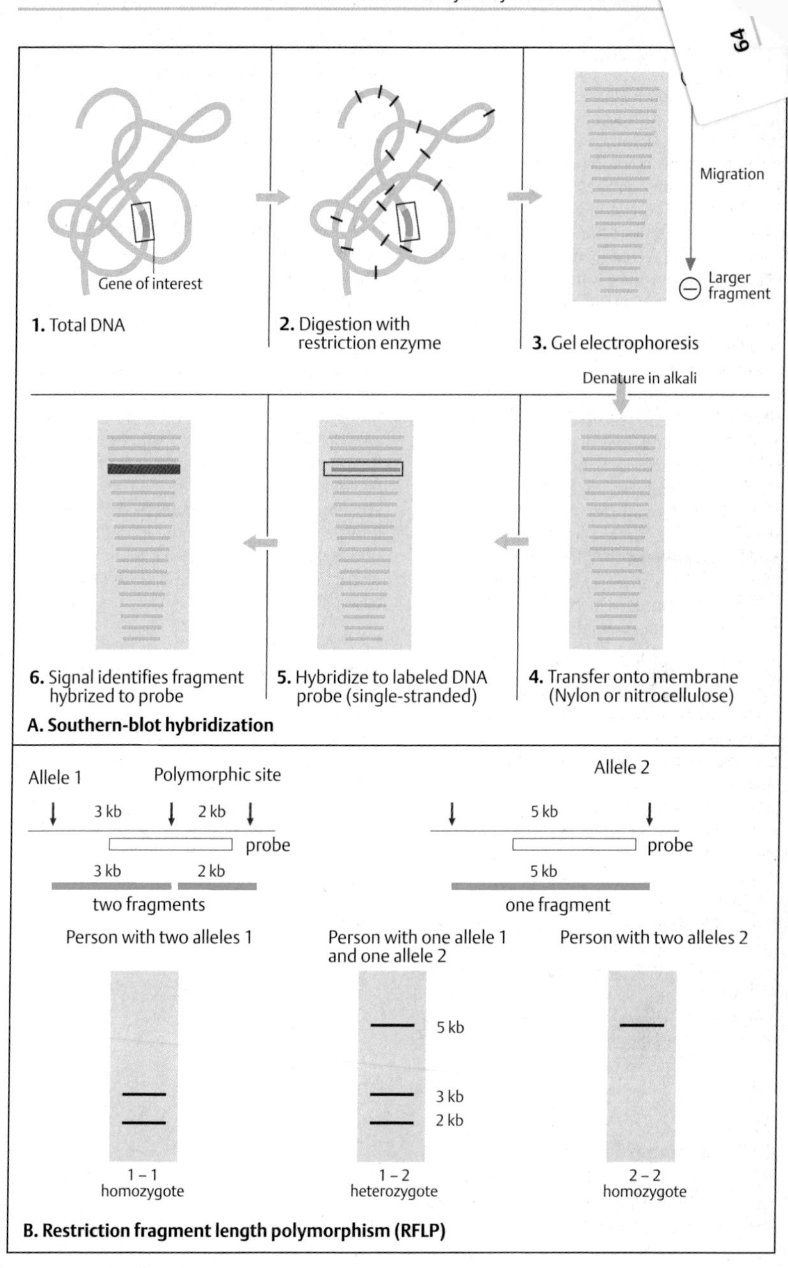


**A. Southern-blot hybridization**

**B. Restriction fragment length polymorphism (RFLP)**

## Restriction Mapping

Restriction endonucleases (restriction enzymes) are DNA-cutting enzymes. They are obtained from bacteria, which produce the enzymes as protection from foreign DNA. A given enzyme recognizes a specific sequence of 4–8 (usually 6) nucleotides (a restriction site) where it cleaves the DNA. The sizes of the DNA fragments produced depend on the distribution of the restriction sites. More than 400 different types of restriction enzymes have been isolated.

### A. DNA cleavage by restriction nucleases

The cleavage patterns (recognition sequences) of three frequently used restriction enzymes, *Eco*RI, *Hin*dIII, and *Hpa*I, are presented. For *Eco*RI and *Hin*dIII the cut is "palindromic," i.e., the cut is asymmetric around an axis on which mirror-image complementary single-stranded DNA segments arise. Each corresponds to its opposite-lying strand in the reverse direction. Therefore, they can be joined to a DNA fragment whose ends contain complementary single-stranded sequences. *Hpa*I cuts both strands so that no single-stranded ends are formed. Frequently cutting and seldom cutting enzymes can be distinguished according to the frequency of occurrence of their recognition sites.

### B. Examples of restriction enzymes

The recognition sequences of some restriction enzymes are shown. The names of the enzymes are derived from those of the bacteria in which they occur, e.g., *Eco*RI from *Escherichia coli* Restriction enzyme I, etc. Some enzymes have a cutting site with limited specificity. In *Hin*dII it suffices that the two middle nucleotides are a pyrimidine and a purine (GTPyPuAC), and it does not matter whether the former is thymine (T) or cytosine (C), and whether the latter is adenine (A) or guanine (G). Such a recognition site occurs frequently and produces many relatively small fragments, whereas enzymes that cut very infrequently produce few and large DNA fragments.

### C. Restriction fragments

In a given DNA segment, the recognition sequence of a restriction enzyme occurs irregularly. Thus, the distances between restriction sites differ. DNA fragments of various sizes (restriction fragments) result from digestion with a restriction enzyme. A given restriction enzyme will cleave a given segment of DNA into a series of DNA fragments of characteristic sizes. This leads to a pattern that can be employed for diagnostic purposes.

### D. Determination of the locations of restriction sites

Since the fragment sizes reflect the relative positions of the cutting sites, they can be used to characterize a DNA segment (restriction map). If a 10-kb DNA segment cut by two enzymes, A and B, yields three fragments, of 2 kb, 3 kb, and 5 kb, then the relative location of the cutting sites can be determined by using enzymes A and B alone in further experiments. If enzyme A yields two fragments of 3 kb and 7 kb, and enzyme B two fragments of 2 kb and 8 kb, then the two cutting sites of enzymes A and B must lie 5 kb apart. To the left of the restriction site of enzyme A are 3 kb; to the right of the restriction site of enzyme B, 2 kb (1 kb = 1000 base pairs).

### E. Restriction map

A given DNA segment can be characterized by the distribution pattern of restriction sites. In the example shown, a DNA segment is characterized by the distribution of the cutting sites for enzymes E (*Eco*RI) and H (*Hin*dIII). The individual sites are separated by intervals defined by the size of the fragments after digestion with the enzyme. A restriction map is a linear sequence of restriction sites at defined intervals along the DNA. Restriction mapping is of considerable importance in medical genetics and evolutionary research.

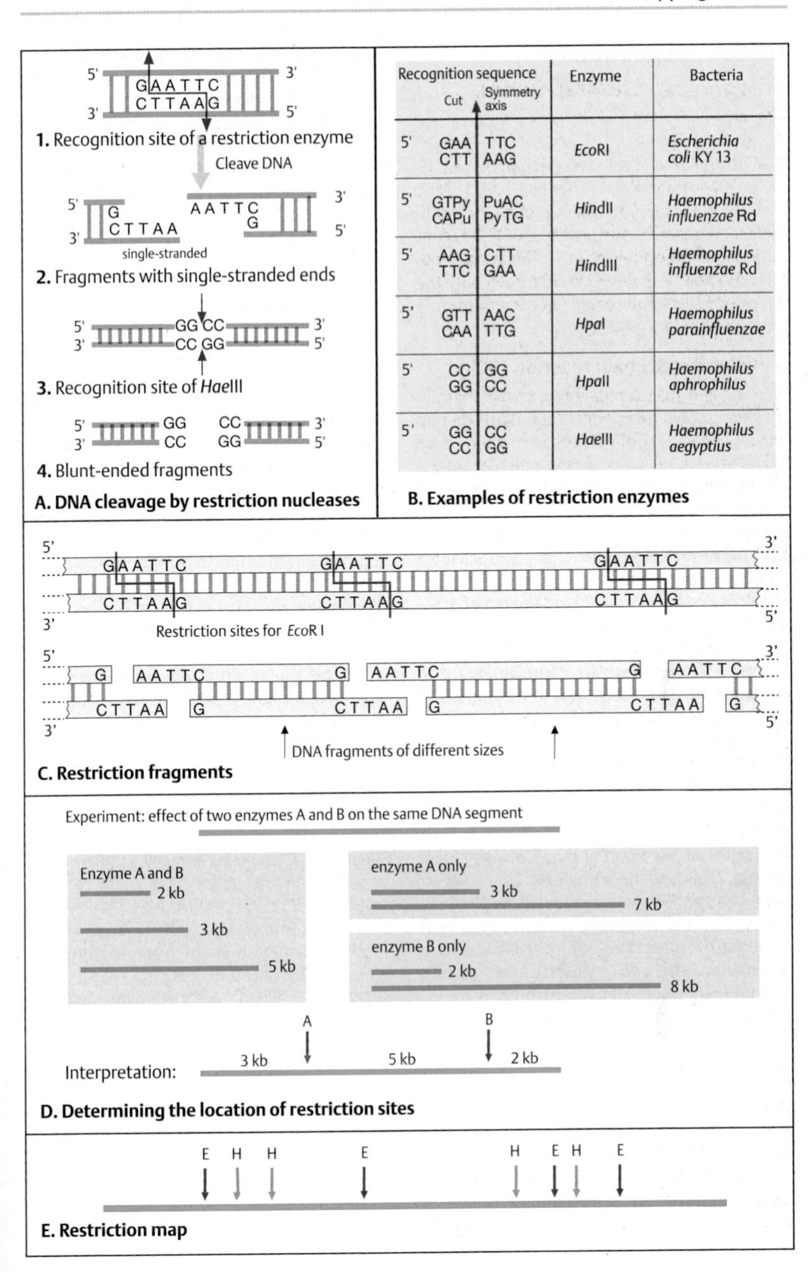


**A. DNA cleavage by restriction nucleases**

**B. Examples of restriction enzymes**

**C. Restriction fragments**

**D. Determining the location of restriction sites**

**E. Restriction map**

# DNA Amplification by Polymerase Chain Reaction (PCR)

The introduction of cell-free methods for multiplying DNA fragments of defined origin (DNA amplification) in 1985 ushered in a new era in molecular genetics (the principle of PCR is contained in earlier publications). This fundamental technology has spread dramatically with the development of automated equipment used in basic and applied research.

## A. Polymerase chain reaction (PCR)

PCR is a cell-free, rapid, and sensitive method for cloning DNA fragments. A standard reaction and a wide variety of PCR-based methods have been developed to assay for polymorphisms and mutations. Standard PCR is an in vitro procedure for amplifying defined target DNA sequences, even from very small amounts of material or material of ancient origin. Selective amplification requires some prior information about DNA sequences flanking the target DNA. Based on this information, two oligonucleotide primers of about 15–25 base pairs length are designed. The primers are complementary to sequences outside the 3′ ends of the target site and bind specifically to these.

PCR is a chain reaction because newly synthesized DNA strands act as templates for further DNA synthesis for about 25–35 subsequent cycles. Theoretically each cycle doubles the amount of DNA amplified. At the end, at least $10^5$ copies of the specific target sequence are present. This can be visualized as a distinct band of a specific size after gel electrophoresis. Each cycle, involving three precisely time-controlled and temperature-controlled reactions in automated thermal cyclers, takes about 1–5 min. The three steps in each cycle are (1) denaturation of double-stranded DNA, at about 93–95 °C for human DNA, (2) primer annealing at about 50–70 °C depending on the expected melting temperature of the duplex DNA, and (3) DNA synthesis using heat-stable DNA polymerase (from microorganisms living in hot springs, such as *Thermophilus aquaticus, Taq* polymerase), typically at about 70–75 °C. At each subsequent cycle the template (shown in blue) and the DNA newly synthesized during the preceding cycle (shown in red) act as templates for another round of synthesis. The first cycle results in newly synthesized DNA of varied lengths (shown with an arrow) at the 3′ ends because synthesis is continued beyond the target sequences. The same happens during subsequent cycles, but the variable strands are rapidly outnumbered by new DNA of fixed length at both ends because synthesis cannot proceed past the terminus of the primer at the opposite template DNA.

## B. cDNA amplification and RT-PCR

A partially known amino acid sequence of a polypeptide can be used to obtain the sequence information required for PCR. From its mRNA one can derive cDNA (see complementary DNA, p. 58) and determine the sequence of the sense and the antisense strand to prepare appropriate oligonucleotide primers (1). When different RNAs are available in small amounts, rapid PCR-based methods are employed to amplify cDNA from different exons of a gene. cDNA is obtained by reverse transcriptase from mRNA, which is then removed by alkaline hydrolysis (2). After a complementary new DNA strand has been synthesized, the DNA can be amplified by PCR (3). Reverse transcriptase PCR (RT-PCR) can be used when the known exon sequences are widely separated within a gene. With rapid amplification of cDNA ends (RACE-PCR), the 5′ and 3′ end sequences can be isolated from cDNA.

## References

Brown, T.A.: Genomes. Bios Scientific Publ., Oxford, 1999.

Erlich, H.A., Gelfand D., Sninsky, J.J.: Recent advances in the polymerase chain reaction. Science **252**:1643–1651, 1991.

Erlich, H.A., Arnheim, N.: Genetic analysis with the polymerase chain reaction. Ann. Rev. Genet. **26**:479–506, 1992.

Strachan, T., Read, A.P.: Human Molecular Genetics. 2nd ed. Bios Scientific Publishers, Oxford, 1999.

Volkenandt, M., Löhr, M., Dicker, A.P.: Gen-Amplification durch Polymerase-Kettenreaktion. Dtsch. Med. Wschr. **17**:670–676, 1990.

White, T.J., Arnheim, N., Erlich, H.A.: The polymerase chain reaction. Trends Genet. **5**:185–189, 1989.

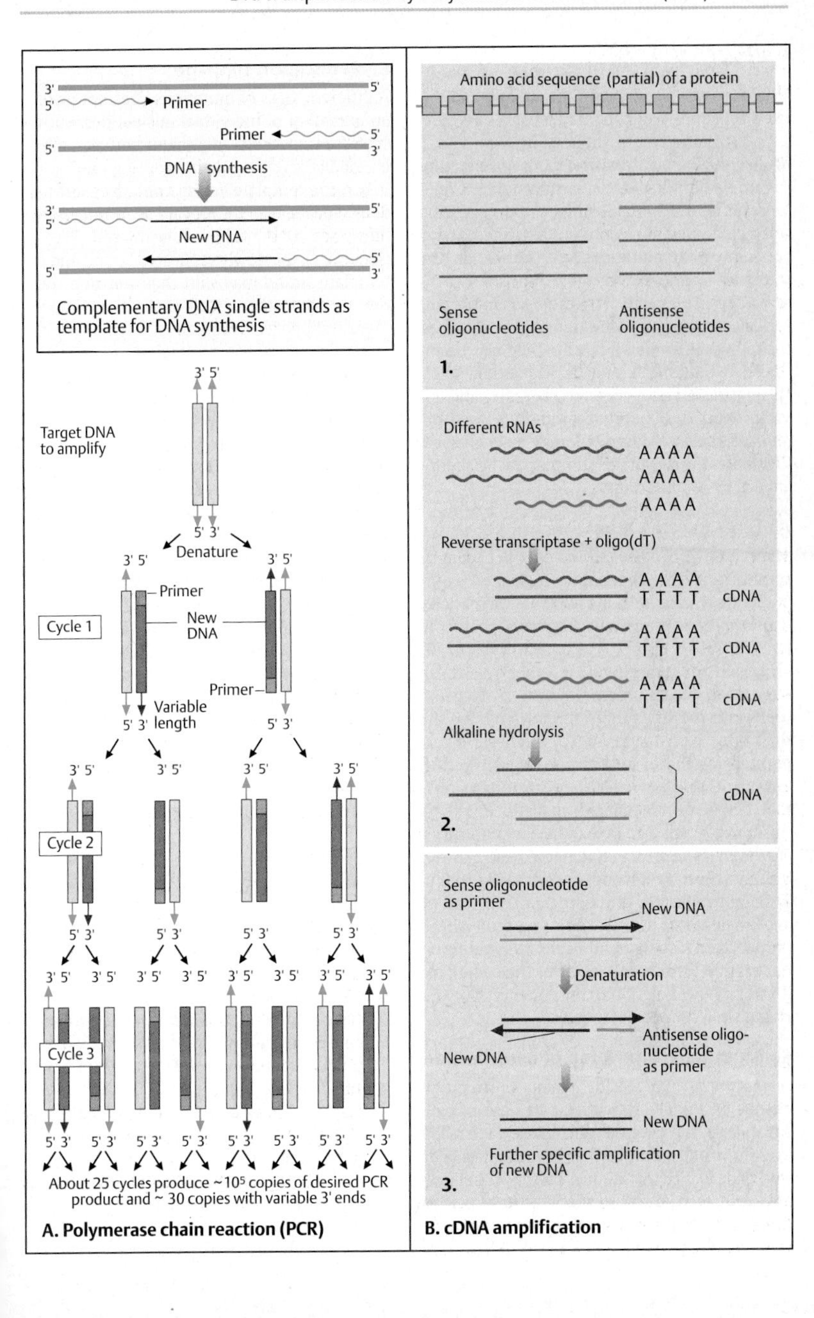


**A. Polymerase chain reaction (PCR)**

**B. cDNA amplification**

# Changes in DNA

When it was recognized that changes (mutations) in genes occur spontaneously (T. H. Morgan, 1910) and can be induced by X-rays (H. J. Muller, 1927), the mutation theory of heredity became a cornerstone of early genetics. Genes were defined as mutable units, but the question what genes and mutations are remained. Today we know that mutations are changes in the structure of DNA and their functional consequences. The study of mutations is important for several reasons. Mutations cause diseases, including all forms of cancer. They can be induced by chemicals and by irradiation. Thus, they represent a link between heredity and environment. And without mutations, well-organized forms of life would not have evolved. The following two plates summarize the chemical nature of mutations.

## A. Error in replication

The synthesis of a new strand of DNA occurs by semiconservative replication based on complementary base pairing (see DNA replication). Errors in replication occur at a rate of about 1 in $10^5$. This rate is reduced to about 1 in $10^7$ to $10^9$ by proofreading mechanisms. When an error in replication occurs before the next cell division (here referred to as the first division after the mutation), e.g., a cytosine (C) might be incorporated instead of an adenine (A) at the fifth base pair as shown here, the resulting mismatch will be recognized and eliminated by mismatch repair (see DNA repair) in most cases. However, if the error is undetected and allowed to stand, the next (second) division will result in a mutant molecule containing a CG instead of an AT pair at this position. This mutation will be perpetuated in all daughter cells. Depending on its location within or outside of the coding region of a gene, functional consequences due to a change in a codon could result.

## B. Mutagenic alteration of a nucleotide

A mutation may result when a structural change of a nucleotide affects its base-pairing capability. The altered nucleotide is usually present in one strand of the parent molecule. If this leads to incorporation of a wrong base, such as a C instead of a T in the fifth base pair as shown here, the next (second) round of replication will result in two mutant molecules.

## C. Replication slippage

A different class of mutations does not involve an alteration of individual nucleotides, but results from incorrect alignment between allelic or nonallelic DNA sequences during replication. When the template strand contains short tandem repeats, e.g., CA repeats as in microsatellites (see DNA polymorphism and Part II, Genomics), the newly replicated strand and the template strand may shift their positions relative to each other. With replication or polymerase slippage, leading to incorrect pairing of repeats, some repeats are copied twice or not at all, depending on the direction of the shift. One can distinguish forward slippage (shown here) and backward slippage with respect to the newly replicated strand. If the newly synthesized DNA strand slips forward, a region of nonpairing remains in the parental strand. Forward slippage results in an insertion. Backward slippage of the new strand results in deletion.

Microsatellite instability is a characteristic feature of hereditary nonpolyposis cancer of the colon (HNPCC). HNPCC genes are localized on human chromosomes at 2p15–22 and 3p21.3. About 15% of all colorectal, gastric, and endometrial carcinomas show microsatellite instability. Replication slippage must be distinguished from unequal crossing-over during meiosis. This is the result of recombination between adjacent, but not allelic, sequences on nonsister chromatids of homologous chromosomes (Figures redrawn from Brown, 1999).

## References

Brown, T.A.: Genomes. Bios Scientific Publ., Oxford, 1999.

Dover, G.A.: Slippery DNA runs on and on and on ... Nature Genet. **10**:254–256, 1995.

Lewin, B.: Genes VII. Oxford University Press, Oxford, 2000.

Rubinstein, D.C., et al.: Microsatellite evolution and evidence for directionality and variation in rate between species. Nature Genet. **10**:337–343, 1995.

Strachan, T.A., Read, A.P.: Human Molecular Genetics. 2nd ed. Bios Scientific Publ., Oxford, 1999.

Vogel, F., Rathenberg, R.: Spontaneous mutation in man. Adv. Hum. Genet. **5**:223–318, 1975.

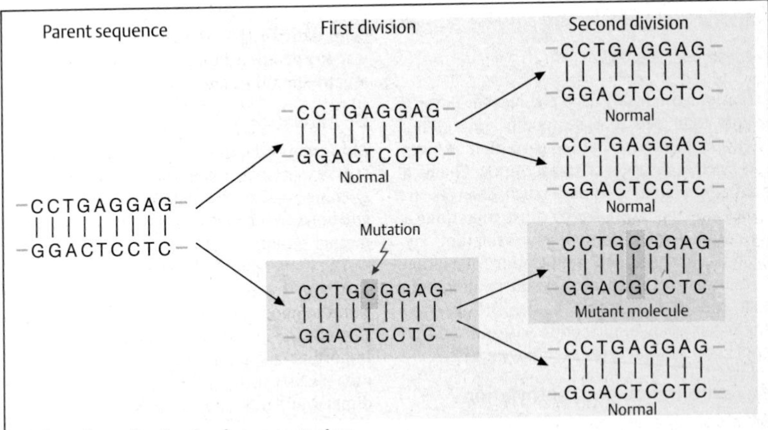


**A. Error in replication leads to a mutation**

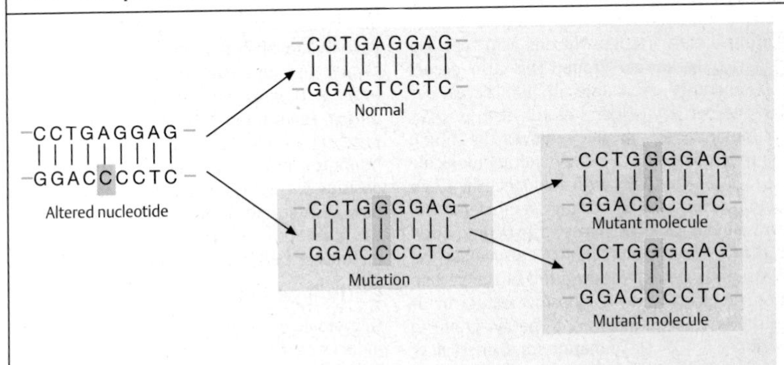


**B. Mutagenic alteration of a nucleotide leads to mismatch resulting in a mutation**

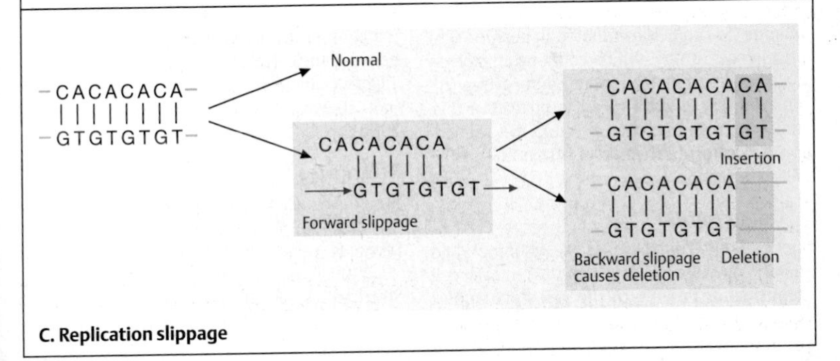


**C. Replication slippage**

## Mutation Due to Different Base Modifications

Mutations can result from chemical or physical events that lead to base modification. When they affect the base-pairing pattern, they interfere with replication or transcription. Chemical substances able to induce such changes are called mutagens. Mutagens cause mutations in different ways. Spontaneous oxidation, hydrolysis, uncontrolled methylation, alkylation, and ultraviolet irradiation result in alterations that modify nucleotide bases. DNA-reactive chemicals change nucleotide bases into different chemical structures or remove a base.

### A. Deamination and methylation

Cytosine, adenine, and guanine contain an amino group. When this is removed (deamination), a modified base with a different base-pairing pattern results. Nitrous acid typically removes the amino group. This also occurs spontaneously at a rate of 100 bases per genome per day (Alberts et al., 1994, p. 245). Deamination of cytosine removes the amino group in position 4 (1). The resulting molecule is uracil (2). This pairs with adenine rather than guanine. Normally this change is efficiently repaired by uracil-DNA glycosylase. Deamination at the RNA level occurs in RNA editing (see Expression of genes). Methylation of the carbon atom in position 5 of cytosine results in 5-methylcytosine, containing a methyl group in position 5 (3). Deamination of 5-methylcytosine will result in a change to thymine, containing an oxygen in position 4 instead of an amino group (4). This mutation will not be corrected because thymine is a natural base. Adenine (5) can be deaminated in position 6 to form hypoxanthine, which contains an oxygen in this position instead of an amino group (6), and which pairs with cytosine instead of thymine. The resulting change after DNA replication is a cytosine instead of a thymine in the mutant strand.

### B. Depurination

About 5000 purine bases (adenine and guanine) are lost per day from DNA in each cell (depurination) owing to thermal fluctuations. Depurination of DNA involves hydrolytic cleavage of the *N*-glycosyl linkage of deoxyribose to the guanine nitrogen in position 9. This leaves a depurinated sugar. The loss of a base pair will lead to a deletion after the next replication if not repaired in time (see DNA repair).

### C. Alkylation

Alkylation is the introduction of a methyl or an ethyl group into a molecule. The alkylation of guanine involves the replacement of the hydrogen bond to the oxygen atom in position 6 by a methyl group, to form 6-methylguanine. This can no longer pair with cytosine. Instead, it will pair with thymine. Thus, after the next replication the opposite cytosine (C) is replaced by a thymine (T) in the mutant daughter molecule. Important alkylating agents are ethylnitrosourea (ENU), ethylmethane sulfonate (EMS), dimethylnitrosamine, and *N*-methyl-*N*-nitro-*N*-nitrosoguanidine.

### D. Nucleotide base analogue

Base analogs are purines or pyrimidines that are similar enough to the regular nucleotide DNA bases to be incorporated into the new strand during replication. 5-Bromodeoxyuridine (5-BrdU) is an analog of thymine. It contains a bromine atom instead of the methyl group in position 5. Thus, it can be incorporated into the new DNA strand during replication. However, the presence of the bromine atom causes ambiguous and often wrong base pairing.

### E. UV-light-induced thymine dimers

Ultraviolet irradiation at 260 nm wavelength induces covalent bonds between adjacent thymine residues at carbon positions 5 and 6. If located within a gene, this will interfere with replication and transcription unless repaired. Another important type of UV-induced change is a photoproduct consisting of a covalent bond between the carbons in positions 4 and 6 of two adjacent nucleotides, the 4–6 photoproduct (not shown). (Figures redrawn from Lewin, 2000).

#### References

Brown, T.A.: Genomes. Bios Scientific Publ., Oxford, 1999.

Lewin, B.: Genes VII. Oxford Univ. Press, Oxford, 2000.

Strachan, T., Read, A.P.: Human Molecular Genetics. 2$^{nd}$ ed. Bios Scientific Publishers, Oxford, 1999.

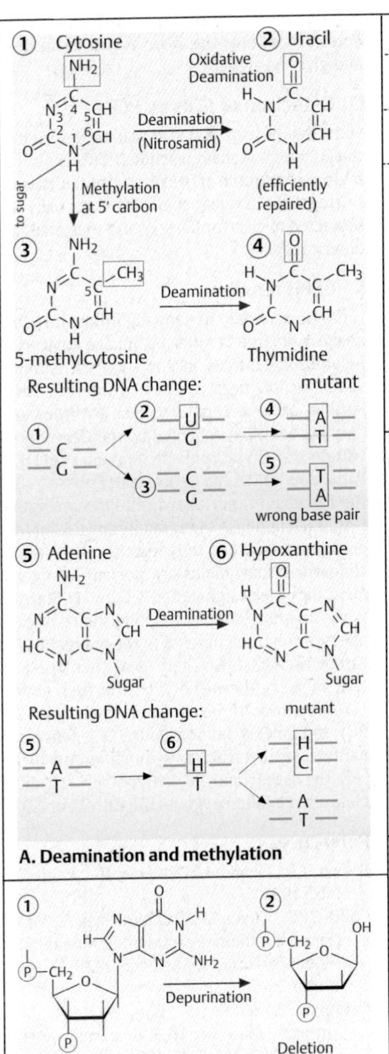


A. Deamination and methylation

B. Depurination

Guanine

6-Methyl-guanine

no hydrogen bond

Alkylation

pairs with cytosine

pairs with thymine

Resulting DNA change:

mutant

C. Alkylation of guanine

Thymine

Adenine

Sugar

change to base analog

BrdU

Guanine

D. Base analog

Ultraviolet irradiation forms thymine dimers with convalent bonds, distorts DNA. Corrected by excision repair

Thymine

Thymine dimer

DNA sugar-phosphate backbone

E. UV-light-induced thymine dimers

# DNA Polymorphism

Genetic polymorphism is the existence of variants with respect to a gene locus (alleles), a chromosome structure (e.g., size of centromeric heterochromatin), a gene product (variants in enzymatic activity or binding affinity), or a phenotype. The term DNA polymorphism refers to a wide range of variations in nucleotide base composition, length of nucleotide repeats, or single nucleotide variants. DNA polymorphisms are important as genetic markers to identify and distinguish alleles at a gene locus and to determine their parental origin.

## A. Single nucleotide polymorphism (SNP)

These allelic variants differ in a single nucleotide at a specific position. At least one in a thousand DNA bases differs among individuals (1). The detection of SNPs does not require gel electrophoresis. This facilitates large-scale detection. A SNP can be visualized in a Southern blot as a restriction fragment length polymorphism (RFLP) if the difference in the two alleles corresponds to a difference in the recognition site of a restriction enzyme (see Southern blot, p. 62).

## B. Simple sequence length polymorphism (SSLP)

These allelic variants differ in the number of tandemly repeated short nucleotide sequences in noncoding DNA. Short tandem repeats (STRs) consist of units of 1, 2, 3, or 4 base pairs repeated from 3 to about 10 times. Typical short tandem repeats are CA repeats in the 5′ to 3′ strand, i.e., alternating CG and AT base pairs in the double strand. Each allele is defined by the number of CA repeats, e.g., 3 and 5, as shown (1). These are also called microsatellites. The size differences due to the number of repeats are determined by PCR. Variable number of tandem repeats (VNTR), also called minisatellites, consist of repeat units of 20–200 base pairs (2).

## C. Detection of SNP by oligonucleotide hybridization analysis

Oligonucleotides, short stretches of about 20 nucleotides with a complementary sequence to the single-stranded DNA to be examined, will hybridize completely only if perfectly matched. If there is a difference of even one base, such as due to an SNP, the resulting mismatch can be detected because the DNA hybrid is unstable and gives no signal.

## D. Detection of STRs by PCR

Short tandem repeats (STRs) can be detected by the polymerase chain reaction (PCR). The allelic regions of a stretch of DNA are amplified; the resulting DNA fragments of different sizes are subjected to electrophoresis; and their sizes are determined.

## E. CEPH families

An important step in gene identification is the analysis of large families by linkage analysis of polymorphic marker loci on a specific chromosomal region near a locus of interest. Large families are of particular value. DNA from such families has been collected by the Centre pour l'Étude du Polymorphisme Humain (CEPH) in Paris, now called the Centre Jean Dausset, after the founder. Immortalized cell lines are stored from each family. A CEPH family consists of four grandparents, the two parents, and eight children. If four alleles are present at a given locus they are designated A, B, C, and D. Starting with the grandparents, the inheritance of each allele through the parents to the grandchildren can be traced (shown here as a schematic pattern in a Southern blot). Of the four grandparents shown, three are heterozygous (AB, CD, BC) and one is homozygous (CC). Since the parents are heterozygous for different alleles (AD the father and BC the mother), all eight children are heterozygous (BD, AB, AC, or CD).

## References


Brown, T.A.: Genomes. Bios Scientific Publ., Oxford, 1999.

Collins, F.S., Guyer, M.S., Chakravarti, A.: Variations on a theme: cataloguing human DNA sequence variation. Science **282**:682–689, 1998.

Deloukas, P., Schuler, G., Gyapay, G., et al.: A physical map of 30,000 human genes. Science **282**:744–746, 1998.

Lewin, B.: Genes VII. Oxford Univ. Press, Oxford, 2000.

Strachan, T., Read, A.P.: Human Molecular Genetics. 2nd ed. Bios Scientific Publishers, Oxford, 1999.

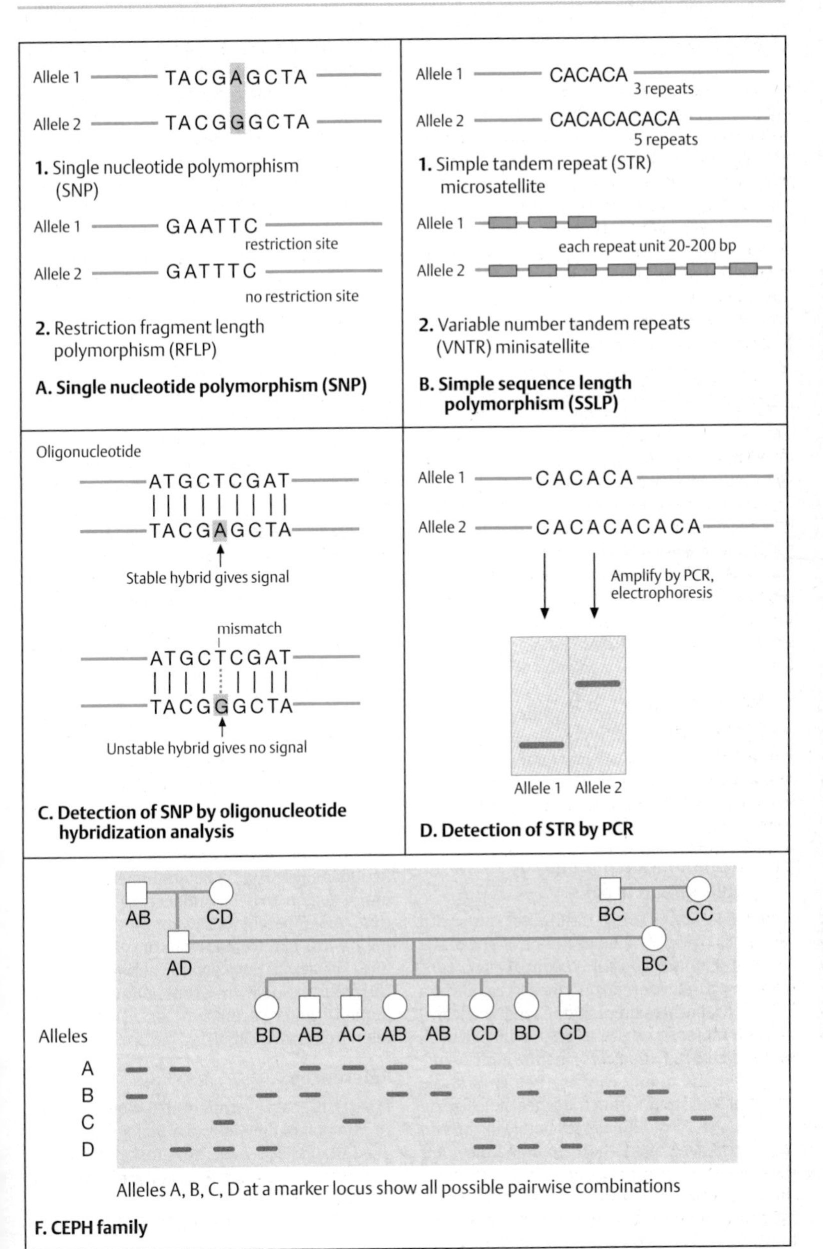


A. Single nucleotide polymorphism (SNP)

B. Simple sequence length polymorphism (SSLP)

C. Detection of SNP by oligonucleotide hybridization analysis

D. Detection of STR by PCR

F. CEPH family

# Recombination

Recombination lends the genome flexibility. Without genetic recombination, the genes on each individual chromosome would remain fixed in their particular position. Changes could occur by mutation only, which would be hazardous. Recombination provides the means to achieve extensive restructuring, eliminate unfavorable mutation, maintain and spread favorable mutations, and endow each individual with a unique set of genetic information. This greatly enhances the evolutionary potential of the genome.

Recombination must occur between precisely corresponding sequences (homologous recombination) to ensure that not one base pair is lost or added. The newly combined (recombined) stretches of DNA must retain their original structure in order to function properly. Two types of recombination can be distinguished: (1) generalized or homologous recombination, which in eukaryotes occurs at meiosis (see p. 116) and (2) site-specific recombination. A third process, transposition, utilizes recombination to insert one DNA sequence into another without regard to sequence homology. Here we consider homologous recombination, a complex biochemical reaction between two duplexes of DNA. The necessary enzymes, which can involve any pair of homologous sequences, are not considered. Two general models can be distinguished, recombination initiated from a single-strand DNA break and recombination initiated from a double-strand break.

## A. Recombination initiated by single-strand breaks

This model assumes that the process starts with breaks at corresponding positions of one of the strands of homologous DNA (same sequences of different parental origin) (1). A nick is made by a single-strand-breaking enzyme (endonuclease) in each molecule at the corresponding site (2), but see below. This allows the free ends of one nicked strand to join with the free ends of the other nicked strand, from the other molecule, to form single-strand exchanges between the two duplex molecules at the recombination joint (3). The recombination joint moves along the duplex (branch migration) (4). This is an important feature because it ensures that sufficient distance for the second nick is present in each of the other strands (5). After the two other strands have joined and gaps have been sealed (6), a reciprocal recombinant molecule is generated (7). Recombination involving DNA duplexes requires topological changes, i.e., either the molecules must be free to rotate or the restraint must be relieved in some other way.

This model has an unresolved difficulty: How is it assured that the single-strand nicks shown in step 2 occur at precisely the same position in the two double helix DNA molecules?

## B. Recombination initiated by double-strand breaks

The current model for recombination is based on initial double-strand breaks in one of the two homologous DNA molecules (1). Both strands are cleaved by an endonuclease, and the break is enlarged to a gap by an exonuclease that removes the new 5′ ends of the strands at the break and leaves 3′ single-stranded ends (2). One free 3′ end recombines with a homologous strand of the other molecule (3). This generates a D loop consisting of a displaced strand from the "donor" duplex. The D loop is extended by repair synthesis until the entire gap of the recipient molecule is closed (4). This displaced strand anneals to the single-stranded complementary homologous sequences of the recipient strand and closes the gap (5). DNA repair synthesis from the other 3′ end closes the remaining gap (6). The integrity of the two molecules is restored by two rounds of single-strand repair synthesis. In contrast to the single-strand exchange model, the double-strand breaks result in heteroduplex DNA in the entire region that has undergone recombination. An apparent disadvantage is the temporary loss of information in the gaps after the initial cleavage. However, the ability to retrieve this information by resynthesis from the other duplex avoids permanent loss. (Figures redrawn from Lewin, 2000).

## References

Alberts, B. et al.: Essential Cell Biology. An Introduction to the Molecular Biology of the Cell. Garland Publishing, New York, 1998.

Brown, T.A.: Genomes. Bios Scientific Publ., Oxford, 1999.

Lewin, B.: Genes VII. Oxford Univ. Press, Oxford, 2000.

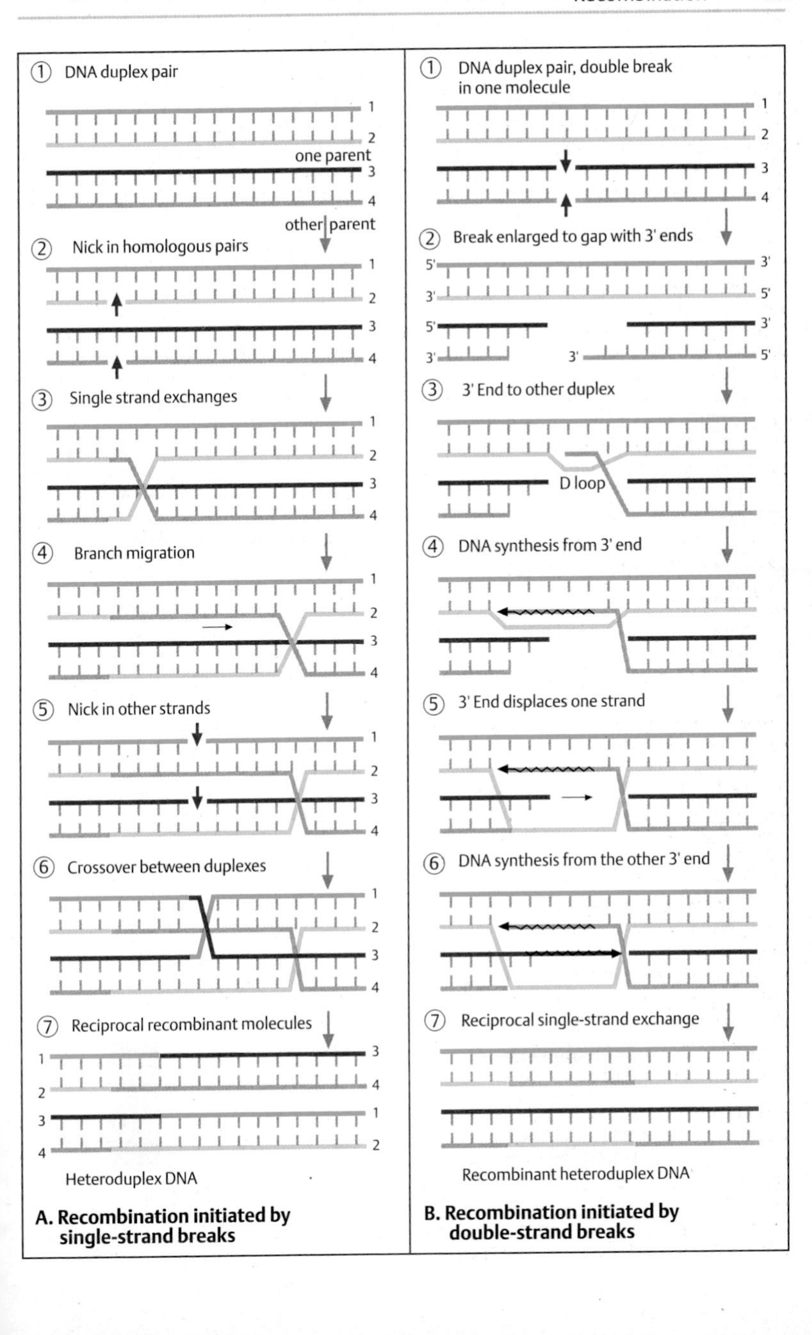


**A. Recombination initiated by single-strand breaks**

**B. Recombination initiated by double-strand breaks**

## Transposition

Aside from homologous recombination, the overall stability of the genome is interrupted by mobile sequences called transposable elements or transposons. There are different classes of distinct DNA sequences that are able to transport themselves to other locations within the genome. This process utilizes recombination but does not result in an exchange. Rather, a transposon moves directly from one site of the genome to another without an intermediary such as plasmid or phage DNA (see section on prokaryotes). This results in rearrangements that create new sequences and change the functions of target sequences. Transposons may be a major source of evolutionary changes in the genome. In some cases they cause disease when inserted into a functioning gene. Three examples are presented below: insertion sequences (IS), transposons (Tn), and retroelements transposing via an RNA intermediate.

### A. Insertion sequences (IS) and transposons (Tn)

A characteristic feature of IS transposition is the presence of a pair of short direct repeats of target DNA at either end. The IS itself carries inverted repeats of about 9–13 bp at both ends and depending on the particular class consists of about 750–1500 bp, which contain a single long coding region for transposase (the enzyme responsible for transposition of mobile sequences). Target selection is either random or at particular sites. The presence of inverted terminal repeats and the short direct repeats of host DNA result in a characteristic structure (1). Transposons carry in addition a central region with genetic markers unrelated to transposition, e.g., antibiotic resistance (2). They are flanked either by direct repeats (same direction) or by inverted repeats (opposite direction, 3).

### B. Replicative and nonreplicative transposition

With replicative transposition (1) the original transposon remains in place and creates a new copy of itself, which inserts into a recipient site elsewhere. Thus, this mechanism leads to an increase in the number of copies of the transposon in the genome. This type involves two enzymatic activities: a transposase acting on the ends of the original transposon and resolvase acting on the duplicated copies.

In nonreplicative transposition (2) the transposing element itself moves as a physical entity directly to another site. The donor site is either repaired (in eukaryotes) or may be destroyed (in bacteria) if more than one copy of the chromosome is present.

### C. Transposition of retroelements

Retrotransposition requires synthesis of an RNA copy of the inserted retroelement. Retroviruses including the human immunodeficiency virus and RNA tumor viruses are important retroelements (see p. 100 and p. 314). The first step in retrotransposition is the synthesis of an RNA copy of the inserted retroelement, followed by reverse transcription up to the polyadenylation sequence in the 3′ long terminal repeat (3′ LTR). Three important classes of mammalian transposons that undergo or have undergone retrotransposition through an RNA intermediary are shown. Endogenous retroviruses (1) are sequences that resemble retroviruses but cannot infect new cells and are restricted to one genome. Nonviral retrotransposons (2) lack LTRs and usually other parts of retroviruses. Both types contain reverse transcriptase and are therefore capable of independent transposition. Processed pseudogenes (3) or retropseudogenes lack reverse transcriptase and cannot transpose independently. They contain two groups: low copy number of processed pseudogenes transcribed by RNA polymerase II and high copy number of mammalian SINE sequences, such as human *Alu* and the mouse B1 repeat families.

### References


Brown, T.A.: Genomes. Bios Scientific Publ., Oxford, 1999.

Lewin, B.: Genes VII. Oxford Univ. Press, Oxford, 2000.

Strachan, T., Read, A.P.: Human Molecular Genetics. 2nd ed. Bios Scientific Publishers, Oxford, 1996.

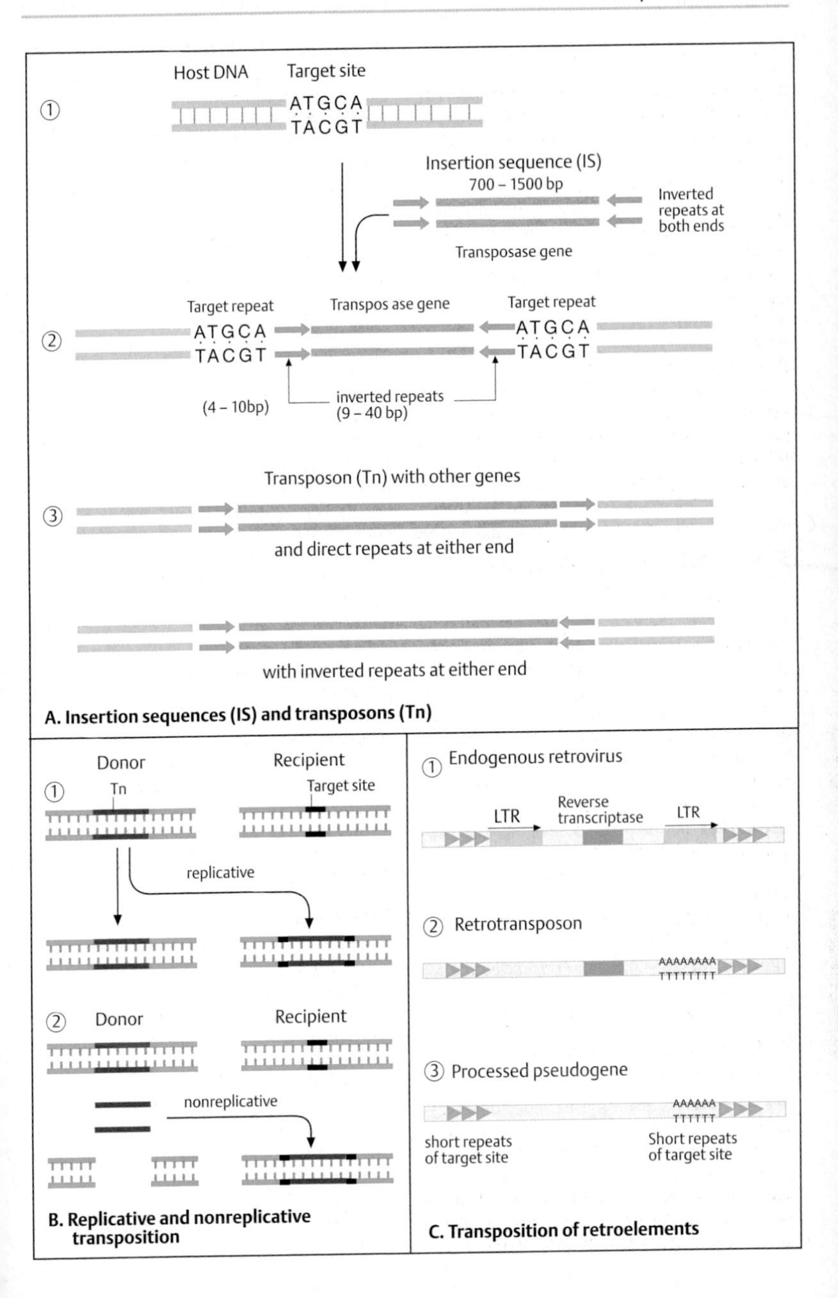


A. Insertion sequences (IS) and transposons (Tn)

B. Replicative and nonreplicative transposition

C. Transposition of retroelements

## Trinucleotide Repeat Expansion

The human genome contains tandem repeats of trinucleotides. Normally they occur in groups of 5–35 repeats. When their number exceeds a certain threshold and they occur in a gene or close to it, they cause diseases. Once the normal, variable length has expanded, the increased number of repeats tends to increase even further when passed through the germline or during mitosis. Thus, trinucleotide expansions form a class of unstable mutations, to date observed in humans only.

### A. Different types of trinucleotide repeats and their expansions

Trinucleotide repeats can be distinguished according to their localization with respect to a gene. Expansions are greater outside genes and more moderate within coding regions. In several severe neurological diseases, abnormally expanded CAG repeats are part of the gene. CAG repeats encode a series of glutamines (polyglutamine tracts). Within a normal number of repeats, which varies according to the gene involved, the gene functions normally (1). However, an expanded number of repeats leads to an abnormal gene product with altered function. Trinucleotide repeats also occur in noncoding regions of a gene (2). Fairly common types are CGG and GCC repeats. The increase in the number of these repeats can be drastic, up to 1000 or more repeats. The first stages of expansion usually do not lead to clinical signs of a disease, but they do predispose to increased expansion of the repeat in the offspring of a carrier (premutation).

### B. Unstable trinucleotide repeats in different diseases

Disorders due to expansion of trinucleotide repeats can be distinguished according to the type of trinucleotide repeat, i.e., the sequence of the three nucleotides, their location with respect to the gene involved, and their clinical features. All involve the central or the peripheral nervous system. Type I trinucleotide diseases are characterized by CAG trinucleotide expansion within the coding region of different genes. The triplet CAG codes for glutamine. About 20 CAG repeats occur normally in these genes, so that about 20 glutamines occur in the gene product. In the disease state the number of glutamines is greatly increased in the protein. Hence, they are collectively referred to as polyglutamine disorders.

Type II trinucleotide diseases are characterized by expansion of CTG, GAA, GCC, or CGG trinucleotides within a noncoding region of the gene involved, either at the 5′ end (GCC in fragile X syndrome type A, FRAXA), at the 3′ end (CGG in FRAXE; CTG in myotonic dystrophy), or in an intron (GAA in Friedreich ataxia). A brief review of these disorders is given on p. 394.

### C. Principle of laboratory diagnosis of unstable trinucleotide repeats

The laboratory diagnosis compares the sizes of the trinucleotide repeats in the two alleles of the gene examined. One can distinguish very large expansions of repeats outside coding sequences (50 to more than 1000 repeats) and moderate expansion within coding sequences (20 to 100–200). The figure shows 11 lanes, each representing one person: normal controls in lanes 1–3; confirmed patients in lanes 4–6; and a family with an affected father (lane 7), an affected son (lane 10), the unaffected mother (lane 11), and two unaffected children: a son (lane 8) and a daughter (lane 9). Size markers are shown at the left. Each lane represents a polyacrylamide gel and the $(CAG)_n$ repeat of the Huntington locus amplified by polymerase chain reaction shown as a band of defined size. Each person shows the two alleles. In the affected persons the band representing one allele lies above the threshold in the expanded region (in practice the bands are somewhat blurred because the exact repeat size varies in DNA from different cells).

### References


Strachan, T., Read, A.P.: Human Molecular Genetics. 2nd ed. Bios Scientific Publishers, Oxford, 1999.

Warren, S.T.: The expanding world of trinucleotide repeats. Science **271**: 1374–1375, 1996.

Rosenberg, R.N.: DNA-triplet repeats and neurologic disease. New Eng. J. Med. **335**: 1222–1224, 1996.

Zoghbi, H.Y.: Spinocerebellar ataxia and other disorders of trinucleotide repeats, pp. 913–920, In: Principles of Molecular Medicine, J.C. Jameson, ed. Humana Press, Totowa, NJ, 1998.

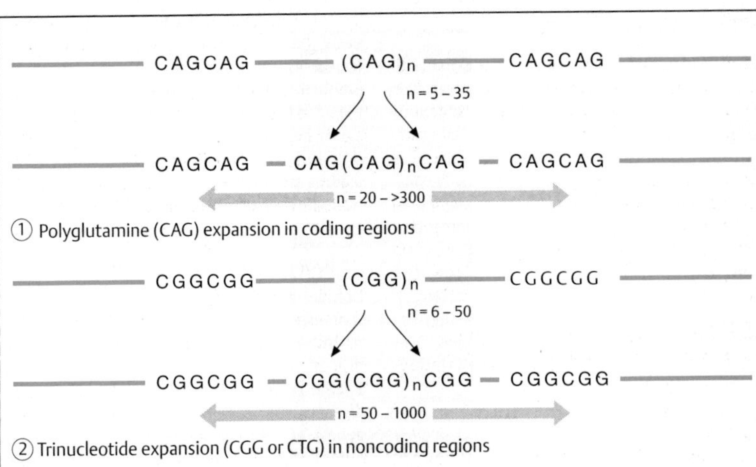


**A. Different types of trinucleotide repeat expression**

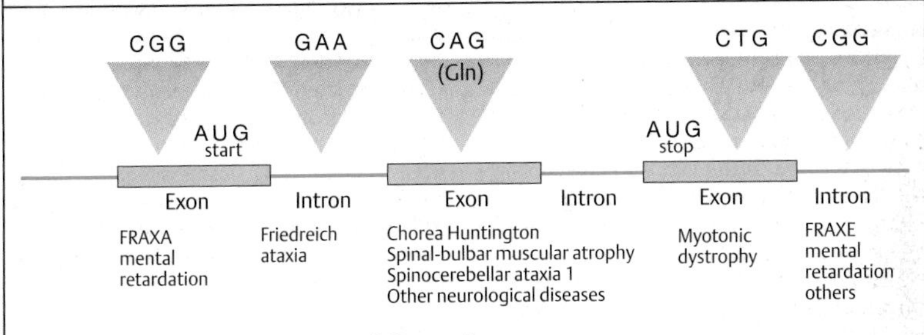


**B. Unstable trinucleotide repeats in different diseases**

Size markers | Three normal control persons | Three affected persons

1 2 3 4 5 6 7 8 9 10 11

Number of trinucleotide repeats: 90, 75, 60, 45, 30, 15

Expanded allele

Normal

**C. Principle of laboratory diagnosis of unstable trinucleotide repeats leading to expansion**

## DNA Repair

Life would not be possible without the ability to repair damaged DNA. Since replication errors, including mismatch, and harmful exogenous factors are everyday problems for a living organism, a broad repertoire of repair genes has evolved in prokaryotes and eukaryotes. The following types of DNA repair can be distinguished by their basic mechanisms: (1) excision repair to remove a damaged DNA site, such as a strand with a thymine dimer; (2) mismatch repair to correct errors of replication by excising a stretch of single-stranded DNA containing the wrong base; (3) repair of UV-damaged DNA during replication; and (4) transcription-coupled repair in active genes.

### A. Excision repair

The damaged strand of DNA is distorted and can be recognized by a set of three proteins, the UvrA, UvrB, and UvrC endonucleases in prokaryotes and XPA, XPB, and XPC in human cells. This DNA strand is cleaved on both sides of the damage by an exonuclease protein complex, and a stretch of about 12 or 13 nucleotides in prokaryotes and 27 to 29 nucleotides in eukaryotes is removed. DNA repair synthesis restores the missing stretch and a DNA ligase closes the gap.

### B. Mismatch repair

Mismatch repair corrects errors of replication. However, the newly synthesized DNA strand containing the wrong base must be distinguished from the parent strand, and the site of a mismatch identified. The former is based on a difference in methylation in prokaryotes. The daughter strand is undermethylated at this stage. *E. coli* has three mismatch repair systems: long patch, short patch, and very short patch. The long patch system can replace 1 kb DNA and more. It requires three repair proteins, MutH, MutL, and MutS, which have the human homologues hMSH1, hMLH1, and hMSH2. Mutations in their respective genes lead to cancer due to defective mismatch repair.

### C. Replication repair of UV-damaged DNA

DNA damage interferes with replication, especially in the leading strand. Large stretches remain unreplicated beyond the damaged site (in the 3′ direction of the new strand) unless swiftly repaired. The lagging strand is not affected as much because Okazaki fragments (about 100 nucleotides in length) of newly synthesized DNA are also formed beyond the damaged site. This leads to an asymmetric replication fork and single-stranded regions of the leading strand. Aside from repair by recombination, the damaged site can be bypassed.

### D. Double-strand repair by homologous recombination

Double-strand damage is a common consequence of $\gamma$ radiation. An important human pathway for mediating repair requires three proteins, encoded by the genes *ATM*, *BRCA1*, and *BRCA2*. Their names are derived from important diseases that result from mutations in these genes: ataxia telangiectasia (see p. 334) and hereditary predisposition to breast cancer (BRCA1 and BRCA2, see p. 328. ATM, a member of a protein kinase family, is activated in response to DNA damage (1). Its active form phosphorylates BRCA1 at specific sites (2). Phosphorylated BRCA1 induces homologous recombination in cooperation with BRCA2 and mRAD5, the mammalian homologue of *E. coli* RecA repair protein (3). This is required for efficient DNA double-break repair. Phosphorylated BRCA1 may also be involved in transcription and transcription-coupled DNA repair (4). (Figure redrawn from Ventikaraman, 1999).

### References


Buermeyer, A.B. et al.: Mammalian DNA mismatch repair. Ann. Rev. Genet. **33**:533–564, 1999.

Cleaver, J.E.: Stopping DNA replication in its tracks. Science **285**:212–213, 1999.

Cortez D., et al.: Requirement of ATM-dependent phosphorylation of Brca1 in the DNA damage response to double-strand breaks. Science **286**:1162–1166, 1999.

Masutani, C., et al.: The XPV (xeroderma pigmentosum variant) gene encodes human DNA polymerase. Nature **399**:700–704, 1999.

Sancar, A.: Excision repair invades the territory of mismatch repair. Nature Genet. **21**:247–249, 1999.

Ventikaraman A.R.: Breast cancer genes and DNA repair. Science **286**:1100–1101, 1999.

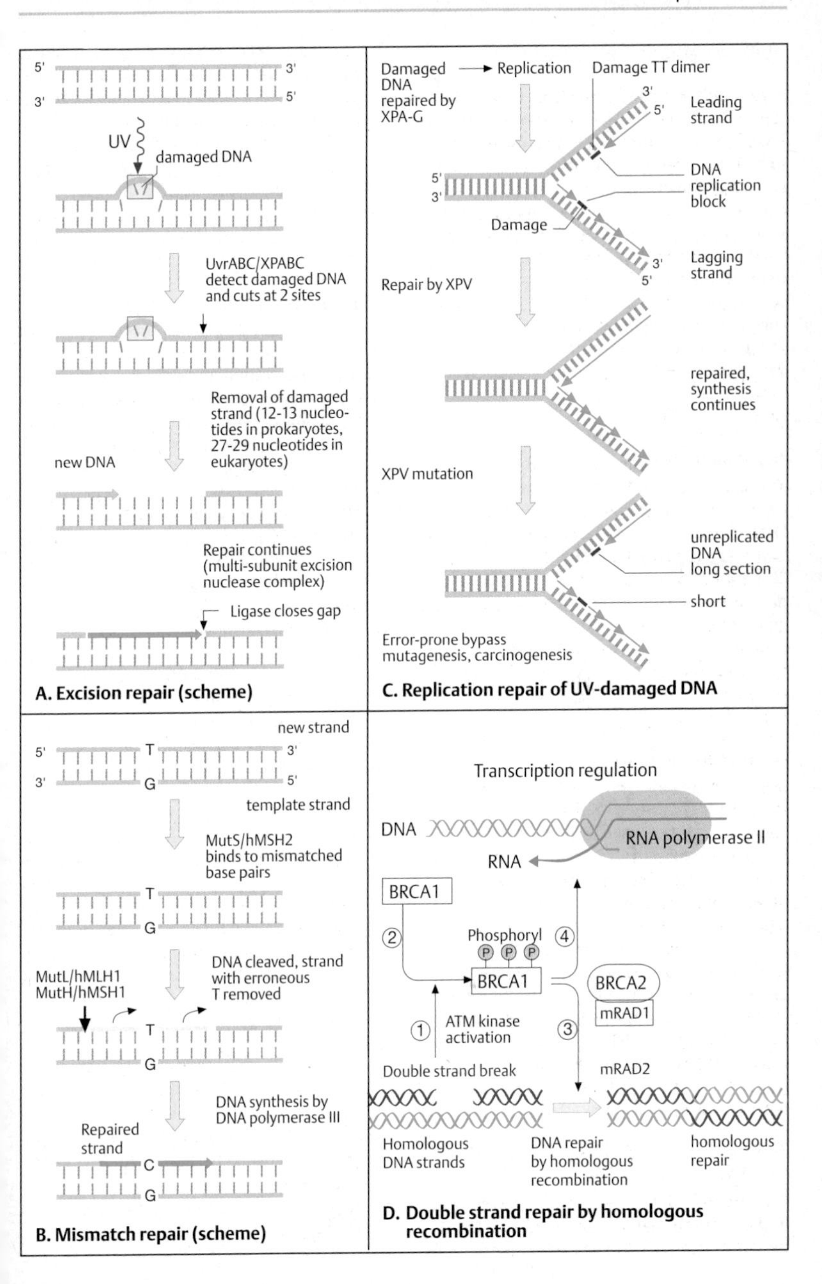


A. Excision repair (scheme)

C. Replication repair of UV-damaged DNA

B. Mismatch repair (scheme)

D. Double strand repair by homologous recombination

## Xeroderma Pigmentosum

Xeroderma pigmentosum (XP) is a heterogeneous group of genetically determined skin disorders due to unusual sensitivity to ultraviolet light. They are manifested by dryness and pigmentation of the exposed regions of skin (xeroderma pigmentosum = "dry, pigmented skin"). The exposed areas of skin also show a tendency to develop tumors. The causes are different genetic defects of DNA repair.
Repair involves mechanisms similar to those involved in transcription and replication. The necessary enzymes are encoded by at least a dozen genes, which are highly conserved in bacteria, yeast, and mammals.

### A. Clinical phenotype

The skin changes are limited to UV-exposed areas (1 and 2). Unexposed areas show no changes. Thus it is important to protect patients from UV light. An especially important feature is the tendency for multiple skin tumors to develop in the exposed areas (3). These may even occur in childhood or early adolescence. The types of tumors are the same as those occurring in healthy individuals after prolonged UV exposure.

### B. Cellular phenotype

The UV sensitivity of cells can be demonstrated in vitro. When cultured fibroblasts from the skin of patients are exposed to UV light, the cells show a distinct dose-dependent decrease in survival rate compared with normal cells (1). Different degrees of UV sensitivity can be demonstrated. The short segment of new DNA normally formed during excision repair can be demonstrated by culturing cells in the presence of [$^3$H]thymidine and exposing them to UV light. The DNA synthesis induced for DNA repair can be made visible in autoradiographs. Since [$^3$H]thymidine is incorporated during DNA repair, these bases are visible as small dots caused by the isotope on the film (2). In contrast, xeroderma (XP) cells show markedly decreased or almost absent repair synthesis. (Photograph of Bootsma & Hoeijmakers, 1999).

### C. Genetic complementation in cell hybrids

If skin cells (fibroblasts) from normal persons and from patients (XP) are fused (cell hybrids) in culture and exposed to UV light, the cellular XP phenotype will be corrected (1). Normal DNA repair occurs. Also, hybrid cells from two different forms of XP show normal DNA synthesis (2) because cells with different repair defects correct each other (genetic complementation). However, if the mutant cells have the same defect (3), they are not be able to correct each other (4) because they belong to the same complementation group. At present about ten complementation groups are known in xeroderma pigmentosum. They differ clinically in terms of severity and central nervous system involvement. Each complementation group is based on a mutation at a different gene locus. Several of these genes have been cloned and show homology with repair genes of other organisms, including yeast and bacteria.

### References


Berneburg, M. et al.: UV damage causes uncontrolled DNA breakage in cells from patients with combined features of XP-D and Cockayne syndrome. Embo J. **19**:1157–1166, 2000.

Bootsma, D.A., Hoeijmakers, J.H.J.: The genetic basis of xeroderma pigmentosum. Ann. Génét. **34**:143–150, 1991.

Cleaver, J.E., et al.: A summary of mutations in the UV-sensitive disorders: xeroderma pigmentosum, Cockayne syndrome, and trichothiodystrophy. Hum. Mutat. **14**:9–22, 1999.

Cleaver, J.E.: Common pathways for ultraviolet skin carcinogenesis in the repair and replication defective groups of xeroderma pigmentosum. J. Dermatol. Sci. **23**:1–11, 2000.

de Boer, J., Hoeijmakers J.H.: Nucleotide excision repair and human syndromes. Carcinogenesis **21**:453–460, 2000.

Hanawalt, P.C.: Transcription-coupled repair and human diseases. Science **266**:1957–1958, 1994.

Sancar, A.: Mechanisms of DNS excision repair. Science **266**:1954–1956, 1994.

Taylor, E.M., et al.: Xeroderma pigmentosum and trichothiodystrophy are associated with different mutations in the XPD (ERCC2). Proc. Natl. Acad. Sci. 94: 8658–8663, 1997.

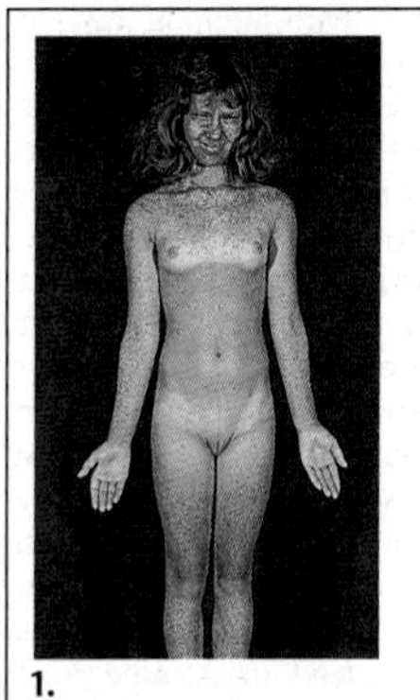
1.

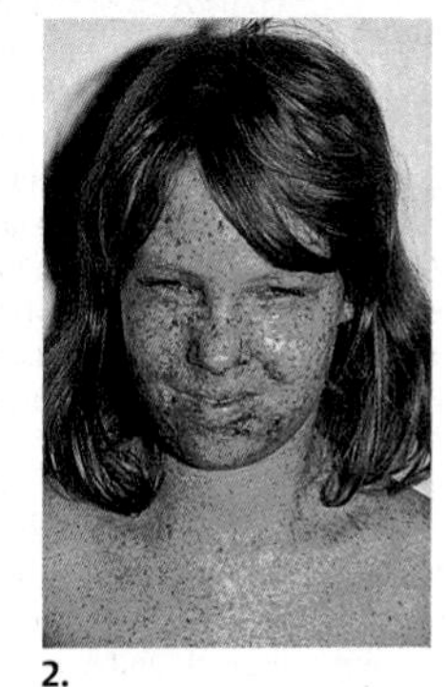
2.

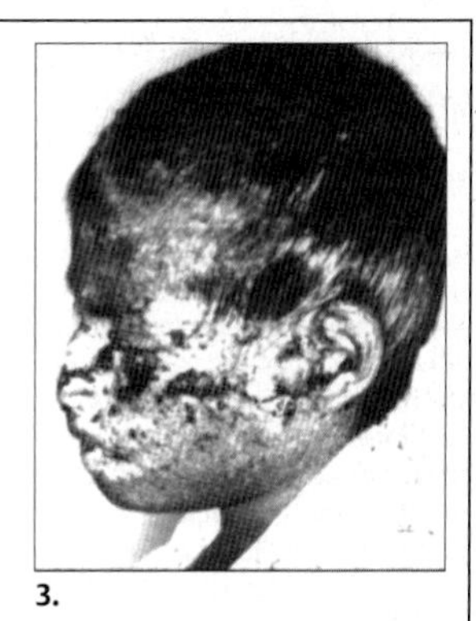
3.

**A. Clinical phenotype**

1.0

0.1

0.01

Surviving cells

Normal

Xp

Dose/UV Light

1. UV sensitivity

XP-A

XP-D

XP-D

Heterokaryon XP-A/XP-D

2. Complementation following fusion of XP-A and XP-D cells

**B. Cellular phenotype**

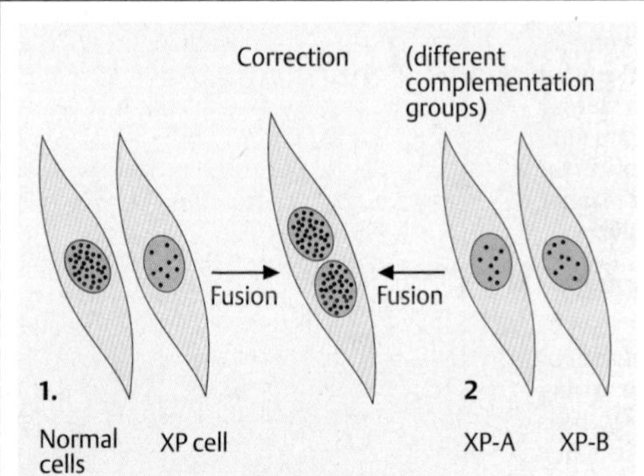


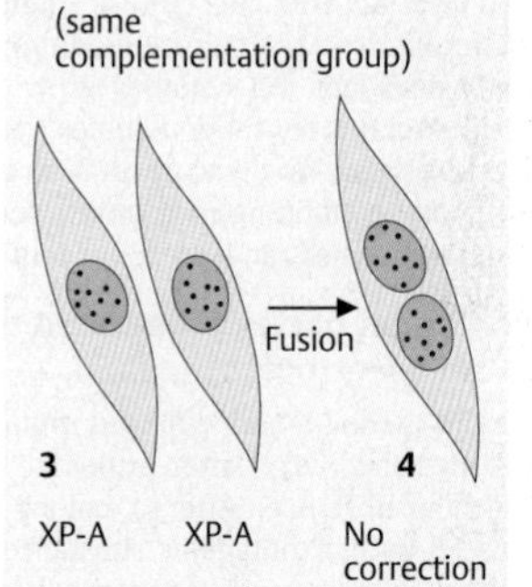


**C. Genetic complementation in cell hybrids**

# Prokaryotic Cells and Viruses

## *Isolation of Mutant Bacteria*

Important advances in genetics were made in the early 1950s through studies of bacteria. As prokaryotic organisms, bacteria have certain advantages over eukaryotic organisms because they are haploid and have an extremely short generation time. Mutant bacteria can be identified easily. The growth of some mutant bacteria depends on whether a certain substance is present in the medium (auxotrophism). Bacterial cultures are well suited for determining mutational events since an almost unlimited number of cells can be tested in a short time. Without great difficulty, it is possible to detect one mutant in $10^7$ colonies. Efficiency to this degree is not possible in the genetic analysis of eukaryotic organisms.

### A. Replica plating to recognize mutants

In 1952, Joshua and Esther Lederberg developed replica plating of bacterial cultures. With this method, individual colonies on an agar plate can be taken up with a stamp covered with velvet and placed onto other culture dishes with media of different compositions. Some mutant bacteria differ from nonmutants in their ability to grow. Here several colonies are shown in the Petri dish of the initial culture. Each of these colonies originated from a single cell. By means of replica plating, the colonies are transferred to two new cultures. One culture (right) contains an antibiotic in the culture medium; the other (left) does not. All colonies grow in normal medium, but only those colonies that are antibiotic resistant owing to a mutation grow in the antibiotic-containing medium. In this manner, mutant colonies can be readily identified.

### B. Mutant bacteria identified through an auxotrophic medium

Here it is shown how different mutants can be distinguished, e.g., after exposure to a mutagenic substance. After a colony has been treated with a mutagenic substance, it is first cultivated in normal nutrient medium. Mutants can then be identified by replica plating. The culture with the normal medium serves as the control. In one culture with minimal medium, from which a number of substances are absent, two colonies do not grow (auxotrophic mutants). Initially, it is known for which of the substances the colonies are auxotrophic. If a different amino acid is added to each of two cultures with minimal medium, e.g., threonine (Thr) to one and arginine (Arg) to the other, it can be observed that one of the mutant colonies grows in the threonine-containing minimal medium, but the other does not. The former colony is dependent on the presence of threonine ($Thr^-$), i.e., it is an auxotroph for threonine. The other culture with minimal medium had arginine added. Only here can the other of the two mutant colonies, an auxotroph for arginine ($Arg^-$), grow. After the mutant colonies requiring specific conditions for growth have been identified, they can be further characterized. This procedure is relatively simple and makes rapid identification of mutants possible. Many mutant bacteria have been defined by auxotrophism. The wild-type cells that do not have special additional growth requirements are called prototrophs (Figures adapted from Stent & Calendar, 1978).

### Reference

Lederberg, J.: Infectious history. Pathways of discovery. Science **288** : 287 – 293, 2000.

Stent, G.S., Calendar, R.: Molecular Genetics. An Introductory Narrative. 2nd ed. W.H. Freeman & Co., San Francisco, 1978.

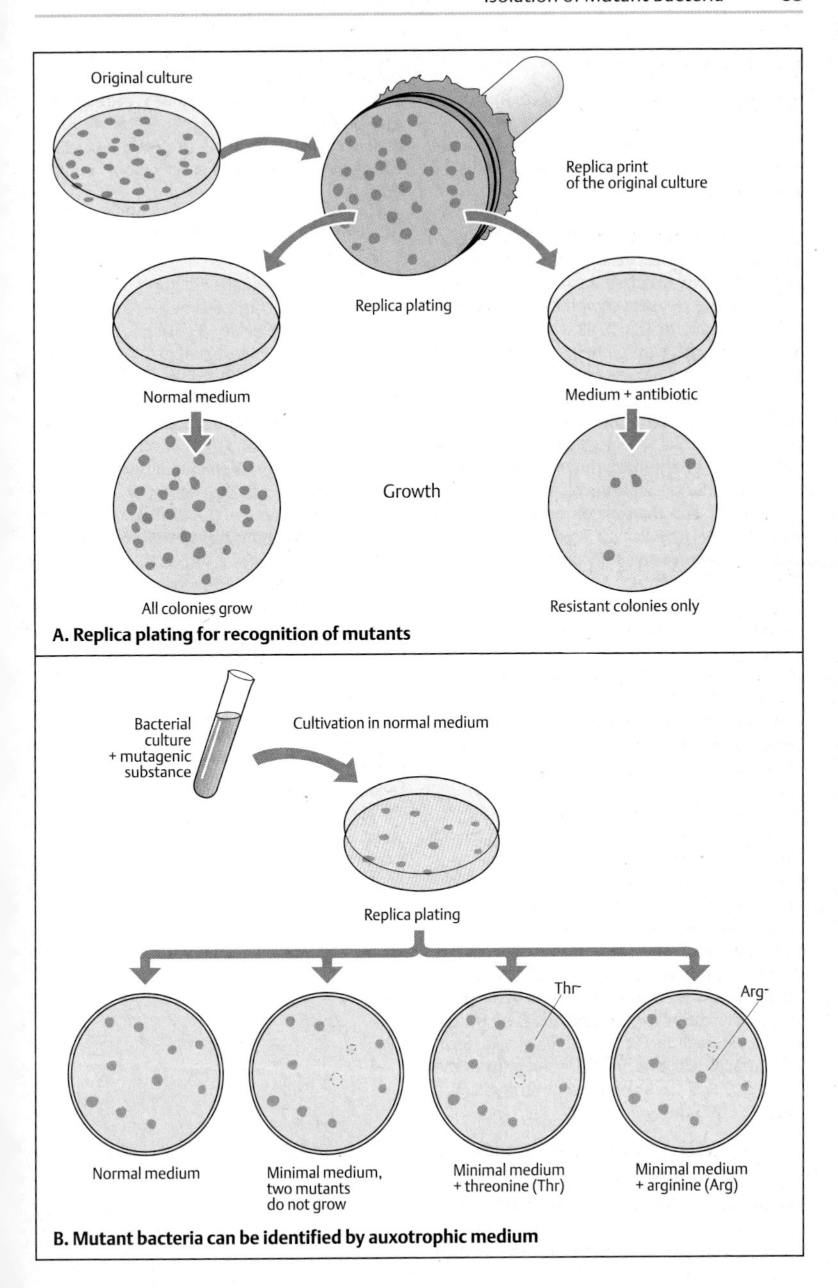


**A. Replica plating for recognition of mutants**

**B. Mutant bacteria can be identified by auxotrophic medium**

## Recombination in Bacteria

In 1946, J. Lederberg and E. L. Tatum first demonstrated that genetic information can be exchanged between different mutant bacterial strains. This corresponds to a type of sexuality and leads to genetic recombination.

### A. Genetic recombination in bacteria

In their classic experiment, Lederberg and Tatum used two different auxotrophic bacterial strains. One (A) was auxotrophic for methionine ($Met^-$) and biotin ($Bio^-$). This strain required methionine and biotin, but not threonine and leucine ($Thr^+$, $Leu^+$), to be added to the medium. The opposite was true for bacterial strain B, auxotrophic for threonine and leucine ($Thr^-$, $Leu^-$), but prototrophic for methionine and biotin ($Met^+$, $Bio^+$). When the cultures were mixed together without the addition of any of these four amino acids and then plated on an agar plate with minimal medium, a few single colonies ($Met^+$, $Bio^+$, $Thr^+$, $Leu^+$) unexpectedly appeared. Although this occurred rarely (about 1 in $10^7$ plated cells), a few colonies with altered genetic properties usually appeared owing to the large number of plated bacteria. The interpretation: genetic recombination between strain A and strain B. The genetic properties of the parent cells complemented each other (genetic complementation). (Figure adapted from Stent & Calendar, 1978).

### B. Conjugation in bacteria

Later, the genetic exchange between bacteria (conjugation) was demonstrated by light microscopy. Conjugation occurs with bacteria possessing a gene that enables frequent recombination. Bacterial DNA transfer occurs in one direction only. "Male" chromosomal material is introduced into a "female" cell. The so-called male and female cells of *E. coli* differ in the presence of a fertility factor (F). When $F^+$ and $F^-$ cells are mixed together, conjugal pairs can form with attachment of a male ($F^+$) sex pilus to the surface of an $F^-$ cell. (Photograph from Science **257**:1037, 1992).

### C. Integration of the F factor into an Hfr⁻ chromosome

The F factor can be integrated into the bacterial chromosome by means of specific crossing-over. After the factor is integrated, the original bacterial chromosome with the sections a, b, and c contains additional genes, the F factor genes (e, d). Such a chromosome is called an Hfr chromosome (Hfr, high frequency of recombination) owing to its high rate of recombination with genes of other cells as a result of conjugation.

### D. Transfer of F DNA from an $F^+$ to an $F^-$ cell

Bacteria may contain the F factor (fertility) as an additional small chromosome, i.e., a small ring-shaped DNA molecule (F plasmid) of about 94000 base pairs (not shown to scale). This corresponds to about 1/40th of the total genetic information of a bacterial chromosome. It occurs once per cell and can be transferred to other bacterial cells. About a third of the $F^+$ DNA consists of transfer genes, including genes for the formation of sex pili. The transfer of the F factor begins after a strand of the DNA double helix is opened. One strand is transferred to the acceptor cell. There it is replicated, so that it becomes double-stranded. The DNA strand remaining in the donor cell is likewise restored to a double strand by replication. Thus, DNA synthesis occurs in both the donor and the acceptor cell. When all is concluded, the acceptor cell is also an $F^+$ cell. (Figures in C and D adapted from Watson et al., 1987).

### References

Lederberg, J., Tatum, E.L.: Gene recombination in *Escherichia coli*. Nature **158**:558, 1946.

Stent, G.S., Calendar, R.: Molecular Genetics. An Introductory Narrative. 2nd ed. W.H. Freeman & Co., San Francisco, 1978.

Watson, J.D. et al.: Molecular Biology of the Gene. 3rd ed. Benjamin/Cummings Publishing Co., Menlo Park, California, 1987.

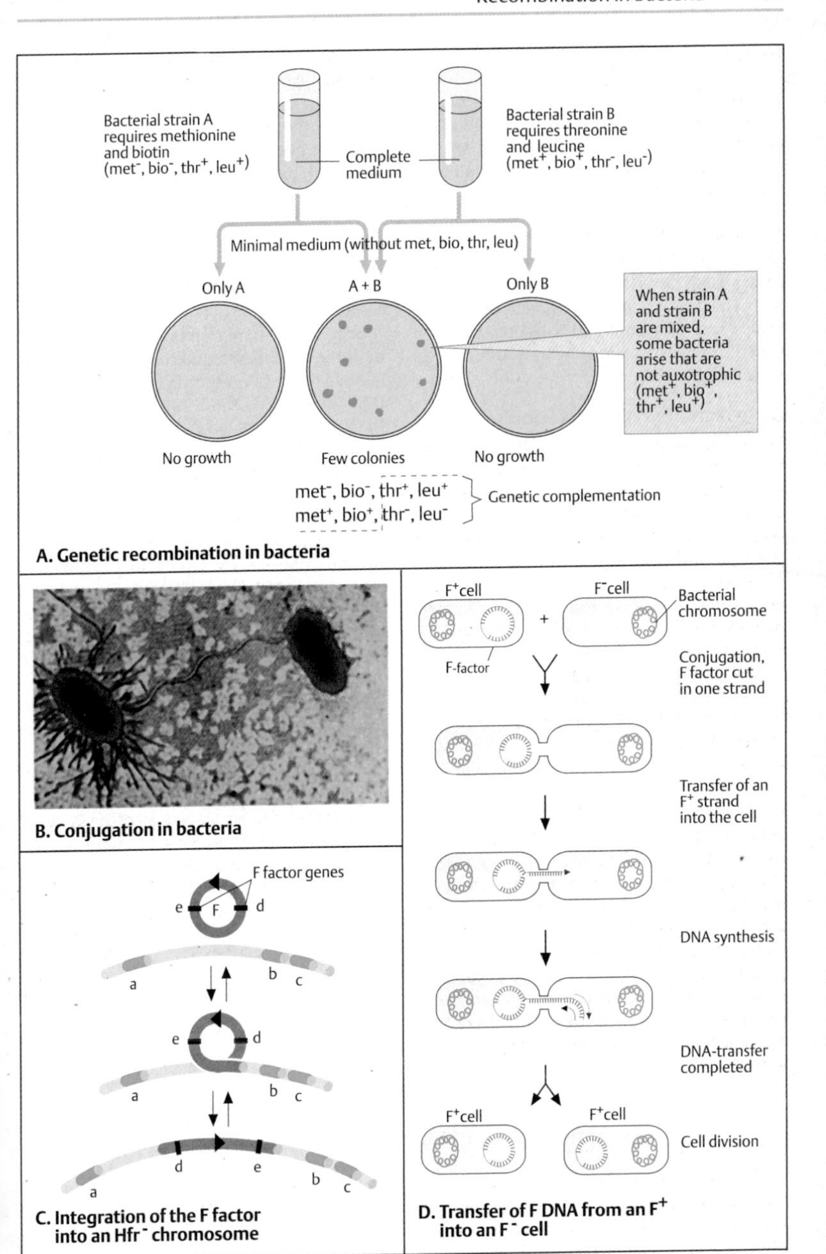


**A. Genetic recombination in bacteria**

**B. Conjugation in bacteria**

**C. Integration of the F factor into an Hfr⁻ chromosome**

**D. Transfer of F DNA from an F⁺ into an F⁻ cell**

# Bacteriophages

The discovery of bacterial viruses (bacteriophages or phages) in 1941 opened a new era in the study of the genetics of prokaryotic organisms. Although they were disappointing in the original hope that they could be used to fight bacterial infections, phages served during the 1950s as vehicles for genetic analysis of bacteria. Unlike viruses that infect plant or animal cells, phages can relatively easily be analyzed in their host cells. Names associated with phage analysis are Max Delbrück, Salvador Luria, and Alfred D. Hershey (the "phage group," see: Cairns et al., 1966).

## A. Attachment of a bacteriophage

Phages consist of DNA, a coat (coat protein) for protection, and a means of attachment (terminal filaments). Like other viruses, phages are basically nothing more than packaged DNA. One or more bacteriophages attach to a receptor on the surface of the outer cell membrane of a bacterium. The figure shows how an attached phage inserts its DNA into a bacterium. Numerous different phages are known, e.g., for *Escherichia coli* and *Salmonella* (phages T1, T2, P1, F1, lambda, T4, T7, phiX174 and others).

## B. Lytic and lysogenic cycles of a bacteriophage

Phages do not reproduce by cell division like bacteria, but by intracellular formation and assembly of the different components. This begins with the attachment of a phage particle to a specific receptor on the surface of a sensitive bacterium. Different phages use different receptors, thus giving rise to specificity of interaction (restriction). The invading phage DNA contains the information for production of coat proteins for new phages and factors for DNA replication and transcription. Translation is provided for by cell enzymes. The phage DNA and phage protein synthesized in the cell are assembled into new phage particles. Finally, the cell disintegrates (lysis) and hundreds of phage particles are released. With attachment of a new phage to a new cell, the procedure is repeated (lytic cycle).

Phage reproduction does not always occur after invasion of the cell. Occasionally, phage DNA is integrated into the bacterial chromosome and replicated with it (lysogenic cycle). Phage DNA that has been integrated into the bacterial chromosome is designated a prophage. Bacteria containing prophages are designated lysogenic bacteria; the corresponding phages are termed lysogenic phages. The change from a lysogenic to a lytic cycle is rare. It requires induction by external influences and complex genetic mechanisms.

## C. Insertion of a lambda phage into the bacterial chromosome by crossing-over

A phage can be inserted into a bacterial chromosome by different mechanisms. With the lambda phage ($\lambda$), insertion results from crossing-over between the *E. coli* chromosome and the lambda chromosome. First, the lambda chromosome forms a ring. Then it attaches to a homologous section of the bacterial chromosome. Both the bacterial and the lambda chromosome are opened by a break and attach to each other. Since the homologies between the two chromosomes are limited to very small regions, phage DNA is seldom integrated. The phage is released (and the lytic cycle is induced) by the reverse procedure. (Figures adapted from Watson et al., 1987).

## References

Cairns J., Stent G.S., Watson J.D., eds. Phage and the Origins of Molecular Biology. Cold Spring Harbor Laboratory Press, New York, 1966.

Watson et al., Molecular Biology of the Gene, 3rd ed. Benjamin/Cummings Publishing Co., Menlo Park, California, 1987.

The Bacteriophage Ecology Group at (http://www.phage.org/).

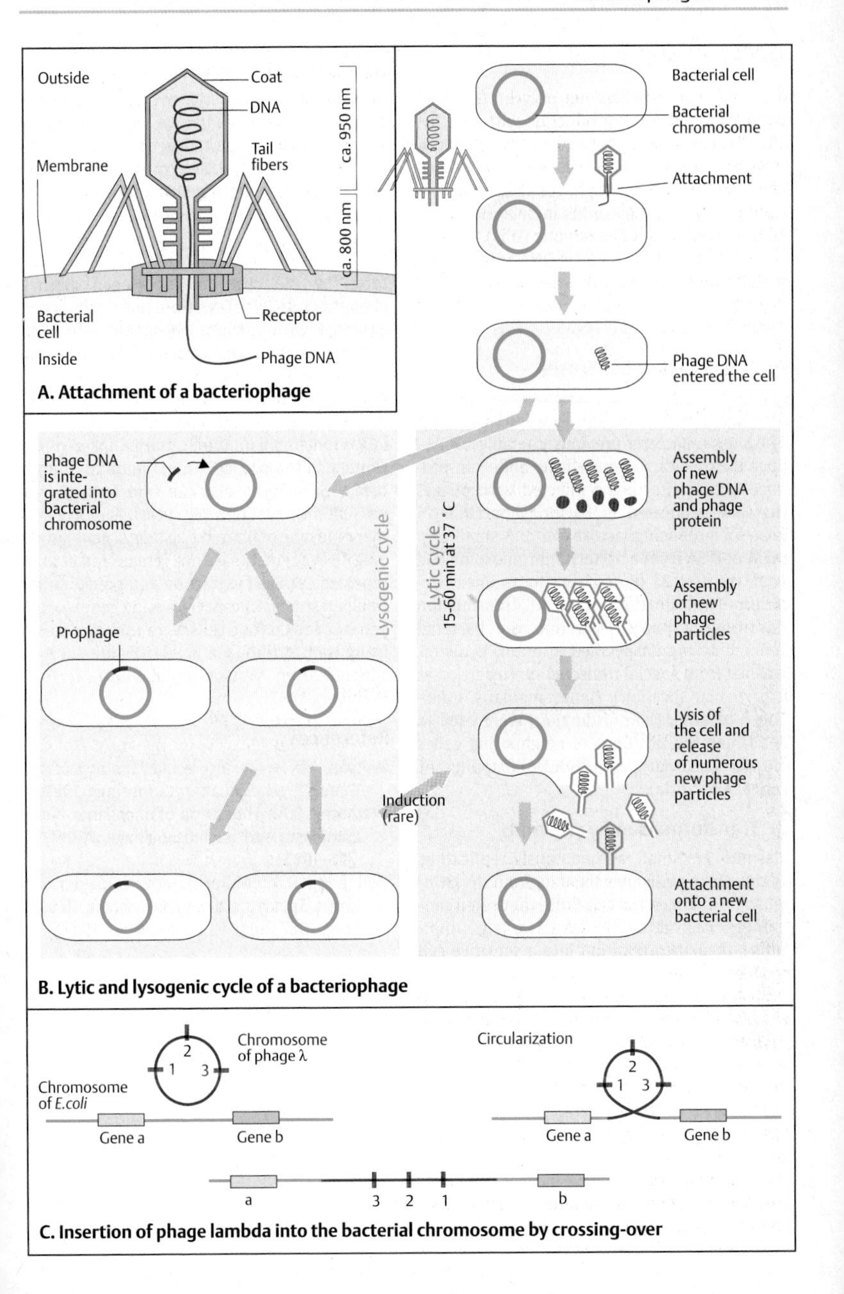


**A. Attachment of a bacteriophage**

**B. Lytic and lysogenic cycle of a bacteriophage**

**C. Insertion of phage lambda into the bacterial chromosome by crossing-over**

## DNA Transfer between Cells

Transfer of DNA occurs not only by fusion of gametes in sexual reproduction but also between other cells of prokaryotic and eukaryotic organisms (conjugation of bacteria, transduction between bacteriophages and bacteria, transformation by plasmids in bacteria, transfection in cultures of eukaryotic cells). Cells altered genetically by taking up DNA are said to be transformed. The term transformation is used in different contexts and refers to the result, not the mechanism.

### A. Transduction by viruses

In 1952, N. Zinder and J. Lederberg described a new type of recombination between two strains of bacteria. Bacteria previously unable to produce lactose (*lac*$^-$) acquired the ability to produce lactose after being infected with phages that had replicated in bacteria containing a gene for producing lactose (*lac*$^+$). A small segment of DNA from a bacterial chromosome had been transferred by a phage to another bacterium (transduction). General transduction (insertion of phage DNA into the bacterial genome at any unspecified location) is distinguished from special transduction (insertion at a particular location). Genes regularly transduced together (cotransduction) were used to determine the positions of neighboring genes on the bacterial chromosome (mapping of genes in bacteria).

### B. Transformation by plasmids

Plasmids are small, autonomously replicating, circular DNA molecules separate from the chromosome in a bacterial cell. Since they often contain genes for antibiotic resistance (e.g., ampicillin), their incorporation into a sensitive cell renders the cell resistant to the antibiotic (transformation). Only these bacteria can grow in culture medium containing the antibiotic (selective medium).

### C. Multiplication of a DNA segment in transformed bacteria

Plasmids are well suited as vectors for the transfer of DNA. A selective medium is used so that only those bacteria that have incorporated a recombinant plasmid containing the DNA to be investigated can grow.

### D. Transfection by DNA

The transfer of DNA between eukaryotic cells in culture (transfection) can be used to examine the transmission of certain genetic traits (transfection assay). Left, a DNA transfer experiment is shown in a culture of mouse fibroblasts; right, in a culture of human tumor cells (Weinberg, 1985, 1987).

The mouse fibroblast culture (see p. 122) is altered by the chemical carcinogen methylcholanthrene (left). DNA from these cells is precipitated with calcium phosphate, extracted, and then taken up by a normal culture (transfection). About 2 weeks later, cells appear that have lost contact inhibition (transformed cells). When these cells are injected into mice that lack a functional immune system (naked mice), tumors develop. DNA from cultured human tumor cells (right) also can transform normal cells after several transfer cycles. The DNA segment must be of limited size (e.g., a gene), since long DNA segments do not remain intact after repeated cycles of extraction and precipitation. Detailed studies of cancer-causing genes (oncogenes) in eukaryotic cells were first carried out using transfection (see p. 90). (Figures in A–C adapted from Watson et al. 1987, D from Weinberg 1987).

### References

Watson, J.D. et al.: Molecular Biology of the Gene, 4th ed., Benjamin/Cummings, 1987.

Weinberg, R.A.: The action of oncogenes in the cytoplasm and nucleus. Science. **230**:770–776, 1985.

Weinberg, R.A.: Molekulare Grundlagen von Krebs. Spektrum der Wissenschaft, Heidelberg, 1987 (engl. in Scient. Amer. 1987).

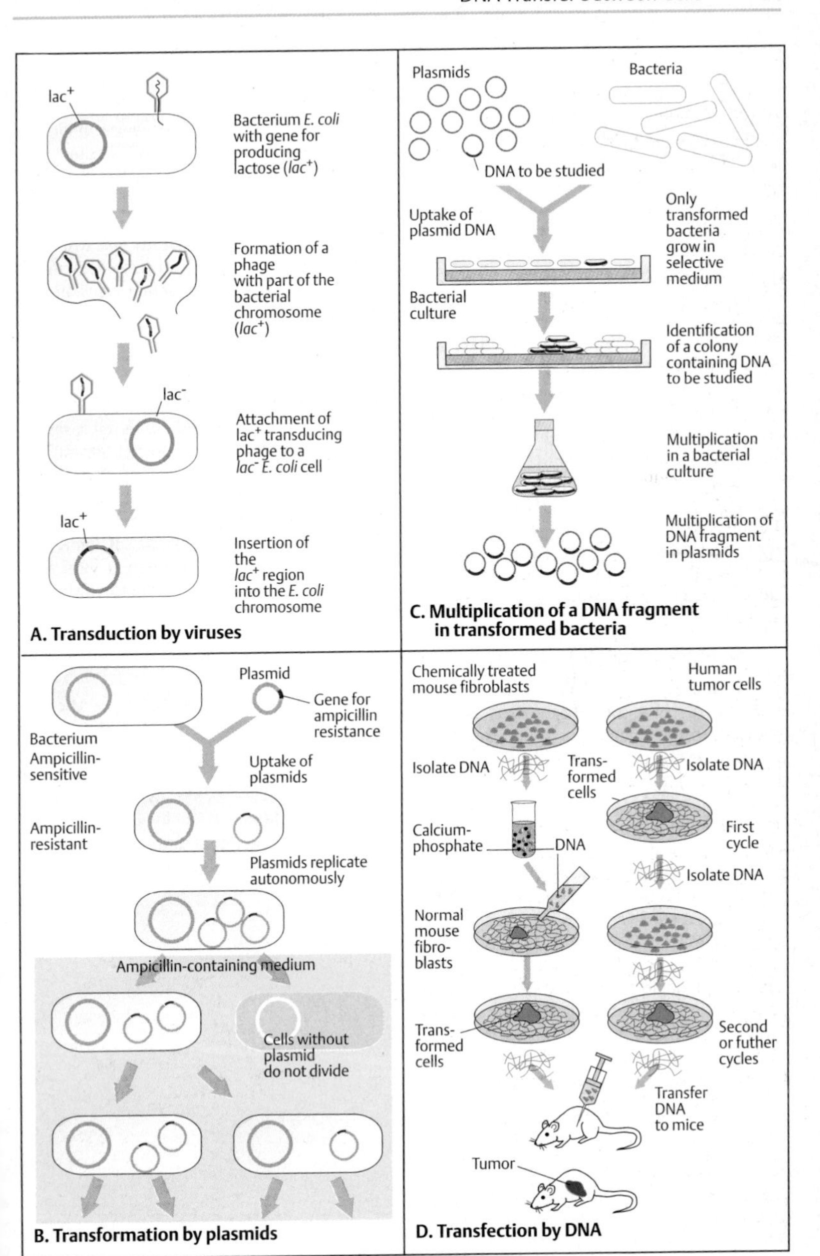
lac+
Bacterium E. coli with gene for producing lactose (lac+)
Formation of a phage with part of the bacterial chromosome (lac+)
lac-
Attachment of lac+ transducing phage to a lac- E. coli cell
lac+
Insertion of the lac+ region into the E. coli chromosome
A. Transduction by viruses
Plasmids
Bacteria
DNA to be studied
Uptake of plasmid DNA
Only transformed bacteria grow in selective medium
Bacterial culture
Identification of a colony containing DNA to be studied
Multiplication in a bacterial culture
Multiplication of DNA fragment in plasmids
C. Multiplication of a DNA fragment in transformed bacteria
Plasmid
Gene for ampicillin resistance
Bacterium
Ampicillin-sensitive
Uptake of plasmids
Ampicillin-resistant
Plasmids replicate autonomously
Ampicillin-containing medium
Cells without plasmid do not divide
B. Transformation by plasmids
Chemically treated mouse fibroblasts
Human tumor cells
Isolate DNA
Trans-formed cells
Isolate DNA
Calcium-phosphate
DNA
First cycle
Isolate DNA
Normal mouse fibro-blasts
Trans-formed cells
Second or futher cycles
Transfer DNA to mice
Tumor
D. Transfection by DNA

# Viruses

Viruses are important pathogens in plants and animals, including man. The complete infectious viral particle is called a virion. Its genome carries a limited amount of genetic information, and it can replicate only in host cells. From analysis of the structure and expression of viral genes, fundamental biological processes such as DNA replication, transcription regulation, mRNA modification (RNA splicing, RNA capping, RNA polyadenylation), reverse transcription of RNA to DNA, viral genome integration into eukaryotic DNA, tumor induction by viruses, and cell surface proteins have been recognized and elucidated. The extracellular form of a virus particle includes a protein coat (capsid), which encloses the genome of DNA or RNA. The capsids contain multiple units of one or a few different protein molecules coded for by the virus genome. Capsids usually have an almost spherical, icosahedral (20 plane surfaces), or occasionally a helical structure. Some viral capsids are surrounded by a lipid membrane envelope.

## A. Classification of viruses

Viruses can be classified on the basis of the structure of their viral coat, their type of genome, and their organ or tissue specificity. The genome of a virus may be enclosed simply in a virus-coded protein coat (capsid) or in the capsid plus an additional phospholipid membrane, which is of cellular origin. The genome of a virus may consist of single-stranded DNA (e.g., parvovirus), double-stranded DNA (e.g., papovavirus, adenovirus, herpesvirus, and poxvirus), single-stranded RNA (e.g., picornavirus, togavirus, myxovirus, rhabdovirus), or double-stranded RNA (e.g., reovirus). Viruses with genomes of single-stranded RNA are classified according to whether their genome is a positive (plus RNA) or negative (minus RNA) RNA strand. Only an RNA plus strand can serve as a template for translation (5′ to 3′ orientation).

## B. Replication and transcription of viruses

Since viral genomes differ, the mechanisms for replicating their genetic material also differ. Viruses must pack all their genetic information into a small genome; thus, one transcription unit (gene) of a viral genome is often used to produce several mRNAs by alternative splicing, and each mRNA codes for a different protein. In some RNA viruses, initially large precursor proteins are formed from the mRNA and subsequently split into several smaller functional proteins. An RNA plus strand can be used directly for protein synthesis. An RNA minus strand cannot be used directly; an RNA plus strand must be formed from the RNA minus strand by a transcriptase before translation is possible. RNA viruses contain a transcriptase to replicate their RNA genomes. RNA viruses in which DNA is formed as an intermediate step (retroviruses) contain a reverse transcriptase. This can form DNA from RNA. The DNA intermediate step in the replication of retroviruses becomes integrated into the host cell. Several RNA viruses have segmented genomes. They consist of individual pieces of RNA genome, each of which codes for one or more proteins (e.g., influenza virus). The exchange of individual pieces of RNA genome of different viral serotypes plays an important role in the formation of new viral strains (e.g., influenza strains). (Figure after Watson et al., 1987).

## References

Brock, T.D., Madigan, M.T.: Biology of Microorganisms. 6th ed. Prentice Hall, Englewood Cliffs, New Jersey, 1991.

Lederberg, J.: Infectious history. Pathways to discovery. Science **288** : 287–293, 2000.

Watson et al.: Molecular Biology of the Gene. 4th ed. The Benjamin/Cummings Publishing Co., Menlo Park California, 1987.

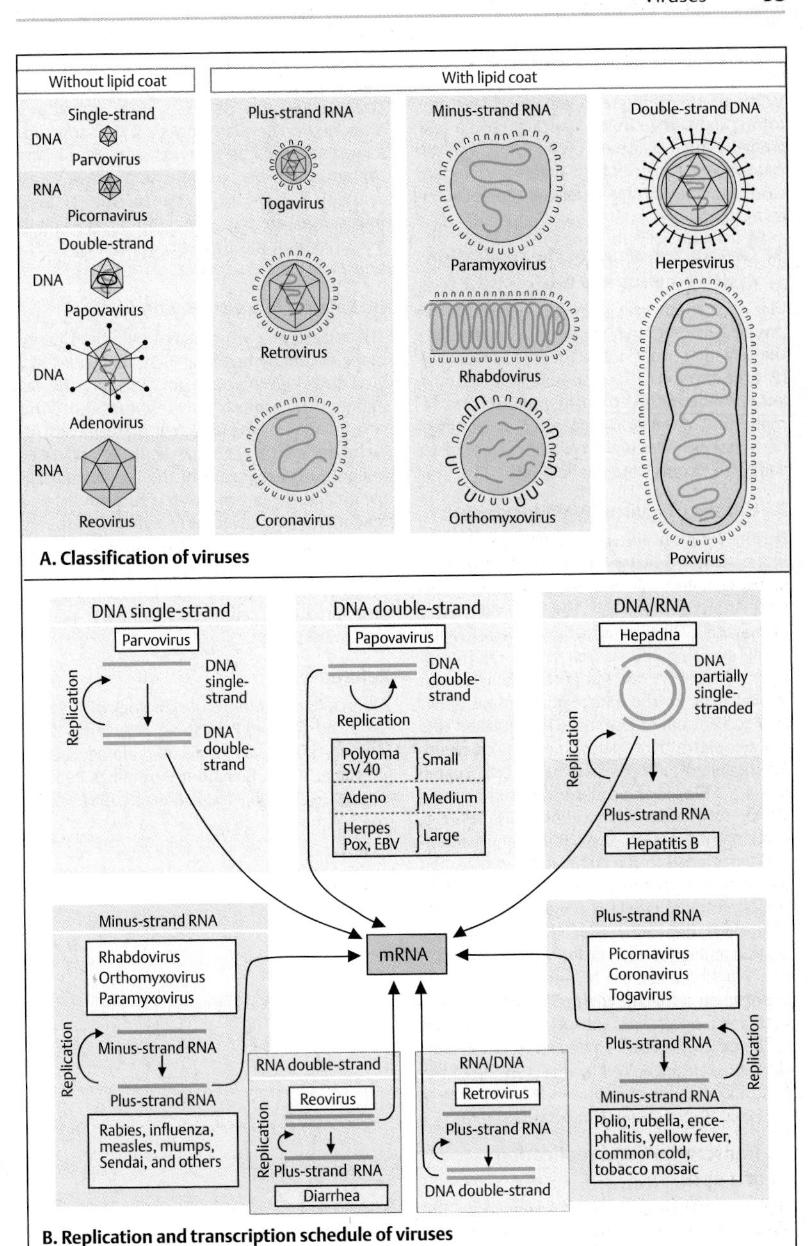


**A. Classification of viruses**

Without lipid coat

Single-strand
DNA — Parvovirus
RNA — Picornavirus

Double-strand
DNA — Papovavirus
DNA — Adenovirus
RNA — Reovirus

With lipid coat

Plus-strand RNA
Togavirus
Retrovirus
Coronavirus

Minus-strand RNA
Paramyxovirus
Rhabdovirus
Orthomyxovirus

Double-strand DNA
Herpesvirus
Poxvirus

**B. Replication and transcription schedule of viruses**

DNA single-strand
Parvovirus
Replication
DNA single-strand
DNA double-strand

DNA double-strand
Papovavirus
DNA double-strand
Replication

| Polyoma SV 40 | Small |
| Adeno | Medium |
| Herpes Pox, EBV | Large |

DNA/RNA
Hepadna
DNA partially single-stranded
Replication
Plus-strand RNA
Hepatitis B

Minus-strand RNA
Rhabdovirus
Orthomyxovirus
Paramyxovirus
Replication
Minus-strand RNA
Plus-strand RNA
Rabies, influenza, measles, mumps, Sendai, and others

RNA double-strand
Reovirus
Replication
Plus-strand RNA
Diarrhea

RNA/DNA
Retrovirus
Plus-strand RNA
DNA double-strand

mRNA

Plus-strand RNA
Picornavirus
Coronavirus
Togavirus
Plus-strand RNA
Replication
Minus-strand RNA
Polio, rubella, encephalitis, yellow fever, common cold, tobacco mosaic

## Replication Cycle of Viruses

With all their different genomic structures, forms, and sizes, viruses basically have a relatively simple replication cycle. While only the genome of a bacteriophage enters a bacterium, the complete virus (genome and capsid) enters a eukaryotic cell.

### A. General sequence of the replication cycle of a virus in a cell

The replication cycle of a virus consists of five principal consecutive steps: (1) entrance into the cell and release of the genome (uncoating), (2) transcription of the viral genes and (3) translation of the mRNAs to form viral proteins, (4) replication of the viral genome, (5) assembly of new viral particles in the cell and release of the complete virions from the host cell (6).

### B. Uptake of a virus by endocytosis

Besides fusion of the lipid membrane of membrane-enclosed viruses with the cell membrane of the host cell, the most frequent mechanism for a virion to enter a cell is by a special form of endocytosis. The virus attaches to the cell membrane by using cell surface structures (receptors), which serve other important functions for the cell, e.g., for the uptake of macromolecules (see p. 352). Like these, the virus is taken into the cytoplasm by a special mechanism, receptor-mediated endocytosis (coated pits, coated vesicles). Within the cell, the virus-containing vesicle fuses with other cellular vesicles (e.g., primary lysosomes). The viral coat is extensively degraded in the endocytotic vesicle, and the viral core (genome, associated with virus-coded proteins) is released into the cytoplasm or nucleus, depending on the viral type. Replication and expression of the viral genome follow. Whether a cell can be infected by a virion depends on a specific interaction between the virus and a cellular receptor. Some viruses, such as the paramyxoviruses (e.g., mumps and Sendai virus), enter the cell by direct fusion of the viral and cellular membranes, mediated by a viral coat glycoprotein (F or fusion protein).

### C. Transcription and replication of a virus

The first viral genes to be expressed after the virus has entered the cell are the early genes of the viral genome. Gene products of these early viral genes regulate transcription of the remaining viral genes and are involved in replicating the viral genome. Synthesis of the capsid proteins begins later (late genes), at the same time as genome replication, when new virions are formed from the genome and capsids (assembly). The virions (nucleocapsids = genome plus capsid) are then released from the cell by one of several mechanisms, depending on the type of virus.

### D. Release of a virus by budding

The release of a virus coated by a lipid membrane occurs by budding. First, molecules of a viral-coded glycoprotein are built into the cell membrane, to which the virus capsid or virus core (containing the viral genome) attaches. Attachment of the genome leads to increased budding of that region of the cell membrane. Eventually, the entire virion is surrounded by a lipid membrane envelope of cellular origin containing viral proteins and is released. Virions can be expelled from the cell continuously and in great numbers without the death of the virus-producing cell. (Figures from J. D. Watson et al., 1987).

### References

Alberts, B. et al.: Molecular Biology of the Cell. 4th ed. Garland Publishing, New York, 1994.

Watson, J.D. et al.: Molecular Biology of the Gene, 3rd ed. Benjamin/Cummings Publishing Co., Menlo Park, California, 1987.

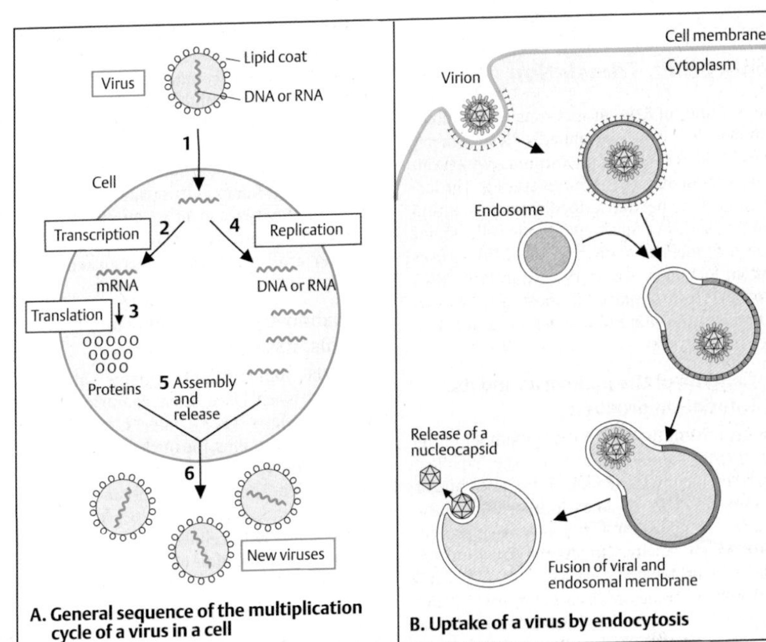


A. General sequence of the multiplication cycle of a virus in a cell

B. Uptake of a virus by endocytosis

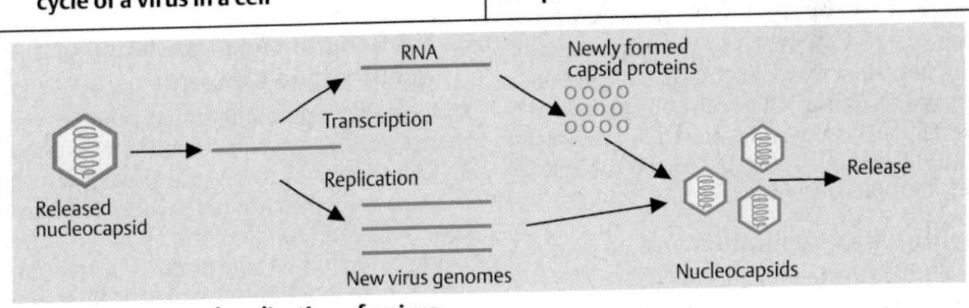


C. Transcription and replication of a virus

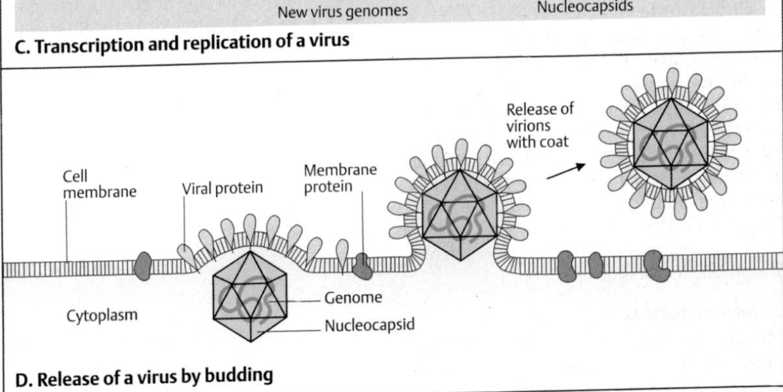


D. Release of a virus by budding

## RNA Viruses: Genome, Replication, Translation

The genomes of RNA viruses consist of double-stranded or single-stranded RNA. Single-stranded RNA can be of plus strands, which can be translated directly, or minus strands. The latter must first be transcribed into plus-strand RNA (sense RNA). Since eukaryotic cells do not have enzymes for copying RNA, RNA viruses contain enzymes necessary to transcribe RNA into DNA (transcriptase). In most of these RNA viruses, transcriptase molecules are included in the virus particle.

### A. Genome of the poliovirus and its translation products

The genome of the poliovirus, a member of the enterovirus family, is one of the smallest genomes. It consists of about 7400 base pairs of plus-strand RNA of known nucleotide base sequence. It codes for a large precursor protein from which smaller, functional proteins are formed by proteolytic cleavage (VPO, 1, 2, 3, and 4), about 60 copies of each being present per virion. VP1 is responsible for attachment to the cellular receptor, which is found only on the epithelial cells, fibroblasts, and nerve cells of primates. Thus, the poliovirus can infect only a very limited range of host cells. The poliovirus is one of the fastest-replicating animal viruses: Within 6–8 hours, an infected cell can release about 10000 new virions.

### B. Togavirus: replication and translation

Togaviruses (e.g., yellow fever, rubella, encephalitis viruses) have an RNA plus strand of about 12000 base pairs as genome. It is enclosed in a capsid and a lipid membrane. Only the replicase protein can be translated from the genomic RNA, since start codons (AUG) for initiating translation lie 3′ and are not recognized before replication. Thus, the capsid proteins are synthesized late after infection, i.e., after replication. This results in a differentiated regulation of viral gene expression.

### C. Influenzavirus

The genome of the influenzavirus consists of eight segments of minus-strand RNA. Each codes for at least one, and some for more than one protein. The lipid membrane contains two virus-coded glycoproteins: hemagglutinin (HA), which recognizes and binds to cell surface receptors, and neuraminidase (NA), a receptor-degrading enzyme. In addition, the virions contain a matrix protein (M) and a nucleocapsid protein (N). If a cell has become infected with two different influenza virus strains, a new type of influenza virus can arise by exchange of genome segments. The recombinant virus is either not at all or only slowly recognized by the immune system.

### D. Rhabdovirus (vesicular stomatitis virus, VSV)

The RNA minus-strand genome of rhabdoviruses is enclosed in a characteristically formed (bullet-shaped) outer membrane; it codes for five proteins, the nucleocapsid protein (N) within the virion, the matrix protein (M) between capsid and outer membrane, a transmembrane viral glycoprotein (G) responsible for interacting with cellular receptors, and two enzymes for replication and mRNA synthesis: protein L (large) and NS (nonstructural). One of the best known types of rhabdoviruses is the rabies virus.

### E. Transcription and translation of a minus-strand RNA virus

RNA viruses with minus-strand genomes (e.g., rhabdoviruses, myxoviruses) must first form an RNA plus strand by means of a virus-coded replicase contained in the viral particle. This serves as a template for the formation of a new genome (replication) and for mRNAs (transcription) for the synthesis of virus-coded proteins (translation). The new viral genomes are then packaged to form virions.
(Figures adapted from Watson et al., 1987).

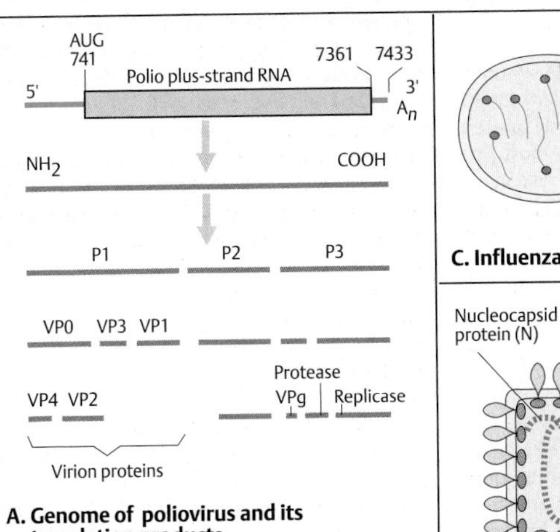


**A. Genome of poliovirus and its translation products**



**B. Togavirus: replication and translation**

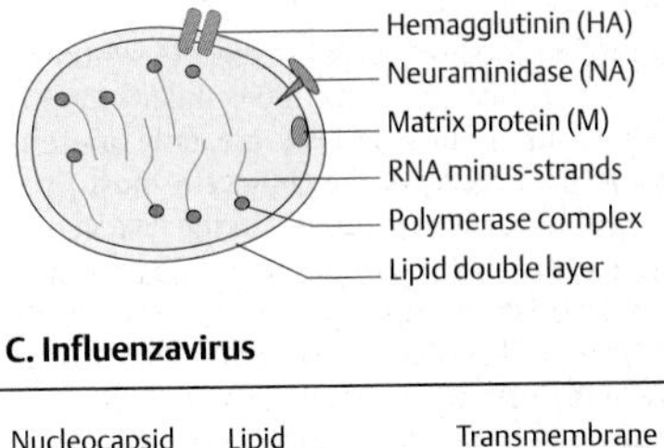


**C. Influenzavirus**

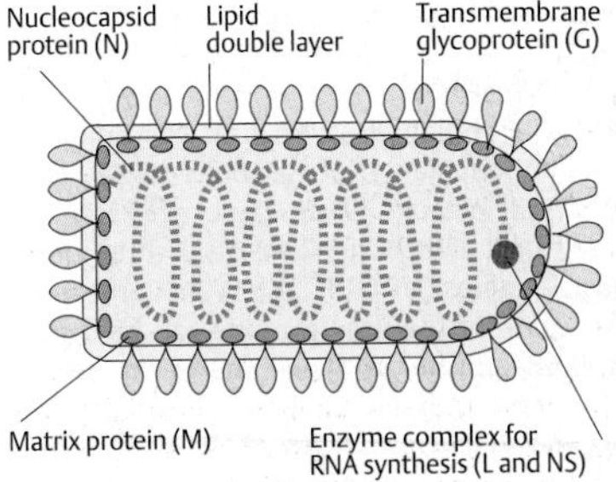


**D. Rhabdovirus (vesicular stomatitis virus, VSV)**

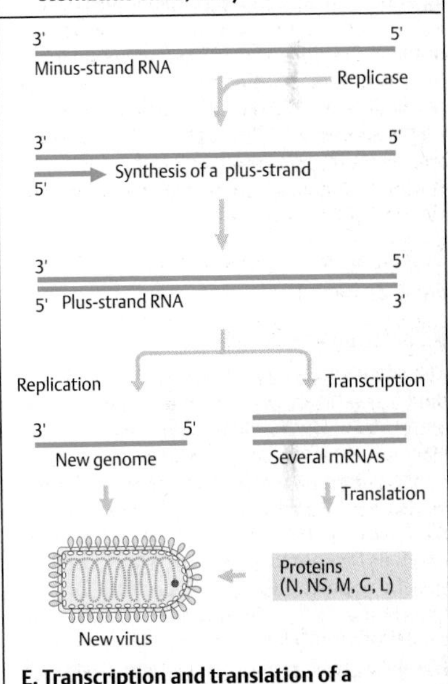


**E. Transcription and translation of a minus-strand RNA virus**

# DNA Viruses

DNA viruses depend on host-cell DNA synthesis for their replication. Since the cellular proteins required for DNA synthesis are only present during the S phase of the cell cycle, most DNA viruses induce DNA synthesis in the host cell. A few contain genes for their own DNA polymerases and other proteins for DNA synthesis. Their genes are expressed in a well-defined chronological pattern. Important DNA viruses include herpesviruses and poxviruses.

## A. SV40 virus

The SV40 (simian virus 40) virus has a double-stranded circular DNA genome (5243 base pairs) and belongs to the papovaviruses. The protein coat of the virion consists of three virus-coded proteins (VP1, VP2, VP3). An early and a late region of the viral genome can be distinguished according to their time of expression. Between them lies a regulative region. Two alternatively spliced mRNAs of the early region code for the large tumor antigen (T, ~90 kilodaltons or kDa) and the small tumor antigen (t, ~20 kDa). They regulate further transcription, the initiation of replication, and the expression of specific cellular genes. The late region, coding for the capsid proteins, is transcribed at the onset of viral DNA synthesis in the opposing DNA strand and in the opposite direction. The late proteins are translated from differently spliced, overlapping mRNAs. A fourth, late gene codes for a small protein (agno protein) of unknown function. Replication of the viral genome DNA begins at a defined point (OR, origin of replication).

## B. Adenovirus

The adenovirus genome (1) consists of a linear double-stranded DNA of about 36000 base pairs (36 kb). Both ends contain a repetitive nucleotide sequence (inverted terminal repeat), which is important for DNA replication. Early transcripts (E), which appear about 2–3 hours after infection and 6–8 hours before the onset of viral DNA replication, are transcribed at specific regions. The region E1a codes for proteins that initiate transcription of all other viral genes and that influence the expression of specific cellular genes. The E2 region codes for proteins directly involved in DNA replication, including a viral DNA polymerase (adenovirus-specific replication mechanism). A single promoter controls transcription of the late region (L), which is transcribed into a large primary RNA (2). From this, at least 20 different mRNAs are produced by alternative splicing. All mRNAs have the same 5′ terminus. Unlike the early transcripts, the late transcripts are encoded by only one DNA strand. The late genes code mainly for viral coat proteins.

## C. Parvovirus

Parvoviruses are small, single-stranded DNA viruses that can replicate only in proliferating cells. Replication of the single DNA strand is initiated (self-priming) by the formation of a specific genomic structure (hairpin structure).

## D. Herpesvirus genome

Herpesviruses and poxviruses are large viruses with DNA genomes of 80–200 kb that contain repetitive sequences. They code for many (50–200) different proteins, including their own DNA polymerase. Other proteins interfere with the regulation of the cell's nucleotide metabolism. The group of herpesviruses includes the Epstein–Barr virus, the varicellazoster virus, and the cytomegalovirus. (Figures after Watson et al., 1987).

## Reference

Watson, J.D. et al.: Molecular Biology of the Gene, 3rd ed. Benjamin/Cummings Publishing Co., Menlo Park, California, 1987.

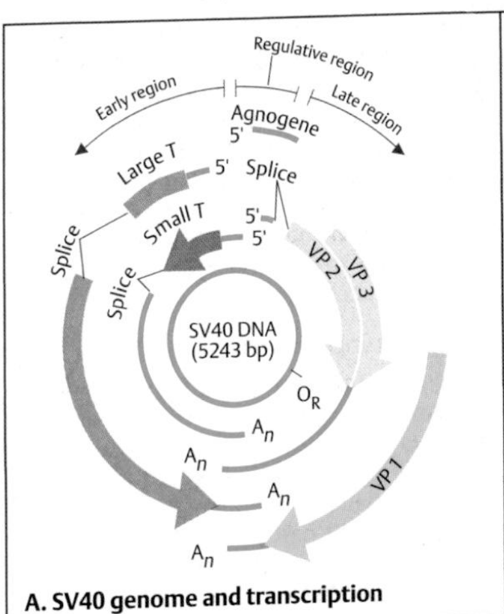


**A. SV40 genome and transcription**

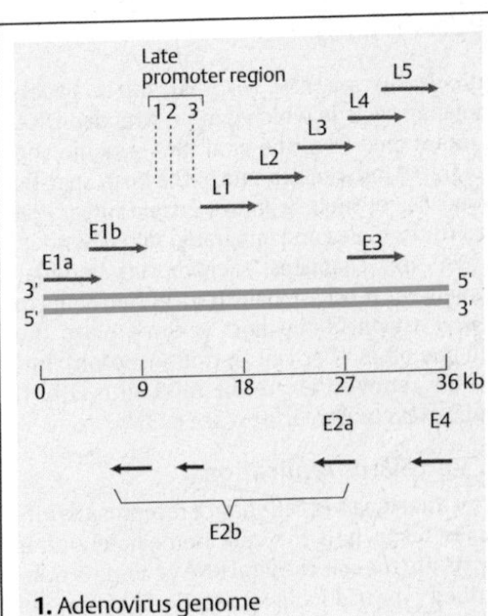


**1.** Adenovirus genome

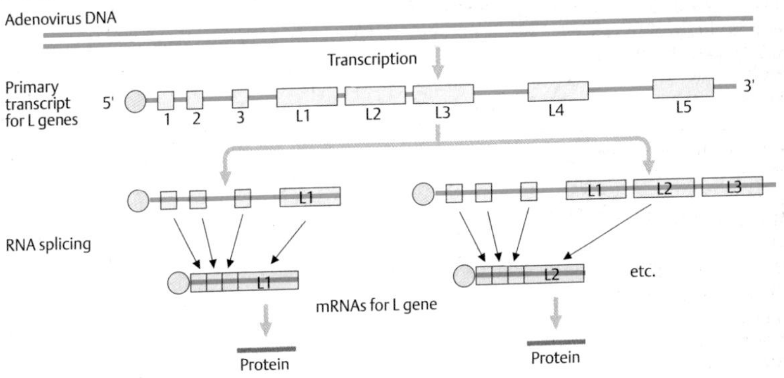


**2.** Adenovirus transcription and RNA splicing

**B. Adenovirus**

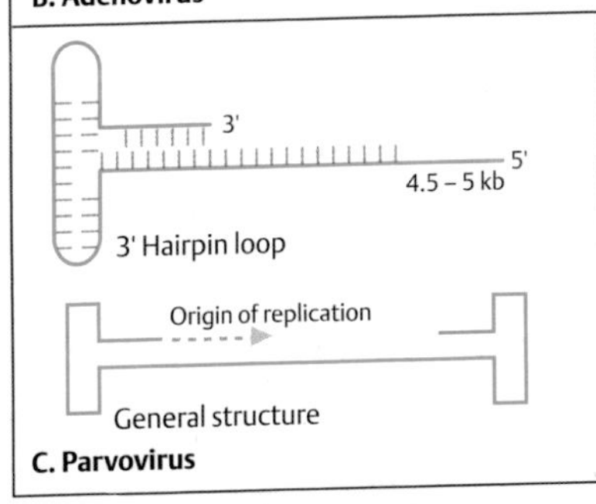


**C. Parvovirus**

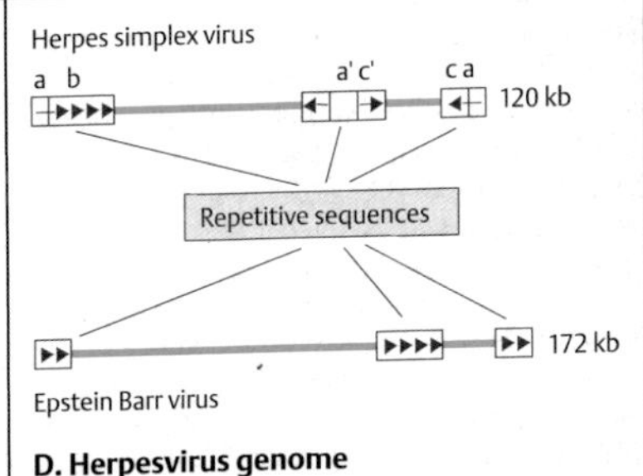


**D. Herpesvirus genome**

# Retroviruses

Retroviruses are RNA viruses with a developmental cycle in which double-stranded DNA is transcribed from the viral RNA genome and integrated into the genome of the host. Specific sequences of their genomes enable integrated proviruses to become integrated into new sites. When this happens, neighboring cellular sequences are occasionally transported along to a new region of the host genome or to the genome of another cell (retrotransposon). Important retroviruses are the AIDS virus (HIV I) and certain tumor viruses (see p. 314).

## A. Retrovirus replication

After attaching to a cell surface receptor, the virion is taken into the cell. Immediately after entry into the cell, the viral RNA genome is transcribed into double-stranded DNA by an enzyme complex called reverse transcriptase, and the new DNA is integrated as a provirus into the DNA of the host cell. The RNAs transcribed from this DNA copy by cellular RNA polymerase II serve either as mRNA for the synthesis of virus proteins or as new virus genomes. Newly formed virions leave the cell by a specific process called exocytosis, without killing the cell.

## B. Genomes of the retroviruses

Several retroviruses cause tumors in mice (mouse leukemia) or chickens (Rous sarcoma). The only retrovirus identified to cause a tumor in humans is HTLV (type 1 and 2) (human T cell leukemia/lymphoma virus). HTLV viruses have genetic similarities with the AIDS virus, which causes acquired immune deficiency syndrome (p. 314).

The RNA genome of a typical retrovirus contains short repetitive (R) and unique (U) sequences at both ends (RU5 at the 5′ end and U3R at the 3′ end). As a rule, three protein-coding genes lie between them: *gag* (group-specific antigen), *pol* (reverse transcriptase), and *env* (a gene that codes for a glycoprotein that is built into the lipid membrane coat of the virion). The 5′ end of the genome contains a nucleotide sequence that is complementary to the 3′ end of a host cell tRNA. This nucleotide sequence binds to tRNA, which serves as a primer for the synthesis by reverse transcriptase of viral DNA from the virus genome RNA. At the 3′ end of the genome, HTLV viruses carry several other genes (*px*, *lor*, *tat*, and others) that are involved in regulating the transcription of the viral genes. The AIDS virus (HIV I/II) contains further genes that do not occur in other retroviruses. These include *vif* (virion infectivity factor, formerly designated *sor*), *rev* (regulator of expression of virion proteins), and *nef* (negative factor, formerly *orf*). The general structure of the genome is more complex than shown here.

## C. DNA synthesis of a retrovirus

Reverse transcriptase transcribes the RNA genome into DNA by RNA-dependent DNA polymerase activity and catalyzes subsequent steps by means of DNA-dependent DNA polymerase activity. Furthermore, the reverse transcriptase has RNAase activity for degrading the RNA of the newly formed RNA/DNA hybrid molecule. The first step in replicating the retrovirus genome is initiated by a primer of host cell tRNA, which is hybridized to the 5′ end of the RNA genome (1). After synthesis of the first DNA strand and removal of the tRNA primer, synthesis of the second DNA strand (plus strand) begins at the RU5 region (2). Here, the previously formed minus DNA strand serves as the primer (3). As DNA synthesis is continued, the remaining RNA is degraded (4), and the DNA plus strand is synthesized to completion (5, 6). The double-stranded DNA copy of the virus contains long terminal repeats (LTR) at both ends. These enable the viral DNA intermediary step to be integrated into the host cell DNA, and they contain the necessary regulatory sequences for transcribing the provirus DNA (see p. 102). (Figures adapted from Watson et al., 1987).

## Reference

Watson, J.D. et al.: Molecular Biology of the Gene, 3rd ed. Benjamin/Cummings Publishing Co., Menlo Park, California, 1987.

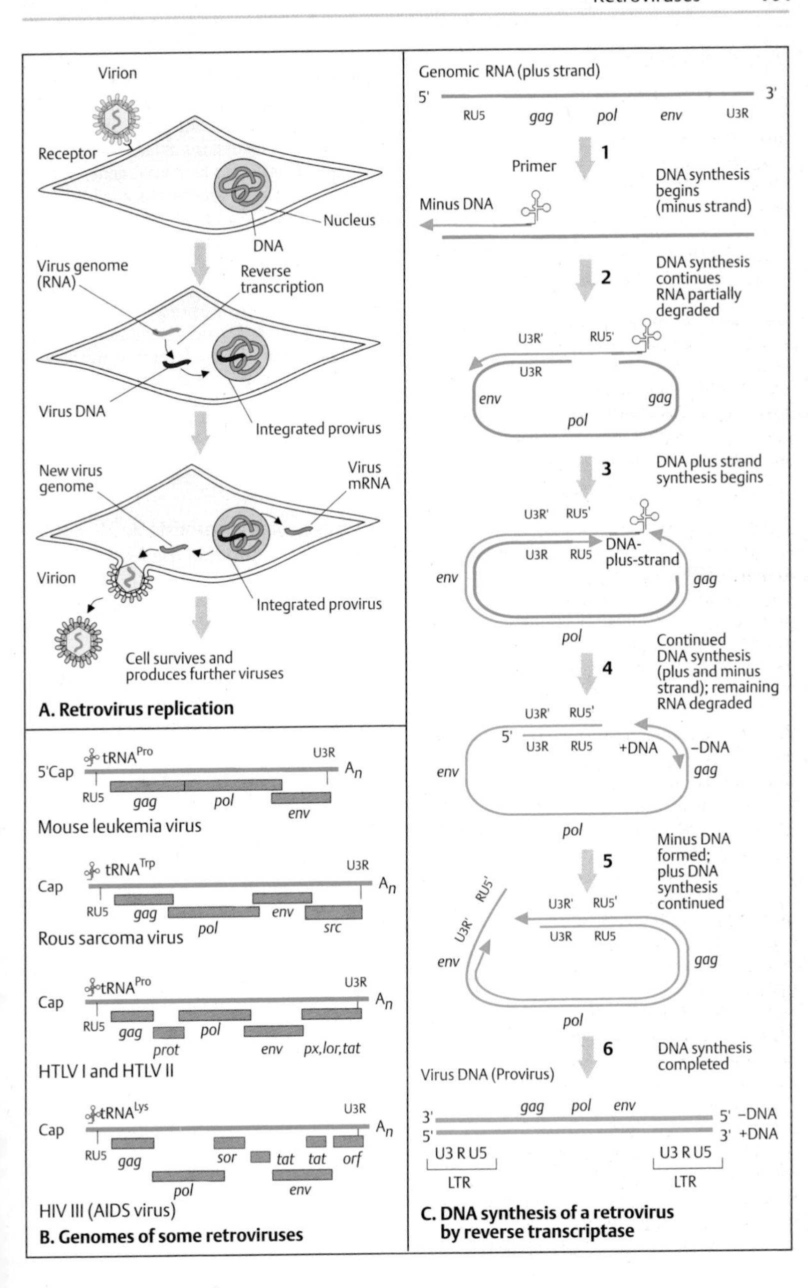


**A. Retrovirus replication**

**B. Genomes of some retroviruses**

**C. DNA synthesis of a retrovirus by reverse transcriptase**

# Retrovirus Integration and Transcription

Integration of the DNA copy of a retrovirus into the host cell DNA occurs at a random location. This may alter cellular genes (insertion mutation). The viral genes of provirus DNA are transcribed by cellular RNA polymerase II. The resulting mRNA serves either for translation or for the production of new RNA genomes, which are packaged into the virion. Some retrovirus genomes may contain an additional viral oncogene (v-*onc*). Viral oncogenes are parts of cellular genes (c-*onc*) previously taken up by the virus. If they enter a cell with the virus, they may change (transform) the host cell so that its cell cycle is altered and the cell becomes the origin of a tumor (tumor virus) (see p. 320).

## A. Retrovirus integration into cellular DNA

In the nucleus of the host cell, the double-stranded DNA (1, virus DNA) produced from virus genome RNA first forms a ring-shaped structure (2) by joining LTRs (long terminal repeats). This is possible because the LTRs contain complementary nucleotide sequences. Recognition sequences in the LTRs and in the cellular DNA (3) allow the circular viral DNA to be opened at a specific site (4) and the viral DNA to be integrated into the host DNA (5). Viral genes can then be transcribed from the integrated provirus (6). As a rule, the provirus remains in the genome of the host cell without disrupting the functions of the cell. One exception is the AIDS virus, which destroys a specific population of lymphocytes (helper T cells) (see p. 310). The genomes of vertebrates (including man) contain numerous DNA sequences that consist of endogenous proviruses. In mice, they may represent up to 0.5% of the total DNA. The genomes of higher organisms also contain LTR-like sequences that are very similar to those of an endogenous retrovirus. These sequences can change their location in the genome (mobile genetic elements or transposons). Since many of them have the fundamental structure of a retrovirus (LTR genes), they are designated retrotransposons (see pp. 76 and 252).

## B. Control of retrovirus transcription

The LTRs are important not only for integration of the virus into cellular DNA, but also because they contain all regulatory signals necessary for efficient transcription of a viral gene. Typical transcription signals are the so-called "CCAAT" and "TATA" sequences of promoters, which are respectively located about 80 and 25 base pairs above the 5′ end of the sequence to be transcribed. Further upstream (in the 5′ direction) are nucleotide sequences that can increase the expression of viral genes (enhancer). Similar regulatory sequences are located at the 5′ ends of eukaryotic genes (see section on gene transcription). Newly synthesized viral RNA is structurally modified at the 5′ end (formation of a cap). Furthermore, numerous adenine residues are added at the 3′ end (polyadenylated, poly(A), see p. 214).

## C. Virus protein synthesis by posttranscriptional modification of RNA

The RNA transcripts synthesized from the provirus by cellular RNA polymerase II serve either for the translation into virus-coded proteins (*gag, pol*) or as the genome for the formation of new virions. Some of the RNA is spliced to form new mRNAs that code for coat proteins (*env*). (Figures after Watson et al., 1987).

## Reference

Watson, J.D. et al.: Molecular Biology of the Gene, 3rd ed. Benjamin/Cummings Publishing Co., Menlo Park, California, 1987.

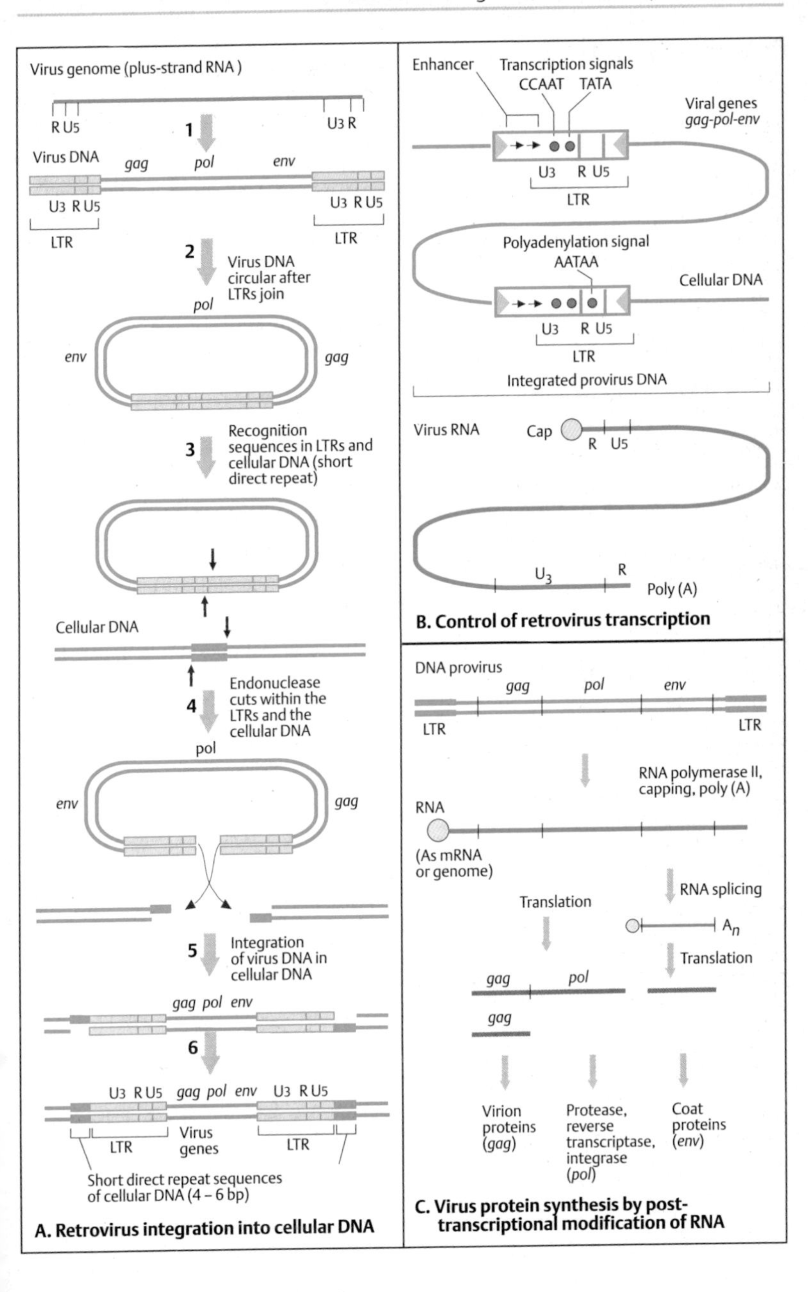


**A. Retrovirus integration into cellular DNA**

**B. Control of retrovirus transcription**

**C. Virus protein synthesis by post-transcriptional modification of RNA**

# Eukaryotic Cells

## Yeast: Eukaryotic Cells with a Diploid and a Haploid Phase

Yeast is a single-celled eukaryotic fungus with a genome of individual linear chromosomes enclosed in a nucleus and with cytoplasmic organelles such as endoplasmic reticulum, Golgi apparatus, mitochondria, peroxisomes, and a vacuole analogous to a lysosome. About 40 different types of yeast are known. Baker's yeast, *Saccharomyces cerevisiae*, consists of cells of about 3 μm diameter that are able to divide every 90 minutes under good nutritional conditions. Nearly half of the proteins known to be defective in human heritable diseases show amino acid similarity to yeast proteins.

The haploid genome of *S. cerevisiae*, contains ca. 6200 genes in $1.4 \times 10^7$ DNA base pairs in 16 chromosomes, about 50% of the genes being without known function (Goffeau, 1996). For about 6200 proteins the following functions have been predicted: cell structure 250 (4%), DNA metabolism 175 (3%), transcription and translation 750 (13%), energy production and storage 175 (3%), biochemical metabolism 650 (11%), and transport 250 (4%). The *S. cerevisiae* genome is very compact compared to other eukaryotic genomes, with about one gene every 2 kb.

### A. Yeast life cycle through a haploid and a diploid phase

Yeast can grow either as haploid or as diploid cells. Haploid cells of opposite types can fuse (mate) to form a diploid cell. Haploid cells are of one of two possible mating types, called **a** and α. The mating is mediated by a small secreted polypeptide called a pheromone or mating factor. A cell-surface receptor recognizes the pheromone secreted by cells of the opposite type, i.e., **a** cell receptors bind only α factor and α cell receptors bind only **a** factor. Mating and subsequent mitotic divisions occur under favorable conditions for growth. Under starvation conditions, a diploid yeast cell undergoes meiosis and forms four haploid spores (sporulation), two of type **a** and two of type α.

### B. Switch of mating type

A normal haploid yeast cell switches its mating type each generation. The switch of mating type (mating-type conversion) is initiated by a double-strand break in the DNA at the MAT locus (recipient) and may involve the boundary to either of the flanking donor loci (HMR or HML). This is mediated by an HO endonuclease through site-specific DNA cleavage.

### C. Cassette model for mating type switch

Mating type switch is regulated at three gene loci near the centromere (cen) of chromosome III of *S. cerevisiae*. The central locus is MAT (mating-type locus) which is flanked by loci HMLα (left) and HMRa (right). Only the MAT locus is active and transcribed into mRNA. Transcription factors regulate other genes responsible for the **a** or the α phenotype. The HMLα and HMRa loci are repressed (silenced). DNA sequences from either the HMLα or the HMLa locus are transferred into the MAT locus once during each cell generation by a specific recombination event called gene conversion. The presence of HMRa sequences at the MAT locus determines the **a** cell phenotype. When HMLα sequences are transferred (switch to an α cassette), the phenotype is switched to α. Any gene placed by recombinant DNA techniques near the yeast mating-type silencer is repressed, even a tRNA gene transcribed by RNA polymerase III, although it uses different transcription factors. Apparently the HML and HMR loci are permanently repressed because they are inaccessible to proteins (transcription factors and RNA polymerase) owing to the condensed chromatin structure near the centromere.

### References

Botstein, D., Chervitz, S. A., Cherry, J.M.: Yeast as a model organism. Science **277**:1259–1260, 1997.

Brown, T.A.: Genomes. Bios Scientific Publishers, Oxford, 1999.

Goffeau, A. et al.: Life with 6000 genes. Science **274**:562–567, 1996.

Haber, J.E.: A locus control region regulates yeast recombination. Trends Genet. **14**:317–321, 1998.

Lewin, B.: Genes VII. Oxford Univ. Press, Oxford, 2000.

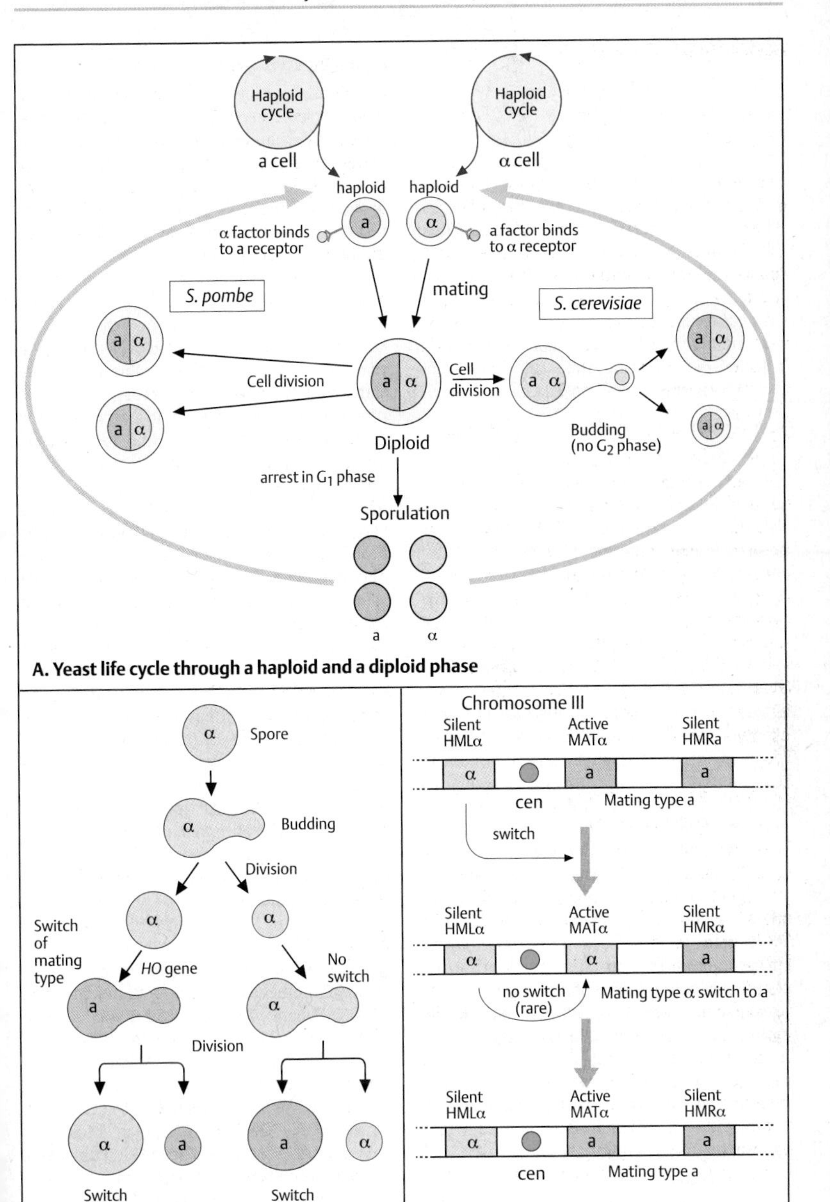


A. Yeast life cycle through a haploid and a diploid phase

B. Switch of mating type

C. Cassette model for mating type switch

# Mating Type Determination in Yeast Cells and Yeast Two-Hybrid System

Yeast cells *(S. cerevisiae)* are unicellular eukaryotes with three different cell types: haploid **a** and α cells and diploid **a**/α cells. Owing to their relative simplicity compared with multicellular animals and plants, yeast serves as a model for understanding the underlying control mechanisms specifying cell types. The generation of many different cell types in different tissues of multicellular organisms probably evolved from mechanisms that determine cell fate in unicellular organisms such as yeast.

## A. Regulation of cell-type specificity in yeast

Each of the three *S. cerevisiae* cell types expresses cell-specific genes. The resulting differences in combinations of DNA-binding proteins determine the cell-type specification. These regulatory proteins are encoded at the MAT locus in combination with a general transcription factor called Mcm1. Mcm1 is expressed in all three cell types. Cells of type **a** express **a**-specific genes, but not α-specific genes. In diploid (**a**/α) cells, diploid-specific genes are expressed. The cell-specific transcription factors are **a**1, α1, and α2, all encoded at the MAT locus. Mcm1 is a dimeric general transcription factor that binds to **a**-specific upstream regulatory sequences (URSs). This stimulates transcription of the **a**-specific genes, but it does not bind too efficiently to the α-specific URSs when α1 protein is absent. In α cells two specific transcription factors, α1 and α2, mediate transcriptional activity. α2-binding sequences associate with MCM helicase and block transcription of **a**-specific genes. α1-binding sequences form a complex with MCM and stimulate transcription of α-specific genes. In diploid cells (**a**/α) the haploid genes are repressed by α2-MCM1 and by α2/**a**1 complexes. In summary, each of the three yeast cell types is determined by a specific combination of transcription factors acting as activators or as repressors depending to which specific regulatory sites they bind.

## B. Yeast two-hybrid system

The problem of determining the function of a newly isolated gene may be approached by determining whether its protein specifically reacts with another protein of known function. Yeast cells can be used in an assay for protein–protein interactions. The two-hybrid method rests on the observation of whether two different proteins, each hybridized to a different protein domain required for transcription factor activity, are able to interact and thereby reassemble the transcription factor. When this occurs, a reporter gene is activated. Neither of the two hybrid proteins alone is able to activate transcription. Hybrid 1 consists of protein X, the protein of interest (the "bait") attached to a transcription factor DNA-binding domain (BD). This fusion protein alone cannot activate the reporter gene because it lacks a transcription factor activation domain (AD). Hybrid 2, consisting of a transcription factor AD and an interacting protein, protein Y (the "prey"), lacks the BD. Therefore, hybrid 2 alone also cannot activate transcription of the reporter gene. Different ("prey") proteins expressed from cDNAs in vectors are tested. Fusion genes encoding either hybrid 1 or hybrid 2 are produced using standard recombinant DNA methods. Cells are cotransfected with the genes. Only cells producing the correct hybrids, i.e., those in which the X and Y proteins interact and thereby reconnect AD and BD to form an active transcription factor, can initiate transcription of the reporter gene. This can be observed as a color change or by growth in selective medium. (Figure adapted from Oliver, 2000, and Frank Kaiser, Essen, personal communication.)

## References

Li, T., et al.: Crystal structure of the MATa1/MAT2 homeodomain heterodimer bound to DNA. Science **270**:262–269, 1995.

Lodish et al.: Molecular Cell Biology. 4th ed. W.H. Freeman & Co., New York, 2000.

Oliver, S.: Guilt-by-association goes global. News & Views. Nature **403**:601–603, 2000.

Strachan, T., Read, A.P.: Human Molecular Genetics. 2nd ed. Bios Scientific Publishers, Oxford, 1999.

Uetz, P., et al.: A comprehensive analysis of protein-protein interaction in *Saccharomyces cerevisiae*. Nature **403**:623–627, 2000 (and at http://www.curatools.curagen.com).

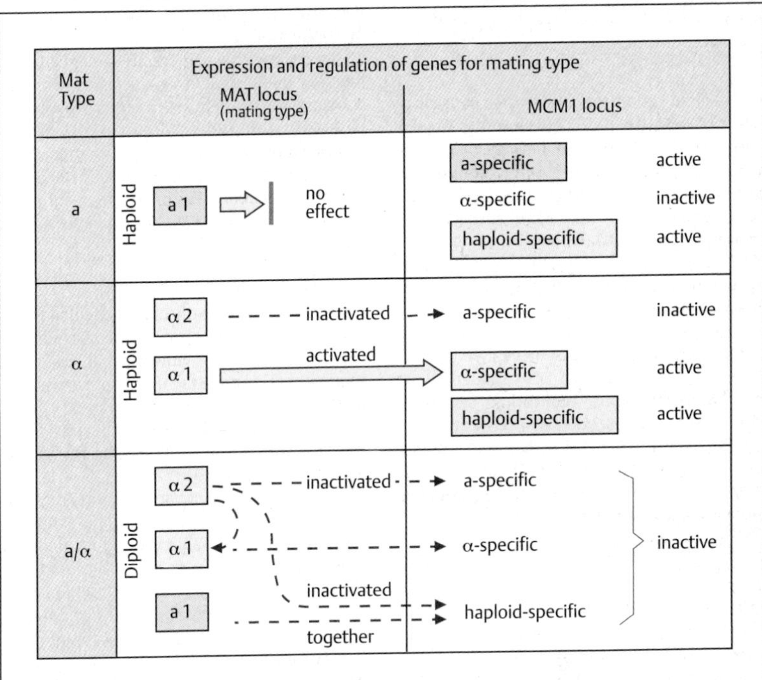


**A. Regulation of cell-type specificity in yeast**

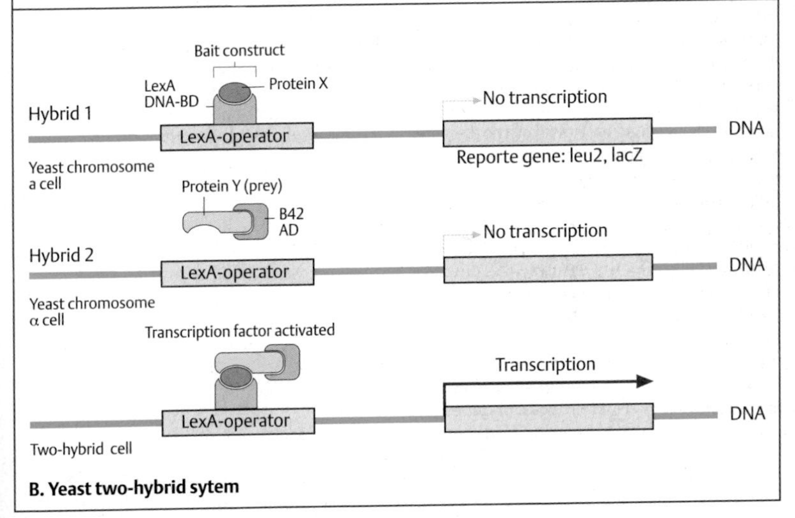


**B. Yeast two-hybrid sytem**

# Functional Elements in Yeast Chromosomes

The haploid genome of *S. cerevisiae* consists of $1.4 \times 10^7$ base pairs in 16 chromosomes. Several functional DNA sequences have been isolated and characterized from yeast chromosomes that are necessary for replication and normal distribution of the chromosomes during mitosis (mitotic segregation). If they lie on the same chromosome, they are designated *cis* elements (as opposed to *trans* on another chromosome). They are genetically stable and can be joined in vitro to DNA of any origin to form stable artificial chromosomes in yeast cells (see p. 104). Three types of functional DNA sequences are known: autonomous replicating sequences (ARS), centromere sequences (CEN), and telomere sequences (TEL). All eukaryotic chromosomes contain these functional elements, but they can be especially well demonstrated in yeast cell chromosomes.

## A. Autonomous replicating sequences (ARS)

Replication proceeds in two directions (bidirectional). The ARS represent the starting points of replication in chromosomes. Their functional significance can be recognized in a transformation experiment: A mutant yeast cell (e.g., with the inability to produce leucine, $Leu^-$) can be transformed into a leucine-producing ($Leu^+$) cell (1) by means of transfection with cloned plasmids containing the gene for leucine formation *(Leu)*. These cells only are able to grow to medium lacking leucine. However, the daughter cells remain leucine-dependent because the plasmid cannot replicate. If the plasmid contains autonomously replicating sequences (ARS) along with the *Leu* gene, a small fraction of the daughter cells will be $Leu^+$ because plasmid replication has occurred (2). But most of the daughter cells remain $Leu^-$ because mitotic distribution (segregation) is defective.

## B. Centromere sequences (CEN)

If the plasmid DNA (1) contains sequences from the centromere (CEN) of the yeast chromosome along with the gene for leucine *(Leu)* and the autonomously replicating sequences (ARS), then normal mitotic segregation takes place (2). This demonstrates that the CEN sequences are necessary for normal distribution of the chromosomes at mitosis (see p. 114).

The centromere sequences are similar but not identical in different yeast chromosomes. They contain three elements with a total of about 220 base pairs (bp), which occur in all chromosomes. Element I is a conserved sequence of 8 bp; element II is AT rich and has about 80 bp; element III has about 25 bp. This segment is nuclease-protected and very important for mitotic stability.

## C. Telomere sequences (TEL)

If the plasmid is linear instead of circular as in B, transformation will occur, but the plasmid does not replicate (2). However, if telomere sequences (TEL) are attached to both ends of the plasmid (3) before it is incorporated (4), then transformation of the yeast cell after incorporating the plasmid is followed by normal replication and mitosis (5).

These observations prove that autonomously replicating sequences (starting point of replication), centromere sequences, and telomere sequences (see p. 180) are necessary for normal chromosome function. These functional elements (ARS, CEN, TEL) can be utilized to produce artifical yeast chromosomes (YACs). (Figures adapted from Lodish et al., 2000, p. 330).

## References

Lodish, H. et al.: Molecular Cell Biology. 4th ed. Scientific American Books, W.H. Freeman & Co., New York, 2000.

Traut, W.: Chromosomen. Klassische und molekulare Cytogenetik. Springer, Heidelberg, 1991.

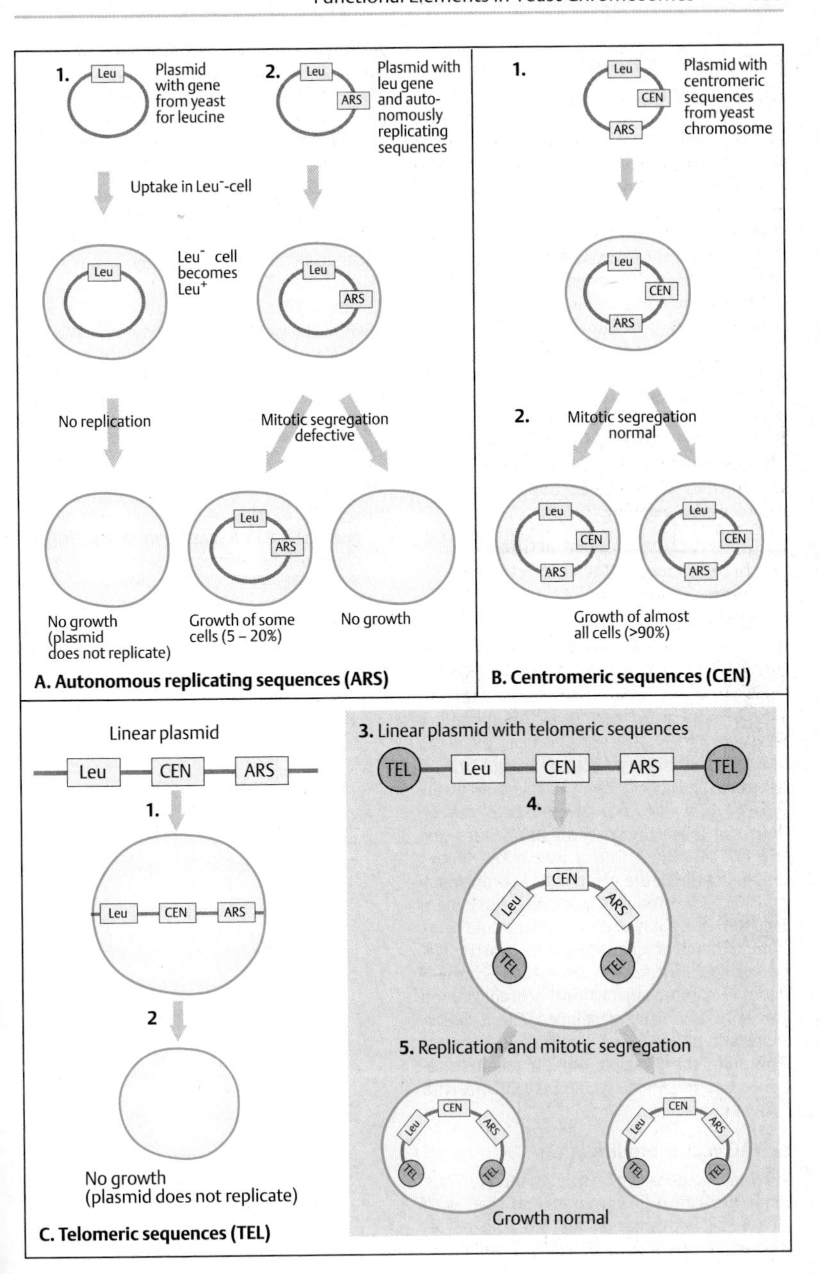
1.
Leu
Plasmid with gene from yeast for leucine
2.
Leu
ARS
Plasmid with leu gene and autonomously replicating sequences
Uptake in Leu⁻-cell
Leu⁻ cell becomes Leu⁺
No replication
Mitotic segregation defective
No growth (plasmid does not replicate)
Growth of some cells (5 – 20%)
No growth
A. Autonomous replicating sequences (ARS)
1.
Leu
CEN
ARS
Plasmid with centromeric sequences from yeast chromosome
2.
Mitotic segregation normal
Growth of almost all cells (>90%)
B. Centromeric sequences (CEN)
Linear plasmid
Leu
CEN
ARS
1.
2
No growth (plasmid does not replicate)
3. Linear plasmid with telomeric sequences
TEL
4.
5. Replication and mitotic segregation
Growth normal
C. Telomeric sequences (TEL)

## Artificial Chromosomes for Analyzing Complex Genomes

The mapping of large and complex genomes (for instance in humans with three billion base pairs) would not be possible if cloned fragments of only a few dozen to a few thousand base pairs were available (with cosmids, up to about 50 kb). Large distances could not be bridged because nonclonable DNA segments would lie between the mapped segments. Since physical mapping of these segments is not possible, attempts were made to develop vectors for cloning large DNA fragments. Artificial chromosomes can be constructed from yeast chromosomes (YACs, yeast artificial chromosomes). These can incorporate DNA fragments of about 100–1000 kb). Since they are stable during growth, they represent an important instrument for gene mapping.

### A. Construction of yeast artificial chromosomes (YACs) for cloning and mapping

The starting material is a linear vector. This consists of telomere sequences (TEL) and selectable markers, here leucine (LEU) and tryptophan (TRP), a starting point for replication (autonomously replicating sequences, ARS), and a centromere (CEN). The vector is incised and the DNA fragment to be replicated (foreign DNA) is inserted. The capacity of a YAC is about 750 kb. Subsequently, the incorporated DNA can be cloned in the artificial chromosome in yeast cells and amplified. In this way, a YAC library containing the entire genome of a complex organism can be produced in a manageable manner. From this, a physical map of large and overlapping fragments (contig) can be constructed. Increasingly, artificial mammalian chromosomes (mammalian artifical chromosomes, MACs) are also being developed. These consist of one arm of a YAC with a selectable marker, a replication starting point, and a telomere attached to a fragment of mammalian DNA with its own telomere.

### B. YAC vector modification

Different methods can be used to modify YACs. Sequences can be incorporated so that, with their help, both ends of the YAC can be recovered as plasmids in *E. coli* (1). As a prerequisite, plasmid sequences with selectable markers (2), here for ampicillin resistance ($Amp^R$), neomycin resistance ($Neo^R$), and onset of plasmid replication (ORI), are included. Digestion with an appropriate restriction enzyme produces fragments of the incorporated foreign DNA and of both vector arms (3). The fragments are formed into a ring and incorporated into *E. coli*. The hybrid plasmids (4) can then be selected for (5) by means of the markers ($Amp^R$ and $Neo^R$).

The use of YACs has considerably simplified the genetic analysis of large genomes. While a complete human-genome library requires about 500 000 clones of lambda phage vectors, a YAC library with 150 kb fragments reduces the number of clones to about 60 000. (Figures adapted from Schlessinger, 1990).

### References

Burke, D.T.: Cloning of large segments of exogenous DNA into yeast by means of artificial chromosome vectors. Science 236:806–812, 1987.

Schlessinger, D.: Yeast artifical chromosomes: tools for mapping and analysis of complex genomes. Trends Genet. 6:248–258, 1990.

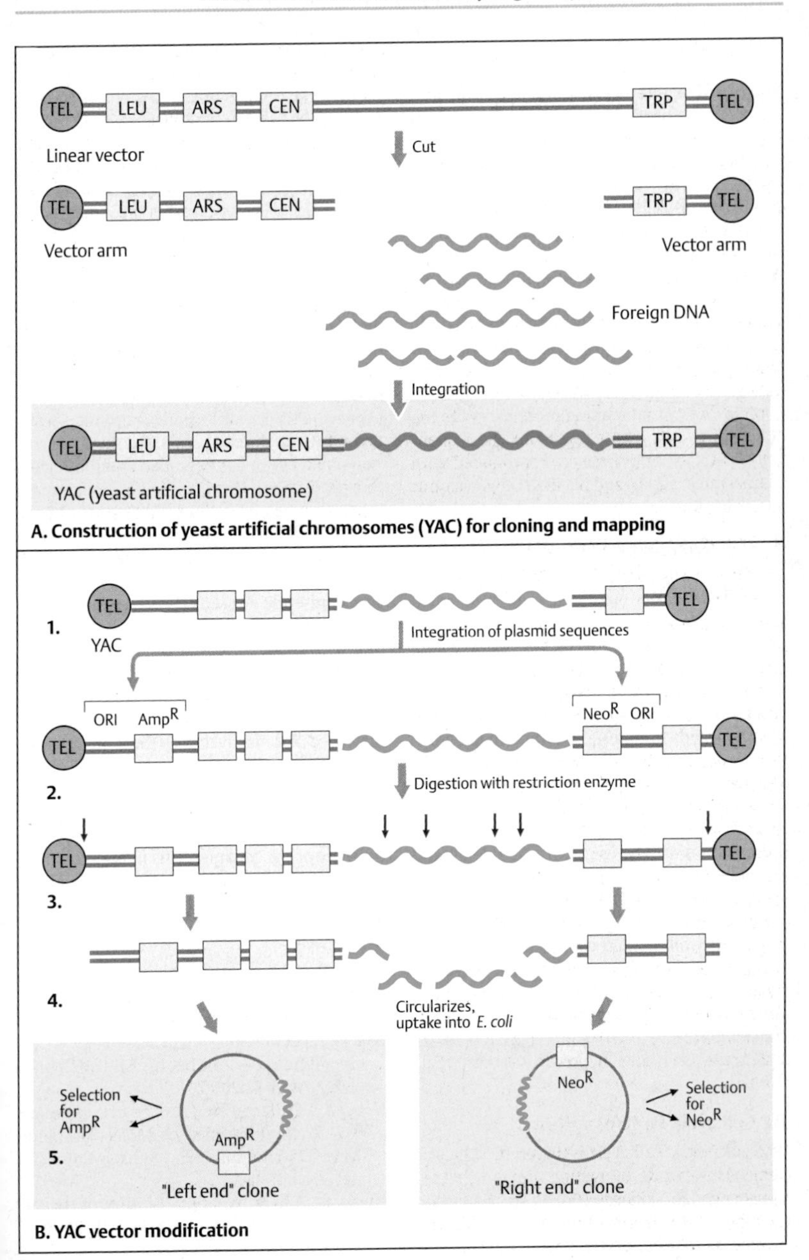


**A. Construction of yeast artificial chromosomes (YAC) for cloning and mapping**

**B. YAC vector modification**

# Cell Cycle Control

The growth of multicellular organisms depends on precise replication of individual cells followed by cell division. During replication, eukaryotic cells go through an ordered series of cyclical events. The time from one cell division to the next is called a cell cycle. The cell cycle has two main phases, interphase and mitosis. Interphase is further divided into three distinct phases: $G_1$ (gap 1), S (DNA synthesis, lasting 6–8 hours in eukaryotic cells, at the end of which the chromosomes have been duplicated), and $G_2$ (gap 2, lasting about 4 hours). Mitosis (M, see p. 114) is the phase of actual division. Cell cycle control mechanisms, which include complex sets of interacting proteins, guide the cell through its cycle by regulating the sequential cyclical events. These are coordinated with extracellular signals and result in cell division at the right time.

## A. Cell division cycle models in yeast

Budding yeast (baker's yeast) divides by mitotic budding to form one large and one small daughter cell. Since a microtubule mitotic spindle forms very early during the S phase, there is practically no $G_2$ phase (1). In contrast, fission yeast *(S. pombe)* forms a mitotic spindle inside the nucleus at the end of the $G_2$ phase, then proceeds to mitosis to form two daughter cells of equal size (2). Unlike in vertebrate cells, the nuclear envelope remains intact during mitosis. An important regulator is cdc2 (cell division cycle, *S. pombe*) (3). Absence of cdc2 activity ($cdc2^-$ mutant) results in cycle delay and prevents entry into mitosis. Thus, too large a cell with only one nucleus results. Increased activity of cdc2 (dominant mutant $cdc^D$) results in premature mitosis and cells that are too small (wee phenotype, from the Scottish word for small). Normally a yeast cell has three options: (a) halt the cell cycle if the cell is too small or nutrients are scarce, (b) mate (see p. 104), or (c) enter mitosis. (Figure adapted from Lodish et al., 2000).

## B. Cell cycle control systems

The eukaryotic cell cycle is driven by cell cycle "engines", a set of interacting proteins, the cyclin-dependent kinases (Cdks). An important member of this family of proteins is cdc2 (also called Cdk1). Other proteins act as rate-limiting steps in cell cycle progression and are able to induce cell cycle arrest at defined stages (checkpoints). The cell is induced to progress through $G_1$ by growth factors (mitogens) acting through receptors that transmit signals to proceed towards the S phase. D-type cyclins (D1, D2, D3) are produced, which associate with and activate Cdks (4 and 6). Other proteins can induce $G_1$ arrest. If these proteins are inactive owing to mutations, cell proliferation becomes uncontrolled as in many forms of cancer. The detection of DNA damage and subsequent cell cycle arrest due to activated p53 is an important mechanism for preventing the cell from entering the S phase.

In early $G_1$ phase cdc2 is inactive. It is activated in late $G_1$ by association with $G_1$ cyclins, such as cyclin E. Once the cell has passed the $G_1$ restriction point, cyclin E is degraded and the cell enters the S phase. This is initiated, among many other activities, by cyclin A binding to Cdk2 and phosphorylation of the RB protein (retinoblastoma protein, see p. 330). The cell passes through the mitosis checkpoint only if no damage is present. Cdc2 (Cdk1) is activated by association with mitotic cyclins A and B to form the mitosis-promoting factor (MPF).

During mitosis, cyclins A and B are degraded, and an anaphase-promoting complex forms (details not shown). When mitosis is completed, cdc2 is inactivated by the S-phase inhibitor Sic1 in yeast. At the same time the retinoblastoma (RB) protein is dephosphorylated. Cells can progress to the next cell cycle stage only when feedback controls have ensured the integrity of the genome. (This figure is an overview only; it omits many important protein transactions).

### References

Hartwell, L., Weinert, T.: Checkpoints: Controls that ensure the order of cell cycle events. Science **246**:629–634, 1989.

Lodish, H., et al.: Molecular Cell Biology. 4th ed. Scientific American Books, F.H. Freeman & Co., New York, 2000.

Nguyen, L.Q., Jameson, J.L.: The cell cycle. pp. 65–72. In: Principles of Molecular Medicine, J.C. Jameson, ed.: Humana Press, Totowa, NJ, 1998.

Nurse, P.: A long twentieth century of the cell cycle and beyond. Cell **100**:71–78, 2000.

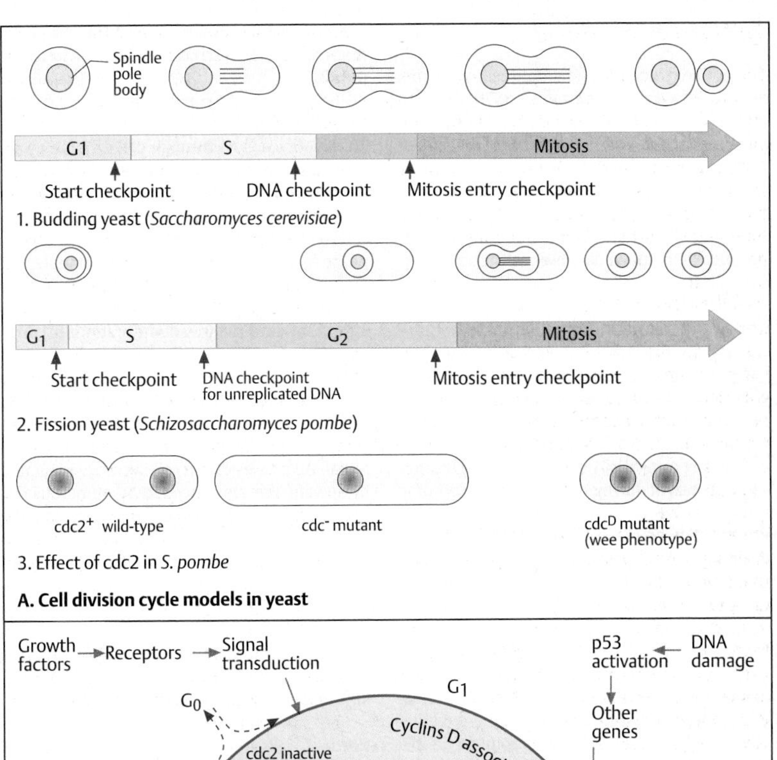
Spindle pole body
G1
S
Mitosis
Start checkpoint
DNA checkpoint
Mitosis entry checkpoint
1. Budding yeast (Saccharomyces cerevisiae)
G1
S
G2
Mitosis
Start checkpoint
DNA checkpoint for unreplicated DNA
Mitosis entry checkpoint
2. Fission yeast (Schizosaccharomyces pombe)
cdc2+ wild-type
cdc- mutant
cdcD mutant (wee phenotype)
3. Effect of cdc2 in S. pombe

**A. Cell division cycle models in yeast**

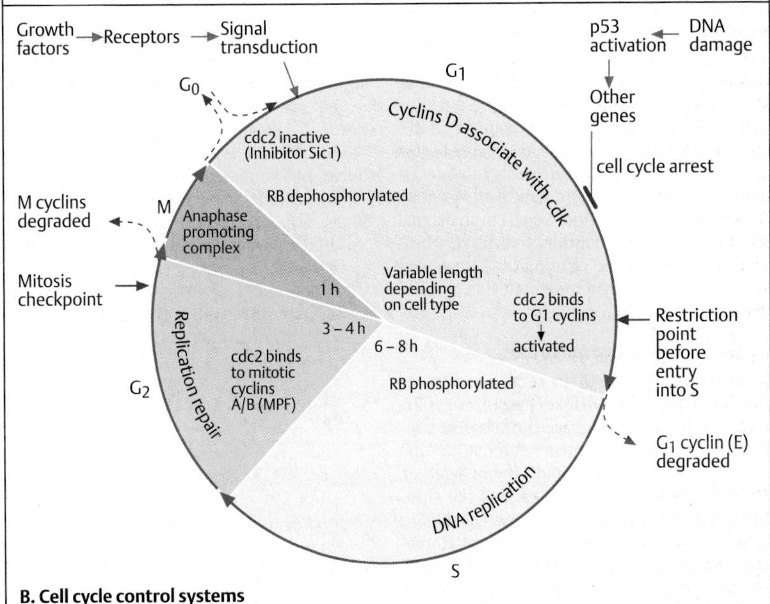
Growth factors
Receptors
Signal transduction
p53 activation
DNA damage
Other genes
cell cycle arrest
G1
G0
Cyclins D associate with cdk
cdc2 inactive (Inhibitor Sic1)
RB dephosphorylated
M cyclins degraded
M
Anaphase promoting complex
Mitosis checkpoint
1 h
Variable length depending on cell type
cdc2 binds to G1 cyclins
activated
Restriction point before entry into S
3–4 h
6–8 h
Replication repair
cdc2 binds to mitotic cyclins A/B (MPF)
G2
RB phosphorylated
G1 cyclin (E) degraded
DNA replication
S

**B. Cell cycle control systems**

## Cell Division: Mitosis

Threadlike structures in dividing cells were first observed by Walter Flemming in 1879. He introduced the term mitosis for cell division. Flemming also observed the longitudinal division of chromosomes during mitosis. In 1884, Strasburger coined the terms prophase, metaphase, and anaphase for the different stages of cell division. A mitosis results in two genetically identical daughter cells.

### A. Mitosis

During the transition from interphase to mitosis, the chromosomes become visible as elongated threads (prophase). In early prophase, each chromosome is attached to a specific site on the nuclear membrane and appears as a double structure (sister chromatids), the result of the foregoing DNA synthesis. The chromosomes contract during late prophase to become thicker and shorter (chromosomal condensation). In late prophase, the nuclear membrane disappears and metaphase begins. At this point, the mitotic spindle becomes visible as thin threads. It begins at two polelike structures (centrioles). The chromosomes become arranged on the equatorial plate, but homologous chromosomes do not pair. In late metaphase during the transition into anaphase, the chromosomes divide also at the centromere region. The two chromatids of each chromosome migrate to opposite poles, and telophase begins with the formation of a nuclear membrane. Finally the cytoplasm also divides (cytokinesis). In early interphase the individual chromosomal structures become invisible. Interphase chromosomes are called chromatin (Flemming 1879), i.e., nuclear structures stainable by basic dyes.

### B. Metaphase chromosomes

Waldeyer (1888) coined the term chromosome for the stainable threadlike structures visible during mitosis. A metaphase chromosome consists of two chromatids (sister chromatids) and the centromere, which holds them together. The centromere may divide each of the chromatids into two chromosome arms. The regions at both ends of the chromosome are the telomeres. The point of attachment to the mitotic spindle fibers is the kinetochore. During metaphase and prometaphase, chromosomes can be visualized under the light microscope as discrete elongated structures, 3–7 μm long (see p. 182).

### C. Cell cycle (mitosis)

Eukaryotic cells go through cell division cycles (cell cycle). In eukaryotic cells, each cell division begins with a phase of DNA synthesis, which lasts about 8 hours (S phase). This is followed by a phase of about 4 hours ($G_2$). During the $G_2$ phase (gap 2), the whole genome is double. Mitosis (M) in eukaryotic cells lasts about an hour (see A). This is followed by a phase, $G_1$ (interphase), of extremely varied duration. It corresponds to the normal functional phase. Cells that no longer divide are in the $G_0$ phase. Every cell must have a "memory" to tell it whether it is in $G_1$ or in $G_2$, since division of the chromosomes before they have been doubled would be lethal. All phases of the cell cycle are regulated by specific proteins encoded by numerous cell division cycle (cdc) genes. In particular, the transition from $G_1$ to S and from $G_2$ to M is regulated by specific cell-cycle proteins (see preceding plate).

### References

Holm, C.: Coming undone. How to untangle a chromosome. Cell 77:956–957, 1994.

Koshland, D.: Mitosis. Back to the basics. Cell 77:951–954, 1994.

North, G.: Regulating the cell cycle. Nature 339:97–98, 1989.

Nurse, P.: The incredible life and times of biological cells. Science **289**:1711–1716, 2000.

Sharp, D.J., Rogers, G.C., Scholey, J.H.: Microtubule motors in mitosis. Nature **407**:41–47, 2000.

Whitehouse, L.H.K.: Towards the Understanding of the Mechanism of Heredity. 3rd ed. Edward Arnold, London, 1973.

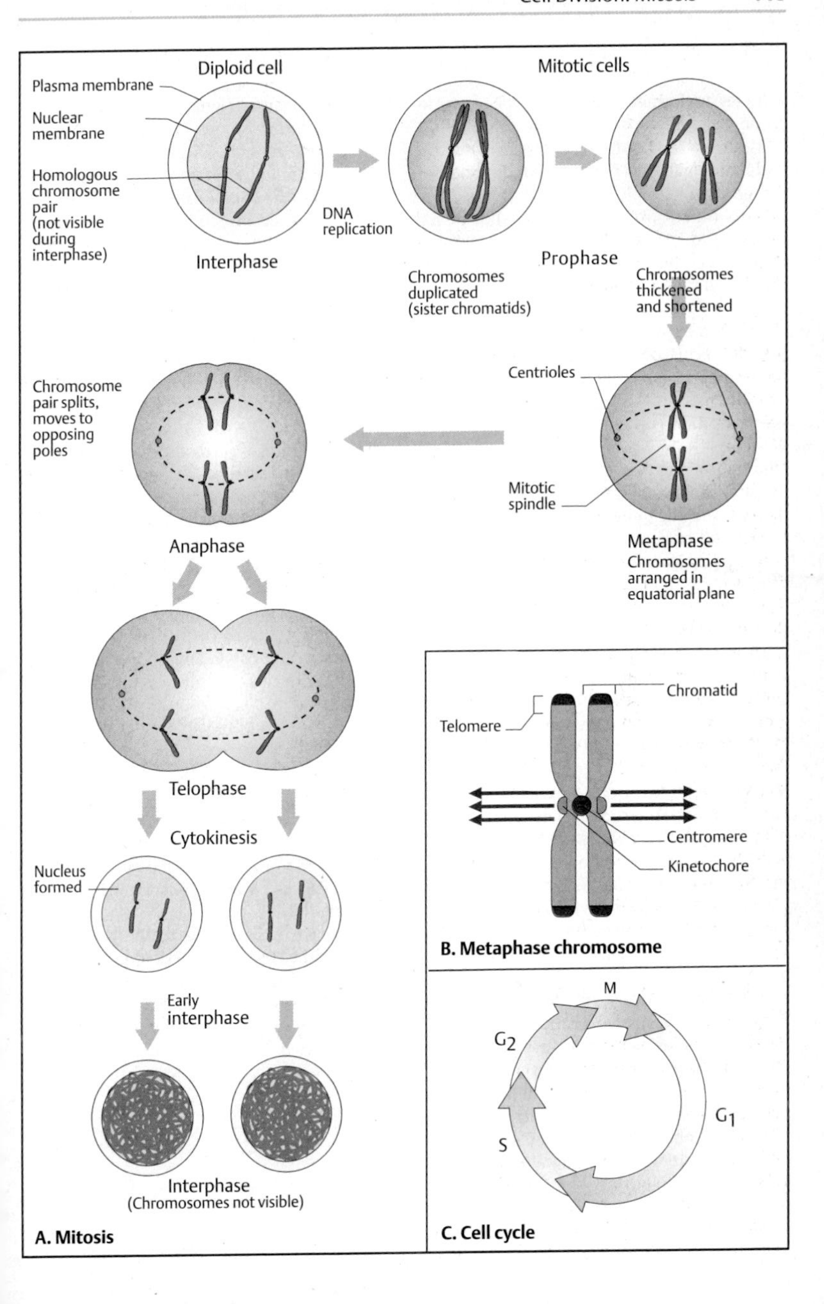
Diploid cell
Mitotic cells
Plasma membrane
Nuclear membrane
Homologous chromosome pair (not visible during interphase)
DNA replication
Interphase
Prophase
Chromosomes duplicated (sister chromatids)
Chromosomes thickened and shortened
Centrioles
Chromosome pair splits, moves to opposing poles
Mitotic spindle
Anaphase
Metaphase
Chromosomes arranged in equatorial plane
Chromatid
Telomere
Telophase
Centromere
Cytokinesis
Kinetochore
Nucleus formed
B. Metaphase chromosome
M
Early interphase
G2
G1
S
Interphase (Chromosomes not visible)
A. Mitosis
C. Cell cycle

## Maturation Division (Meiosis)

Gametes are formed by a special type of cell division that differs from somatic cell division. Strasburger (1884) introduced the term meiosis (reduction division or maturation division) for this process.

Meiosis differs from mitosis in fundamental genetic and cytological respects. Firstly, homologous chromosomes pair up. Secondly, exchanges between homologous chromosomes occur regularly (crossing-over). This results in chromosome segments with new constitutions (genetic recombination). Thirdly, the chromosome complement is halved during the first cell division (meiosis I). Thus, the daughter cells resulting from this division are haploid (reduction division).

Meiosis is a complex cellular and biochemical process. The cytologically observable course of meiosis and the genetic consequences do not correspond exactly in time. A genetic process occurring in one phase is usually not cytologically manifest until a subsequent phase.

### A. Meiosis I

A complete meiosis consists of two cell divisions, meiosis I and meiosis II. The relevant genetic events, genetic recombination by means of crossing-over and reduction to the haploid chromosome complement, occur in meiosis I.

Meiosis begins with chromosome replication. Initially the chromosomes in late interphase are visible only as threadlike structures, as in mitosis. At the beginning of prophase I, the chromosomes are doubled, but this is not visible until a later period of prophase I (see p. 118). Subsequently, the pairing of homologous chromosomes can also be visualized. The pairing makes an exchange between homologous chromosomes (crossing-over) possible by juxtapositioning homologous chromatids (chiasma formation). The result of crossing-over is an exchange of material between two chromatids of homologous chromosomes (genetic recombination). This exchange has occurred by the time the cell has entered metaphase I. After the homologous chromosomes migrate to opposite poles, anaphase I begins.

### B. Meiosis II

Meiosis II consists of longitudinal division of the doubled chromosomes (chromatids) and a further cell division. Each daughter cell contains one chromosome of a chromosome pair and is therefore haploid. Because of the recombination that occurred in prophase I, the chromosomes of the resulting haploid cell differ from those of the original cells. Thus, unlike in mitosis, the chromosomes of the daughter cells are not genetically identical with those of the original cell. On each chromosome, recombinant and nonrecombinant sections can be identified. The genetic events relevant to these changes have occurred in the prophase of meiosis I (see p. 118).

The distribution of chromosomes during meiosis explains the segregation (separation or splitting) of traits according to Mendelian laws (1 : 1 segregation, cf. p. 134).

Recombination is the most striking event in meiosis. The molecular mechanisms of recombination are complex. Occasionally, recombination may occur during mitosis (mitotic recombination), e.g., during DNA repair. Gene conversion designates a unilateral, nonreciprocal exchange. With this, one allele is lost in favor of another.

### References


Carpenter, A.T.C.: Chiasma function. Cell **77**:959–962, 1994.

McKim, K.S., Hawley, R.S.: Chromosomal control of meiotic cell division. Science **270**:1595–1601, 1995.

Moens, P.B., ed.: Meiosis. Academic Press, New York, 1987.

Whitehouse, L.H.K.: Towards the Understanding of the Mechanism of Heredity. 3rd ed. Edward Arnold, London, 1973.

Zickler, D., Kleckner, N.: Meiotic chromosomes: Integrating structure and function. Ann. Rev. Genet. **33**:603–754, 1999.

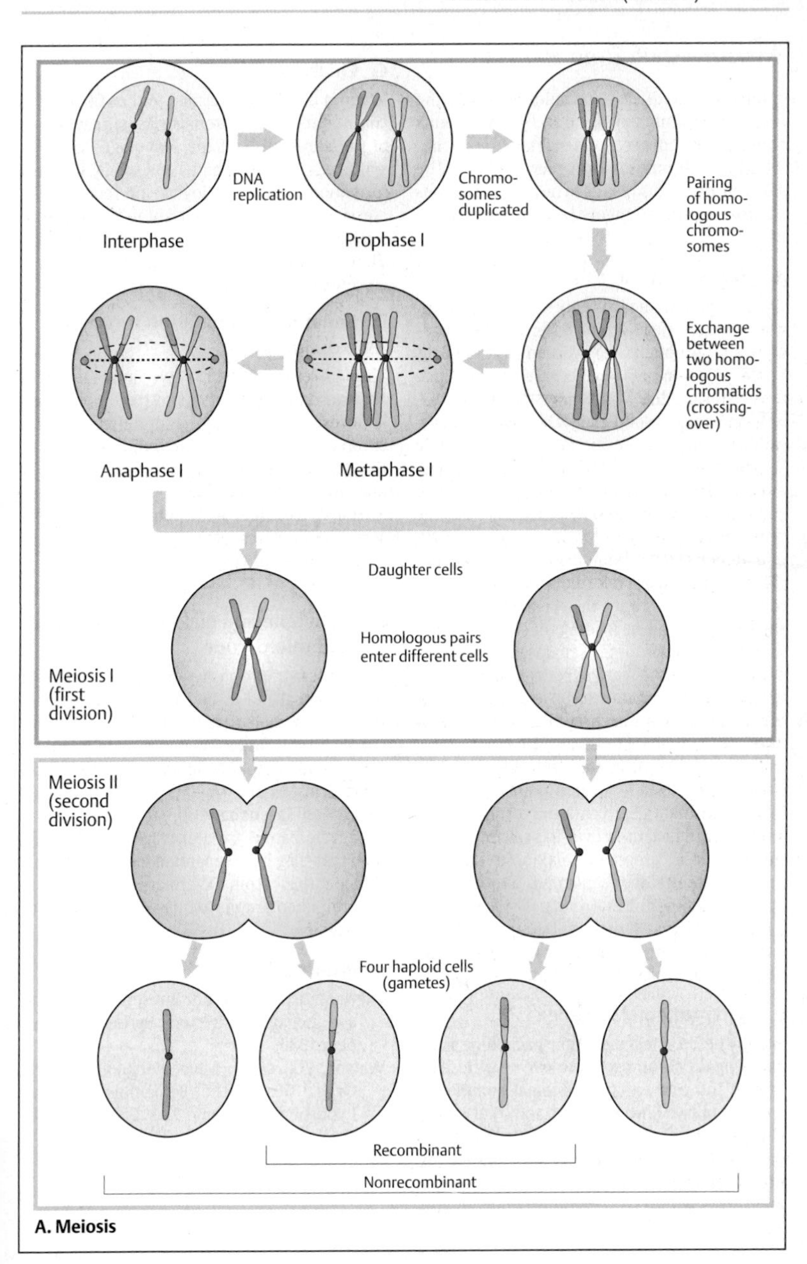
DNA replication
Chromo-somes duplicated
Pairing of homo-logous chromo-somes
Interphase
Prophase I
Exchange between two homo-logous chromatids (crossing-over)
Anaphase I
Metaphase I
Daughter cells
Homologous pairs enter different cells
Meiosis I (first division)
Meiosis II (second division)
Four haploid cells (gametes)
Recombinant
Nonrecombinant

**A. Meiosis**

# Crossing-Over in Prophase I

Prophase of meiosis I is a complex period during which important cytological and genetic events occur. In this phase, exchanges between homologous chromosomes occur regularly by crossing-over. This results in new combinations of chromosome segments (genetic recombination).

## A. Prophase of meiosis I

The prophase of meiosis I goes through a number of stages that can be differentiated schematically, although they proceed continuously. In the leptotene stage, the chromosomes are first visible as fine threadlike structures (in A only one chromosome pair is shown schematically). In zygotene the chromosomes are visible as paired structures. By this time, every chromosome has been doubled and consists of two chromatids, which are held together at the centromere (each chromatid contains a DNA double helix). Two homologous chromosomes that have paired are referred to as a bivalent. In the pachytene stage, the bivalents become thicker and shorter. In diplotene the two homologous chromosomes separate for the most part, but still remain attached to each other at a few points (chiasmata). In subsequent stages of diplotene, each of the chromosome pairs separates even more extensively (diakinesis), especially at the centromere region, but not yet at one or more distally located points of attachment (chiasmata). Each chiasma corresponds to a region in which crossover has taken place. In the last stage of prophase I, diakinesis, the chromosomes are widely separated, although still attached at their distal ends. The chiasmata have shifted distally (terminalization). At the end of diakinesis, the nuclear membrane disappears and the cell enters metaphase I.

## B. Synaptonemal complex

Shortly before the onset of the pachytene stage, homologous chromosomes move very close together and form a synaptonemal complex. It consists of two chromatids (1 and 2) of maternal origin (mat) and two chromatids (3 and 4) of paternal origin (pat). This initiates chiasma formation and is the prerequisite for crossing-over and subsequent recombination. (Diagram after Watson et al., 1987).

## C. Chiasmata

When a chiasma is formed, either of the two chromatids of one chromosome pairs with one of the chromatids of the homologus chromosome (e.g., 1 and 3, 2 and 4 and so on). Chiasma formation is the cytological prerequisite for crossing-over and is important in the definitive separation (segregation) of the chromosomes. The centromere (Cen) plays an important role in chromosome pairing.

## D. Genetic recombination through crossing-over

Through crossing-over, new combinations of chromosome segments arise (recombination). As a result, recombinant and nonrecombinant chromosome segments can be differentiated. In the diagram, the areas A–E (shown in red) of one chromosome and the corresponding areas a–e (shown in blue) of the homologous chromosome become respectively a–b–C–D–E and A–B–c–d–e in the recombinant chromosomes.

## E. Pachytene and diakinesis under the light microscope

In the micrograph, pachytene chromosomes are readily visualized as bivalents (a). An unusual structure in pachytene is formed by the X and the Y chromosomes. They appear to be joined end-to-end. Actually, short segments of the short arms in the regions with homologous sequences (pseudoautosomal region, see p. 390) have paired. In later stages (b), it can be seen that they have separated for the most part. (Photographs from Therman, 1986). Today, electron micrographs are usually used for meiotic studies.

## References

Therman, E.: Human Chromosomes: structure and behaviour. 2nd ed. Springer, Heidelberg, 1986.

Watson, J.D., et al.: Molecular Biology of the Gene. 3rd ed. The Benjamins/Cummings Publishing Co., Menlo Park, California, 1987

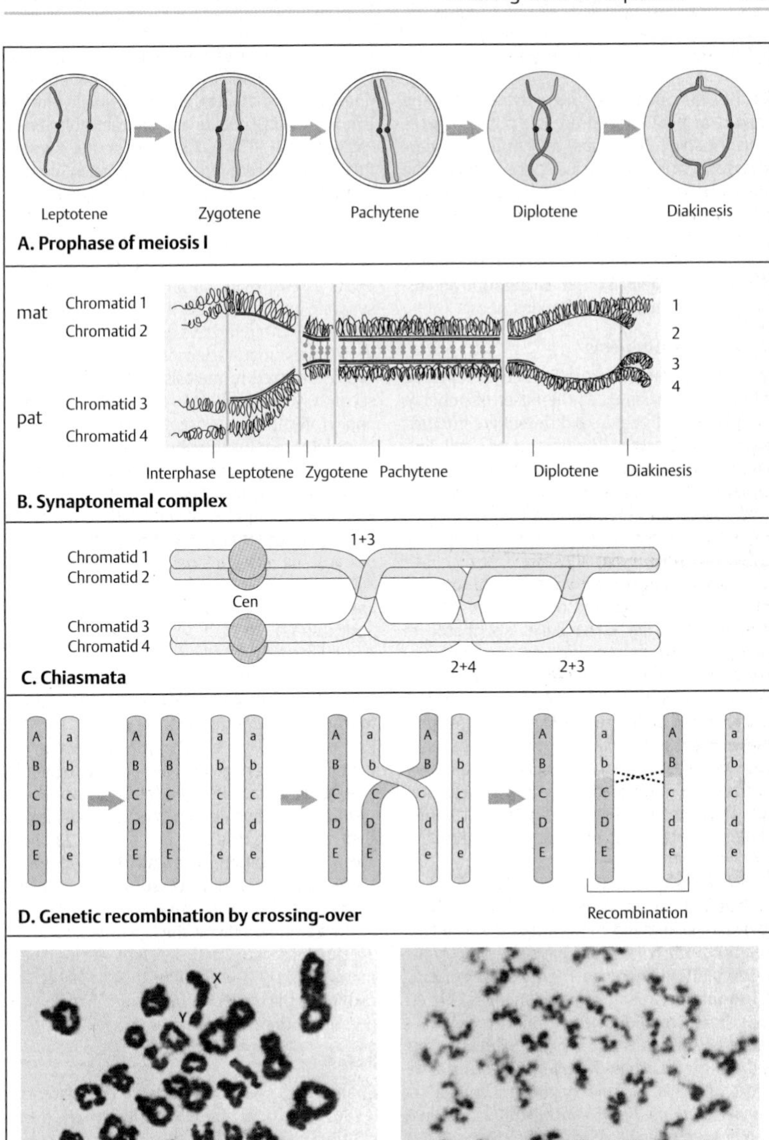


E. Photographs of pachytene and diakinesis under the light microscope

## Formation of Gametes

Germ cells (gametes) are produced in the gonads. In females the process is called oogenesis (formation of oocytes) and in males, spermatogenesis (formation of spermatozoa). The primordial germ cells, which migrate to the gonads during early fetal development, increase in number by mitotic division. The actual formation of germ cells (gametogenesis) begins with meiosis. Meiosis differs in duration and results between males and females.

### A. Spermatogenesis

Diploid spermatogonia are formed by repeated mitotic cell divisions. At the onset of puberty, some of the cells begin to differentiate into primary spermatocytes. The first meiotic cell division occurs in these cells. At the completion of meiosis I, a primary spermatocyte has given rise to two secondary spermatocytes, each of which has a haploid set of duplicated chromosomes (recombination is not illustrated here). Each chromosome consists of two sister chromatids, which become separated during meiosis II. In meiosis II, each secondary spermatocyte divides to form two spermatids. Thus, one primary spermatocyte forms four spermatids, each with a haploid chromosome complement. The spermatids differentiate into mature spermatozoa. Male spermatogenesis is a continuous process. In human males, the time lapse between differentiation into a primary spermatocyte at the onset of meiosis I and the formation of mature spermatocytes is about 6 weeks.

### B. Oogenesis

Oogenesis (formation of oocytes) differs from spermatogenesis in timing and in the result. At first the germ cells, which have migrated to the ovary, multiply by repeated mitosis (formation of oogonia). In human females, meiosis I begins about 4 weeks before birth. Primary oocytes are formed. However, meiosis I is arrested in a stage of prophase designated dictyotene. The primary oocyte persists in this stage until ovulation. Only then is meiosis I continued (recombination is not shown here).

In females, the cytoplasm divides asymmetrically in both meiosis I and meiosis II. The result each time is two cells of unequal size: a larger cell that will eventually form the egg and a small cell, called a polar body. When the primary oocyte divides, the haploid secondary oocyte and polar body I are formed. When the secondary oocyte divides, again unequally, the result is a mature oocyte and another polar body (polar body II). The polar bodies do not develop further, but degenerate. On rare occasions when this does not occur, a polar body may become fertilized. This can give rise to an incompletely developed twin.

In the secondary oocyte, each chromosome still exists as two sister chromatids. These do not separate until the next cell division (meiosis II), when they enter into two different cells. In most vertebrates, maturation of the secondary oocyte is arrested in meiosis II. At ovulation the secondary oocyte is released from the ovary, and if fertilization occurs, meiosis is then completed. Faulty distribution of the chromosomes (nondisjunction) may occur in meiosis I as well as in meiosis II (see p. 116).

The maximal number of germ cells in the ovary of the human fetus at about the 5th months is $6.8 \times 10^6$. By birth this has been reduced to $2 \times 10^6$, and by puberty to about 200,000. Of these, about 400 are ovulated (Connor & Ferguson-Smith, 1993).

The long period between meiosis I and ovulation is presumably a factor in the relatively frequent nondisjunction of homologous chromosomes in older mothers.

The difference in time in the formation of gametes during oogenesis and spermatogenesis is reflected in the difference in germline cell divisions. In the female there are 22 cell divisions before meiosis, resulting in a total of 23 chromosome replications. In contrast, 610 chromosome replications have taken place in the ancestral cells of spermatozoa produced in a male at the age of 40 (380 at the age of 30), yielding 25 times as many cell divisions during spermatogenesis (Crow, 2000). This probably accounts for the higher mutation rate in males, especially with increased paternal age.

#### References

Connor, J.M., Ferguson-Smith, M.A.: Essential Medical Genetics. 4th ed. Blackwell Scientific, London, 1993.

Crow, J.F.: The origins, patterns and implications of human spontaneous mutation. Nature Reviews **1** :40–47, 2000.

Hurst, L.D., Ellegren, H.: Sex biases in the mutation rate. Trends Genet. **14**:446–452, 1998.

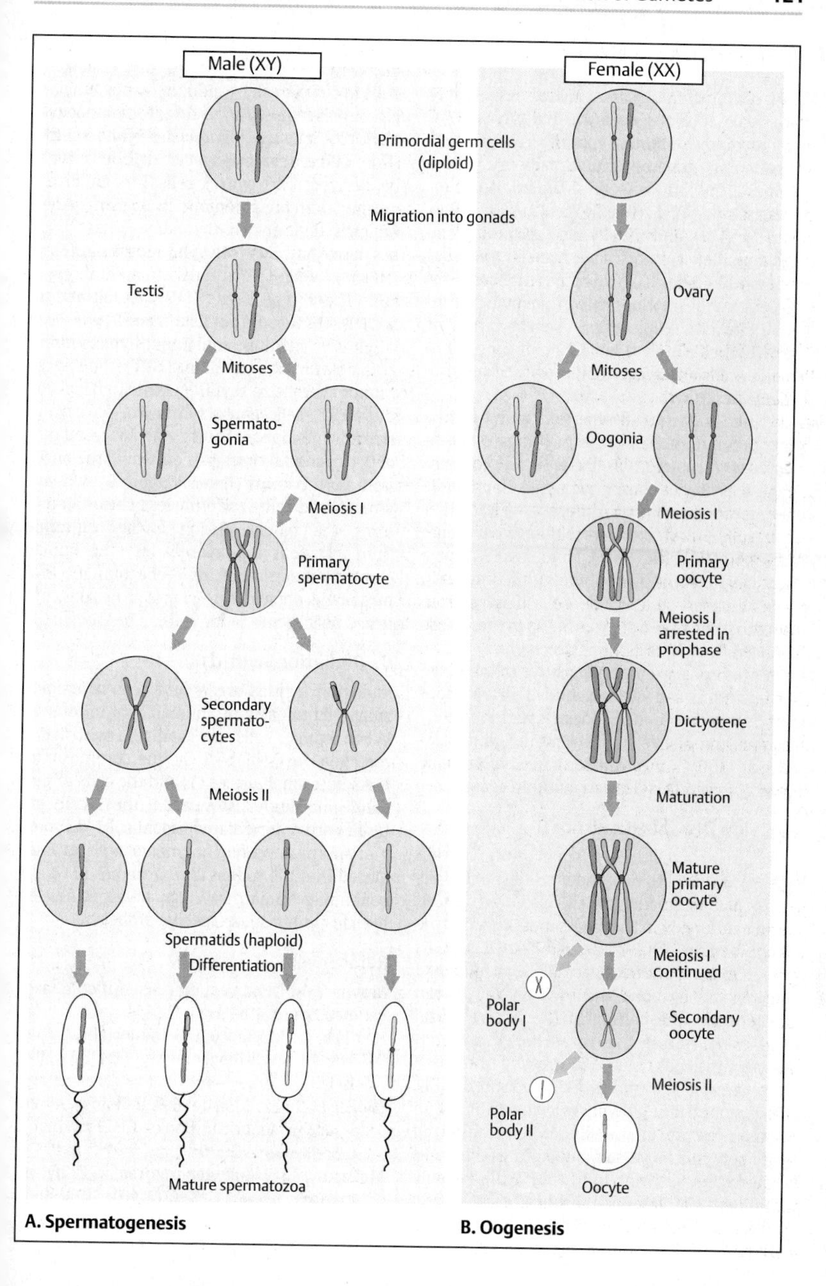


**A. Spermatogenesis**

**B. Oogenesis**

# Cell Culture

Cells of animals and plants can live and multiply in a tissue-culture dish (as a cell culture) at 37 °C in a medium containing vitamins, sugar, serum (containing numerous growth factors and hormones), the nine essential amino acids for vertebrate animals (His, Ile, Leu, Lys, Met, Phe, Thr, Tyr, Val), and usually also glutamine and cysteine. Cell cultures have been in wide use since 1965 and have become the basis for genetic studies not possible in the living mammals *(somatic cell genetics)*. A great variety of growth media are available for culturing mammalian cells and accommodating their requirements for growth.

The predominant cell type that grows from a piece of mammalian tissue in culture is the fibroblast. Although fibroblasts secrete proteins typical of fibrous connective tissue, in principle they retain the ability to differentiate into other cell types. Cultured skin fibroblasts have a finite life span (Hayflick, 1997). Human cells have a capacity for about 30 doublings until they reach a state called senescence. Cells derived from adult tissues have a shorter life span than those derived from fetal tissues.

Cultured cells are highly sensitive to increased temperature and do not survive above about 39 °C, whereas under special conditions they can be stored alive in vials kept in liquid nitrogen at −196 °C. They can be thawed after many years or even decades and cultured again.

## A. Skin fibroblast culture

To initiate a culture, a small piece of skin (2×4 mm) is obtained under sterile conditions and cut into smaller pieces, which are placed into a culture dish. The pieces must attach to the bottom of the dish. After about 8–14 days, cells begin to grow out from each piece and begin to multiply. They grow and multiply only when adhering to the bottom of the culture vessel (adhesion culture due to anchorage dependency of the cells). When the bottom of the culture vessel is covered by a dense layer of cells, they stop dividing owing to contact inhibition (this is lost in tumor cells). When transferred into new culture vessels (subcultures), the cells will resume growing until they again become confluent. By a series of subcultures, several millions of cells can be obtained for a given study.

## B. Hybrid cells for study

Cells in culture can be induced by polyethylene glycol or Sendai virus to fuse. If parental cells from different species are fused, interspecific (from different animal species) hybrid cells can be derived. The hybrid cells can be distinguished from the parental cells by using parental cells deficient in thymidine kinase ($TK^-$) or hypoxanthine phosphoribosyltransferase ($HPRT^-$). When cell cultures of parental type A ($TK^-$, 1) and type B ($HPRT^-$, 2) are cultured together, cells that did not fuse (3 and 5) will die in a selective medium containing hypoxanthine, aminopterin, and thymidine (HAT) (the $TK^-$ cell cannot synthesize thymidine monophosphate; the $HPRT^-$ cell cannot synthesize purine nucleoside monophosphates), whereas fused cells of both parental types (4) containing the nuclei from both parents (heterokaryon, 6) will survive. Hybrid cells randomly lose chromosomes from one parent during further culturing (human/rodent cell hybrids lose the human chromosomes). When only one human chromosome is present, the genetic properties conferred by its genes to the cell can be studied (8).

## C. Radiation hybrids

Radiation hybrids are rodent cells containing small fragments of human chromosomes. When human cells are irradiated with lethal roentgen doses of 3–8 Gy, the chromosomes break into small pieces (1) and the cells cannot divide in culture. However, if these cells are fused with nonirradiated rodent cells (2), some human chromosome fragments will be integrated into the rodent chromosomes (3). Cells containing human DNA can be identified by human chromosome-specific probes.

### References

Brown, T.A.: Genomes. Bios Scientific Publishers, Oxford, 1999.

Hayflick, L.: Mortality and immortality at the cellular level. Biochemistry **62**:1180–1190, 1997.

Lodish, H., et al.: Molecular Cell Biology. 4th ed. Scientific American Books, F.H. Freeman & Co., New York, 2000.

McCarthy, L.: Whole genome radiation hybrid mapping. Trends Genet. **12**:491–493, 1996.

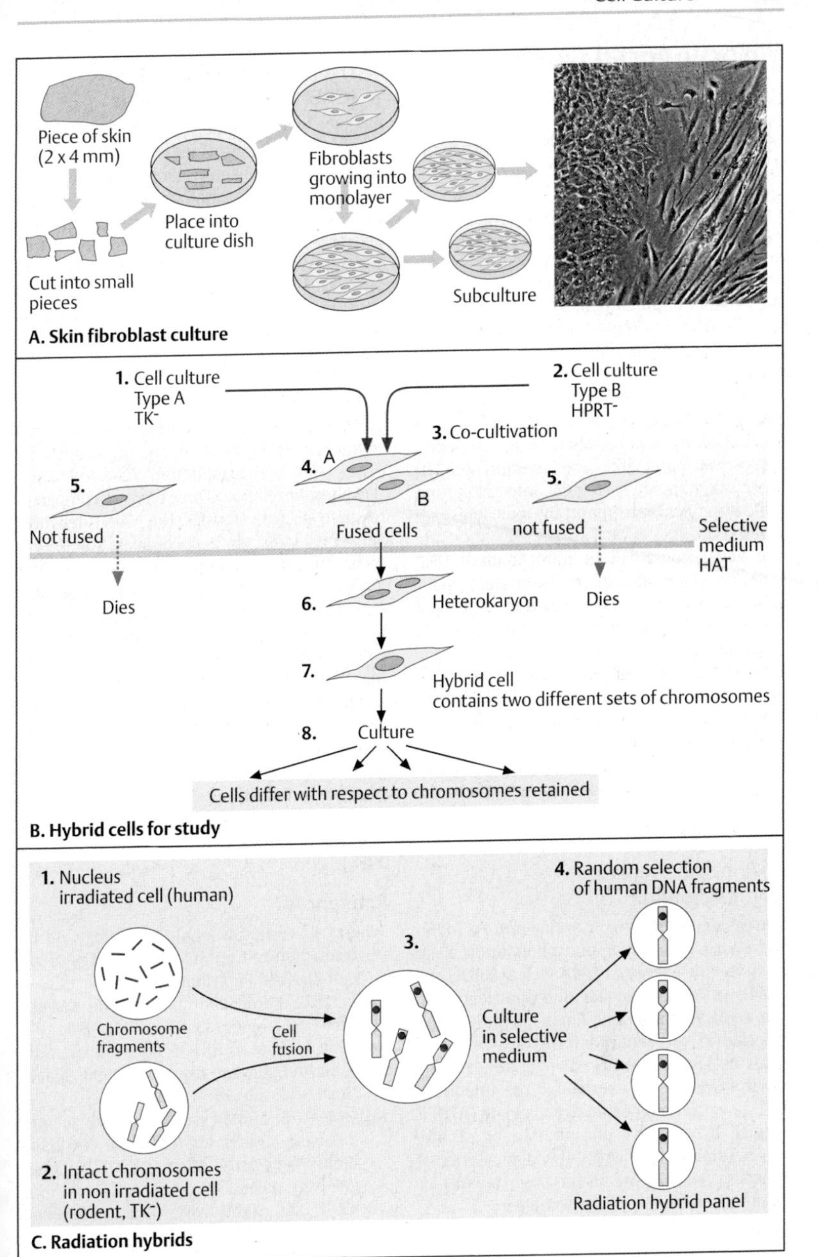
Piece of skin
(2 x 4 mm)
Cut into small pieces
Place into culture dish
Fibroblasts growing into monolayer
Subculture
A. Skin fibroblast culture
1. Cell culture
Type A
TK⁻
2. Cell culture
Type B
HPRT⁻
3. Co-cultivation
4. A
B
Fused cells
5.
Not fused
5.
not fused
Selective medium HAT
Dies
Dies
6. Heterokaryon
7. Hybrid cell
contains two different sets of chromosomes
8. Culture
Cells differ with respect to chromosomes retained
B. Hybrid cells for study
1. Nucleus irradiated cell (human)
Chromosome fragments
Cell fusion
2. Intact chromosomes in non irradiated cell (rodent, TK⁻)
3.
Culture in selective medium
4. Random selection of human DNA fragments
Radiation hybrid panel
C. Radiation hybrids

# Mitochondrial Genetics

## Genetically Controlled Energy-Delivering Processes in Mitochondria

Eukaryotic organisms contain essential genetic information separate from the nuclear DNA, in extrachromosomal genomes called mitochondria. The mitochondria of all eukaryotes and the chloroplasts of green plants and algae contain circular DNA molecules (mitochondrial DNA, mtDNA). Each eukaryotic cell contains $10^3 - 10^4$ copies. Mitochondria and chloroplasts are the site of essential energy-delivering processes and photosynthesis. Each mitochondrion is surounded by two highly specialized membranes, the outer and inner membranes. The inner membrane is folded into numerous cristae and contains important molecules (see C).

The genes contained in mitochondrial DNA code for 13 proteins of the respiratory chain, subunits of the ATPase complex, and subunits of the NADH-dehydrogenase complex (ND); 22 other genes code for transfer RNA (tRNA) and two rRNAs. A number of diseases due to mutations and deletions in mtDNA are known in humans. Sequence homologies indicate evolutionary relationships. In particular, evolutionary transfer of DNA segments from chloroplasts to mitochondria, and from chloroplasts to nuclear DNA of eukaryotic organisms, has been demonstrated.

### A. Principal events in mitochondria

The essential energy-conserving process in mitochondria is oxidative phosphorylation. Relatively simple energy carriers such as NADH and $FADH_2$ (nicotinamide–adenine dinucleotide in the reduced form and flavin adenine dinucleotide in the reduced form) are produced from the degradation of carbohydrates, fats, and other foodstuffs by oxidation. The important energy carrier adenosine triphosphate (ATP) is formed by oxidative phosphorylation of adenosine diphosphate (ADP) through a series of biochemical reactions in the inner membrane of mitochondria (respiratory chain).

### B. Oxidative phosphorylation (OXPHOS) in mitochondria

Adenosine triphosphate (ATP) plays a central role in the exchange of energy in biological systems. ATP is a nucleotide consisting of adenine, a ribose, and a triphosphate unit. It is energy-rich because the triphosphate unit contains two phospho-anhydride bonds. Energy (free energy) is released when ATP is hydrolyzed to form ADP. The energy contained in ATP and bound to phosphate is released, for example, during muscle contraction.

### C. Electron transfer in the inner mitochondrial membrane

The genomes of mitochondria and chloroplasts contain genes for the formation of the different components of the respiratory chain and oxidative phosphorylation. Three enzyme complexes regulate electron transfer: the NADH-dehydrogenase complex, the b–$c_1$ complex, and the cytochrome oxidase complex (C). Intermediaries are quinone (Q) derivatives such as ubiquinone and cytochrome *c*. Electron transport leads to the formation of protons ($H^+$). These lead to the conversion of ADP and $P_i$ (inorganic phosphate) into ATP (oxidative phosphorylation). ATP represents a phosphate-bound reservoir of energy, which serves as an energy supplier for all biological systems. Thus it is understandable that genetic defects in mitochondria become manifest primarily as diseases with reduced muscle strength and other degenerative signs. (Figures adapted from Bruce Alberts et al., 1998).

### References

Alberts, B., et al.: Essential Cell Biology. An Introduction to the Molecular Biology of the Cell. Garland Publishing, New York, 1998.

Johns, D.R.: Mitochondrial DNA and disease. New Eng. J. Med. **333**:638–644, 1995.

Kogelnik, A.M., et al.: MITOMAP: a human mitochondrial genome database—1998 update. Nucl. Acids Res. **26**:112–115, 1998.

MITOMAP: A human mitochondrial genome database: Center for Molecular Medicine, Emory University, Atlanta, GA, USA, 2000. (Website see p. 130)

Turnball, D.M., Lighttowlers, R.N.: An essential guide to mtDNA maintenance. Nature Genet. **18**:199–200, 1998.

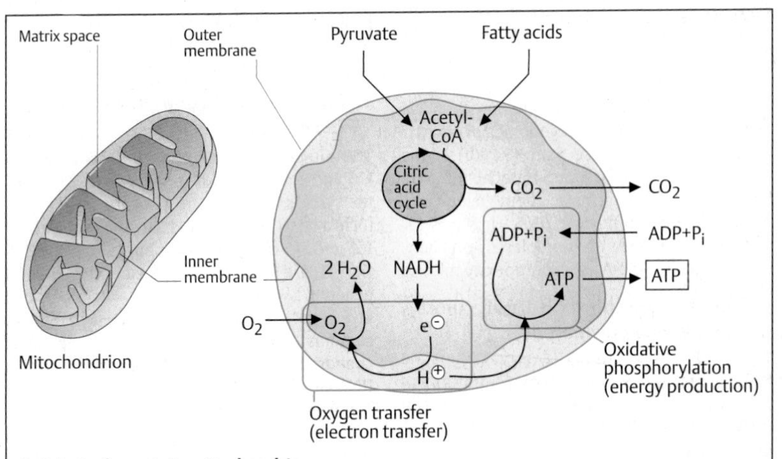


**A. Principal events in mitochondria**

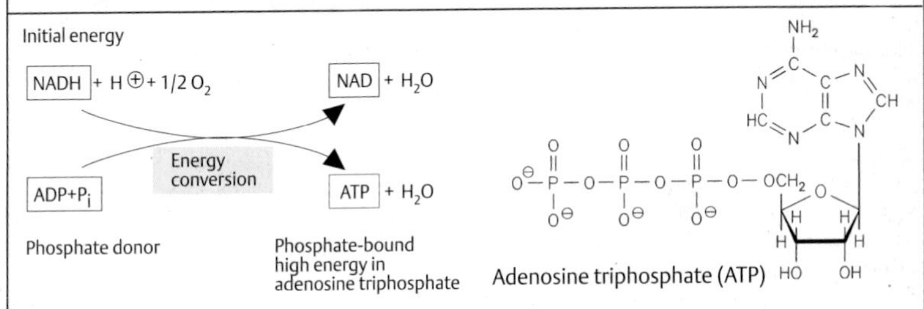


**B. Oxidative phosphorylation (OXPHOS) in mitochondria**

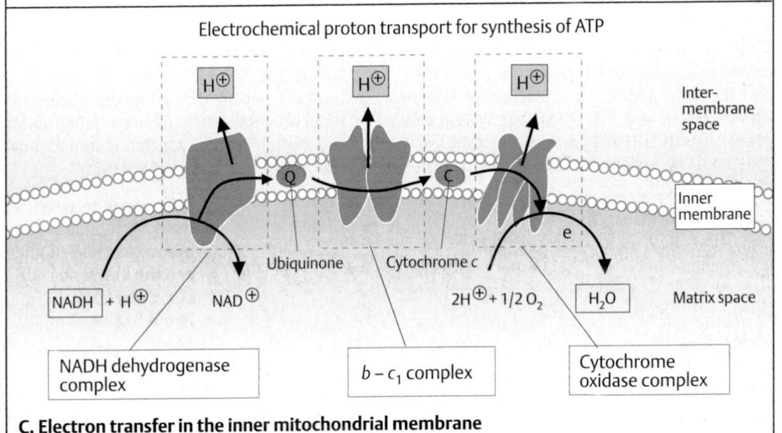


**C. Electron transfer in the inner mitochondrial membrane**

# The Genome in Chloroplasts and Mitochondria

Chloroplasts of higher plants and mitochondria of eukaryotic cells contain genomes of circular DNA. About 12000 base pairs (12 kb) of the genomes of chloroplasts and mitochondria are homologous. Furthermore, homologous regions are found in nuclear DNA. Thus it is assumed that the DNAs of chloroplasts, mitochondria, and nuclear DNA are evolutionarly related.

## A. Genes in the chloroplasts of a moss (*Marchantia polymorpha*)

The genome in chloroplasts (ctDNA) is large: 121 kb in the moss *M. polymorphia* and 155 kb in the tobacco plant. Yet the organization and number of their genes are similar. Protein synthesis shows certain similarities with that of bacteria. Many of the ribosomal proteins are homologous with those of *E. coli.* Genes in the chloroplast genomes are interrupted and contain introns. Each chloroplast contains about 20–40 copies of ctDNA and there about are 20–40 chloroplasts per cell. Among these are genes for two copies each of four ribosomal RNAs (16S rRNA, 23S rRNA, 4.5S rRNA, and 5S rRNA). The genes for ribosomal RNA are located in two DNA segments with opposite orientation (inverted repeats), which are characteristic of chloroplast genomes. An 18–19-kb segment with short single gene copies lies between the two inverted repeats. The genome of chloroplasts contains genetic information for about 30 tRNAs and about 50 proteins. The proteins belong to photosystem I (two genes), photosystem II (seven genes), the cytochrome system (three genes), and the $H^+$-ATPase system (six genes). The NADH dehydrogenase complex is coded for by six genes; ferredoxin by three genes; and ribulose by one gene. Twenty-nine genes have not been identified to data. (Data after Lewin, 2000).

## B. Mitochondrial genes in yeast (*S. cerevisiae*)

The mitochondrial genome of yeast is large (120 kb) and contains introns. It contains genes for the tRNAs, for the respiratory chain (cytochrome oxidase 1, 2, and 3; cytochrome *b*), for 15S and 21S rRNA, and for subunits 6, 8, and 9 of the ATPase system. The yeast mitochondrial genome is remarkable because its ribosomal RNA genes are separated. The gene for 21S rRNA contains an intron. About 25% of the mitochondrial genome of yeast contains AT-rich DNA without a coding function.

The genetic code of the mitochondrial genome differs from the universal code in nuclear DNA with respect to usage of some codons. The nuclear stop codon UGA codes for tryptophan in mitochondria, while the nuclear codons for arginine (AGA and AGG) function as stop codons in mammalian mitochondria.

## References


Alberts, B., et al.: Molecular Biology of the Cell. 3 rd ed. Garland Publishing, New York, 1994.

Alberts, B., et al.: Essential Cell Biology. An Introduction to the Molecular Biology of the Cell. Garland Publishing, New York, 1998.

Lewin, B.: Genes VII. Oxford Univ. Press, Oxford, 2000.

Ohyama, K., et al.: Chloroplast gene organization deduced from complete sequence of liverwort *M. polymorpha* chloroplast DNA. Nature **322**:572–574, 1986.

Shinozaki, K., et al.: The complete nucleotide sequence of the tobacco chloroplast genome: its gene organization and expression. EMBO J. **5**:2043–2049, 1986.


| Codon | Nuclear DNA | Mitochondrial genome | | | |
|---|---|---|---|---|---|
| | | Mammals | Drosophila | Yeast | Plants |
| UGA | Stop | Trp | Trp | Trp | Stop |
| AUA | Ile | Met | Met | Met | Ile |
| CUA | Leu | Leu | Leu | Thr | Leu |
| AGA } AGG } | Arg | Stop | Ser | Arg | Arg |

Differences between the genetic code of the mitochondrial genome and the universal code of nuclear DNA (From B. Alberts et al., 1994).

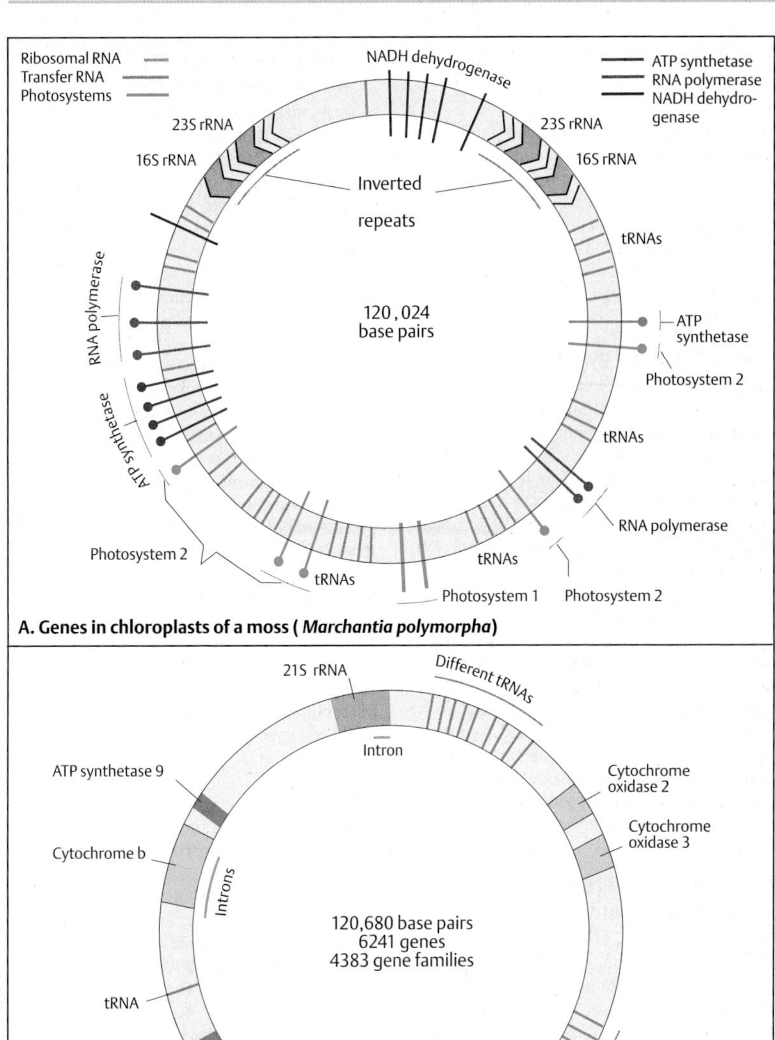


**A. Genes in chloroplasts of a moss (*Marchantia polymorpha*)**

**B. Mitochondrial genes in yeast (*S. cerevisiae*)**

# The Mitochondrial Genome of Man

The mitchondrial genome in mammals is small and compact. It contains no introns, and in some regions the genes overlap, so that practically every base pair belongs to a gene. The mitochondrial genomes of humans and mice have been sequenced and contain extensive homologies. Each consists of about 16.5 kb, i.e., is considerably smaller than a yeast mitochondrial genome.

## A. Mitochondrial genes in man

Each mitochondrion contains 2–10 DNA molecules. The mitochondrial genome of humans contains 13 protein-coding regions: genes for the cytochrome *c* oxidase complex (subunits 1, 2, and 3), for cytochrome *b*, and for subunits 6 and 8 of the ATPase complex. Unlike that of yeast, mammalian mitochondrial DNA contains seven subunits for NADH dehydrogenase (ND1, ND2, ND3, ND4L, ND4, ND5, and ND6). Of the mitochondrial coding capacity, 60% is taken by the seven subunits of NADH reductase (ND). A heavy (H) and a light (L) single strand can be differentiated by a density gradient. Most genes are found on the H strand. The L strand codes for a protein (ND subunit 6) and 8 tRNAs. From the H strand, two RNAs are transcribed, a short one for the rRNAs and a long one for mRNA and 14 tRNAs. A single transcript is made from the L strand. A 7S RNA is transcribed in a counterclockwise manner close to the origin of replication (ORI) (not shown).

## B. Cooperation between mitochondrial and nuclear genome

Some mitochondrial proteins are aggregates of gene products of nuclear and mitochondrial genes. These gene products are transported into the mitochondria after nuclear transcription and cytoplasmic translation. In the mitochondria, they form functional proteins from subunits of mitochondrial and nuclear gene products. This explains why a number of mitochondrial genetic disorders show Mendelian inheritance, while purely mitochondrially determined disorders show exclusively maternal inheritance.

## C. Evolutionary relationship of mitochondrial genomes

Mitochondrial DNA has a mutation rate tentimes higher than that of nuclear DNA. Mutations are generated during OXPHOS (p. 124) through pathways involving reactive oxygen molecules. Mutations accumulate because effective DNA repair and protective histones are lacking. At birth most mtDNA molecules are identical (homoplasmy); later they differ as a result of mutations accumulated in different mitochondria (heteroplasmy).

Mitochondria probably evolved from independent organisms that were integrated into cells. They replicate, transcribe, and translate their DNA independently of the nuclear DNA. Since mitochondria are present in oocytes only, and not spermatozoa, their genes are exclusively inherited through the mother. The maternal inheritance can be used effectively in studies of evolutionary relationships. An evolutionary tree of related populations can be reconstructed by comparing mtDNA variants, called haplotypes, based on sequence differences. A phylogenetic tree reconstructed from 147 humans of different geographic regions thoughout the world suggested that an ancestral sequence existed about 200000 years ago (dubbed "Eve"). Mitochondrial DNA can be studied from ancient biological specimens such as Neandertals (see p. 258).

## References

Anderson, S., et al.: Sequence and organization of the human mitochondrial genome. Nature **290**:457–474, 1981.

Lang, B.F., et al.: Mitochondrial genome evolution and the origin of eukaryotes. Ann. Rev. Genet. **33**:351–397, 1999.

Singer, M., Berg, P.: Genes and Genomes. Blackwell Scientific Publishers, Oxford, 1991.

Suomalainen, A., et al.: An autosomal locus predisposing to deletions of mitochondrial DNA. Nature Genet. **9**:146–151, 1995.

Wallace, D.C.: Mitochondrial diseases: genotype versus phenotype. Trends Genet. **9**:128–133, 1993.

Wallace, D.C.: Mitochondrial DNA sequence variation in human evolution and disease. Proc. Nat. Acad. Sci. **91**:8739–8746, 1994.

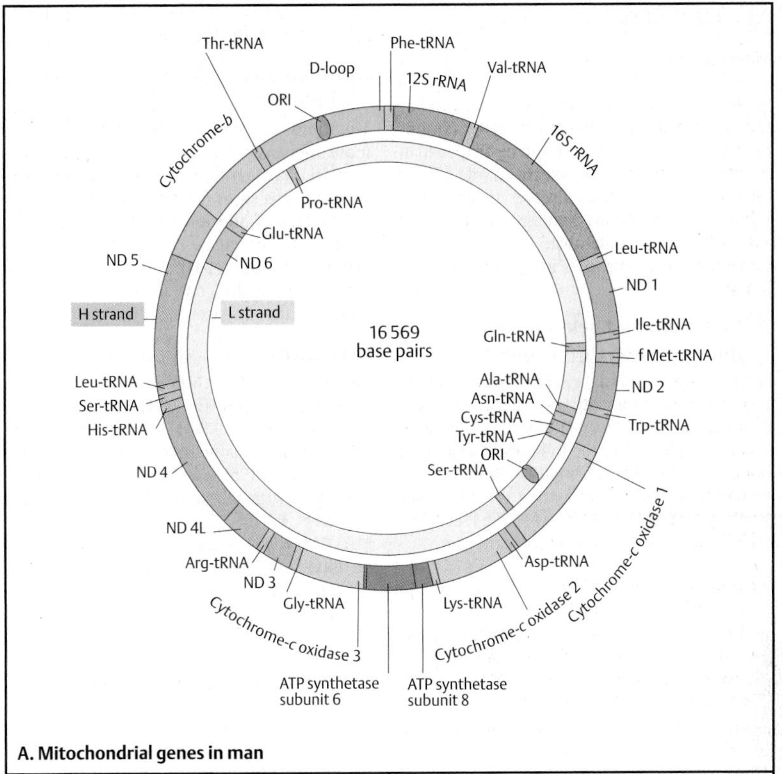


**A. Mitochondrial genes in man**

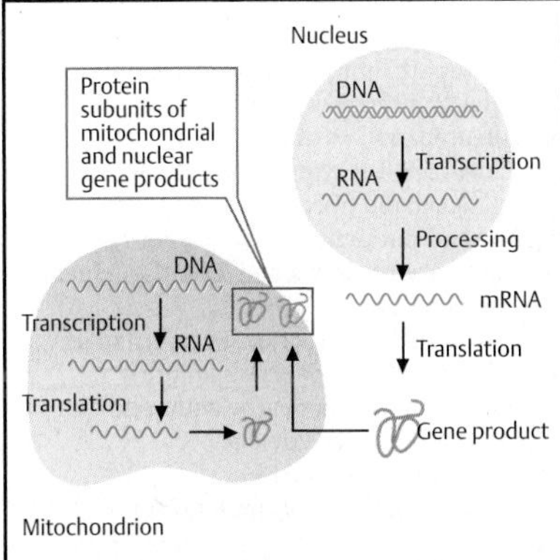


**B. Cooperation between mitochondrial and nuclear genome**

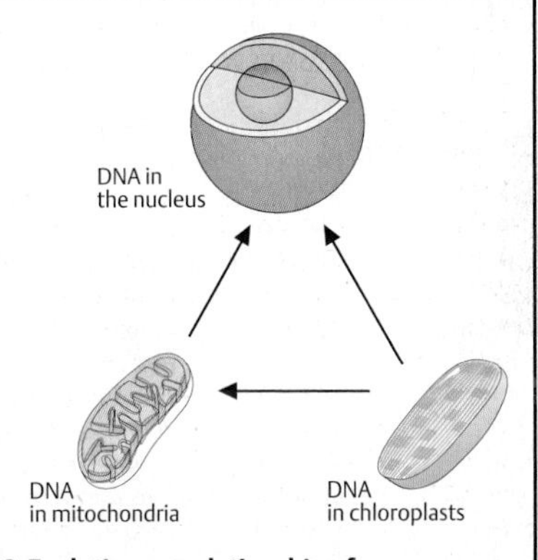


**C. Evolutionary relationship of mitochondrial genomes**

## Mitochondrial Diseases

Mitochondrial function can be disrupted by genetic changes involving one or several of the respiratory chain proteins or the tRNAs or rRNAs. In addition, the interaction of mitochondrial and nuclear genes can be disturbed in many ways. The clinical spectrum and age of onset of mitochondrial diseases is extremely wide. Organs with high energy requirements are particularly vulnerable: the brain, heart, skeletal muscle, eye, ear, liver, pancreas, and kindey.

### A. Mutations and deletions in mitochondrial DNA in man

Both deletions and point mutations are causes of mitochondrial genetic disorders. Some are characteristic and recur in different, unrelated patients. Panel A and the table show examples of important mutations and deletions, and mitochondrial diseases. (Figure adapted from Wallace, 1999; MITOMAP; and Marie T. Lott and D. C. Wallace, personal communication).

### B. Maternal inheritance of a mitochondrial disease

Hereditary mitochondrial diseases are transmitted only through the maternal line, since spermatozoa contain hardly any mitochondria. Thus the disease will not be transmitted from an affected man to his children.

### C. Heteroplasmy for mitochondrial mutations

Mutations or deletions in mitochondria are more frequently limited to a single tissue (mitochondrial cytopathy) than are germline mutations. In such cases, the cells contain different proportions of affected mitochondria (heteroplasmy). The proportion of defective mitochondria varies after repeated cell divisions. This contributes to the considerable variability of mitochondrial diseases.

### References


Chinnery, P.E., et al.: Mitochondrial genetics. J. Med. Genet. **36**:425–436, 1999.

Estivill, X., et al.: Familial progressive sensorineural deafness is mainly due to the mtDNA: A 1555 G and is enhanced by treatment with aminoglycosides. Am. J. Hum. Genet. **62**:27–35, 1998.

MITOMAP: A human mitochondrial genome database: (http://www.gen.emory.edu/mitomap.html) Center for Molecular Medicine, Emory University, Atlanta, GA, USA, 2000.

Wallace, D.C.: Mitochondrial diseases in man and mice. Science **283**:1482–1488, 1999.


Examples of diseases due to mutations or deletions in mitochondrial DNA

| Abbreviation | McKusick No. | Name |
|---|---|---|
| KSS | 530000 | Kearns–Sayre syndrome (ophthalmoplegia, pigmentary degeneration of the retina, and cardiomyopathy) |
| LHON | 535000 | Leber hereditary optic neuropathy |
| MELAS | 540000 | Encephalomyopathy, Lactic-acidosis with strokelike symptoms |
| MERRF | 545000 | Myoclonic epilepsy and ragged red fibers |
| MMC | 590050 | Mitochondrial myopathy and cardiomyopathy |
| NARP | 551500 | Neurogenic muscle weakness with ataxia and retinitis pigmentosa |
| CEOP | 555000 | Chronic external ophthalmoplegia |
| MNGIE | 550900 | Myoneurogastrointestinal pseudoobstruction, Encephalopathy |
| PEAR | 557000 | Pearson syndrome (bone marrow and pancreas failure) |
| ADMIMY | 157640<br>550000 | Autosomal dominant mitochondrial myopathy with mitochondrial deletions in the D-loop (type Zeviani) |
| None | 515000 | Chloramphenicol induced toxicity |
| None | 580000 | Streptomycin-induced ototoxicity (A1555G mutation) |
| None | 520000 | Diabetes mellitus–deafness |

MIM: Mendelian Inheritance in Man, McKusick's Catalog of Human Genes and Genetic Disorders. 12th ed., Johns Hopkins University Press, Baltimore, 1998; MITOMAP; and *http://www.mips.biochem.mpg.de/proj/medgen/mitop/*

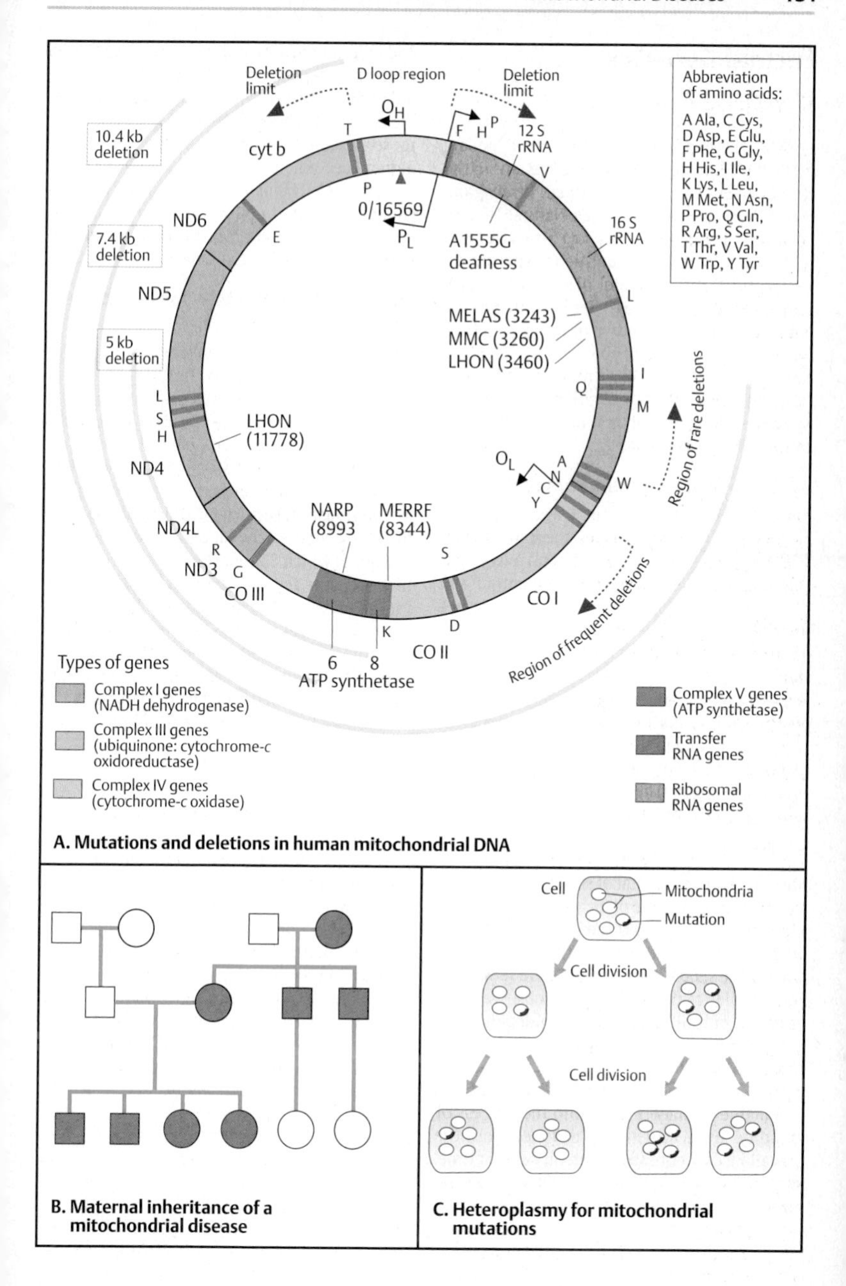


**A. Mutations and deletions in human mitochondrial DNA**

**B. Maternal inheritance of a mitochondrial disease**

**C. Heteroplasmy for mitochondrial mutations**

# Formal Genetics

## The Mendelian Traits

In 1865, the Augustinian monk Gregor Mendel published some remarkable observations in the *Berichte der Naturgeschichtlichen Vereinigung von Brünn* (Proceedings of the Natural History Society of Brno), which received little attention at the time. In this work, entitled "Experiments on Hybrid Plants," Mendel observed that certain traits in garden peas *(Pisum sativum)* are inherited independently of one another. Moreover, Mendel described certain regularities in the pattern of occurrence of individual traits in consecutive generations. His experimental system, the plants, and the observed traits will be presented here.

### A. The pea plant (*Pisum sativum*)

The plant consists of a stem, leaves, blossoms, and seedpods. In the blossom, the (female) pistil (comprising the stigma, style, and ovary) and the (male) stamen, comprising the anther and filament, can be differentiated. The garden pea normally reproduces by self-fertilization. Pollen from the anther falls onto the stigma of the same blossom.
However, it is relatively easy to carry out cross-fertilization. To do this, Mendel opened a blossom and removed the anther before pollen could escape and used pollen from another blossom instead. The resulting seedpod contains the seeds from which new plants develop.

### B. The observed traits (phenotypes)

Mendel observed a total of seven characteristic traits: (1) height of the plants, (2) location of the blossoms on the stem of the plant, (3) the color of the pods, (4) the form of the pods, (5) the form of the seeds, (6) the color of the seeds, and (7) the color of the seed coat. He observed that each pair of traits was inherited independently from all other pairs of traits. Mendel's main observation was that independent traits are inherited in certain predictable patterns. This was a fundamental new insight into the process of heredity. Since it distinctly deviated from the prevailing concepts about heredity at that time, its significance was not immediately recognized.

Today it is known that genetically determined traits are independently inherited (segregation) only when they are located on different chromosomes or far enough apart on the same chromosome to be separated each time by recombination (i.e., when there is no genetic linkage; see pp. 144). This is true of the genes investigated by Mendel. In recent years, some of these genes have been cloned and their molecular structures have been characterized.

### Deviation from the Mendelian pattern of inheritance

Mendelian traits do not always occur in the expected proportions (see p. 134). As a result of the phenomenon of meiotic drive, one trait may occur much more frequently than others. Examples exist in the t complex of the mouse (about 99% instead of 50% of offspring of heterozygous t/+ male mice are also heterozygous) and Segregation Distorter (SD) in *Drosophila.*
In 1993, a mouse population in Siberia was described in which 85% and 65% of offspring were heterozygous for an inversion. Homozygosity for the inversion leads to reduced fitness and is a selective disadvantage. The shift was not due to early embryonal death. Possibly, deviations from Mendelian laws are more frequent than previously assumed. Further deviations occur with genomic imprinting (see p. 226) and germline mosaicism.

### References


Brink, R.A., Styles, E.D.: Heritage from Mendel. Univ. of Wisconsin Press, Madison, 1967.

Corcos, A.F., Monaghan F.V.: Gregor Mendel's Experiments on Plant Hybrids. Rutgers Univ. Press, New Brunswick, 1993.

Mendel, G.: Versuche über Pflanzenhybriden. Verh. naturf. Ver. Brünn **4**:1–47, 1866.

Pomiankowski, A., Hurst, D.L.: Siberian mice upset Mendel. Nature 363: 396–397, 1993.

Weiling, F.: Johann Gregor Mendel: Der Mensch und Forscher. II Teil. Der Ablauf der Pisum Versuche nach der Darstellung. Med. Genetik. 2: 208–222, 1993.

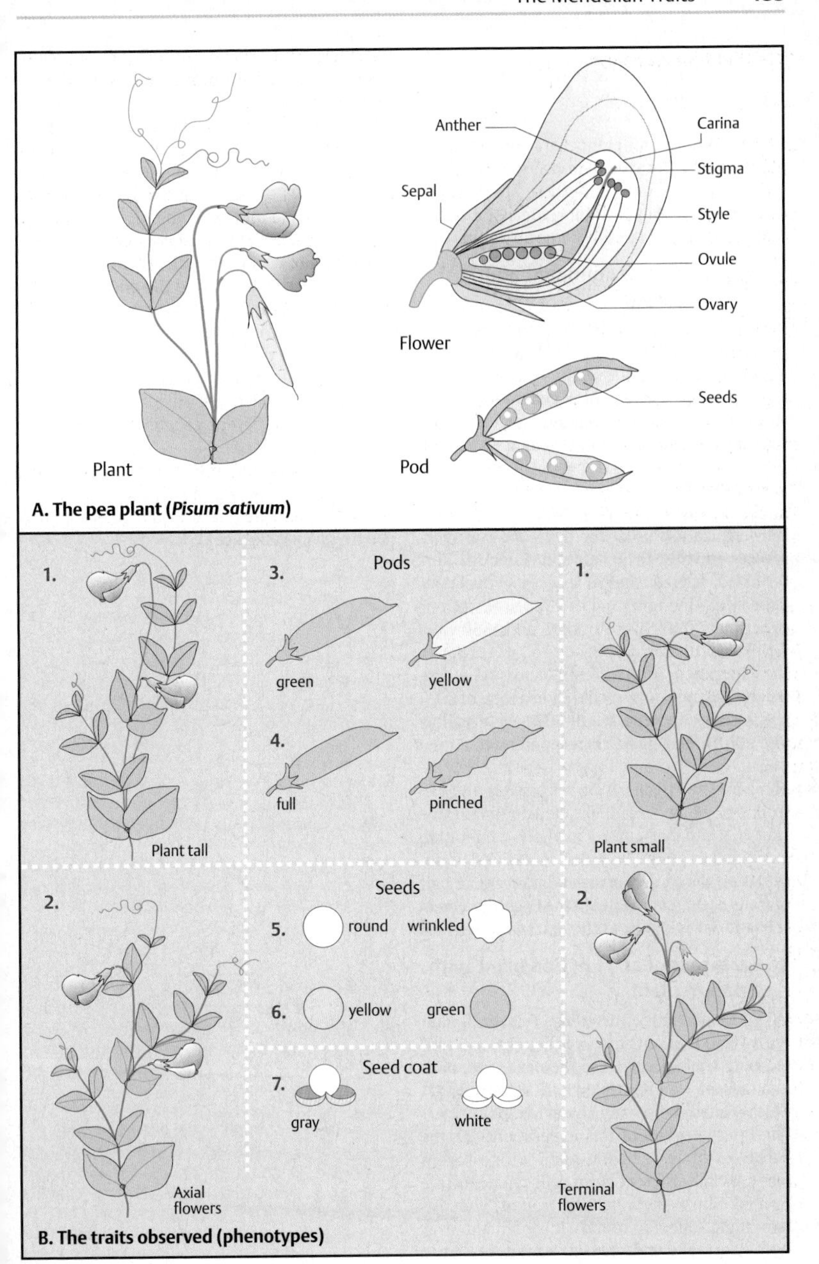


**A. The pea plant (*Pisum sativum*)**

**B. The traits observed (phenotypes)**

# Distribution (Segregation) of Mendelian Traits

Mendel observed different characteristics of the pea plant and followed their occurrence in consecutive generations. From these observations, certain regularities in their patterns of occurrence (Mendelian laws) became apparent.

## A. Segregation of dominant and recessive traits

In two different experiments, Mendel observed the shape (smooth or wrinkled) and the color (yellow or green) of the seeds. When he crossed the plants of the parental generation P, i.e., smooth × wrinkled or yellow × green, he observed that in the first filial (daughter) generation, $F_1$, all seeds were respectively smooth and yellow.

In the next generation, $F_2$, which arose by the self-fertilization usual for peas, the traits observed in the P generation (smooth and wrinkled, or green and yellow, respectively) reappeared. Among a total of 7324 seeds of one experiment, 5474 were smooth and 1850 were wrinkled. This corresponded to a ratio of 3:1. In the experiment with different colors (green vs. yellow), Mendel observed that in a total of 8023 seeds of the $F_2$ generation, 6022 were yellow and 2001 green, again corresponding to a ratio of 3:1.

The trait the $F_1$ generation exclusively showed (round or yellow), Mendel called dominant; the trait that did not appear in the $F_1$ generation (wrinkled or green) he called recessive. His observation that a dominant and a recessive pair of traits occur (segregate) in the $F_2$ generation in the ratio 3:1 is known as the first law of Mendel.

## B. Backcross of an $F_1$ hybrid plant with a parent plant

When Mendel backcrossed the $F_1$ hybrid plant with a parent plant showing the recessive trait (1), both traits occurred in the next generation in a ratio of 1:1 (106 round and 102 wrinkled). This is called the second law of Mendel.

The interpretation of this experiment (2), the backcross of an $F_1$ hybrid plant with a parent plant, is that different germ cells (gametes) are formed. The $F_1$ hybrid plant (round) contains two traits, one for round (R, dominant over wrinkled, r) and one for wrinkled (r, recessive to round, R). This plant is a hybrid (heterozygote) and therefore can form two types of gametes (R and r).

In contrast, the other plant is homozygous for wrinkled (r). It can form only one type of gamete (r, wrinkled). Half of the offspring of the heterozygous plant receive the dominant trait (R, round), the other half the recessive trait (r, wrinkled). The resulting distribution of the observed traits is a ratio of 1:1, or 50% each.

The observed trait is called the phenotype (the observed appearance of a particular characteristic). The composition of the two factors (genes) R and r, (Rr) or (rr), is called the genotype. The alternative forms of a trait (here, round and wrinkled) are called alleles. They are the result of different genetic information at one given gene locus.

If the alleles are different, the genotype is heterozygous; if they are the same, it is homozygous (this statement is always in reference to a single, given gene locus).

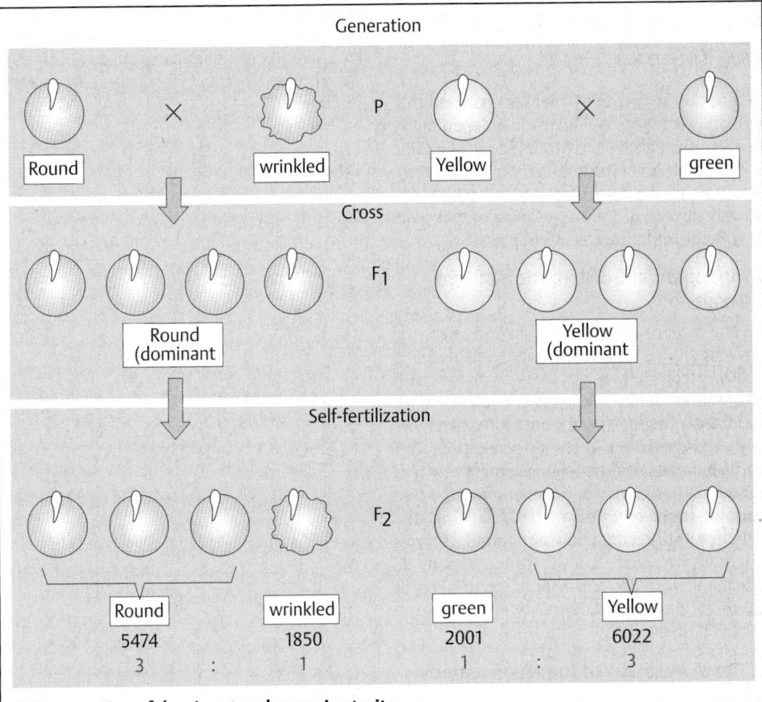


**A. Segregation of dominant and recessive traits**

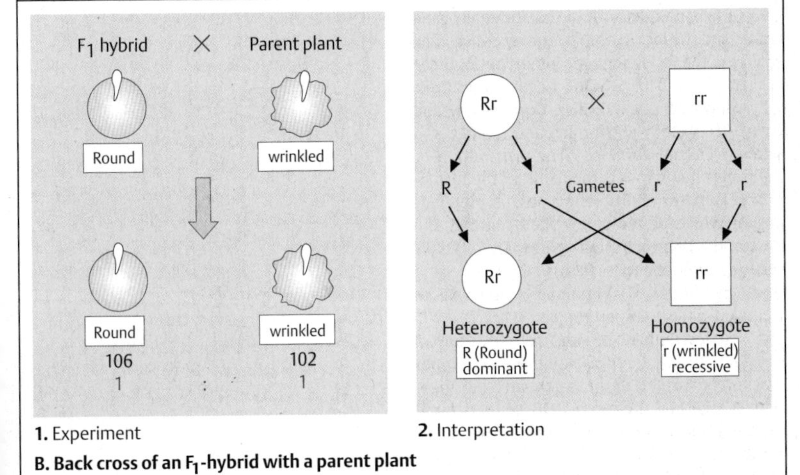


**1.** Experiment

**2.** Interpretation

**B. Back cross of an $F_1$-hybrid with a parent plant**

## Independent Distribution of Two Different Traits

In a further experiment, Mendel observed that two different traits are inherited independently of each other. Each pair of traits shows the same 3:1 distribution of the dominant over the recessive trait in the $F_2$ generation as he had previously observed. The segregation of two pairs of traits again followed certain patterns.

### A. Independent distribution of two traits

In one experiment, Mendel investigated the crossing of the trait pairs round/wrinkled and yellow/green. When he crossed plants with round and yellow seeds with plants with wrinkled and green seeds, only round and yellow seeds occurred in the $F_1$ generation. This corresponded with the original experiments, as shown on p. 132. In the $F_2$ generation, the two pairs of traits occurred in the following distribution: 315 seeds yellow and round, 108 yellow and wrinkled, 101 green and round, 32 green and wrinkled, corresponding to a segregation ratio of 9:3:3:1. This is referred to as the third Mendelian law.

### B. Interpretation of the observations

If we assign the capital letter **G** to the dominant gene *yellow*, a lowercase **g** to the recessive gene *green*, the capital letter **R** to the dominant gene *round*, and the lowercase **r** to the recessive gene *wrinkled*, the following nine genotypes of these two traits can occur: **GGRR, GGRr, GgRR, GgRr** (all *yellow* and *round*); **GGrr, Ggrr** (*yellow* and *wrinkled*); **ggRR, ggRr** (*green* and *round*); and **ggrr** (*green* and *wrinkled*). The distribution of the traits shown in A is the result of the formation of gametes of different types, i.e., depending on which of the genes they contain. The ratio of the dominant trait yellow (**G**) to the recessive trait green (**g**) is 12:4, or 3:1.
Also, the ratio of dominant round (**R**) to wrinkled (**r**) seeds was 12:4, i.e., 3:1.
The square (Punnett square, named after an early geneticist) shows the different genotypes that can be formed in the zygote after fertilization. Altogether there are 9/16 yellow round seeds (**GRGR, GRGr, GrGR, GRgR, gRGR, GRgr, GrgR, gRGr, grGR**), 3/16 green round (**gRgR, gRgr, grgR**), 3/16 yellow wrinkled (**GrGr, Grgr, grGr**), and 1/16 green wrinkled seeds (**grgr**). Each of the two traits (dominant yellow versus recessive green or dominant round versus recessive wrinkled) occurs in a 3:1 ratio (dominant vs recessive).

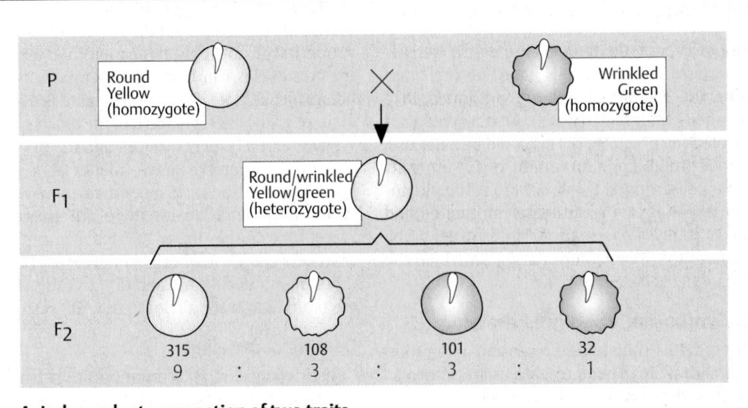


**A. Independent segregation of two traits**

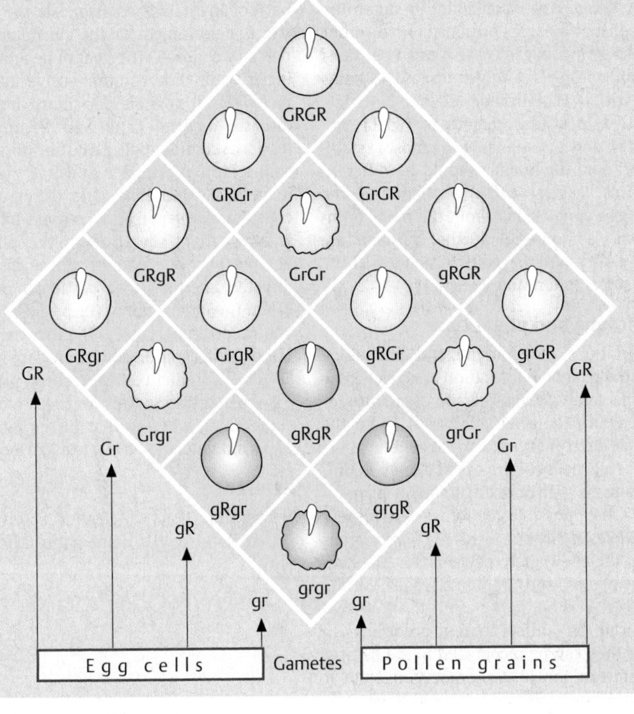


**B. Interpretation of the observation**

# Phenotype and Genotype

Formal genetic analysis examines the genetic relationship of individuals based on their kinship. These relationships are presented in a pedigree (pedigree analysis). An observed trait is called the phenotype. This could be a disease, a blood group, a protein variant, or any other attribute determined by observation. The phenotype depends to a great degree on the method and accuracy of observation. The term genotype refers to the genetic information on which the phenotype is based.

## A. Symbols in a pedigree drawing

The symbols shown here represent a common way of drawing a pedigree. Males are shown as squares, females as circles. Individuals of unknown sex (e.g., because of inadequate information) are shown as diamonds. In medical genetics, the degree of reliability in determining the phenotype, e.g., presence of a disorder, should be stated. In each case it must be stated which phenotype (e.g. which disease) is being dealt with. Established diagnoses (data complete), possible diagnoses (data incomplete), and questionable diagnoses (statements or data doubtable) should be differentiated. False assignment of a phenotype can lead to false conclusions about the mode of inheritance. A number of further symbols are used, e.g., for heterozygous females with X-chromosomal inheritance (see p. 142).

## B. Genotype and phenotype

The definitions of genotype and phenotype refer to the genetic information at a given gene locus. The gene locus is the site on a chromosome at which the genetic information for the given trait lies. Different forms of genetic information at a gene locus are called alleles. In diploid organisms – all animals and many plants – there are three possible genotypes with respect to two alleles at any one locus: (1) homozygous for one allele, (2) heterozygous for the two different alleles, and (3) homozygous for the other allele.

Alleles can be differentiated according to whether they can be recognized in the heterozygous state or only in the homozygous state. If they can be recognized in the heterozygous state, they are called dominant. If they can be recognized in the homozygous state only, they are recessive. The concepts dominant and recessive are an attribute of the accuracy in observation and do not apply at the molecular level. If the two alleles can both be recognized in the heterozygous state, they are designated codominant (e.g., the alleles A and B of the blood group system AB0; 0 is recessive to A and B).

If there are more than two alleles at a gene locus, there will be correspondingly more genotypes. With three alleles there are six genotypes; for example, in the AB0 blood group system there are AA, A0 (both phenotype A), BB, B0 (both phenotype B), AB, and 00 (actually, there are more than three alleles in the AB0 system).

## Genetic counseling

Genetic counseling is a communication process relating to the diagnosis and the possible occurrence of a genetically determined disease in a family and in more distant relatives. On the basis of an established diagnosis, the individual risk is determined for the consultand or that person's children. The goal of genetic counseling is to provide comprehensive information, including all possible decisions, course of the disease, medical care, and treatment. Professional confidentiality must be observed. The counselor makes no decisions.

In genetic counseling, one distinguishes the consultand, the person who seeks information, and the patient, since they are very often different persons. The patient whose disease first directed attention to a particular pedigree is called the index patient (or proposita if female and propositus if male). The increasing availability of information about a disease based on a DNA test (predictive DNA testing) prior to disease manifestation calls for the utmost care in establishing whether it is in the interest of a given individual to have a test carried out.

## Reference

Harper, P.S.: Practical Genetic Counselling, 5th ed. Butterworth-Heinemann, Oxford, 1998.

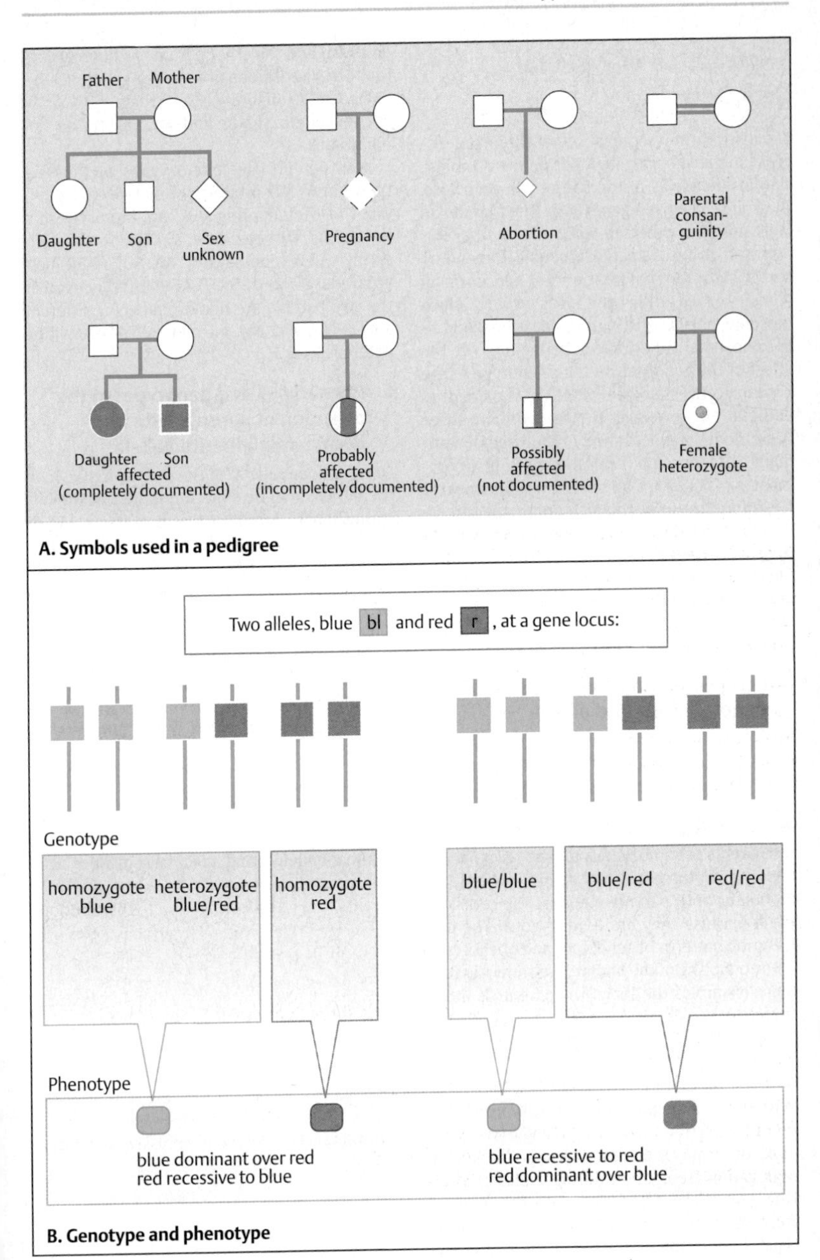
Father
Mother
Daughter
Son
Sex unknown
Pregnancy
Abortion
Parental consanguinity
Daughter Son affected (completely documented)
Probably affected (incompletely documented)
Possibly affected (not documented)
Female heterozygote
A. Symbols used in a pedigree
Two alleles, blue bl and red r, at a gene locus:
Genotype
homozygote blue
heterozygote blue/red
homozygote red
blue/blue
blue/red
red/red
Phenotype
blue dominant over red
red recessive to blue
blue recessive to red
red dominant over blue
B. Genotype and phenotype

# Segregation of Parental Genotypes

The distribution (segregation) of parental genotypes in the offspring depends on the combination of the alleles in the parents. In each case, they refer to a given gene locus. The Mendelian laws state the expected combination of alleles in the offspring of a parental couple. Depending on the chromosomal location, i.e., a locus on an autosome (see p. 184) or a locus on the X chromosome, one can distinguish autosomal and X-chromosomal inheritance. Depending on the effect of the genotype on the phenotype in the heterozygous state, an allele is classified as dominant or recessive. Hence, there are three basic modes of inheritance: (1) autosomal dominant, (2) autosomal recessive, and (3) X-chromosomal. For genes on the X chromosome it is not important to distinguish dominant and recessive (see below). Since genes on the Y chromosome are always transmitted from the father to all sons and the Y chromosome bears very few disease-causing genes, Y-chromosomal inheritance can be neglected when considering monogenic inheritance.

## A. Possible mating combinations of the genotypes for two alleles

For a gene locus with two alleles there are six possible combinations of parental genotypes (1–6). Here two alleles, blue (bl) and red (r), are shown, blue being dominant over red. In three of the parental combinations (1, 3, 4) neither of the parents is homozygous for the recessive allele (red). In three parental combinations (2, 5, 6), one or both parents manifest the recessive allele because they are homozygous. The distribution patterns of genotypes and phenotypes in the offspring of the parents are shown in B. In these examples the sex of the parents is interchangable.

## B. Distribution pattern in the offspring of parents with two alleles A and a

With three of the parental mating types for the two alleles **A** (dominant over **a**) and **a** (recessive to **A**), there are three combinations that lead to segregation (separation during meiosis) of allelic genes. These correspond to the parental combinations 1, 2, and 3 shown in A. In mating types 1 and 2, one of the parents is a heterozygote (**Aa**) and the other parent is a homozygote (**aa**). The distribution of observed genotypes expected in the offspring is 1: 1; that is 50% (0.50) are **Aa** heterozygotes and 50% (0.50), **aa** homozygotes.

If both parents are heterozygous **Aa** (mating type 3 in A), the proportions of expected genotypes of the offspring (**AA, Aa, aa**) occur in a ratio of 1:2:1. In each case, 25% (0.25) of the offspring will be homozygous **AA,** 50% (0.50) heterozygous **Aa,** and 25% (0.25) homozygous **aa.** If the two parents are homozygotes for different alleles (**AA** and **aa**), all their offspring will be heterozygotes.

## C. Phenotypes and genotypes in the offspring of parents with one dominantly inherited allele

One dominant allele (in the first pedigree, **A**, in the father) is to be expected in 50% of the offspring. If both parents are heterozygous, 25% of the offspring will be homozygous **aa**. If both parents are homozygous, one for the dominant allele **A**, the other for the recessive allele **a**, then all offspring are obligate heterozygotes (i.e., must necessarily be heterozygotes). It should be emphasized that the figures are percentages for expected distributions of the genotypes. The actual distribution may deviate form the expected, especially with small numbers of children.

Expected distribution of genotypes for parents with different combinations of genotypes for a dominant allele A and a recessive allele a

| Parents | Offspring | Distribution |
|---|---|---|
| AA × AA | AA | 1 |
| AA × Aa | AA, Aa | 1 : 1 |
| Aa × Aa | AA, Aa, aa* | 1 : 2 : 1 |
| AA × aa | Aa | 1 |
| Aa × aa | Aa, aa | 1 : 1 |
| aa × aa | aa | 1 |

* Dominant phenotype to recessive phenotype 3 : 1

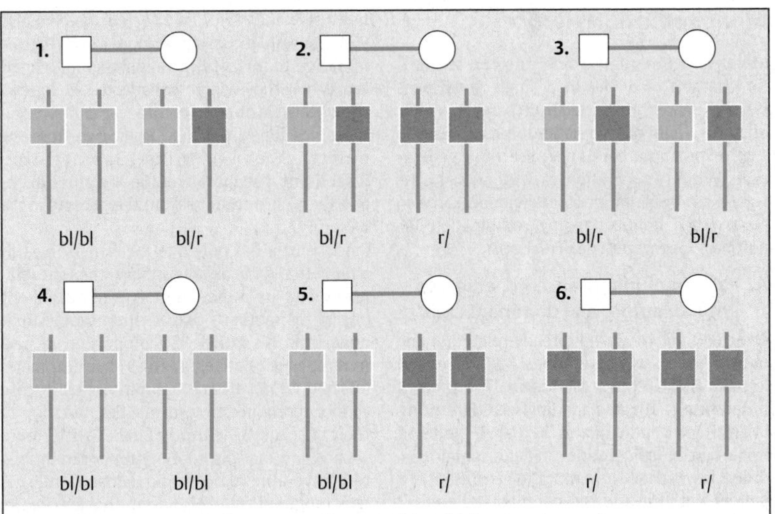


A. Possible mating types of genotypes for two alleles (bl dominant over r)

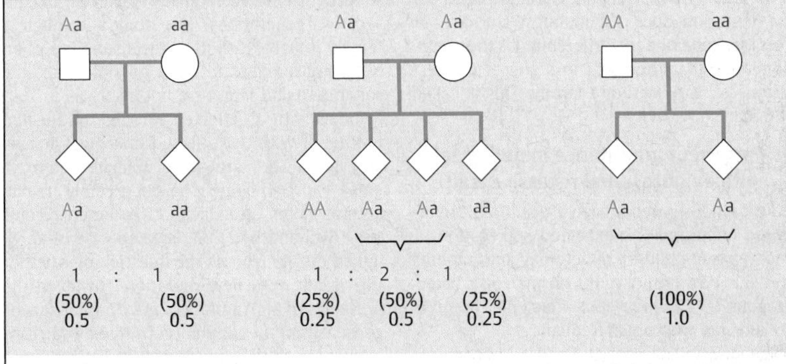


B. Expected distribution of genotypes in offspring of parents with two alleles, A and a

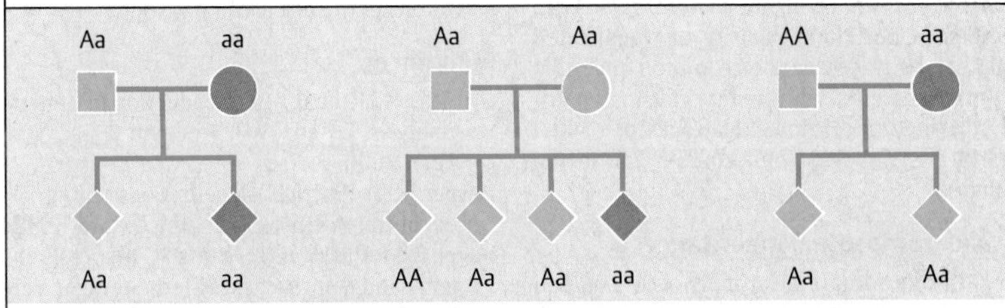


C. Phenotypes and genotypes in the offspring of parents with a dominant allele A and a recessive allele a

# Monogenic Inheritance

Monogenic inheritance refers to the segregation pattern of alleles at a single gene locus when they are transmitted from parents to their offspring. In medical genetics it is customary to assign Roman numerals to consecutive generations. Within a generation, each individual is assigned an Arabic numeral. Individuals can also be assigned nonoverlapping combinations of numbers for computer calculations.

## A. Pattern of inheritance in pedigrees with an autosomal dominant trait

Affected individuals are directly related in one or more successive generations. Both males and females are affected in a 1:1 ratio. The expected proportion of affected and unaffected offspring of an affected individual is 50% (0.50) each. An important consideration in autosomal disorders is whether a new mutation is present in a patient without affected parents. Pedigrees 2 and 3 show a new mutation in generation II. In some autosomal dominant disorders a carrier of the mutation does not manifest the disease. This is called nonpenetrance, but it is the exception rather than the rule. The degree of manifestation can vary within a family. This is called variable expressivity.

## B. Pattern of inheritance in pedigrees with an autosomal recessive trait

Heterozygous parents have a risk of 25% of affected offspring. The expected segregation of genotypes of children of heterozygous patients is 1:2:1 (25% homozygous normal, 50% heterozygous, 25% homozygous affected). The sexes are affected with equal frequency.
The unaffected parents (II-3 and II-4 of pedigree 1, I-1 and I-2 of pedigree 2, and III-1 and III-2 of pedigree 3) are obligate heterozygotes. In pedigree 3, the homozygosity of the affected child can be traced back to common ancestors of the two parents, who are first cousins. Parental consanguinity (blood relationship) of III-1 and III-2 is indicated by a double line in the pedigree.

## C. X-chromosomal inheritance

An X-chromosomal trait usually occurs only in males because they are hemizygous for all genes located on their X chromosome. A female heterozygous for an X-chromosomal mutation has a risk of 50% for an affected son. She also will transmit the X chromosome carrying the mutation to 50% of her daughters, but as heterozygotes they will not be affected. Pedigree 1 shows the distribution of three parental X chromosomes (one from the father, two from the mother) in the offspring. Panel 2 shows the corresponding distribution of the X chromosomes and the Y chromosome from the parents to the offspring.
Since a male has only one X chromosome, his daughter(s) will always inherit this chromosome with its mutation(s). A son will inherit one of his mother's two X chromosomes, but none from his father. The proportion of new mutations is relatively high (3) because males affected with a severe X chromosomal disease cannot reproduce to transmit the mutation to their offspring. One third of males with severe X-chromosomal disorders without family history have a new mutation (Haldane's rule). A new mutation could also occur in a female (4). Typical X-chromosomal inheritance (5) is easy to recognize. Affected males may occur in consecutive generations, but always through a female lineage. Father–son transmission of an X-chromosomal trait is not possible since the son inherits his father's Y, not his X.
Females with an affected son and an affected brother or with two affected sons must be heterozygous. They are called obligate heterozygotes. Those who might be, but possibly are not, heterozygotes are facultative heterozygotes (e.g., III-5 and IV-2). An isolated case of an X-linked disorder poses the question of whether this is due to a new mutation. However, the presence of the mutation in a proportion of germ cells of the mother (germline mosaicism) may increase the apparent risk. The situation can only be resolved on the basis of previous observations of the disease in question.

### References

Griffiths, A.J.F., et al.: An Introduction to Genetic Analysis. 7th ed. W.H. Freeman & Co., New York, 2000.

Harper, P.S.: Practical Genetic Counselling. 5th ed. Butterworth-Heinemann, Oxford, 1998.

Vogel, F., Motulsky, A.G.: Human Genetics. Problems and Approaches. 3rd ed. Springer Verlag, Heidelberg–New York, 1997.

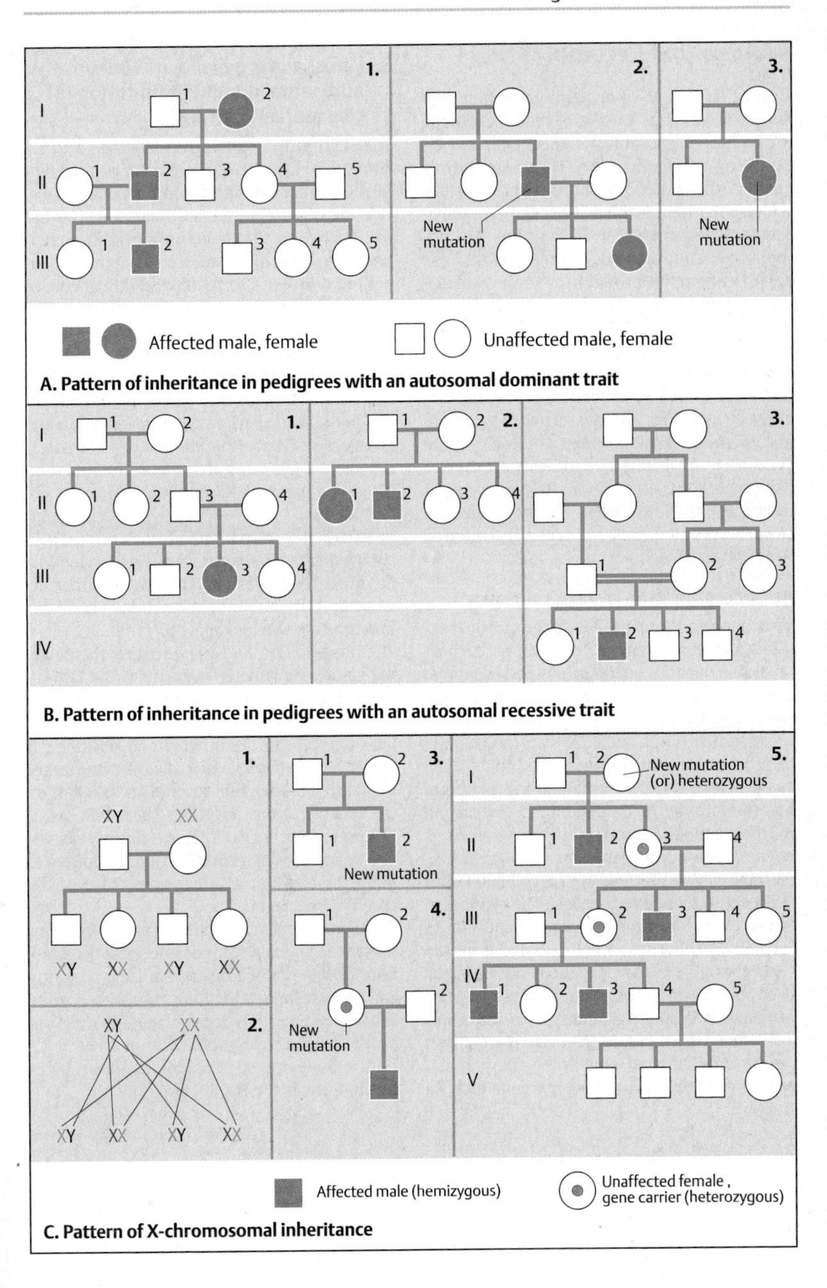


A. Pattern of inheritance in pedigrees with an autosomal dominant trait

B. Pattern of inheritance in pedigrees with an autosomal recessive trait

C. Pattern of X-chromosomal inheritance

## Linkage and Recombination

Linkage refers to two or more genes being inherited together as a result of their location on the same chromosome. This depends on the distance between their loci. The closer they lie next to each other, the more frequently they will be inherited together (linked). Recombination due to crossing-over between the loci (breakage and reunion during meiosis, see p. 116) leads to the formation of a new combination of linked genes. When the loci are very close together, recombination is rare; when they lie farther apart, recombination is more frequent. In fact, the frequency of recombination can be used as a measure of the distance between gene loci. Linkage relates to gene loci, not to specific alleles. Alleles at closely linked gene loci that are inherited together are called a haplotype. If this occurs more frequently or less frequently than expected by the individual frequencies of the alleles involved, it is referred to as linkage disequilibrium (p. 158).

### A. Recombination by crossing-over

Whether neighboring genes on the same parental chromosome remain together or become separated depends on the cytological events (1) during meiosis. If there is no crossing-over between the two gene loci **A** and **B,** having the respective alleles **A, a** and **B, b**, then they remain together on the same chromosome (linked). The gamete chromosomes formed during meiosis in this case are not recombinant and correspond to the parental chromosomes. However, if crossing-over occurs between the two gene loci, then the gametes formed are recombinant with reference to these two gene loci. The cytological events (1) are reflected in the genetic result (2). For two neighboring gene loci **A** and **B** on the same chromosome, the genetic result is one of two possibilities: not recombinant (gametes correspond to parental genotype) or recombinant (new combination). The two possibilities can be differentiated only when the parental genotype is informative for both gene loci (**Aa** and **Bb**).

### B. Linkage of a gene locus with an autosomal dominant mutation (B) to a marker locus (A)

The segregation of two linked gene loci in a family is shown here. There are two possibilities: 1, no recombination and 2, recombination. One locus (**B**) represents an autosomal dominant mutation that leads to a certain phenotype, e.g., that of an autosomal dominant inherited disorder. The father and three children (red symbols in the pedigree) are affected. The other locus (**A**) is a neighboring marker locus. All three affected children have inherited the mutant allele **B** as well as the marker allele **A** from their father. The three unaffected individuals have inherited the normal allele **b** and the marker allele a from their father. The paternal allele **a** indicates absence of the mutation (i.e., **B** not present). Recombination has not occurred (1).

In situation 2, recombination has occurred in two (indicated) persons: An affected individual has inherited alleles **a** and **B** from the father, instead of **A** and **B.** An unaffected individual has inherited allele **A** and allele **b.**

The precondition for differentiating the paternal genotypes is heterozygosity at the father's loci. In the case presented, the alleles **A** and **B** lie on one of the father's chromosomes, and the alleles **a** and **b** on the other (in *cis* position). It would also be possible that allele **A** in the father would lie on one chromosome and allele **B** on the other (in *trans* position). These two possibilities represent two different linkage phases. The recognition of recombination as opposed to nonrecombination assumes knowledge of the parental linkage phase.

Segregation analysis of linked genes is very important in medical genetics because the presence or absence of a disease-causing mutation can be determined without directly knowing the type of mutation (indirect gene analysis). In order to reduce the probability of recombination, closely linked, flanking markers (DNA polymorphisms, see p. 72) are used.

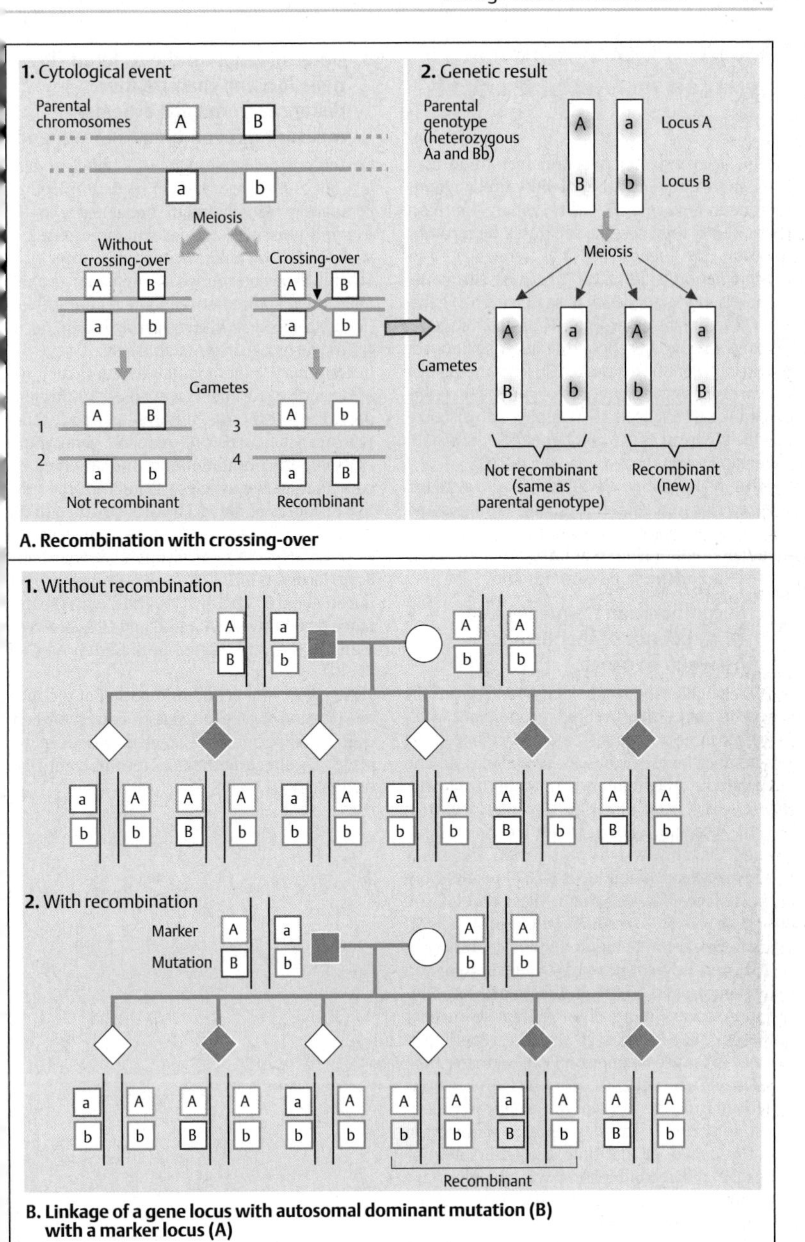


A. Recombination with crossing-over

B. Linkage of a gene locus with autosomal dominant mutation (B) with a marker locus (A)

# Genetic Distance between Two Gene Loci and Recombination Frequency

The closer together two gene loci are located, the more frequently they are inherited together (genetic linkage); the farther apart, the more frequently they become separated by recombination. The highest possible frequency of recombination is 50% (0.50), because this corresponds to the frequency of segregation of genes on different chromosomes. Thus, the frequency of recombination reflects the distance between two loci (genetic distance). This distance can be expressed as the frequency of genetic recombination (as opposed to the physical distance, which is given as the number of DNA base pairs lying between the two loci, see p. 240).

Synteny (H. J. Renwick, 1971) refers to gene loci being located on the same chromosome, whether or not they are linked. Thus the term synteny also includes unlinked, widely separated loci on the same chromosome.

## A. Recombination frequency as a consequence of the distance between two loci

Two neighboring gene loci **A** and **B** in the parents may either become recombinant or remain nonrecombinant (see p. 144). If one of the parents is heterozygous for two alleles **Aa** and **Bb**, but the other homozygous for both, then homozygosity at only locus **A** (1) or only locus **B** (2) in the offspring will be the result of recombination. The observed recombination frequency between locus **A** and locus **B** (3%) results from the distance between them. These two loci are said to be 0.03 recombination units (3 cM) apart. One recombination unit is a centimorgan (cM), and 1 cM corresponds to a recombination frequency of 1% (0.01). In mammals, recombination occurs more often in female meiosis than in male meiosis, so that the genetic distance in females is about 1.5 times greater than in males (see p. 240). The term morgan is derived from the name of the American geneticist who in 1911 first described recombination in *Drosophila*. At that time, the observation of linkage and recombination was an important argument for genes being linearly arranged along the chromosomes.

## B. Determination of the order of three gene loci and their relative distances from each other by measuring recombination frequency

Not only the relative distances between gene loci but also their order can be determined by comparing recombination frequencies. In the example presented, the order of three gene loci **A, B,** and **C** of unknown distance from each other is to be determined (1). In plants and animals, the distance between any two of the loci (locus **A** from locus **B,** locus **B** from locus **C,** locus **A** from locus **C**) can be established.

In traditional experimental genetics, such hybridization experiments were used for this purpose. For a test cross, differing parental genotypes can be used. One parental genotype is crossed with a different one. The observed recombination frequencies (2) indicate the relative distances of the loci from each other. In the example presented, the distance from locus **A** to locus **C** is 0.08 (8%); the distance between locus **B** and locus **C** is 0.23 (23%); and the distance between **A** and **B** is 0.31 (31%). Thus, with the distance 0.08 between **A** and **C** and 0.23 between **C** and **B,** they are located in the order **A–C–B** (3).

This type of indirect determination of the relative location of gene loci in their correct order in classical experimental genetics has been replaced by direct methods of recombinant DNA technology.

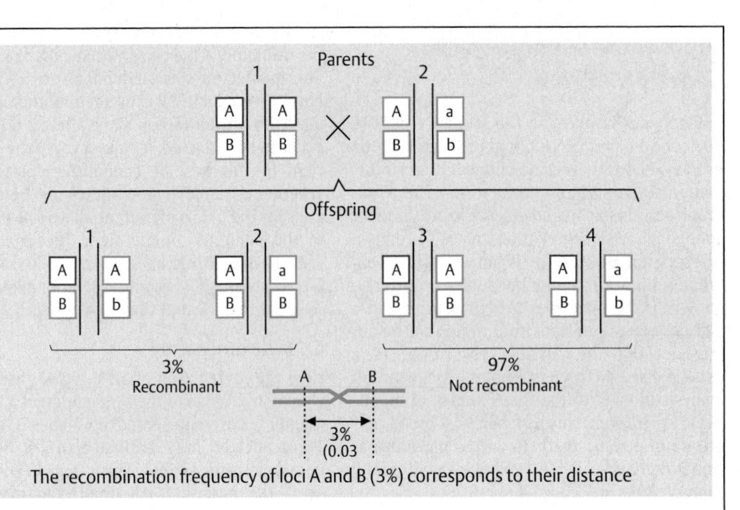


**A. Recombination frequency as a consequence of the distance of two loci**

**1.** Gene loci A, B, C of unknown distance

A B C

**2.** Test cross of homozygous parental genotypes

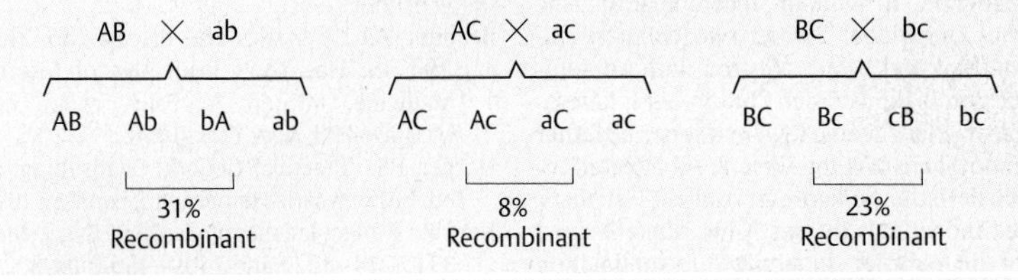


**3.** Relative distance

A ← 0.08 → C ← 0.23 → B

← 0.31 →

**B. Determination of the relative distance and sequence of three gene loci by measuring the frequency of recombination**

## *Segregation Analysis with Genetic Markers*

Individual differences in the sequence of DNA nucleotide bases (DNA polymorphism) can be used as genetic markers to obtain indirect information about the presence or absence of a mutation at a closely linked disease locus (indirect gene diagnosis). Segregation analysis of linked markers and a disease locus can be applied even when a mutation cannot be found. By using the known linkage between a DNA marker and a disease locus, important information can be obtained about the possible occurrence of a genetic disease. Three examples of segregation analysis in each of the three modes of monogenic inheritance are given below. A broad repertoire of genetic markers with high information content is available (see microsatellites and minisatellites, p. 72).

### A. Autosomal dominant

Two pedigrees are presented that were studied by marker analysis, one without recombination between disease locus and marker (1), one with recombination (2). Above, the pedigree is shown; below, a diagram of a Southern blot analysis or some other marker system that distinguishes two alleles 1 and 2. A given individual will be homozygous (1-1, or 2-2) or heterozygous (2-1) at this locus (marker locus). In pedigree 1, without recombination, the mother (individual 2) and two children (individuals 4 and 5) are affected with an autosomal dominant disorder. The mother is heterozygous 1-2 for the two DNA markers. The father is homozygous 2-2 for allele 2. All affected individuals in this pedigree carry allele 1, all unaffected individuals do not. Thus, allele 1 must carry the mutation. In family 2, recombination has occurred in individual 5. The frequency of recombination determines the frequency of false predictions. For this reason, very closely linked markers with very low recombination frequencies are used, preferably marker loci that flank the disease locus on both sides.

### B. Autosomal recessive

The two affected individuals (individuals 4 and 7) in the left-hand pedigree are homozygous for allele 1 (1-1), inherited from each of their parents. Thus one allele 1 of the father and allele 1 of the mother represent the allele that carries the mutation at the closely linked disease locus. The unaffected sibs (individuals 3, 5, and 6) have received allele 2 from their mother and allele 1 from their father. Since allele 2 does not occur in the affected, it cannot carry the mutation. In this case it cannot be determined whether the unaffected children are heterozygous for the paternal mutant allele 1 or not.
In the pedigree on the right, recombination must have occurred because this child is not affected, although it is homozygous for allele 2 (2-2) as are the affected sibs (individuals 3 and 5).

### C. X chromosomal

Here the segregation of an X-linked disease is shown in relation to the segregation of a linked marker locus represented by alleles 1 and 2. Three affected male individuals (5, 8, 10) are hemizygous for allele 1. Thus, this allele represents the mutation. All unaffected male individuals are hemizygous for allele 2. The female individuals 2 and 6 are heterozygous 1-2. They must be considered obligate heterozygotes for the mutation. Recombination must have occurred in individual 13 because he is hemizygous for allele 2 but is affected.
The examples show how information about the presence or absence of a mutation can be derived from segregation analysis of closely linked markers (indirect DNA diagnosis).

### References

Beaudet, A.L.: Genetics and disease. pp. 365–394. In: Harrison's Principles of Internal Medicine. 14th ed. A.S. Fauci, et al., eds. McGraw-Hill, New York, 1998.
Harper, P.S.: Practical Genetic Counselling. 5th ed. Butterworth-Heinemann, London, 1998.
Korf, B.: Molecular diagnosis. New Eng J. Med. **332**:1218–1220 and 1499–1502, 1995.
Muller, R.F., Young, I.D.: Emery's Elements of Medical Genetics. 10th ed. Harcourt Brace–Churchill Livingstone, Edinburgh, 1998.
Richards, C.S., Ward, P.A.: Molecular diagnostic testing. pp. 83–88. In: Principle of Molecular Medicine. J.L. Jameson, ed. Humana Press, Totowa, New Jersey, 1998.
Strachan, T., Read, A.P.: Human Molecular Genetics. 2nd ed. Bios Scientific Publishers, Oxford, 1999.

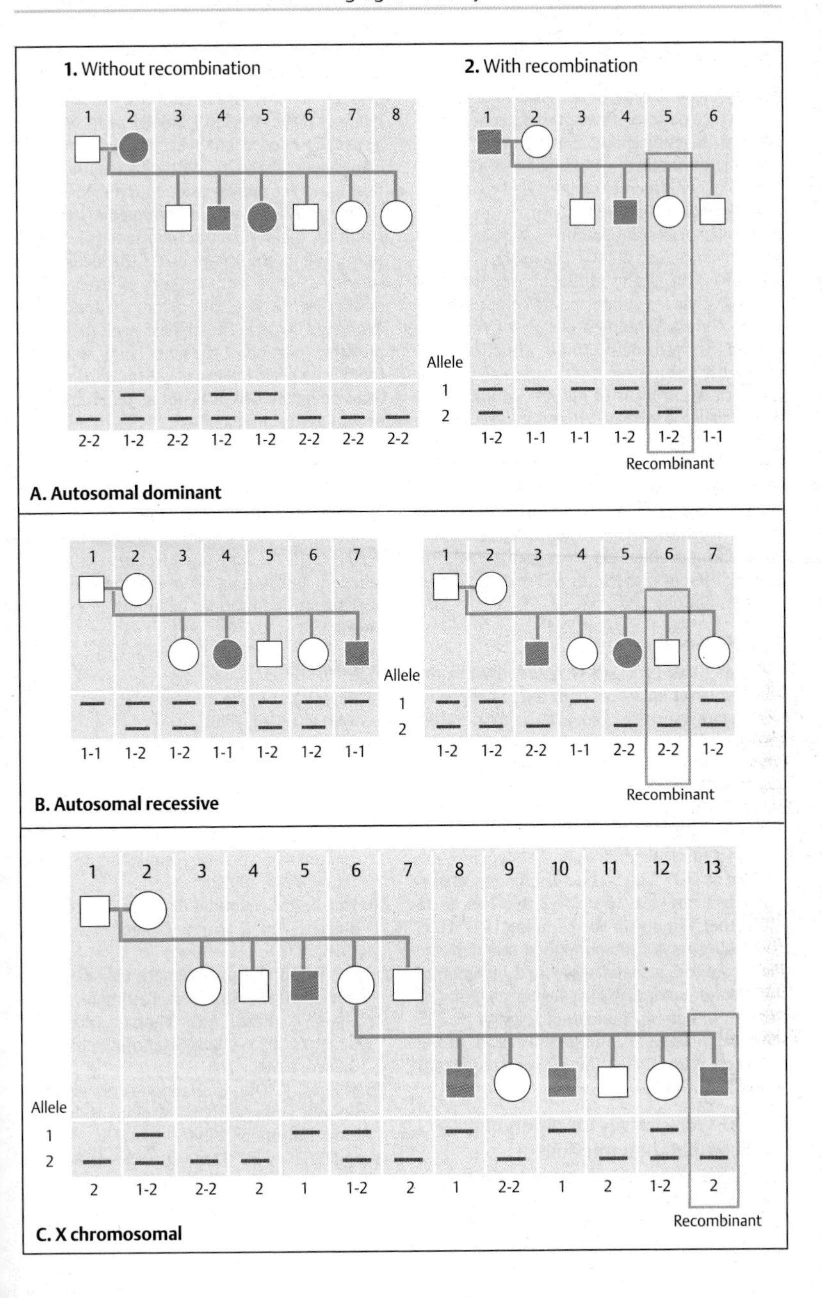
1. Without recombination
2. With recombination
1 2 3 4 5 6 7 8
2-2 1-2 2-2 1-2 1-2 2-2 2-2 2-2
Allele
1
2
1 2 3 4 5 6
1-2 1-1 1-1 1-2 1-2 1-1
Recombinant
A. Autosomal dominant
1 2 3 4 5 6 7
1-1 1-2 1-2 1-1 1-2 1-2 1-1
Allele
1
2
1 2 3 4 5 6 7
1-2 1-2 2-2 1-1 2-2 2-2 1-2
Recombinant
B. Autosomal recessive
1 2 3 4 5 6 7 8 9 10 11 12 13
Allele
1
2
2 1-2 2-2 2 1 1-2 2 1 2-2 1 2 1-2 2
Recombinant
C. X chromosomal

# Linkage Analysis

Linkage analysis is a test designed to detect the inheritance pattern of two or more gene loci located sufficiently close to each other on the same chromosome. The alleles present on linked loci will tend to be inherited together if recombination between them is rare. The frequency of recombination is a measure of their distance (see p. 146). The analysis of linkage rests on many different sophisticated methods that are supported by computer programs. Here a simplified outline of the background is provided. Three situations are frequently encountered: (i) linkage analysis of two loci, e.g., a locus of interest, mutations of which cause a disease (disease locus) and a locus characterized by a detectable DNA polymorphism (marker locus); (ii) linkage analysis of several loci (multilocus analysis); and (iii) linkage analysis involving the entire genome by means of DNA markers (microsatellites) along each chromosome (genome scan). Each requires different procedures and rests on different assumptions.

## A. LOD scores

Linkage of two gene loci is assumed when the probability for linkage as opposed to the probability against linkage is equal to or greater than the ratio of 1000:1 ($10^3$:1). The logarithm of this ratio (odds) is called the LOD score (logarithm of the odds). A LOD score of 3 corresponds to an odds ratio of 1000:1. The likelihood of linkage as opposed to nonlinkage for different recombination fractions (recombination frequencies), expressed as LOD scores, is determined by observations in families. The closer two loci lie to each other, the higher the resulting LOD score. The table shows (in a simplified manner) the LOD score for close linkage with a recombination fraction under 0.05 (a), high probability of linkage with a recombination fraction of 0.15 (b), weak linkage (c), and no linkage (d). If the LOD score is less than 0, linkage is excluded (not shown in the table).

## B. LOD score curves for different recombination fractions

The diagram shows a simplified form of the LOD score curves for the values in the table in A. (Figure after Emery, 1986).

## C. Multilocus analysis

Because of the great number of available markers at the DNA level, linkage analysis today is usually carried out with multiple markers (multilocus analysis). With the chromosomal position of the marker taken into consideration, the localization score is determined as the logarithm of the probability quotient (likelihood ratio). The localization score of the locus being sought is noted with respect to each of the marker loci (A, B, C, D). Each of the four peaks expresses linkage. The highest peak marks the probable location of the gene being sought. If there is no peak, linkage is not present, and the locus being studied does not map to the region tested (exclusion mapping).

In contrast to linkage, which refers to the genetic distance of gene loci, the term association refers to the co-occurrence of alleles or phenotypes. If one particular allele occurs more frequently in people with a disease than expected from the individual frequencies, this is called an association. There may be many reasons for an association, including nongenetic causes.

## References


Byerley, W.F.: Genetic linkage revisited. Nature **340**:340–341, 1989.

Emery AEH.: Methodolgy in Medical Genetics. 2nd ed. Churchill Livingstone, Edinburgh, 1986.

Lander, E.S., Kruglyak, L.: Genetic dissection of complex traits: guidelines for interpreting and reporting linkage results. Nature Genet. **11**:241–247, 1995.

Morton, N.E.: Sequential tests for detection of linkage. Am. J. Hum. Genet. **7**:277–318, 1955.

Ott J.: Analysis of Human Genetic Linkage. Johns Hopkins University Press, Baltimore, 1991.

Strachan, T., Read, A.P.: Human Molecular Genetics. 2nd ed. Bios Scientific Publishers, Oxford, 1999.

Terwilliger, J., Ott, J.: Handbook for Human Genetic Linkage. Johns Hopkins University Press, Baltimore, 1994.

Wang, D.G., et al.: Large-scale identification, mapping and genotyping of single nucleotide polymorphisms in the human genome. Science **280**:1077–1082, 1998.

Recombination fraction

| | <0.05 | 0.05 | 0.10 | 0.15 | 0.20 | 0.30 | 0.35 | 0.40 | 0.45 | 0.50 |
|---|---|---|---|---|---|---|---|---|---|---|
| a | 3 | 0.7 | 0.3 | 0.2 | 0.01 | 0 | | | | |
| b | | 0 | 0.1 | 3 | 0.2 | 0 | | | | |
| c | 0 | 0.2 | 0.7 | 1.6 | 1.0 | 0 | | | | |
| d | | 0 | 0.1 | 0.2 | 0.3 | 0.2 | 0.1 | 0.1 | 0 | |

**A. LOD scores**

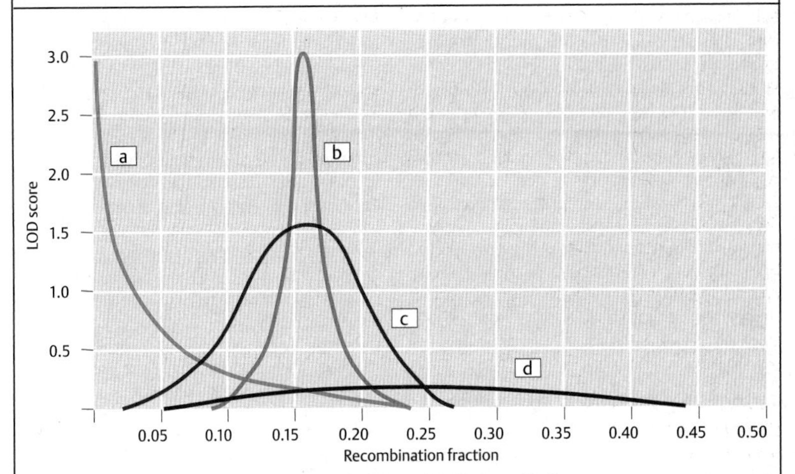


**B. LOD score curves for close (a), probable (b), and no linkage (c,d)**

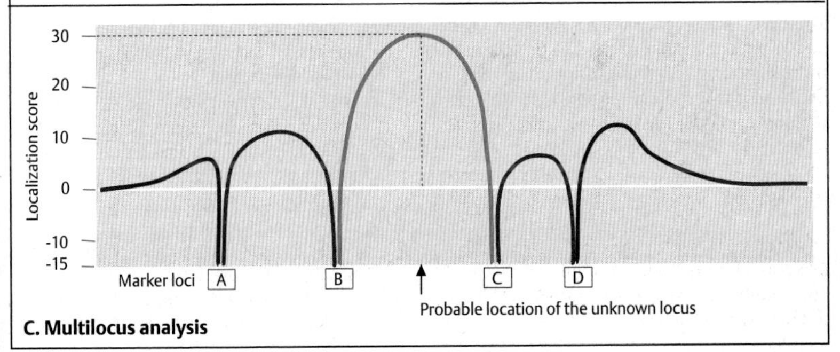


**C. Multilocus analysis**

# Quantitative Differences in Genetic Traits

Most variation among organisms is not qualitative but quantitative. Height, weight, shades of color in flowers or skin, behavioral patterns, learning capacity, metabolic rate, and blood pressure are examples. They cannot be grouped into separate, easily distinguished categories as can monogenic traits displaying a Mendelian mode of inheritance. The term quantitative genetics was introduced in 1883 by Francis Galton to refer to the variation of complex traits. R. A. Fisher in 1918 and other researchers later laid the foundations to assess the genetics of such traits. Distinguishing genetic and enviromental contributions to the trait is a primary goal. As this is usually not possible for individuals, quantitative traits have to be defined for populations. This poses considerable difficulties in defining a trait (question of cut-off point for a trait to be considered abnormal) and finding an appropriate control (comparison with another population). The underlying genetic variation (genotypic variation) is interchangeably called polygenic (many genes), multigenic (several genes), or multifactorial (many factors including nongenetic contributions to the variation). It is important to understand that, at least in theory, a quantitative trait is the result of distinct genotypes at each of the gene loci involved.

## A. Length of the corolla in *Nicotiana longiflora*

When parent plants with average corolla lengths of 40 cm and 90 cm are crossed, the next generation ($F_1$) shows corolla length distribution that is longer than that of the short parent and shorter than that of the tall parent. In the next generation ($F_2$) the distribution spreads at both ends towards long and short. If plants from the short, middle, and long varieties are used (crossed to breed a new population, $F_3$), the mean resulting distribution corresponds to the corolla length of the parental plants, i.e., short (shown on the left), long (shown on the right), or average (shown in the middle). This can be explained by a difference in the distribution of genes contributing to the variation in the trait, in this example length of the corolla. (Figure adapted from Ayala and Kiger, 1984).

## B. Distribution of frequency in the $F_2$ generation with different numbers of involved gene loci

A variable characteristic due to different alleles at a single gene locus can be defined by observable differences in the two types of homozygotes **aa** and **AA** (column 1 in the figure). As the number of loci contributing to the phenotype increases from two (column 2) to three (column 3) or to many loci (column 4), the $F_1$ generation tends to deviate towards a mean. In the next generation ($F_2$) the distribution of the frequency of occurrence of the trait spreads. It is no longer possible to use the phenotype to infer the genotype. As the number of loci increases, a smooth distribution of the quantitative trait in the population becomes apparent. Even with just three loci contributing to the phenotype (column 3), the distribution curve is relatively smooth, fitting into a Gaussian (normal) curve as seen with many loci (column 4). Each individual's phenotype is the result of the contribution of each allele at the loci involved (their identity generally remains unknown). The variance of the phenotype ($V_P$) is the sum of the genetic variance ($V_G$) and the environmental variance ($V_E$). However, in many cases these types of variance cannot be distinguished, especially in humans. A locus contributing to a quantitative characteristic is called a quantitative trait locus (QTL). There are many approaches to assessing QTLs (see *Trends Genet.* 11: 463–524, 1995 Special Issue).

## References


Ayala, F.J., Kiger, J.A.: Modern Genetics. 2nd ed. Benjamin/Cummings Publishing Co., Menlo Park, California, 1984.

Burns, G.W., Bottinger, P.J.: The Science of Genetics. 6th ed. Macmillan Publ. Co., New York–London, 1989.

Falconer, DS.: Introduction to Quantitative Genetics. 2nd ed. Longman, London, 1981.

Griffith, A.J.F., et al.: An Introduction to Genetic Analysis. 5th ed. W.H. Freeman & Co., New York, 2000.

King. R., Rotter, J. Motulsky, A.G., eds.: The Genetic Basis of Common Disorders. Oxford Univ. Press, Oxford, 1992.

Vogel, F., Motulsky, A.G.: Human Genetics and Approaches. 3rd ed. Springer Verlag, Heidelberg–New York, 1997.

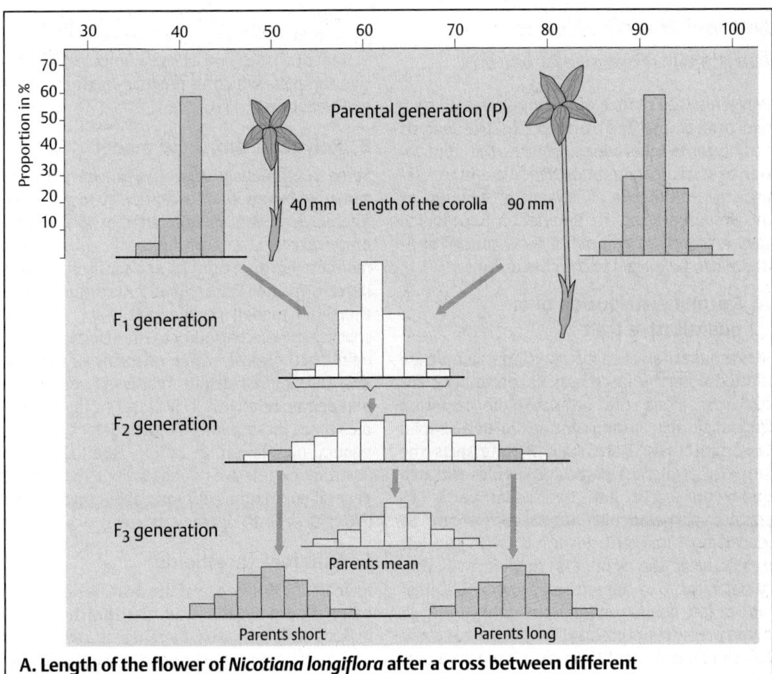


**A. Length of the flower of *Nicotiana longiflora* after a cross between different types of parental plants**

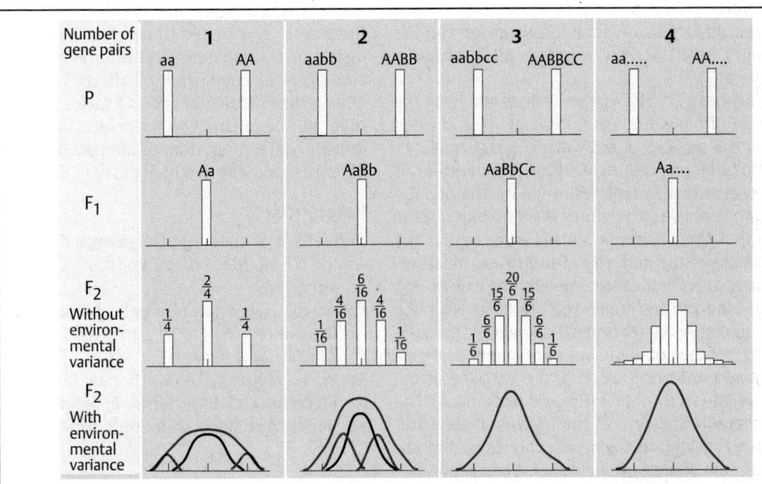


**B. Distribution of frequency in the $F_2$ generation with a different number of gene loci**

## Normal Distribution and Polygenic Threshold Model

Polygenic inheritance of quantitative traits occurs in all animal and plant species. The analysis of polygenic inheritance requires the application of statistical methods. These aid in assessing the difference between a sample of measurements and the population from which they are derived and define how much confidence can be placed in the conclusions.

### A. Normal distribution of a quantitative trait

When quantitative data from a large sample are plotted along the abscissa and the number of individuals along the ordinate, the resulting frequency distribution assumes a bell-shaped Gaussian curve. The mean ($\bar{x}$) intersects the curve at its highest point and divides the area under the curve into two equal parts (1). Further perpendicular intersections can be placed one standard deviation (*s*) to the left ($-1s$) and to the right ($+1s$) to yield two additional areas, *c* to the left (e.g., low blood pressure or low height) and *d* to the right (e.g., high blood pressure or increased height). Areas *a* and *b* each comprise 34.13% of the total area under the curve (2). Further partitioning with perpendiculars to the abscissa at two and three standard deviations ($-2s$ and $-3s$ to the left and $+2s$ and $+3s$ to the right) results in further subsections (3).

The mean ($\bar{x}$) of a sample is determined by the sum of individual measurements ($\Sigma x$) divided by the number of individuals (*n*) (Formula 1). Many measurements yield the frequency $f_x$ of observed individuals (Formula 2). The population variance ($\sigma^2$) defines the variability of the population (Formula 3). It is expressed as the square of the sum ($\Sigma$) of individual measurements (x) minus the population mean ($\mu$), divided by the number of individuals in the population (*N*) (Formula 3). However, the variance of the population usually cannot be determined directly. Therefore, the variance of the sample ($s^2$) must be estimated (Formula 4). The square of the sum of the individual measurements (x) minus the mean ($\bar{x}$) is divided by the number of measurements (*n*). A correction factor $n/n-1$ is introduced because the number of independent measurements is $n-1$, resulting in a simplified formula 5. The standard deviation (*s*, formula 6), the square root of the sample variance ($s^2$), rests on a large number of successive samples, which in practice might be difficult to obtain.

### B. Polygenic threshold model

Some continuously variable characters are assumed to represent a susceptibility to a disease. Therefore, it was postulated that if susceptibility exceeds a threshold, the disease is manifest as, for example, shown in B for cleft lip/cleft palate (Fraser, 1980). According to the threshold model, populations differ with respect to the susceptibility to the disease (1). The liability to disease in the offspring of affected individuals (first-degree relative, 2) or in second degree relatives (3) is closer to the threshold than in unrelated individuals from the general population. In general, the liability of first-degree relatives is about half that of the general population ($^1/_2\,\bar{x}$) and about one quarter ($^1/_4\,\bar{x}$) for second-degree relatives.

### C. Different thresholds

Instead of a difference of the distribution of liability for a polygenic trait, the threshold may differ. This is the case when there is a sex difference in the frequency of the disease (1). For example, if the disease occurs more frequently in males than in females, then the susceptibility to the disease is assumed to be greater in males owing to a lower sex-specific threshold. Accordingly, the proportion of affected persons among the offspring of affected males is lower than in the offspring of females (2). Consequently, the proportion of affected offspring of females exceeds that of males (3).

### References

Burns, G.W., Bottino, P.J.: The Science of Genetics. 6th ed. Macmillan Publishing Co., New York, 1989.

Fraser, F.C.: Evolution of a palatable multifactorial threshold model. Am. J. Hum. Genet. **32**:796–813, 1980.

Vogel, F., Motulsky, A.G.: Human Genetics—Problems and Approaches. 3rd ed. Springer Verlag, Heidelberg–New York, 1997.

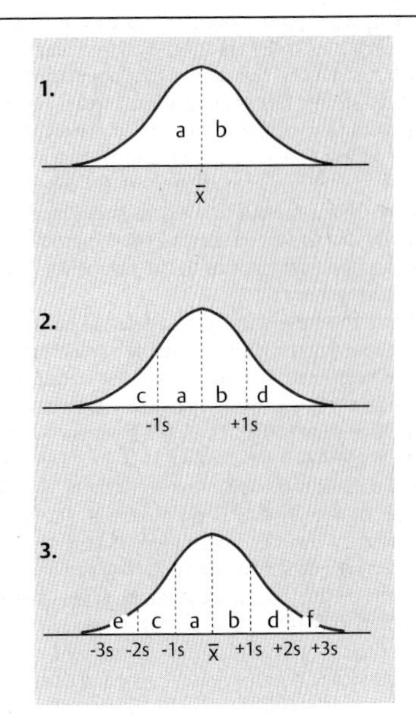


1 $$\bar{x} = \frac{\sum x}{n} \quad \text{(Mean)}$$

2 $$\bar{x} = \frac{\sum fx}{n}$$

3 $$\sigma^2 = \frac{\sum (x-\mu)^2}{N} \quad \text{Population variance}$$

4 $$s^2 = \left[\frac{\sum (x-\bar{x})^2}{n}\right]\left[\frac{n}{n-1}\right]$$

5 $$s^2 = \frac{\sum (x-\bar{x})^2}{(n-1)}$$

6 $$s = \sqrt{\frac{\sum f (x-\bar{x})^2}{n-1}}$$

(Standard deviation)

**A. Normal distribution of a trait**

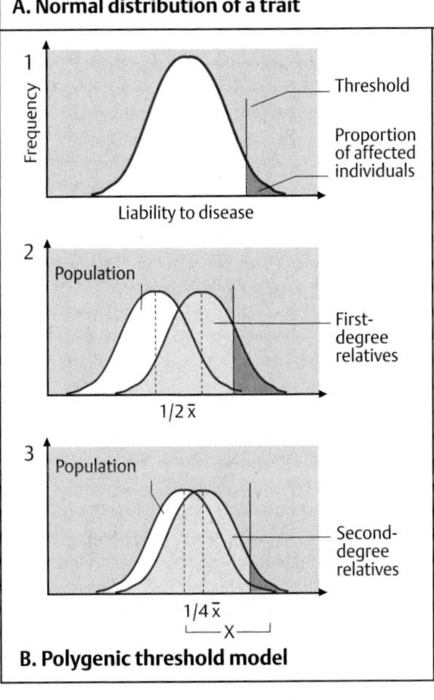


**B. Polygenic threshold model**

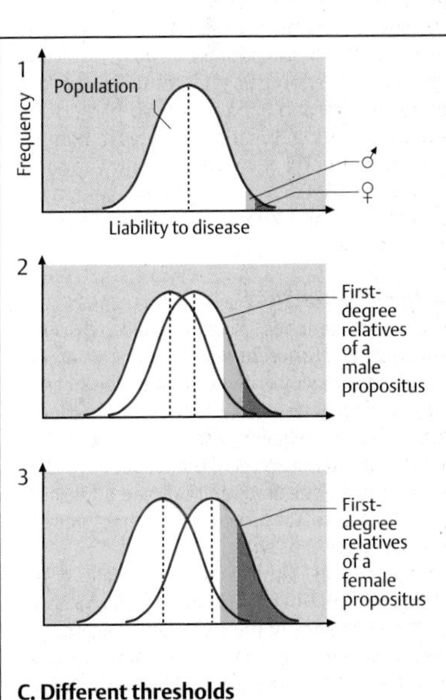


**C. Different thresholds**

# Distribution of Genes in a Population

A population can be characterized by its distribution of genes, i.e., the frequency of alleles at different gene loci. The field of population genetics is concerned with the distribution patterns of alleles and the factors that alter or maintain their frequencies. A detrimental allele that causes disease in an individual might be expected to disappear sooner or later from a population. For the reasons given below, this is usually not the case. The frequency of an allele in a population is related to the distribution of genotypes among the offspring of different parental genotypes.

## A. Expected frequency of genotypes in the children of parents with various genotypes

As shown on p. 140, a gene locus with two alleles (a dominantly inherited allele **A** and a recessively inherited allele **a**) can show six possible combinations (1–6). Each of these has an expected distribution of genotypes in their offspring according to the Mendelian laws. If one parent is homozygous and the other heterozygous, as in combinations 2 and 4, a 1:1 distribution of genotypes will result in the offspring (0.5 **AA** and 0.5 **Aa**). If both parents are heterozygous (combination 3), three genotypes will occur among their children in a ratio of 1:2:1 (0.25 **AA**, 0.5 **Aa**, and 0.25 **aa**). With combinations 1, 5, and 6 the respective offspring are of one genotype.

## B. Allele frequency

Population genetics is concerned with the frequencies of different genotypes in a population. The frequency with which an allele is present at a given gene locus in a given population is called allele frequency (or also, gene frequency). The concept of allele frequency refers exclusively to the frequency of the allele in a population, and not directly to the frequency of the individual genotypes.

For a gene locus with two possible alleles **A** and **a**, the only possible genotypes are **AA** or **Aa** or **aa.** The frequency of the two alleles together ($p$ the frequency of **A** and $q$ the frequency of **a**) must be 100% (1.0). If two alleles **A** and **a** are equally frequent (each 0.5), they have the frequency of $p=0.5$ for the allele **A** and $q=0.5$ for the allele **a** (1); i.e., $p+q=1$. The frequency distribution of the two alleles in a population follows a simple binomial relationship: $(p+q)^2=1$. Accordingly, the distribution of genotypes in the population corresponds to $p^2+2pq+q^2=1.0$. The expression $p^2$ corresponds to the frequency of the genotype **AA;** the expression $2pq$ corresponds to the frequency of the heterozygotes **Aa;** and $q^2$ corresponds to the frequency of the homozygotes **aa.**

When the frequency of an allele is known, the frequency of the genotype in the population can be determined. For instance, if the frequency $p$ of allele **A** is 0.6 (60%), then the frequency $q$ of allele **a** is 0.4 (40%). Thus, the frequency of the genotype **AA** is 0.36; of **Aa** it is $2\times0.24=0.48$; and of **aa** it is 0.16 (2). And conversely, if genotype frequency has been observed, the allele frequency can be determined. If only the homozygotes **aa** are known (e.g., they can be identified owing to an autosomal recessive inherited disease), then $q^2$ corresponds to the frequency of the disorder. From $p=1-q$, the frequency of heterozygotes ($2pq$) and of normal homozygotes ($p^2$) can also be determined.

| | Genotype of parents | Genotype of offspring |
|---|---|---|
| 1 | AA and AA | 1.0 AA |
| 2 | AA and Aa | 0.50 AA<br>0.50 Aa |
| 3 | Aa and Aa | 0.25 AA<br>0.50 Aa<br>0.25 aa |
| 4 | Aa and aa | 0.50 Aa<br>0.50 aa |
| 5 | AA and aa | 1.0 Aa |
| 6 | aa and aa | 1.0 aa |

**A. Expected frequency of genotypes in children of parents with different genotypes**

| Parents | 0.5 A | 0.5 a |
|---|---|---|
| 0.5 A | AA 0.25 | Aa 0.25 |
| 0.5 a | Aa 0.25 | aa 0.25 |
| | Offspring | |

1

$p = 0.50$ (Frequency of A)
$q = 0.50$ (Frequency of a)

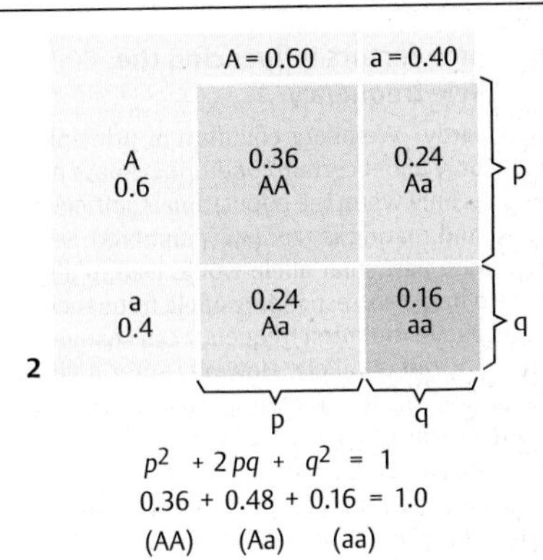


**B. Allele frequency**

# Hardy-Weinberg Equilibrium

When an allele in the homozygous or in the heterozygous state leads to a disorder that owing to its severity and the time of onset prevents the individual from reproducing, it represents a selective disadvantage for that individual. Although it could be assumed that such a mutant allele would eventually disappear from a population, this is not the case. Rather, a frequency equilibrium develops according to principles and conditions elucidated in 1908 by the English mathematician G. F. Hardy and the German physician W. Weinberg (Hardy-Weinberg equilibrium principle).

## A. Constant allele frequency (Hardy–Weinberg equilibrium)

An allele (here referred to as allele **a**) that leads to a severe disorder in the homozygous state is mainly present in a population in the undetectable heterozygous state. Only homozygotes (**aa**) come to attention as a result of illness. The heterozygotes (**Aa**) remain unrecognized and can not be distinguished by phenotype from the homozygotes **AA.** The frequency of affected individuals (homozygotes **aa**) is determined by the frequency of allele **a** (corresponding to $q$). The frequency of the three genotypes is determined by the binomial relationship $(p+q)^2 = 1$, where $p$ represents the frequency of allele **A,** and $q$ the frequency of allele **a.** The homozygous alleles **(aa)** eliminated in one generation by illness are replaced by new mutations. An equilibrium is reached between the elimination due to illness and the mutation frequency.

## B. Some factors influencing the allele frequency

The Hardy–Weinberg equilibrium principle is valid only under certain conditions: Above all, it applies only when the population is sufficiently large and mating is random (panmixia). Selection for a particular allele would lead to an increase in the corresponding allele frequency.
A change in mutation frequency can change the frequency of an allele. However, after a single alteration of the mutation frequency, equilibrium will occur at a new level within one generation.
In a small population, the proportion of genotypes may be shifted owing to casual fluctuations (genetic drift). An allele that is very rare or not present in a population can be introduced by migration and spread out as the population grows (founder effect).
Finally, a population may experience a drastic reduction in size, followed by an increase in the number of individuals. An allele that was previously rare in this population may by chance subsequently become relatively common as the population expands again.
Linkage disequilibrium is a deviation of the frequency of the co-occurrence of certain alleles at two or more neighboring (linked) loci from that expected from their individual frequencies in the population. Linkage disequilibrium may result from a selective advantage of one allele or from a geographic difference with respect to the time of occurrence of a mutation in one population compared to another. In such a case one finds a disease-causing mutation preferably on one haplotype of closely linked marker loci, but rarely or not at all on another (see hemoglobin disorders as an example). (Photographic detail from "Coney Island, 1938, photograph by Weegee," hilma hilcox, Fotofolio).

## References


Cavalli-Sforza, L.L., Bodmer, W.F.: The Genetics of Human Populations. W.H. Freeman & Co., San Francisco, 1971.

Cavalli-Sforza, L.L., Menozzi, P., Piazza, A.: The History and Geography of Human Genes. Princeton Univ. Press, Princeton, New Jersey, 1994.

Eriksson, A.W., et al., eds.: Population Structure and Genetic Disorders. Academic Press, London, 1980.

Jorde, L.: Linkage disequilibrium and the search for complex diseases. Genome Res. **10**:1435–1444, 2000.

Kimura, M., Ohta, T.: Theoretical Aspects of Population Genetics. Princeton Univ. Press, Princeton, New Jersey, 1971.

Kruglyak, L.: Prospects for whole-genome linkage disequilibrium mapping of common disease genes. Nature Genet. **22**:139–144, 1999.

Terwilliger, J., Ott, J.: Handbook for Human Genetic Linkage. Johns Hopkins Univ. Press, Baltimore, 1994.

Vogel, F., Motulsky, A.G.: Human Genetics. Problems and Approaches. 3rd ed. Springer Verlag, Heidelberg–New York, 1997.

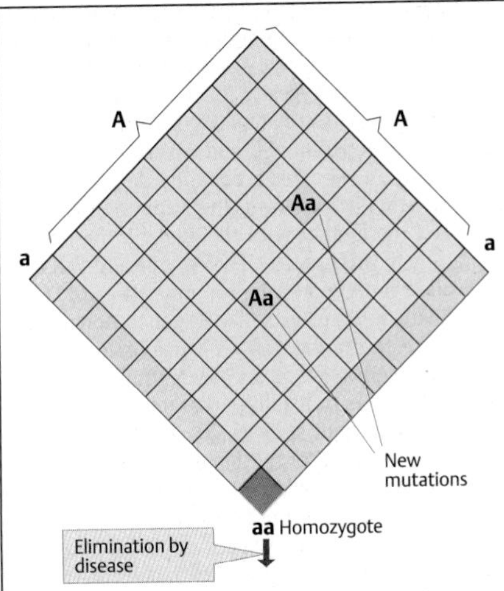
A
A
a
a
Aa
Aa
New mutations
aa Homozygote
Elimination by disease

Alleles **a** are eliminated from the population owing to severe illness in homozygotes **aa**, but are replaced by new mutation. An equilibrium results.

| Genotypes | | Frequency |
|---|---|---|
| Homozygote | **AA** | $p^2$ |
| Heterozygote | **Aa** | $2pq$ |
| Homozygote | **aa** | $q^2$ |

**A. Constant allele frequency (Hardy Weinberg equilibrium)**

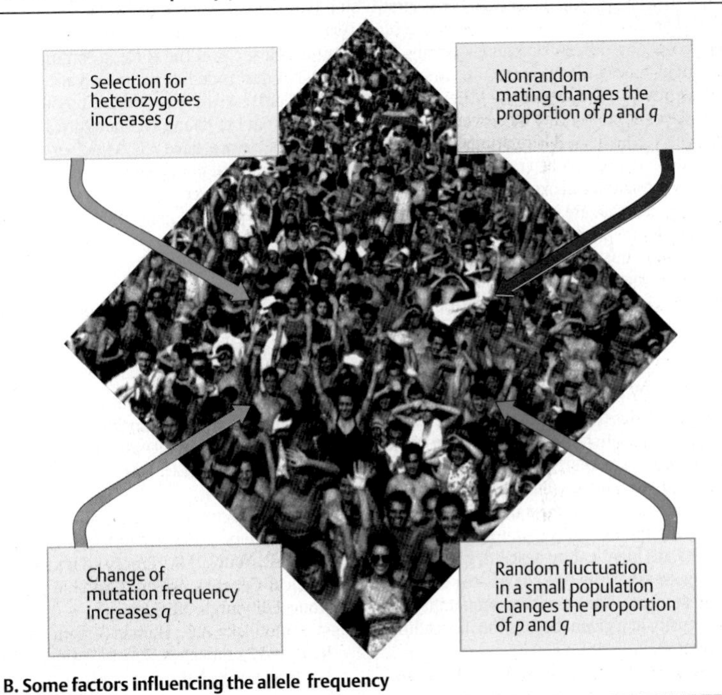
Selection for heterozygotes increases q
Nonrandom mating changes the proportion of p and q
Change of mutation frequency increases q
Random fluctuation in a small population changes the proportion of p and q

**B. Some factors influencing the allele frequency**

## Consanguinity and Inbreeding

Consanguinity (being "of the same blood") of the parents of an individual (parental consanguinity) means that they are genetically related because they have one or more ancestors in common. Consanguinity increases the probability that two related individuals carry an identical mutation as a recessive allele inherited from a common ancestor.

### A. Simple types of consanguinity

A mating between closely related individuals is called inbreeding. The degree of inbreeding depends on the closeness of their relationship. A mating between brother and sister or father and daughter is called incest (1). Sibs share half their genes, corresponding to a coefficient of relatedness of $^1/_2$. The chance of homozygosity by descent at a given locus in their offspring is $^1/_4$ ($^1/_2$ each from C and D to E). Two of the most frequent types of consanguineous unions in man are of first (2) and second cousins. First cousins share $^1/_8$ of their genes. The chance of homozygosity by descent in their offspring is $^1/_{16}$. An uncle–niece union is another example of consanguinity (3). They share $^1/_4$ of their genes. The chance of homozygosity by descent in their offspring is $^1/_8$. The degree of relationship of two individuals is expressed as coefficient of relationship *r*. The value of *r* corresponds to the proportion of genes that are identical by descent in the two persons ($^1/_2$ for sibs, $^1/_8$ for first cousins, $^1/_{32}$ for second cousins, $^1/_4$ for aunt–nephew or uncle–niece). The (generally) remote possibility that the unrelated individuals E or F transmit a mutant (allozygous, see below) allele at this locus can usually be disregarded.

### B. Identity by descent (IBD)

If related individuals mate, the chance increases that a particular allele present in one of their ancestors will have been inherited by both individuals (identity by descent, IBD). This results in an extra chance that their offspring will be homozygous for this allele (autozygous since the alleles are identical by descent, in contrast to allozygous when the two alleles are not identical by descent). The resulting probability of homozygosity at a given locus is the inbreeding coefficient, *F.*

The pedigree in panel B shows a consanguineous mating between individuals III and IV (first cousins) who share two common ancestors, I and II (their grandparents). The risk for an autosomal recessive disease in their offspring V due to homozygosity for a mutant allele at a given gene locus depends on the probability that both parents (III and IV) are heterozygous for this allele. This probability depends on the genetic distance from their common ancestor and the frequency of the allele in the population. In this case, the path of one and the same allele **A** present in one common ancestor I to individual III consists of two generational steps. Since the probability of transmission at each step (equivalent to one generation) is one half (50%), the probability that transmission occurred at both steps is $^1/_2 \times ^1/_2$, or $^1/_4$. However, homozygosity for the allele **A** in individual V requires that individual IV also has inherited allele **A** from the same common ancestor (I) as individual III. The number of steps required is also $^1/_2 \times ^1/_2$. The joint probability that allele **A** has descended from individual I to both individuals III and IV, therefore, is $(^1/_2)^4$ or $^1/_{16}$. Since there are two common ancestors and we do not know from which ancestor the allele in common to III and IV has descended (the common allele could also be **C** or **D**), we have to add this possibility to the path from I to III and IV. Therefore, for both paths possible we have $(^1/_2)^4+(^1/_2)^4$ or $^1/_8$. The final probability of homozygosity of V resulting from identity by descent (IBD) is one half of the proportion of parental genes that are identical by descent ($^1/_{16}$).

### References


Bittles, A.H., Neel, J.V.: The costs of human inbreeding and their implications for variations at the DNA level. Nature Genet. **8**:117–121, 1994.

Emery, A.E.H.: Methodology in Medical Genetics. Churchill Livingstone, Edinburgh, 1986.

Griffith, A.J.F., et al.: An Introduction to Genetic Analysis. 7th ed. W.H. Freeman & Co., New York, 2000.

Harper, P.S.: Practical Genetic Counselling. 5th ed. Butterworth-Heinemann, Oxford, 1998.

Mueller, R.F., Young, I.D.: Emery's Elements of Medical Genetics. 9th ed. Churchill Livingstone, Edinburgh, 1995.

Vogel, F., Motulsky, A.G.: Human Genetics. Problems and Approaches. 3rd ed. Springer Verlag, Heidelberg–New York, 1997.

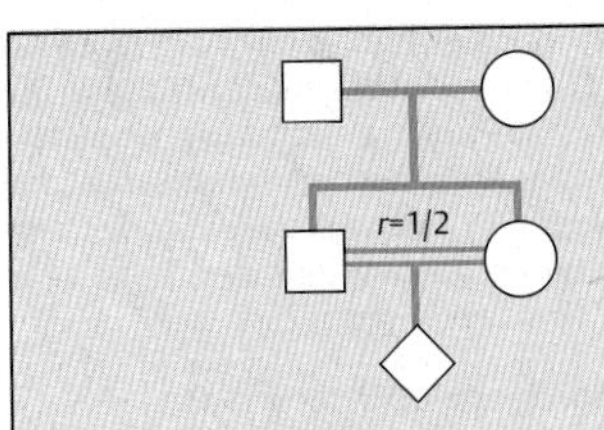


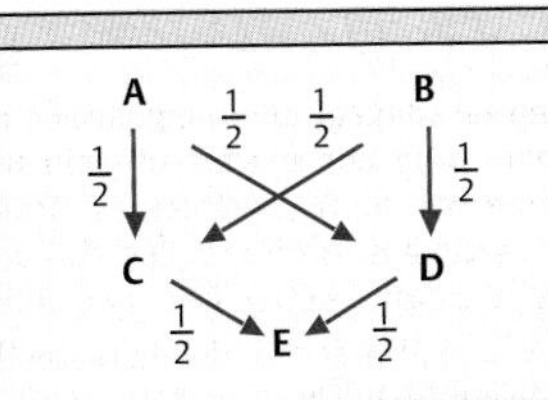


1. Brother/sister mating

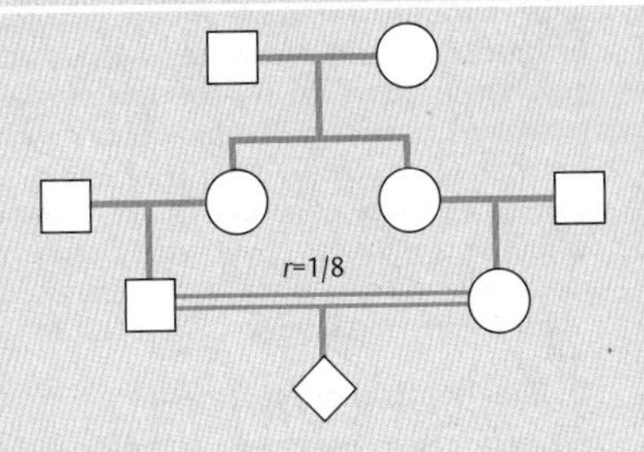


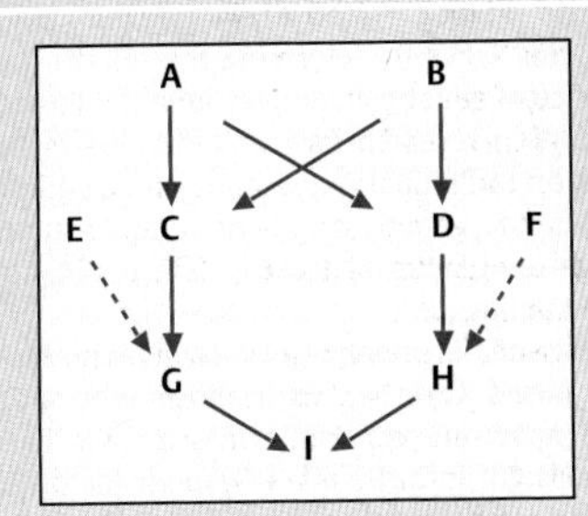


2. First cousins

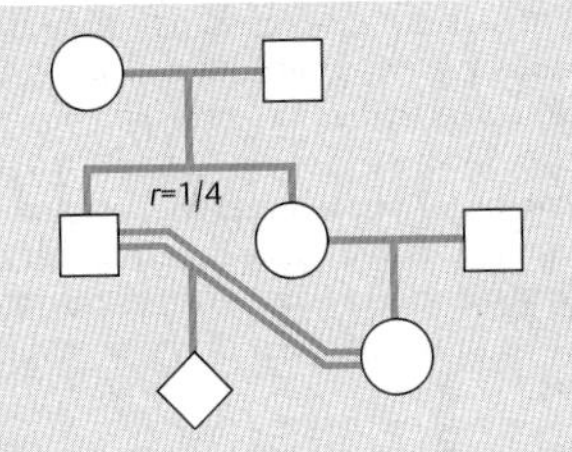


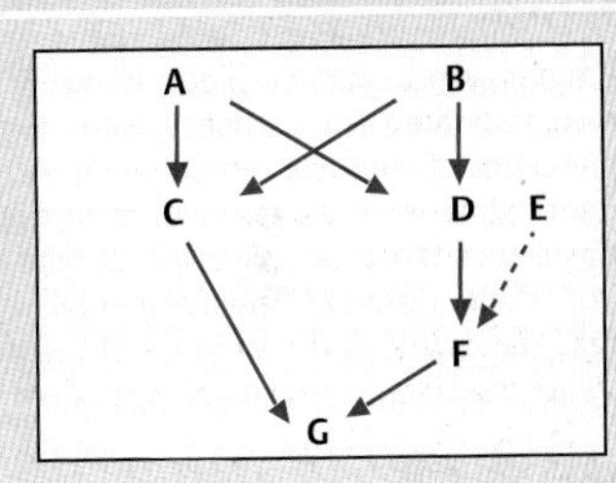


3. Uncle/niece

**A. Simple types of consanguinity**

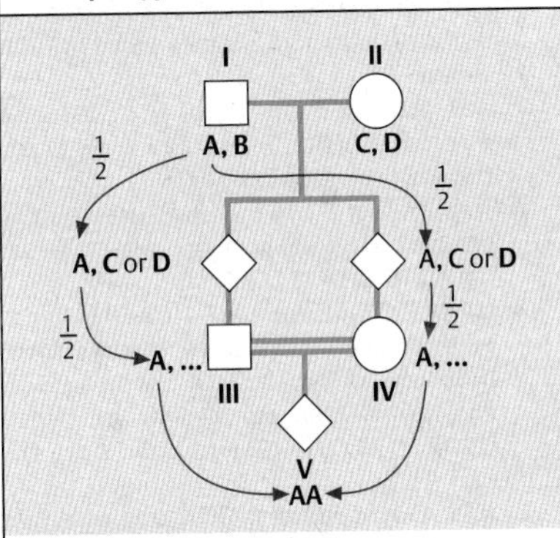


$$\mathrm{I} \longrightarrow \mathrm{III} = \left(\tfrac{1}{2}\right)^2$$

$$\mathrm{I} \longrightarrow \mathrm{IV} = \left(\tfrac{1}{2}\right)^2$$

$$\mathrm{I} \longrightarrow \mathrm{III\ and\ IV} = \left(\tfrac{1}{2}\right)^4$$

$$\mathrm{III\ and\ IV} \longrightarrow \mathrm{V} = \left(\tfrac{1}{2}\right)^2$$

$$\mathrm{IBD}: \left[\left(\tfrac{1}{2}\right)^4 + \left(\tfrac{1}{2}\right)^4\right] \times \tfrac{1}{2} = \tfrac{1}{16}$$

Probability of descent from a common ancestor

**B. Identity by descent (IBD)**

## Twins

Twin, triplet, or quadruple pregnancies regularly occur in many species of animals. In humans, about 1 in 80 pregnancies results in twins. They can arise from a single fertilized egg (monozygotic twins, MZ) or from two different eggs (dizygotic twins, DZ). Monozygotic twins are genetically identical.

The study of genetic differences between monozygotic (identical) and dizygotic (nonidentical) twins seeks to determine the relative importance of genetic and environmental factors. While twin research has limited value in the search for individual genes, it may shed light on the possible genetic origin of complex traits, such as multifactorial diseases (see p. 154) and human behavior.

Comparisons of monozygotic twins who have been raised together with those who were raised apart are especially informative. They allow insight into the role of genetic factors as opposed to environmental influences. In a large study, the Minnesota Twin Study (Bouchard, 1990), monozygotic and dizygotic twins who had been separated since infancy were compared. A series of complex tests demonstrated that monozygotic twins raised apart, compared with those raised together, were very similar in spite of their different environmental influences. About 70% of the variance of mental ability was associated with genetic variation.

### A. Types of twins

Monozygotic twins arise by separation at a very early stage of development. They always have a common placenta (monoplacental). They may lie in two amniotic cavities (diamniotic) or in a common amniotic cavity (monoamniotic). Dizygotic twins (DZ) always have their own amniotic cavity (diamniotic). They may each have their own placenta (diplacental), or they may share a common placenta.

### B. Pathological conditions in twins

Incomplete separation (so-called Siamese twins) or connections between the blood circulations lead to pathological conditions in twins. A relatively frequent form of incomplete separation is thoracopagus (1), in which the twins are joined to various extents at the thoracic region. The blood circulation of twins can be connected via the common placenta (formation of a shunt) (2). This can lead to one twin receiving less blood or even bleeding to death. Especially severe malformations result from incompletely formed organs, e.g., absence of the heart in one twin (acardius) (3). If one of the twins dies very early in pregnancy, it may disappear by the time the other twin is born (vanishing twin).

### C. Concordance or discordance

When twins show the same trait, they are said to be concordant; when they differ, discordant. Comparisons of the rate of concordance allow conclusions about the relative role of genetic factors in the etiology of complex traits such as some malformations, predisposition to diseases, height, blood pressure, etc. However, definitive conclusions about genetic causes can rarely be reached from concordance rates alone. (After Connor and Ferguson-Smith, 1991.)

### D. Biochemical differences

Dizygotic and monozygotic twins differ also biochemically. Due to genetic differences, many chemical substances used in therapy are metabolized or excreted at different rates, owing to different activities of corresponding enzymes (see p. 166 and 372). An example is the rate of excretion of phenylbutazone, as studied by E. S. Vesell. This substance is excreted at the same rate in identical twins, whereas the rates of excretion differ between dizygotic twins or siblings.

### References

Bouchard, T.J. et al.: Sources of human psychological differences: The Minnesota study of twins reared apart. Science **250**:223–228, 1990.

Connor, J.M., Ferguson-Smith, M.A.: Essential Medical Genetics, 3rd ed. Blackwell Scientific Publishers, Oxford, 1991.

McGregor A.J., et al.: Twins. Novel uses to study complex traits and genetic diseases. Trends Genet. **16**:131–134, 2000.

Phelan, M.C.: Twins. pp. 1047–1079. In: Human Malformations and Related Anomalies. Vol. II. Stevenson, R.E., Hall, J.G., Goodman, R.M., eds. Oxford Monographs on Medical Genetics **37**:1047–1079, Oxford Univ. Press, Oxford, 1993.

Segal, N.: Entwined Lives: Twins and what they tell us about human behavior. Dutton Books, New York, 1999.

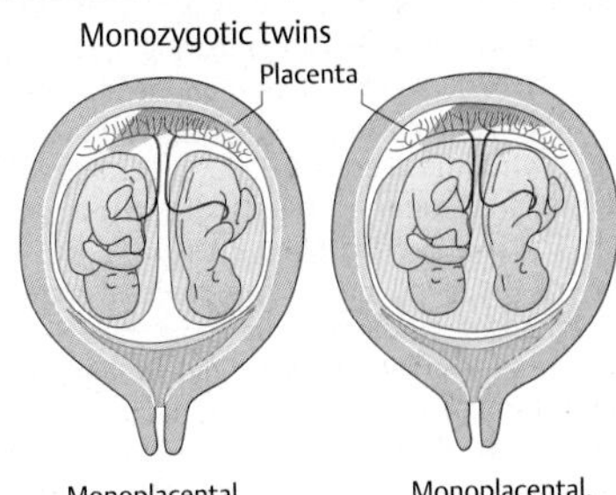


Dizygotic twins

Diplacental, diamniotic

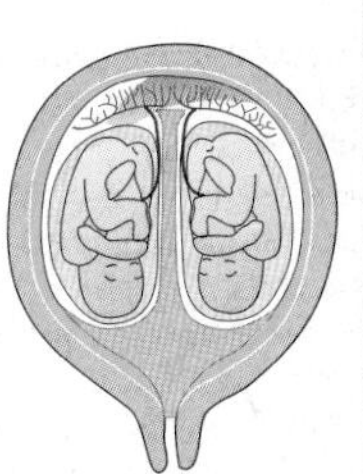


**A. Types of twins**

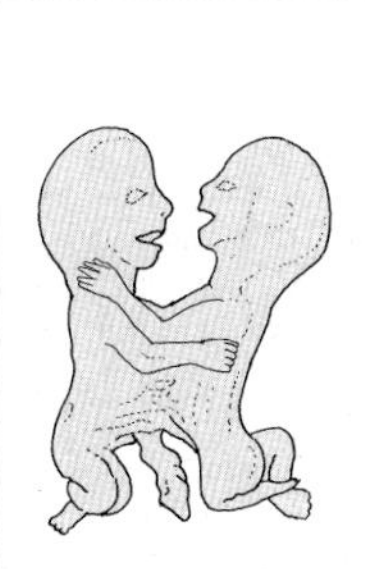

1. Thoracopagus

2. Dizygotic twins connected by a shunt

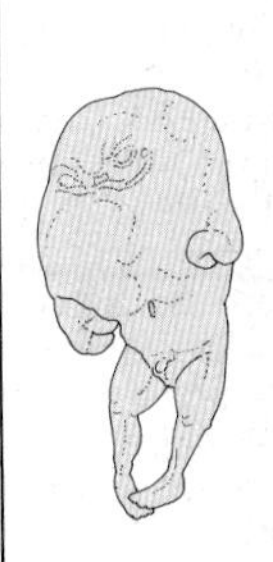

3. Acardius

**B. Pathological conditions in monozygotic twins**

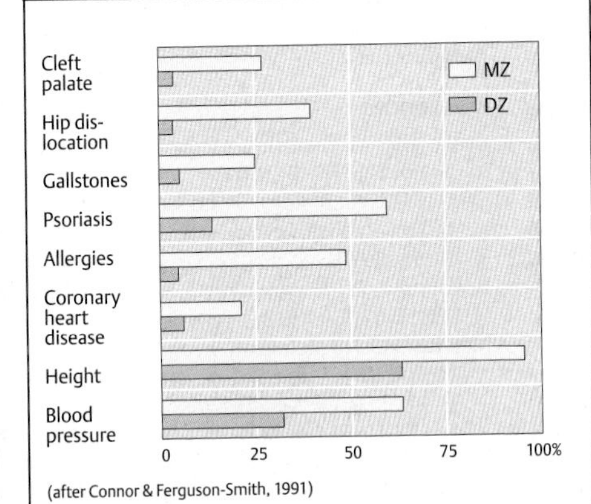


(after Connor & Ferguson-Smith, 1991)

**C. Concordance of some traits in monozygotic (MZ) and dizygotic (DZ) twins**

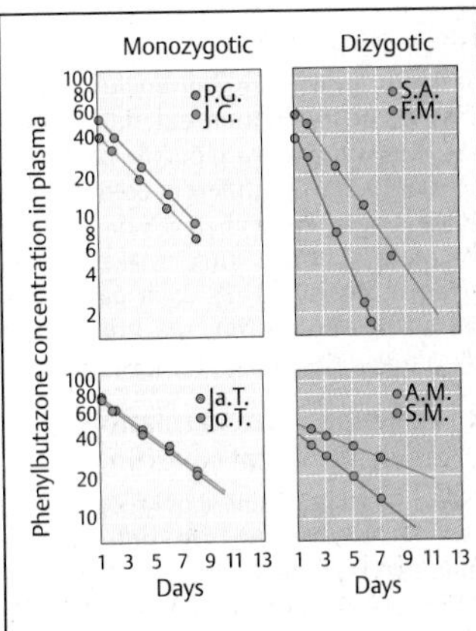


**D. Excretion rate of a drug (data of E. S. Vesell)**

# Polymorphism

In nature, organisms of the same species usually differ in some aspects of their appearance. The differences are genetically determined and are referred to as polymorphisms. Genetic polymorphism is defined as the occurrence in a population of two or more genetically determined alternative phenotypes due to different alleles, whereby the least frequent allele cannot be maintained by repeated mutation alone. A gene locus is defined as polymorphic if the rare allele(s) has (have) a frequency of at least 0.01 (1%), and as a result, heterozygotes for such alleles occur with a frequency of at least 2%. Polymorphism can be observed at the level of the whole individual (phenotype), in variant forms of proteins and blood group substances (biochemical polymorphism), in morphological features of chromosomes (chromosomal polymorphism), or at the level of DNA in differences of nucleotides (DNA polymorphism, p. 72). A polymorphism is considered as neutral if the presence or absence of a certain allele does not confer any advantage or disadvantage. A polymorphism may represent an advantage for the population, because it is likely to contain individuals who are better prepared for certain environmental conditions or changes thereof than if there were genetic uniformity.

## A. Polymorphism of the phenotype

An impressive example of phenotypic polymorphism is the color pattern on the wing sheaths of the Asian ladybug *(Harmonia axyridis)* (1). In the area of distribution, extending from Siberia to Japan, multiple variants can be distinguished (Ayala, 1978). The different color combinations are due to different alleles of the same gene (1). In the California king snake *(Lampropeltis getulus californiae)* (2), color patterns differ to such an extent within the same species that they would seem to represent different species.

## B. Polymorphism in relation to environmental conditions

Most living organisms show geographical variation; that is, there are differences in the frequencies of alleles among populations of the same species in different areas. The differences may be gradual and reflect adaption to environmental conditions. For example, the average height of the yarrow plant on the slopes of the Sierra Nevada in California decreases as the altitude increases (1). By comparing the growth of plants whose seeds were obtained at different altitudes and sown in one garden, it can be shown that the average height of plants at the different altitudes is genetically determined. (Illustration after Campbell, 1990.)

One of the most impressive cases of natural selection for a polymorphic color pattern has been represented by the English peppered moth *Biston betularia.* In the midlands of England it was observed in two varieties, a light gray *(typica),* with spots of pigments, and a rare dark gray variety *(carbonaria)* (2). It was thought that the black variety was rare because it was easily seen by predators and therefore selected against. In the late 1880s, the light variety became rare and the black type prevailed. It was assumed that this was due to selection for *carbonaria,* which was camouflaged on the dark bark of trees after the light-colored lichens covering trees and rocks had been killed by the beginning industrial pollution. After pollution was reduced in the 1950s, the light type reappeared and again became common. Unfortunately, serious doubt has been cast on this beautiful story of natural selection by genetic adaptation. Majerus (1998) pointed out unconfirmed observations that make natural selection untenable in this case (see Nature, 5 November 1998, p. 35). What a pity. (Origin of illustrations: F. J. Ayala (A1), J. H. Tashjian, San Diego Zoo (A2), N.A. Campbell, Biology, 1990 (B1), Laßwitz and K. Jäkel, Okapia, Frankfurt (B2)).

## References

Ayala, F.J.: Mechanisms of evolution. Scient. Amer. **329**:48–61, 1978.

Campbell, N.A.: Biology. 2nd ed. Benjamin/Cummings Publishing Co., Menlo Park, California, 1990.

Lewontin, R: Adaptation. Scient. Amer. **239**:156–169, 1978.

Majerus, M.E.N.: Melanism: Evolution in Action. Oxford Univ. Press, Oxford, 1998.

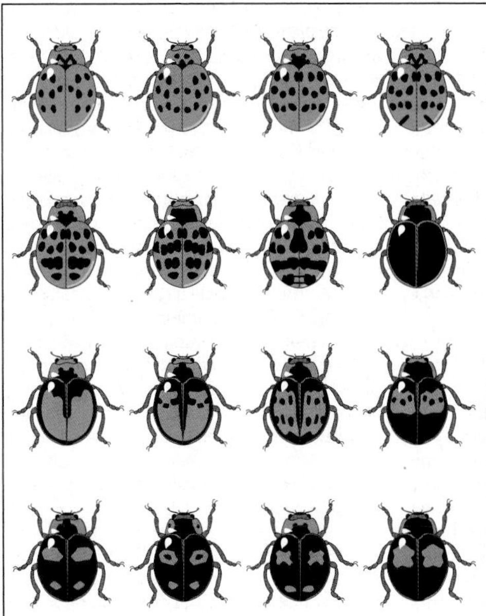

1. Asiatic beetle *Harmonia axyridis*

2. California king snake *Lampropeltis getulus californiae*

**A. Polymorphism of the phenotype**

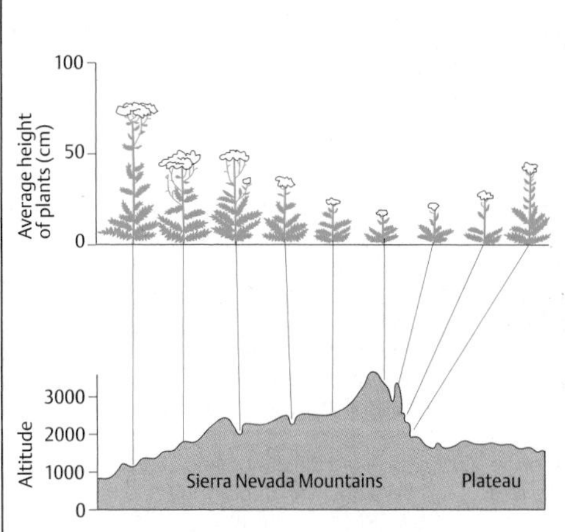


1. Yarrow plant in the Sierra Nevada Mountains

2. Moth *Biston betularia*, light and dark type

**B. Polymorphism in relation to environmental conditions**

# Biochemical Polymorphism

Biochemical polymorphism is detected by laboratory methods. It results from individual differences in the sequences of nucleotide bases of DNA. If a difference leads to a change in a codon, a different amino acid may be incorporated at the corresponding site of the protein. This can be demonstrated by analyzing the gene product.
Subtle differences in the many biochemical reactions between individuals was recognized early last century as the basis of genetic individuality by Archibald Garrod (Bearn, 1993).

## A. Recognition of polymorphism by means of gel electrophoresis

Polymorphism of a protein (gene product) can be demonstrated by gel electrophoresis when one variant form differs from the others by the presence of an amino acid with a different electric charge. In this case, the allelic forms of the gene product can be distinguished owing to their different speeds of migration in an electrical field (electrophoresis). When a gene product consists of two subunits (dimeric protein), heterozygous individuals will show three bands. A polymorphism that does not lead to a change in electric charge cannot be identified in this manner.

## B. Demonstration of polymorphism in gene products

Here, polymorphism is shown to be frequent in three typical gene products, the enzymes phosphoglucomutase, malate dehydrogenase, and acid phosphatase, in different species of *Drosophila*. Each of the diagram sections 1, 2, and 3 shows a starch gel electrophoresis of 12 fruit flies, each gel being specifically stained for the respective protein. With phosphoglucomutase (1) different migration speeds are observed in homozygous individuals as opposed to heterozygous (2, 4, 10). Malate dehydrogenase is a dimeric protein, so that heterozygous individuals (4, 5, 6, 8) show three bands. Acid phosphatase (3) shows a complex pattern because four alleles are involved. (Figure after Ayala and Kieger, 1984).

## C. Frequency of polymorphism

In a study of average heterozygosity in *Drosophila willistoni* (1), 17.7% of 180 gene loci were found be heterozygous. The average heterozygosity is the proportion of heterozygous individuals in a population with reference to the number of analyzed loci. This is determined by summing the proportion of heterozygotes per total number of individuals per gene locus. The best insight into the considerable frequency of polymorphism is obtained from direct analysis of DNA. Hypervariable regions are found in many areas of the genome. Here, digestion with a restriction enzyme produces several different patterns due to individual differences in the size of the DNA fragments. In (2), the polymorphism at one locus near the H chain of the immunoglobulin J region in 16 individuals is shown (White et al., 1986). Hypervariable regions lead to an individual pattern for each person (DNA fingerprint).

## D. Genetic diversity and evolution

A population of genetically relatively homogeneous individuals is less likely to be able to adapt to changing environmental conditions than is a genetically diverse species. Two populations of *Drosophila serrata* kept for 25 generations in separate closed bottles with limited availability of space and nourishment were compared. The population with the greater genetic diversity adjusted better to the environmental conditions. (Figure after Ayala & Kieger, 1984).

### References

Ayala, F.J., Kieger, J.A.: Modern Genetics. 2nd ed. Benjamin/Cummings Publishing Co., Menlo Park, California, 1984.

Beaudet, A.L., et al.: Genetics, biochemistry, and molecular basis of variant human phenotypes. pp. 53–118 In: C.R. Scriver, et al., eds. The Metabolic and Molecular Bases of Inherited Disease. 7th ed. McGraw-Hill, New York, 1995.

Bearn, A.: Archibald Garrod and the Individuality of Man. Clarendon Press, Oxford, 1993.

White, R., et al.: Construction of human genetic linkage images. I. Progress and perspectives. Cold Spring Harbor Symp. Quant Biol. **51**:29, 1986.

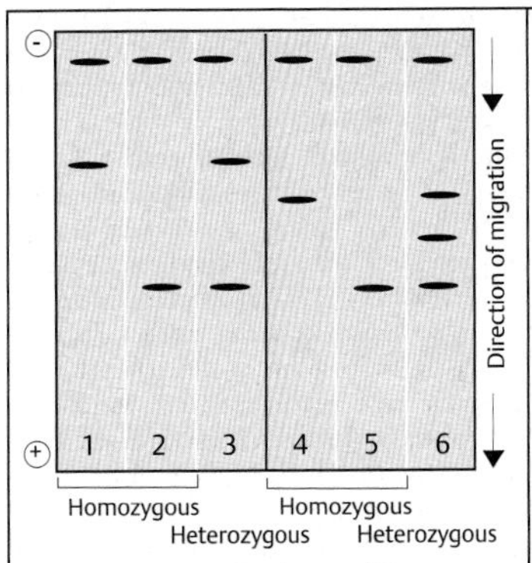


**A. Recognition of polymorphism by gel electrophoresis**

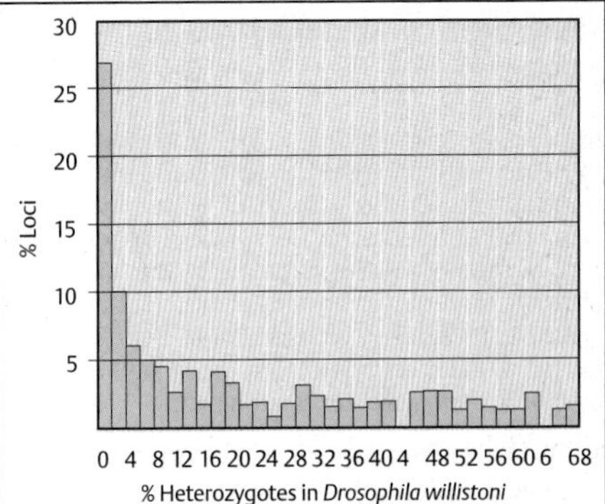


**1.** High-average heterozygosity (17.7 % of 180 loci)

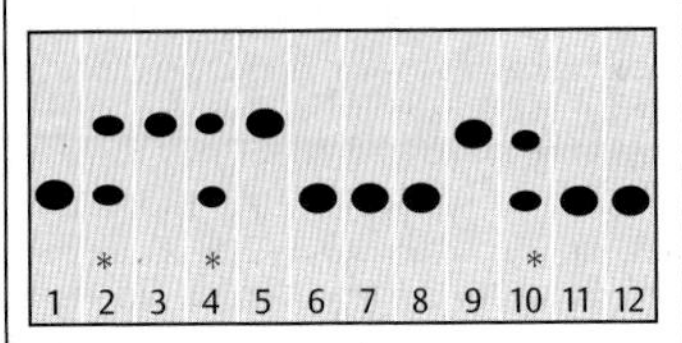

**1.** Phosphoglucomutase in *Drosophila pseudoobscura*

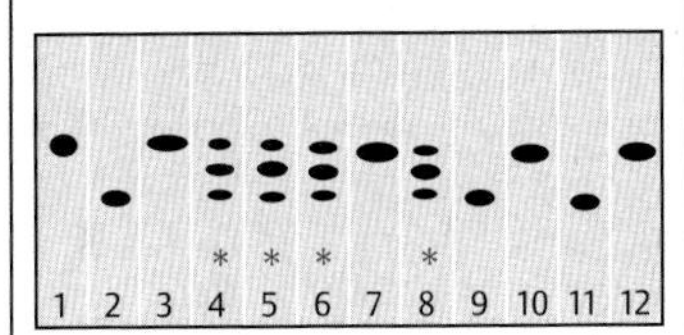

**2.** Malate dehydrogenase in *Drosophila equinoxialis*

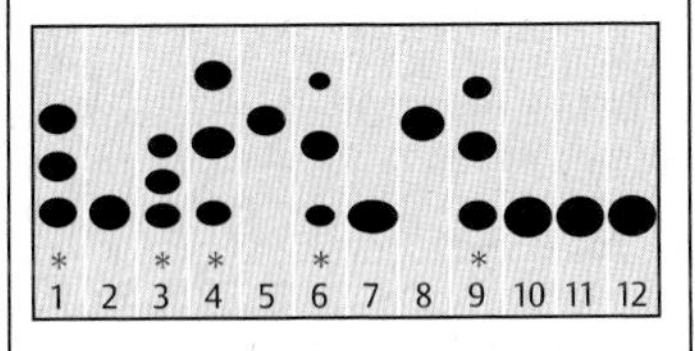

**3.** Acid phosphatase in *Drosophila equinoxialis*

(∗ = Heterozygous fruit flies)

**B. Demonstration of polymorphism in gene products**

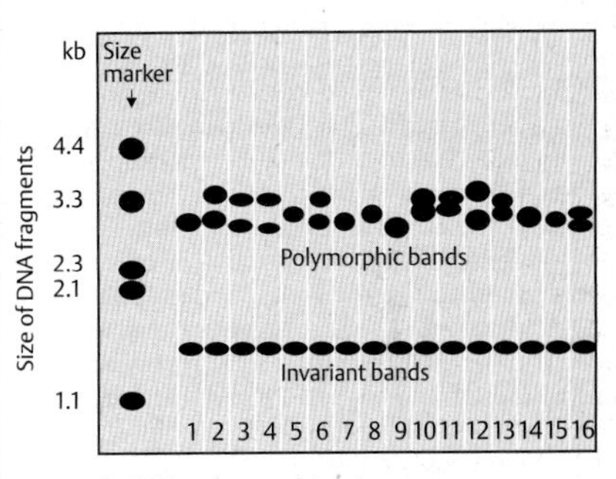


**2.** DNA polymorphism in man (16 individuals)

**C. Frequency of polymorphism**

Two populations of Drosophila serrata

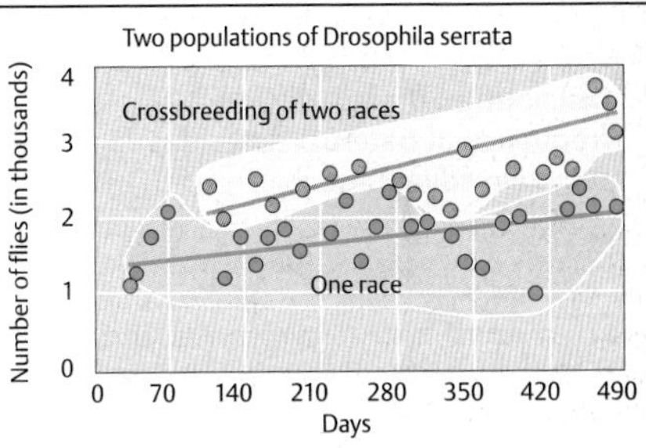


Reduced food and space conditions

**D. Genetic diversity and evolution**

# Geographical Distribution of Genes

Within a species, alleles at a given gene locus may occur with different frequencies in different geographical regions. This can be due either to random differences in frequencies (genetic drift) or to selection. Selection means that one allele has conferred an advantage over another (selective advantage), with the result that the individual has a slightly higher probability of survival in that particular environment.

## A. Different frequencies of hereditary disorders, e.g., in Finland

Genetically determined diseases in humans (hereditary diseases) occur with different frequencies in different regions of the world. For example, the distribution of the grandparents of persons affected with three autosomal recessive inherited diseases in Finland shows an increased regional clustering because each mutation has arisen in a different region of Finland: (1) congenital flat cornea, in the western part of the country; (2) the Finnish type of congenital nephrosis (a severe renal disorder), in the southern part; and (3) diastrophic skeletal dysplasia, in other southern regions. There is no known basis for a selective advantage of these mutant alleles in the heterozygotes to explain their frequencies. The different distribution merely reflects the place and relative point in time of the mutation. The mutations must have arisen after Finland was settled, since they are unusually frequent there but rare elsewhere.
Similar examples are found in many other regions of the world and in other populations.

## B. Distribution of malaria and frequency of heterozygotes due to selective advantage

The distribution of malaria in the Mediterranean region, West Africa, Sub-Saharan Africa, and East Africa including the Nile Valley and Asia (1) is similar to the frequency of heterozygotes for three forms of genetic disease: sickle cell anemia, thalassemia (see p. 340 ff.), and glucose-6-phosphate dehydrogenase deficiency (a red-cell disorder). Heterozygotes for these three genetically determined diseases have a selective advantage here. They are relatively protected because conditions in the blood of heterozygotes are less favorable for the malaria parasite than in the blood of normal homozygotes (see p. 340).
Glucose-6-phosphate dehydrogenase deficiency (2) is an X-chromosomal disorder that leads to severe illness in affected males (hemizygotes). In the same way, homozygotes for sickle cell anemia (3) and different forms of a disorder of hemoglobin formation (thalassemia, 4) are very severely affected, whereas heterozygotes are not or are only mildly affected. This has led to a striking increase of heterozygotes for the mutant allele in areas infested with malaria.
Protection for the heterozygotes occurs at the cost of affected homozygotes, but has a positive effect for the population as a whole, since the total frequency of affected individuals is reduced. Nevertheless, the price of this advantage is considerable. More than 400 million people become ill with malaria each year in Africa, Asia, and South America. About 1 to 3 million die. It is hoped that sequencing the malaria-causing parasite, *plasmodium falciparum*, and other approaches will invigorate the fight against malaria (Malaria New Focus, Science 20 October 2000, p. 428–441).
The sickle cell mutant has arisen independently at least four times in different regions and has become established there due to selective advantage. The situation is similar for thalassemia (see p. 344).

### References

Cavalli-Sforza, L.L., Menozzi, P., Piazza, A.: The History and Geography of Human Genes. Princeton Univ. Press, Princeton, 1994.

Norio, R., Nevanlinna, H.R., Perheentupa, J.: Hereditary diseases in Finland. Ann. Clin. Res. **5**:109–141, 1973.

Norio, R.: Diseases of Finland and Scandinavia. pp. 359–415. In: Rothschild, H.R., ed. Biocultural Aspects of Disease. Academic Press, New York, 1981.

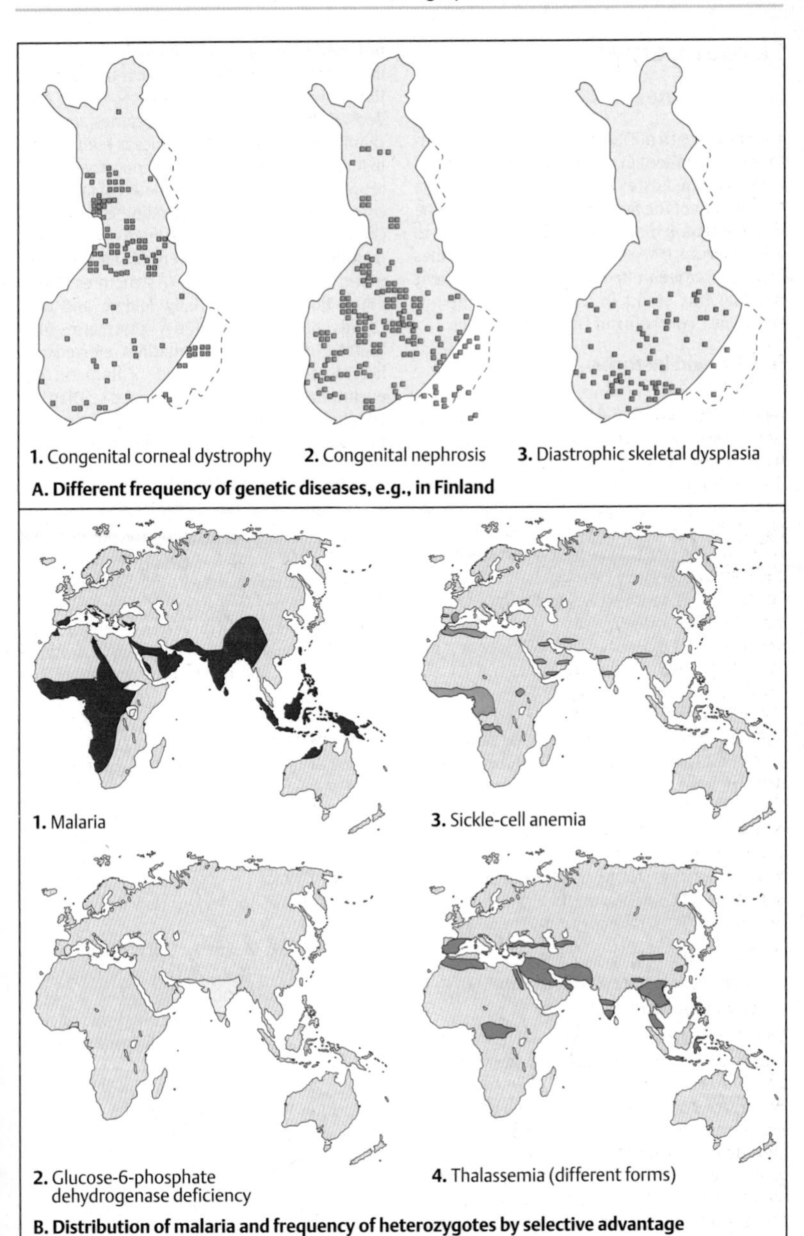


**1.** Congenital corneal dystrophy    **2.** Congenital nephrosis    **3.** Diastrophic skeletal dysplasia

A. Different frequency of genetic diseases, e.g., in Finland

**1.** Malaria

**3.** Sickle-cell anemia

**2.** Glucose-6-phosphate dehydrogenase deficiency

**4.** Thalassemia (different forms)

B. Distribution of malaria and frequency of heterozygotes by selective advantage

# Chromosomes

## Nucleosomes

The total length of DNA in the haploid chromosome complement in nondividing cells of mammals corresponds to about 1 meter. Since the total length of the human haploid chromosome complement in metaphase cells corresponds to about 115 μm, DNA in metaphase chromosomes must be shortened 10 000-fold by very efficient packing. This occurs in discrete units, the nucleosomes, consisting of DNA and histones.

### A. DNA and histones

Histone proteins (histones) are proteins with a high proportion of positively charged amino acids (lysine and arginine), which enable them to bind firmly to the negatively charged DNA double helix. There are five types of histone molecules: H1, H2 A, H2 B, H3, and H4. Except for H1, they occur in exactly equal numbers. Histones are evolutionarily highly conserved. There is an especially close spatial relationship between DNA and histones in the A+T-rich regions of the minor grooves. Here, the DNA double helix is most easily bent. (It is bent more than shown here).

### B. Nucleosome

The nucleosome is the fundamental subunit of DNA and histones in interphase chromosomes, or chromatin (see next plate). A nucleosome consists of about 146 bp of DNA wound twice around an octamer of histones, two copies each of H2 A, H2 B, H3, and H4, the core histones. The histone octamer forms a cylinder of 11 nm diameter and 6 nm height (the figure in B is not to scale and is highly schematic; for details see Lewin, 2000).

### C. Three-dimensional structure of a nucleosome

The ribbon diagram of the nucleosome based on the X-ray structure at high resolution shows one DNA strand in green, the other in brown. The histone are shown in different colors in a view from above (figure adapted from Luger et al., with kind permission by T. J. Richmond).

### D. Chromatin structures

Chromatin occurs in condensed (tightly folded), less condensed (partially folded), and extended unfolded form. When extracted from cell nuclei in isotonic buffers, most chromatin appears as fibers of about 30 nm in diameter. In less condensed chromatin fibers, it becomes apparent that DNA between nucleosomes is bound to H1 histones. H1 histones vary among tissues and represent about half of the amount of individual core histones. The corresponding electron-microscopic photographs obtained by different techniques show the condensed (folded) chromatin as compact 300–500 Å structures (top), a 250 Å fiber when partially folded, and as a "beads on a string" 100 Å chromatin fiber (bottom). Condensed chromatin is referred to as heterochromatin, extended chromatin as euchromatin (see following pages). (Diagram adapted from Alberts et al. 1994; the electronmicrographs are from Thoma and Koller, 1979).

### E. Chromatin segments

The chromatin of the nucleus consists of segments of nucleosomes condensed to different degrees. DNA wound around a nucleosome is inactive and unreactive. In the DNA segments lying between series of nucleosomes, sequence-specific DNA-binding proteins are found. They regulate the expression (activity) of genes (see pp. 204 ff). (Figures after Alberts et al. 1994).

### References

Alberts, B., et al.: Molecular Biology of the Cell. 3rd ed. Garland Publishing, New York, 1994.

Kornberg, R.D., Lorch, Y.: Twenty-five years of the nucleosome, fundamental particle of the eukaryote chromosome. Cell **98**:285–294, 1999.

Lewin, B.: Genes VII. Oxford Univ. Press, Oxford, 2000.

Lodish, H., et al.: Molecular Cell Biology (with an animated CD-ROM). 4th ed. W.H. Freeman & Co., New York, 2000.

Luger, K., et al.: Crystal structure of the nucleosome core particle at 2.8 Å resolution. Nature **389**:251–260, 1997.

Thoma, F., Koller, Th., Klug, A.: Involvement of histone H1 in the organization of the nucleosome and of the salt dependent superstructures of chromatin. J. Cell Biol. **83**:403–427, 1979.

Watson, J.D., et al.: Molecular Biology of the Gene. 3rd ed. Benjamin/Cummings Publishing Co., Menlo Park, California, 1987.

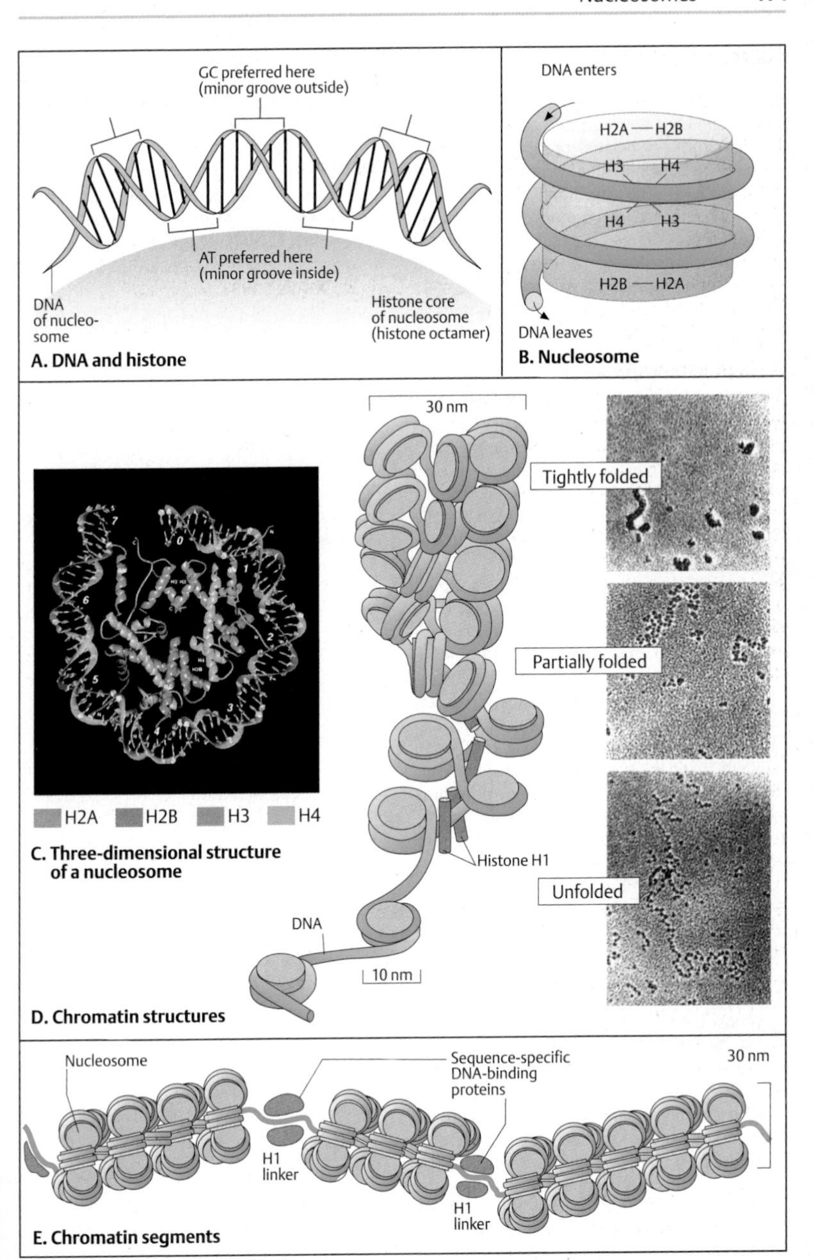


**A. DNA and histone**

**B. Nucleosome**

**C. Three-dimensional structure of a nucleosome**

**D. Chromatin structures**

**E. Chromatin segments**

# DNA in Chromosomes

In eukaryotic organisms the DNA is contained in chromosomes. Chromosomes consist of DNA and its closely associated proteins, which are histones and nonhistone proteins in about equal amounts. The chromatin of the cell nucleus is composed of chromosomes in interphase. Chromosomes are visible as individual structures only during mitosis. The haploid genome in humans contains 3 x $10^9$ base pairs in 22 autosomes and two sex chromosomes (X and Y).

## A. DNA in metaphase chromosomes

A gene is a nucleotide sequence that serves as the functional unit for the formation of a complementary RNA molecule (transcription unit). A chromosome consists of a single, very long DNA molecule that contains a series of genes. It is folded and coiled in a complex manner. The spatial arrangement is secured by the nucleosomes.

In eukaryotic cells, the DNA is closely associated with histones to form characteristic structures, the nucleosomes. In addition, there are hundreds of sequence-specific DNA-binding proteins that recognize short DNA segments. Their binding is an important means of regulating gene activity (gene expression, p. 204 ff.).

Chromosomal DNA is folded and packed in an efficient manner. Schematically, several levels of packing of DNA in a metaphase chromosome can be differentiated. The figure shows (at the top) a segment of DNA double helix from part of a chromatin segment. The chromatin segment consists of several nucleosomes connected by so-called linker stretches of DNA. The folding of a chromatin segment produces a chromatin fiber of 30 nm diameter, visible on electron microscopy. It consists of a series of tightly packed nucleosomes. These in turn form a part of a chromosome segment of about 300 nm diameter in the extended stage. A further packing level is represented by a thickened segment of a metaphase chromosome. This is called a condensed chromosomal segment. This segment corresponds to only a small part of one chromatid of a metaphase chromosome. As described on p. 182, each metaphase chromosome consists of two chromatids (sister chromatids), one of these chromatids being the result of DNA replication in the S phase. No transcription occurs in the chromatin of mitotic chromosomes. RNA synthesis is halted with the onset of chromosome condensation. (Figure after Alberts et al., 1989).

### Reference

Alberts, et al.: Molecular Biology of the Cell. 3rd ed. Garland Publishing, New York, 1994.

Tyler-Smith, C., Willard, H.F.: Mammalian chromosome structure. Curr. Opin. Genet. Dev. **3**:390–397, 1993.

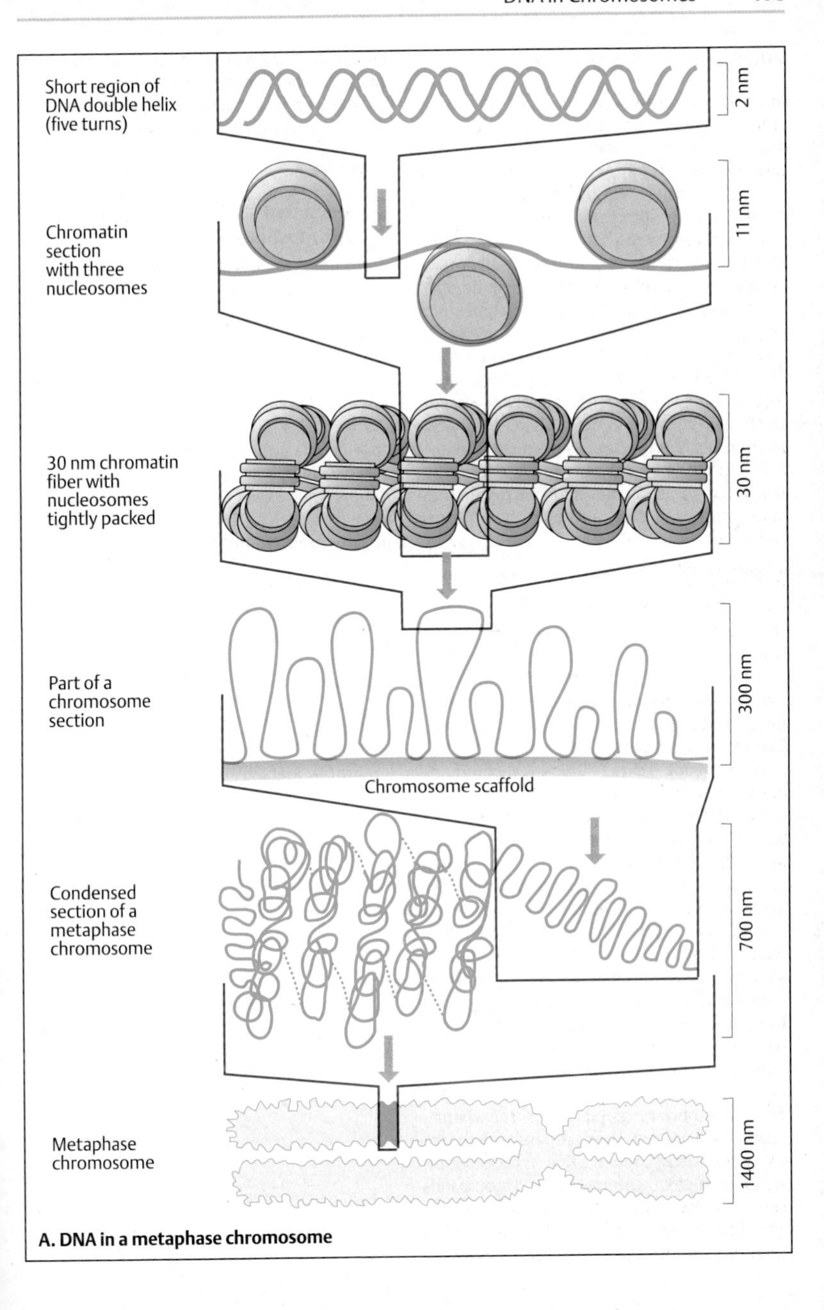
Short region of DNA double helix (five turns)
2 nm
Chromatin section with three nucleosomes
11 nm
30 nm chromatin fiber with nucleosomes tightly packed
30 nm
Part of a chromosome section
300 nm
Chromosome scaffold
Condensed section of a metaphase chromosome
700 nm
Metaphase chromosome
1400 nm

**A. DNA in a metaphase chromosome**

# Polytene Chromosomes

In certain cells of some insects, chromatin structures are visible under the light microscope as individual looplike domains. These cells and the chromosomes in them are greatly enlarged by prior DNA synthesis without cell division. That is, they contain about a thousand times more DNA than usual. The multiple copies of homologous chromosomes remain side by side in these cells and form a gigantic polytene chromosome. If the multiple copies of a polytene chromosome did not occur side by side, but rather as discrete chromosomes, the result would be referred to as polyploidy. Since there is a direct transition from polytene to polyploid chromosomes, the polytene chromosomes probably correspond in structure to normal chromosomes. Polytene chromosomes are readily visible under the light microscope because they are so large and lie precisely next to each other.

## A. Polytene chromosomes in the salivary glands of *Drosophila* larvae

Under the light microscope, polytene chromosomes show readily recognized patterns of alternating dark and light bands. The light bands are called interbands. A polytene chromosome results from ten cycles of replication without division into daughter chromosomes. Thus, there are about 1024 ($2^{10}$) identical chromatid strands, which lie strictly side by side. About 80% of the DNA in polytene chromosomes is located in bands, and about 15% in interbands. The chromatin in the darkly stained band is condensed to a much greater degree than the chromatin in the interbands. This can be attributed to an especially high degree of folding. Depending on the size, an individual band contains between 3000 and 300 000 nucleotide base pairs. Altogether, the Drosophila genome contains about 5000 bands and the same number of interbands. They have been numbered to produce a polytene chromosome map. A micrographic detail of a polytene chromosome from a *Drosophila* salivary gland shows the characteristic band pattern. The dark bands are the result of chromatin condensation in the large interphase polytene chromosomes. These bands must be differentiated from those visible only in metaphase after special staining techniques in the banding patterns of the karyotypes of eukaryotic organisms (see p. 184). Intensely stained chromosomal segments correspond to a high degree of packing and are genetically inactive (heterochromatin); less tightly packed segments stain less distinctly and correspond to segments with genetic activity (euchromatin).

## B. Functional stages in polytene chromosomes

Polytene chromosomes form structures that correlate with the functional state: During the larval development of drosophila, a series of expansions (puffs) appear in temporal stages in the polytene chromosomes. Chromosome puffs are decondensed, expanded segments that represent active chromosomal regions, i. e., regions that are being transcribed. The location and duration of the puffs reflect different stages of larval development (1). The incorporation of radioactively labeled RNA has been used to demonstrate that RNA synthesis, a sign of gene activity (transcription), occurs in these regions (2).

(Figures after Alberts et al., 1994 and Watson et al., 1987).

### References

Alberts, et al.: Molecular Biology of the Cell. 3rd ed. Garland Publishing, New York, 1994.

Watson, J.D., et al.: Molecular Biology of the Gene. 3rd ed. Benjamin/Cummings Publishing Co., Menlo Park, California, 1987.

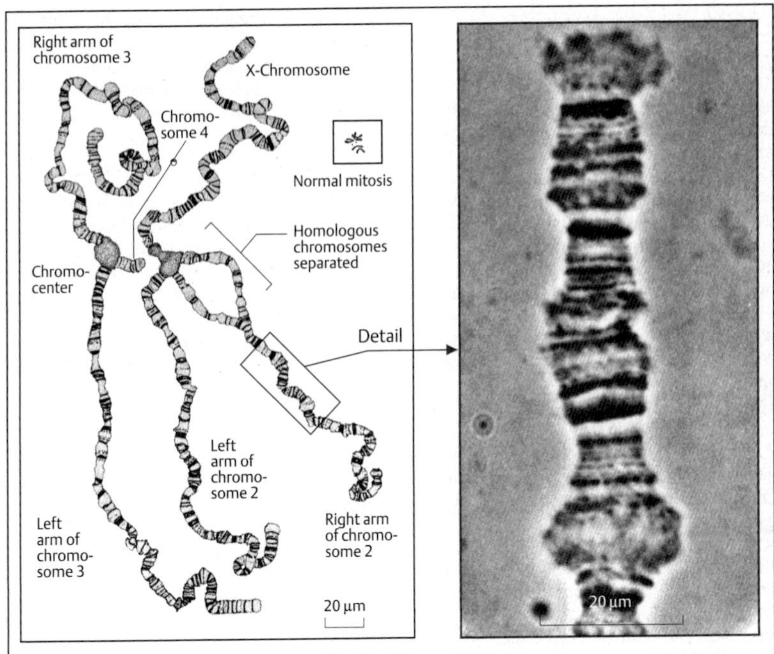


**A. Polytene chromosomes in salivary glands of *Drosophila* larvae**

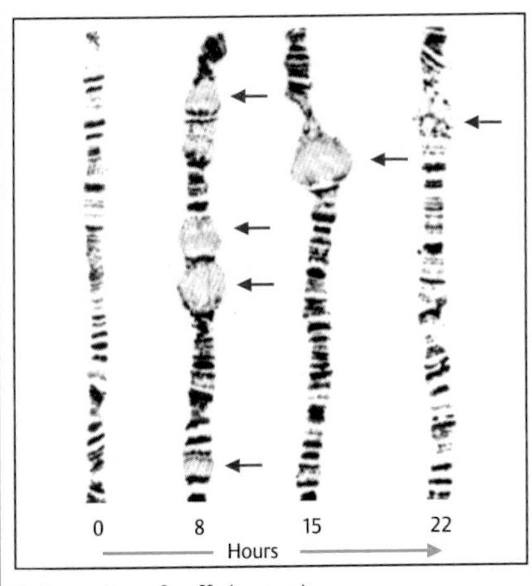


**1.** Formation of puffs (arrows)

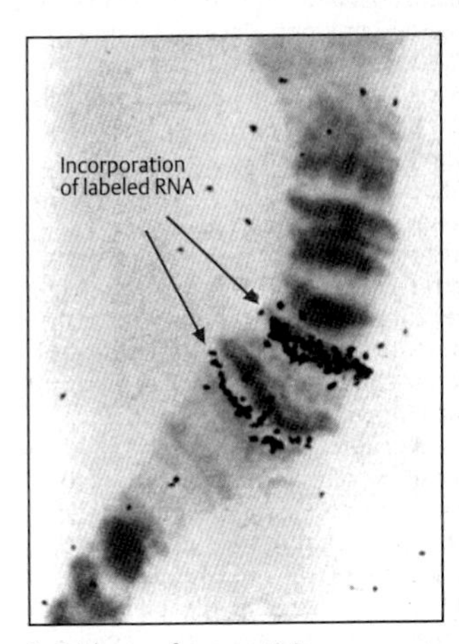


**2.** Evidence of gene activity

**B. Functional stages in polytene chromosomes**

# DNA in Lampbrush Chromosomes

Usually, the chromatin in interphase nuclei is so densely packed that single chromatin threads are not directly visible. An exception is in the interphase nuclei of certain cells, especially immature amphibian oocytes with paired meiotic chromosomes. These chromosomes show RNA synthesis and form unusual chromatin loops. They are covered with newly transcribed RNA and are packed in dense RNA–protein complexes. These are visible under the light microscope as the so-called lampbrush chromosomes.

## A. Chromosome structure in amphibian oocytes (lampbrush chromosomes)

The paired chromosomes of oocytes in meiosis consist of numerous chromatin loops arranged along an axis (1). Chiasma formation is visible at various locations.

Each segment of a lampbrush chromosome consists of a series of chromatin loops, originating from an axis (2). There, a condensed structure, the chromomere, is visible. Any given chromatin loop always contains the same DNA sequence and is formed in the same manner as the oocyte grows. Each loop corresponds to a particular unit of unfolded chromatin, which has unraveled and is transcriptionally active. Each cell contains four copies of a loop, two per chromosome (four-strand structure).

Transcription occurs either along the whole loop or at parts of a loop. The chromatin loops of a chromosome are paired, mirror-image structures (3). Each corresponds to the loop of a sister chromatid. The chromomere at the base of the loops consists of dense chromatin of the two sister chromatids, while the chromatin loops themselves are not condensed.

Lampbrush chromosomes in amphibian oocytes are unusually large compared with mitotic chromosomes (about 400 μm long, as opposed to at the most 10 μm) as shown in the micrograph (4, phase contrast exposure, Gall 1956). At the beginning of meiosis, when DNA replication is complete, the homologous pairs lie immediately next to each other and form characteristic structures composed of four chromatids (p. 116).

Lampbrush chromosomes are distinguished by an especially high rate of RNA transcription. Most of the RNA transcripts are longer than in other chromosomes. (Figures redrawn from Alberts et al., 1994).

### References

Alberts, B., et al.: Molecular Biology of the Cell. 3rd ed. Garland Publishing, New York, 1994.

Callan, H.G.: Lampbrush chromosomes. Proc. R. Soc. Lond. B. Biol Sci. **214**:417–448, 1982.

Gall, J.G.: On the submicroscopic structure of chromosomes. Brookhaven Symp. Biol. **8**:17–32, 1956.

Traut, W.: Chromosomen. Klassische und molekulare Cytogenetik. Springer, Heidelberg, 1991.

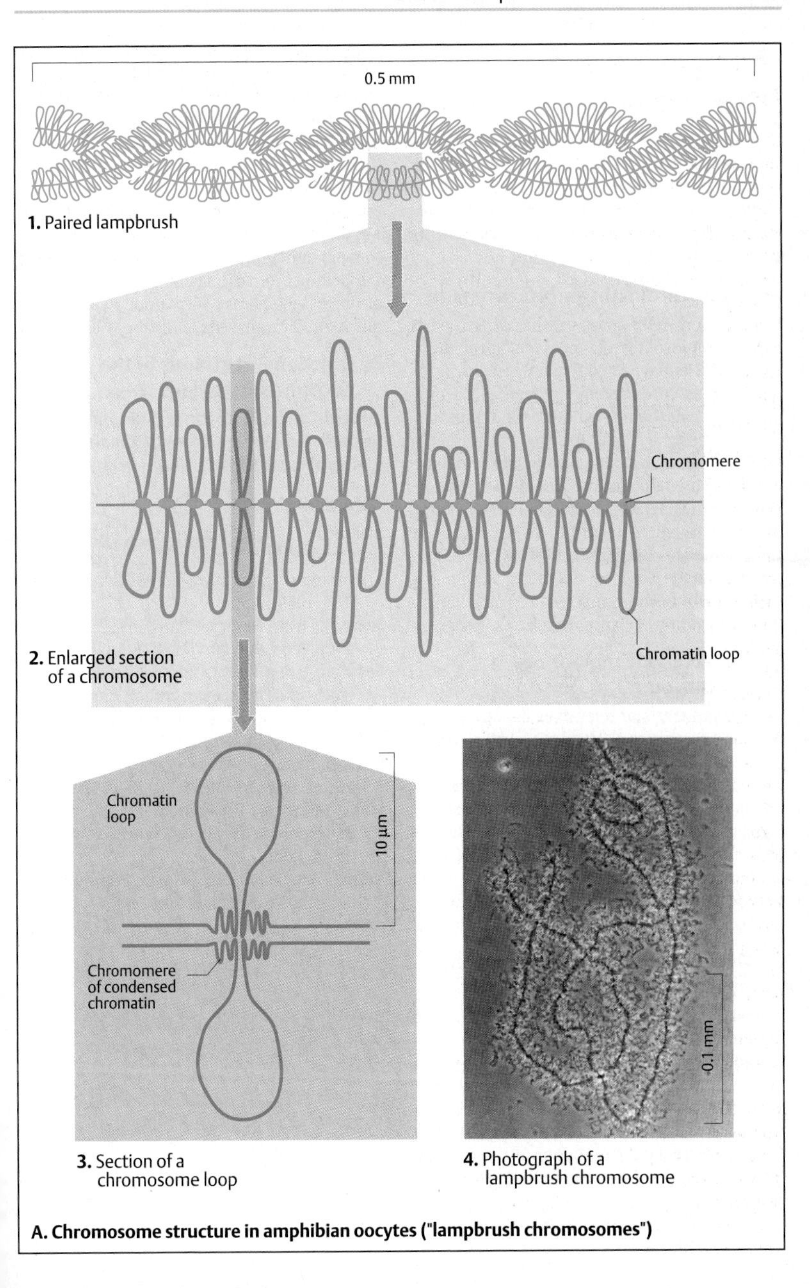


**A. Chromosome structure in amphibian oocytes ("lampbrush chromosomes")**

# Correlation of Structure and Function in Chromosomes

In a chromosome, regions with special structures and functions can be differentiated. The centromere is the point of attachment of the microtubuli of the mitotic spindle (kinetochore). The telomeres at the ends contain no genes and have a special structure.

## A. Heterochromatin and euchromatin

In 1928 Emil Heitz observed that certain parts of the chromosomes of a moss (*Pellia epiphylla*) remain thickened and deeply stained during interphase, as chromosomes otherwise do only during mitosis. He named these structures heterochromatin, as opposed to euchromatin, which becomes invisible during late telophase and subsequent interphase. Functionally, heterochromatin is defined as a region in which few or no active genes lie and in which repetitive DNA sequences occur. When active genes become located close to the heterochromatin, they usually become inactivated (position effect–variegation). (Figure from Heitz, 1928).

## B. Characteristic regions of a chromosome

The centromere and telomeres contain repetitive DNA sequences. They are evolutionarily conserved because they are important for chromosome stability. The segments located between the telomeres and centromere consist of trypsin-sensitive light and trypsin-resistant dark G-bands. The light G-band areas of DNA form loops in which the protein-coding genes lie. The DNA loops are bound to a protein matrix at special attachment sites.

## C. Model of a chromosome segment in interphase

A three-dimensional model of a chromosome segment shows that the constitutive heterochromatin (C-band) in the centromere region is very tightly wound. In the light G-bands, the euchromatin is relatively loosely packed, and in the dark G-bands, somewhat more tightly packed. (With kind permission of the author, from Manuelidis, 1990, copyright 1990 by the AAAS).

## D. Constitutive heterochromatin (C-bands) in the centromeric region

The constitutive heterochromatin in the centromeric region can be specifically stained (C bands). The distal half of the long arm of the Y chromosome is also C-band positive. The centromeric heterochromatin in chromosomes 1, 9, and 16 and in the long arm of the Y chromosome in humans is polymorphic. The lengths of the heterochromatic segments in one or more of these regions may vary among different individuals. (From Verma and Babu, 1989).

## E. Functional attributes of the euchromatin regions

The light and dark G-bands differ in functional respects. An average G-band contains around 1.5 megabases (Mb) of DNA.

### References


Bickmore, W.A., Sumner, A.T.: Mammalian chromosome banding: an expression of genome organization. Trends Genet. **5**:144–148, 1989.

Heitz, E.: Das Heterochromatin der Moose. I Jahrb.Wiss. Bot. **69**:762–818, 1928.

Passarge, E.: Emil Heitz and the concept of heterochromatin: Longitudinal chromosome differentiation was recognized fifty years ago. Am. J. Hum. Genet. **31**:106–115, 1979.

Manuelidis, L.: View of metaphase chromosomes. Science **250**:1533–1540, 1990.

Pluta, A.F., et al.: The centomere: Hub of chromosomal activities. Science **270**:1591–1594, 1995.

Verma, A.S., Babu, A.: Human Chromosomes. Pergamon Press, New York, 1989.

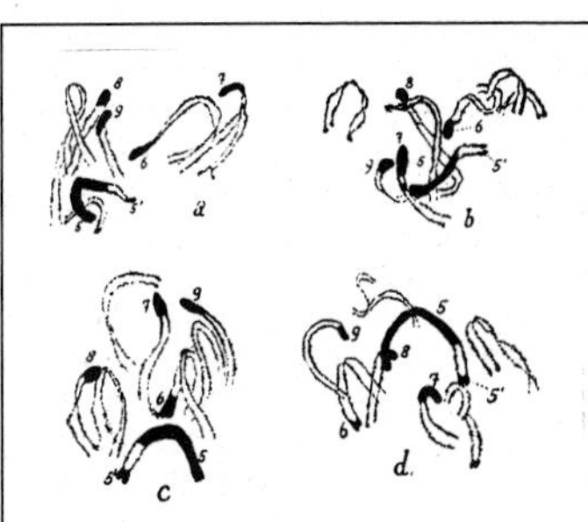

A. Heterochromatin and euchromatin

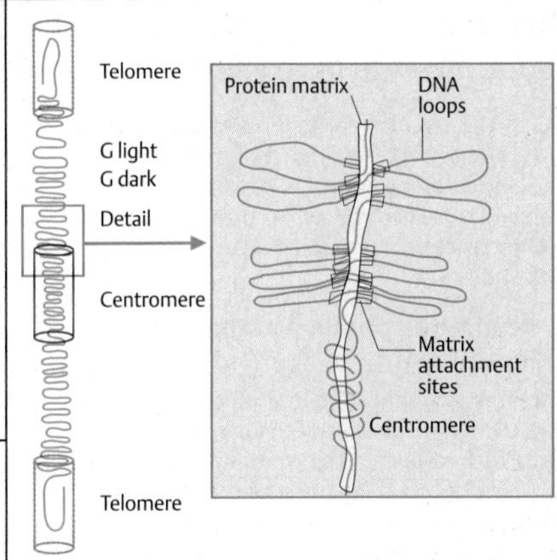


B. Special parts of a chromosome

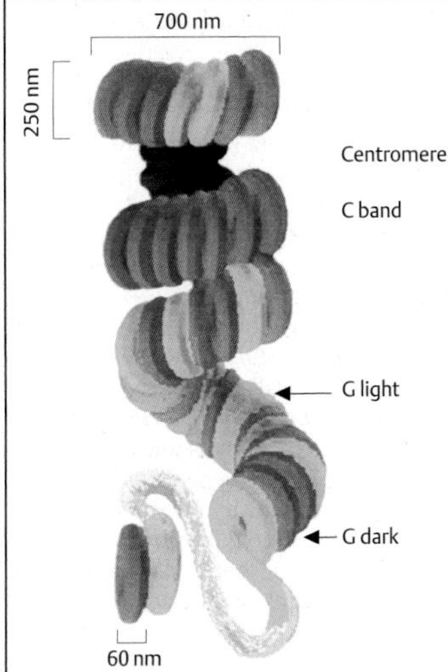


C. Model of a chromosomal segment in interphase

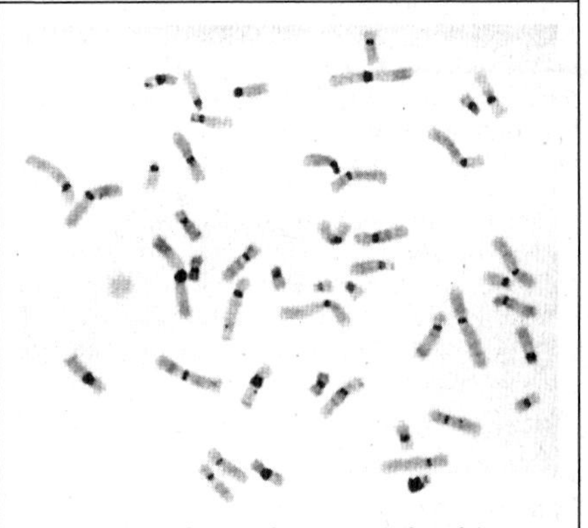

D. Constitutive heterochromatin (C bands) at the centromeres

| Light G bands | Dark G bands |
|---|---|
| GC-rich | AT-rich |
| Fluorescence with G-specific fluorochromes, e.g., olivomycin | Fluorescence with AT-specific fluorochromes, e.g., quinacrine |
| Early replicating | Late replicating |
| Gene-rich | Gene-poor |
| Alu repeats | LINE repeats (long interspersed repetitive sequences) |
| SINE repeats (short interspersed repetitive DNA sequences) | HMG-1 nonhistone proteins bound to AT-rich areas |
| Z-DNA conformation possible | Minisatellites |

E. Functional properties in euchromatin

# Special Structure at the Ends of a Chromosome: the Telomere

Unlike the circular chromosomes of bacteria, bacteriophages, plasmids, and mitochondrial DNA, the chromosomes of eukaryotes are linear. Each end is "sealed" by a specialized region, the telomere. Telomeres stabilize chromosomes at both ends.

## A. Replication problem at the ends of linear DNA

Since DNA is synthesized in the 5' to 3' direction only, the two templates of the parent molecule differ with respect to the continuity of synthesis. On the 5' to 3' template strand, synthesis occurs in the reverse direction relative to the fork movement (lagging strand synthesis). There, DNA is synthesized in short fragments about 1000 – 2000 nucleotides long in bacteria and about 200 nucleotides in eukaryotes (Okazaki fragments, see DNA replication, p. 42). However, 8 – 12 bases at the end of the lagging strand template cannot be synthesized by DNA polymerase because the primer it requires cannot be attached beyond the end of the template strand. Hence, at each round of replication before cell division, these 8 – 12 nucleotides will be lost at the chromosome ends. Some organisms compensate for this loss by adding telomeric repeats to the ends of the chromosome during the replication cycle.

## B. G-rich repetitive sequences at the telomeric region

DNA at the telomeres consists of G-rich tandem sequences (5'-TTAGGG-3' in vertebrates, 5'-TGTGGG-3' in yeast, 5'-TTGGGG-3' in protozoa). The G-strand overhangs are important for telomeric protection by formation of a duplex loop (see C).

## C. Telomerase activity and stabilization by a loop

Two features characterize the telomere: telomerase activity to compensate for replication-related loss of nucleotides at the chromosome ends and telomeric DNA loop formation to stabilize the chromosome ends. Telomerase is a modified reverse transcriptase consisting of protein and about 450 nucleotides of RNA. Near the RNA 5' end are sequences complementary to telomeric DNA repeat sequences. A short nucleotide sequence of this RNA pairs with terminal DNA sequences. The adjacent RNA nucleotides provide the template for adding nucleotides to the 3' end of the chromosome. After telomerase has extended the 3' (G-rich) strand, a new Okazaki fragment can be synthesized at the 5' strand by DNA polymerase. Griffith et al. (1999) have shown that telomeric duplex DNA forms a loop (t-loop), thus avoiding the "sticky end" problem. The loop formation is mediated by the two related proteins TRF1 (telomeric repeat-binding factor) and TRF2, which bind to mammalian telomere repeats, and the loop is anchored by the insertion of the G-strand overhang (see B) into a proximal segment of duplex telomeric DNA.

## D. General structure of a telomere

In the terminal 6 – 10 kb of a chromosome, telomeric sequences and telomere-associated sequences can be differentiated (1). The telomere-associated sequences contain autonomously replicating sequences (ARS). The telomere sequences consist of about 250 to 1500 G-rich repeats (≈ 9 kb). They are highly conserved among different species (2). Telomerase activity is essential for survival in protozoans and yeast. In vertebrates it occurs mainly in germ cells, and no telomerase activity is found in somatic tissues. The cell division-dependent decrease of telomere length is viewed as being related to aging and death of cells because, ultimately, functional DNA will be lost. Unlike normal cells, many tumors have telomerase activity.

### References

Blackburn, E.H.: Telomere states and cell fate. Nature **408**:53 – 56, 2000.

Buys, C.H.C.M.: Telomeres, telomerase, and cancer. New Eng. I. Med. **342**:1282 – 1283, 2000.

Griffith, J.D., et al.: Mammalian telomeres end in a large duplex loop. Cell **97**:503 – 514, 1999.

Hodes, R.J.: Telomere length, aging, and somatic cell turnover. J. Exp. Med. **190**:153 – 156, 1999.

Lewin, B.: Genes VII. Oxford Univ. Press, Oxford, 2000.

Shay, J.W.: At the end of the millenium, a view of the end. Nature Genet. **23**:382 – 383, 1999.

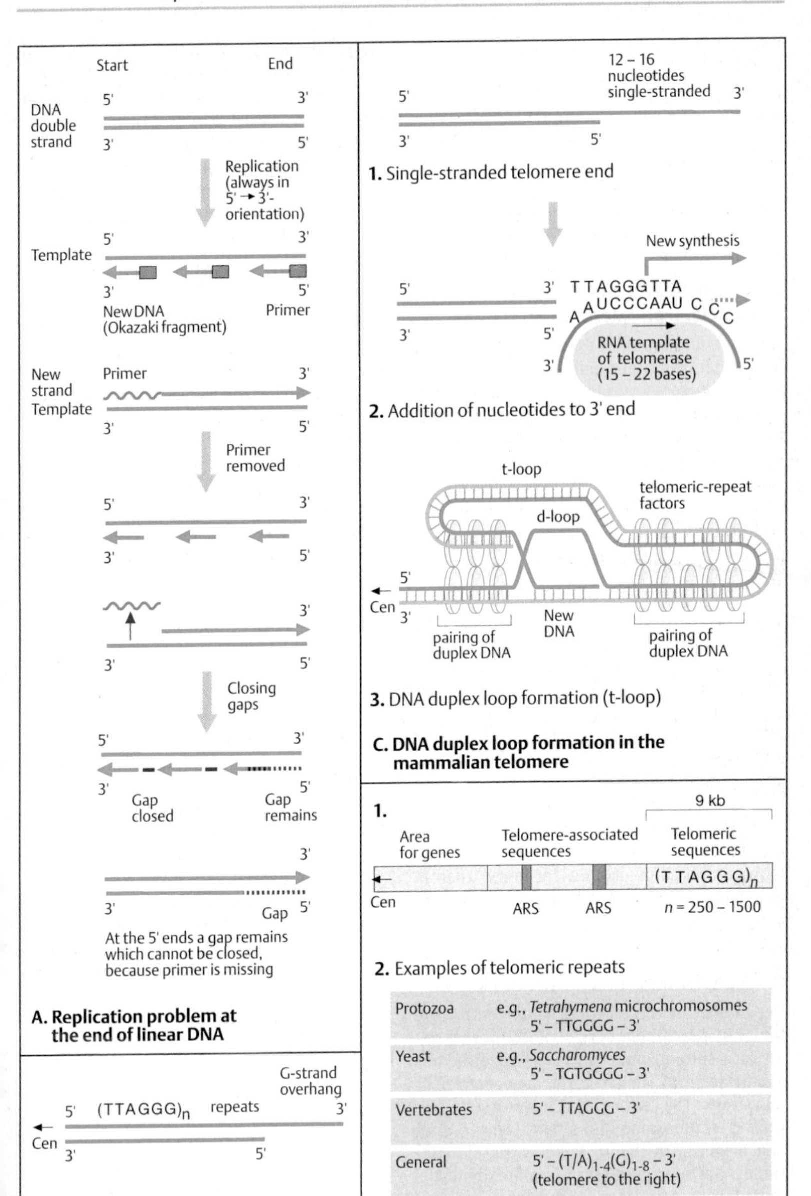


**A. Replication problem at the end of linear DNA**

**B. Telomeric DNA has G-rich repetitive repeats in the 3' end**

**C. DNA duplex loop formation in the mammalian telomere**

2. Examples of telomeric repeats

| | |
|---|---|
| Protozoa | e.g., *Tetrahymena* microchromosomes 5' – TTGGGG – 3' |
| Yeast | e.g., *Saccharomyces* 5' – TGTGGGG – 3' |
| Vertebrates | 5' – TTAGGG – 3' |
| General | 5' – $(T/A)_{1-4}(G)_{1-8}$ – 3' (telomere to the right) |

**D. General structure of a telomere**

## Metaphase Chromosomes

Chromosomes are visible as separate structures only during mitosis. In interphase, the chromosomes in chromatin cannot be individually differentiated. In metaphase chromosomes, DNA is packed about 10 000 times more densely than in interphase. Electron–microscopic studies of metaphase chromosomes have yielded some insight into chromosomal structure.

### A. A histone-free chromosome under the electron microscope

When certain proteins, especially histones, are removed from chromosomes, the chromosomal skeleton becomes visible under the electron microscope (1). Such a structure is surrounded by numerous darkly stained threads. A higher magnification (2) shows that this is a single continuous thread. It corresponds to the DNA double helix. (Photographs from Paulson and Laemmli, 1977.)

### B. The microscopic appearance of metaphase chromosomes of man

With an approximately 1000-fold magnification, the metaphase chromosomes of man and other vertebrates can readily be recognized under the light microscope as individual rodlike structures. A metaphase is shown here at about 2800-fold magnification. The chromosomes differ from each other in length, in the size and arrangement of their transverse light and dark bands (banding pattern), and in the point of attachment of the spindle (centromere), which is recognizable as a constriction. In prometaphase, the chromosomes are longer than in metaphase and show more bands. Thus, for certain purposes chromosomes are also studied in prometaphase.

### C. Types of metaphase chromosomes

Depending on the location of its centromere (point of attachment of the spindle during mitosis), a chromosome can be distinguished as submetacentric, metacentric, acrocentric, or telocentric. The centromere divides a submetacentric chromosome into a short arm (p arm) and a long arm (q). In metacentric chromosomes, the short and long arms are about the same length. Acrocentric chromosomes show a dense appendage called a satellite (not to be confused with satellite DNA) at the end of the short arm. Satellite size differs for each acrocentric chromosome of an individual (chromosomal polymorphism). Telocentric chromosomes have neither a short arm nor a satellite. None of the human chromosomes are telocentric, whereas all chromosomes are telocentric in the house mouse, *Mus musculus*. However, it is debatable whether telocentric chromosomes actually exist as defined.

### D. Simple structural aberrations

A functionally relevant deviation from the normal structure is called a structural aberration. This is to be differentiated from chromosomal polymorphism. Loss (deletion) or doubling (duplication) of a particular segment may occur. A deletion may occur at the end of a chromosome (terminal deletion) or within a chromosomal segment (interstitial deletion). Prerequisite for a terminal deletion is one break; for an interstitial deletion, two breaks. A segment that has been doubled is called a duplication. In metaphase chromosomes, an aberration is seen in both chromatids because as a rule it has occurred before the S phase. Duplications and deletions represent opposite, and in some respects complementary, aberrations of chromosomal structure (p. 404).

### References


Paulson, J.R., Laemmli, U.K.: The structure of histone-depleted metaphase chromosomes. Cell **12**:817–828, 1977.

Rooney, D.E., Czepulkowski, B.H., eds.: Human Cytogenetics. A Practical Approach. Vol. I. Constitutional Analysis. 2nd ed. The Practical Approach Series. IRL Press, Oxford, 1992.

Rooney, D.E., Czepulkowski, B.H., eds.: Human Cytogenetics. A Practical Approach. Vol. II. Malignancy and Acquired Abnormalities. 2nd ed. The Practical Approach Series. IRL Press, Oxford, 1992.

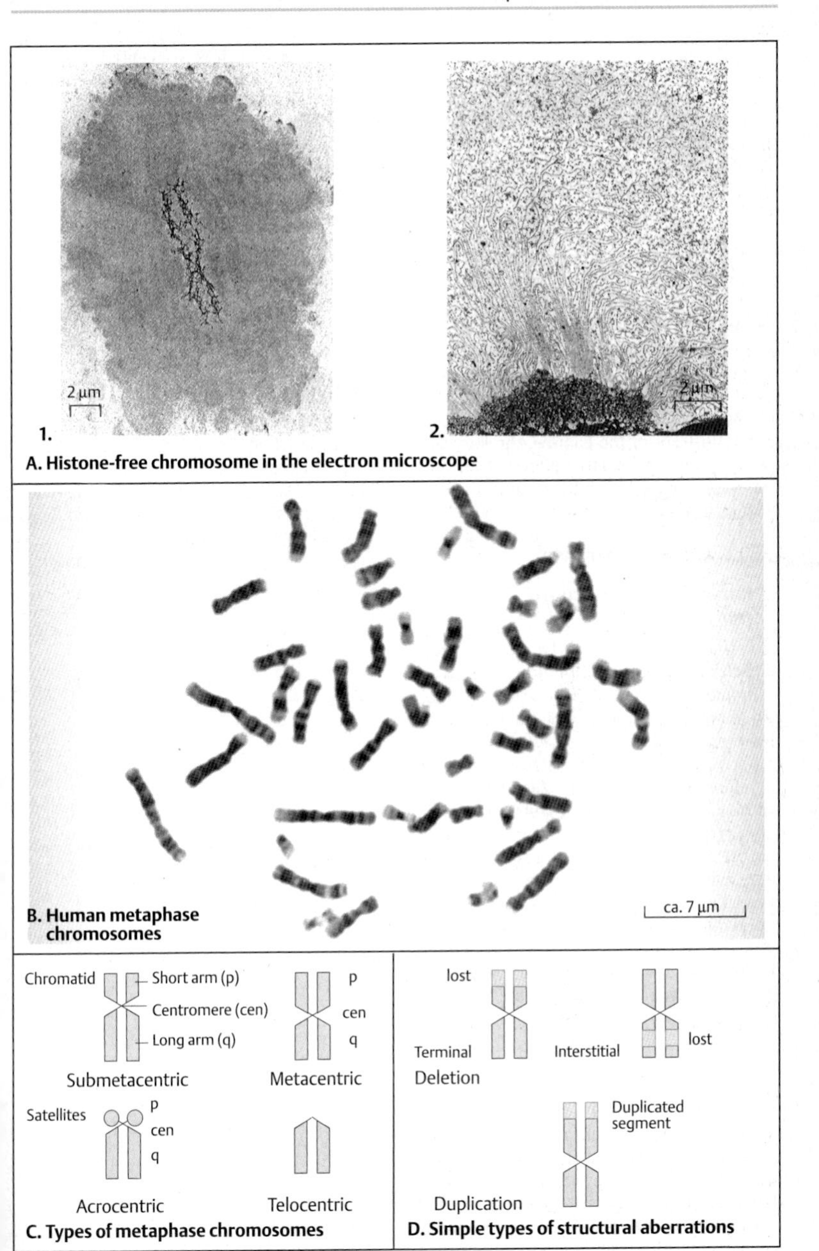


A. Histone-free chromosome in the electron microscope

B. Human metaphase chromosomes

C. Types of metaphase chromosomes

D. Simple types of structural aberrations

# Karyotype

Karyotype refers to the arrangement of chromosomes in homologous pairs. They are arranged and numbered according to a convention. The basis for the arrangement is size of a chromosome, position of the centromere, and the chromosome-specific banding pattern. The karyotype is characteristic for each species. However, the term karyotype can also be applied to an individual or to a single cell.

## A. The karyotype of man

Man *(Homo sapiens)* has 22 pairs of chromosomes (autosomes) and in addition either two X chromosomes, in females, or an X and a Y chromosome, in males (karyotype resp. 46,XX or 46,XY). In front of the comma, the karyotype formula gives the total number of chromosomes present, and after the comma, the composition of the sex chromosomes. The 22 pairs of autosomes in man are divided into seven groups (A–G) (see p. 186–190).

## B. Karyotype of the mouse *(Mus musculus)*

The standard karyotype of the mouse consists of 19 chromosome pairs in addition to the X and Y chromosomes. All chromosomes except the X and Y are telocentric and of similar size (1). However, they differ in their banding patterns, characteristic for each chromosome pair, and therefore are individually distinct. Certain strains of mice may show variants of the karyotype (2). These variants arise from fusion of certain of the chromosomes. In the example shown here, only chromosome pairs 1, 15, 19, and X correspond with those of the standard karyotype, while the others consist of fused chromosomes, e. g., chromosomes 4 and 2, chromosomes 8 and 3, etc. Structural rearrangements of the karyotype occurred with the separation of different species in evolution. (Figure from Traut, 1991; photographs by H. Winking, Lübeck.)

## C. Flow cytometry karyotype in man

Because of their different lengths, metaphase chromosomes can also be presented in a flow-cytometry-based karyotype. With this method, individual chromosomes, stained and passed by a laser light source, give signals corresponding to their sizes. Although there are overlaps, e. g., between the similarly sized human chromosomes 9–12, or with chromosomes 1 and 2, a distribution pattern of light impulses based on individual chromosome sizes is obtained. The size of the X chromosome lies between those of chromosomes 8 and 7; the size of the Y chromosome as a rule corresponds to that of a chromosome 22, although Y chromosome sizes may differ considerably. Because of the technical expenditure required and the unsharp resolution, flow cytometry is not frequently used for a practical diagnosis. However, it is of advantage in certain studies. (Figure from Connor and Ferguson-Smith, 1991.)

### References

Buckle, V.J., Kearney, L.: New methods in cytogenetics. Curr. Opin. Genet. Develop. **4**:374–382, 1994.

Connor, J.M., Ferguson-Smith, M.A.: Essential Medical Genetics. 3rd ed. Blackwell Scientific Publications, London, 1991.

Miller, O.J., Therman, E.: Human Chromosomes. 4th ed., Springer-Verlag, New York, 2001.

Traut, W.: Chromosomen. Klassische und molekulare Cytogenetik. Springer, Heidelberg, 1991.

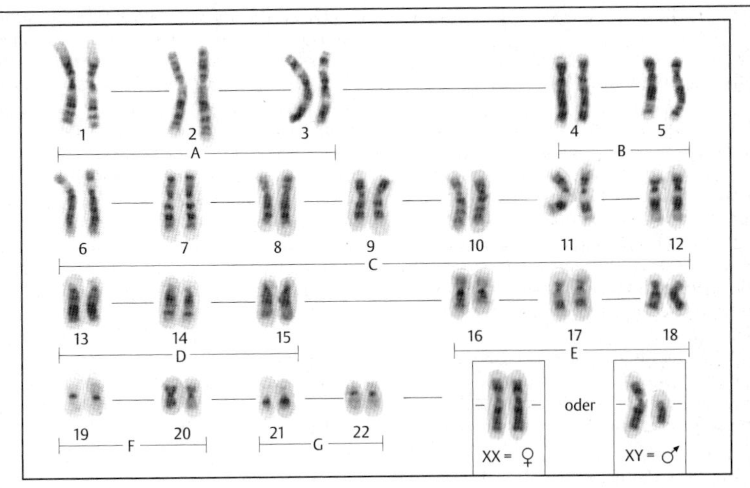


A. Karyotype of man

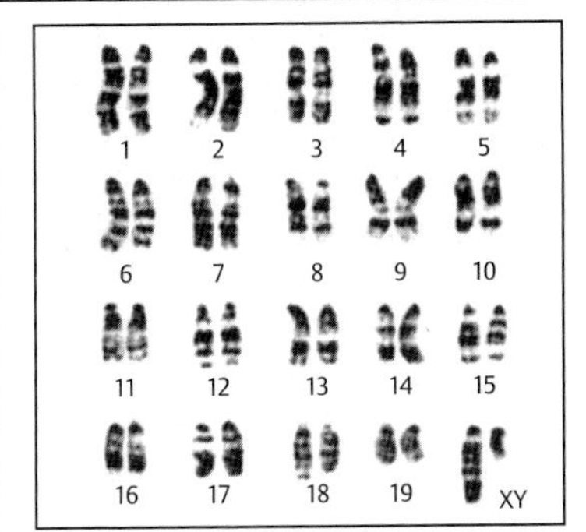


1. Standard

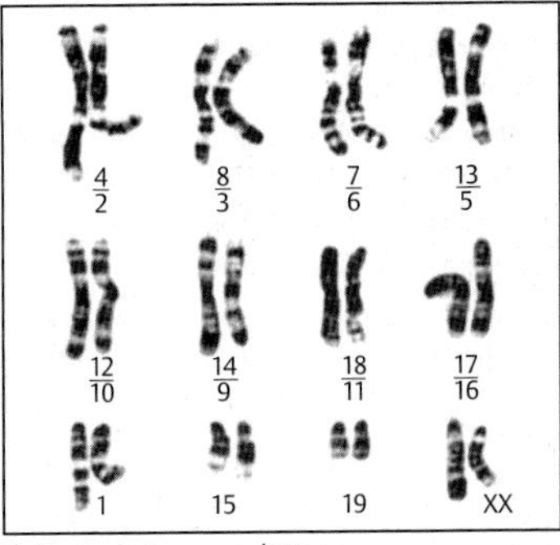


2. Variants in a population with fused chromosomes

B. Karyotype of the mouse (*Mus musculus*)

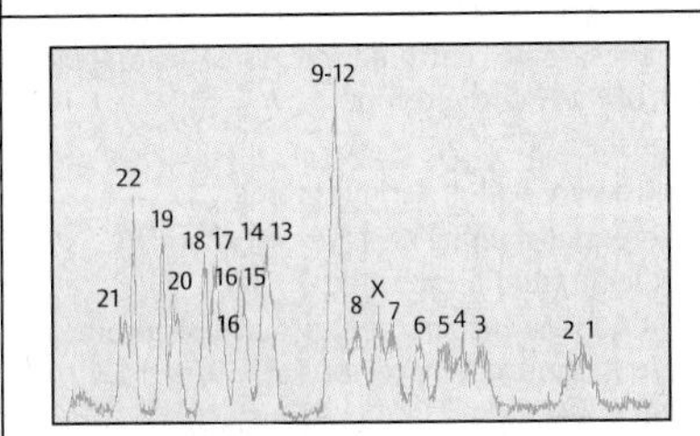


1. Normal female

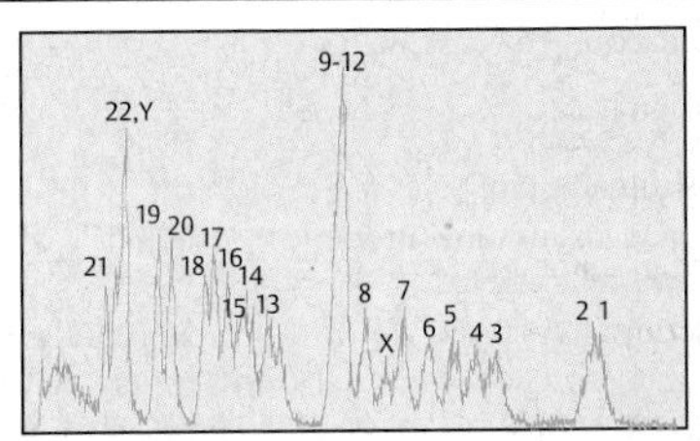


2. Normal male

C. Flow cytometry karyotype

## *The G- and R-Banding Patterns of the Human Metaphase Chromosomes*

With certain preparation and staining procedures, it is possible to distinguish light and dark bands in metaphase and prometaphase chromosomes under the light microscope. Each chromosome pair and each chromosome segment of sufficient length has a recognizable band pattern, specific for the respective chromosome. The basic types of bands are G-bands (Giemsa stain-induced) and R-bands (reverse bands). A procedure used for fluorescence microscopy produces quinacrine-induced bands (Q-bands). Their pattern corresponds to that of G-bands. Pages 187 and 189 show photographs of human metaphase chromosomes with G-bands (left) and R-bands (right). Between them, the correspoding G-band and R-band patterns are presented diagrammatically. Starting at the centromere, each chromosome is divided into defined regions and bands. Only the G-band-system numbering is shown. About 550 individually distinguishable bands of the haploid chromosome complement can be recognized in the illustration (550-band stage).

### A. The metaphase chromosomes of man (pairs 1 – 22, X and Y)

The 22 pairs of autosomes (1 – 22) are divided into seven groups. Chromosome pairs 1 – 3 form group A (chromosomes 1 and 3 are metacentric; chromosome 2 is submetacentric); chromosomes 4 and 5 form group B; 6 – 12 group C; and so on. The acrocentric chromosomes in humans belong to groups D and G. The chromosomes of group F are metacentric. The possibility of identifying every chromosome region and many bands makes it possible to establish breakpoints and to state the localization of genes according to region and band number on each chromosome. In good preparations subdivisions can be defined by decimals. For instance, band 31 includes bands 31.1, 31.2, and 31.3. (Figures correspond to Figure 5 from Harnden and Klinger, 1985, in Birth Defects: Original Article Series, vol. 21, no. 1, March of Dimes Birth Defects Foundation, New York, 1985. The G-band photographs are from Francke, 1981; and the R-band photographs are from Camargo and Cervenka, 1982.)

### References


Camargo M., Cervenka J.: Am J. Hum. Genet. **34**:757 – 780, 1982.

Francke, U., Cytogenet. Cell Genet. **31**:24 – 32, 1981.

Mitelman, F., ed.: ISCN 1995, An International System for Human Cytogentic Nomenclature. Cytogenetics and Cell Genetics, Karger, Basel, 1995.


The main types of chromosome bands

| Banding methods | Type | Principal use |
|---|---|---|
| Trypsin-induced Giemsa stain | G | Differentiates light and dark bands |
| AT-specific fluorochrome (quinacrine, Hoechst 3325B) | Q | Light fluorescence in the region of dark G-bands, some centromere regions, distal long arm of the Y chromosome |
| Reverse bands | R | Opposite of G |
| Centromere stain | C | Centromere region darkly stained |
| Bromodeoxyuridine (BrdU) for two cell cycles | SCE | Differential staining of sister chromatids (sister chromatid exchanges) |
| Distamycin A – DAPI | DA/DAPI | Light fluorescence in the short arm of chromosome 15, centromere regions of 1, 9, and 16; distal long arm of Y |
| Silver nitrate stain | NOR | Short arms of all acrocentric chromosomes |
| Giemsa 11 | G11 | Centromere of chromosome II |

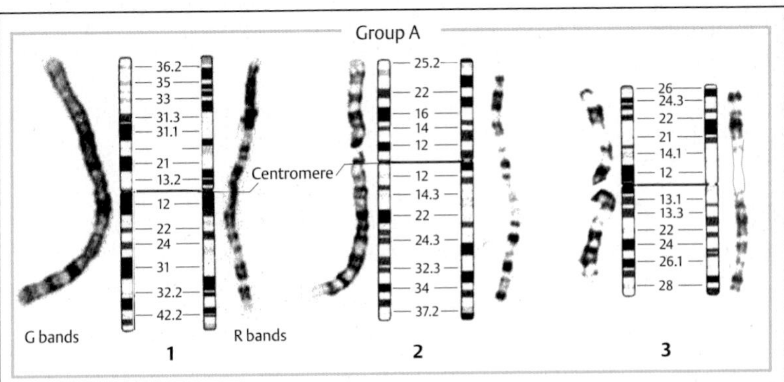


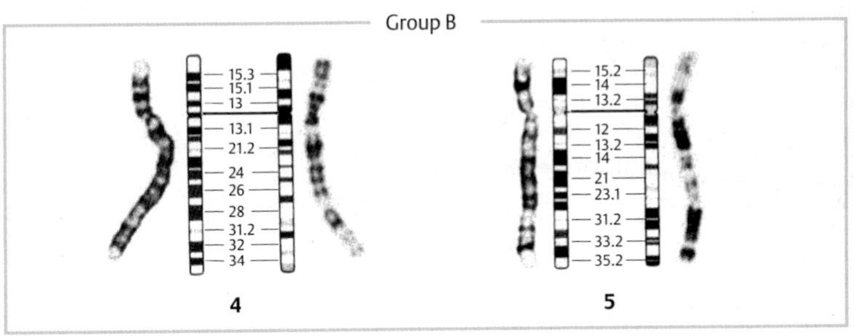


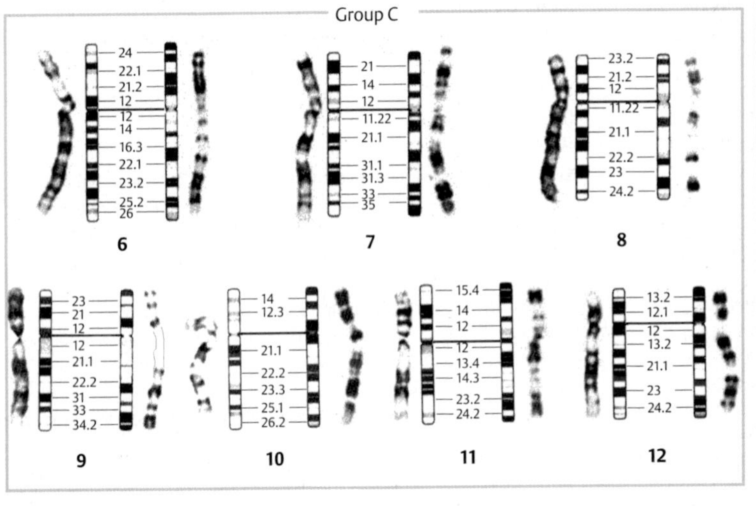


**A. Metaphase chromosomes of man (pairs 1 – 12)**

## Designation of Chromosomal Aberrations

The following list of chromosome abbrevations includes the main types of chromosomal aberrations in man, abbreviated according to the International System for Human Cytogenetic Nomenclature (ISCN 1995).

| | |
|---|---|
| 46,XX | Normal female karyotype with 46 chromosomes (two X chromosomes) |
| 46,XY | Normal male karyotype with 46 chromosomes (an X and a Y chromosome) |
| 47,XXY | Karyotype with 47 chromosomes (two X chromosomes, a Y chromosome) |
| 47,XY,+21 | Karyotype with 47 chromosomes (X and Y chromosomes); an additional chromosome 21 (trisomy 21) |
| 13p | Short arm of chromosome 13 |
| 13q | Long arm of chromosome 13 |
| 13q14 | Region 1, band 4 of the long arm of chromosome 13 |
| 13q14.2 | Subband 2 of 13q14 |
| 2q– | Shortening of the long arm of chromosome 2 |
| del(2) | Deletion in chromosome 2 |
| del(2)(q21–qter) | Deletion in chromosome 2 of region 2, band 1 (2q21) of the long arm to the end (telomere) of the long arm (qter) |
| inv(4) | Inversion in chromosome 4 |
| inv(4)(p11q21) | Inversion in chromosome 4, 4p11 to 4q21 (pericentric inversion) |
| dup(1) | Duplication in chromosome 1 |
| inv dup(1) | Inverted duplication in chromosome 1 |
| inv dup(2)(p23–p24) | Inverted duplication of the bands p23 to p24 in chromosome 2) |
| r(13) | Ring-shaped chromosome 13 (deletion implied) |
| i(Xq) | Isochromosome for the long arm of an X chromosome |
| dic(Y) | Dicentric Y chromosome |
| idic(X) | Isodicentric X |
| t(2;5) | Reciprocal translocation between a chromosome 2 and a chromosome 5 |
| t(2q–;5q+) | Reciprocal translocation between a chromosome 2 and a chromosome 5; the long arm of 2 has been shortened and that of 5 has become longer |
| t(2;5)(q21;q31) | Reciprocal translocation with the breakpoints in q21 of chromosome 2 and q31 of chromosome 5 |
| der(2) | A ("derivative") chromosome derived from chromosome 2 |
| t(13q14q) | Translocation of the centric fusion type, of the long arms of a chromosome 13 and a chromosome 14. It results in a single chromosome |
| ins(5) | Insertion into chromosome 5 |
| ins(5;2)(p14;q22;q32) | The segment q22 to q32 of a chromosome 2 has been inserted into region p14 of a chromosome 5 |
| fra((X)(q27.3) | X chromosome with a fragile site at position q27.3 |

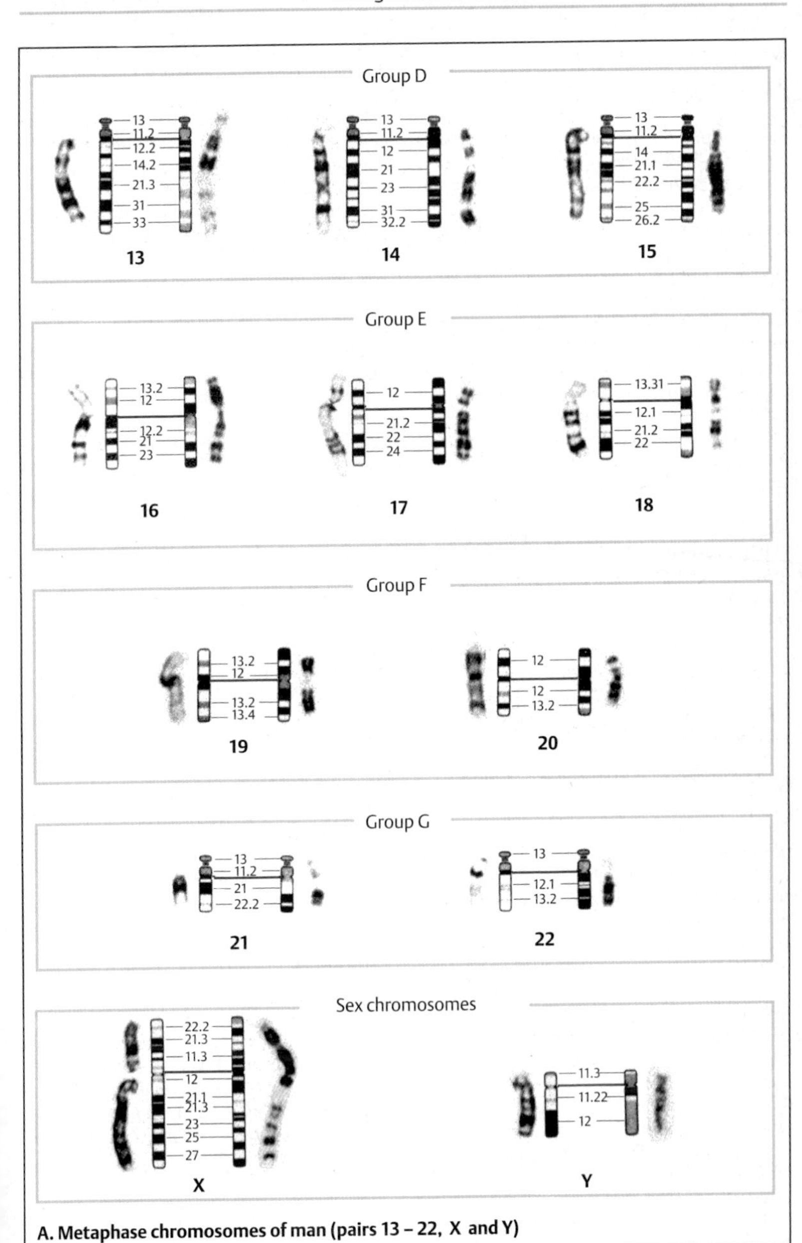


A. Metaphase chromosomes of man (pairs 13 – 22, X and Y)

# Preparation of Metaphase Chromosomes for Analysis

Since chromosomes are visible as individual structures under the light microscope only during metaphase (or under special conditions and for certain purposes, during prometaphase), every chromosome analysis requires dividing cells. In vivo, only bone marrow cells contain a sufficient proportion of cells in mitosis. Thus, in-vivo chromosome analysis of cells is limited to bone marrow. All other procedures for analyzing chromosomes in mitosis require culturing of suitable cells (cell culture). Most commonly, lymphocytes from blood are cultured for chromosome preparations.

Peripheral blood lymphocytes stimulated by phytohemagglutinin grow in a suspension culture. Their life span is limited to a few cell divisions. However, by exposing the culture to Epstein-Barr virus they can be transformed into a lymphoblastoid cell line with permanent growth potential. Such cultures are widely used because they are much easier to handle than adhesion cultures (p. 122).

In addition, fibroblasts from a piece of skin can be propagated in cell culture for analysis (see p. 122). However, since this procedure is somewhat elaborate and time-consuming, it is used only for certain purposes.

## A. Chromosome analysis from blood

For the cell culture, either peripheral blood is used directly or lymphocytes are isolated from peripheral blood (T lymphocytes). A sample of about 2 ml of peripheral blood is needed. The blood is prevented from clotting by use of a heparinized syringe, since clumping of the blood cells precludes culturing (the proportion of heparin to blood is about 1 : 20). Peripheral blood or isolated lymphocytes are placed in a vessel with culture medium. The cells are generally stimulated with phytohemagglutinin, a protein from plants that unspecifically stimulates lymphocytes to divide. The culture requires about 72 hours at 37 °C for cells to divide. Lymphocyte cultures are suspension cultures; i.e., the cells divide in culture medium without attaching to the culture vessel. Cell division is arrested and the culture is terminated by adding a suitable concentration of a colchicine derivative (colcemid) two hours prior to harvest. Colcemid interrupts mitosis during metaphase, so that a relative enrichment of cells in metaphase results.

Cell preparation is carried out as follows: the culture solution is centrifuged; the cell sediment is placed in a hypo-osmolar KCl solution (0.075 molar), incubated for about 20 minutes, and centrifuged again. The resulting cell sediment is placed in fixative. The fixing solution is a mixture of methyl alcohol and glacial acetic acid in a ratio of 3 : 1. Usually the fixative is changed two to three times with subsequent centrifugation. After that, the fixed cells are taken up in a pipette and dropped onto a slide. The preparation is stained, and the slide is covered with a cover glass.

At this point the cells are ready for analysis. Suitable metaphases are located under the microscope with about 100 x magnification and are subsequently examined at about 1250 x magnification. During direct analysis with the microscope, the number of chromosomes and the presence or absence of all chromosomes and recognizable chromosome segments are noted. Since the preparation procedure itself may induce deviations from the normal chromosome number or structure in some cells, more than one cell must be analyzed. Depending on the purpose of the analysis, between 5 and 100 metaphases (usually 10–15) are examined. Some of the metaphases are photographed with the microscope and can subsequently be cut out of the photograph (karyotyping). In this way a karyotype can be obtained from the photograph of a metaphase. The time needed for a chromosome analysis varies depending on the problem, but is usually 3–4 hours. Analysis and karyotyping time can be shortened by computer procedures.

### References

Miller, O.J., Therman, E.: Human Chromosomes. 4th ed., Springer-Verlag, New York, 2001.

Therman, E., Susman, M.: Human Chromosomes. Structure, Behavior, and Effects. 3rd ed. Springer, New York–Heidelberg, 1993.

Verma, R.S., Babu, A.: Human Chromosomes. Manual of basic techniques. Pergamon Press, New York, 1989.

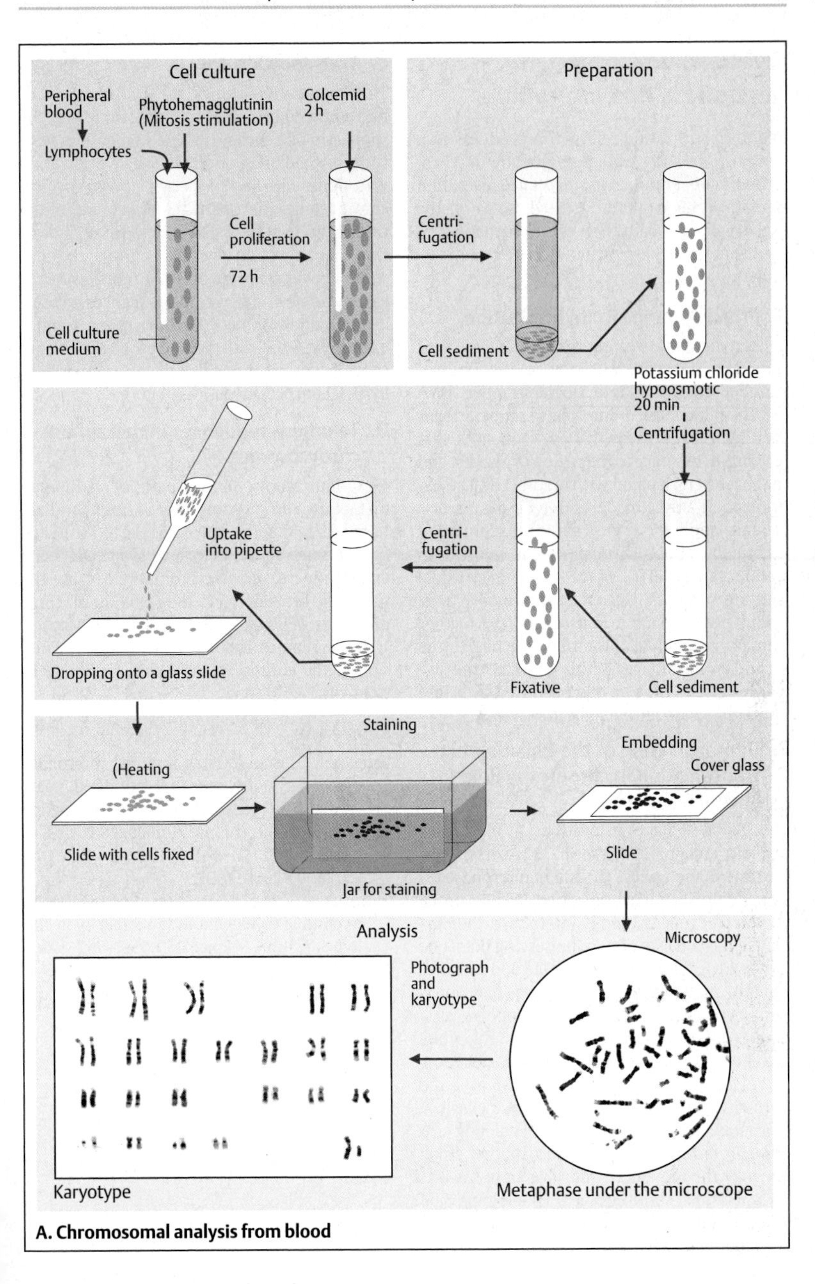


**A. Chromosomal analysis from blood**

# In Situ Hybridization in Metaphase and Interphase

In situ hybridization refers to procedures that demonstrate DNA sequences directly on chromosome preparations (in situ). Since resolution is relatively good (about 12 x $10^7$ base pairs), the exact regional localization of a sequence on its corresponding chromosome can be determined.

## A. Principle of in situ hybridization

Cells in metaphase or interphase are fixed on a slide and denatured to change the double-stranded DNA (1 a) into single-stranded DNA (2). The metaphase or interphase preparation is then hybridized (3) with DNA sequences that are complementary to the region of interest and that have been labeled with biotin (1 b). The hybridization site is made visible by means of a primary antibody against biotin; this antibody is bound to a fluorochrome (4), e. g., fluorescein isothiocyanate (FITC). Since the primary signal is quite weak, a secondary antibody (e.g., avidin) bound to biotin is attached (5). A further primary antibody can then be attached to the secondary antibody (6). This amplifies the signal, which can then be demonstrated by bright fluorescence under the light microscope.

## B. Demonstration of the Philadelphia translocation in chronic myeloid leukemia (CML)

The Philadelphia translocation (1) in chronic myeloid leukemia (CML, see p. 332) can be demonstrate in metaphase (2) and in interphase (3) by means of in situ hybridization. When a probe for the *BCR* gene is used in interphase, the normal signal consists of two fluorescing dots, one dot on each chromosome 22. (On good preparations of metaphase chromosomes, one dot is seen over each chromatid and appears as a double dot on a chromosome.) When the probe includes the breakpoint of the translocation, three signals are visible: the largest over the normal chromosome 22, a small one over the *BCR* sequences remaining in the distal long arm of a chromosome 22 (22 q), and another small one over the sequences translocated to the distal long arm of chromosome 9. (photographs by kind permission of T. Cremer, München; from Lengauer et al. 1992).

## C. Translocation 4;8

This preparation shows the translocation of part of the long arm of a chromosome 8 to the short arm of a chromosome 4 in a patient with Langer–Giedion syndrome. The hybridization was done with a 170-kb YAC (yeast artifical chromosome) that spans the breakpoint of the translocation in the 8 q24 region. Three fluorescent signals result: over the normal chromosome 8, over the part of 8 q24 translocated to chromosome 4, and over the sequences remaining on chromosome 8. Chromosomes 4 and 8 were hybridized with alphoid probes, which are specific for the centromere region. (Preparation by H. J. Lüdecke, Essen.)

## D. Telomere sequences in metaphase chromosomes

This illustration shows part of a human metaphase chromosome after in situ hybridization with telomere sequences (see p. 180). Each chromosome shows four signals, one over each end (telomere) of each chromatid, because the telomeric sequences are the same for all chromosomes. (Photographs by Robert M. Moyzes, 1991, Los Alamos, Laboratory, with kind permission of the author, Scient. Amer. August 1991, p. 34–41.)

### References

Lengauer, C., et al.: Metaphase and interphase cytogenics with Alu-PCR-amplified yeast artificial chromosome clones containing the BCR gene and the proto-oncogenes c-raf-1, c-fms, and c-erbB-2. Cancer Research **25**:2590–2596, 1992.

Tkachuk, D.C., et al.: Detection of *bcr-abl* fusion in chronic myelogenous leukemia by in situ hybridization. Science **250**:559–562, 1990.

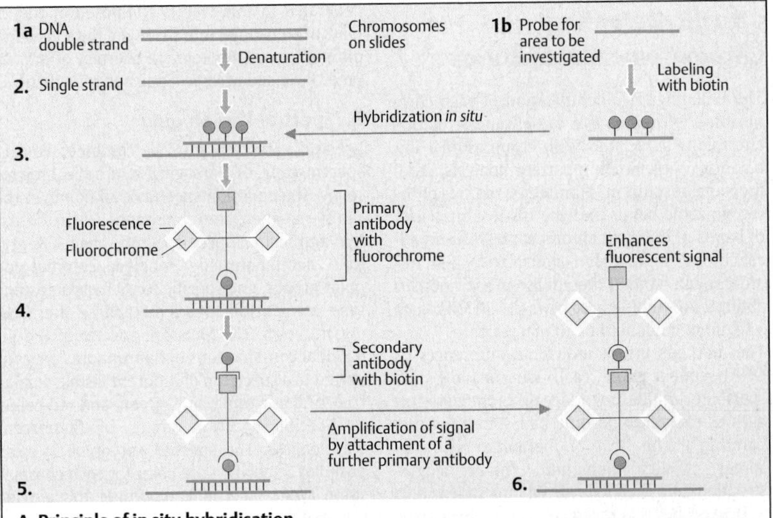


A. Principle of in situ hybridisation

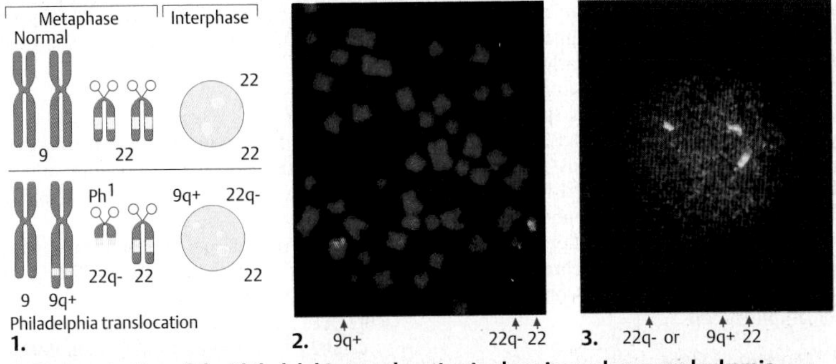


B. Demonstration of the Philadelphia translocation in chronic myelogenous leukemia

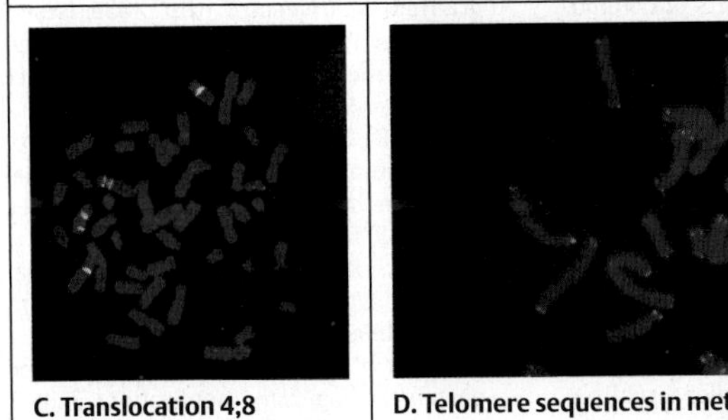

C. Translocation 4;8

D. Telomere sequences in metaphase chromosomes

## Specific Metaphase Chromosome Identification

The unambiguous identification of each chromosome is prerequisite to defining structural alterations associated with chromosomal imbalance. With banding pattern analysis, chromosomal resolution is limited by the size of the recognizable bands and the relative similarity of bands of different chromosomes. The smallest chromosomal region detectable by light microscopy in banded metaphase preparations is about 5–10 million base pairs (5–10 Mb). Such a segment could harbor 10–50 genes.
The methods utilize individual differences of DNA sequences of each chromosome and special techniques to induce different color images of each chromosome pair ("chromosome painting"). The DNA of metaphase chromosomes is first denatured (made single-stranded) and then hybridized on a slide in situ to DNA probes that make up a large collection of DNA fragments from one particular chromosome. Probes for all 24 human chromosomes are hybridized to the metaphase. Two approaches have proved particularly useful: multiplex fluorescence in situ hybridization (M-FISH, Speicher et al., 1996) and spectral karyotyping (SKY, Schröck et al., 1996). Other approaches and modifications exist; for example, the use of artificially extended DNA or chromatin fibers. In comparative genome hybridization (CGH), the quantitative difference in allelic sequences on homologous chromosomes is compared to detect deletions or duplications and amplifications.

### A. Multiplex FISH

With this approach (M-FISH), sets of chromosome-specific DNA probes, each labeled with its own combination of DNA-binding fluorescent dyes, are hybridized to metaphase chromosomes. For each chromosome type, a specific multicolor bar code is constructed by using different YAC clones (yeast artificial chromosomes) containing the DNA probes to be applied. Only five different fluorophores are necessary for image analysis by epifluorescence microscopy using a charge-coupled device (CCD) camera to generate a composite image of each chromosome in a pseudocolor visualized by appropriate software. The karyotype generated in this way is composed of the 22 pairs of autosomes and the X and the Y each in a different color. (Photograph courtesy of Drs. Sabine Uhrig and Michael Speicher, München).

### B. Spectral karyotyping

Spectral karyotyping (SKY) combines Fourier spectroscopy, CCD imaging, and optical microscopy. The emission spectra of all points in the sample are measured simultaneously in the visible and near-infrared spectral range. Twenty-four combinatorially labeled chromosome paint probes, one specific to each chromosome type, are hybridized to a metaphase after DNA denaturation. The emission spectra of the individual combinations of fluorophores are converted to a spectrum of different visible display colors by assigning blue, green, and red colors to specific spectral ranges of fluorescent wavelengths. The spectral karyotype is composed of a specific false color for each chromosome type. Spectral karyotyping has a wide range of diagnostic applications in the analysis of constitutional structural chromosome aberrations and cancer cytogenetics. (Photographs courtesy of Dr. Evelyn Schröck, Bethesda, Maryland).

### References

Haaf, T., Ward, D.C.: High resolution ordering of YAC contigs using extended chromatin and chromosomes. Hum. Mol. Genet. **3**:629–633, 1994.

Heiskanen, M., Peltonen, L., Palotie, A.: Visual mapping by high resolution FISH. Trends Genet. **12**:379–384, 1996.

Lichter, P.: Multicolor FISHing: what's the catch? Trends Genet. **13**:475–479, 1997.

Ried, T., et al.: Chromosome painting: a useful art. Hum. Mol. Genet. **7**:1619–1626, 1998.

Schröck, E., et al.: Multicolor spectral karyotyping of human chromosomes. Science **273**:494–497, 1996.

Speicher, M.R., Ballard, S.G., Ward, D.C.: Karyotyping human chromosomes by combinatorial multi-fluor FISH. Nature Genet. **12**:368–375, 1996.

Strachan, T., Read, A.P.: Human Molecular Genetics. 2nd ed. Bios Scientific Publishers, Oxford, 1999.

Uhrig, S., et al.: Multiplex-FISH for pre- and postnatal diagnostic application. Am. J. Hum. Genet. **65**:448–462, 1999.

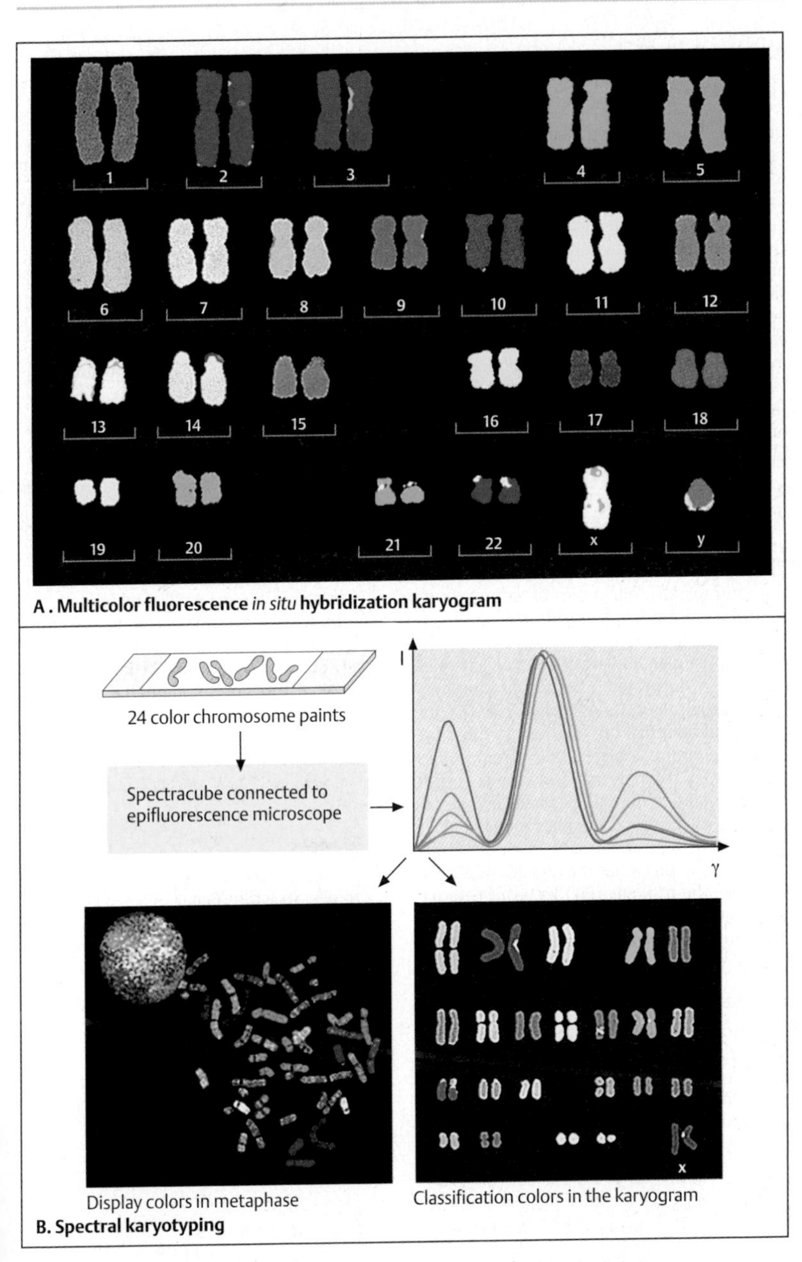


**A. Multicolor fluorescence *in situ* hybridization karyogram**

Display colors in metaphase

Classification colors in the karyogram

**B. Spectral karyotyping**

# Numerical Chromosome Aberrations

Deviation from the normal chromosome number in a single pair of chromosomes is referred to as aneuploidy. In humans, numerical chromosome aberrations occur in about 1 in 400 newborns. An abnormality of the number of chromosomes occurs as a result of their abnormal distribution (nondisjunction) during meiosis I or II (meiotic nondisjunction). With meiotic nondisjunction, the aberration occurs in all cells of a resulting organism. Nondisjunction in meiosis I and meiosis II can be differentiated (see pp. 116). Abnormal chromosomal distribution during mitosis leads to an aberration in only a proportion of the cells (chromosomal mosaicism).

## A. Triploidy

Triploidy refers to a deviation from the normal number of chromosomes in which each chromosome is present threefold instead of twofold. With tetraploidy, four copies of each chromosome are present. Triploidy arises when an abnormal oocyte with a double (46,XX) chromosome complement instead of a haploid complement (23,X) is formed. After fertilization by a normal spermatocyte, triploidy (69,XXX or 69,XXY) of maternal origin arises. In this case, two of the three complete sets of chromosomes are maternal. Triploidy may also arise as a result of abnormalities during spermatogenesis, resulting in an abnormal spermatozoon that does not contain the normal haploid chromosome complement, but rather the diploid (46,XY). In this case, the triploidy (69,XXY) is of paternal origin (see p. 402). A further cause of triploidy is dispermy, or fertilization of a normal egg by two normal sperm.

## B. Aneuploidy

In a trisomy (1), only one of the chromosomes is present threefold; all other chromosome pairs are normal. Rarely, two different trisomies occur, for two different chromosomes (double aneuploidy). If one chromosome of a pair is missing, it is referred to as monosomy (2).

## C. Origin of trisomy and monosomy

The result of normal meiosis (see p. 116), consisting of two cell divisions (not shown here), is a normal distribution and a haploid chromosome complement. With abnormal distribution (nondisjunction either in meiosis I or in meiosis II), one gamete is formed with an additional chromosome, whereas the other is missing a chromosome. After fertilization, the respective zygote contains either three copies of one chromosome (trisomy) or only a single chromosome of a pair (monosomy). Abnormal chromosome distribution can occur during oogenesis (maternal nondisjunction) or during spermatogenesis (paternal nondisjunction).

## D. Abnormalities of chromosome number in humans

In humans, the following autosomal trisomies may occur in liveborn infants: trisomy 13 with a frequency of 1 in 5000 newborns; trisomy 18 with 1 in 3000; and trisomy 21 with about 1 in 650 newborns (1). Additional X or Y chromosomes occur in about 1 in 800 newborns, much more frequently than the autosomal trisomies (2). But unlike the autosomal trisomies, they usually do not lead to defined clinical pictures. Triple X (47,XXX) or an additional Y chromosome (47,XYY) are generally not clinically apparent. On the other hand, monosomy X (3) leads to the clinical picture of Turner syndrome (see p. 402), and XXY causes Klinefelter syndrome.

## References

Therman, E., Susman, M.: Human Chromosomes: Structure, Behaviour, and Effects. 3rd ed. Springer, Heidelberg, 1993.

Traut, W.: Chromosomen. Klassische und molekulare Genetik. Springer, Heidelberg, 1991.

Origin of triploidy

| Oocyte abnormal | Spermatozoon normal | Oocyte normal | Spermtozoon abnormal |
|---|---|---|---|
| 46, XX | 23, X or 23, Y | 23, X | 46, XY |
| 69, XXX or 69, XXY | | 69, XXY | |
| Maternal origin | | Paternal origin | |

**A. Triploidy (all chromosomes threefold)**

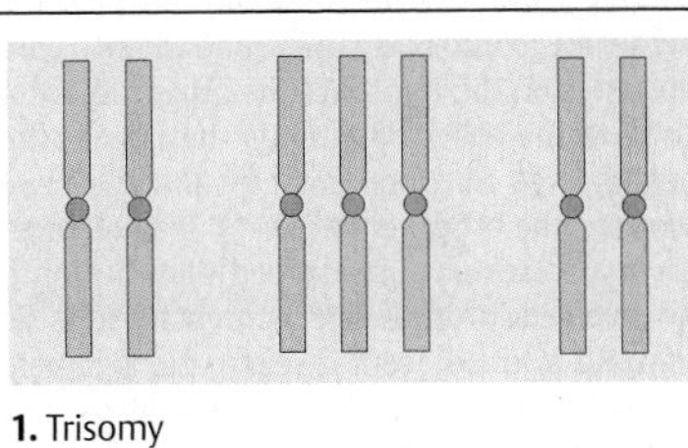

**1.** Trisomy

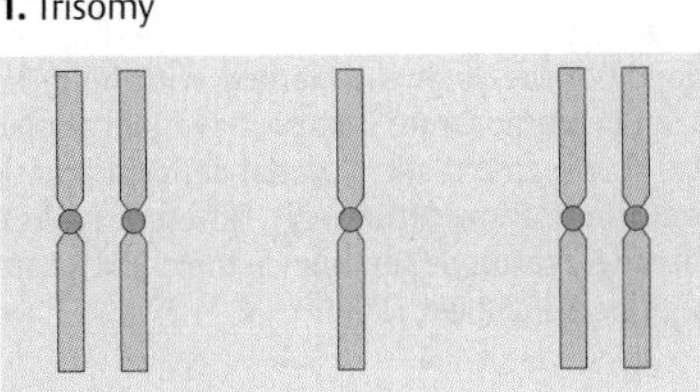

**2.** Monosomy

**B. Aneuploidy**

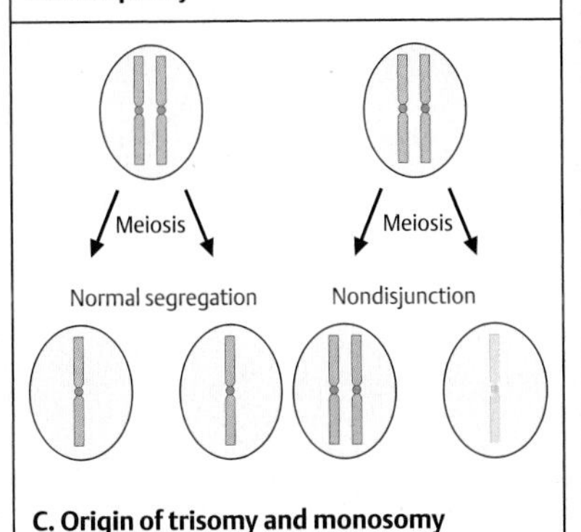


**C. Origin of trisomy and monosomy**

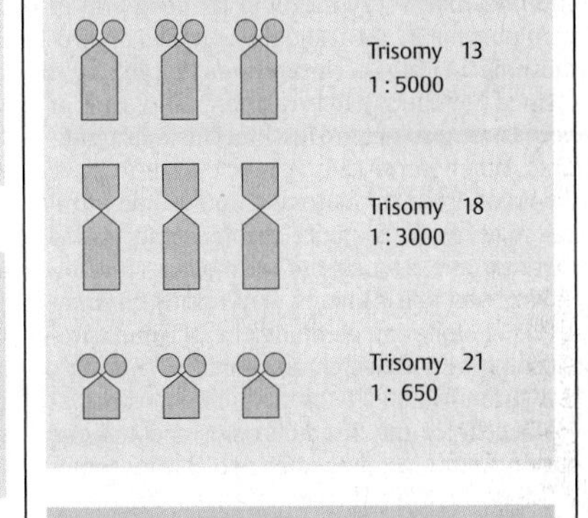


**2.** Additional X or Y chromosome

| | |
|---|---|
| XXX | 1 : 800 |
| XXY | 1 : 800 |
| XYY | 1 : 800 |

**3.** Monosomy X 1 : 3000

**D. Numerical aberrations occurring in liveborn infants**

## Translocation

Translocation refers to an exchange of chromosome segments. A translocation can arise by centric fusion of two acrocentric chromosomes (Robertsonian translocation) or by exchange between two chromosomes (reciprocal translocation). With centric fusion, two complete chromosomes are involved; with reciprocal translocation, only a part of each of the two chromosomes is exchanged. In a translocation it is important to determine the breakpoints in each of the chromosomes involved.

### A. Centric fusion of acrocentric chromosomes

Chromosome 14 and chromosome 21 (1) are the most frequently involved in fusions (about 1 in 1000 newborns). By fusion of the long arm of chromosome 21 (21q) and the long arm of chromosome 14 (14q), a chromosome t (14q21q) is formed (2). The satellite-carrying short arms of both chromosomes are lost, but this is insignificant. When germ cells (gametes) are formed, deviation from the normal chromosome number may result (3). Since chromosome 14 and chromosome 21 pair during meiosis, the following possible gametes may result: chromosome 14 alone (no chromosome 21), one chromosome 14 and one chromosome 21 (normal), the chromosome 14 fused to chromosome 21 (balanced), or the fused chromosome and one chromosome 21 (altogether two chromosomes 21).

After fertilization, the corresponding zygotes contain either only one chromosome 21 (unviable monosomy 21), a normal chromosome complement, a balanced chromosome complement with the fused chromosome, or three chromosomes 21 (trisomy 21). In the latter case, the clinical disorder Down syndrome (formerly called mongolism) results (p. 400).

### B. Reciprocal translocation

A reciprocal translocation is an exchange of chromosomal material between two chromosomes. Since usually no chromosomal material is lost or added with a reciprocal translocation, it does not cause clinical signs (i.e., it is balanced). However, carriers of a reciprocal translocation may form gametes with unbalanced chromosome complements. During meiosis, the chromosomes involved in the reciprocal translocation take part as usual in the homologous pairing of meiosis I. Each of the chromosomes not involved in the translocation pairs with its homologous partner that is involved in the translocation. This leads to the formation of a characteristic quadriradial configuration of the involved chromosomes. When these four chromosomes separate (segregation) during anaphase of meiosis (see p. 116), one of three possibilities may occur: With alternate segregation, one gamete receives the two normal chromosomes, and the other gamete the chromosomes involved in the translocation, i.e., it is balanced.

With nonalternate segregation (neighboring or adjacent chromosomes), the two chromosomes on the left go into one gamete and the two chromosomes on the right into the other (adjacent-2). With the other possibility, the upper chromosomes go into one gamete, and the lower two into the other (adjacent-1). In each of the last two cases, an unbalanced distribution of the involved chromosome segments results. For example, after adjacent-2 segregation, gametes receive a partial duplication of the chromosome segment marked with red and a partial deficiency of the segment marked with blue (left pair of chromosomes) or a partial duplication of the blue segment and a partial deficiency of the red (duplication/deficiency). Different types of disorders result depending on the chromosome segments involved.

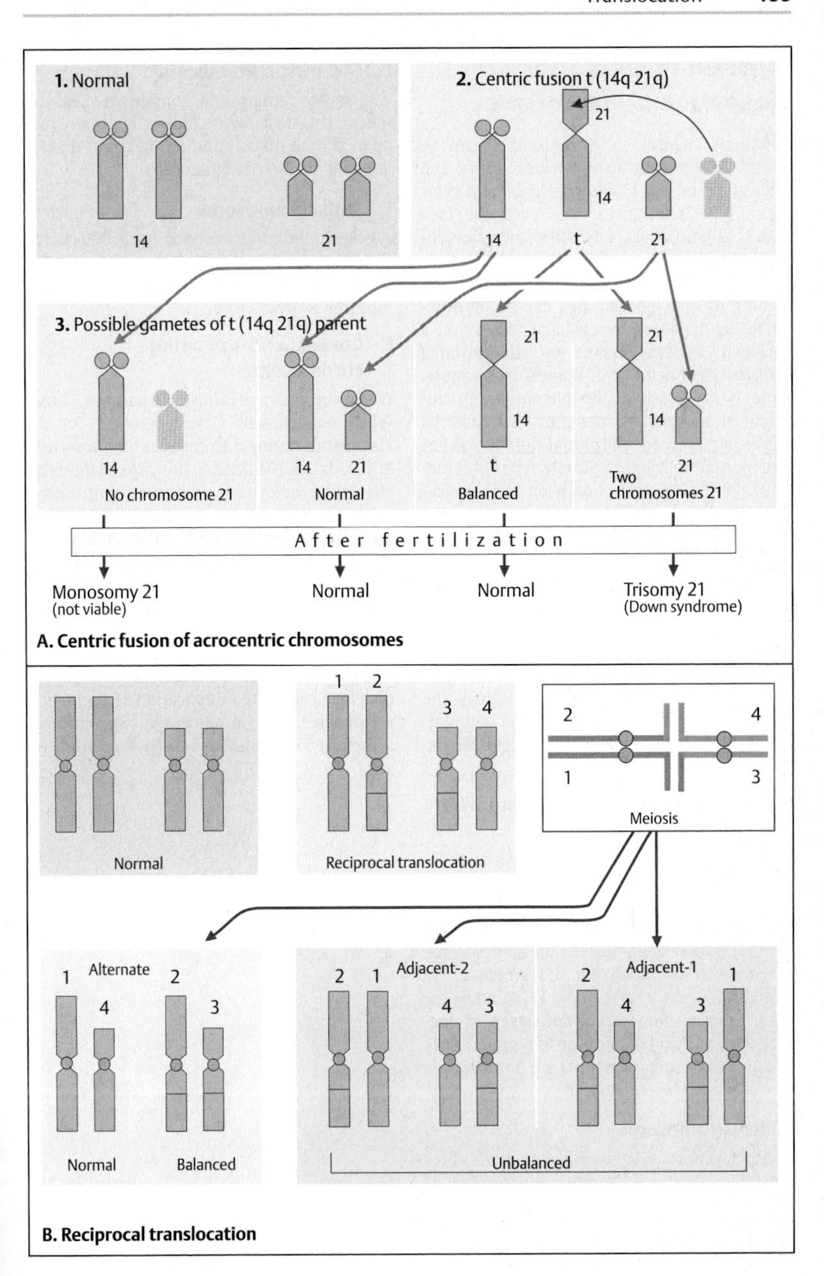


A. Centric fusion of acrocentric chromosomes

B. Reciprocal translocation

# *Different Types of Structural Chromosomal Aberrations*

Structural changes in chromosomes can be classified according to cytological types and their effect on the phenotype. The main cytological types are translocation (exchange) (see p. 198), deletion (loss, see p. 182), inversion, insertion, isochromosome, dicentric chromosome, and ring chromosome (see below). According to their effects, they can be differentiated as balanced or unbalanced. With a balanced rearrangement, no chromosomal material has been lost or gained. In this case, there is no effect on the phenotype. In unbalanced aberrations, chromosomal material has either been added (partial duplication) or lost (partial deficiency). Simple types of structural aberration, such as deletion and duplication are shown on p. 182.

## A. Inversion

An inversion is a 180-degree change in direction of a chromosomal segment. Prerequisite for every inversion is a break at two different sites, followed by reunion of the inverted segment. Depending on whether the centromere is involved, a pericentric inversion (when the centromere lies within the inverted segment) and a paracentric inversion can be differentiated.

## B. The consequences of crossing-over in the inverted region

With homologous pairing during meiosis, an inversion loop is formed in the region of the inversion (1). When the inverted segment is relatively large, crossing-over may occur in this region (2). In the daughter cells, one chromosome may show a duplication (e. g., of segments A and B) and a deficiency (of segment F) (3), while the other chromosome shows deficiency of segments A and B and duplication of segment F (4). These chromosome segments are not balanced (aneusomy by recombination).

## C. Isochromosome

An isochromosome arises when a normal chromosome (1) divides transversely instead of longitudinally, so that it is composed of two long arms (2) or of two short arms (3). In each case, the other arm is missing.

## D. Dicentric chromosome

A dicentric chromosome contains two centromeres. It is unstable because it is usually torn apart during mitosis and its parts are divided between the two daughter cells.

## E. Ring chromosome

A ring chromosome arises after two breaks followed by a joining of the two opposite ends. The distal segments are lost. Therefore, a ring chromosome is unbalanced.

## F. Consequences of a ring chromosome

A ring chromosome is unstable because a break with reattachment ("crossing-over") of the chromatids during the prophase of mitosis usually leads to difficulties. In this case, a large ring-shaped chromosome with two centromeres arises during metaphase and telophase. Since the centromeres migrate in different directions during anaphase, the ring becomes disrupted. If this does not occur strictly symmetrically, two daughter cells will result with certain segments either missing (deficiency) or duplicated (duplication). In the example, one daughter cell with a deficiency of segment 4 and one daughter cell with a duplication of segment 4 are formed. Not infrequently, ring chromosomes are lost completely and a monosomy results.

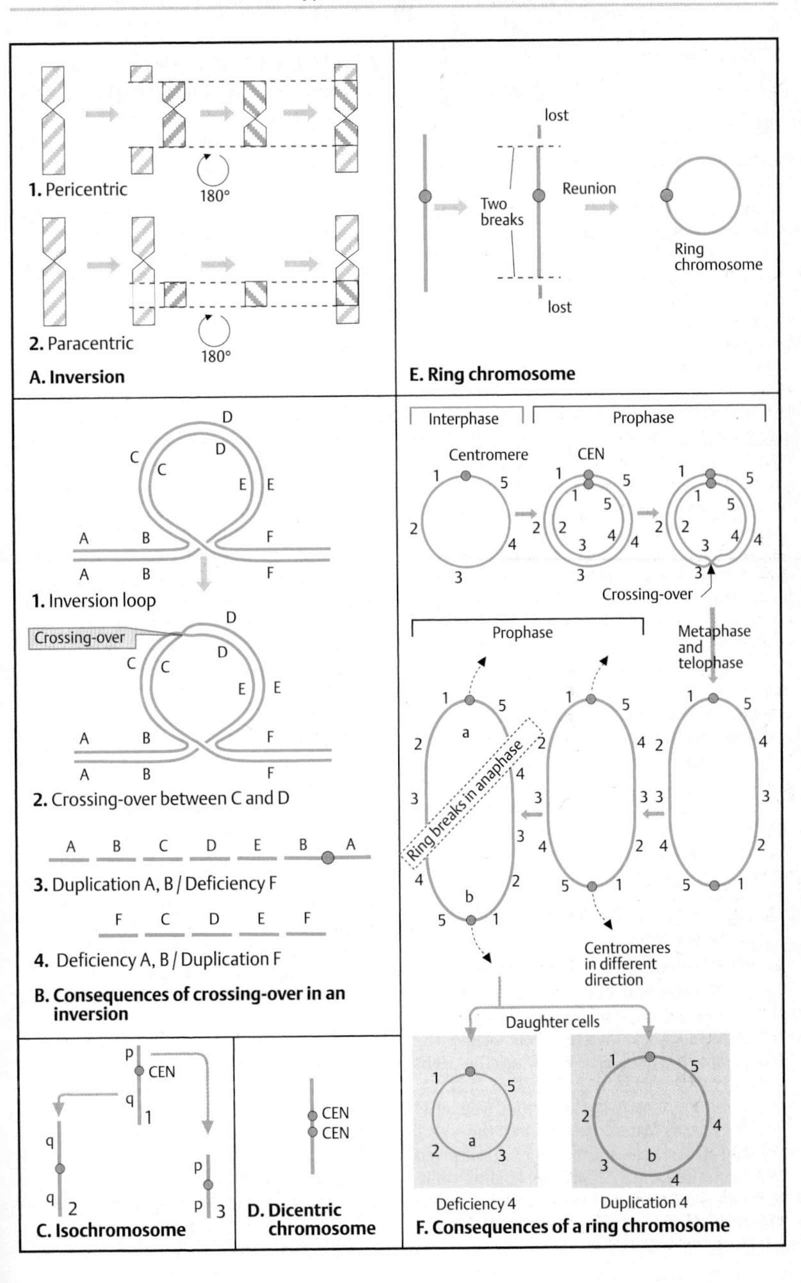


A. Inversion

B. Consequences of crossing-over in an inversion

C. Isochromosome

D. Dicentric chromosome

E. Ring chromosome

F. Consequences of a ring chromosome

# Detection of Structural Chromosomal Aberrations by Molecular Methods

Structural chromosomal rearrangements occur with a frequency of about 0.7–2.4 per 1000 mentally retarded individuals. Small supernumerary chromosomes are observed about once in 2500 prenatal diagnoses. In both situations it is mandatory to identify the chromosome or chromosomes involved. Many small changes cannot be identified with conventional light-microscopic chromosome analysis, even in the best preparations using one of the banding techniques. The proportion of identifiable aberrations can be greatly enhanced by molecular cytogenetics (see p. 192) using fluorescence in situ hybridization (FISH).

A great variety of methods is available for identifying small rearrangements. Single-copy probes hybridizing to specific sites on individual chromosomes can be used to identify specific locations on a chromosome. Numerous probes from one chromosome can be applied to identify a whole chromosome (chromosome painting). In comparative genomic hybridization, the genomic DNA of a cell population, for example tumor cells, is hybridized to normal metaphase chromosomes. DNA segments that are overrepresented or underrepresented in the tumor tissue owing to duplication or deletion will appear as increased or decreased signals. The preparation of extended DNA fibers (fiber FISH) increases the resolution. The following are selected examples of the use of multiplex fluorescence in situ hybridization.

## A. Derivative chromosome 1 with extra material: 46,XX,der(1)t(1:12)(q43;p13.3)

In the conventional G-banded karyogram (shown on the left) a small amount of additional chromosomal material is not visible. The karyogram by multiplex FISH on the right shows a small extra band at the end of one chromosome (1), shown by an arrow. The FISH analysis reveals that the extra band at the end of the long arm of a chromosome 1 (1 q) is derived from a chromosome 12. The breakpoints could be determined to be in chromosome 1 in region 4, band 3 (1q43) and chromosome 12 in region 1, band 3.3 (12p13.3).

## B. Additional isodicentric chromosome 15: 47,XY,+psu idic (15)(q11)

A small additional chromosome is present in the metaphase on the left (arrow) and in the corresponding karyotype on the right. The FISH karyotype reveals that the extra chromosome belongs to chromosome 15. It consists of a small isodicentric chromosome with a duplication of the proximal long arm 15 q11 (region 1, band 1 of 15 q).

## C. Additional derivative chromosome 21

The metaphase on the left and the karyotype on the right show a small additional chromosome (*arrow*) with fluorescence in two colors. Detailed analysis of many metaphases revealed a compound chromosome consisting of material from a chromosome 18 and a centromere from a chromosome 21. Thus, the karyotype is unbalanced for additional material from a chromosome 18 (partial trisomy 18), which is a cause of developmental retardation. The origin of such a compound chromosome usually remains obscure. (All photographs kindly provided by Drs. Sabine Uhrig and Michael Speicher, München).

## References


Chudoba, I., et al.: High resolution multicolor-banding: A new technique for refined FISH analysis of human chromosomes. Cytogenet. Cell Genet. **84**:156–160, 1999.

Crolla, J.A.: FISH and molecular studies of autosomal supernumerary marker chromosomes excluding those derived from chromosome 15. II. Review of the literature. Am. J. Med. Genet. **75**:367–381, 1998.

Haddad, R.R., et al.: Identification of de novo chromosomal markers and derivatives by spectral karyotyping. Hum. Genet. **103**:619–625, 1998.

Jalal, S.M., Law, M.E.: Utility of multicolor fluorescent in situ hybridization in clinical cytogenetics. Genetics in Medicine **1**:181–186, 1999.

Uhrig, S., et al.: Multiplex-FISH for pre- and postnatal diagnostic application. Am. J. Hum. Genet. **65**:448–462, 1999.

Viersbach, R., et al.: Delineation of supernumerary marker chromosomes in 38 patients. Am. J. Med. Genet. **76**:351–358, 1998.

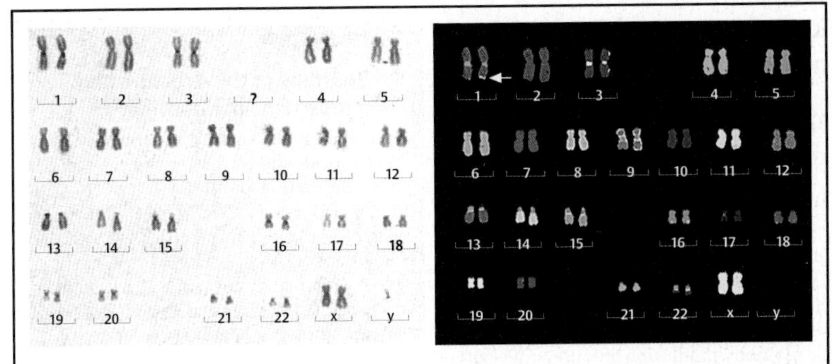


**A. Derivative chromosome 1 with extra material (arrow): 46,XX, der (1)t(1;12)(q43;p13.3)**

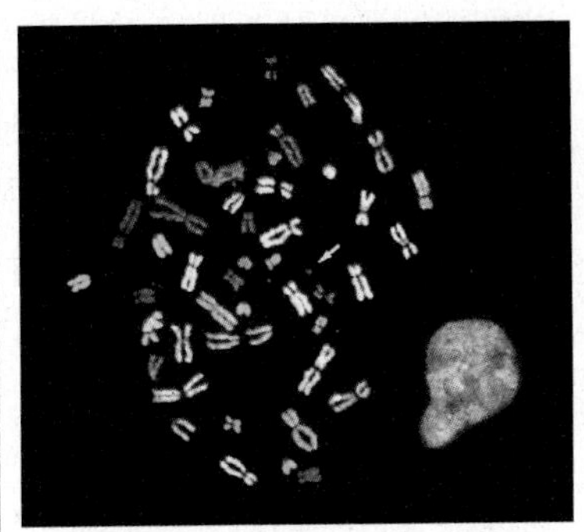

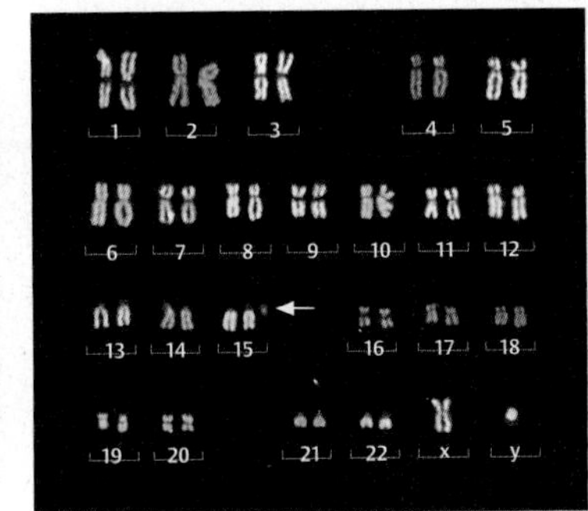


**B. Additional isodicentric chromosme 15 (arrow): 47,XY,+psu idic(15)(q11)**

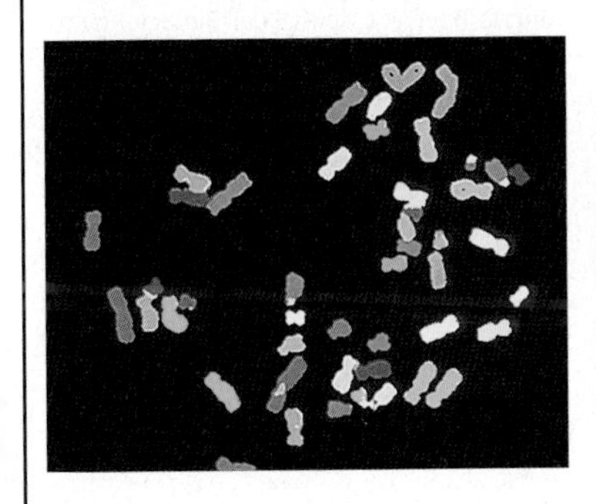

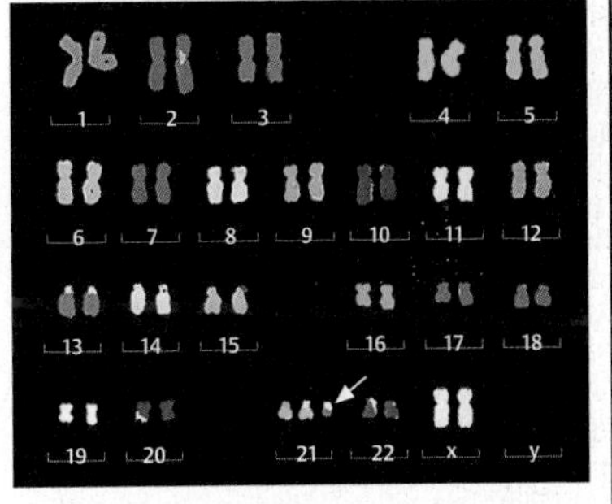


**C. Additional derivative chromosome 21 (arrow): 47,XX,+der(21)t(18;21)(p11.2;q11.1)**

# Regulation and Expression of Genes

## *The Cell Nucleus and Ribosomal RNA*

The cell nucleus is the main center from which cell functions are regulated. A highly important function of cells is to produce endogenous proteins (protein synthesis). The proteins, in turn, are required for innumerable vital processes, such as the catalyzation of complex biochemical reactions, production of energy, transport of molecules, etc. Cells from different tissues differ with respect to the genes that are expressed.

### A. Cell nucleus and protein synthesis

Transcription and processing of the primary transcripts (RNA splicing) occur in the cell nucleus. RNA in the nucleus is bound to nuclear RNA-binding proteins for stabilization. The mature RNA is then released from the nucleus into the cytoplasm. For translation the mRNA must be bound to ribosomes. Ribosomes are complex protein structures made up of numerous subunits, which in turn are the products of individual genes (ribosomal genes).

### B. Nucleolus and the synthesis of ribosomes

The nucleolus is a morphologically and functionally specific region in the cell nucleus in which ribosomes are synthesized. In man, the rRNA genes (200 copies per haploid genome) are transcribed by RNA polymerase I to form 45 S rRNA molecules. After the 45 S rRNA precursors have been produced, they are quickly packaged with ribosomal proteins (from the cytoplasm). Before they are transferred from the nucleus to the cytoplasm, they are cleaved to form three of the four rRNA subunits. These are released into the cytoplasm with the separately synthesized 5 S subunit. Here they form functional ribosomes. The sizes of ribosomes, their subunits, and different types of ribosomal RNAs (rRNA) are given in Svedberg units (S). This is the rate at which a molecule sediments in a solvent. The S values are not additive. A functional ribosome consists of a small and a large subunit. In eukaryotes, the small contains 18 S rRNA, and the large subunit 5.8 S, 5 S, and 28 S rRNA.

### C. Overview of the structure and components of ribosomes

Ribosomes are the centers of protein synthesis. They provide the workplace and the tools. The 70 S ribosome in prokaryotes consists of two subunits of 30 S and 50 S. The 50 S subunit consists of a large (23 S) and a small (5 S) rRNA of ~2900 and 120 nucleotides, respectively, and 33–35 different proteins; the 30 S subunit contains a large 16 S rRNA and 21 proteins. The 50 S subunit provides peptidyltransferase activity, while the 30 S subunit is the site where genetic information is decoded. The 30 S subunit also has a proofreading mechanism to minimize errors in translation. The whole ribosome has a molecular weight of 2.5 million daltons (MDa) and a sedimentation coefficient of 70 S. The eukaryotic ribosome is much larger (4.2 MDa and 80 S), with 60 S and 40 S subunits, which contain an array of rRNAs and proteins as shown in the figure. Recent observations of bacterial 30 S and 50 S ribosomal subunit structures at 5 Å resolution have helped to elucidate the details of ribosomal structure and function. (Figures adapted from Alberts et al., 1998).

### References

Agalarov, S. C., et al.: Structure of the S15, S6, S18-rRNA complex: assembly of the 30*S* ribosome central domain. Science **288**:107–112, 2000.

Alberts, B., et al.: Molecular Biology of the Cell. 3rd ed. Garland Publishing, New York, 1994.

Alberts, B., et al: Essential Cell Biology. An Introduction to the Molecular Biology of the Cell. Garland Publishing, New York, 1998.

Ban, N., et al.: Placement of protein and RNA structures into a 5 Å-resolution map of the 50 S ribosomal subunit. Nature **400**:841–847, 1999.

Garrett, R.: Mechanics of the ribosome. Nature **400**:811–812, 1999.

Wimberly, B.T., et al.: Structure of the 30 S ribosomal subunit. Nature **407**:327–339, 2000.

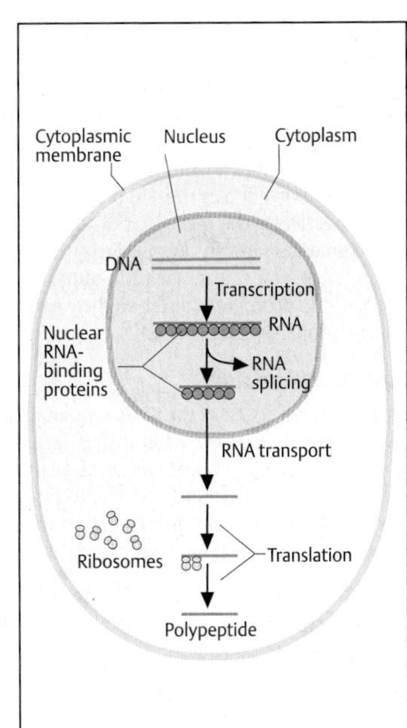


**A. Nucleus and protein synthesis**

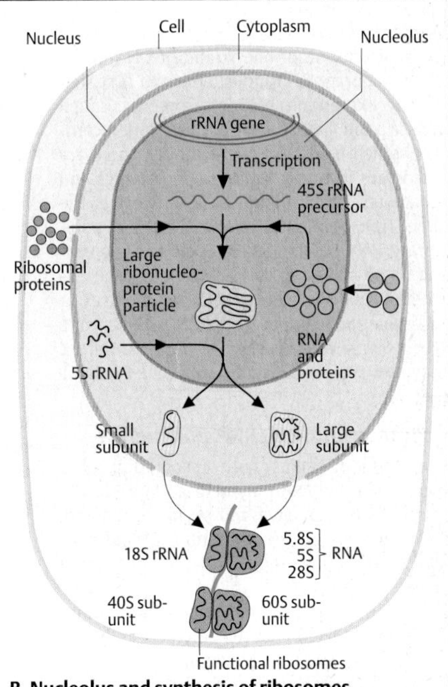

**B. Nucleolus and synthesis of ribosomes**

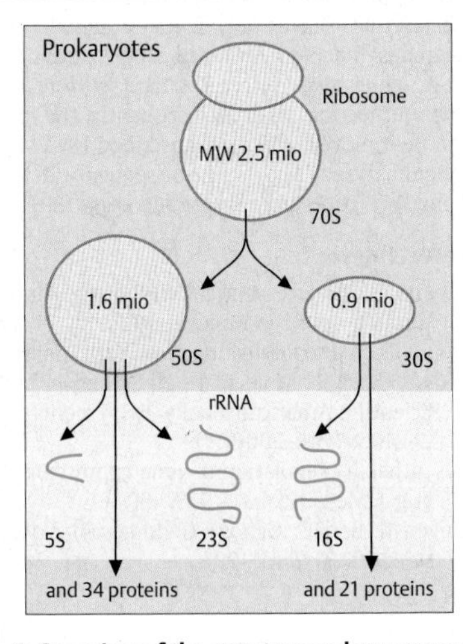


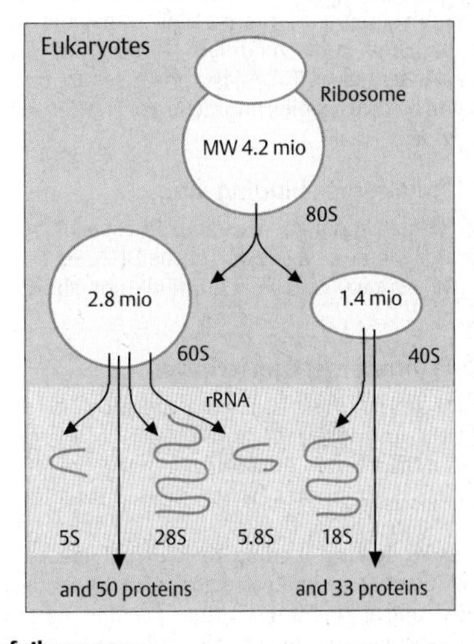


**C. Overview of the structure and components of ribosomes**

# Transcription

The transcription of one strand of DNA into a complementary RNA molecule is the first step in gene expression. Multiple proteins (called transcription factors) form a transcription complex, which binds to DNA. Although there are differences in transcription in prokaryotes and eukaryotes, much of the basic process is the same. Transcription is catalyzed by RNA polymerase. RNA polymerase in *E. coli* has five subunits (two $\alpha$, two $\beta$, and one sigma), each encoded by its own genes. RNA polymerases in eukaryotes are complex (see p. 214). Eukaryotic RNA polymerase consists of three different enzymes, which transcribe different types of genes.

## A. Transcription by RNA polymerase

Transcription begins with recognition of a specific site by RNA polymerase (1). Here the double helix is opened and begins to unwind. RNA synthesis begins (initiation, 2) and continues with elongation (3). As the polymerase moves along the DNA, mRNA is synthesized. The DNA that has been transcribed rewinds into the double helix behind the polymerase. At termination (4), the RNA polymerase is removed from the DNA. At this point, the formation of the unstable primary transcript is completed. Since it is unstable, it is immediately translated in prokaryotes and modified (processed) in eukaryotes (see p. 50). All the processes are mediated by the complex interaction of a variety of enzymes.

## B. Polymerase binding site

Bacterial RNA polymerase binds to a specific region of about 60 base pairs of the DNA. Several active centers can be identified (not shown here).

## C. Promoter of transcription

Transcription must begin at a specific position of DNA, just upstream (at the 5′ end) of a gene. This transcription initiation site is called a promoter. A promoter is a short nucleotide sequence of DNA that regulates the onset of transcription by binding to RNA polymerase. Two distinct promoter regions can be recognized above the transcription starting point. These sequences are evolutionarily highly conserved (consensus sequences). In prokaryotes, a promoter with a consensus sequence consisting of six base pairs, TATAAT (also called a Pribnow box after its discoverer) is located 10 base pairs above the starting point; another region of conserved sequences, TTGACA, is located 35 base pairs above the gene (at the 5′ end). These sequences are referred to as the –10 box and the –35 box respectively (the term "box" is derived from the sequence identity or similarity in all genes). In eukaryotes, the location and the sequences of the promoters differ slightly from those of the prokaryotes (see p. 212).

## D. A transcription unit

A transcription unit is all of the DNA sequences in a given segment that are used in transcription. It begins at the promoter and ends at the terminator. The region around the promoter at the 5′ end is designated proximal; that around the terminator at the 3′ end is designated distal.

## E. Determination of the starting point of transcription

One way to identify an active gene is to determine the starting point of transcription. This can be done by comparing the RNA formed and the DNA template. After transcription, the RNA formed (single-stranded) is hybridized to a complementary single strand of DNA (RNA/DNA hybridization). An endonuclease (S1 nuclease) that cleaves only single-stranded DNA degrades the nonhybridized single strand of DNA, while the hybridized strand is protected (RNA protection assay). Subsequently, the RNA can be removed and the transcribed DNA segment analyzed (e.g., its size or sequence determined). (Figures after Singer and Berg, 1991).

### References

Alberts, B., et al.: Essential Cell Biology. An Introduction to the Molecular Biology of the Cell. Garland Publishing, New York, 1998.

Lodish, H., et al.: Molecular Cell Biology. 4th ed. Scientific American Books, F.H. Freeman & Co., New York, 2000.

Rosenthal, N. Regulation of gene expression. N Eng. J. Med. **331**:931–933, 1994.

Singer, M., Berg, P.: Genes & Genomes. Blackwell Scientific, Oxford, 1991.

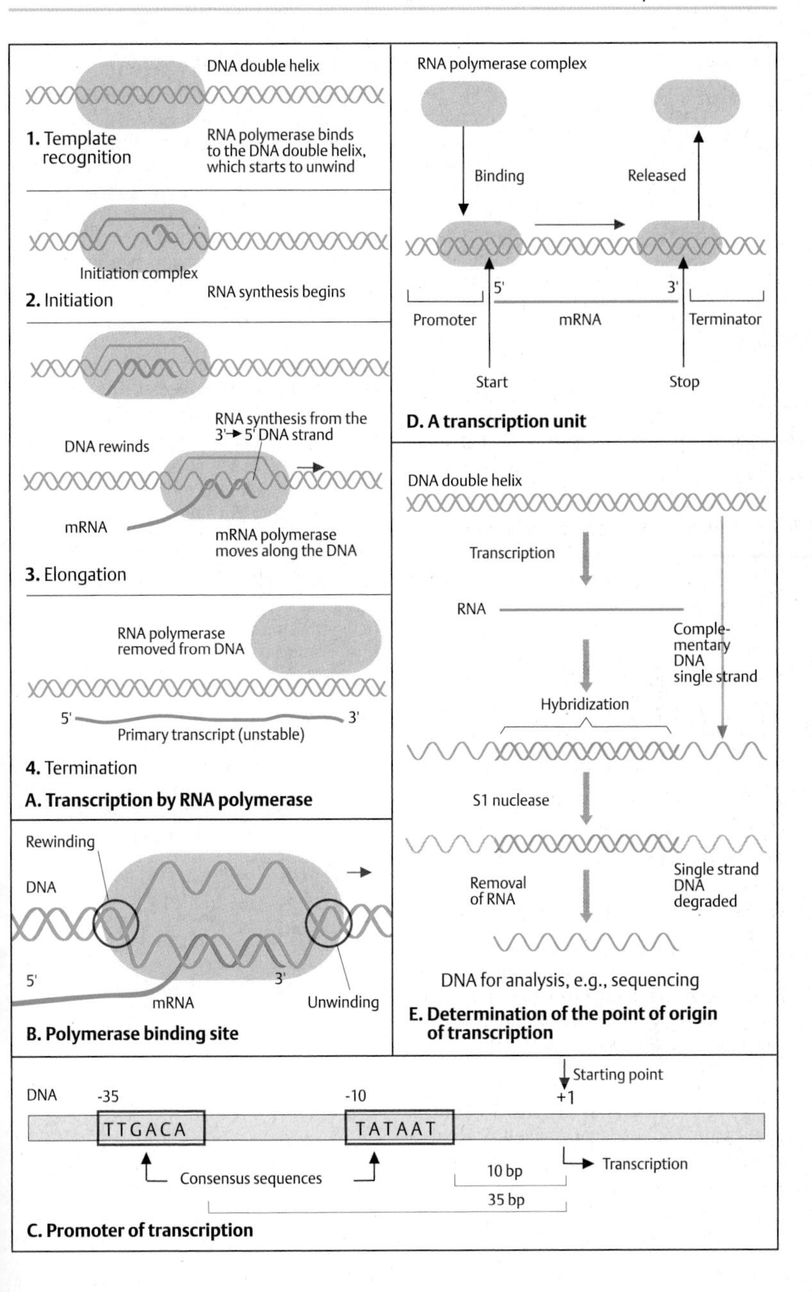


A. Transcription by RNA polymerase

B. Polymerase binding site

C. Promoter of transcription

D. A transcription unit

E. Determination of the point of origin of transcription

# Control of Gene Expression in Bacteria by Induction

The regulation of gene expression is a basic function of prokaryotic and eukaryotic organisms. Prokaryotic organisms rely entirely on their ability to adapt rapidly to changes in external conditions. Substances usually present in the nutrient medium need not be synthesized by the bacterium itself. On the other hand, substances not present must be synthesized by the cell. The control of gene expression occurs at different levels. Regulator proteins may act as repressors (suppressing RNA polymerase activity) or as activators (inducing RNA polymerase activity). Control of prokaryotic genes is often facilitated in that functionally related genes usually lie together and therefore can be regulated together (operon) (see p. 234).

## A. Induction of enzymes in bacteria

The presence of certain substances in the nutrient medium induces the synthesis of enzymes for their utilization in bacteria. An example in *Escherichia coli (E. coli)* is the activation of three enzymes for lactose catabolism by lactose. Within 10 minutes after lactose is added to the nutrient medium, the enzymes β-galactosidase, β-galactoside permease, and β-galactoside transacetylase increase manyfold (1). β-Galactosidase is the enzyme that splits lactose into galactose and glucose (2).

## B. The lactose operon in *E. coli*

A series of genes whose regulation is coordinated is called an operon. Three structural genes that code for the synthesis of lactose-degrading enzymes (genes *lacZ, lacY,* and *lacA*) form the lactose operon (*lac* operon). These three genes are regulated by a promoter at the 5′ end and are transcribed into a common mRNA (polycistronic transcipt). Normally they show little activity because a *lac* repressor inhibits (*lac*i) *lac* mRNA snythesis. The former is the gene product of the *lacI* regulator gene. *E. coli* can use lactose as its sole source of carbon and energy because large amounts of β-galactosidase can be synthesized within a short time.

## C. Control of the *lac* operon

The three structural genes of the *lac* operon, *lacZ, lacY,* and *lacA,* are controlled by means of a repressor protein that binds to the promoter/operator region (P–O). When the repressor is bound to the P–O region, RNA polymerase cannot bind to the promoter region. Transcription is blocked (1), and the three gene products are not formed.The *lac* operon is activated when a β-galactoside molecule binds to one of the subunits of the repressor (2) and inactivates it. RNA polymerase can then bind to the promoter region and transcription can begin.

## D. Gene-regulating nucleotide sequence of the *lac* operator

The activity of a gene is mediated by gene regulatory proteins that bind to DNA at specific sites. The repressor of the lactose operon (*lac* repressor) was the first such protein to be isolated (in 1966 by Gilbert and Müller-Hill). The repressor is a tetramer of identical 37-kDa subunits. Each has a binding site for the inducer. In the absence of the inducer, the repressor binds very tightly to the DNA in the operator region. The recognition sequence of the repressor is a short sequence of 28 nucleotide base pairs that are related by an axis of symmetry (shown as colored boxes). Symmetry matching is an important principle of protein–DNA interaction at a genetic switch. Gene-regulatory proteins can be dinstinguished by the specificity of the DNA sequences they recognize (compare the recognition sequences of restriction enzymes, p. 64).

## References

Stryer, L.: Biochemistry. 4th ed. W.H. Freeman, New York, 1995.

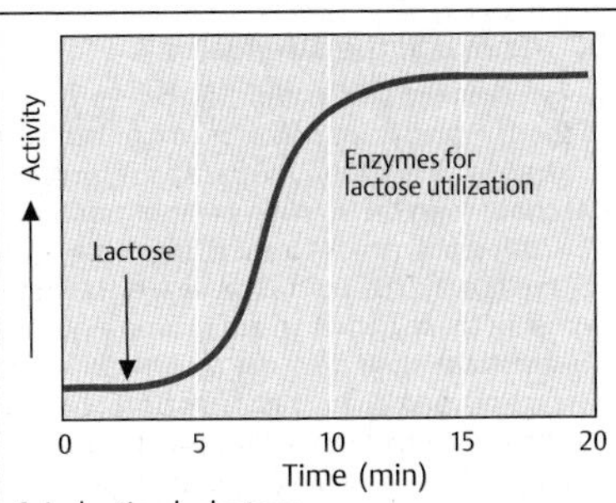


1. Induction by lactose

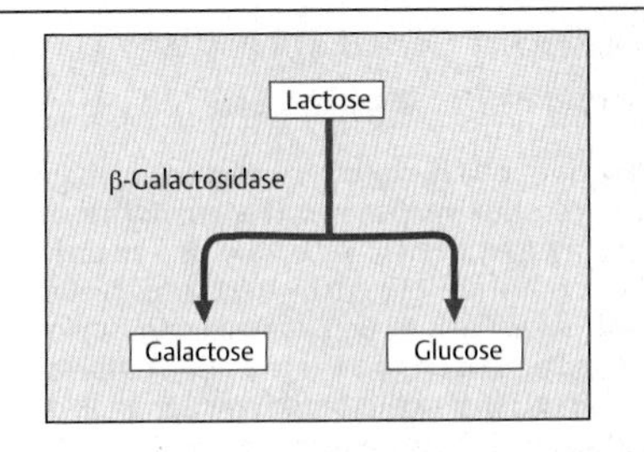


2. Enzymatic lactose degradation

**A. Induction of enzymes in bacteria**

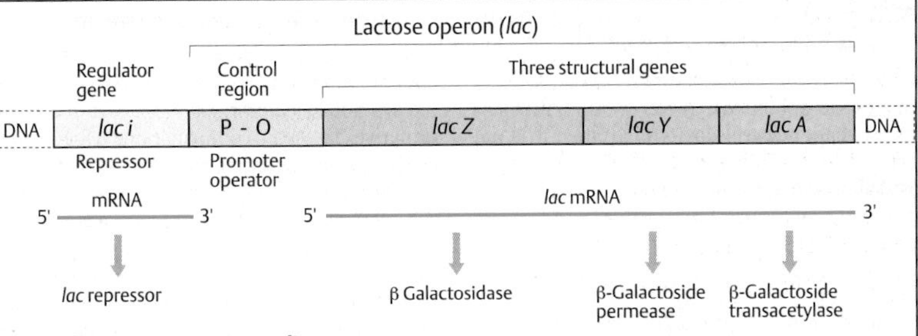


**B. The lactose operon in *E. coli***

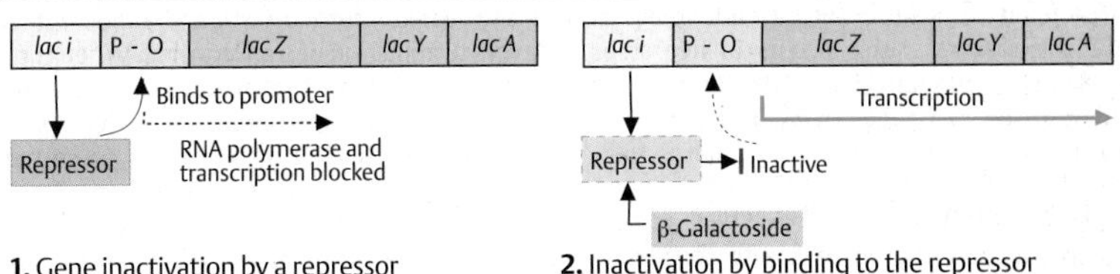


1. Gene inactivation by a repressor

2. Inactivation by binding to the repressor

**C. Control of the *lac* operon**

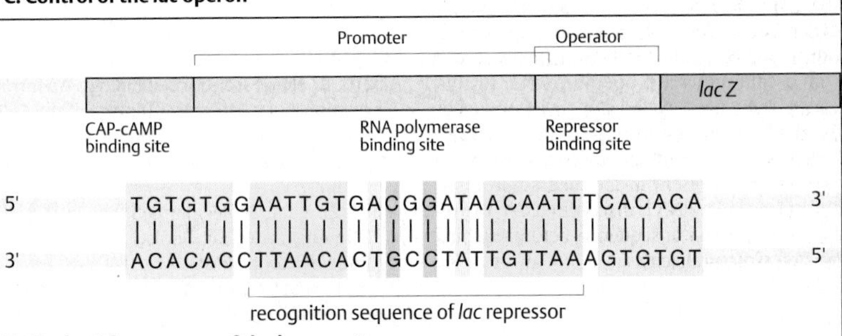


**D. Nucleotide sequence of the *lac* operator**

# Control of Gene Expression in Bacteria by Repression

If a gene is usually expressed (is active), it is said to be constitutive. Gene expression in bacteria can vary considerably depending on the presence or absence of certain substances in the nutrient medium. An important mechanism for controlling transcription is a signal (termination signal) that can terminate transcription or translation. It lies between the promoter and the beginning of the first structural gene and is called the attenuator (attenuation of translation).

## A. Regulation of synthesis of the amino acid tryptophan in *E. coli*

Tryptophan is an essential amino acid in eukaryotic organisms. Bacteria can synthesize tryptophan, but will do so only when it is not present in the nutrient medium (2). If tryptophan is added to the medium, enzyme activity for tryptophan biosynthesis decreases within about 10 minutes (1).

## B. Biosynthesis of tryptophan by means of five enzymes and five genes

Tryptophan is synthesized from chorismate via four intermediates; this occurs in five steps, regulated by five enzymes. The enzymes are coded for by five genes (*trpA–E*) (CdRP = carboxyphenylamino-deoxyribulose-phosphate).

## C. Tryptophan operon in *E. coli*

The tryptophan operon in *E. coli* consists of these five genes and their regulating sequences. The latter include promoter and operator, a leader sequence, and attenuator sequences. Translation of the five structural genes results from a continuous trp operon mRNA. In this, leader sequences coded for by L-sequence genes are connected in series. The attenuator sequences are part of the L-sequences. When tryptophan is present in the medium, translation of *trp* leader RNA is discontinued in the region of an attenuator sequence, before reaching the first structural gene.

## D. The role of the attenuator

The weakening (attenuation) of the expression of the tryptophan operator is controlled by a sequence of about 100–140 base pairs (in the 3′ direction) from the starting point of transcription (tryptophan mRNA leader). In the presence of tryptophan, the *trp* mRNA leader is interrupted in the region of an attenuator sequence (1), and translation does not take place. In the absence of tryptophan, translation is continued. The *trp* leader peptide contains two tryptophan residues (2). When tryptophan is deficient, translation is delayed and a stop signal weakened.

## E. Attenuation of the *trp* operon

In *E. coli*, attenuation is often mediated by a tight relation of transcription and translation. The *trp* mRNA leader region can exist in two alternative base-pair conformations. One allows transcription, the other not. When tryptophan ist present (1), ribosomes can synthesize the complete leader peptide. The ribosome closely follows the RNA polymerase transcribing the DNA template (not shown). The ribosome has passed region 1 and prevents complementary regions 2 and 3 from forming a hairpin by base pairing. Instead, part of complementary region 3 and region 4 form a stem and a loop, which favors termination of transcription. When tryptophan is deficient (2), the ribosome stalls at the two UGG *trp* codons owing to deficiency of tryptophanyl-tRNA. This alters the conformation of the mRNA so that regions 2 and 3 pair. The stem structure favoring termination is not formed, region 4 remains single-stranded, and transcription continues. (Figure adapted from Stryer, 1995).

## References

Alberts, B., et al: Essential Cell Biology. An Introduction to the Molecular Biology of the Cell. Garland Publishing Co., New York, 1998.

Lewin, B.: Genes VII. Oxford Univ. Press, Oxford, 2000.

Lodish, H., et al.: Molecular Cell Biology. 4th ed., Scientific American Books, F.H. Freeman & Co., New York, 2000.

Stryer, L.: Biochemistry. 4th ed. W.H. Freeman & Co, New York, 1995.

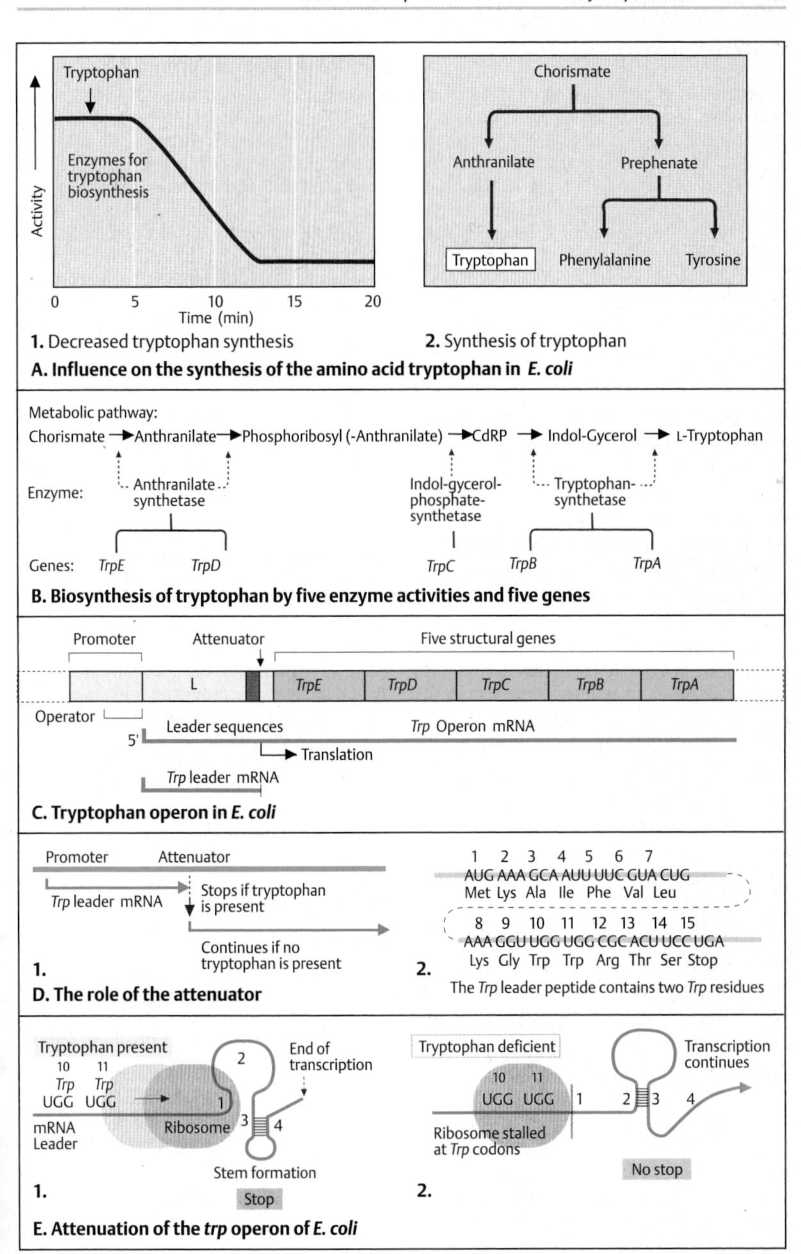


**A. Influence on the synthesis of the amino acid tryptophan in *E. coli***

**B. Biosynthesis of tryptophan by five enzyme activities and five genes**

**C. Tryptophan operon in *E. coli***

**D. The role of the attenuator**

**E. Attenuation of the *trp* operon of *E. coli***

# Control of Transcription

Transcription is controlled at promoters and other DNA sequences (enhancers) outside of the actual coding region. Transcription control in prokaryotes and eukaryotes differ in some respects and correspond in others. An important region for controlling gene expression, the promoter region, lies upstream (5′ direction) of the coding sequence.

## A. Promoter region

In prokaryotes, two important areas in the promoter region are 35 and 10 nucleotide base pairs upstream of (in the 5′ direction) the starting point of transcription. Mutations in the region of the regulative sequences (promoter region) in certain regions are extremely sensitive to base substitution (mutation), an indication of their importance. (Figure redrawn from Watson et al., 1987).

## B. Assembly of general transcription factors to initiate transcription

The activation of polymerase II (Pol II) to transcribe most eukaryotic genes (1) requires an initiation complex assembled at the promoter. It consists of general transcription factors (TF) that associate in an ordered sequence. In the first step (2), TFIID (transcription factor D for polymerase II) binds to the TATA region. The TATA box is recognized by a small, 30-kDa TATA-binding protein (TBP), which is part of one of the many subunits of TFIID (the bending of the DNA by TBP is not shown here).

Following this, TFIIB can bind to the complex (3). Subsequently, other transcription factors (TFIIH, followed by TFIIE) and Pol II escorted by TFIIF join the complex and assure that Pol II is attached to the promoter (4). The binding of TFIIE extends the polymerase binding sites further downstream in the 3′ direction. Pol II is then released from the complex and transcription can begin. A key step is phosphorylation of Pol II by a subunit of TFIIH, which is a protein kinase. Other activities of TFIIH involve a helicase and an ATPase. The site of phosphorylation is a polypeptide tail, composed in mammals of 52 repeats of the amino acid sequence YSPTSPS in which the serine (S) and threonine (T) side chains are phosphorylated. (The figure is a simplified scheme, redrawn from p. 421 in Alberts et al., 1994 and p. 631 in Lewin, 2000).

## C. RNA polymerase promoters

Eukaryotic cells contain three RNA polymerases (Pol I, Pol II, and Pol III). Pol I is located in the nucleolus, synthesizes ribosomal RNA, and accounts for about 50 – 70% of the relative activity. Pol II and Pol III are located in the nucleoplasm (the part of the nucleus excluding the nucleolus). Pol II represents 20 – 40% of cellular activity. It is responsible for the synthesis of heterogeneous nuclear RNA (hnRNA), the precursor of mRNA. Pol III, responsible for the synthesis of tRNAs and other small RNAs, contributes only a minor activity of about 10%. Each of the large eukaryotic RNA polymerases (500 kDa or more) is more complex, with 8 – 14 subunits, than the single prokaryotic RNA polymerase. Each RNA polymerase uses a different type of promoter. RNA polymerase II cannot initiate transcription without a complex of general transcription factors (see B) that bind to a single upstream promoter (1). RNA polymerase I (2) has a bipartite promoter, one 170 to 180 bp upstream (5′ direction) and another from about 45 bp upstream to 20 bp downstream (3′ direction). The latter is called the core promoter. Pol I requires two ancillary factors, UBF1 and SL1. SL1 consists of four proteins including a TBP (see B) that cannot bind directly to the promoter. It binds to UBF1, after which Pol I can bind to the core promoter to initiate transcription (2). RNA polymerase III uses either upstream promoters or two internal promoters downstream of the transcription start site (3). Three transcription factors are required with internal promoters, TFIIIA (a zinc finger protein, see p. 218), TFIIIB (a TBP and two other proteins), and TFIIIC (a large, more than 500 kDa protein).

### References

Alberts, B., et al.: Molecular Biology of the Cell. 3rd ed. Garland Publishing Co., New York, 1994: Chapter 8, p. 365.

Lewin, B.: Genes VII. Oxford Univ. Press, Oxford, 2000: Chapter 28, p. 811.

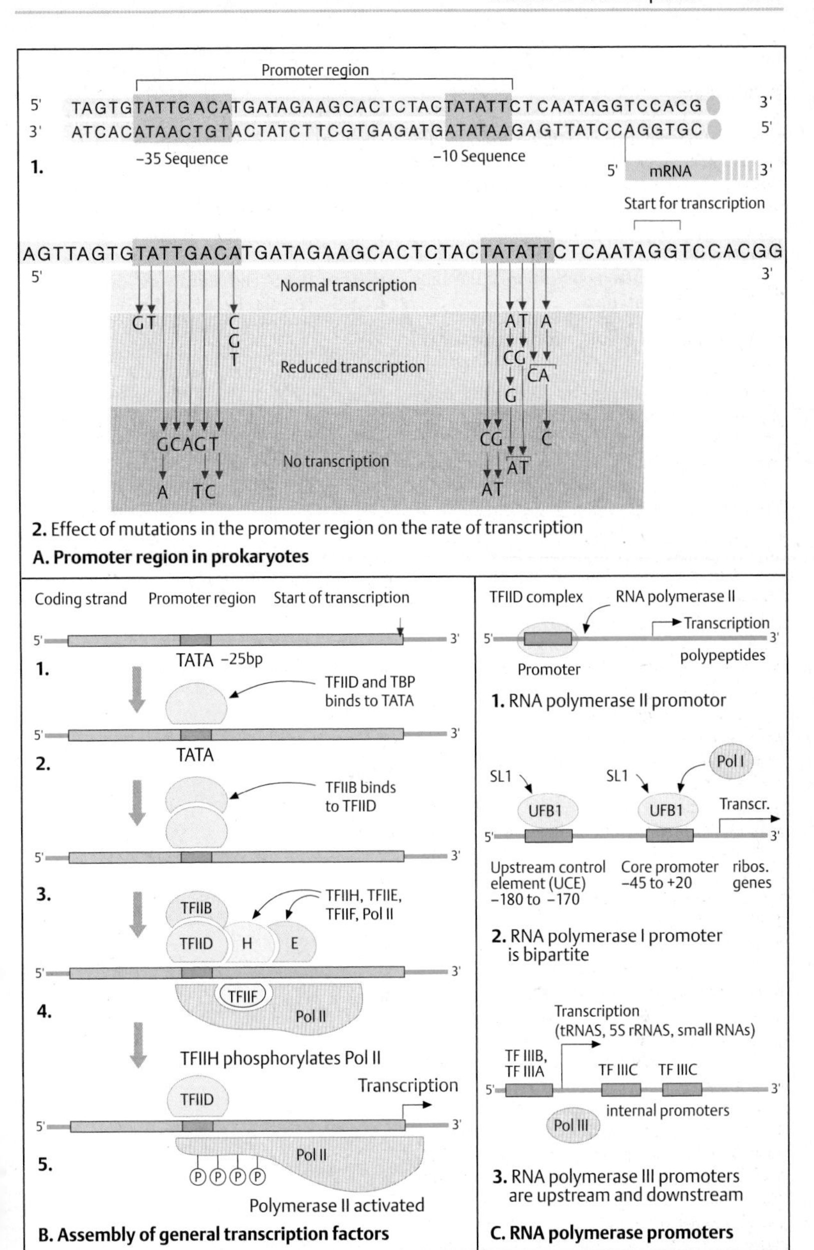


**1.**

5' TAGTGTATTGACATGATAGAAGCACTCTACTATATTCTCAATAGGTCCACG 3'
3' ATCACATAACTGTACTATCTTCGTGAGATGATATAAGAGTTATCCAGGTGC 5'

Promoter region

−35 Sequence          −10 Sequence

5' ▭ mRNA ▥ 3'

Start for transcription

**2.** Effect of mutations in the promoter region on the rate of transcription

AGTTAGTGTATTGACATGATAGAAGCACTCTACTATATTCTCAATAGGTCCACGG

Normal transcription

Reduced transcription

No transcription

**A. Promoter region in prokaryotes**

**1.**
Coding strand    Promoter region    Start of transcription
TATA −25bp

**2.**
TATA
TFIID and TBP binds to TATA

**3.**
TFIIB binds to TFIID

**4.**
TFIIB  TFIID  H  E    TFIIH, TFIIE, TFIIF, Pol II
TFIIF    Pol II

**5.**
TFIIH phosphorylates Pol II
TFIID    Transcription
Pol II
(P)(P)(P)(P)
Polymerase II activated

B. Assembly of general transcription factors

**1.** RNA polymerase II promotor
TFIID complex    RNA polymerase II
Promoter
Transcription
polypeptides

**2.** RNA polymerase I promoter is bipartite
SL1    SL1    Pol I
UFB1    UFB1    Transcr.
Upstream control element (UCE) −180 to −170
Core promoter −45 to +20
ribos. genes

**3.** RNA polymerase III promoters are upstream and downstream
Transcription (tRNAS, 5S rRNAS, small RNAs)
TF IIIB, TF IIIA    TF IIIC    TF IIIC
Pol III    internal promoters

C. RNA polymerase promoters

# Transcription Control in Eukaryotes

Transcription in eukaryotes differs from that in prokaryotes in two main respects. In eukaryotes, one gene codes for a single polypeptide (monocistronic transcription unit) and the initial transcript is processed into mature messenger mRNA. This involves intron splicing (see p. 50) and substantial modification of the ends of the primary transcript.

## A. Prototype of a eukaryotic structural gene

A structural gene is a gene that codes for a polypeptide gene product. It can be divided into sections involved in transcription (transcription unit) and regulatory sequences. Regulatory sequences are located both upstream (the 5′ direction) and downstream (the 3′ direction) of the gene. In addition, internal regulatory sequences may occur in introns. Some regulatory sequences are located far from the gene. Together with the promoter (see p. 206), they are required to regulate transcription.

## B. Prototype of mature eukaryotic mRNA

Mature eukaryotic mRNA is produced from its precursor RNA by the removal of introns, addition of a 5′ cap at the 5′ end, and addition of numerous adenine nucleotides at the 3′ end (polyadenylation). A noncoding sequence (5′ leader) is located in front of the translation start signal (AUG), and a trailer sequence, at the 3′ end in back of the translation stop signal (UAA). Both addition of the 5′ cap and polyadenylation involve enzymatic reactions.

## C. 7-Methyl-guanosine cap

The translation of eukaryotic mRNA is similar to that of prokaryotic mRNA, with two distinct differences: (1) transcription and translation occur at different locations in the eukaryotic cell: transcription occurs in the cell nucleus, and translation in the cytoplasm; (2) the 5′ and 3′ ends of eukaryotic mRNA have special structures. The structure at the 5′ end is called a cap. Through the action of guanosine-7-methyltransferase, guanosine is bound by a triphosphate bridge to the first and second ribose groups of the precursor mRNA chain. The guanosine is methylated in position 7, as are the two initial ribose residues at the beginning of the RNA chain. Except for the mRNAs transcribed by DNA viruses, eukaryotic mRNA usually contains a single protein-coding sequence (monocistronic messenger).

## D. Polyadenylation at the 3′ end

Eukaryotic termination signals have been less well recognized than the regulators of gene activity at the 5′ end. Eukaryotic primary transcripts are split by a specific endonuclease shortly after the sequence AAAUAA. Subsequently, about 100 – 250 adenine nucleotides are attached to the 3′ end of the transcript by means of a poly(A)-polymerase (polyadenylation). The poly(A) end binds to a protein. All mRNAs, except those that code for histone proteins, possess a poly(A) terminus.
(Figures after Singer and Berg, 1991).

## References

Singer, M., Berg, P.: Genes & Genomes. Blackwell Scientific, Oxford, 1991.

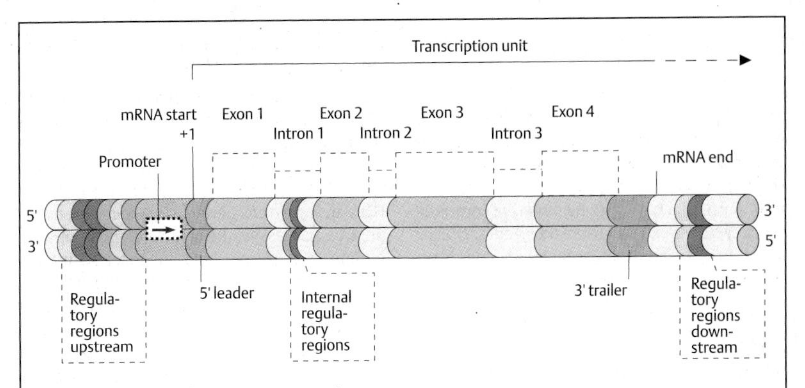


A. Prototype of a eukaryotic structural gene

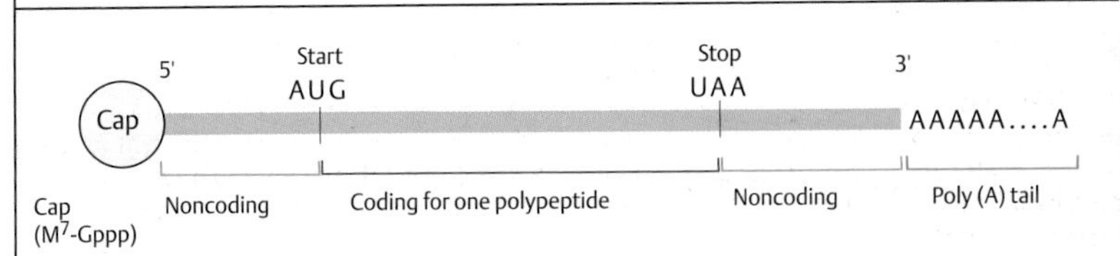


B. Prototype of a mature eukaryotic mRNA

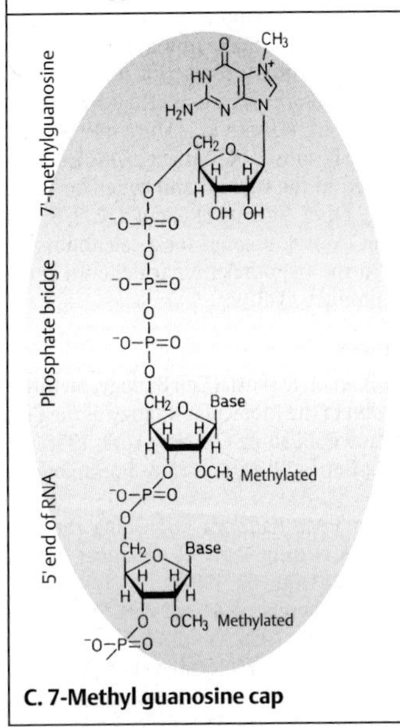


C. 7-Methyl guanosine cap

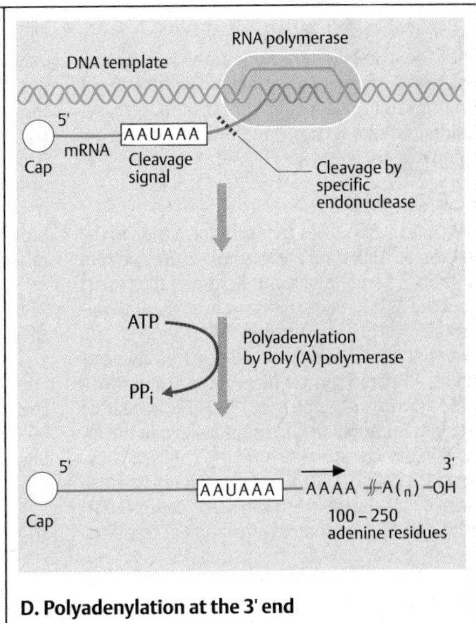


D. Polyadenylation at the 3' end

# Regulation of Gene Expression in Eukaryotes

Precisely regulated gene expression is a prerequisite for producing and maintaining the many different types of cells and tissues of a multicellular organism. Cells differentiate into their particular cell types by means of combinations of expressed and repressed genes. During differentiation the tightly regulated genes function in the order, usually sequential, required for a particular cell fate (developmental pathways). Many regulator genes and their proteins have been identified (cf. part III, Genetics and Medicine). The following outlines some important principles of the specific control of gene expression in eukaryotic cells.

## A. Levels of control of eukaryotic gene expression

In principle, expression can be regulated at four distinct levels. The first and by far the most important is primary control of transcription. Processing to mature RNA can be regulated at the level of the primary RNA transcript. A frequently observed process is alternative splicing (see D). Translation can be varied by RNA editing (see B for an example). At the protein level, posttranslational modifications can determine the activity of a protein. The cleavage of preproinsulin to form mature insulin, glycosylation or hydroxylation, and protein folding are some examples (see p. 32, 362).

## B. RNA editing

RNA editing modifies genetic information at the RNA level. An important example is the apolipoprotein-B gene involved in lipid metabolism. It encodes a protein of 4538 amino acids, apolipoprotein B. This is synthesized in the liver and secreted into the blood, where it transports lipids. A related shorter form of the protein with 2153 amino acids, Apo B-48 (250 kDa, instead of 512 kDa for Apo B-100), is synthesized in the intestine. An intestinal deaminase converts a cytosine in codon 2158, CAA (glutamine), to uracil (UAA). This change results in a stop codon (UAA) and thereby terminates translation at this site.

## C. Long-range gene activation by an enhancer

Enhancers control gene activity at a distance. An enhancer is a distant site involved in initiation of transcription (see p. 206). It may be located either upstream or downstream of the same DNA strand (*cis*-acting) or on a different DNA strand (*trans*-acting). Enhancer elements provide tissue-specific or time-dependent regulation. It is unclear how enhancers can exert their effect from a considerable distance. One model suggests that DNA forms a loop between enhancer and promoter. Activator proteins bound to the enhancer, e.g., a steroid hormone, could then come into contact with the general transcription factor complex at the promoter. Others might function as repressors (cf. transcription control in prokaryotes, p. 210).

## D. Alternative RNA Splicing

A DNA segment can code for different forms of mRNA when different introns are removed from the primary transcript (alternative splicing). By means of alternative gene splicing, a gene can code for different, albeit similar gene products. This allows a high degree of functional flexibility. Numerous examples of differential RNA splicing are known for mammalian genes. For example, the primary transcript for the calcitonin gene contains six exons. They are spliced into two different types of mature mRNA. One, consisting of exons 1–4 (but not exons 5 and 6), is produced in the thyroid and codes for calcitonin. The other consists of exons 1, 2, 3, 5, and 6, but not exon 4. It codes for a calcitonin-like protein in the hypothalamus (calcitonin gene-related product, CGRP).

## References


Alberts, B., et al: Essential Cell Biology. An Introduction to the Molecular Biology of the Cell. Garland Publishing Co., New York, 1998.

Lewin, B.: Genes VII, Oxford Univ. Press, Oxford, 2000.

Blackwood E.M., Kadonga J.F.: Going the distance: A current view of enhancer action. Science **281**:60–63, 1998.

Stryer, L.: Biochemistry. 4th ed. W.H. Freeman & Co, New York, 1995.

Watson J.D., et al.: Molecular Biology of the Gene. 4th ed. Benjamin/Cummings Publishing Co., Menlo Park, California, 1987.

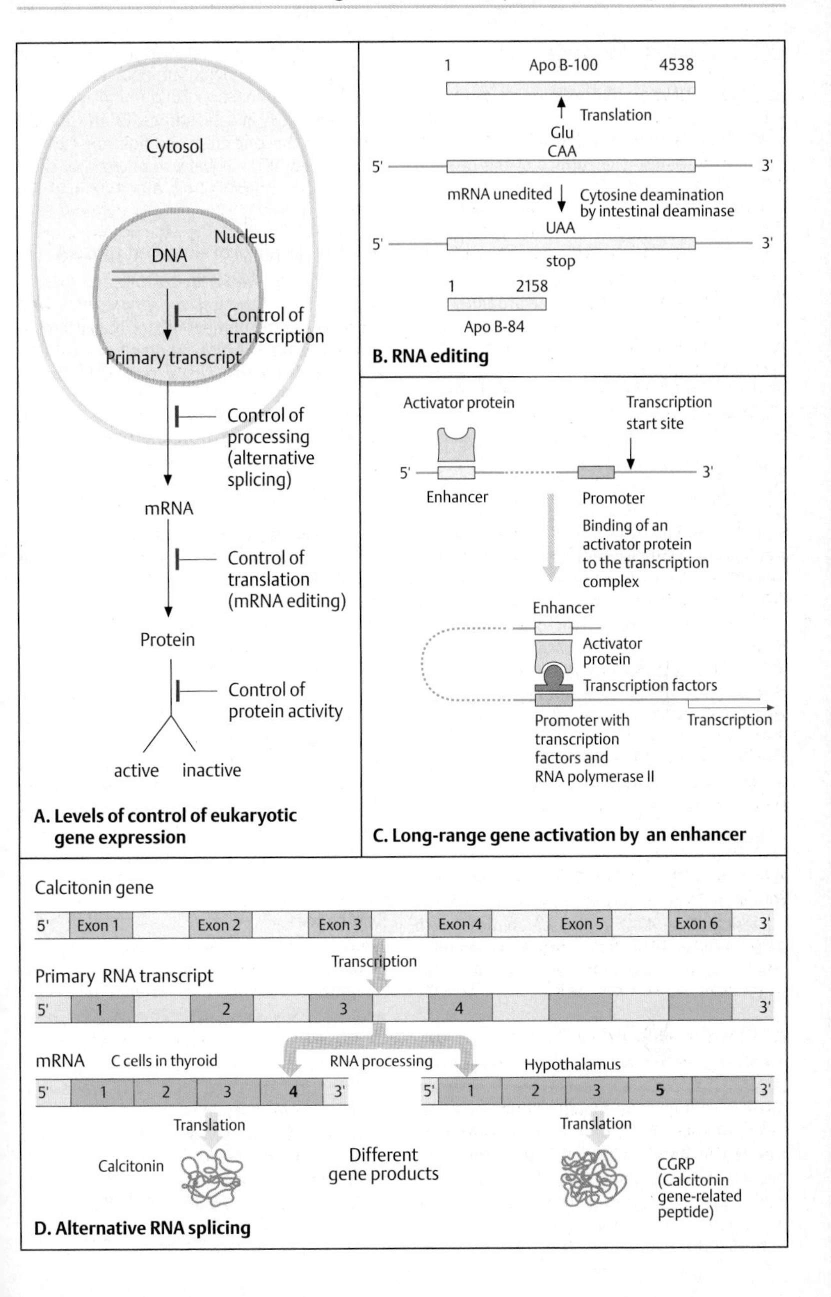
Cytosol
Nucleus
DNA
Control of transcription
Primary transcript
Control of processing (alternative splicing)
mRNA
Control of translation (mRNA editing)
Protein
Control of protein activity
active
inactive
A. Levels of control of eukaryotic gene expression
1
Apo B-100
4538
Translation
Glu
CAA
5'
3'
mRNA unedited
Cytosine deamination by intestinal deaminase
UAA
stop
1
2158
Apo B-84
B. RNA editing
Activator protein
Transcription start site
Enhancer
Promoter
Binding of an activator protein to the transcription complex
Enhancer
Activator protein
Transcription factors
Promoter with transcription factors and RNA polymerase II
Transcription
C. Long-range gene activation by an enhancer
Calcitonin gene
Exon 1
Exon 2
Exon 3
Exon 4
Exon 5
Exon 6
Transcription
Primary RNA transcript
mRNA
C cells in thyroid
RNA processing
Hypothalamus
Translation
Calcitonin
Different gene products
CGRP (Calcitonin gene-related peptide)
D. Alternative RNA splicing

# DNA-Binding Proteins

Regulatory DNA sequences interact with proteins to exert proper functional control. Regulatory proteins can recognize specific DNA sequences because the surface of the proteins fits precisely onto the DNA surface. Three basic groups of regulatory DNA sequences can be distinguished: (1) sequences that establish the exact beginning of translation; (2) DNA segments that regulate the end, or termination; and (3) DNA sequences near the promoter that have specific effects on gene activity (repressors, activators, enhancers, and others).

## A. Binding of a regulatory protein to DNA

Gene regulatory proteins can recognize DNA sequence information without having to open the hydrogen bonds within the helix. Each base pair represents a distinctive pattern of hydrogen bond donors (example shown in red) and hydrogen acceptors (example shown in green). These proteins recognize the major groove of DNA, where binding takes place. Here a single contact of an asparagine (Asn) of a gene-regulatory protein with a DNA base adenine (A) is shown. A typical area of surface-to-surface contact involves 10–20 such interactions. (Figure redrawn from Alberts et al., 1998, p. 276).

## B. An α helix inserts into a major groove of operator DNA

One part of the protein, an α helix (the sequence-reading or recognition helix) is inserted into the major groove of DNA. Here the sequence Q-Q-Q-S-T (glutamine Q, serine S, threonine T) in the recognition sequence of the bacteriophage 434 repressor bonds with specific bases in a major groove of operator DNA. (Figure redrawn from Lodish et al., 2000, p. 351).

## C. Zinc finger motif

Another group of proteins are called zinc fingers because they resemble fingers (see D). They are involved in important functions during embryonic development and differentiation. The basic zinc finger motif consists of a zinc atom connected to four amino acids of a polypeptide chain. Here, two histidine (H) and two cysteine (C) residues are shown in the schema on the left. The three-dimensional structure on the right consists of an antiparallel β sheet (amino acids 1–10), an α helix (amino acids 12–24), and the zinc connection. Four amino acids, cysteines 3 and 6 and histidines 19 and 23, are bonded to the zinc atom and hold the carboxy (COOH) end of the α helix to one end of the β sheet. (Figure redrawn from Alberts et al., 1994, p. 411).

## D. Zinc finger proteins bind to DNA

The interaction with DNA is strong and specific. Each protein recognizes a specific DNA sequence. As the number of zinc fingers can be varied, this type of DNA-binding has great evolutionary flexibility. (Figure redrawn from Alberts et al. 1994).

## E. Binding to a response element

Many hormones and growth factors activate cell-surface receptors. In contrast, steroid hormones enter the target cells and interact there with a specific receptor protein in the cytosol. The hormone–receptor complex then migrates to specific sites of DNA. The hormone-binding domain will prevent binding to DNA unless the hormone is present. Activated receptors bind to specific DNA sequences called hormone response elements (HREs). Each polypeptide chain of the receptor contains a zinc atom bound to four cysteines (1). The skeletal model shows the two DNA-binding domains binding to the HRE in two adjacent major grooves of the target DNA (2). The space-filling model shows how tightly the recognition helix of each dimer of this protein fits into the major groove of DNA shown in red and green (3). (Figure redrawn from Stryer, 1995, p. 1002).

### References

Alberts, B., et al.: Molecular Biology of the Cell. 3rd ed. Garland Publishing Co., New York, 1994.

Alberts, B., et al.: Essential Cell Biology. Garland Publishing Co., New York, 1998.

Lodish, H., et al.: Molecular Cell Biology. 4th ed. Scientific American Books, F.H. Freeman & Co., New York, 2000.

Stryer, L.: Biochemistry. 4th ed. W.H. Freeman & Co., New York, 1995.

Tjian, R.: Molecular machines that control genes. Sci. Am. **272**:38–45, 1995.

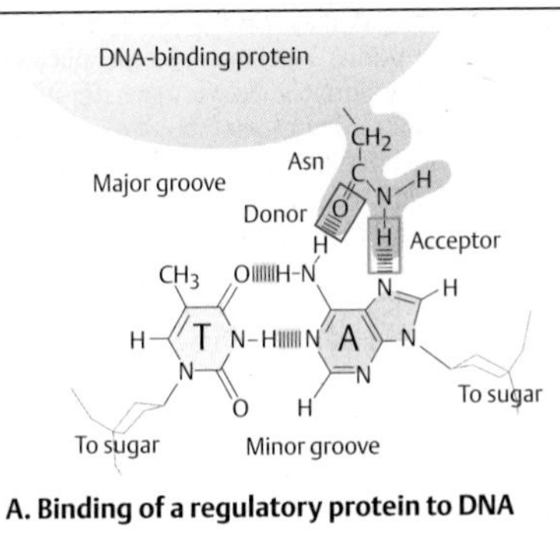


**A. Binding of a regulatory protein to DNA**

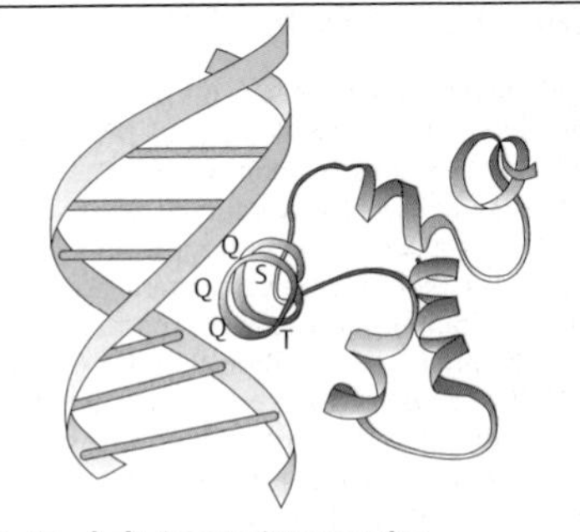


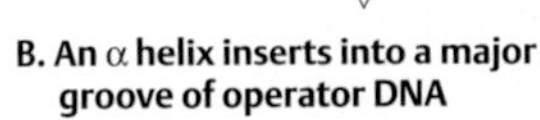

**B. An α helix inserts into a major groove of operator DNA**

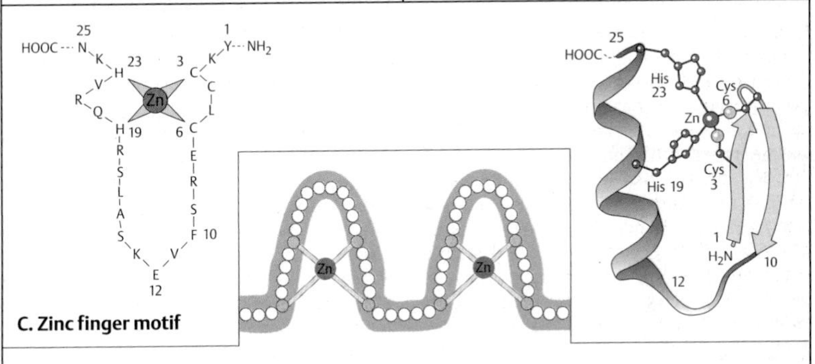


**C. Zinc finger motif**

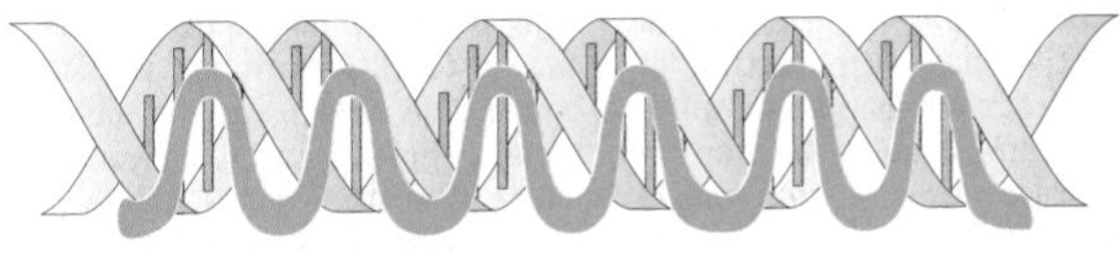

**D. A zinc finger protein binds to DNA**

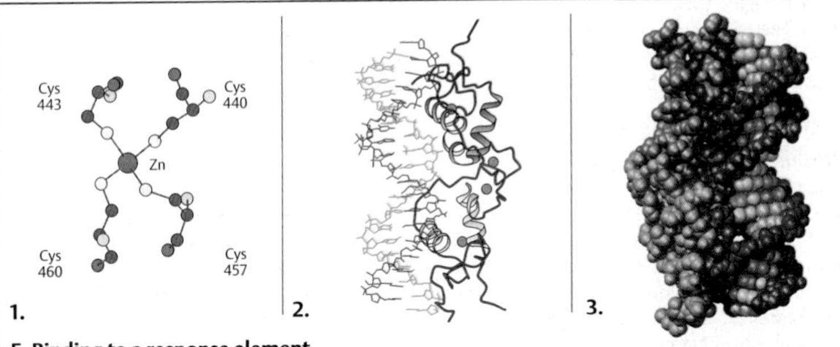


**E. Binding to a response element**

# Other Transcription Activators

Transcription activators are dimeric proteins with distinct functional domains: a DNA-binding domain and an activation domain. The DNA-binding domain interacts with specific regulatory DNA sequences. The activation domain interacts with other proteins that stimulate transcription. Transcription activators participate in the assembly of the initiation complex, for example, by stimulating the binding of transcription factor IID (TFIID, see p. 212) to the promoter. Other activators may interact with general transcription factors. They provide a second level of transcriptional control.

## A. Leucine zipper dimer

Most DNA-binding regulatory proteins recognize specific sites as dimers. One part of the molecule serves as the recognition molecule, the other stabilizes the structure. A particularly striking example is given by proteins with a leucine zipper motif. The name is derived from the basic structure. Two α helices are joined like a zipper by periodically repeated leucine residues located at the interface of the two helices. The two helices separate, form a Y-shaped structure, and extend into the major groove of the DNA (1). Leucine zipper proteins may be homodimers with identical subunits (2, 3) or heterodimers with different albeit similar subunits (4). The ability to form unlike dimers (heterodimerization) greatly expands the spectrum of specificites. The use of combinations of different proteins to control cellular functions is called combinatorial control. (Figure redrawn from Alberts et al., 1994).

A DNA-binding motif related to the leucine zipper is the helix–loop–helix (HLH) motif (not shown). The HLH motif consists of one short α helix and one longer α helix. The two α helices are connected by a flexible loop of protein.

## B. Activation by steroid hormone binding

Transcriptional enhancers are regulatory regions of DNA that increase the rate of transcription. Their spacing and orientation vary relative to the starting point of transcription. An enhancer is activated by binding to a hormone–receptor complex. This activates the promoter, and transcription begins (active gene). Numerous important genes in mammalian development are regulated by steroids (steroid-responsive transcription). The latter include glucocorticoids and mineralocorticoids, the steroids of glycogen and mineral metabolism; sex hormones, which function in embryonic sex differentiation and control of reproduction; and others. Normal bone development and function are under the control of steroidlike vitamin D. Another steroidlike hormone is retinoic acid, an important regulator of differentiation during embryogenesis (morphogen). These hormones initiate their physiological effects by association with corresponding steroid-specific transcellular receptors (hormone–receptor complex).

## C. Evidence of a protein-binding region in DNA

Protein-binding regions in DNA represent regulatory areas; thus, their analysis can yield some insights into gene regulation. Protein-binding DNA regions can be demonstrated in several ways. With band-shift analysis (1), protein-bound and non-protein-bound DNA fragments are differentiated using gel electrophoresis in direction towards the small fragments, a DNA fragment that is part of a DNA-protein complex migrates more slowly than a free DNA fragment of the same size. The DNA-protein complex is found at a different position ("band shift"). DNA footprinting (2) is another procedure for identifying protein-binding sites on DNA. The principle of DNA footprinting is that a protein-bound DNA region, e.g., the polymerase-promoter complex, is protected from the effects of a DNA-cleaving enzyme (DNAase I). Previously isolated DNA is cut into different fragments by DNAase, and the fragments are sorted according to size by gel electrophoresis. Since the DNA protein-binding region is protected from cleavage by DNAase I (DNAase I protection experiment), DNA bands from the binding region are missing ("footprint").

## References

Alberts, B., et al.: Molecular Biology of the Cell. 3rd ed. Garland Publishing Co., New York, 1994.

Alberts, B., et al.: Essential Cell Biology. Garland Publishing, Co., New York, 1998.

Lodish, H., et al.: Molecular Cell Biology. 4th ed. Scientific American Books, F.H. Freeman & Co., New York, 1999.

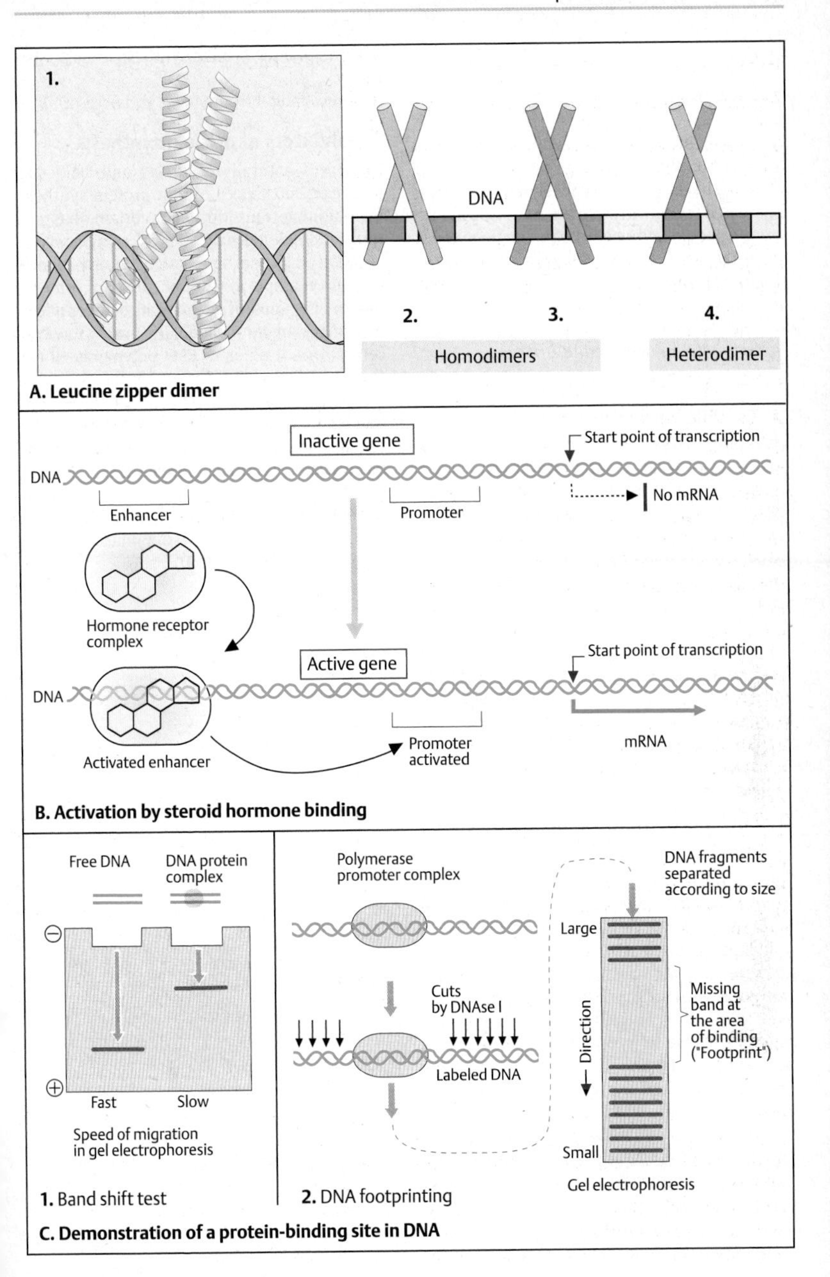


A. Leucine zipper dimer

B. Activation by steroid hormone binding

C. Demonstration of a protein-binding site in DNA

# Inhibitors of Transcription and Translation

A number of natural and artificial substances are able to inhibit transcription or translation. They can be used to treat cancer or as antibiotics to treat infections. Although most substances are unspecific and not suitable for therapeutic purposes, some are very specific and therefore important for the understanding of transcription and translation or for therapy. Basically, one can distinguish whether an agent interferes with transcription or with translation.

## A. Insertion of actinomycin D between a GC base pair

Actinomycin D is a complex polypeptide produced by a species of streptomyces bacteria. It consists of a phenoxazone ring with two symmetrical side chains (1). It acts by becoming intercalated between two neighboring GC base pairs in double-stranded DNA. Viewed from the side (2), the inserted actinomycin D molecule is seen very distinctly within the DNA double helix. In the view from above (3) the actinomycin D molecule forms a narrow layer within the DNA double helix, bound by the two neighboring GC base pairs. The degree of inhibition by actinomycin D varies greatly. High concentrations of actinomycin D block replication, whereas low concentrations suffice to inhibit transcription.

## B. Puromycin imitates an aminoacyl tRNA

Puromycin, a polypeptide from *Streptomyces alboniger,* blocks polypeptide synthesis in the ribosomes of prokaryotes and eukaryotes. Its action is based on the structural similarity with an aminoacyl tRNA. An aminoacyl tRNA is a tRNA molecule with an amino acid attached to its 3′ end. Normally a peptide bond is formed by peptidyltransferase between the amino group of the incoming aminoacyl tRNA at the A (aminoacyl) position and the carboxyl group of the peptidyl tRNA at the P (peptidyl) position. The structure of puromycin resembles that of aminoacyl tRNA, but lacking an interaction with the codon it cannot be attached to the A position in the ribosome. The resulting polypeptidyl–puromycin adduct is removed from the ribosome, and the protein synthesis ends prematurely.

(Figures A and B from Singer and Berg, 1991).

## C. Inhibitors of protein synthesis

Numerous naturally occurring and artificially produced substances inhibit protein synthesis by inhibiting transcription or certain phases of translation. Some have clinical significance as antibiotics; others are toxicologically significant. An example for the specificity of some inhibitors is α-amanitin, a dicyclic octapeptide of the fungus *Amanita phalloides.* In very low concentrations, it binds to RNA polymerase II and thereby blocks the formation of precursor mRNA in eukaryotes. In contrast, RNA polymerase I is insensitive to this toxin, and polymerase III binds to it only in higher concentrations. (Data after Singer and Berg, 1991).

## References

Singer, M., Berg, P.: Genes & Genomes. Blackwell Scientific, Oxford, 1991.

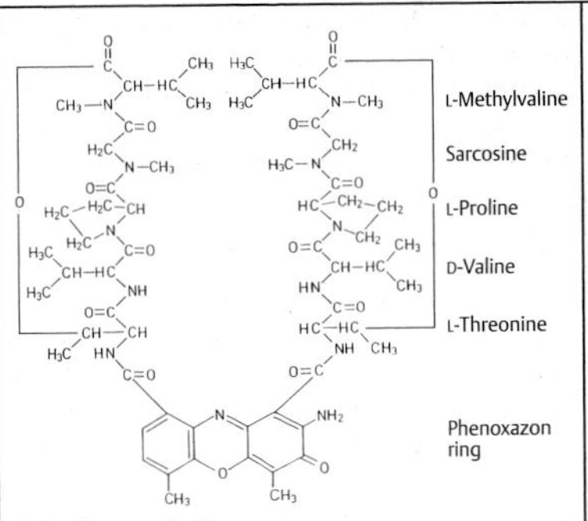


**1.** Actinomycin D

Ribosome

Peptidyl-puromycin

Puromycin at the A site

Peptidyl-tRNA at the P site

**B. Puromycin imitates an aminoacyl-tRNA**

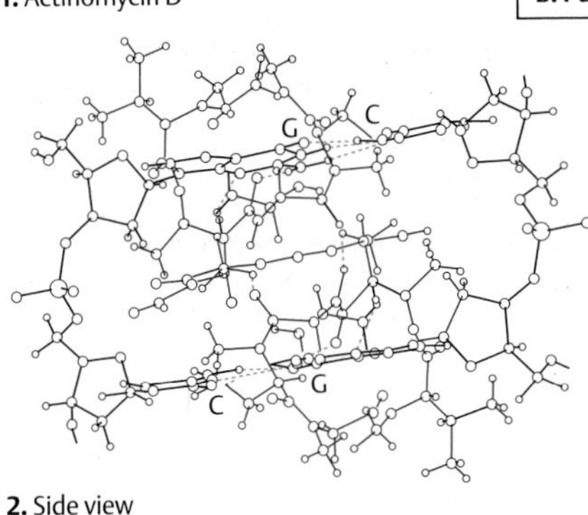


**2.** Side view

G C C G

**3.** View from above

**A. Intercalation of actinomycin D between a GC base pair**

| In Prokaryotes (Examples): | |
|---|---|
| Actinomycin | binding between adjacent G - C bases in DNA (lintercalation) |
| Chloramphenicol | inhibits peptidyltransferase of the 70S ribosome |
| Erythromycin | binds to the 50S particle and arrests synthesis of the 70S ribosome |
| Neomycin | binds to the 30S ribosomal subunit and inhibits binding of a tRNA |
| Puromycin | premature chain termination |
| Rifamycin | inhibits RNA synthesis by binding to the β subunit of the RNA polymerase holoenzyme |
| Streptomycin | as erythromycin |
| Tetracycline | inhibits binding of tRNA to the 30S ribosomal subunit |

| In Eukaryotes (Examples): | |
|---|---|
| α-Amanitin | inhibits polymerase II |
| Chloramphenicol | inhibits peptidyltransferase of mitochondrial ribosomes |
| Cycloheximide | inhibits peptidyltransferase |
| Diphtheria toxin | inhibits initiation factor 2 and translocation |

**C. Inhibitors of protein synthesis**

# DNA Methylation

Methylation of cytosine residues in DNA plays an important role in gene regulation. DNA methylation is required for normal embryonic development. Genomic imprinting, X chromosome inactivation, chromatin modification, and silencing of endogenous retroviruses all depend on establishing and maintaining proper methylation patterns. DNA methylation is gene-specific and occurs genome-wide. Two types of methyltransferase can be distinguished by their basic functions: maintenance methylation and *de novo* methylation.

## A. Maintenance methylation

This type of methylation is responsible for adding methyl groups to the newly synthesized DNA strand after replication and cell division. The methylated sites in the parental DNA (1) after replication (2) serve as template for correct methylation of the two new strands. This ensures that the previous methylation pattern is correctly maintained (3). It results in both daughter strands being methylated at the same sites as the parental DNA. The enzyme responsible for this is Dnmt1 (DNA methylase 1). Its function is essential. Mice deficient in this enzyme die as a result of genome-wide demethylation.

## B. Recognition of a methylated DNA segment

Certain restriction enzymes do not cleave DNA when their recognition sequence is methylated (1). The enzyme *Hpa*II cleaves DNA only when its recognition sequence 5′-CCGG-3′ is not methylated (2). *Msp*I recognizes the same 5′-CCGG-3′ sequence irrespective of methylation and cleaves DNA at this site every time. This difference in cleavage pattern results in DNA fragments of different sizes serves to distinguish the methylation pattern of the DNA.

## C. DNA methylation *de novo*

This is the second type of DNA methylation. Here methyl groups are added at new positions of both strands of DNA, not just in the hemimethylated strand as in maintenance methylation shown in A. Two genes for different methyltransferases with overlapping functions in global remethylation have been identified recently: *Dnmt3a* and *Dnmt3b*. Unmethylated DNA (1) is methylated by their enzymes (2) in a site-specific and tissue-specific manner (3). Targeted homozygous disruption of the mouse *Dnmt3a* and *Dnmt3b* genes results in severe developmental defects. Double homozygous mutants die before day 11.5 of the 21-day embryonic development.

## D. Human *DNMT3B* gene

Mutations in the human gene *DNMT3B* encoding type 3B *de novo* methyltransferase cause a distinctive disease called ICF syndrome (*i*mmunodeficiency, *c*entromeric chromosomal instability, and *f*acial anomalies, McKusick catalog no. 242860). The centromeres of chromosomes 1, 9 and 16, where satellite DNA types 2 and 3 are located, are unstable. Classical satellite DNA is grossly undermethylated in all tissues. The human gene (1) consists of 23 exons spanning 47 kb of genomic DNA. Six exons are subject to alternative splicing. The protein (2) consists of 845 amino acids with five DNA methyltransferase motifs (I, IV, V, IX, X) in the C-terminal region. The arrows point to six different mutations. The mutation at position 809 (3), a change of A to G in codon 809, i.e., GAC (Asp) to GGC (Gly), leads to the replacement of asparagine (Asn) by glycine (Gly). Both parents are heterozygous for this mutation.
(Figure adapted form Xu et al., 1999).

## References


Bird, A: DNA methylation de novo. Science **286**:2287 – 2288, 1999.

Hansen, R. S., et al.: *DNMT3B* DNA methyltransferase gene is mutated in the ICF immunodeficiency syndrome. Proc. Nat. Acad. Sci. **96**:14412 – 14417, 1999.

Okano, M., et al.: DNA Methyltransferases Dnmt3a and Dnmt3b are essential for de novo methylation and mammalian development. Cell **99**:247 – 257, 1999.

Reik, W., Kelsey, G., Walter, J.: Dissecting *de novo* methylation. Nature Genet. **23**: 380 – 382, 1999.

Robertson, K.D., Wolffe, A.P.: DNA methylation in health and disease. Nature Reviews **1**:11 – 19, 2000.

Xu, G., Bestor, T., et al.: Chromosome instability and immunodeficiency syndrome caused by mutations in a DNA methyltransferase gene. Nature **402**:187 – 191, 1999.

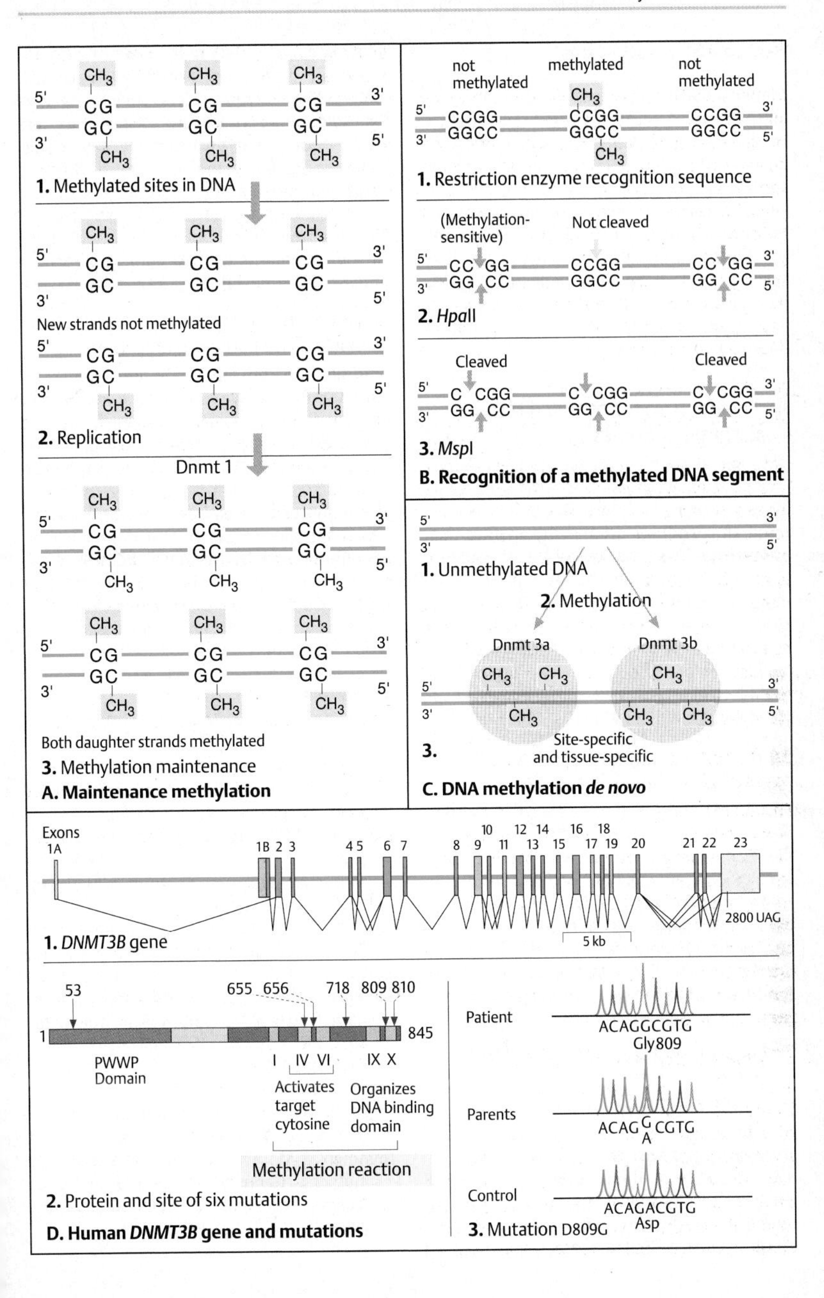
CH3
5'
CG
GC
3'
1. Methylated sites in DNA
New strands not methylated
2. Replication
Dnmt 1
Both daughter strands methylated
3. Methylation maintenance
A. Maintenance methylation
not methylated
methylated
CCGG
GGCC
1. Restriction enzyme recognition sequence
(Methylation-sensitive)
Not cleaved
2. HpaII
Cleaved
3. MspI
B. Recognition of a methylated DNA segment
1. Unmethylated DNA
2. Methylation
Dnmt 3a
Dnmt 3b
3.
Site-specific and tissue-specific
C. DNA methylation de novo
Exons
1A
1B 2 3
4 5
6 7
8 9 10 11 12 13 14 15 16 17 18 19 20
21 22 23
2800 UAG
5 kb
1. DNMT3B gene
53
655 656
718 809 810
1
845
PWWP Domain
I IV VI
IX X
Activates target cytosine
Organizes DNA binding domain
Methylation reaction
2. Protein and site of six mutations
Patient
ACAGGCGTG
Gly809
Parents
ACAG G/A CGTG
Control
ACAGACGTG
Asp
3. Mutation D809G
D. Human DNMT3B gene and mutations

## Genomic Imprinting

Genetic contributions from both the maternal and the paternal sets of chromosomes are necessary for a mammalian zygote to develop normally. The reason lies in a parent-of-origin-specific expression of certain genes. The genome contains defined regions where only the maternal gene copy is expressed, and not the paternal copy, and vice versa. This allele-specific gene expression, depending on the parental origin, results from so-called genomic imprinting. Genomic imprinting has important implications for human genetic disease (see p. 398).

### A. The importance of two different parental genomes

Different developmental results are observed when the female pronucleus is removed from a mouse zygote (1) before the pronuclei have fused and is replaced either by another male pronucleus (2) or by a control (i.e., the same as removed) (3), or when the male pronucleus is removed and replaced by a female pronucleus (4) or a control. If the female pronucleus is replaced by a male pronucleus (2), the zygote first appears normal. However, after implantation should ensue, nearly all androgenotes fail to complete preimplantation (2). Those few that, rarely, reach postimplantation develop completely abnormally. Such embryos do not develop beyond the 12-somite stage.

When a male pronucleus is replaced by a female pronucleus (4), a gynogenetic zygote, which differs markedly from the androgenote. Here, about 85% of gynogenotes reach normal preimplantation development. But although the embryo at first develops fairly well, the extra-embryonic membranes are absent or underdeveloped, and the gynogenote dies before reaching the 40-somite stage. (Figure redrawn from Sapienza and Hall, 1995).

### B. Human embryonic development depends on the presence of a maternal and a paternal genome

A naturally occurring human androgenetic zygote is a hydatidiform mole (1). This is an abnormal placental formation containing two sets of paternal chromosomes and none from the mother. No embryo develops although implantation takes place. The placental tissues develop many cysts (2). When only maternal chromosomes are present, an ovarian teratoma with many different types of fetal tissues develops (3). No placental tissue is present in this naturally occurring gynogenetic zygote. In triploidy, a relatively frequent global chromosomal lethal human disorder (see p. 402), extreme hypoplasia of the placenta and fetus is observed when the additional chromosomal set is of maternal origin (4). (Photographs kindly provided by Professor Helga Rehder, Marburg.)

### C. Genomic imprinting is established in early embryonic development

The imprint pattern present in somatic cells (1), with one allele active only either, the maternal or the paternal, propagated through all mitotic divisions. However, in primordial germ cells the imprint is erased (2). During gametogenesis the imprint is reset (3). In the male germline all gametes receive the paternal imprint and in the female germline all gametes receive the maternal imprint. After fertilization, the correct imprint pattern is present in the zygote (4). It is maintained through all subsequent cell divisions under the control of a regional imprinting center (see p. 398).

### References

Barlow, D.P.: Gametic imprinting in mammals, Science **270**:1610–1613, 1995.

Horsthemke, B.: Genomisches Imprinting beim Menschen: Grundlagen und klinische Relevanz. Biospektrum **4**:23–26, 1998.

Morrison, I.M., Reeve, A.E.: Catalogue of imprinted genes and parent-of-origin effects in humans and animals. Hum. Mol. Genet. **7**:1599–1609, 1998.

Reik, W., Surani, A., eds.: Genomic Imprinting. IRL Press at Oxford University Press, Oxford, 1997.

Sapienza, C., Hall, J.G.: Genetic imprinting in human disease. pp. 437–458. In: The Metabolic and Molecular Bases of Inherited Disease. 7th ed. C.R. Scriver, et al., eds. McGraw-Hill, New York, 1995.

Surani, A.: Imprinting and the initiation of gene silencing. Cell **93**:309–312, 1998.

Tilghman, S.M.: The sins of the fathers and mothers: Genomic imprinting in mammalian development. Cell **96**:185–193, 1999.

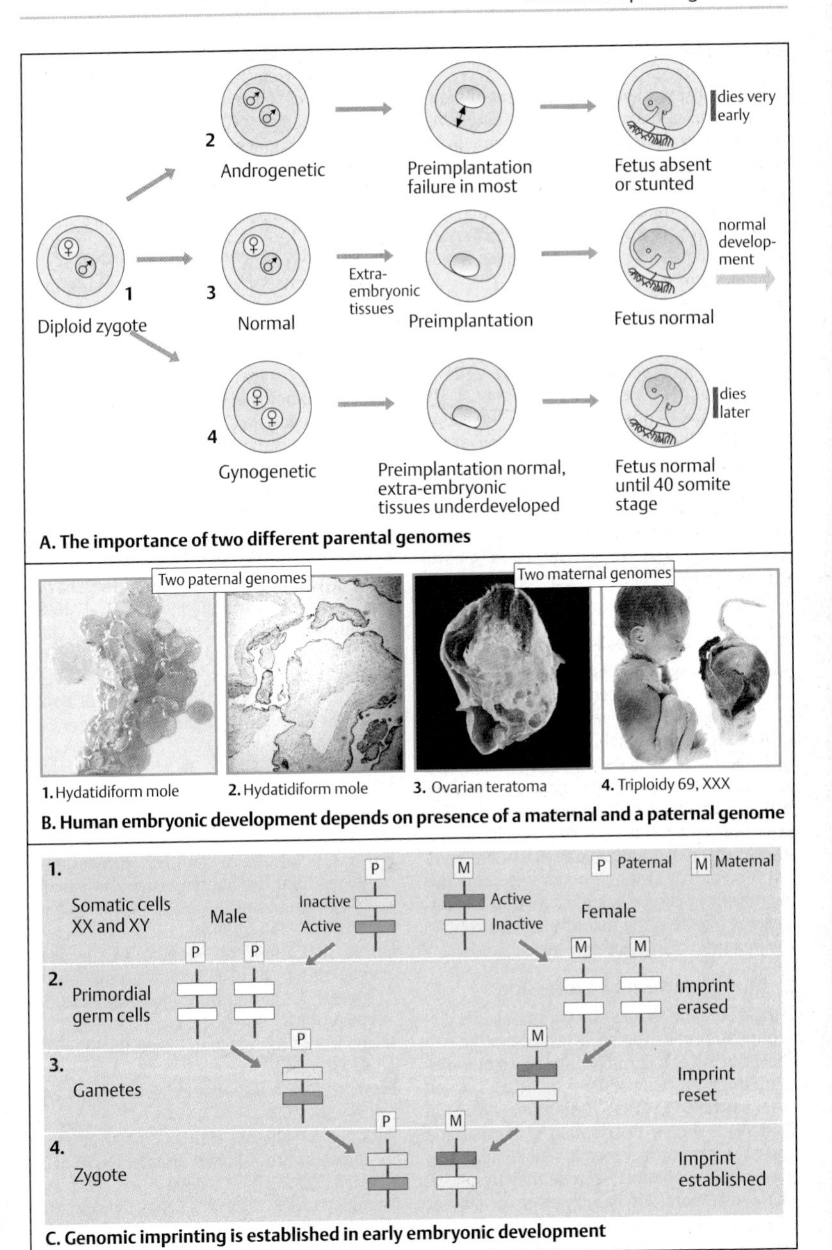


**A. The importance of two different parental genomes**

1. Hydatidiform mole 2. Hydatidiform mole 3. Ovarian teratoma 4. Triploidy 69, XXX

**B. Human embryonic development depends on presence of a maternal and a paternal genome**

**C. Genomic imprinting is established in early embryonic development**

# X-Chromosome Inactivation

During early embryonic development of mammalian females, one of the two X chromosomes becomes inactivated. The inactivation is induced by a gene (*XIST*, X-inactivation-specific transcript) on the proximal long arm of the X chromosome, transcribed from the inactive X. X inactivation is a mechanism to balance X-chromosomal gene expression between female and male cells.

## A. X chromatin

In 1949, Barr und Bertram observed a stainable appendage in the nucleus of nerve cells of female (1 and 3) but not of male cats (2). The authors named this structure as "sex chromatin". Similar structures were found drumsticks in peripheral blood leukocytes (4) and small peripheral bodies in the nuclei of fibroblasts and oral mucosa cells (5) in humans. Each of these structures represents one of the two X chromosomes and are referred to as X chromatin. (Figures 1–3 from Barr and Bertram, 1949).

## B. Scheme of X inactivation

Random inactivation of most of the genes of one of the two X chromosomes in female cells occurs early in embryogenesis, at about day 21 in humans. In a given cell, it involves the X chromosome of either maternal or paternal origin. The inactivation pattern is normally irreversible and stably transmitted to all daughter cells. The expected distribution is usually 1 : 1. In rare instances this may be skewed toward a preferential type of inactivation. In extreme cases this may result in clinical manifestation in a heterozygous female if the majority of cells contain the mutation on the active X chromosome.

## C. Mosaic pattern of expression

Female somatic tissues show a mosaiclike distribution of cells expressing just one of the two alleles. In mice, X-chromosomal coat-color mutants show a mosaic of light- and dark-colored coat patches (1, after Thompson, 1965). In humans, a similar distribution of normal and absent sweat pores is seen in female heterozygotes for hypohidrotic ectodermal dysplasia. The sweat pores of hemizygotes are absent (hypohidrosis). Fingerprints of female heterozygotes show areas with normal sweat pores (black points) and areas with absent sweat pores (2a and 2b, from Passarge and Fries, 1973). In cell cultures from female heterozygotes for X-chromosomal HGPRT deficiency (hypoxanthine–guaninephosphoribosyltransferase) the colonies are either $HGPRT^-$ or $HGPRT^+$ (3). (From Migeon, 1971, with kind permission of the author and publisher).

## D. Exceptions from X-inactivation

Transcriptional silencing of X-chromosomal genes in mammalian female cells is not complete. Some genes escape inactivation and are expressed from both the active and the inactive X chromosome. The majority of these genes have a homologue on the Y chromosome, reflecting a common evolutionary origin. Panel D shows genes that are expressed on the inactive human X chromosome. Most are located at the ends of the X chromosomes. (Figure adapted from Brown et al., 1997).

## E. X-inactivation profile

An analysis of 224 X-linked transcripts showed that 34 escape inactivation (Carrel et al., 1999). Of these, 31 map to the short arm of the X chromosome. The expressed genes are open circles, the inactivated genes filled circles. Several genes in the pseudoautosomal region of Xp are shown as diamonds. (Figure adapted from Carrel et al., 1999).

## References

Barr, M.L., Bertram, L.F.: Nature **163**:676–677, 1949.

Brown, C.J., Carrel, L, Willard, H.F.: Expression of genes from the human active and inactive X-Chromosomes. Am. J. Hum. Genet. **60**:1333–1343, 1997.

Carrel, L., Cottle, A.A., Goglin, K.C.: A first-generation X inactivation profile of the human X Chromosome. Proc. Nat. Acad. Sci. **96**:14440–14444, 1999.

Migeon, B.R.: Am. J. Hum. Genet. **23**:199–200, 1971.

Passarge, E., Fries, E.: Nature New Biology **245**: 58–59, 1973.

Puck J.M., Willard, H.F.: X inactivation in females with X-linked disease. New Engl. J. Med. **338**:325–327, 1998.

Thompson, M.W.: Canad. J. Genet. Cytol. **7**:202–213, 1965.

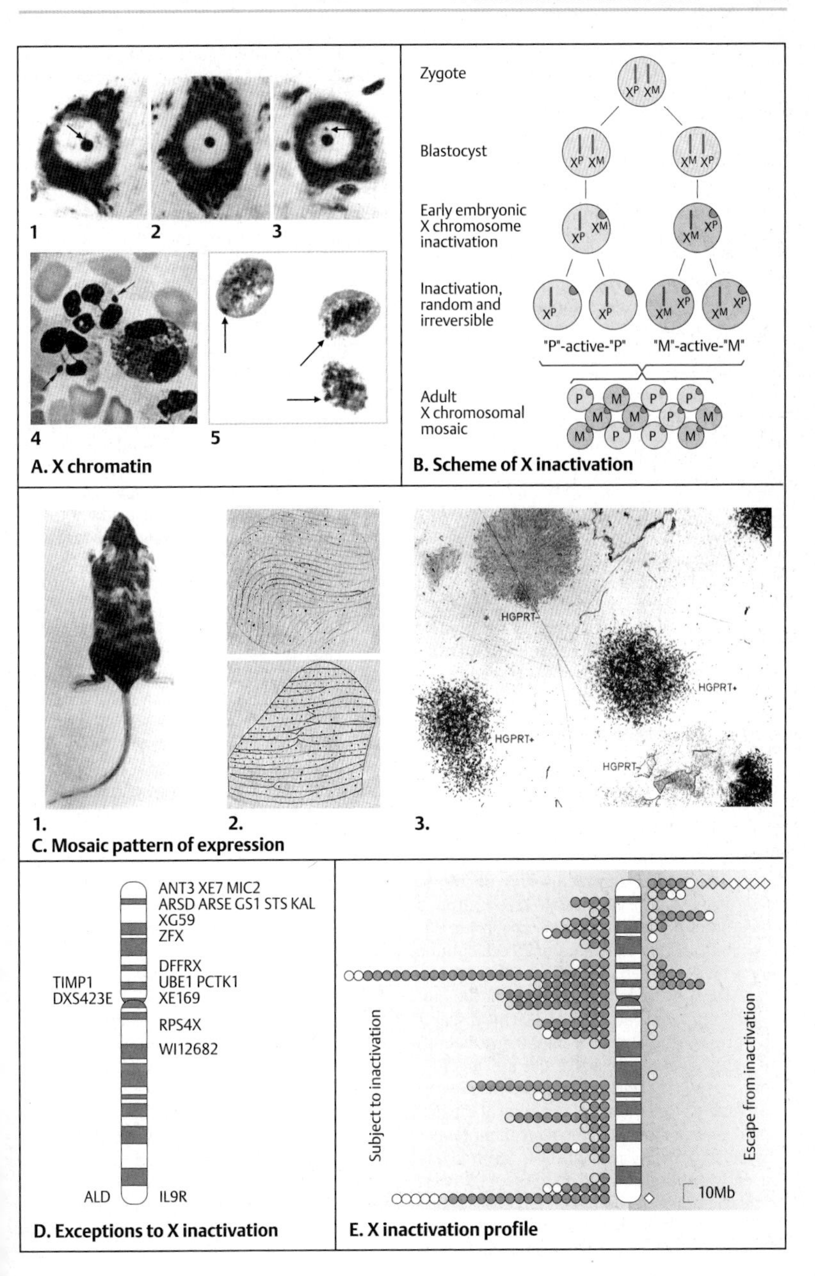


A. X chromatin

B. Scheme of X inactivation

C. Mosaic pattern of expression

D. Exceptions to X inactivation

E. X inactivation profile

## Targeted Gene Disruption in Transgenic Mice

The specific experimental inactivation of a gene affords an opportunity to study its normal function by comparing it with that of the inactive state. This will yield information about the role of a particular gene in development or other functions that would not be available otherwise. In the approach described here, a normal gene is replaced by a mutant allele by disrupting the normal gene (targeted gene disruption). The effects can be studied in different embryonic stages of mouse development and after birth. Subsequently the results can be compared to the effects of mutations in homologous human genes as seen in human genetic diseases. The method requires the use of mouse embryonic stem (ES) cells. ES cells are pluripotential, i.e., they are capable of giving rise to different kinds of cells but not to an entire organism. These cells can be grown in culture through many generations and yet retain the potential to be integrated into a mouse blastocyst. Here they participate in embryonic development allowing mice to produce that are homozygous for a mutation introduced into a specific gene (knockout mice).

### A. Transgenic mouse derived from ES cells with a disrupted gene

Embryonic stem cells (ES) from a mouse blastocyst are isolated (1) after 3.5 days of gestation (of a total of 19.5 days) and transferred to a cell culture (2). Here they will grow on a layer of irradiated cells that are themselves unable to grow (feeder layer). Target DNA (see B) from a mouse homozygous for a marker coat color, e.g., dominant black, is added to the ES cell culture. Very few cells or perhaps just one may take up target DNA by recombination with the homologous gene in the ES cell (3). This disrupts the normal gene. These recombinant ES cells can be grown in a selective culture medium (nonrecombinant cells will not grow, see B). Recombinant ES cells containing a copy of the disrupted gene are injected into a recipient mouse blastocyst (4). These cells will be integrated into the early mouse embryo (5). The blastocyst partly containing recombinant ES cells is transferred into a pseudopregnant mouse (6). After birth (6), mice derived from normal and recombinant cells (chimeric mice) can easily be identified by spots of black coat color (7). When adult chimeric mice are mated to normal mice homozygous for another coat color allele (e.g., white, 8), the birth of black progeny indicates that the targeted gene is present in the germ line (9). The mating of such heterozygous mice (not shown) will produce mice that are homozygous for the disrupted gene (knockout mice). (Figure redrawn from Alberts et al., 1998).

### B. Double selection for ES cells containing the disrupted gene

The isolation of mouse ES cells with a gene-targeted disruption requires positive and a negative selection. A bacterial gene conferring resistance to neomycin (*neo*$^R$) is added (2) into DNA cloned from the target gene (1). DNA containing the thymidine kinase gene (*tk*$^+$) from herpes simplex virus is then also added to the vector outside the region of homology. These two selectable modifications (*neo*$^R$ *and tk*$^+$) are part of the replacement vector (3). The vector is introduced into ES cells, which are grown in culture. All ES cells that do not take up the vector (the majority) are sensitive to neomycin and will not grow in a medium containing the antibiotic (positive selection, not shown). Cells that take up the vector (about 1%) do so either by nonhomologous insertion at random sites (4) or by homologous recombination at the target site (5). These cells can be distinguished when grown in a selective medium containing gancyclovir, a nucleotide analogue toxic to cells containing *tk*$^+$. Only cells containing the gene-targeted insertion (through homologous recombination) will grow because they do not contain *tk*$^+$ (negative selection). (Figure redrawn from Lodish et al., 2000).

### References

Alberts, B. et al.: Essential Cell Biology. An Introduction to the Molecular Biology of the Cell. Garland Publishing Co., New York, 1998.

Capecchi, M.R.: Targeted gene replacement. Scient. Amer., March 1994, pp. 52–59.

Lodish, H. et al.: Molecular Cell Biology. 4$^{th}$ ed. Scientific American Books, F.H. Freeman & Co., New York, 2000.

Majzoub, J.A., Muglia, L.J.: Knockout mice. Molecular Medicine. New Eng. J. Med. **334**:904–907, 1996.

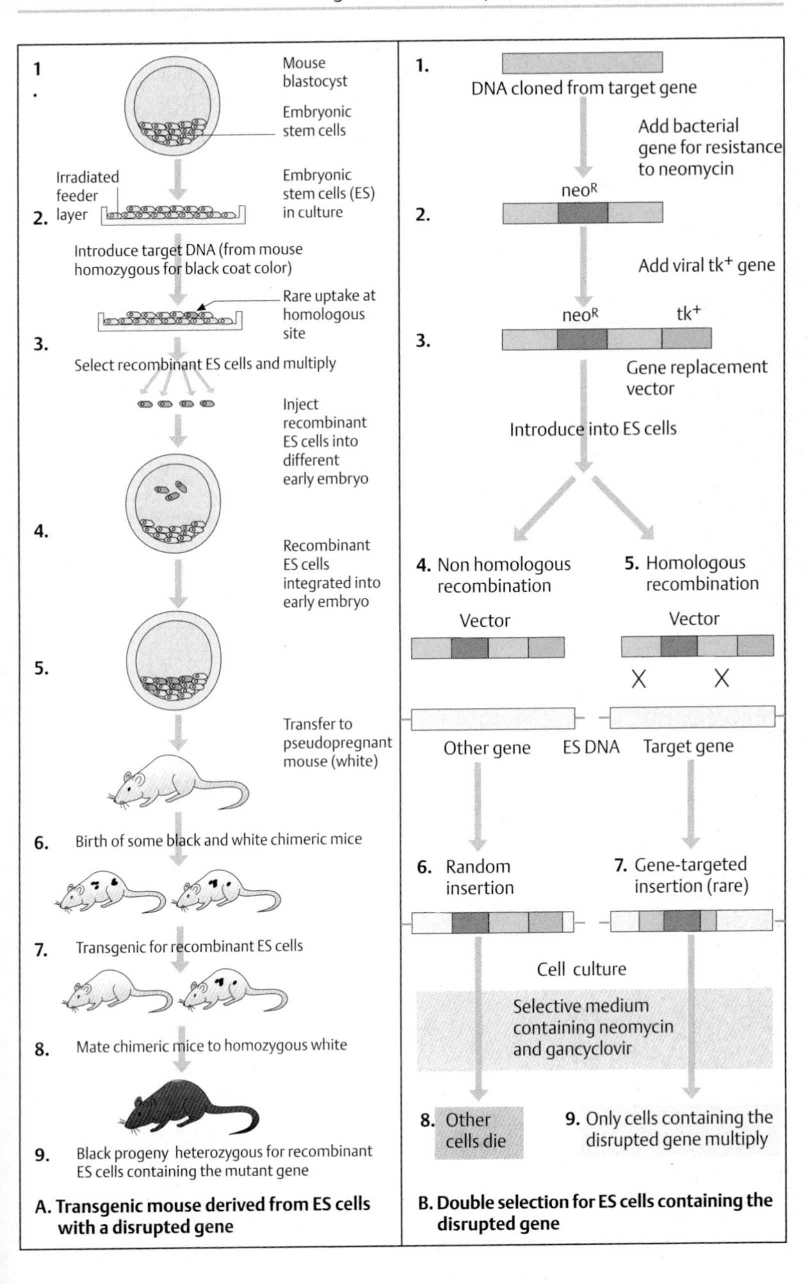


**A. Transgenic mouse derived from ES cells with a disrupted gene**

**B. Double selection for ES cells containing the disrupted gene**

# Genomics

# Genomics, the Study of the Organization of Genomes

A genome contains all biological information required for life and/or reproduction.
The term genomics for the study of genomes was introduced in 1987 by V. A. McKusick and F. H. Ruddle at the suggestion of T. H. Roderick of the Jackson Laboratory, Bar Harbor, Maine, USA. The term *genomics* extends beyond *genetics.* While the latter mainly deals with heredity and its mechanism and consequences, the term genomics encompasses many aspects relating to molecular and cell biology: the different types of genomic maps; nucleic acid sequencing; assembly, storage, and management of data; gene identification; functional analysis (functional genomics), evolution of genomes; and other interdisciplinary areas relating to the wide variety of genomes in different organisms. A eukaryotic genome, contained in the chromosomes, is many times larger than a prokaryote genome. A prokaryotic genome consists of a circular chromosome with compactly arranged genes.
Important insights about the functions, evolutionary relationships, antibiotic resistance, and other metabolic aspects required for the development of new strategies for therapy are gained from sequencing the genome of microorganisms (reviewed by Fraser et al., 2000.)

## A. The genome of a small bacteriophage

The genome of a bacteriophage usually consists of double-stranded DNA, although some phage genomes consist of single-stranded DNA or of RNA. The size of the genome of phages ranges from 1.6 kb to over 150 kb, representing anywhere from a few to over 200 genes. One of the smallest phages is ΦX174. F. Sanger and coworkers sequenced the genome completely and found that several of the ten genes of ΦX174 overlap. (Figure adapted from Watson et al., 1987, and Sanger et al., 1977).

## B. Overlapping genes in ΦX174

The genes A and B, B and C, and D and E partially overlap. In these overlapping regions of the genome, a different reading frame is used differently by different genes. Gene E begins with the start codon ATG, the first two nucleotides of which (AT) are the last two nucleotides of codon TAT for tyrosine in gene D. Similarly, the stop codon TGA for the E gene is part of codons GTG (valine) and ATG (methionine) of the D gene.

## C. Genome of *Escherichia coli*

*E. coli* is an important microorganism. It colonizes the lower intestines of mammals including man in a symbiotic relationship. Pathogenic strains of *E. coli* cause gastrointestinal, urinary, pulmonary, and nervous system infections in humans. The *E. coli* genome has 4639221 bp. A total of 2657 protein-coding genes with known function (62% of all genes) and 1632 genes (38%) without known function have been identified. The simplified figure shows eight genes (A–H), the origin of replication (ORI), and the genes for DNA polymerase and methionine. Four operons are shown: the operons for lactose consisting of three genes, for galactose with four genes, for tryptophan with five genes, and for histidine with nine genes. About a fourth of all genes in *E. coli* are organized into 75 different operons. Most genes of the *E. coli* genome are present as a single copy; only the genes for ribosomal RNA (rRNA) are present in multiple copies. As a result, the bacteria can double their protein content every 20 minutes during cell division.
(Figure adapted from Watson et al., 1987).

## References


Adams, M.D., et al.: The genome sequence of *Drosophila melanogaster*. Science **287**:285–2195, 2000.

Brent, R.: Genomic biology. Cell **100**:169–183, 2000.

Brown, T.A.: Genomes. Bios Scientific Publ., Oxford, 1999.

Fraser, C.M., Eisen, J.A., Sulzberg, S.L.: Microbial genome sequencing. Nature **406**:799–803, 2000.

Lander, E.S., Weinberg, R.A.: Genomics: Journey to the center of biology. Science **287**:1777–1782, 2000.

Sanger, F., et al.: Complete sequence of the bacteriophage Φ174. Nature **265**:687–695, 1977.

Watson, J.D., et al., eds.: Molecular Biology of the Gene. Vol. I, 4th ed. Benjamin/Cummings Publishing Co., Menlo Park, California, 1987.

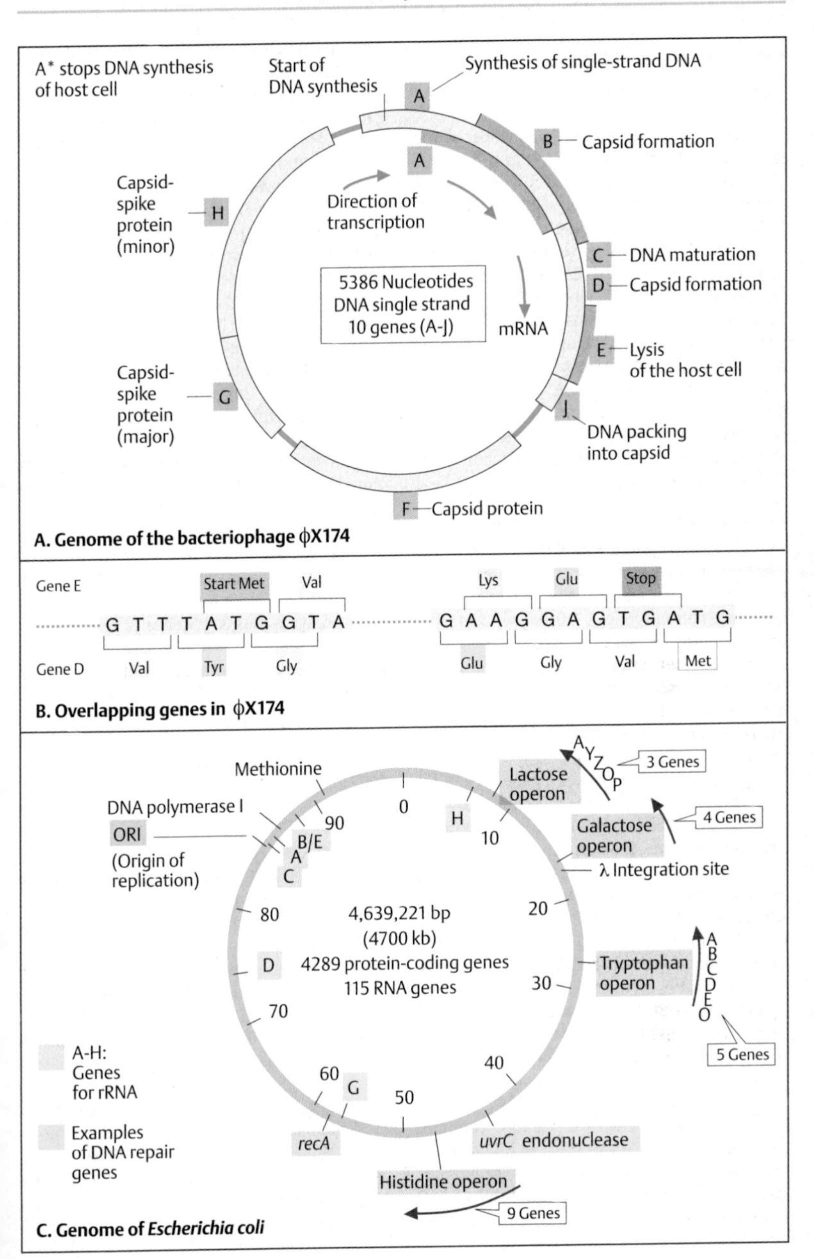


**A. Genome of the bacteriophage ϕX174**

**B. Overlapping genes in ϕX174**

**C. Genome of *Escherichia coli***

# *The Complete Sequence of the Escherichia coli Genome*

The report of the complete sequence of the genome of the *E. coli* K-12 strain in 1997 (Blattner et al., 1997) with a full map of the 4289 protein-coding genes of the 4 639 221-base pair genome is presented here as an example of one of many sequenced microorganism genomes (see Entrez Genomes at *http://www.ncbi.nlm.nih.gov/*, and search for sequenced microorganisms). General insights into the overall structure of prokaryote genomes can be derived as outlined below.

Genome sequences can be correlated with other information to identify regulatory sites, operons, mobile genetic elements, repetitive sequences, deduced functions, and homology to other organisms.

## A. Overall structure and comparison with other genome sequences

The figure shows a small section of about 80 kb of the entire genome from the original publication. Base pair numbers 3 310 000 to ~3 345 000 are shown in the first row (top), and 3 339 000 to 4 025 000 in the second row. The top double line shows color-coded genes of *E. coli* encoding a protein on either of the two strands of the DNA double helix. Six other completed genomes are shown for comparison.

The average distance between genes is 118 bp. The protein-coding genes (87.8% of the genome) can be assigned to 22 functional groups (see gene function color code at the bottom of the figure). Among these are 45 genes with recognized regulatory functions (1.05% of the total); 243 genes for energy metabolism (5.67%); 115 genes for DNA replication, recombination, and repair (2.68%); 255 genes for transcription, RNA synthesis, and metabolism (5.94%); 182 genes for translation (4.24%); 131 genes for amino acid biosynthesis and metabolism (3.06%); and 58 genes for nucleotide biosynthesis and metabolism (1.35%).

Homologies between *E. coli* proteins and those of other organisms with sequenced genomes were detected in 1703 proteins of *H. influenzae* and 468 proteins of *Mycoplasma genitalium*. Yeast (*S. cerevisiae*) has 5885 proteins that match proteins in *E. coli*. 60% of *E. coli* proteins have no match in the other prokaryotes with completed genomes as of 1997.

How many genes are required for a microorganism? The smallest known cellular genome is that of *Mycoplasma genitalium*, with only 480 protein-coding genes and 37 RNA genes in 580 kb of DNA. However, not all of these genes are essential. Hutchison et al. (1999) determined by global transposon mutagenesis that 1354 of 2209 insertions into genes were not lethal to the organism. The results suggested that 265–350 of the 480 genes are essential under laboratory conditions. The limited capacity for metabolism in *M. genitalium* is compensated for by a greater dependence on the transport of molecules from the extracellular environment into the cell, mainly by ABC transporters. An ABC transporter is a heterotrimeric transport system made up of a specific ligand-binding subunit, a permease, and an ATP-binding protein.

The total of protein families required for an organism is called a *proteome*. Its study is called *proteomics*. Since many genes are required to encode the different proteins of a given metabolic or signal pathway, genes of related functions are grouped into families. Their number is considerably smaller than the total number of genes. The CAI (Codon Adaption Index) reflects the preferred codon usage of an organism. (The figure is adapted from a small part of the complete map published by Blattner et al., 1997).

## References


Blattner, F.R., et al.: The complete genome sequence of Escherichia coli K-12. Science **277**:1453–1474, 1997.

Brown, T.A.: Genomes. Bios Scientific Publ., Oxford, 1999.

Fraser, C.M., Eisen, J.A., Sulzberg, S.L.: Microbial genome sequencing. Nature **406**:799–803, 2000.

Hutchison, C.A., III., et al.: Global transposon mutagenesis and a minimal mycoplasma genome. Science **286**:2165–2169, 1999.

Neidhardt, F.C., et al., eds.: *Escherichia coli* and *Salmonella*. Cellular and Molecular Biology. ASM Press, Washington, D.C., 1996.

Wren, B.W.: Microbial genome analysis: insights into virulence, host adaptation and evolution. Nature Reviews **1**:30–39, 2000.

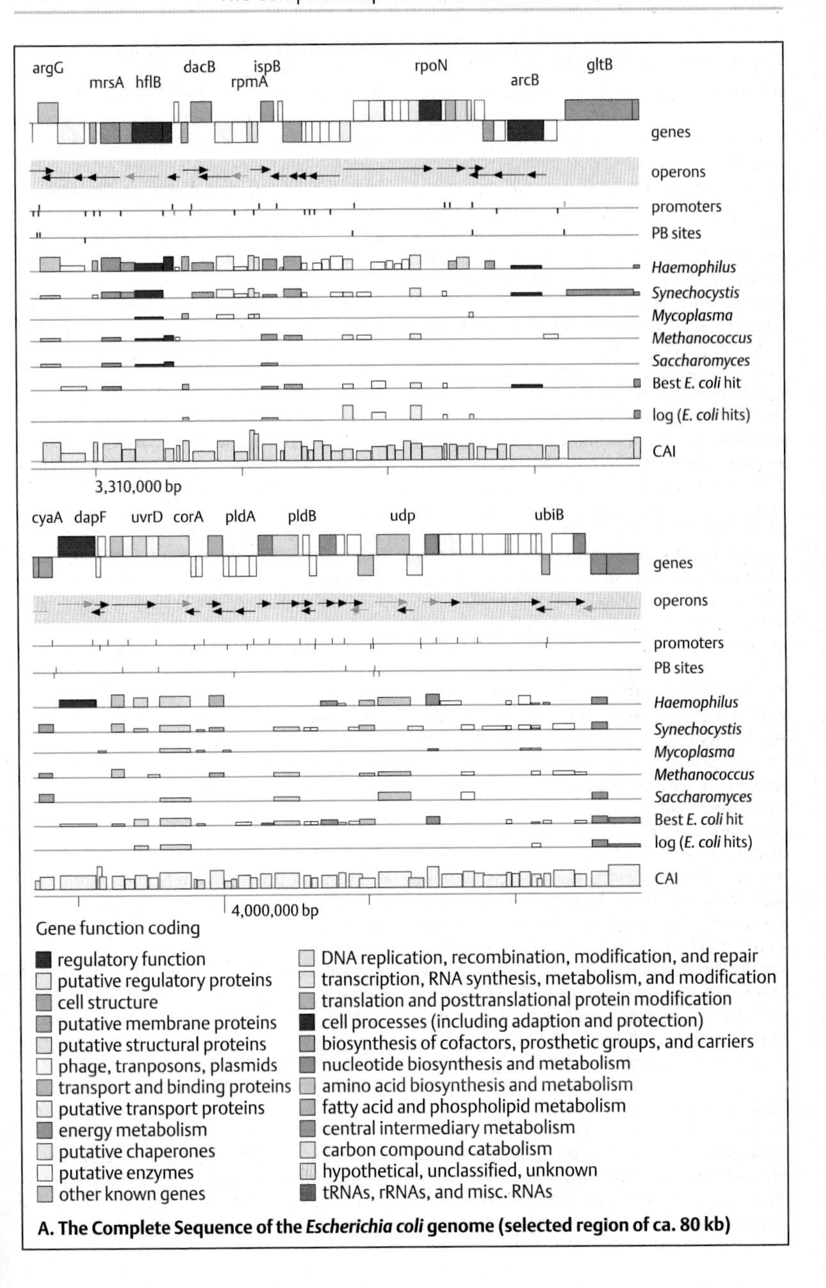


**A. The Complete Sequence of the *Escherichia coli* genome (selected region of ca. 80 kb)**

## Genome of a Plasmid from a Multiresistant Corynebacterium

Plasmids are double-stranded circular DNA molecules in bacteria but separate from the bacterial chromosome. They are self-replicating and occur in a symbiotic or parasitic relationship with the host cell. The number of plasmids per bacterial cell varies from a few to thousands. Their sizes range from a few thousand base pairs to more than 100 kb. Plasmids usually confer a benefit to the host cell, often because they contain genes encoding enzymes that inactivate antibiotics. Drug-resistant plasmids pose a major threat to successful antibiotic therapy. Since many plasmids also contain transfer genes encoding proteins that form a macromolecular tube, or pilus, through which a copy of plasmid DNA can be transferred to other bacteria, antibiotic resistance can spread very rapidly. The following example provides new insights into the origin and evolution of a multiresistant plasmid composed of DNA segments derived from bacteria of very different origins (soil bacteria and plant, animal, and human pathogens).

### A. The multiresistant plasmid pTP10

A large, 51 409 bp plasmid from the multiresistant Gram-positive *Corynebacterium striatum* M82B containing genes encoding proteins renders its host bacteria resistant to 16 antimicrobial agents from six structural classes (Tauch et al., 2000). This is the largest plasmid to have been sequenced to date. It contains DNA segments from a plasmid-encoded erythromycin (Em, shown inside the circular genome diagram) resistance region from the human pathogen *Corynebacterium diphtheriae*, a chromosomal DNA region from *Mycobacterium tuberculosis* containing tetracycline (Tc) and oxacillin resistance, a plasmid-encoded chloramphenicol (Cm) resistance region from the soil bacterium *Corynebacterium glutamicum*, and a plasmid-encoded aminoglycoside resistance to kanamycin (Km), neomycin, lividomycin, paramomycin, and ribostamycin from the fish pathogen *Pasteurella piscicida*. In addition, the plasmid contains five transposons and four insertion sequences (IS*1249*, IS*1513*, *IS1250*, and IS*26*) at eight different sites. Altogether eight genetically distinct DNA segments of different evolutionary origin were identified.

### B. Genetic map of plasmid pTP10

Plasmid pTP10 from the Gram-positive opportunistic human pathogen *Corynebacterium striatum* M82B has 47 open reading frames (ORFs). They can be assigned to eight different DNA segments forming a contiguous array of subdivided stretches in the linear representation shown (I, II, VIIb, III, VIIa, VIII, IVa, Va, VI, Vb, IVb, and VIIc).

Segment I (shown in green) consists of five ORFs comprising the composite resistance transposon Tn*5432*. The insertion sequences IS*1249b* (ORF1) and IS*1249a* (ORF5) flank the erythromycin resistance gene region *ermCX* (ORF3) and *ermLP* (ORF4). An identical copy of IS*1249* occurs in ORF29 (segment VIII). ORF3, the central region of Tn*5432*, encodes a 23 S rRNA methyltransferase preceded by a short leader peptide probably involved in the regulation of erythromycin-inducible translational attenuation. This region is virtually identical to the antibiotic resistance gene region (erythromycin, clindamycin) of plasmid pNG2 from *C. diphtheriae* S601.

Segment II (ORFs 6 – 14), located downstream of Tn*5432*, contains the tetracycline resistance genes *tetA* (ORF6) and *tetB* (ORF7). This segment (ORFs 6 – 14) is very similar to ATP-binding cassette (ABC) transporters identified in a mycobacterium *(M. smegmatis)* chromosome. The tandemly arranged genes *tetA and tetB* also mediate resistance to the β-lactam antibiotic oxacillin, although it is structurally and functionally unrelated to tetracycline. Presumably this results from TetAB protein heterodimerization and subsequent export of the antibiotics out of the bacterial cell. Similarly, the other segments have been delineated.

(Figures adapted from Tauch et al., 2000. I thank Professor Alfred Pühler, University of Bielefeld, for providing the figures and advice).

#### References

Tauch, A., Krieft, S., Kalinowski, J., Pühler, A.: The 51,409-bp R-plasmid pTP10 from the multiresistant clinical isolate *Corynebacterium striatum* M82B is composed of DNA segments initially identified in soil bacteria and in plant, animal, and human pathogens. Mol. Gen. Genet., **263**:1 – 11, 2000.

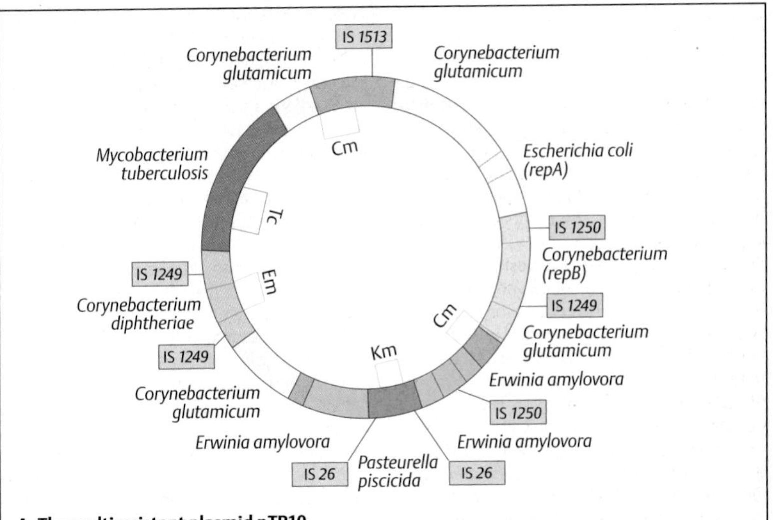


**A. The multiresistant plasmid pTP10**

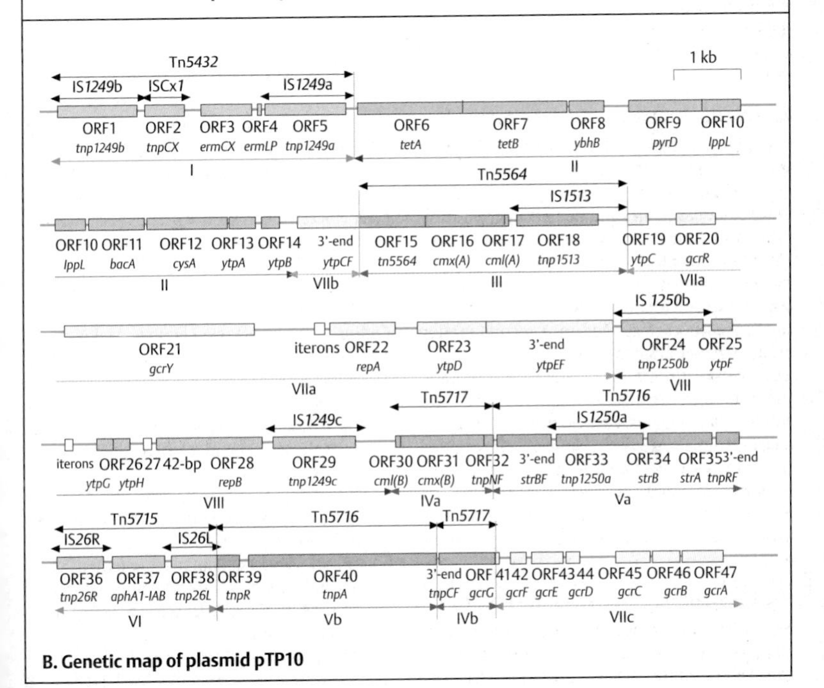


**B. Genetic map of plasmid pTP10**

## Genome Maps

A genome map is a linear representation of genomic landmarks (genes and markers). It refers either to a chromosome (cytogenetic map) or to a stretch of DNA. A map provides knowledge of the position of a particular genomic landmark and its relation to others. Unraveling the human genome in some respect resembles the mapping of new continents five hundred years ago.

A *genetic map* expresses the positions of genes relative to each other without a physical anchor on the chromosome. This is the type of map that was first used by A. H. Sturtevant in 1913 when working with T. M. Morgan (who started studying *Drosophila* in 1910). Here, the distance between markers is determined by the frequency of recombination during meiosis, which is in turn determined by the relative distance between the loci (see p. 116). A *physical map* provides knowledge of the exact position of a gene or marker. Its distance to another locus on the same chromosome is expressed by number of base pairs (bp), a physical equivalent. A variety of methods are employed to arrive at a physical map.

### A. Physical and genetic gene maps

The physical map gives the position of a gene locus and its distance from other genes on the same chromosome in absolute values, expressed in base pairs and related to given positions along the chromosome. The genetic map gives the relative position of gene loci according to the frequency of recombination, expressed as recombination units or centimorgans (cM). One centimorgan corresponds to a recombination frequency of 1%. Since recombination occurs almost twice as often in oocytes as in spermatocytes, the genetic map in females is about 40% longer than in males. Each gene locus has an official designation with a defined abbreviation using the letter D (for DNA), the number of the chromosome, and the number of the marker, preceded by an S for single-copy DNA, e.g., D1S77. (Diagram adapted from Watson et al., 1992).

### B. STS mapping from a clone library

STS mapping plays a major role in genome mapping. An STS (sequence-tagged site) is a short stretch (60–1000 bp) of a unique DNA nucleotide sequence, An STS has a specific location and can be analyzed by PCR (see p. 66). The relevant information, i.e., the sequence of the oligonucleotide primers used for the PCR reaction and other data can be stored electronically and does not depend on biological specimens.

One can start with a clone library containing DNA fragments in unknown order (1). Each end of the chromosomal fragment is characterized by a pattern of restriction sites (see p. 64). The DNA fragments are ordered by determining which ends overlap, then assembling them as a contiguous array of overlapping fragments into a clone contig (2). These are linearly arranged. This establishes a map that shows the location and the physical distance of the landmarks, here A, B, C, etc. (3). Sequence-tagged sites (STSs) are generated from the two ends of the overlapping clones. This involves sequencing 100–300 bp of DNA (4). More than 40000 STSs were characterized in the human genome in 1999 (see GenBank at *http://www.ncbi.nlm.nih.gov*).

### C. EST mapping

ESTs (expressed sequence tags) are short DNA sequences obtained from cDNA clones (complementary DNA, see p. 58). Each EST represents part of a gene. Their location is determined by hybridizing an assembly of different cDNAs (1) to genomic DNA (2). Thus, the locations of defined sequences of expressed genes can be determined (3). These can be mapped to a location on a chromosome to establish an EST map.

### References

Collins, F.S.: Shattuck Lecture—Medical and societal consequences of the human genome project. New Eng. J. Med. **341**:28–37, 1999.

Green, E.D., Cox D.R., Myers R.M.: The human genome project and its impact on the study of human disease, p. 401–436. In: Scriver, C.R., et al., eds. The Metabolic and Molecular Bases of Inherited Disease. 7th ed. McGraw-Hill, New York, 1995.

Sturtevant, A.H.: The linear arrangement of six sex-linked factors in Drosophila, as shown by their mode of association. J. Exp. Zool. **14**:43–59, 1913.

Watson, J.D., et al.: Recombinant DNA. 2nd ed. Scientific American Books, W.H. Freeman & Co., New York, 1992.

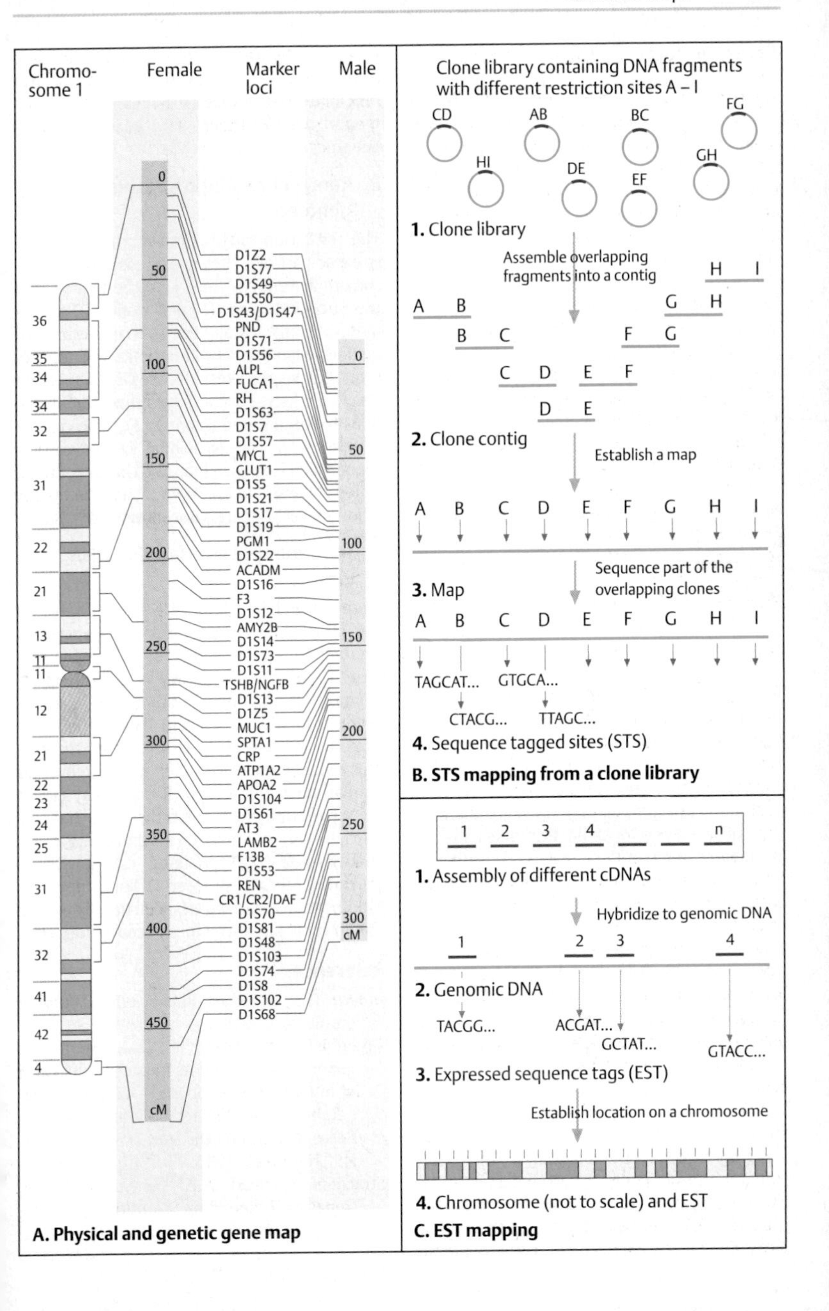

Chromosome 1
Female
Marker loci
Male
D1Z2
D1S77
D1S49
D1S50
D1S43/D1S47
PND
D1S71
D1S56
ALPL
FUCA1
RH
D1S63
D1S7
D1S57
MYCL
GLUT1
D1S5
D1S21
D1S17
D1S19
PGM1
D1S22
ACADM
D1S16
F3
D1S12
AMY2B
D1S14
D1S73
D1S11
TSHB/NGFB
D1S13
D1Z5
MUC1
SPTA1
CRP
ATP1A2
APOA2
D1S104
D1S61
AT3
LAMB2
F13B
D1S53
REN
CR1/CR2/DAF
D1S70
D1S81
D1S48
D1S103
D1S74
D1S8
D1S102
D1S68
cM
Clone library containing DNA fragments with different restriction sites A – I
CD AB BC FG HI DE EF GH
1. Clone library
Assemble overlapping fragments into a contig
2. Clone contig
Establish a map
3. Map
Sequence part of the overlapping clones
TAGCAT... GTGCA...
CTACG... TTAGC...
4. Sequence tagged sites (STS)
1. Assembly of different cDNAs
Hybridize to genomic DNA
2. Genomic DNA
TACGG... ACGAT... GCTAT... GTACC...
3. Expressed sequence tags (EST)
Establish location on a chromosome
4. Chromosome (not to scale) and EST

**A. Physical and genetic gene map**

**B. STS mapping from a clone library**

**C. EST mapping**

# Approach to Genome Analysis

The approach to genome analysis encompasses several goals. Of primary interest is the number, type, and distribution of genes. Knowing all genes and their positions and structures in a eukaryotic genome will provide the basis for understanding their function. The size of a genome needs to be taken into account for a systematic study.

Two basic approaches to sequencing a genome can be distinguished: clone-by-clone sequencing and the so-called shotgun approach. In the former, individual DNA clones of known relation to each other are isolated, arranged in their proper alignment, and sequenced. The shotgun approach breaks the genome into millions of fragments of unknown relation. The individual DNA clones, for which prior knowledge of their precise origin is lacking, are sequenced. Subsequently, they are aligned by high-capacity computers. The two approaches complement each other.

## A. Sizes of genomes and cloning vectors

The sizes of genomes of different organisms vary considerably. In general, genome size reflects the complexity of the organism. A mammalian genome (human and mouse are known best) contains $3 \times 10^9$ base pairs (bp) or 3000 Mb. If each nucleotide pair were represented by a 1-mm-wide letter, the text would be more than 3000 km long or take up more than ten sets of the Encyclopaedia Britannica or 750 megabytes of computer capacity. Thus, finding all genes, mapping their position, and determining their structure and function is an enormous task (see Human Genome Project).

By comparison, the genome of important model organisms such as *Drosophila*, the nematode *C. elegans*, yeast, and bacteria are much smaller. The genomes of some important plants such as maize, rice, and wheat are even larger (5000–17000 Mb) than mammalian genomes.

Since the size of DNA fragments that can be isolated and multiplied in cloning vectors for analysis is relatively small, a huge cloning capacity is necessary for analysis of a large genome. Yeast artificial chromosomes (YAC) can accommodate about 1.4 Mb, bacterial artificial chromosomes (BAC) about 0.5 Mb, whereas bacteriophages and cosmids (a phage cloning vector consisting of the λ COS site inserted into a phage) take up only 25–50 kb of foreign DNA for cloning. Thus, vectors that clone large fragments are desirable for analyzing large genomes.

## B. Range of resolution within the genome

The resolution ranges from a whole chromosome or part of a chromosome isolated from a somatic hybrid cell line (1) to the sequence of the nucleotide pairs (5) and cloned DNA fragments (2). Each fragment is characterized by distinct landmarks (restriction sites or sequence-tagged sites, STS, see Genome Maps). They are aligned according to their contiguous linear orientation in a contig (3), which can be mapped (4). The individual clones can be sequenced (sequence map, 5). This approach is called a clone-by-clone approach in contrast to "shotgun sequencing" (not shown).

## C. Alignment of overlapping DNA clones

From a stretch of genomic DNA to be characterized by schematic segments A–E (1), a series of overlapping clones is derived (2). In the first step a radiolabeled clone is hybridized to genomic DNA (clone 1) from a DNA library. A probe (probe A) is used to find the adjacent clone by hybridizing to a new clone (clone 2). This establishes that clones 1 and 2 overlap. Similar steps follow. DNA farther away can be identified until one reaches the clone that contains the gene of interest. This procedure, called chromosome (DNA) walking, can start from several points and proceed in both directions. Modifications are used to speed up this process and to cover large stretches of contiguous DNA.

## References


Brown, T.A.: Genomes. Bios Scientific Publishers, Oxford, 1999.

Green, E.D., Cox, D.R. Myers, R.M.: The human genome project and its impact on the study of human disease, pp. 401–436. In: Scriver, C.R., et al., eds. The Metabolic and Molecular Bases of Inherited Disease. 7th ed., McGraw-Hill, New York; 1995.

Strachan, T., Read, A.P.: Human Molecular Genetics. 2nd ed. Bios Scientific Publishers, Oxford, 1999.

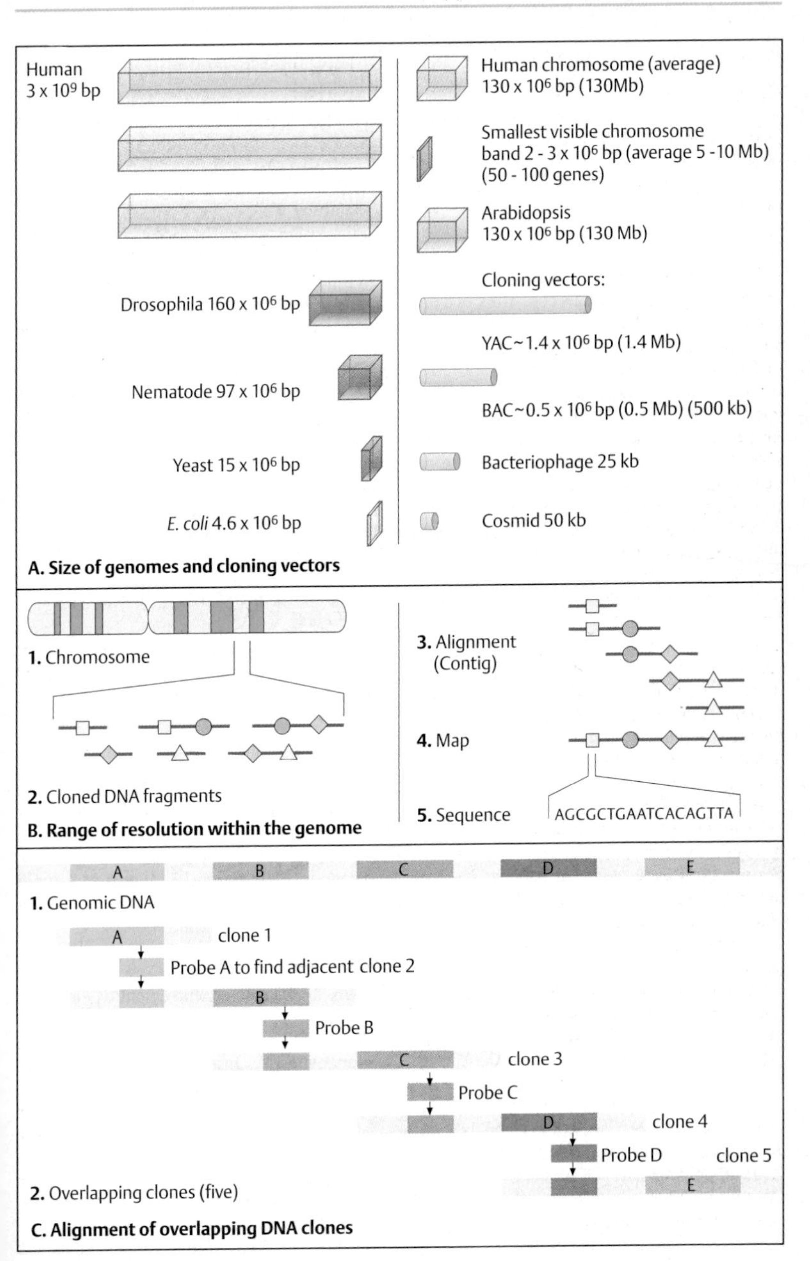


**A. Size of genomes and cloning vectors**

**B. Range of resolution within the genome**

**C. Alignment of overlapping DNA clones**

# Organization of Eukaryotic Genomes

The genomes of higher organisms contain considerably more DNA than required for genes since most of their DNA does not have coding functions.

## A. The components of the nuclear human genome

The human genome is huge, consisting of 3 billion base pairs ($3\times10^9$ bp or 3000 Mb) per haploid set of chromosomes. Only 30% of mammalian DNA is related to genes (900 Mb), whereas 70% of the DNA is not (2100 Mb). Coding DNA in genes accounts for only 3% (90 Mb) of the total amount of DNA. The bulk of DNA (70%) consists of sequences that are repeated many times (repetitive DNA). Characteristic types of repetitive DNA are tandem repeats. Depending on their size and pattern, different types are distinguished: classic satellite DNA, minisatellites, and microsatellites. Together they constitute 14% of the total DNA (420 Mb). More than half of human DNA (56%) consists of repeats interspersed throughout the genome. The most important types are long terminal repeats (LTRs), LINEs (long interspersed nuclear elements), SINEs (short interspersed nuclear elements), and transposons. (Figure adapted from Strachan & Read, 1999).

## B. Satellite DNA

When fractionated human DNA is centrifuged in a cesium chloride density gradient, the main portion of DNA (1) forms a band at buoyant density 1.701 g · $cm^{-3}$. Three additional bands (satellites) appear at 1.687, 1.693, and 1.697 g · $cm^{-3}$, respectively. These are less dense because their CG content differs from that of the main DNA. One distinguishes classic satellite DNA (2) made up of repeats of 100–6500 bp, minisatellites (3) of 10–20 bp repeats, and microsatellites (4) of 2–5 bp repeats. AT-rich and CG-rich segments can be recognized. Microsatellites are the most frequent form of repetitive DNA. Their general structure is $(CA)_n$ where n equals about 2–10. The human genome contains 50000–100000 polymorphic $(CA)_n$ blocks.

## C. Long interspersed nuclear elements (LINEs)

Long interspersed repeat sequences (LINEs) are mammalian retrotransposons that in contrast to retroviruses lack long terminal repeats (LTRs). They account for up to 70% of the human genome by weight. They consist of repetitive sequences up to 6500 bp long that are adenine-rich at their 3′ ends. They may contain one or two open reading frames (ORF), although they are usually shorter and contain no ORF. At the 5′ end and at the 3′ end they have an untranslated region (5′ UTR and 3′ UTR). LINE elements are thought to have arisen by transposition (see p. 76). Mammalian genomes contain 20000–60000 copies of LINE sequences. The major human LINE element is the L1 sequence, a segment that spans up to 6.4 kb. Approximately 100 000 L1 elements are dispersed throughout the human genome. This can result in genetic disease if one is inserted into a gene (e.g. hemophilia, p. 366).

## D. Short interspersed nuclear elements (SINEs)

Short interspersed repeat sequences (SINEs) consist of midsize repetitive segments of similar nucleotide sequences with an average of 300 bp. Their basic structure is a tandem duplication of CG-rich segments separated by adenine-rich segments. The most frequent SINE sequences in humans are the Alu family (Alu sequences). With about 500 000 copies, they make up about 3–6% of the total genome of humans. An Alu sequence consists of two 130-bp tandem duplications with A-rich sections between them. The 3′ side ("right side") contains an insertion of 32 bp.

## References


Brown, T.A.: Genomes. Bios Scientific Publ., Oxford, 1999.

Kazazian, H.H. Jr.: An estimated frequency of endogenous insertional mutations in humans. Nature Genet. **22**:130, 1999.

Kazazian, H.H. Jr.: L1 retrotransposons shape the mammalian genome. Science **289**:1152–1153, 2000.

Lodish, H., et al.: Molecular Cell Biology. 4th ed., Scientific American Books, F.H. Freeman & Co., New York, 2000.

Strachan, T., Read, A.P.: Human Molecular Genetics. 2nd ed. Bios Scientific Publishers, Oxford, 1999.

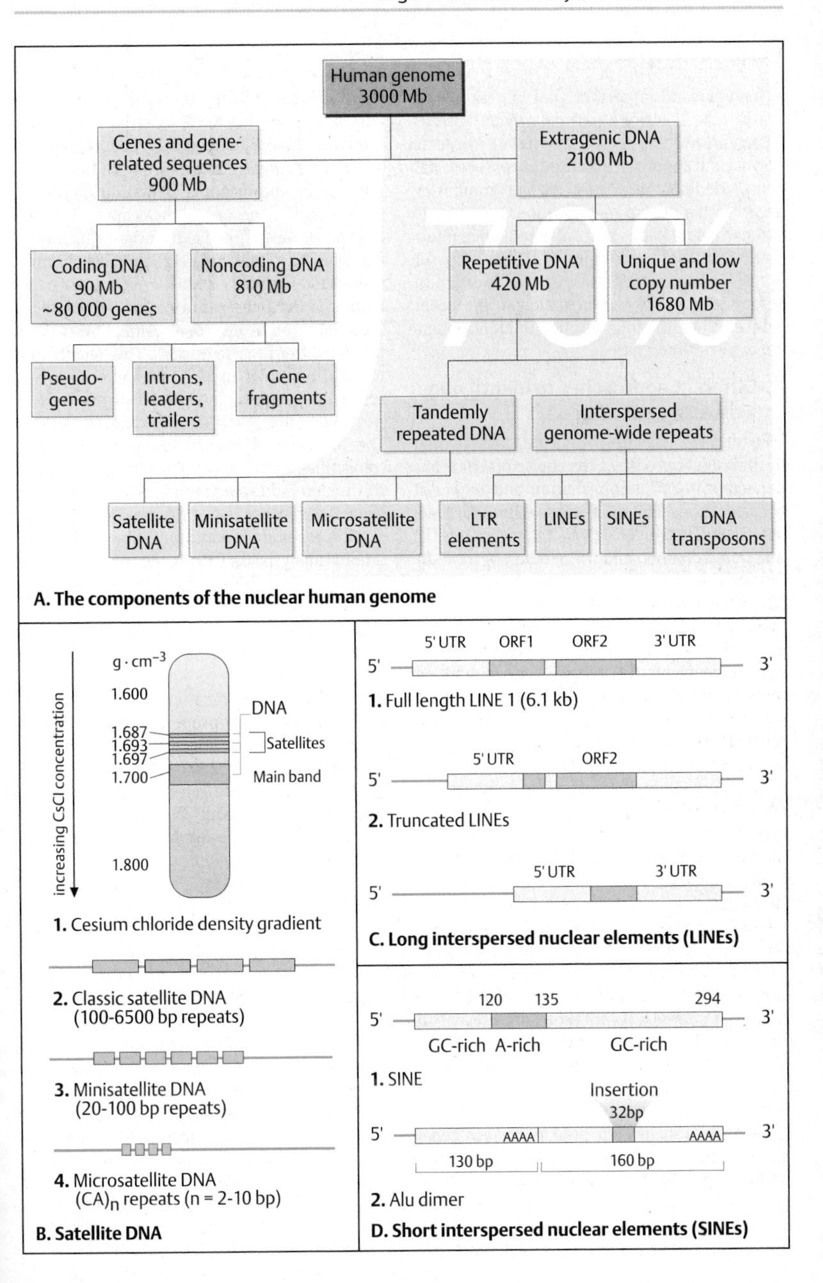


A. The components of the nuclear human genome

B. Satellite DNA

C. Long interspersed nuclear elements (LINEs)

D. Short interspersed nuclear elements (SINEs)

## Gene Identification

A frequent and important goal in genome research is to identify a particular gene with its structure and expression and relate these to normal and abnormal function. This process has been aided by the ever-increasing number of genes with a known map location, by access to extensive data banks that store sequence information, by the availability of dense maps of marker loci, by comparative data from different organisms, and by information from other sources. The principles applied in identifying a gene are outlined here.

### A. Different approaches to identifying a disease-related gene

Disease genes have been identified and isolated by different approaches, the choice of which has depended on prior information and technical considerations. The three approaches that have been useful to date are (i) positional cloning, (ii) functional cloning, and (iii) cloning of a candidate gene.

Positional cloning has been used to identify more than 20 human disease genes. The first and crucial step is clinical identification of the phenotype. Next, the putative gene is mapped to a chromosomal position with defined limits (see B). From here the gene can be isolated (cloned). In the next step, mutations in the gene are demonstrated in patients and are shown not to be present in unaffected family members and normal controls.

Functional cloning requires prior knowledge of the function of the gene. As this information is rarely available at the outset, this approach is very limited.

The candidate gene approach utilizes independent paths of information. If a gene with a function relevant to the disorder is known and has been mapped, mutations of this gene can be sought in patients. If mutations are present in the candidate gene of patients, this gene is likely to be causally related to the disease.

### B. Principal steps in gene identification

To identify a suspected human disease gene, clinical and family data together with blood samples for DNA must be collected from patients with the same particular disorder. The disorder may follow one of the three modes of monogenic inheritance (autosomal recessive, autosomal dominant, X-chromosomal; 1) or multigenic complex inheritance (not shown). Genetic heterogeneity must be ruled out. A chromosome region likely to harbor the gene is then identified by one of the genetic mapping techniques (linkage analysis or physical mapping using a chromosomal structural aberration such as a deletion or a translocation) (2). The map position is refined and the gene location is narrowed to a small area, e.g., a visible chromosome band within 2 – 3 Mb (3). A contig of overlapping DNA clones from a YAC or BAC (bacterial artificial chromosome) or a cosmid library is established from the region (4). This results in a refined molecular map (5) characterized by a set of localized polymorphic DNA marker loci (restriction map, STS map, EST map, see p. 242). The presence of open reading frames (ORF), transcripts, exons, and polyadenylated sites in this region indicates that one is dealing with a gene. Such stretches of DNA most likely representing genes are isolated and subjected to mutational analysis (6). Those genes not containing mutations in patients can be excluded (7). When mutations are found and a polymorphism is excluded, the correct gene has been identified (8). For confirmation, the expression pattern is analyzed, its structure (exons, introns) and size are determined, the transcript is analyzed, and these properties are compared with those of similar genes in other organisms ("zoo blot", see p. 250). Finally, the DNA sequence of the entire gene can be determined and conclusions about the function drawn. From here on, a gene-directed diagnostic procedure can be developed.

### References


Ballabio, A.: The rise and fall of positional cloning? Nature Genet. **3**:277, 1993.

Brown, T.A.: Genomes. Bios Scientific Publishers, Oxford, 1999.

Collins, F.S.: Positional cloning: Let's not call it reverse anymore. Nature Genet. **1**:3, 1992.

Green, E.D., Cox, D.R., Myers, R.M.: The human genome project and its impact on the study of human disease, pp. 401 – 436. In: Scriver, C.R., et al., eds. The Metabolic and Molecular Bases of Inherited Disease. 7th ed. McGraw-Hill, New York; 1995.

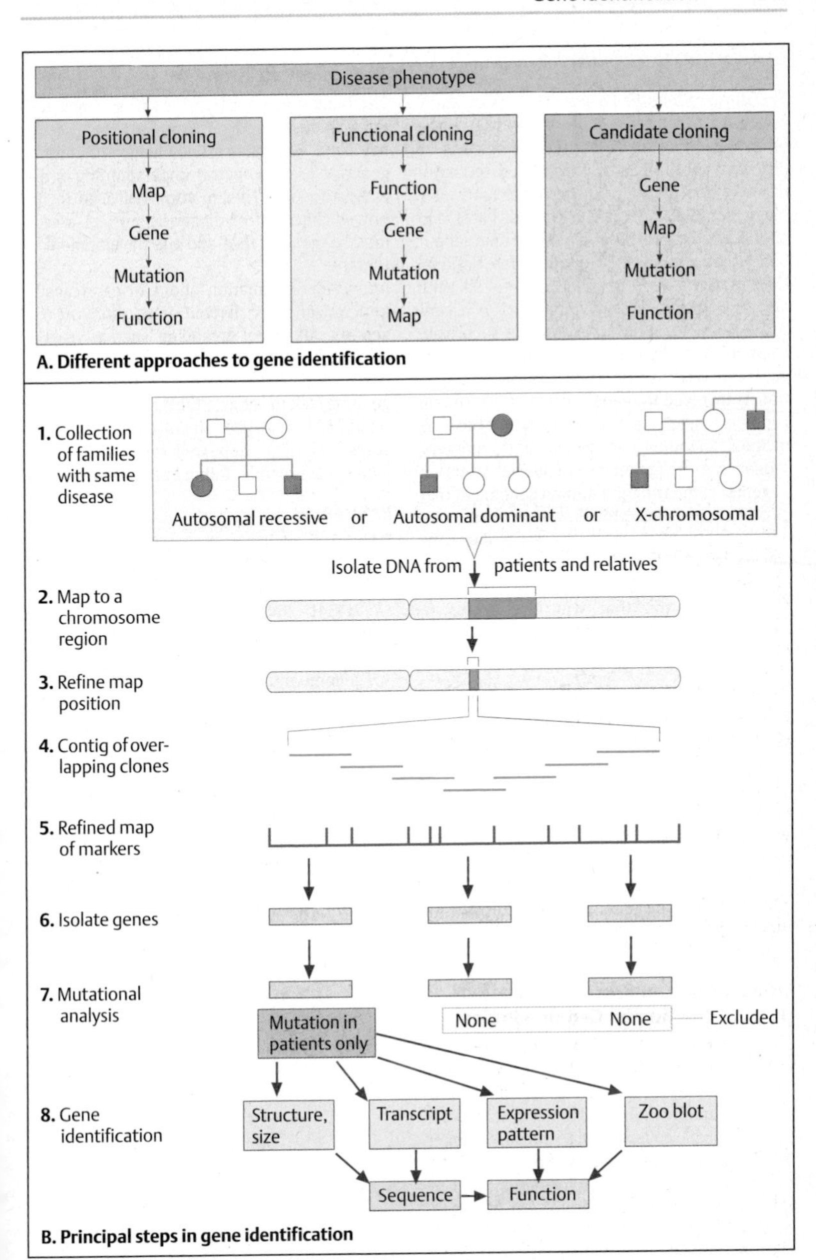


**A. Different approaches to gene identification**

**B. Principal steps in gene identification**

# The Human Genome Project

The Human Genome Project (HGP) is an international cooperative effort to investigate the human genome in its entirety. It consists of different approaches, each directed toward a different goal. Many countries participate. Five major centers, four in the United States and one in the United Kingdom, contribute about 85% of the data. The United States National Human Genome Research Institute (NHGRI: *http://www.nhgri.nih.gov/*) functions as a central agency. In addition, a number of private companies participate. The Human Genome Organization (HUGO), an international organization, is involved in many aspects of the human genome project. The principal goals of the HGP include mapping of genes and markers, sequencing the genome with special attention to genes, comparing the human genome to that of other organisms, providing new techniques, and handling of the vast amount of data. The speed of progress precludes an up-to-date presentation here. Therefore, the opposite page contains information that will give access to various aspects of the human genome project via the Internet.

## A. The Human Genome Project (HGP) and related information online

Information about the Human Genome Project is best gleaned from the internet. Here, several websites provide access to detailed information much beyond the limited space available here. The opposite page (p. 249) provides information about the main areas, the HGP itself, genes and disease, gene maps, networks of databases, education, and the genomes of other organisms than man. For microbial genomes see page 234–236.

## Ethical, legal, and social implications (ELSI) of the Human Genome Project

From the outset, the ethical, legal, and social implications of the human genome project have been an important consideration. About 3% of the total funding is directed to the ELSI Research program. ELSI covers a wide range of issues. These include confidentiality and fairness in the use of individual genetic information, prevention of genetic discrimination, use of genetic methods in clinical diagnostics, public and professional education, and other related issues.

## Medical implications

The human genome project carries important implications for the theory and practice of medicine. Complete knowledge of human genes will lead to better understanding of disease processes. This, in turn, will lead to improved precision of diagnosis, correct assessment of genetic risk, and the development of therapy.

Important information about disease-causing human genes is derived from studying the genomes of model organisms such as yeast, *C. elegans*, *Drosophila*, and the mouse. The recently published sequence of the *Drosophila* genome (Adams et al., 2000) revealed that of a set of 289 human genes involved in causing diseases, 177 (62%) appeared to have an orthologue in *Drosophila* (Rubin et al., 2000).

## References

Borsani, G., Ballabio, A., Banfi, S.: A practical guide to orient yourself in the labyrinth of genome databases. Hum. Mol. Genet. **7**:1641–1648, 1998.

Collins, F.S.: Shattuck Lecture—Medical and societal consequences of the human genome project. New Eng. J. Med. **341**:28–37, 1999.

Collins, F.S. et al.: New goals for the U.S. human genome project: 1998–2003. Science **282**:682–689, 1998.

Goodman, L.: The human genome project aims for 2003. Genome Res. **8**:997–999, 1998.

Rubin, G.M., et al.: Comparative genomics of the eukaryotes. Science **287**:2204–2215, 2000.

New Human Genome Website: (*http://www.ncbi.nlm.nih.gov/genome/central*).

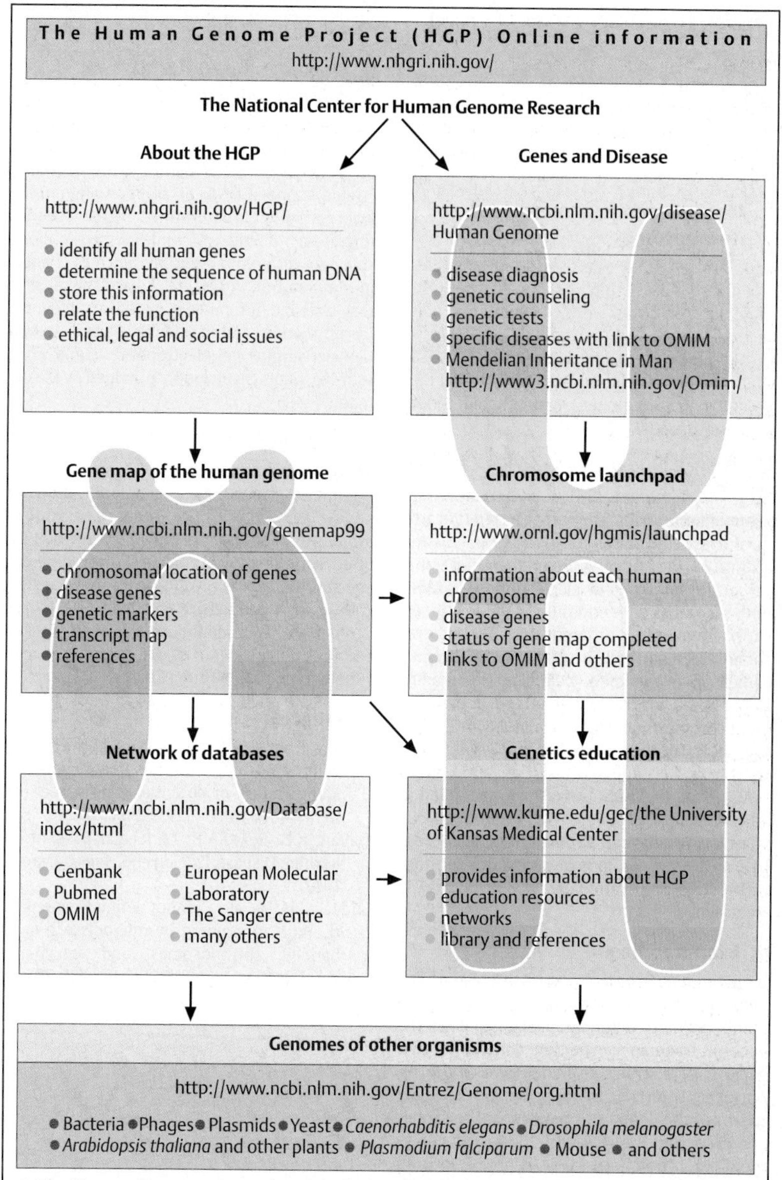

**The Human Genome Project (HGP) Online information**
http://www.nhgri.nih.gov/

**The National Center for Human Genome Research**

**About the HGP**

http://www.nhgri.nih.gov/HGP/

- identify all human genes
- determine the sequence of human DNA
- store this information
- relate the function
- ethical, legal and social issues

**Genes and Disease**

http://www.ncbi.nlm.nih.gov/disease/
Human Genome

- disease diagnosis
- genetic counseling
- genetic tests
- specific diseases with link to OMIM
- Mendelian Inheritance in Man
  http://www3.ncbi.nlm.nih.gov/Omim/

**Gene map of the human genome**

http://www.ncbi.nlm.nih.gov/genemap99

- chromosomal location of genes
- disease genes
- genetic markers
- transcript map
- references

**Chromosome launchpad**

http://www.ornl.gov/hgmis/launchpad

- information about each human chromosome
- disease genes
- status of gene map completed
- links to OMIM and others

**Network of databases**

http://www.ncbi.nlm.nih.gov/Database/index/html

- Genbank
- Pubmed
- OMIM
- European Molecular
- Laboratory
- The Sanger centre
- many others

**Genetics education**

http://www.kume.edu/gec/the University of Kansas Medical Center

- provides information about HGP
- education resources
- networks
- library and references

**Genomes of other organisms**

http://www.ncbi.nlm.nih.gov/Entrez/Genome/org.html

- Bacteria • Phages • Plasmids • Yeast • *Caenorhabditis elegans* • *Drosophila melanogaster* • *Arabidopsis thaliana* and other plants • *Plasmodium falciparum* • Mouse • and others

**A. The Human Genome Project (HGP) and related information online**

# Identification of a Coding DNA Segment

Numerous methods are available to identify a gene of interest that do not require large segments of DNA to be sequenced. Some examples are presented here.

## A. Microdissection of metaphase chromosomes

If the chromosomal location of a gene of interest is known, this region can be cut out of a metaphase chromosome by means of microdissection (arrow). As first applied by B. Horsthemke and co-workers (Lüdecke et al., 1989) this method has the advantage that all other chromosomal segments are eliminated. (Photo from Buiting et al., 1990).

## B. Artificial yeast chromosomes (YACs)

Large DNA fragments (200 – 300 kb) can be replicated in yeast cells. They are inserted into artificial yeast chromosomes (YAC, see p. 110) and replicated with them. A photograph of a transverse alternating field electrophoresis (TAFE) with nine lanes after ethidium bromide staining shows fragments of different sizes. These correspond to the naturally occurring yeast chromosomes. Six of the lanes (2 – 7) contain an additional band, which corresponds to an artificial yeast chromosome. They are marked with a yellow point: the lowest band in lane 2 (YAC9), the third band from the bottom in lanes 3, 4, and 6 (YAC41, YAC45, YAC51), and the lowest band in lane 7 (YAC52). In lane 5, YAC50 is masked by a yeast chromosome (third fragment from below). Lane 1 contains the standards for fragment size (preparation by K. Buiting and B. Horsthemke, Essen).

## C. Exon trapping

In an unidentified segment of DNA, a gene can be recognized by the occurrence of coding segments (exons). To find and isolate an exon, the genomic fragment can be cloned in a vector that consists of a strong promoter gene and a reporter gene. An exon that is present is cut out by means of the donor and acceptor splice signals and expressed together with the genomic fragment (exon trapping). cDNA is produced from the mRNA and replicated by means of PCR. Finally, the trapped exon can be sequenced or characterized by other means. (Diagram after Davies and Read, 1992.)

## D. Single-strand conformation polymorphism (SSCP)

This procedure helps establish a difference in the nucleotide base sequence due to mutation or polymorphism. Whether DNA segments (single-stranded DNA) of common origin differ from each other is determined by their speed of migration in a polyacrylamide gel electrophoresis under different conditions, such as changes in temperature, in pH, etc. A base substitution may lead to a difference in spatial arrangement (conformation) and in mobility (lane 4, arrow). (Polyacrylamide gel electrophoresis and silver staining, photograph kindly provided by D. Lohmann, Essen).

## E. "Zoo blot"

The cross hybridization of DNA across species boundaries ("zoo blot") is an indication for coding sequences, since genes have similar structures in different organisms. A zoo blot is a Southern blot of genomic DNA from different species. Owing to evolutionary relationship, coding DNA sequences will hybridize to the same probe. This can be taken as evidence that the DNA probed is part of a functioning gene. (Photograph of K. Buiting, Essen).

## References


Buiting, K., et al.: Microdissection of the Prader-Willi syndrome chromosome region and identification of potential gene sequences. Genomics 6 : 521 – 527, 1990.

Davies, K.E., Read, A.P.: Molecular Basis of Inherited Disease. 2nd ed. IRL Press, Oxford, 1992.

Lüdecke, H.J. et al.: Cloning defined regions of the human genome by microdissection of banded chromosomes and enzymatic amplification. Nature **338**:348 – 350, 1989.

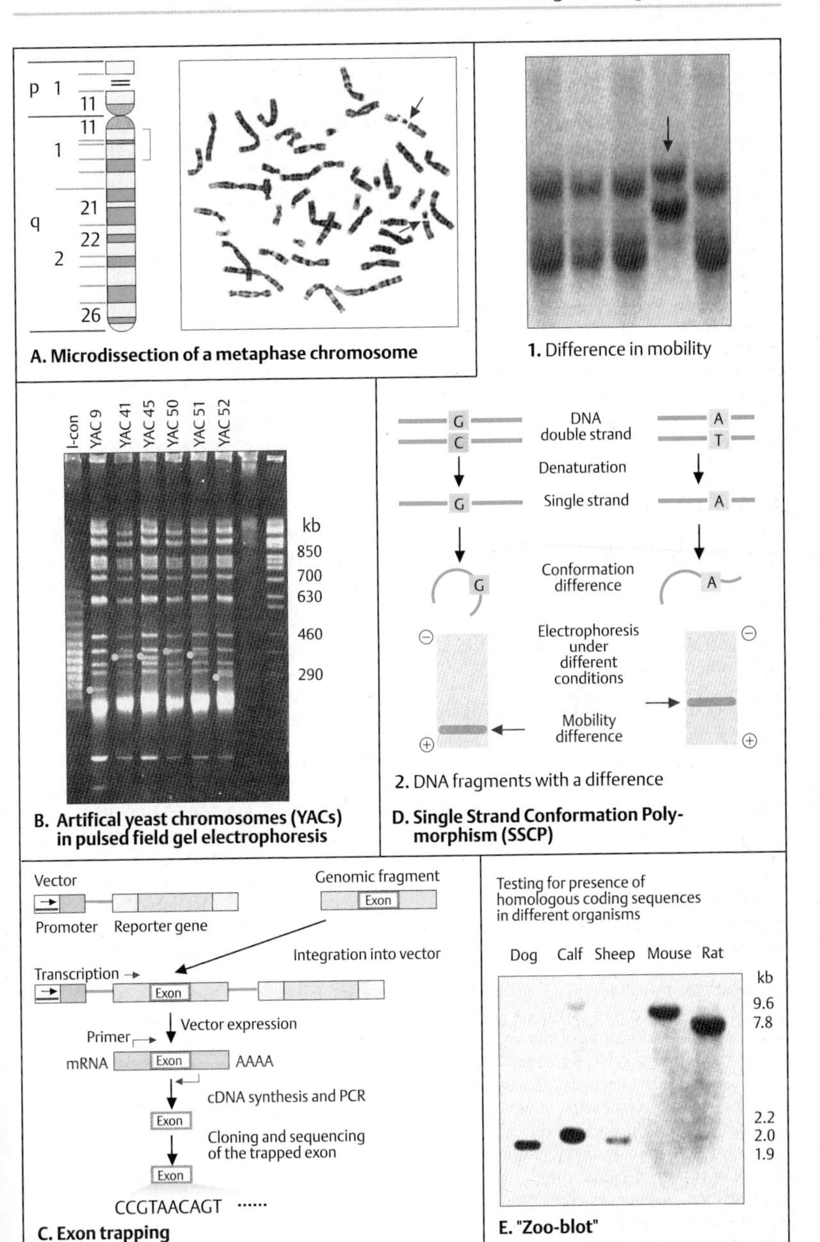


**A. Microdissection of a metaphase chromosome**

**B. Artifical yeast chromosomes (YACs) in pulsed field gel electrophoresis**

**C. Exon trapping**

**D. Single Strand Conformation Polymorphism (SSCP)**

**E. "Zoo-blot"**

# The Dynamic Genome: Mobile Genetic Elements

In the 1950s Barbara McClintock observed an unusual phenomenon during her genetic investigations of Indian corn (maize, *Zea mays*), namely, "jumping genes" (mobile genetic elements). A mobile element causes a break in a chromosome at the site of an insertion and causes a gene locus to move to a different location in the chromosome (transposition). During the last 25 years, mobile elements have been found in every organism in which they have been sought: in bacteria, in *Drosophila*, in the nematode *C. elegans*, and in mammals, including man. These observations have resulted in the concept of a dynamic genome that is by no means fixed and unchangeable. Mutations due to mobile elements inserted into genes have also been demonstrated in man.

## A. Stable and unstable mutations in Indian corn

McClintock observed not only stable mutations (e.g., violet corn kernels) but also fine or somewhat coarser pigment spots (variegation) in some kernels due to unstable mutations.

## B. Effect of mutation and transposition

A gene at the C locus produces a violet pigment of the aleurone in cells of Indian corn. When a mobile element *(Ds)* inactivates this gene locus, the corn is colorless. If *Ds* is removed by transposition, C-locus function is restored and small pigmented spots appear.

## C. Insertion and removal *(Ds)*

As defined by McClintock, an activator (*Ac* locus) is an element that can activate another locus, dissociation *(Ds)*, and cause a break in the chromosome (1). While *Ac* can move independently (autonomous transposition), *Ds* can only move to another location of the chromosome under the influence of *Ac* (nonautonomous transposition). The C locus is inactivated by the insertion of *Ds* (2). Under the influence of *Ac*, *Ds* is then removed from some of the cells, and the C locus is returned to normal function. Since the cells of corn are of clonal origin, the time of transposition influences the phenotype. If transposition occurs early in development, the pigmented spots are relatively large; if it occurs late, the spots are small.

## D. Transposons in bacteria

Mobile genetic elements are classified according to their effect and molecular structure: simple insertion sequences (IS) and the more complex transposons (Tn). A transposon contains additional genes, e.g., for antibiotic resistance in bacteria.

Transposition is a special type of recombination by which a DNA segment of about 750 bp to 10 kb is able to move from one position to another, either on the same or on another DNA molecule. The insertion occurs at an integration site (1) and requires a break (2) with subsequent integration (3). The sequences on either side of the integrated segment at the integration site are direct repeats. At both ends, each IS element or transposon carries inverted repeats whose lengths and base sequences are characteristic for different IS and Tn elements. The expression "direct" signifies that two copies of a sequence are oriented in the same direction (e.g., TTAG on each side of the integrated transposon). Direct and inverted repeats are evidence of the presence of a mobile genetic element. One *E. coli* cell contains on average about ten copies of such sequences. They have also been demonstrated in yeast, drosophila, and other eukaryotic cells.

(Photographs from N.V. Fedoroff, 1984).

## References

Fedoroff, N.V.: Transposable genetic elements in maize. Sci. Am. **250**:65–74, 1984.

Fedoroff, N.V.: Maize transposable elements. Perspect. Biol. Med. **35**:2–19, 1991.

Fedoroff, N.V., Botstein, D., eds.: The Dynamic Genome: Barbara McClintock's Ideas in the Century of Genetics. Cold Spring Harbor Laboratory Press, New York, 1992.

Fox-Keller, E.: A Feeling for the Organism: The Life and Work of Barbara McClintock. W.H. Freeman & Co., San Francisco, 1983.

McClintock, B.: Introduction of instability at selected loci in maize. Genetics **38**:579–599, 1953.

McClintock, B.: Controlling genetic elements. Brookhaven Symp. Biol. **8**:58–74, 1955.

McClintock, B.: The significance of responses of the genome to challenge. Science **226**:792–801, 1984.

Schwartz, R.S.: Jumping genes. New Engl. J. Med. **332**:941–944, 1995.

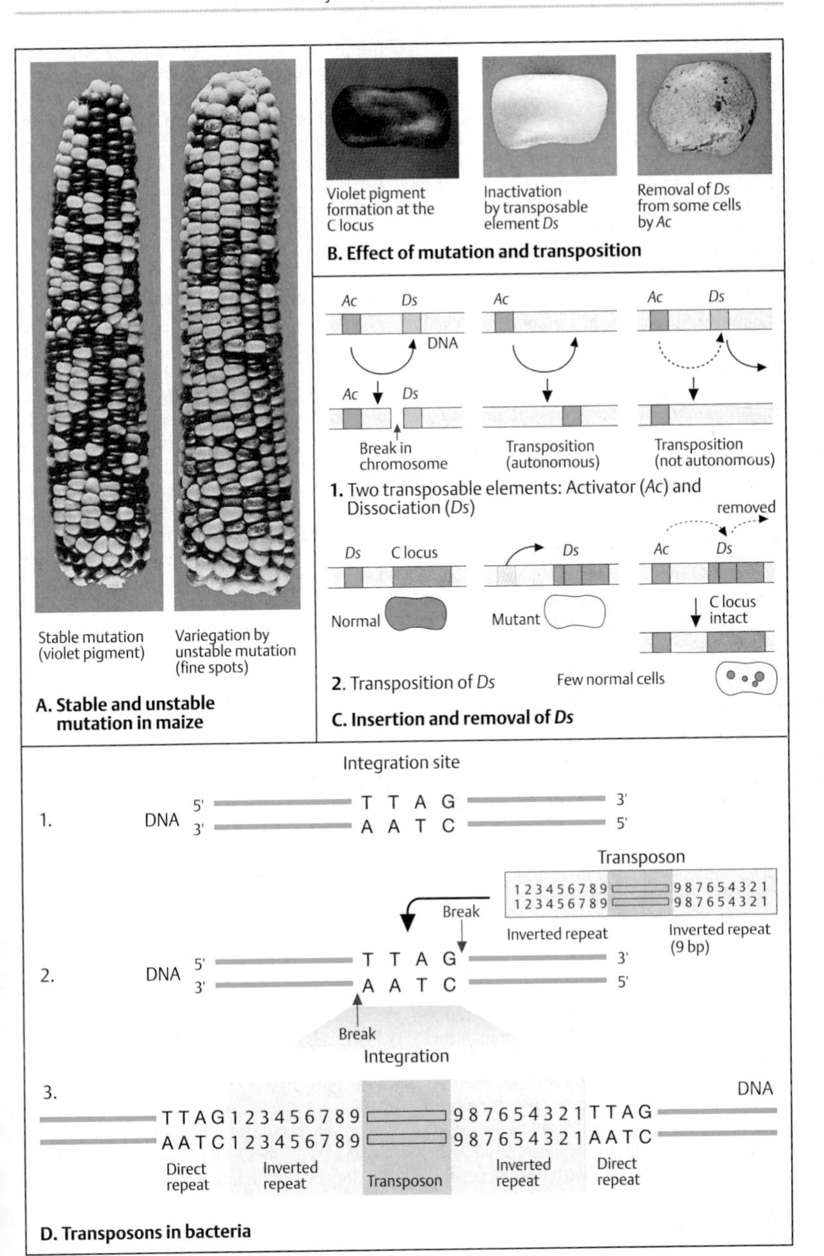

Stable mutation (violet pigment)
Variegation by unstable mutation (fine spots)
A. Stable and unstable mutation in maize
Violet pigment formation at the C locus
Inactivation by transposable element Ds
Removal of Ds from some cells by Ac
B. Effect of mutation and transposition
Ac
Ds
DNA
Break in chromosome
Transposition (autonomous)
Transposition (not autonomous)
1. Two transposable elements: Activator (Ac) and Dissociation (Ds)
removed
C locus
Normal
Mutant
C locus intact
2. Transposition of Ds
Few normal cells
C. Insertion and removal of Ds
Integration site
1.
DNA
5'
3'
T T A G
A A T C
Transposon
1 2 3 4 5 6 7 8 9
9 8 7 6 5 4 3 2 1
Break
Inverted repeat
Inverted repeat (9 bp)
2.
Integration
3.
T T A G 1 2 3 4 5 6 7 8 9
A A T C 1 2 3 4 5 6 7 8 9
9 8 7 6 5 4 3 2 1 T T A G
9 8 7 6 5 4 3 2 1 A A T C
Direct repeat
Inverted repeat
Transposon
D. Transposons in bacteria

# Evolution of Genes and Genomes

Genes and genomes existing today are the cumulative result of events that have taken place in the past. The classical theory of evolution as formulated by Charles Darwin in 1859 states that (i) all living organisms today have descended from organisms living in the past; (ii) the organisms living during earlier times differed from those living today; (iii) the changes were more or less gradual, with only small changes at a time; (iv) the changes usually led to divergent organisms, with the number of ancestral types of organisms being much smaller than the number of types today; and (v) all changes result from causes that continue to exist today and can thus be studied today.

## A. Gene evolution by duplication

Observations on different genomes indicate that different types of duplication must have occurred: of individual genes, of parts of genes (exons), subgenomic duplications, and rare duplications of the whole genome. Duplication of a gene relieves the selective pressure on this gene. After a duplication event, the gene can accumulate mutations without compromising the original function, provided the duplicated gene attains a separate regulatory control. (Figure adapted from Strachan and Read, 1999).

## B. Gene evolution by exon shuffling

The exon/intron structure of eukaryotic genes lends great evolutionary versatility. New genes can be created by placing parts of existing genes into a new context, using functional properties in a new combination. (Figure adapted from Strachan and Read, 1999).

## C. Evolution of chromosomes

Evolution occurs also by structural rearrangements of the genome at the chromosomal level. Related species, e.g., mammals, differ by the number of their chromosomes and chromosomal morphology, but not by the number of genes, which often are conserved to a remarkable degree. The human chromosomes 2 and 5 are shown in comparison with the corresponding chromosomes in three closely related primates, chimpanzee, gorilla, and orangutan. The human chromosome 2 appears to have evolved from fusion of the two primate chromosomes. The differences in chromosome 5 are much more subtle. The orangutan chromosome 5 differs from that of man and the other primates by a pericentric inversion. The banding pattern of all primate chromosomes is remarkably similar. This reflects their close evolutionary relationship. (Figure adapted from Yunis and Prakash, 1982).

## D. Molecular phylogenetics and evolutionary tree reconstruction

A convincing evolutionary relationship can be established by reconstructing past events. A phylogenetic tree may be based on different types of evidence: on fossils, on differences in proteins, on immunological data, on DNA–DNA hybridization, and on DNA sequence similarity. One determines the number of events that must have taken place to explain the diversity observed today. In the path from an ancestral gene (1), different events of divergence can be distinguished (two events are shown schematically here). However, trees are not always as simple as shown here. Since the term homology does not distinguish whether the common evolutionary origin is within or between species, the terms *paralogy* and *orthology* are used (2). Genes or nonallelic chromosomal segments or DNA sequences are called *paralogous* if they have evolved *within* a species (for example the two α-globin loci in humans). If they evolved *between* species prior to diversion, they are called *orthologous* (for example the α-globin and the β-globin genes).

## References


Brown, T.A.: Genomes. Bios Scientific Publ., Oxford, 1999.

Cavalli-Sforza, L.L., Menozzi, P., Piazza, A.: The History and Geography of Human Genes. Princeton Univ. Press, Princeton, New Jersey, 1994.

Moran, J.V., De Berardinis, J., Kazazian, H.H.: Exon shuffling by L1 retrotransposition. Science **283**:1530–1534, 1999.

Ohno, S.: Evolution by Gene Duplication. Springer Verlag, Heidelberg, 1970.

Strachan, T., Read, A.P.: Human Molecular Genetics. 2nd ed. Bios Scientific Publishers, Oxford, 1999.

Yunis, J.J., Prakash, O.: The origin of man: A chromosomal pictorial legacy. Science **215**:1525–1530, 1982.

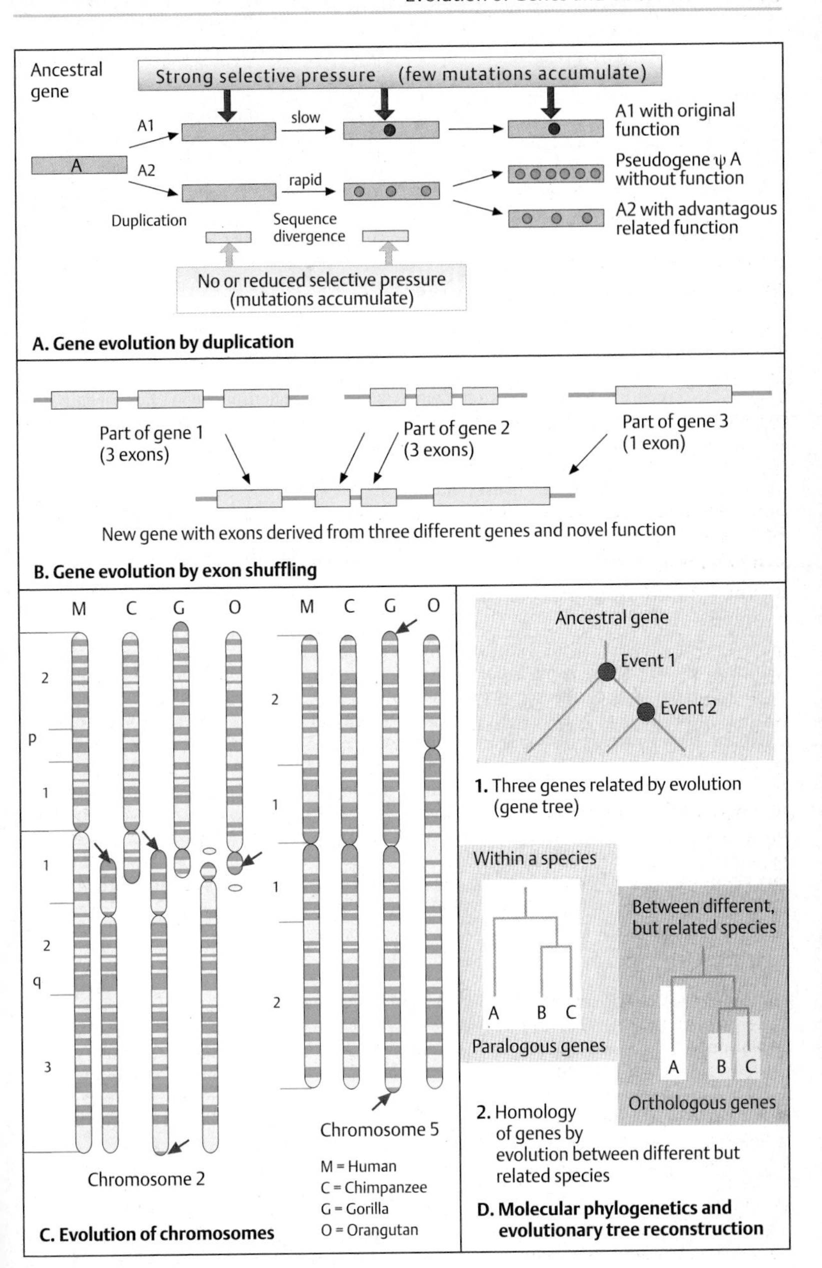
Ancestral gene
Strong selective pressure (few mutations accumulate)
A1
slow
A1 with original function
A
A2
rapid
Pseudogene ψ A without function
A2 with advantagous related function
Duplication
Sequence divergence
No or reduced selective pressure (mutations accumulate)
A. Gene evolution by duplication
Part of gene 1 (3 exons)
Part of gene 2 (3 exons)
Part of gene 3 (1 exon)
New gene with exons derived from three different genes and novel function
B. Gene evolution by exon shuffling
M C G O
p
q
Chromosome 2
Chromosome 5
M = Human
C = Chimpanzee
G = Gorilla
O = Orangutan
C. Evolution of chromosomes
Ancestral gene
Event 1
Event 2
1. Three genes related by evolution (gene tree)
Within a species
A B C
Paralogous genes
Between different, but related species
Orthologous genes
2. Homology of genes by evolution between different but related species
D. Molecular phylogenetics and evolutionary tree reconstruction

## Comparative Genomics

In comparing the genomes of different organisms it is interesting to determine the minimal set of gene families required to encode all proteins required for the overall function of the organism, the so-called core proteome. This is one of the goals of comparative genomics. An example for four different organisms is given in the following table.

### A. Comparison of gene loci in man, mouse, cat, and cow

The conservation of certain groups of genes during mammalian evolution can be demonstrated by comparing reference loci in four different orders of mammals. Many of the loci of humans (representing the order Primates) are found in the same order, although often on different chromosomes, in the house mouse (*Mus musculus*, representing Rodentia), the house cat (representing Carnivora), and the cow (representing Artiodactyla).

### B. The X chromosomes of man and mouse

Many chromosomal segments with the same sequences of two or more homologous loci are syntenic (on the same chromosome) in mouse and man. Gene loci sequences on the X chromosomes of eutherian mammals are especially similar owing to the conserving effect of X inactivation. In the X chromosomes of humans and mice, there are five blocks of gene loci that have the same sequences, although they are in different regions and in part oriented in different directions; these have been retained for about 80 million years, since their divergence from a common ancestor. (After Lavale and Boyd, 1993).

### C. Sequence homologies of the X and the Y chromosomes of man

During the evolution of vertebrates, the X and the Y chromosomes were derived from an autosome. While the Y chromosome has become considerably smaller, the X chromosome has for the most part retained the gene loci of originally autosomal character. In five regions (I – V), the X chromosome and Y chromosome of man contain sequence homologies. The homologous sequences on the extreme distal ends of both chromosomes (I and II) correspond to a pseudoautosomal region. In region I, homologous pairing and crossing-over occur regularly during meiosis. (After Wolf et al., 1992).

### References

Green E.D., Cox, D.R., Myers, R.M.: The human genome project and its impact on the study of human disease, pp. 401 – 436. In: C.R. Scriver, et al., eds. The Metabolic and Molecular Bases of Inherited Disease. 7th ed. McGraw-Hill, New York, 1995.

Lavale, S.H., Boyd, Y. Novel sequences conserved on the human and mouse X chromosomes. Genomics 15 : 483 – 491, 1993.

O'Brien, S.J., et al.: Anchored references loci for comparative genome mappings in mammals. Nature Genet. 3 : 103 – 112, 1993.

Rubin, G.M., et al.: Comparative genomics of the eukaryotes. Science 287 : 2204 – 2215, 2000.

Wolf, U., Schempp, W., Scherer, G.: Molecular biology of the human Y chromosome. Rev. Physiol. Biochem. Pharmacol. **121** : 148 – 213, 1992.

Comparative genomics: Total number of genes and the "core proteome"

| Organism | Number of Genes | Genes Duplicated | Gene Families (Core Proteome) | Type of Organism |
|---|---|---|---|---|
| *H. influenzae* | 1709 | 284 | 1425 | Bacterium |
| *C. cerevisiae* | 6241 | 1858 | 4383 | Yeast |
| *C. elegans* | 18424 | 8971 | 9453 | Nematode |
| *Drosophila* | 13601 | 5536 | 8065 | Insect |

Data from G. M. Rubin et al.; 2000.

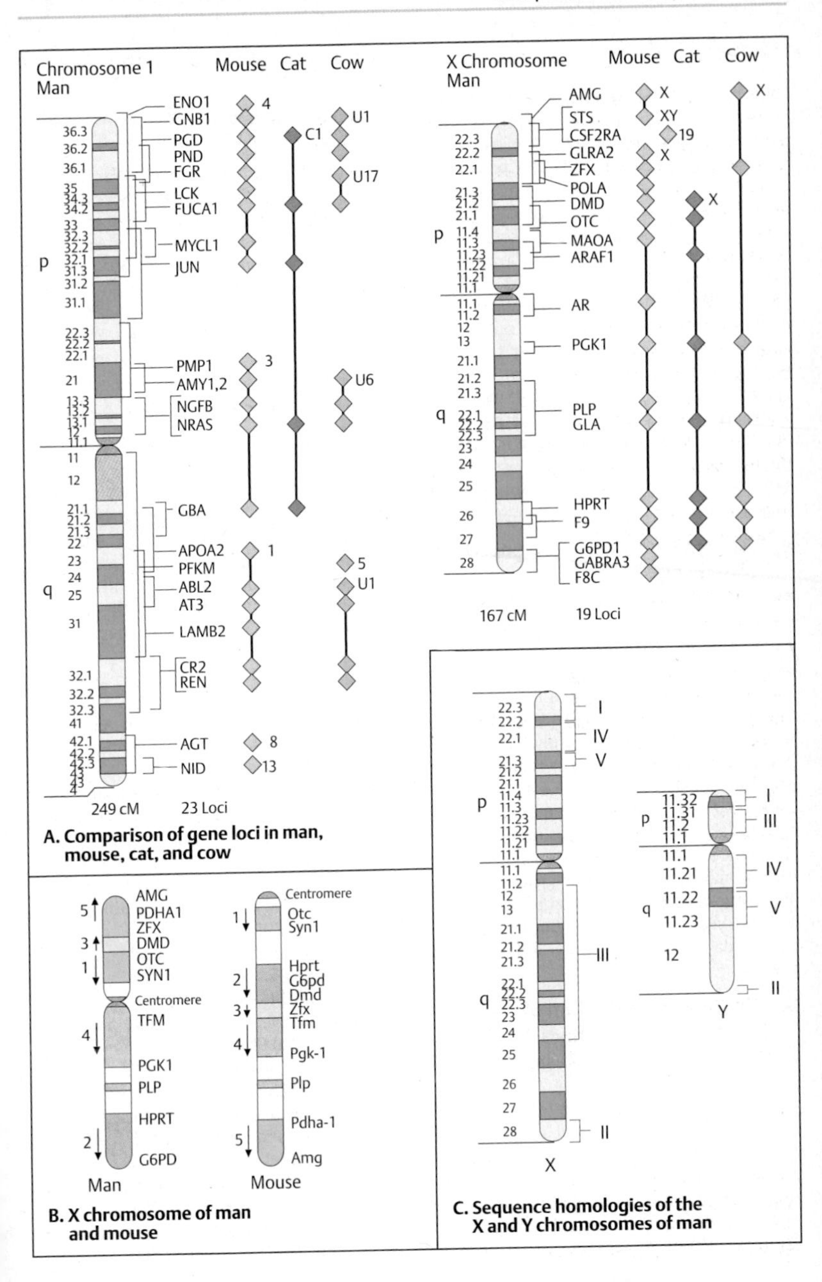


**A. Comparison of gene loci in man, mouse, cat, and cow**

**B. X chromosome of man and mouse**

**C. Sequence homologies of the X and Y chromosomes of man**

## Human Evolution

Human beings belong to the family of Hominidae. Within this family they are a single living species, *Homo sapiens*. Studies of DNA sequences obtained from different parts of the human genome from populations living in different regions of the world indicate that the genetic diversity of humans living today is surprisingly limited. The origin of today's humans can be traced back to about 100000 to 300000 years ago in Africa. From here, an anatomically modern *H. sapiens* ancestor spread out over the earth and diversified. About 90% of the overall genetic variation in humans is among individuals and only 10% among different ethnic groups. There are no races of *H. sapiens;* rather, different ethnic groups have developed in different geographic regions during the past 30000–40000 years (see glossary for genetic definition of race).

### A. Hominid family tree with uncertain relationships

The oldest identified hominid skeletal remains have been found in Eastern Africa. They are attributed to an extinct genus, *Australopithecus*. Several different species originated about 4.5 to 2 million years ago during the Pliocene epoch (5.3 to 1.6 million years ago). During this time fundamental changes in morphology and behavior occurred, presumably to adapt to changing the habitat from the forest to the plains: bipedalism, a dramatic increase of brain volume, and anatomic changes in the pharynx to allow speech were accompanied by tool making and other complex behavior.

The earliest fossils of *Homo sapiens* are from 100000 years ago. Modern humans as they exist today date back about 30000–40000 years.

### B. Geographic distribution of important hominid finds

Two hypotheses for the origin of modern humans have been put forward. The first is that evolution was multiregional, that is, early humans migrated to different geographic regions, resulting in parallel evolution throughout the world. The other hypothesis assumes that all modern humans moved through the middle east into Europe and Asia between 50000 and 100000 years ago and replaced the descendents of the earlier *Homo erectus*. (Figure adapted from Wehner and Gehring, 1995).

### C. Relationship of modern humans and Neandertals

Modern humans and Neandertals coexisted about 30000–40000 years ago. Recent studies indicate that Neandertals did not contribute mitochondrial DNA to modern humans. At two different locations about 2000 km apart, mtDNA from Neandertal specimens at Feldhofer Cave, Neandertal, and Mezmaiskaya Cave in the northern Caucasus showed only 3.48% sequence divergence. (Figures adapted from Krings et al., 1997).

### D. Phylogenetic tree reconstruction from mitochondrial DNA

Genetic studies are consistent with the Out of Africa hypothesis. The strongest evidence came from an analysis of mitochondrial DNA from 147 modern humans of African, Asian, Australian, New Guinean, and European origin. The genealogical tree could be traced back to an ancestral haplotype presumably 200000 years old ("mitochondrial Eve"). (Figure adapted from Cann et al., 1987).

### References

Cann, R.L., Stoneking, M., Wilson A.C.: Mitochondrial DNA and human evolution. Nature **325**:31–36, 1987.

Kaessmann, H., et al.: DNA sequence variation in a non-coding region of low recombination on the human X chromosome. Nature Genet. **22**:78–81, 1999.

Krings, M., et al.: Neanderthal DNA sequences and the origin of modern humans. Cell **90**:19–30, 1997.

Krings, M., Capelli, C., Tschentscher, F., et al.: Neanderthal mtDNA diversity. Nature Genet. **26**:144–146, 2000.

Ovchinnikov, I.V., et al.: Molecular analysis of Neanderthal DNA from the northern Caucasus. Nature **404**:490–493, 2000.

Tattersall, I., Schwartz, J.H.: Hominids and hybrids: The place of Neanderthals in human evolution. Proc. Natl. Acad. Sci. **96**:7117–7119, 1999.

Tattersall, I., Matternes, J.H.: Once we were not alone. Sci. Am. **282**:38–44, 2000.

Wehner, R., Gehring, W.: Zoologie. 23rd ed. Thieme Verlag, Stuttgart, 1995.

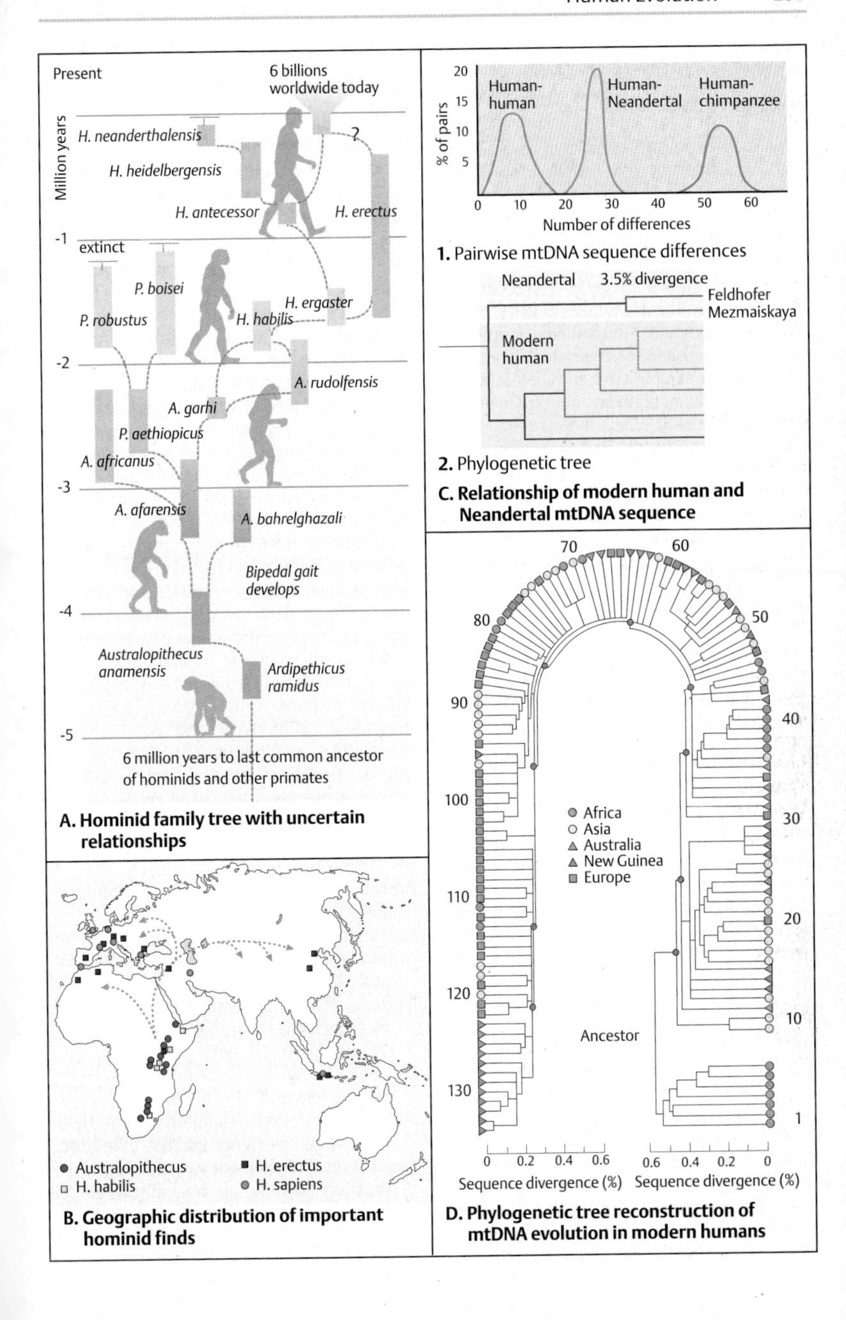


**A. Hominid family tree with uncertain relationships**

**B. Geographic distribution of important hominid finds**

**1.** Pairwise mtDNA sequence differences

**2.** Phylogenetic tree

**C. Relationship of modern human and Neandertal mtDNA sequence**

**D. Phylogenetic tree reconstruction of mtDNA evolution in modern humans**

## Genome Analysis by DNA Microarrays

A microarray or DNA chip is an assembly of oligonucleotides or other DNA probes, e.g., cDNA clones, fixed on a fine grid of surfaces. It is used to analyze the expression states of a series of genes represented in cDNA prepared from mRNA (expression screening) or to recognize sequence variations in genes (screening for DNA variation). The advantages of using microarrays are manyfold: simultaneous large-scale analysis of thousands of genes at a time, automation, small sample size, and easy handling given the right equipment. Several manufacturers offer highly efficient microarrays that can accomodate 300 000 DNA probes on a high-density glass slide of small size (e.g. 1.28 cm × 1.28 cm).

Two basic types of DNA microarrays can be distinguished, although many variations are being developed: (i) microarrays of previously prepared DNA clones or PCR products that are attached to the surface, arranged in a high-density gridded array in two-dimensional linear coordinates; (ii) microarrays of oligonucleotides synthesized in situ on a suitable surface. Both types of DNA arrays can be hybridized to labeled DNA probes in solution.

### A. Gene expression profile by cDNA array

This figure shows a microarray of 1500 different cDNAs from the human X chromosome. The cDNAs were obtained from lymphoblastoid cells of a normal male (XY) and a normal female (XX). The cDNAs of the male cells were labeled with the fluorochrome Cy3 (green) and the cDNAs of the female cells were labeled with Cy5 (red). The inactivation of most genes in one of the two X chromosomes in female cells leads to a 1 : 1 ratio of cDNAs from the expressed genes in the male and the female X chromosomes (yellow signal at most sites because red fluorescence (female) and green fluorescence (male) signals are superimposed owing to the similar expression levels in male and female cells). An exception is the *XIST* gene, which regulates X inactivation (see p. 228). It is expressed on the inactive X chromosome only. It lights up bright red because it is not expressed on the male X chromosome and is not superimposed by a green signal. (Photograph courtesy of G. M. Wieczorek, U. Nuber, and H. H. Ropers, Max-Planck Institute for Molecular Genetics, Berlin).

### B. Gene expression patterns in human cancer cell lines

Microarrays analyzing the pattern of expression in cancer cells can be expected to have a great impact on diagnosis, surveillance of therapy, and screening for anticancer drugs. Approximately 8000 genes among 60 cell lines derived from different types of cancer have been studied by Ross et al. (2000). A consistent relationship between gene expression patterns and tissue of origin was detectable.

Panel 1 shows the cell-line dendrogram relating the patterns of gene expression with respect to the tissue of origin of the cell lines as derived from 1161 cDNAs in 64 cell lines. Panel 2 shows a colored microarray representation of the data using Cy5-labeled (red) cDNA reverse-transcribed from mRNA isolated from the cell lines compared with Cy3-labeled (green) cDNA derived from reference mRNA. The columns (1161 genes) and the rows (60 cell lines) show in red clusters of increased gene expression at several locations. These observations forecast the future of analysis of altered gene expression patterns in tumor cells. (Figure adapted from Ross, et al. 2000 with kind permission by the authors and Nature Genetics; the data can be visualized at the autors' websites *http://genome-www.stanford.edu/nci60* and *http://discover.nci.nih.gov*).

### References


Brown, T.A.: Genomes. Bios Scientific Publ., Oxford, 1999.

Gaasterland, T., Bekiranov, S.: Making the most of microarray data. Nature Genet. **24**:204–206, 2000.

Lipshutz, R.J., et al.: High density synthetic oligonucleotide arrays. Nature Genet. **21**(suppl.):20–24, 1999.

Marshall, E.: Do-it-yourself gene watching. Science **286**:444–447, 1999.

Pinkel, D.: Cancer cells, chemotherapy and gene clusters. Nature Genet. **24**:208–209, 2000.

Ross, D.T., et al.: Systematic variation in gene expression patterns in human cancer cell lines. Nature Genet. **24**:227–235, 2000.

Strachan, T., Read, A.P.: Human Molecular Genetics. 2nd ed. Bios Scientific Publishers, Oxford, 1999.

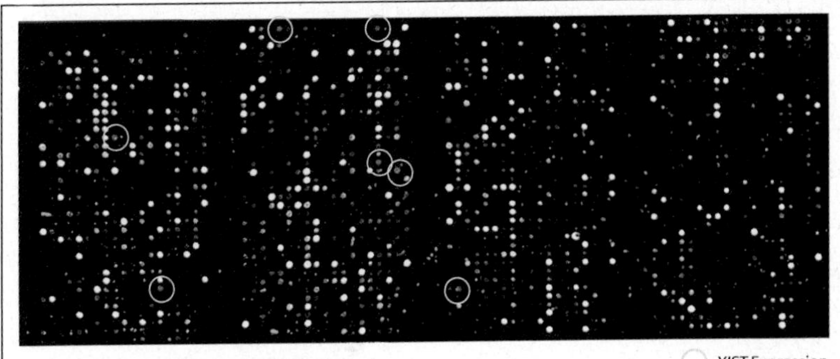

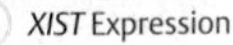


**A. Gene expression profile by cDNA array**

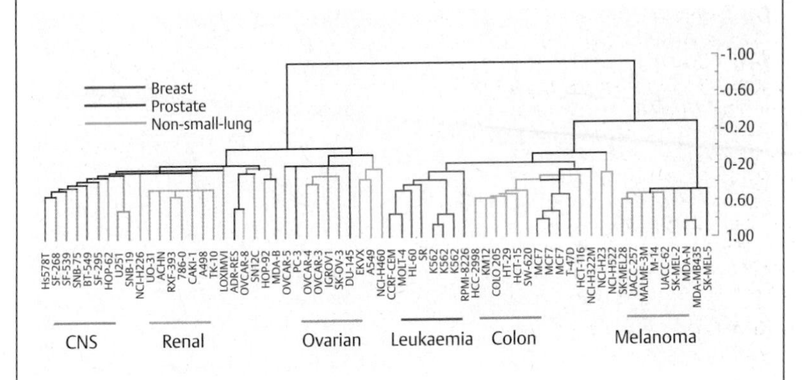


**1.** Cell line dentogram

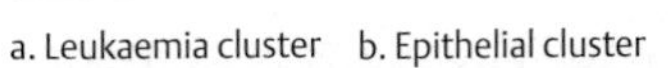

d. CNS cluster

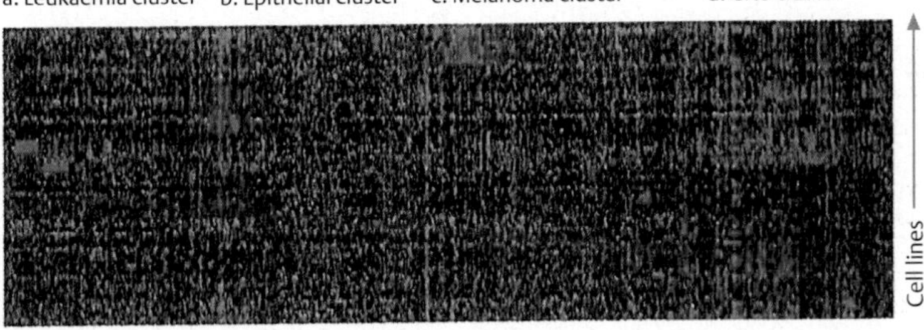


**2.** Microarray data

**B. Gene expression patterns in human cancer cell lines**

# Genetics and Medicine

# Cell-to-Cell Interactions

## Intracellular Signal Transduction Systems

Multicellular organisms depend on communication between cells to assure growth, differentiation, specific functions in different types of cells, and proper response to external stimuli. Specific cell–cell interactions between different types of cells have evolved. A common leitmotif is the specific binding of an extracellular signaling molecule (ligand) to a specific receptor of the target cell to trigger a specific functional response. The vast variety of molecules involved in the many different types of cells can be classified into families of related structure and function (see Lodish et al., 2000; Alberts et al., 1994). Two areas are selected here: the main intracellular functions controlling growth and the receptor tyrosine kinases.

### A. Main intracellular functions controlling growth

Growth factors are a large group of different extracellular molecules that bind with high specificity to cell surface receptors (1). Their binding to the receptor (2) activates intracellular signal transduction proteins (3). This initiates a cascade of events resulting in activation of other proteins (often by phosphorylation) that act as second messengers (4). Hormones of different types are a heterogeneous class of signaling molecules (5). They enter the cell either by diffusion through the plasma membrane or by binding to a cell surface receptor (6). Some hormones require an intranuclear receptor (7). Eventually the signal cascade results in activation or inactivation of transcription factors (8). Before transcription and translation ensue, an elaborate system of DNA damage recognition and repair systems (9) make sure that cell proliferation is safe (cell cycle control, 10). In the event that faults in DNA structure have not been repaired prior to replication, an important pathway sacrifices the cell by apoptosis (cell death, 11). (Figure adapted from Lodish et al., 2000.)

### B. Receptor tyrosine kinase family

Like the G protein-coupled receptors (GPCRs, see p. 268) and their effectors, the receptor tyrosine kinases (RTKs) are a major class of cell surface receptors. Their ligands are soluble or membrane-bound growth factor proteins. RTK signaling pathways involve a wide variety of other functions. Mutations in RTKs may send a proliferative signal even in the absence of a growth factor, resulting in errors in embryonic development and differentiation (congenital malformation) or cancer. Of the more than twenty different RTK families, five examples are selected here: the epidermal growth factor receptor (EGFR); insulin receptor (IR); fibroblast growth factor receptor (FGFR) types 1, 2, and 3; platelet-derived growth factor (PDGFR); and RET (*re*arranged during *t*ransformation).
These receptors share structural features, although they differ in function. All have a single transmembrane domain and an intracellular tyrosine kinase domain of slightly varied size. The extracellular domains consist of evolutionarily conserved motifs: cystein-rich regions, immunoglobulin (Ig)-like domains, fibronectin repeats in the tyrosine kinase with Ig and the EGF. RTK mutations cause a group of important human diseases and malformation syndromes. The phenotypes of the mutations differ according to the particular type of RTK involved and the type of mutation.

#### References

Alberts, B., et al.: Molecular Biology of the Cell. 3rd ed. Garland Publishing Co., New York, 1994.

Cohen, M.M.: Fibroblast growth factor receptor mutations, pp. 77–94, In: M.M. Cohen Jr., R.E. MacLean, eds., Craniosynostosis, Diagnosis, Evaluation, and Management. 2nd ed. Oxford University Press, Oxford, 2000.

Lodish, H., et al.: Molecular Cell Biology (with an animated CD-ROM). 4th ed. W.H. Freeman & Co., New York, 2000.

Muenke, M., et al.: Fibroblast growth factor receptor–related skeletal disorders: craniosynostosis and dwarfism syndromes, pp. 1029–1038, In: J.L. Jameson, ed., Principles of Molecular Medicine. Humana Press, Totowa, New Jersey, 1998.

Münke, M., Schell, U.: Fibroblast-growth-factor receptor mutations in human skeletal disorders. Trends Genet. 11:308–313, 1995.

Roberton, S.C., Tynan, J.A., Donoghue, D.J.: RTK mutations and human syndromes: when good receptors turn bad. Trends Genet. 16:265–271, 2000.

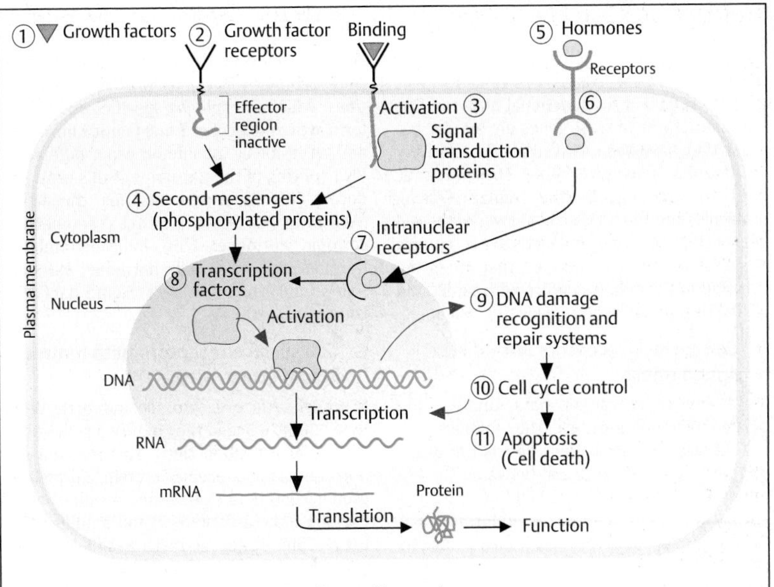


A. Main intracellular functions controlling cell growth

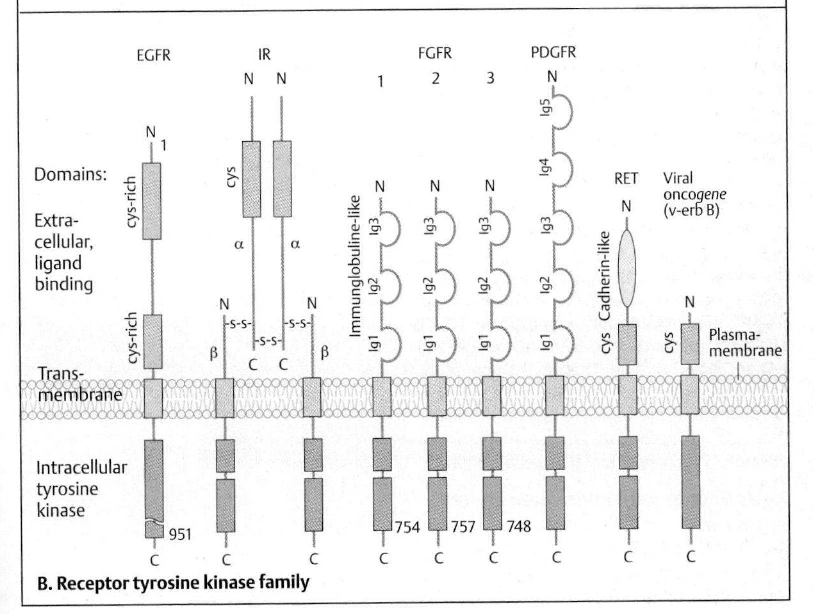


B. Receptor tyrosine kinase family

# Types of Cell Surface Receptors

Specific receptors on cell surfaces (and in the nucleus or cytosol) convey cell-to-cell signals into the cells and the functional answers. The basic structures of their genes are similar because they have been derived from a relatively small group of ancestral genes. They way they bind to the ligand (the signal-releasing molecule) and the functional answer of the cell are specific. When a ligand binds to a receptor, a series of reactions is initiated that alters the function of the cell. Receptors with direct and indirect ligand effects can be distinguished.

## A. Cell surface receptors with direct ligand effect

Many hormones cannot pass through the plasma membrane; instead, they interact with cell surface receptors. Their effects are direct and very rapid. With ligand-activated (or ligand-gated) ion channels (1), binding of the ligand to the receptor changes the conformation of the receptor protein. This causes an ion-specific channel in the receptor protein to open. The resulting flow of ions changes the electric charge of the cell membrane. Receptors with ligand-activated protein kinase (2) further activate a substrate protein. Most protein kinases phosphorylate tyrosine (tyrosine kinase), serine, or threonine by transferring a phosphate residue from adenosine triphosphate (ATP), which is then converted to adenosine diphosphate (ADP). Other receptors mediate the removal of phosphate from a phosphorylated tyrosine side chain by means of their phosphatase activity (3). With one important type of receptor, ligand binding activates guanylate cyclase (4), which catalyzes the formation of cyclic guanosine monophosphate (cGMP) from guanosine triphosphate (GTP). The cGMP functions as a second messenger and brings about a rapid change of activity of enzymes or nonenzymatic proteins. Removal or degradation of the ligand reduces the concentration of the second messenger and ends the reaction. (Diagrams after Lodish et al., 2000.)

## B. Hormones with immediate effects on cells

Important examples of hormones that function as ligands are amino acid derivates, arachidonic acid derivatives, and many peptide hormones. Epinephrine, norepinephrine, and histamine act directly and very rapidly. Peptide hormones such as insulin or adrenocorticotropic hormone (ACTH) initially occur as precursor polypeptides, which are split by specific proteases to form active molecules. Some peptide hormones are coded for by a common gene; differential RNA splicing of the transcript of this gene produces different precursors for translation. (Abbreviations used: ACTH, adrenocorticotropic hormone; FSH, follicle-stimulating hormone; LH, leutinizing hormone; TSH, thyroid-stimulating hormone.) (Figure data after Lodish et al., 2000.)

## C. Cell surface receptors with indirect ligand effect

Many cell surface receptors act indirectly. When they bind to a ligand they induce a series of intracellular activation steps. This reaction system consists of a receptor protein, a protein (G protein) bound to a guanosine residue, and an enzyme to be activated. Ligand binding alters the receptor protein and activates the G protein (2). This moves to the effector, e.g., an enzyme complex (3), and activates it (4). In this way, a second messenger is formed that triggers further reactions in the cell, e.g., cyclic adenosine monophosphate (cAMP) by means of the enzyme adenylate cyclase (see p. 268).

### References

Lodish, H. et al.: Molecular Cell Biology. 4th ed. Scientific American Books, New York, 2000.
Watson, J.D., et al.: Recombinant DNA. 2nd ed. Scientific American Books, New York, 1992.

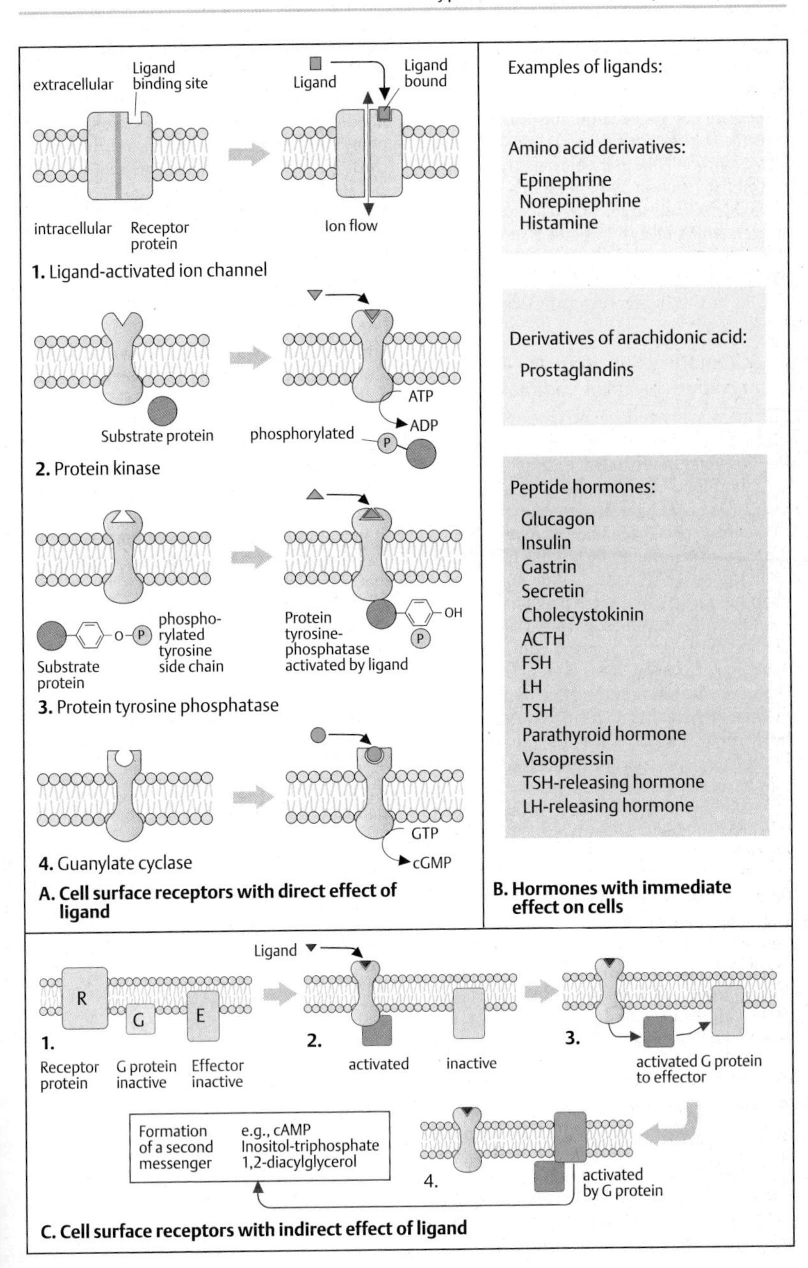


**A. Cell surface receptors with direct effect of ligand**

**B. Hormones with immediate effect on cells**

**C. Cell surface receptors with indirect effect of ligand**

# G Protein-coupled Receptors

The indirect transmission of signals into the cell is mediated by transmembrane proteins, which traverse the cell membrane. A first messenger, e.g., a hormone like epinephrine, triggers an intracellular reaction by binding to a specific receptor. This leads to activation of a second messenger, which in turn initiates a series of reactions that result in a change of cell function. Many of the genes for the different proteins involved in the indirect transmission of signals are known.

## A. Stimulatory G protein ($G_s$) and hormone–receptor complex

There are many endogenous messengers (hormones) with their own specific receptors. First the hormone binds to the receptor (formation of a hormone–receptor complex). The intracellular transmission of signals is mainly carried out by special guanine-nucleotide-binding proteins, or G proteins. By binding to guanosine triphosphate (GTP, a nucleotide composed of guanine, a sugar, and three phosphate groups), the G protein becomes activated and initiates further reactions. G proteins consist of three subunits: $\alpha$, $\beta$, and $\gamma$. The $\alpha$ subunit (stimulatory G protein, $G_s$) binds to the effector protein. Immediately thereafter, $G\alpha$ is inactivated (GTPase) by hydrolysis of GTP to GDP (guanosine diphosphate). This transforms the G protein back into an inactive form ($G_i$).
Several human diseases due to defective G protein or a defective G protein receptor are known. (Clapham, 1993.)

## B. Four hormone classes

Four principal classes of hormones can be differentiated: (1) amino acid derivatives such as epinephrine and epinephrine derivatives; (2) polypeptides such as glucagon; (3) steroids such as cortisol and its derivatives; and (4) fatty acid derivatives such as the prostaglandins.

## C. Formation and hydrolysis of cAMP

The key reaction is the formation of cyclic adenosine monophosphate (cAMP) from adenosine triphosphate (ATP) by means of adenylate cyclase. Intracellular cyclic AMP transmits the activation initiated by the hormone–receptor complex without a molecule having passed through the plasma membrane. cAMP is responsible for many physiological reactions. It becomes inactivated when converted into adenosine monophosphate (AMP) by phosphodiesterase. cGMP (cyclic guanosine monophosphate) functions in the same manner as cAMP to initiate an intracellular reaction.

## D. G protein cycle to activate adenylate cyclase

When a hormone binds to its specific receptor, a structural change occurs (1). This activates the $\alpha$ subunit of the G protein, which separates from the $\beta$ and $\gamma$ subunits (2). The stimulatory G protein ($G_s$-$\alpha$) binds to the effector protein, usually adenylate cyclase, and activates it (3). cAMP is then formed from ATP, while GTP is hydrolyzed to GDP at the G-$\alpha$ subunit. This inactivates the effector protein and the formation of cAMP is terminated. Thus, the signal is of very short duration, and the initial conditions are rapidly restored. Several toxins exert their activity by interrupting this cycle. For example, cholera toxin inhibits inactivation of the $G_s$-$\alpha$ protein so that adenylate cyclase remains activated and large amounts of sodium and water are lost through the intestinal mucous membranes. (Figures adapted from Watson et al., 1992).

### References


Bourne, H.R., Sanders, D.A., McCormick, F.: The GTPase superfamily: conserved structure and molecular mechanism. Nature **349**:11–127, 1991.

Clapham, D.E.: Mutations in G protein-linked receptors: novel insights on disease. Cell **75**:1237–1239, 1993.

Linder, M.E., Gilman, A.G.: G-Proteins, Sci. Am. 36–43, 1992.

Watson, J.D., et al.: Recombinant DNA. 2nd ed. W.H. Freeman, Scientific American Books, New York, 1992.

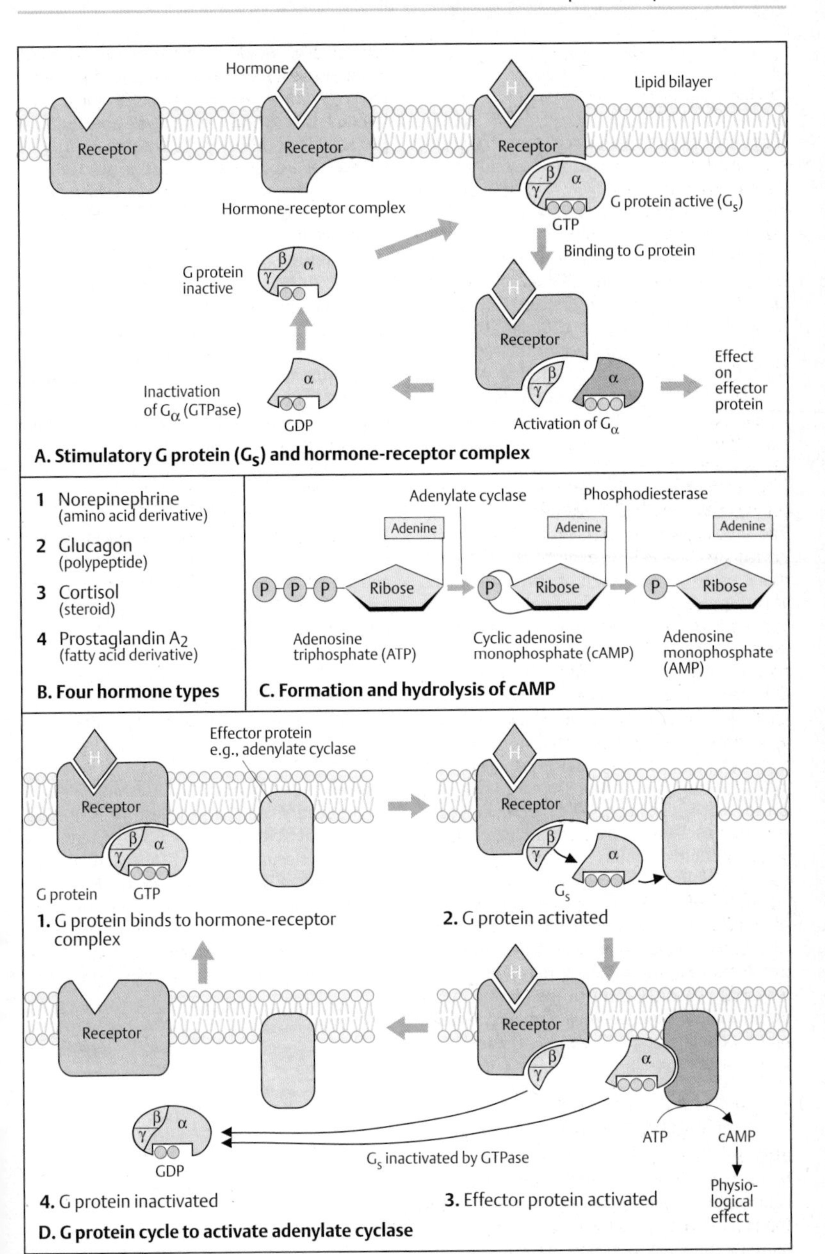


A. Stimulatory G protein ($G_S$) and hormone-receptor complex

B. Four hormone types

C. Formation and hydrolysis of cAMP

D. G protein cycle to activate adenylate cyclase

# Transmembrane Signal Transmitters

Functional signals between cells are received by transmembrane proteins as signal transmitters. During evolution, relatively simple precursor genes for such proteins gave rise to multiple structurally and functionally related genes. Their corresponding proteins serve to transmit ions (sodium, potassium, calcium, chloride, and others), as neurotransmitters, and for perception of light and odors, etc. Cloning of these genes has yielded insight into the variety of functions of transmembrane signal transmitters. Their general structure can be traced back to an evolutionarily conserved ancestral molecule.

## A. Transmembrane structure of voltage-gated ion channels

The direct flow of ions across the cell membrane is regulated by ion channels. The transmembrane proteins, composed of several domains, are arranged so that they form pores that can be opened and closed. The simplest model is the potassium channel (1). This membrane-bound polypeptide contains six transmembrane domains. The amino and the carboxy ends of the protein lie within the cell. Changes in cell membrane potential or voltage cause the channel to open (or close) in order to initiate (or terminate) a brief flow of ions. Domain 4, which is composed of polar amino acids, is crucial for the flow of ions. Sodium and calcium ion channels consist of four subunits (2) of similar structure, each resembling a potassium channel. The similarity is due to the common evolutionary origin of their genes. The four subunits of the sodium channel (3) are positioned to form a very narrow porelike passage, much narrower than a potassium channel, through the plasma membrane. Ion transport is brought about by membrane depolarization (3 and 4). (Figure after Watson et al., 1992.)

## B. Seven-helix structure of transmembrane signal transmitters

Indirect transmission of signals is more frequent than the direct transport of ions or ligand-gated impulse transmission. Here, the transmembrane protein is involved only in the first step of signal transmission. Further steps follow. An especially common structural motif is a transmembrane protein containing seven α helices within the plasma membrane. The amino end is extracellular; the carboxy end is intracellular. Different oligosaccharide side chains are usually bound to the extracellular domains. The intracellular domains have binding sites for other molecules involved in signal transmission. The seven-helix motif is the characteristic structure of G protein-binding receptors (p. 268). As the G proteins themselves, these receptors and their genes form a large family with a long evolutionary history. In yeast, they serve to discern the pheromones of the mating types (p. 186); in higher organisms they are the basis for transmitting signals of vision, smell, and taste (p. 278–286). (Figure redrawn from Stryer, 1995).

## C. A receptor with two transmembrane protein chains, α and β

The receptor for γ-aminobutyric acid (GABA) utilizes two transmembrane protein subunits, α and β. Both the amino and the carboxy ends are extracellular. The two chains are coded for by different genes. Several oligosaccharide side chains are present on the extracellular side. The β chain contains a phosphorylation site for cAMP-dependent protein kinase.
(Figures adapted from Watson et al., 1992).

## References

Sabatini, D.D., Adesnik, M.B.: The biogenesis of membranes and organelles. pp. 459–553, In: Scriver, C.R., et al., eds., The Metabolic and Molecular Bases of Inherited Disease. 7th ed. McGraw-Hill, New York, 1995.

Stryer, L.: Biochemistry, 4th ed. Freeman Publ., San Francisco, 1995.

Watson, J.D. et al.: Recombinant DNA, 2nd ed., Scientific American Books, New York, 1992.

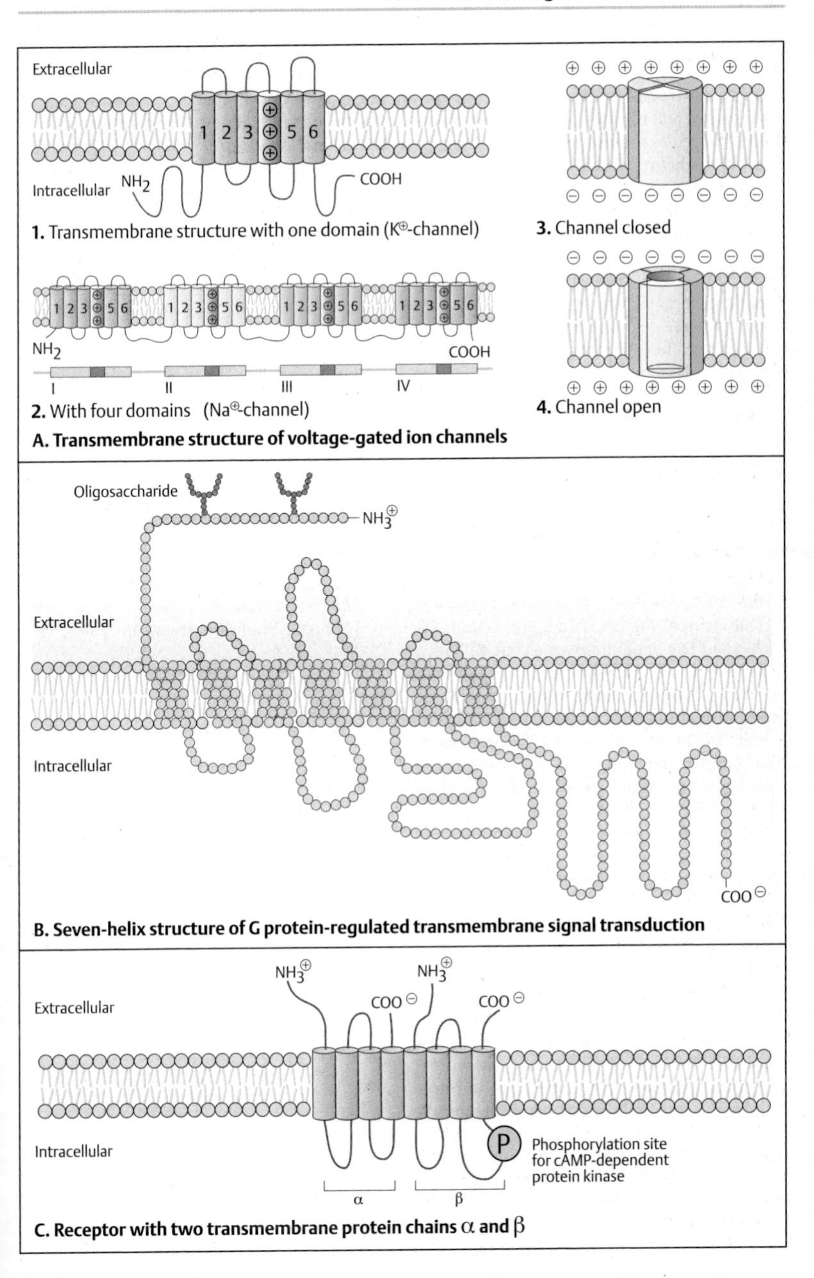


**A. Transmembrane structure of voltage-gated ion channels**

**B. Seven-helix structure of G protein-regulated transmembrane signal transduction**

**C. Receptor with two transmembrane protein chains α and β**

## Receptors of Neurotransmitters

Impulses are relayed between nerve cells or between nerve and muscle cells by various transmitter molecules (neurotransmitters). Their effects are further relayed by receptors in the cell membrane. Receptors can be differentiated according to their structure, which in turn determines their specificity.

### A. Acetylcholine as a neurotransmitter

Cholinergic synapses convey the nerve impulse from one nerve cell to another or from a nerve cell to a muscle cell (motor endplate). Acetylcholine leads to postsynaptic depolarization through the release of potassium ions ($K^+$) and the uptake of sodium ions ($Na^+$). This process is regulated by an acetylcholine receptor.

### B. Acetylcholine receptors

The acetylcholine receptors are of two genetically and functionally different types. Pharmacologically they can be differentiated according to the effects of nicotine and muscarine. The nicotine-sensitive acetylcholine receptor is an ion channel for potassium and sodium. It consists of five subunits: two α, one β, one γ, and one δ (1). Acetylcholine binds as a ligand to the two α subunits. Each subunit consists of four transmembrane domains (2). Each subunit is encoded by its own gene (3). These genes have similar structures and nucleotide base sequences. The ligand-gated ion channel is an example of direct transport without an intermediate carrier. A mutation in the second transmembrane region has been shown to change the ion selectivity from cations to anions (Galzi, 1992.)

The muscarine-sensitive type of acetylcholine receptor is a protein that contians seven transmembrane subunits (4). Since each exists in the form of an α helix, it is referred to as a seven-helix transmembrane protein (p. 270). The amino end ($NH_2$) lies extracellularly; the carboxy end (COOH), intracellularly. The transmembrane domains are connected by intracellular and extracellular polypeptide loops (4). Different domains of the whole protein are distinguished (5) according to location and the relative proportion of hydrophilic and hydrophobic amino acids. The amino end and the carboxy end each form a domain just like the intracellular (a–c), and extracellular portions (d–f). The transmembrane domains located within the plasma membrane (1–7) consist for the most part of hydrophobic amino acids. The structure of the gene product corresponds to the general structure of the gene (6). The different domains are coded for by individual exons. The DNA nucleotide sequences within functionally similar domains are similar.

The seven-helix transmembrane motif occurs in many receptors. The general structures of the genes and of the gene products are very similar, but they differ in their specificity of binding to other functionally relevant molecules (G proteins). They play a role not only as neurotransmitters but also in the transmission of light, odors, and taste. (Figures based on Watson et al., 1992.)

### References

Galzi, G.L.: Mutations in the channel domain of a neuronal nicotinic receptor convert ion selectivity from cationic to anionic. Nature **359**:500–505, 1992.

Watson, J.D., Gilman, M., Witkowski, J., Zoller, M.: Recombinant DNA. 2nd ed. W.H. Freeman, Scientific American Books, New York, 1992.

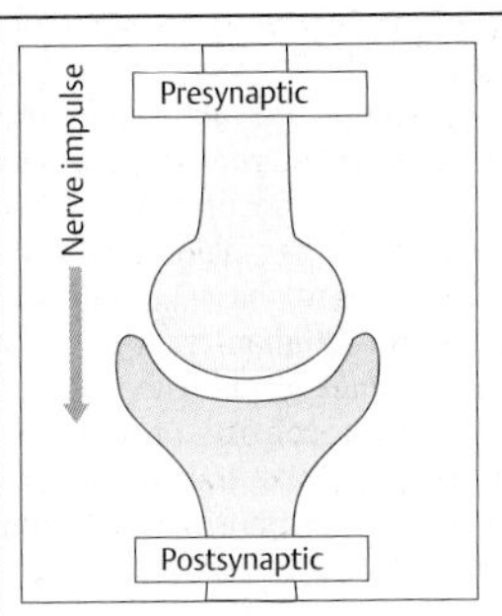


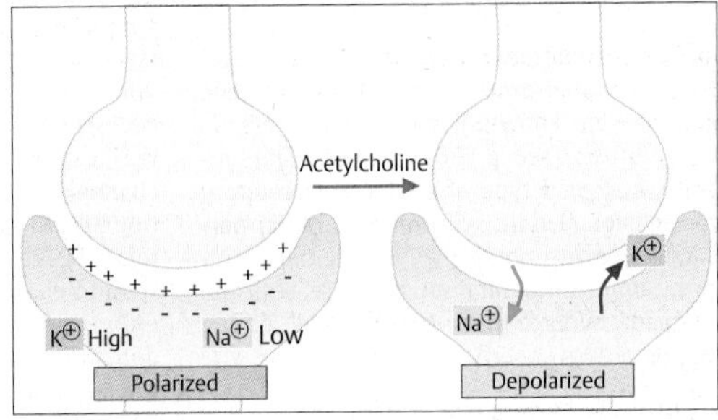


Cholinergic synapse (nerve/nerve or nerve/muscle)

**A. Acetylcholine as neurotransmitter**

Two types of acetylcholine receptors

Cation-specific channel in muscle of vertebrates (nicotine-sensitive)

Seven-helix transmembrane protein bound to G proteins (muscarine-sensitive)

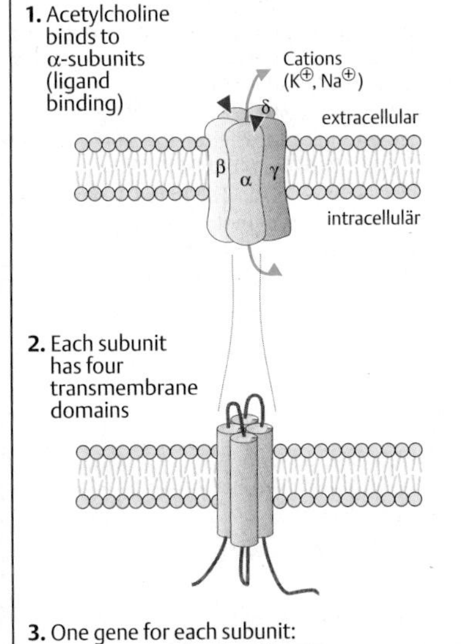


**3.** One gene for each subunit:

2 for α-subunits
1 for β-subunit
1 for γ-subunit
1 for δ-subunit

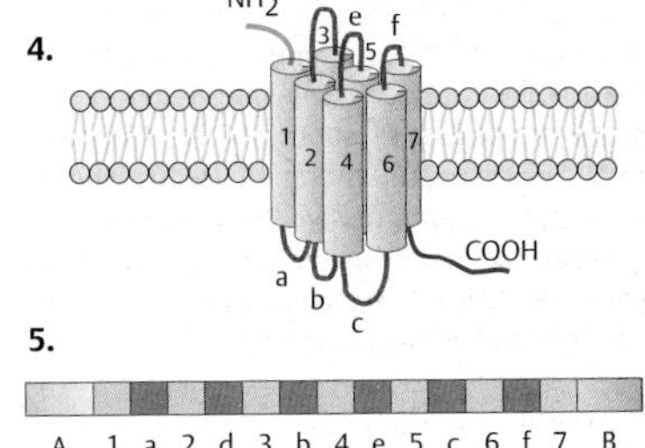


**5.**

A 1 a 2 d 3 b 4 e 5 c 6 f 7 B

A Amino end
B Carboxy end
a - c intracellular domains
d - f extracellular domains
1 - 7 transmembrane domains (hydrophobic)

**6.** Gene structure (diagram)

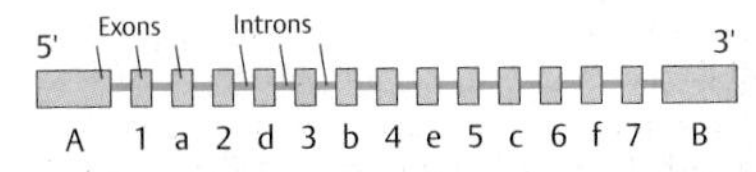


The different domains are encoded by individual exons

**B. Acetylcholine receptor**

## Genetic Defects in Ion Channels

More than 20 different disorders due to defective ion channel proteins resulting from gene mutations are known. Such disorders include cystic fibrosis (see p. 276), the long-QT syndrome, a special type of deafness, hereditary hypertension (Liddle syndrome), familial persistant hyperinsulinemic hypoglycemia of infancy, some hereditary muscle diseases, and malignant hyperthermia (see p. 372), among other disorders.

### A. Long-QT syndrome, a genetic cardiac arrhythmia

Congenital long-QT syndrome is characterized by a prolonged QT interval in the electrocardiogram (more than 460 ms, corrected for heart rate), sudden attacks of missed heart beats (syncopes) or series of rapid heart beats (torsades de pointes), and an increased risk for sudden death from ventricular fibrillation in children and young adults.

### B. Different molecular types of long-QT syndrome

Prolongation of the QT interval in the electrocardiogram results from an increase in the duration of the cardiac action potential (1). The normal potential lasts about 300 ms (phases 1 and 2). The resting membrane potential (phase 3) is reached by progressive inactivation of calcium currents and increasing depletion of potassium currents, which repolarize the cell. In phase 0 the cell is quickly depolarized by activated sodium currents following an excitatory stimulus.

LQT1 accounts for about half of the patients with long-QT syndrome. The gene for LQT2 encodes a 1195-amino-acid transmembrane protein responsible for the other major potassium channel that participates in phase 3 repolarization (*HERG* stands for (*h*uman-*e*ther-*r*-go-go-related gene, a *Drosophila* homologue). LQT3, a sodium channel protein, consists of four subunits, each containing six transmembrane domains and a number of phosphate-binding sites. Homozygosity for LQT1 (*KVLQT1* gene) or LQT5 (*KCNE1* gene) causes a form of long-QT syndrome associated with deafness, the Jervell and Lange-Nielsen syndrome. (Figure adapted from Ackerman and Clapham, 1997.) It is important to distinguish the different types because the choice of medication differs.

### References


Ackerman, M.J., Clapham, D.D.: Ion channels—Basic science and clinical disease. New Eng. J. Med. **336**: 1575–1586, 1997.

Keating, M.T., Sanguinetti, M.C.: Molecular genetic insights into cardiovascular disease. Science **272**:681–685, 1996.

Schulze-Bahr, E., et al.: *KCNE1* mutations cause Jervell and Lange-Nielsen syndrome. Nature Genet. **17**:267–268, 1997.

Schulze-Bahr, E., et al.: The long-QT syndrome. Current status of molecular mechanisms. Z. Kardiol. **88**:245–254, 1999.

Viskin, S.: Long QT syndromes and torsades de pointes. Lancet **354**:1625–1633, 1999.


Examples of Diseases due to Genetic Ion Channel Defects

| Disease | Inheritance | Ion Channel Gene | Locus |
|---|---|---|---|
| Cystic fibrosis | AR | *CFTR* (chloride) | 7q31 |
| Long-QT syndrome (6 types) | AD | *KVLQT1* (cardiac potassium) | 11p15.5 |
| | | *HERG* | 7q35–36 |
| | | *SCNA5* (cardiac sodium) | 3p21 |
| | | Three other types (see text) | |
| Malignant hyperthermia | AD | (Ryanodine, muscle calcium) | 19q13.1 |
| | | (skeletal-muscle sodium) | 17q23–25 |

AD, autosomal dominant; AR, autosomal recessive.

(About 18 others not listed, adapted from Ackerman and Clapham, 1997.)

Prolonged QT interval in the electrocardiogram

Syncope

Sudden death

Autosomal dominant

Six genes involved (LQT1 - LQT6)

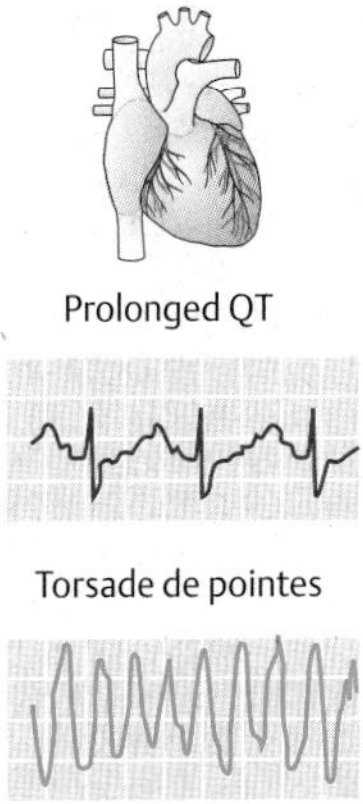

**1.** Main features

Prolonged QT

Torsade de pointes

**2.** Electrocardiogram

Romano Ward syndrome (Long-QT syndrome)

| Type | Locus | Gene |
|---|---|---|
| LQT1 | 11p15.5 | *KCNQ1 (KVLQT1)* |
| LQT2 | 7q35-36 | *HERG* |
| LQT3 | 3p21-24 | SCNA5 |
| LQT4 | 4q25-27 | unknown |
| LQT5 | 21q22.1 | *KCNE1* |
| LQT6 | 21q21.1 | *KCNE2* |

Long-QT and Deafness (Jervell and Lange-Nielsen) due to allelic mutations at LQT1 and LQT5 (autosomal recessive)

**3.** Genetics

**A. Long-QT syndrome, a genetic cardiac arrhythmia**

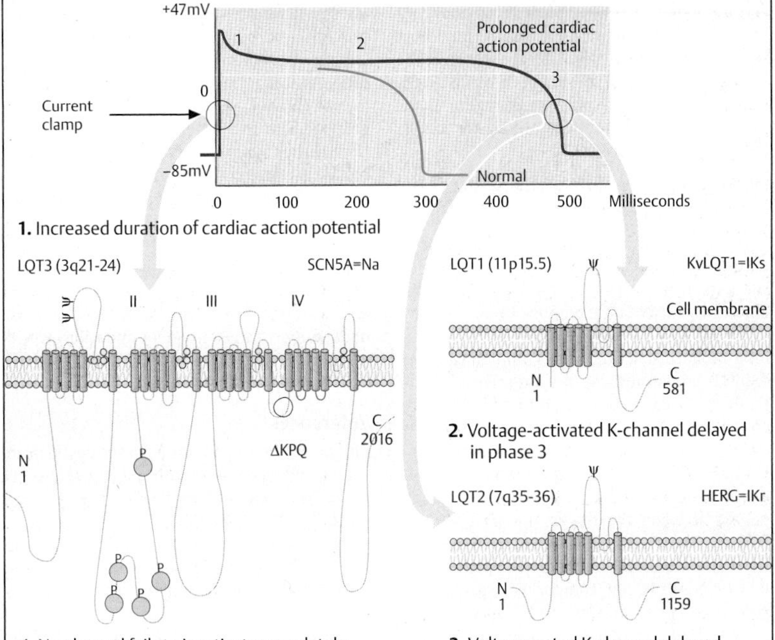

+47mV

1

2

Prolonged cardiac action potential

3

0

Current clamp

–85mV

Normal

0 100 200 300 400 500 Milliseconds

**1.** Increased duration of cardiac action potential

LQT3 (3q21-24) SCN5A=Na

ψ ψ II III IV

N 1

P

ΔKPQ

C 2016

4. Na-channel fails to inactivate completely during phase 0

LQT1 (11p15.5) ψ KvLQT1=IKs

Cell membrane

N 1

C 581

**2.** Voltage-activated K-channel delayed in phase 3

ψ

LQT2 (7q35-36) HERG=IKr

N 1

C 1159

**3.** Voltage-gated K-channel delayed in phase 3

**B. Different molecular types of long-QT syndrome**

# Chloride Channel Defects: Cystic Fibrosis

Cystic fibrosis (mucoviscidosis) is a highly variable multisystemic disorder due to mutations of the cystic fibrosis transmembrane conduction regulator gene *(CFTR)*. Cystic fibrosis (CF) is one of the most frequent autosomal recessive hereditary diseases in populations of European origin (about 1 in 2500 newborns). The high frequency of heterozygotes (1:25) is thought to result from their selective advantage due to reduced liability to epidemic diarrhea (cholera).

## A. Cystic fibrosis: clinical aspects

The disease primarily affects the bronchial system and the gastrointestinal tract. Viscous mucus formation leading to frequent, recurrent bronchopulmonic infections and eventually chronic oxygen deficiency characterize the common, severe form of the disease. The average life expectancy in typical CF is about 30 years. The disease may take a less severe, almost mild course. Congenital bilateral absence of the vas deferens (CBAVD) occurs in 95% of patients with CF. It may be the only manifestation in individuals with different mutant allelic combinations at the CF locus.

## B. Positional cloning of the gene for cystic fibrosis (CF)

The *CFTR* gene was isolated on the basis of its chromosomal location (positional cloning) on the long arm of chromosome 7 at q31 (7q31). Since the gene could be mapped to the long arm of chromosome 7 near a marker locus D7S15, a long-range restriction map comprising about 1500 kb containing the presumptive CF locus flanked by two marker loci, MET and D7S8, was constructed. From there a region of 250 kb was isolated by a combination of chromosome walking and chromosome jumping. Several genes were located in this region (candidate genes) between the marker loci D7S340 and D7S424. The gene sought was identified by the finding of mutations of this gene in patients but not in controls, by comparing with similar genes in other organisms (evolutionary conservation), by determining its exon/intron structure, by sequencing it, and by determining the expression pattern of the gene in different tissues.

## C. The *CFTR* gene and its protein

The *CFTR* gene is large, extending over 250 kb of genomic DNA, and is organized into 27 exons (24 are shown in the diagram) encoding a 6.5 kb transcript with several alternatively spliced forms of mRNA. The CFTR protein has 1480 amino acids. It is a membrane-bound chloride ion channel regulator with several functional domains: two nucleotide-binding domains (encoded by exons 9–12 and 19–23), a regulatory domain (exons 12–14a), and two transmembrane-spanning domains (exons 3–7 and 14b–18). Each of the two transmembrane regions consists of six transmembrane segments. The nucleotide-binding domain 1 (NBD1) confers cAMP-regulated chloride channel activity. The most common mutation (occurring in 66% of patients), a deletion of a phenylalanine codon in position 508 (ΔF508), is located here. The protein is a member of the ATP-binding cassette (ABC) family of transporters. The R domain contains putative sites for protein kinase A and protein kinase C phosphorylation. *CFTR* is widely expressed in epithelial cells. The more than 800 different mutations observed in the *CFTR* gene (see *http://www.genet.sickkids.on.ca/cftr*) can be grouped into five different functional classes: (i) abolished synthesis of full-length protein, (ii) block in protein processing, (iii) reduced chloride channel regulation, (iv) reduced chloride channel conductance, (v) reduced amount of normal CFTR protein. The underlying genetic defects include missense mutations, nonsense mutations, RNA splicing mutations, and deletions. Aside from ΔF508 in about 66% of patients, the most frequent mutations worldwide are G542X (2.4%), G551D (1.6%), N1303K (1.3%), and W1282X (1.2%).

### References

Chillon, M., et al.: Mutations in the cystic fibrosis gene in patients with congenital absence of the vas deferens. N. Engl. J. Med. **332**:1475–1480, 1995.

Rosenbluth, D.B., Brody, S. L.: Cystic fibrosis, pp. 329–338, In: J.L. Jameson, ed., Principles of Molecular Medicine. Human Press, Totowa, N.J., 1998.

Rosenstein, B.J., Zeitline, P.C.: Cystic fibrosis. Lancet **351**:277–282, 1998.

Tsui, L.C.: The spectrum of cystic fibrosis mutations. Trends Genet. **8**:392–398, 1992.

Cystic fibrosis (Mucoviscidosis)

Severe progressive disease of the bronchial system and gastro-intestinal tract

Disturbed function of a chloride ion channel by mutations of the *CFTR* gene

Autosomal recessive

Gene locus 7q31.3

Disease incidence approx. 1:2500

Heterozygote frequency approx. 1:25

Mutation ΔF508 in approx. 70%

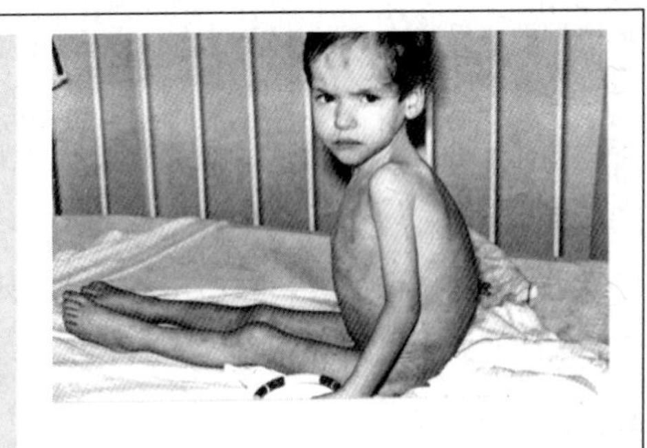

**A. Cystic fibrosis, a very frequent recessive disease in Europe and North America**

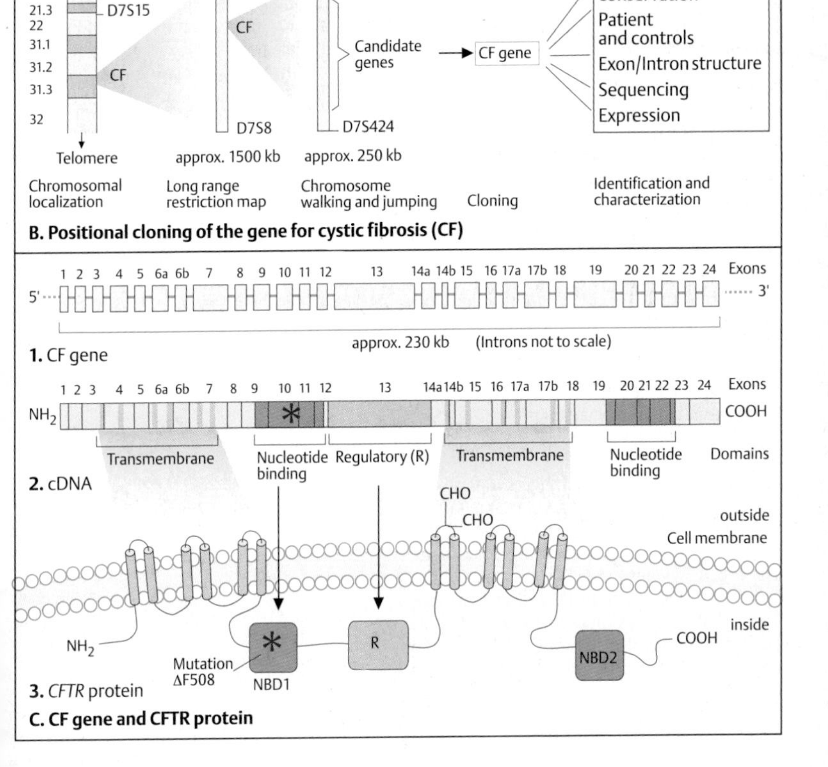


**B. Positional cloning of the gene for cystic fibrosis (CF)**

**C. CF gene and CFTR protein**

# Rhodopsin, a Photoreceptor

The human retina contains about 110 million rod cells for vision in the dark and 6 million cone cells for color vision in the light. These cells contain photoreceptors that convert light into a nerve impulse. Rhodopsin is the photoreceptor for weak light. The light-transmitting system consists of numerous components coded for by genes that are similar in structure and function to genes for other transmembrane signal-transmitting molecules.

## A. Rod cells

A rod cell consists of an outer segment with a photoreceptor region and an inner segment comprising cell nucleus and cytoplasm with endoplasmic reticulum, Golgi apparatus, and mitochondria. The outer segment contains about 1000 disks with rhodopsin molecules in the membrane. In the periphery, the approximately 16 nm thick disks are folded by the protein peripherin. (Diagram after Stryer, 1995).

## B. Photoactivation

In 1958, George Wald and co-workers discovered that light isomerizes 11-*cis*-retinal (1) very rapidly into all-*trans*-retinal (2), a form that practically does not exist in the dark (~1 molecule/1000 years). The light-induced structural change is so great that the resulting atomic motion can trigger a reliable and reproducible nerve impulse. The absorption spectrum of rhodopsin (3) corresponds to the spectrum of sunlight, with an optimum at a wavelength of 500 nm. Although vertebrates, arthropods, and mollusks have anatomically quite different types of eyes, all three phyla use 11-*cis*-retinal for photoactivation.

## C. Light cascade

Photoactivated rhodopsin triggers a series of enzymatic steps (light cascade). First, a signal-transmitting protein of visualization, transducin, is activated by photoactivated rhodopsin. Transducin belongs to the G protein family, i.e., it can assume an inactive GDP and an active GTP form. GTP activates phosphodiesterase. This very rapidly hydrolyzes cGMP and lowers the cGMP concentration in cytosol, which leads to closure of the sodium ion channels. Immediately thereafter, phosphodiesterase is inactivated by means of a G protein cycle.

## D. Rhodopsin

Rhodopsin is a seven-helix transmembrane protein with binding sites for functionally important molecules such as transducin, rhodopsin kinase, and arrestin on the cytosol side. The binding site of the light-sensitive molecule (chromophore) is lysine in position 296 of the seventh transmembrane domain. The light-absorbing group consists of 11-*cis*-retinal. The amino end of rhodopsin is located in the disk interspaces, and the carboxy end on the cytosol side. About half of the molecule is contained in the seven transmembrane hydrophobic domains, one-fourth in the disk interspaces and one-fourth on the cytosol side.

## E. cGMP as transmitter in the vizualization process

The light cascade ends with rapid hydrolysis of cGMP, the internal transmitter in visualization. This leads to rapid closure of the sodium ion channels and hyperpolarization of the membrane to initiate nerve impulse, which is transmitted as a signal to the brain.

## References


Palczewski, K. et al.: Crystal structure of rhodopsin: A G protein-coupled receptor. Science **289**:739–745, 2000.

Schoenlein, R.W., et al.: The first step in vision: Femtosecond isomerization of rhodopsin. Science **254**:412–415, 1991.

Stryer, L.: Biochemistry. 4th ed. W.H. Freeman, New York, 1995.

Stryer, L.: Molecular basis of visual excitation. Cold Spring Harbor Symp Quant Biol. **53**:28–294, 1988.

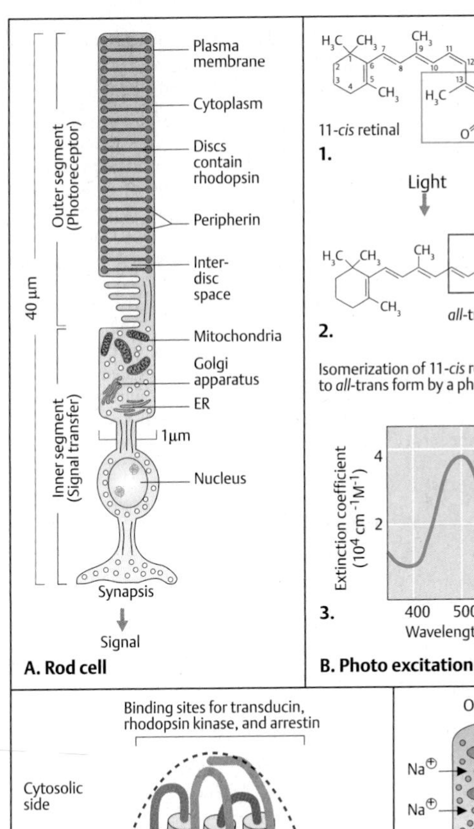
Plasma membrane
Cytoplasm
Discs contain rhodopsin
Peripherin
Inter-disc space
Mitochondria
Golgi apparatus
ER
1µm
Nucleus
Synapsis
Signal
Outer segment (Photoreceptor)
Inner segment (Signal transfer)
40 µm

**A. Rod cell**

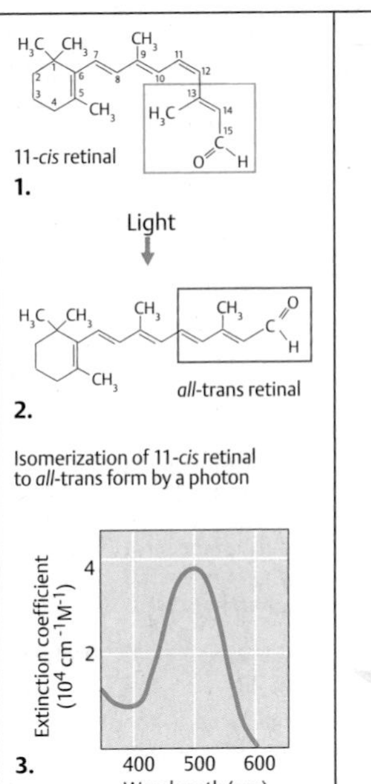
11-cis retinal
1.
Light
all-trans retinal
2.
Isomerization of 11-cis retinal to all-trans form by a photon
Extinction coefficient ($10^4$ cm$^{-1}$M$^{-1}$)
3.
400
500
600
Wavelength (nm)

**B. Photo excitation**

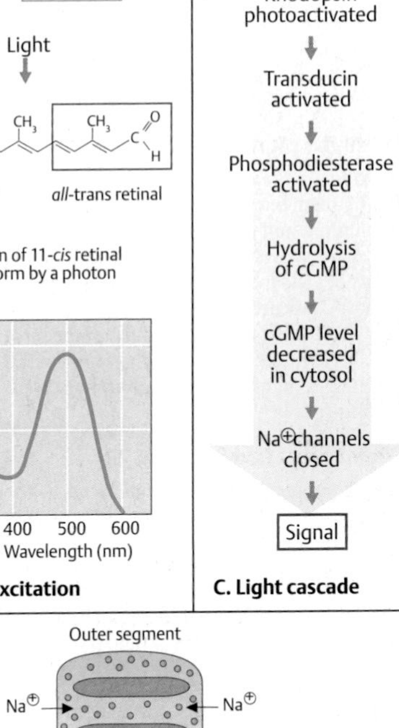
Light
Rhodopsin photoactivated
Transducin activated
Phosphodiesterase activated
Hydrolysis of cGMP
cGMP level decreased in cytosol
Na⊕channels closed
Signal

**C. Light cascade**

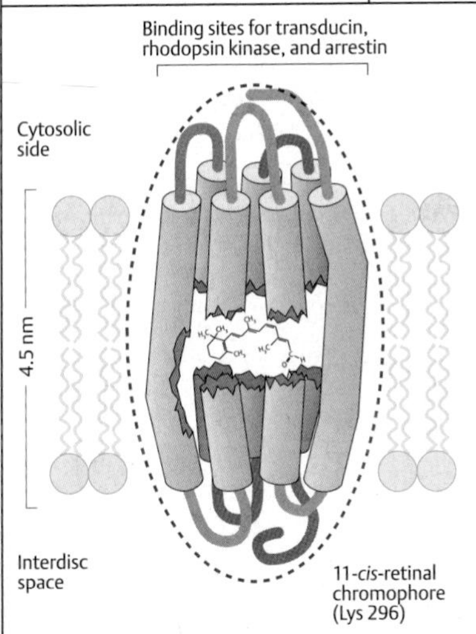
Binding sites for transducin, rhodopsin kinase, and arrestin
Cytosolic side
4.5 nm
Interdisc space
11-cis-retinal chromophore (Lys 296)

**D. Rhodopsin**

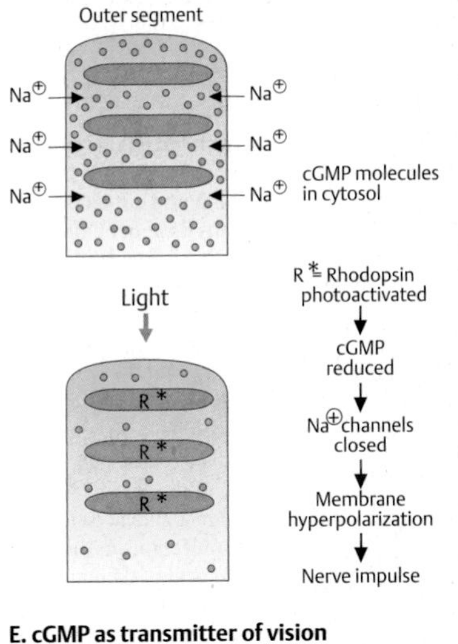
Outer segment
Na⊕
cGMP molecules in cytosol
Light
R* = Rhodopsin photoactivated
cGMP reduced
Na⊕channels closed
Membrane hyperpolarization
Nerve impulse

**E. cGMP as transmitter of vision**

# Mutations in Rhodopsin

Retinitis pigmentosa (RP) is a genetically heterogeneous group of diseases that lead to pigmental degeneration of the retina and progressive blindness. Numerous mutations in the rhodopsin gene have been shown to be the cause of different forms of RP. Mutations in other genes coding for proteins of the light cascade may also cause retinitis pigmentosa.

## A. Retinitis pigmentosa

The fundus of the eye shows distinct displacement of pigmentation, with irregular hyperpigmentation and depigmentation. The papilla (optic disk) shows waxy yellow discoloration. The loss of vision, especially in dim light (night blindness), proceeds from the periphery to the center at different rates depending on the form of the disease, until only a very narrow central visual field remains. (Photograph from E. Zrenner, Tübingen.)

## B. Point mutation in codon 23

The first point mutation demonstrated in the rhodopsin gene (Dryja et al., 1990) was a transversion from cytosine to adenine in codon 23. This changed the codon CCC for proline (Pro) into CAC for histidine (His). Since the proline in position 23 occurs in more than ten related G protein receptors, it must be very important for normal function.

## C. Mutations in rhodopsin

The gene locus for rhodopsin (RHO) in man lies on the long arm of chromosome 3 in region 2, band 1.4 (3q21.4). Dominant and autosomal recessive inherited mutations have been demonstrated in humans. Most mutations lead to the exchange of an amino acid, although deletions may also occur. Of the 348 amino acids of rhodopsin, 38 are identical (invariant) at various positions in vertebrates. More than 100 different mutations are known for autosomal dominant inherited RP. An increasing number of mutations are recognized to cause autosomal recessive RP. In addition, mutations in several other gene loci have been recognized to lead to retinitis pigmentosa, e.g. mutations in the gene for peripherin on the short arm of chromosome 6 in humans (6p) and a locus in the centromeric region of chromosome 8. Other photoreceptor gene disease loci are the $\alpha$ and $\beta$ subunits of phosphodiesterase (PDE).

## D. Demonstration of a mutation in codon 23 by means of oligonucleotides after PCR

This pedigree (1) with autosomal dominant inherited retinitis pigmentosa due to mutation in codon 23 includes 13 affected individuals in three generations (affected females, black circles; affected males, black squares). Using polymerase chain reaction (PCR) (see p. 166), Dryja et al. (1990) demonstrated the mutation in amplified fragments of exon 1 (2). The normal oligonucleotide corresponds to the normal sequence between codons 26 and 20. The mutant sequence of the oligomere RP contains the mutant sequence CAC. All affected individuals gave a hybridization signal with the RP oligomer (2) (II-2, II-12, and III-4 were not examined), whereas unaffected individuals did not (see p. 408 for demonstration of a point mutation with oligonucleotides).

## References

Barkur, S. S.: Retinitis pigmentosa and related disorders. Am J Med Genet. **52**:467–474, 1994.

Dryja, T.P.: Retinitis pigmentosa, pp. 4297–4309, In: C.R. Scriver, et al., eds., The Metabolic and Molecular Bases of Inherited Disease. 7th ed. McGraw-Hill, New York, 1995.

Dryja, T.P., et al.: A point mutation of the rhodopsin gene in one form of retinitis pigmentosa. Nature **343**:364–366, 1990.

McInnes, R.R., Bascom, R.A.: Retinal genetics: a nullifying effect for rhodopsin. Nature Genetics **1**:155–157, 1992.

Wright, A.F.: New insights into genetic eye disease. Trends Genet. **8**:85–91, 1992.

A group of hereditary diseases with degeneration of the retinal pigment

Night blindness

Progressive loss of vision

Frequency about1:3500

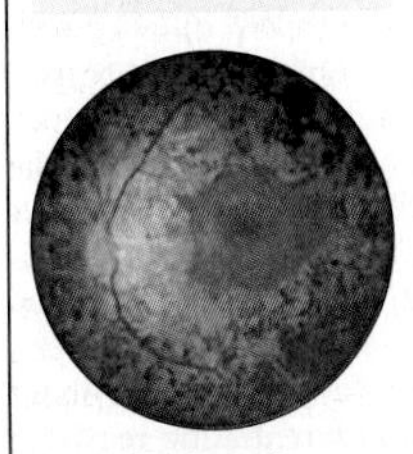

Typical fundus with pigment changes, narrow vessels, and pale, waxy optical nerve

Frequency of the different genetic forms

25% autosomal dominant
20% autosomal recessive
8% X-chromosomal
47% Mode of inheritance uncertain in an individual patient

Important diagnostic signs

Fundus:
narrow vessels
pale optic nerve
macular changes
widened light reflex
pigment epithelium changes
electroretinogram silent

Secondary changes in the anterior chamber:
vitreous body changes

Cataract
Myopia

**A. Retinitis pigmentosa**

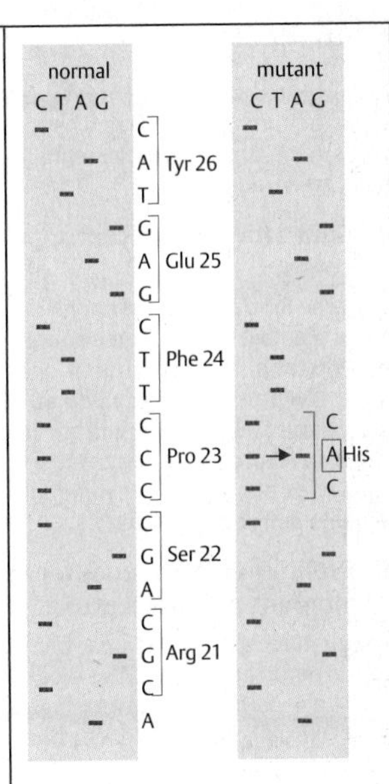


**B. Point mutations in codon 23**

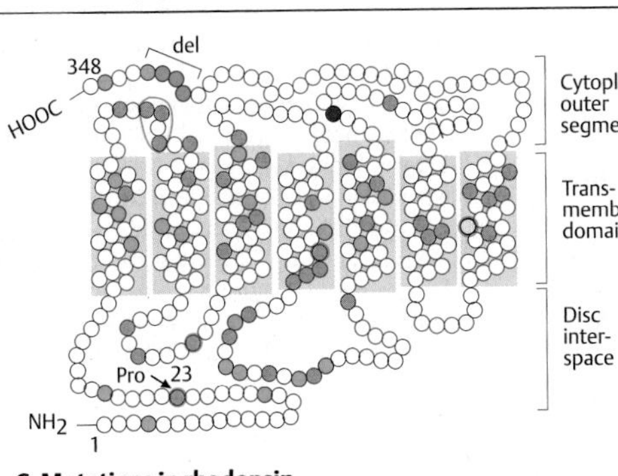


● = Invariant amino acid in different vertebrates
○ = Retinal binding site
● = Point mutation *
● = Deletions *
● = Point mutation in autosomal recessive RP
— = gt → tt Intron 4-donor splice site mutation

* in autosomal dominant retinitis pigmentosa

**C. Mutations in rhodopsin**

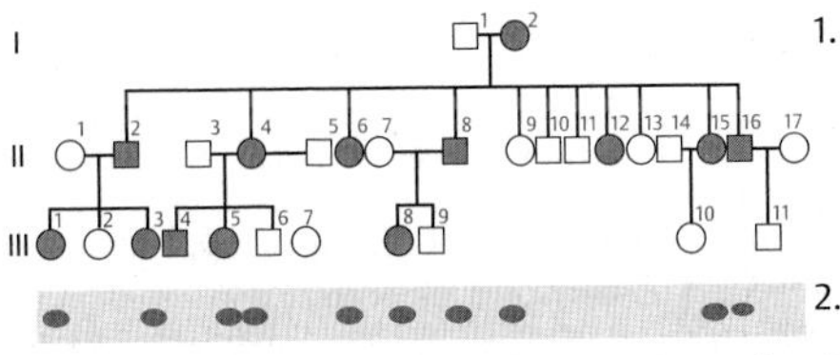


1. Pedigree with autosomal dominant retinitis pigmentosa due to mutation in codon 23 (P23H)

2. Autoradiogram of hybridization of amplified DNA fragments in codon 23 with oligomer 3'-CATGAGCTTCACCGACGCA-5' for the mutant sequence

**D. Demonstration of mutation P23H in codon 23 by oligonucleotides after PCR**

## Color Vision

As suggested by Thomas Young in 1802, color vision in humans is mediated by three receptor types in the cone cells of the retina, one each for blue, green, and red.

### A. Genes for photoreceptor proteins in cones

The gene for the blue receptor is autosomal; the genes for the red and green receptors are X chromosomal. The absorption spectra of the three receptors show maxima of 426 nm for blue, about 530 for green, and about 550 for red. The red receptor was discovered to be polymorphic, with two somewhat different absorption maxima at 552 and 557 nm.

### B. Evolution of the genes for visual pigment photoreceptors

The photoreceptor genes arose from a single ancestral gene (protogene). The rhodopsin–transducin pair is found in invertebrates and is at least 700 million years old. The blue receptor is almost as old as rhodopsin, about 500 million years. The separation into a receptor for green and one for red must have occurred only about 30 million years ago, after the Old World and New World apes separated, since man and the Old World apes have three cone pigments whereas New World apes have two.

### C. Structural similarity of the visual pigments

In 1986, J. Nathans and co-workers sequenced the genes for color photoreceptors and observed marked structural similarities, especially of the green and red receptor genes. Here the gene products (the receptors) are shown and their similarities are compared. The dark dots indicate variant amino acids; the light dots are identical amino acids, given in percentages.

### D. Polymorphism in the photoreceptor for red

A. G. Motulsky and co-workers (Winderickx et al., 1992) demonstrated variant codons in three regions of the red receptor gene (1). Serine was found at position 180 in 60% of the investigated males; alanine in 40%. Position 230 showed polymorphism of isoleucine (Ile) and threonine (Thr); position 233 of alanine (Ala) and serine (Ser) (2). Differences in red color perception could be demonstrated by the color-mixing test procedure of Raleigh (3).

### E. Normal and defective red–green vision

One gene for red and one to three genes for green lie close together on the long arm of the X chromosome in humans (1). Since the sequences of these genes are very similar, unequal crossing-over is not infrequent (2). Intergenic crossing-over leads to loss (green blindness) or duplication; intragenic crossing-over leads to a hybrid gene (red blindness). Green blindness results from loss of a gene for the green receptor; red blindness, from a defective or absent red receptor. With red–green blindness, neither a normal red nor a normal green receptor is present. About 1% of all men are red–green blind and about 2% are green blind. About 8% show weakness in differentiating red from green.

### References


Kohl, S., et al.: Total colour blindness is caused by mutations in the gene encoding the α-subunit of the cone photoreceptor cGMP-gated cation channel. Nature Genet. **19**:257–259, 1998.

Motulsky, A.G., Deeb, S. S.: Color vision and its genetic defects, pp. 4275–4295. In: C.R. Scriver, et al., eds., The Metabolic and Molecular Bases of Inherited Disease. 7th ed. McGraw-Hill, New York, 1995.

Nathans, J., Thomas, D., Hogness, D.S.: Molecular genetics of human color vision: the genes encoding blue, green, and red pigments. Science **232**:193–202, 1986.

Neitz, M., Neitz, J.: Numbers and ratios of visual pigment genes for normal red-green color vision. Science **267**:1013–1016, 1995.

Winderickx, J., et al.: Polymorphism in red photopigment underlies variation in colour matching. Nature **356**:431–433, 1992.

Wissinger, B., Sharpe, L.T.: New aspects of an old theme: The genetic basis of human color vision. Am. J. Hum. Genet. **63**:1257–1262, 1998.

Wissinger, B., et al.: Human rod monochromacy: linkage analysis and mapping of a cone photoreceptor expressed candidate gene on chromosome 2q11. Genomics 51:325–331, 1998.

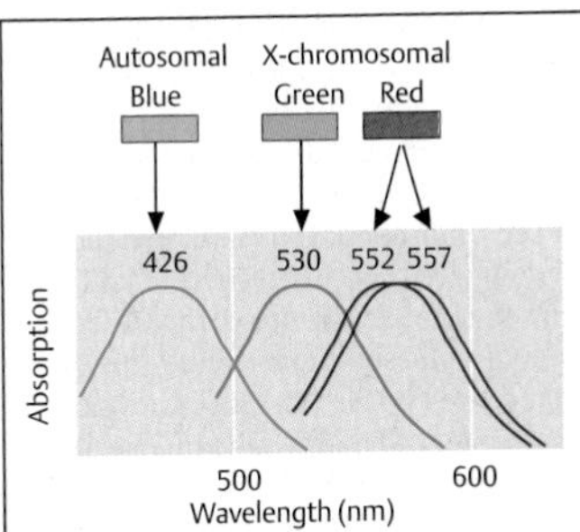


**A. Genes for photoreceptor proteins in rods**

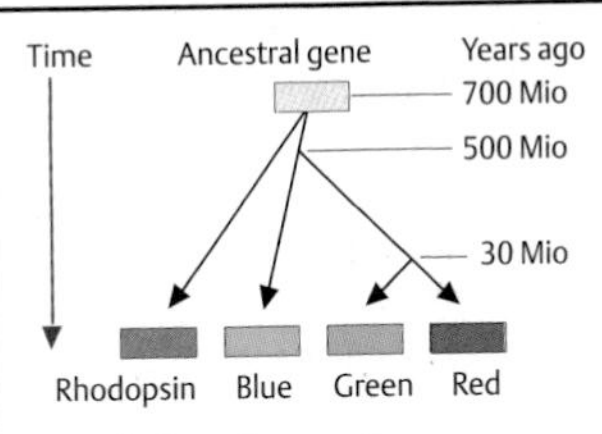


**B. Evolution of genes for visual pigment photoreceptors**

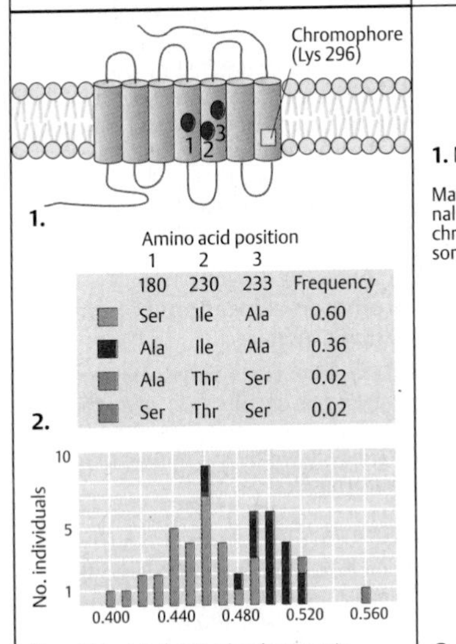


| | 1<br>180 | 2<br>230 | 3<br>233 | Frequency |
|---|---|---|---|---|
| | Ser | Ile | Ala | 0.60 |
| | Ala | Ile | Ala | 0.36 |
| | Ala | Thr | Ser | 0.02 |
| | Ser | Thr | Ser | 0.02 |

**D. Polymorphism in the photoreceptor red**

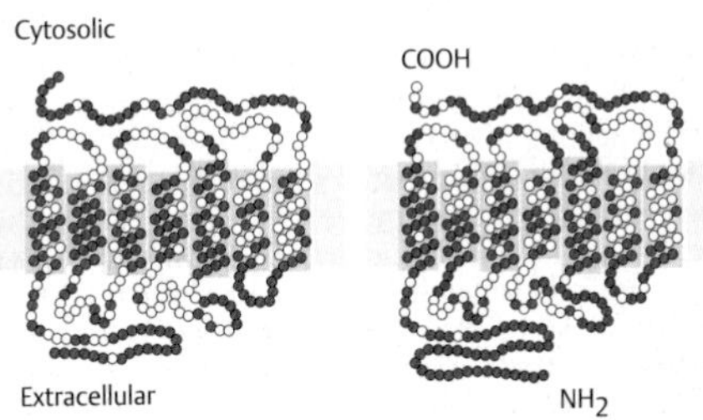


**1.** Blue/Rhodopsin 75% **2.** Green/Rhodopsin 41%

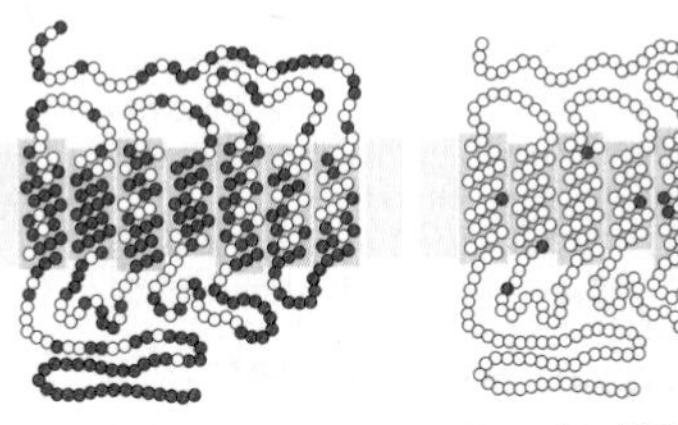

**3.** Green/Blue 44% **4.** Green/Red 96%

**C. Similar structure of visual pigments**

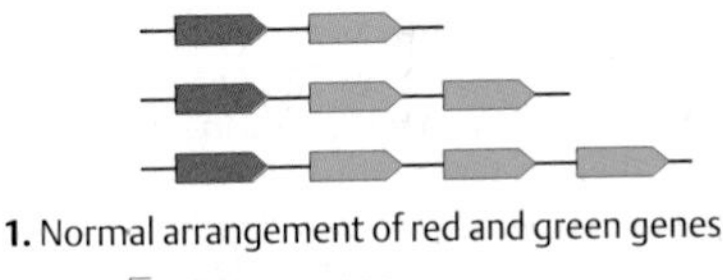

**1.** Normal arrangement of red and green genes

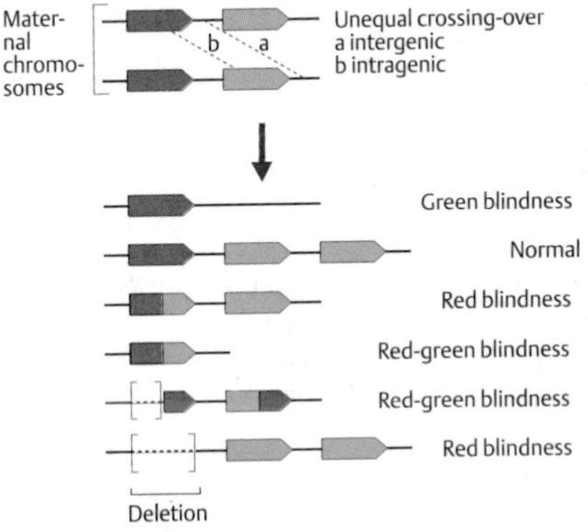


**2.** Examples of different consequences of unequal crossing-over

**E. Normal and abnormal red/green vision**

## Hearing and Deafness

Acoustic signals are essential for an animal's ability to respond appropriately to its environment. Hearing is orchestrated by a large ensemble of proteins acting in concert. Specialized sensory cells in the cochlea of the inner ear process the incoming sound waves, converting them into cellular information that is relayed to the brain via the acoustic nerve. A missing or defective protein involved in the hearing process results in hearing loss. Hearing loss is common in humans. One out of 1000 newborns lacks the ability to hear or has severely impaired hearing. Two categories of genetic hearing loss can be distinguished: nonsyndromic and syndromic. In the former category the genetic defect is limited to the ear; in the latter the ear is one of several organ systems affected.
The types of genes implicated when defective as the cause of nonsyndromic hearing loss include those encoding proteins involved in cytoskeletal structure, transcription factors, ion channels (potassium channel), and intercellular gap channels composed of junction connexins.

### A. The main components of the ear

The auditory system consists of the outer ear, the middle ear, and the inner ear. Sound waves are funneled through the outer ear (auricle) and transmitted through the external ear canal to the tympanic membrane, which they cause to vibrate. These vibrations are transmitted through the tympanic cavity of the middle ear by a chain of three movable bones, the malleus, the incus, and the stapes. Three major cavities form the inner ear: the vestibule, the cochlea, and the semicircular canals. The chochlea is the site where auditory signals are processed. The cochlea contains a membranous labyrinth filled with a fluid, the endolymph. The vestibular apparatus includes three semicircular canals oriented at 90° degree angles to each other. They respond to rotatory and linear acceleration. Signals received here are transmitted via the vestibular nerve, which fuses with the cochlear nerve to form the acoustic nerve. The latter transmits the information to the brain.

### B. The cochlea

The cochlea contains the cochlear duct, which forms the organ of Corti. The organ of Corti converts sound waves in the endolymph of the cochlea into intracellular signals. These are transmitted to auditory regions of the brain. The organ of Corti contains two types of sensory cells: one row of inner hair cells and three rows of outer hair cells. The inner hair cells are pure receptor cells. Vibrations induced by sound lead to slight deflections of the stereocilia and open potassium channels at the tips of the stereocilia. The influx of potassium ions at the tips of the cilia of the hair cells (see C) causes a change in membrane potential that results in a nerve impulse, which is transmitted as an auditory signal to the auditory cortex of the brain. Potassium ions are recycled to the supporting cells and the spiral ligament into the endolymph of the scala media. The tectorial membrane amplifies the sound waves as a resonator. (Figure adapted and redrawn from Willems, 2000.)

### C. The outer hair cell

The outer hair cells combine sensory function with the ability to elongate and contract when acoustically stimulated. The apical pole of a hair cell carries an array of about 100 cylindrical stereocilia in a V-shaped arrangement. Each stereocilium contains an actin molecule, which enables it to elongate or to contract. The tips of the stereocilia are connected by tip links. The potassium channels are formed by the KCNQ4 protein (yellow) and by connexins (red). Important for the structural integrity and dynamics of the hair cells is a cytoskeleton involving actin, myosin 7A, myosin 15, and the protein diaphanous. (Figure adapted and redrawn from Willems, 2000.)

### D. Chromosomal locations of human deafness genes

Almost every human chromosome harbors at least one gene involved in nonsyndromic monogenic hearing loss. The diagrammatic presentation shown here is limited to nonsyndromic hearing loss.

### References

Robertson N.D., Morton, R.N.D.: Beginning of a molecular era in hearing and deafness. Clin. Genet. **55**:149–159, 1999.

Steel, K.P., Bussoli, T.J.: Deafness genes. Expressions of surprise. Trends Genet. **15**:207–211, 1999.

Willems, P.J.: Genetic causes of hearing loss. New Engl. J. Med. **342**:1101–1109, 2000.

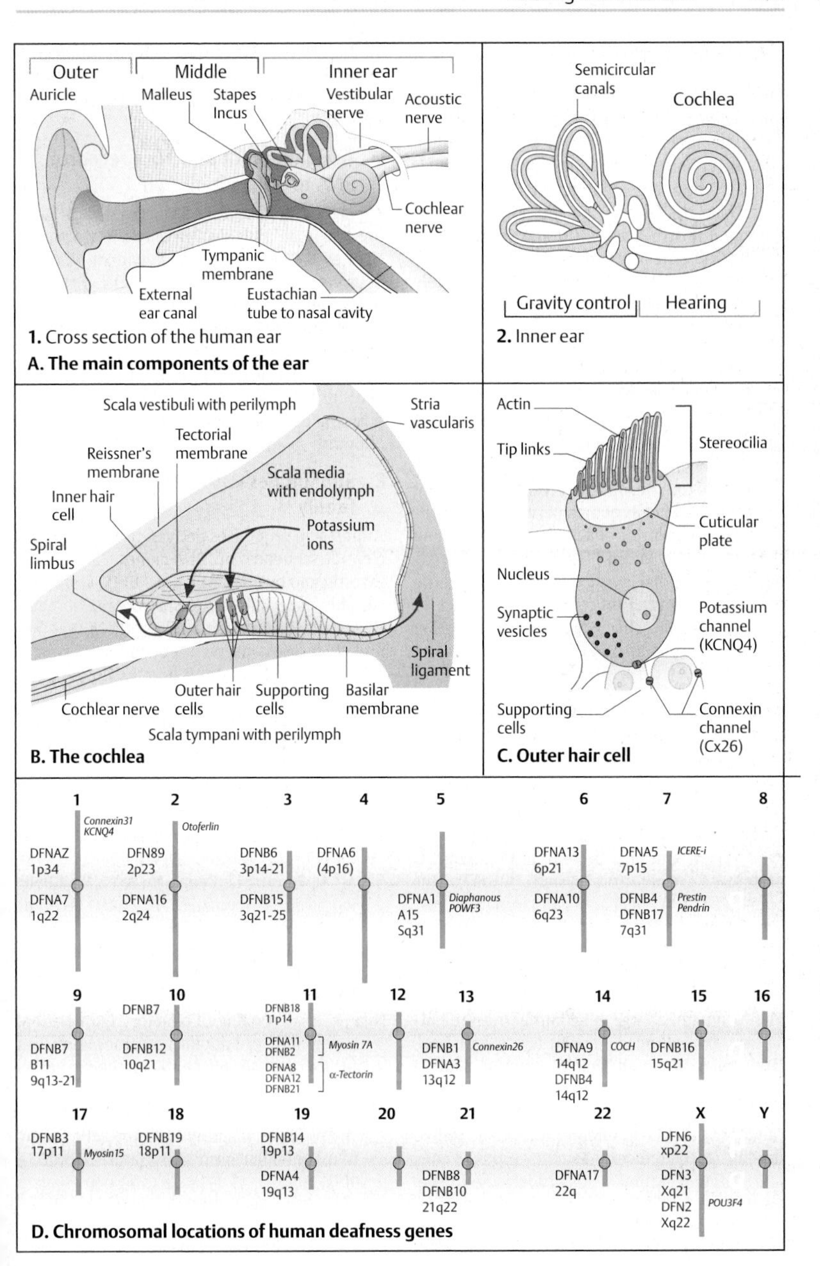


**1.** Cross section of the human ear

**2.** Inner ear

**A. The main components of the ear**

**B. The cochlea**

**C. Outer hair cell**

**D. Chromosomal locations of human deafness genes**

## Odorant Receptor Gene Family

Vertebrates can differentiate thousands of individual odors. Although their ability to distinguish differences in color is based on only three classes of photoreceptors, their sense of smell is regulated by a large multigene family of receptors that are highly specific for individual odorants. In fish, about 100 and in mammals about 1000 genes code for specific olfactory receptors. These genes are expressed exclusively in the olfactory epithelium of the nasal mucous membrane.

### A. Olfactory nerve cells in the nasal mucous membrane

The peripheral olfactory neuroepithelium of the nasal mucous membrane consists of three cell types: olfactory sensory neurons, whose axons lead to the olfactory bulb, supporting cells, and basal cells, which serve as stem cells for the formation of olfactory neurons during the individual's entire lifespan. Each olfactory neuron is bopolar, with olfactory cilia in the lumen of the nasal mucous membrane and a projection to the olfactory bulb, the first relay station of the olfactory system on the way to the brain.

### B. Odor-specific transmembrane receptors and GTP-binding protein ($G_s$[olf])

Each receptor in the cilia of the olfactory neurons binds specifically to one odorant. Binding of the receptor activates adenylate cyclase via a specific GTP-binding protein (stimulatory G protein of the olfactory system, $G_s$[olf]). This opens a sodium ion channel and initiates a cascade of intracellular signals that result in a nerve signal, which is transmitted to the brain.

### C. Olfactory receptor protein

The cloning of a large gene family from the olfactory epithelium of the rat (Buck and Axel, 1991), demonstrated that a receptor protein contains seven transmembrane regions and shows marked structural homology with rhodopsin and β-adrenergic receptors. Unlike rhodopsin, the olfactory receptor proteins contain variable amino acids, especially in the fourth and fifth transmembrane domains. The third intracellular loop between transmembrane domains V and VI is relatively short (17 amino acids), in contrast to other receptor proteins. It is assumed that contact with the various G proteins takes place here (Buck and Axel, 1991).

### D. Assignment of olfactory receptor RNA to neurons

A gene for the receptor of a given odorant is expressed in an individual in only a few neurons. Ngai et al. 1993 classified individual olfactory neurons in the olfactory epithelium of the catfish *(Ictalurus punctatus)*. Only 0.5–2% of all olfactory neurons recognize a given receptor probe such as probe 202 (1) or 32 (2). Odors are distinguished in the brain according to which neurons are stimulated. The topographical position of each neuron is specific for each odorant.

### E. Subfamilies within the multigene family

Amino acid sequences derived from partial nucleotide sequences of cDNA clones (F2–F24) (1) investigated by Buck and Axel (1991) were very variable, especially in transmembrane domains III and IV. Within subfamilies, there was homology due to conserved sequences (2). For example, F12 and F13 differ in only 4 of 44 positions (91% identical).
(Figures adapted from Buck & Axel, 1991, and Ngai et al., 1993).

### References

Buck, L., Axel, R.: A novel multigene family may encode odorant receptors: a molecular basis for odor recognition. Cell **65**:175–187, 1991.

Chess, A., Femon, I.I, Cedar, H., Axel, R.: Allelic inactivation regulates olfactory receptor gene expression. Cell **78**:823–834, 1994.

Ngai, J., et al.: The family of genes encoding odorant receptors in the channel catfish. Cell **72**:657–666, 1993.

Ngai, J., et al.: Coding of olfactory information: topography of odorant receptor expression in the catfish olfactory epithelium. Cell **72**:667–680, 1993.

Parmentier, M., et al.: Expression of members of the putative olfactory receptor gene family in mammalian germ cells. Nature **355**:453–455, 1992.

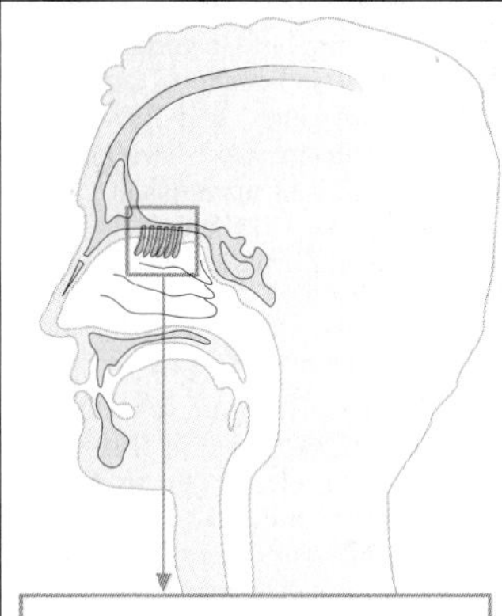

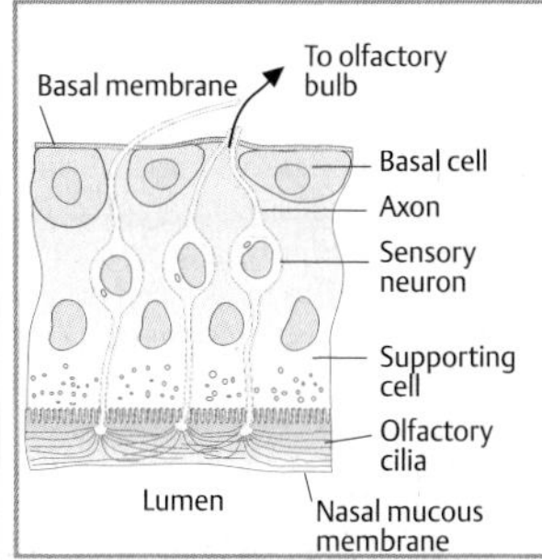


**A. Olfactory nerve cells in the nasal mucous membrane**

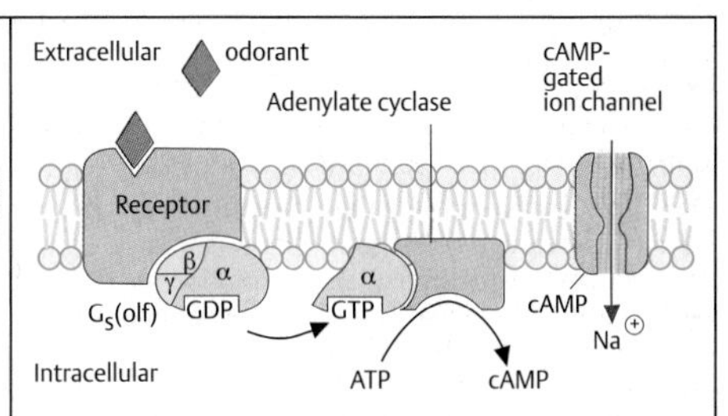


**B. Odor-specific transmembrane receptor and GTP-binding protein [$G_s$(olf)]**

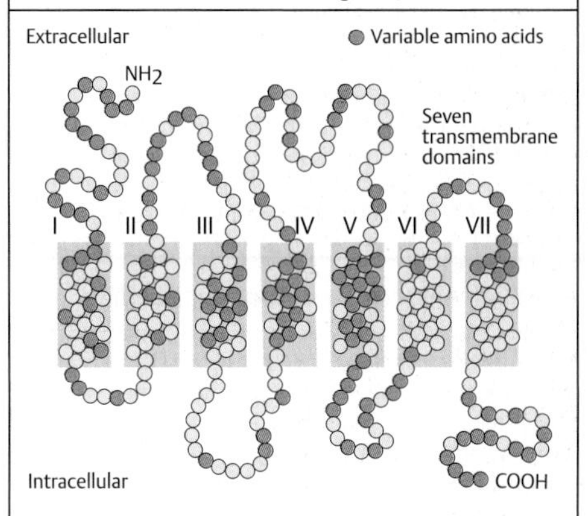


**C. Olfactory receptor protein**

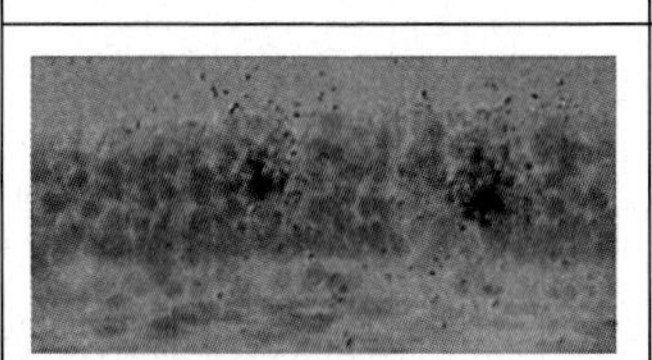

**1.** Receptor probe 202

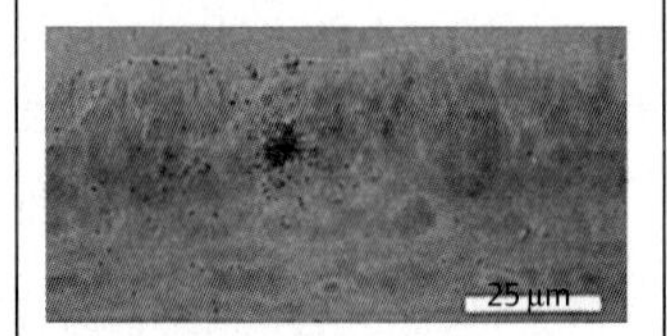


**2.** Receptor probe 32

**D. Correspondence of olfactory receptor RNA to neurons**

| cDNA clones | | Transmembrane domain | |
|---|---|---|---|
| F2 | RVNE | VVIFIVVSLFLVLPFALIIMSYV | RIVSSILKVPSSQGIYK |
| F3 | FLND | LVIYFTLVLLATVPLAGIFYSYF | KIVSSICAISSVHGKYK |
| F5 | HLNE | LMILTEGAVVMVTPFVCILISYI | HITCAVLRVSSPRGGWK |
| F6 | QVVE | LVSFGIAFCVILGSCGITLVSYA | YIITTIIKIPSARGRHR |
| F7 | HVNE | LVIFVMGGIILVIPFVLIIVSYV | RIVSSILKVPSARGIRK |
| F8 | FPSH | LTMHLVPVILAAISLSGILYSYF | KIVSSIRSMSSVQGKYK |
| F12 | FPSH | LIMNLVPVMLAAISFSGILYSYF | KIVSSIHSISTVQGKYK |
| F13 | FPSH | LIMNLVPVMLAAISFSGILYSYF | KIVSSIRSVSSVKGKYK |
| F23 | FLND | VIMYFALVLLAVVPLLGILYSYS | KIVSSIRAISTVQGKYK |
| F24 | HEIE | MIILVLAAFNLISSLLVVLVSYL | FILIAILRMNSAEGRRK |

**1.** Variable amino acid sequences

| | | | |
|---|---|---|---|
| F12 | FPSH | LIMNLVPVMLAAIISFSGILYSYF | KIVSSIHSISTVQGKYK |
| F13 | FPSH | LIMNLVPVMLAAIISFSGILYSYF | KIVSSIRSVSSVKGKYK |
| F8 | FPSH | LTMHLVPVILAAIISLSGILYSYF | KIVSSIRSMSSVQGKYK |
| I12 | FPSH | LIMNLVPVMLGAIISLSGILYSYF | KIVSSVRSISSVQGKHK |
| F23 | FLND | VIMYFALVLLAVVVPLLGILYSYS | KIVSSIRAISTVQGKYK |
| F3 | FLND | LVIYFTLVLLATVVPLAGIFYSYF | KIVSSICAISSVHGKYK |

**2.** Homology within subfamilies

**E. Subfamilies within the multigene family**

## Mammalian Taste Receptor Gene Family

Aside from the main olfactory system, mammals have evolved two other chemosensory systems, the taste receptor gene family (for bitter taste) and the mammalian pheromone receptor gene family. Five different types of taste can be perceived: salty, sour, bitter, sweet, and umani (the taste of monosodium glutamate, present in Asian food). Salty and sour tastes involve direct effects due to the entry of $H^+$ and $Na^+$ ions through specialized membrane channels. In contrast, bitter, sweet, and umani tastes are mediated via a G protein-coupled receptor (GPCR) signaling pathway system. A sweet taste may herald a desirable carbohydrate content, whereas a bitter taste is associated with potentially toxic substances such as alkaloids, cyanides, or other detrimental aromatic compounds.

### A. Mammalian chemosensory epithelia

The oral and nasal cavities of mammals contain three distinct chemosensory epithelia: (i) the main olfactory epithelium (MOE) containing sensory cells with odorant receptors in the nose (see previous page), (ii) the taste sensory epithelium of the taste buds of the tongue, soft palate, and epiglottis, and (iii) the vomeronasal organ (VOM, also called Jacobson's organ), a tubular structure in the nasal septum containing sensory cells with pheromone receptors. The main olfactory bulb (MOB) relays signals from the MOE to the olfactory cortex of the brain. The accessory olfactory bulb (AOB) relays signals from the VOM to areas of the amygdala and hypothalamus.

### B. Mammalian chemosensory systems

The receptor cells are organized in three corresponding molecular and cellular chemosensory systems. Each neuron of the main olfactory sensory system (1) expresses one of the different olfactory receptor genes and sends axons to specific glomeruli of the main olfactory bulb (mitral cells). The odorant receptor (OR) gene family comprises about 1000 members, each encoding a seven-transmembrane cyclic nucleotide-gated channel with distinct odorant specificity (G-olfactory proteins, $G_{olf}$). The bitter taste sensory system (2) connects axonal projections of receptor cells in the taste sensory epithelium of the taste buds to gustatory nuclei of the brain stem. Two families of taste receptors, the TIRs (two genes) and T2Rs (50–80 genes of the gustducin class) have been described. Two families of mammalian putative pheromone receptors (V1Rs and V2Rs) are encoded by 30–50 and over 100 genes, respectively (3).

### C. Taste receptor gene family

The two novel taste receptor gene families, T1R1 and T1R2, are expressed in distinct subsets of taste receptor cells. The figure shows an alignment of the predicted amino acid sequences of 23 different T2 receptors (T2Rs) of human (h), rat (r), and mouse (m) origin between the first (TM1) and the third (TM3) transmembrane domain. Dark blue indicates identity in at least half of the aligned sequences; light blue represents conserved substitutions; and the remainder are divergent regions. They reflect the ability to bind many structurally different ligands. The T2R genes cluster on a few chromosomes, human chromosomes 5, 7, and 12 and mouse chromosomes 6 and 15.

### D. Expression of many taste receptor genes in the same cell

Unlike olfactory system receptor cells, individual taste receptor cells express multiple T2R receptors. Up to ten T2R probes hybridize to only a few cells, shown darkened (1). Double-label fluorescent in-situ hybridization shows that different receptor genes (2, T2R-3 in green and T2R-7 in red, 3) are expressed in the same taste receptor cell. The T2Rs confer high sensitivity for bitter substances at low concentrations but do not distinguish between them. (Figures adapted from Dulac, 2000, and Adler et al., 2000.)

### References

Adler, E., et al.: A novel family of mammalian taste receptors. Cell **100**:693–702, 2000.

Buck, L.B.: The molecular architecture of odor and pheromone sensing in mammals. Cell **100**:611–618, 2000.

Chandrashekar, J., et al.: T2 Rs function as bitter taste receptors. Cell **100**:703–711, 2000.

Dulac, C.: The physiology of taste, vintage 2000. Cell **100**:607–610, 2000.

Malnic, B., et al.: Combinatorial receptor codes for odors. Cell **96**:713–723, 1999.

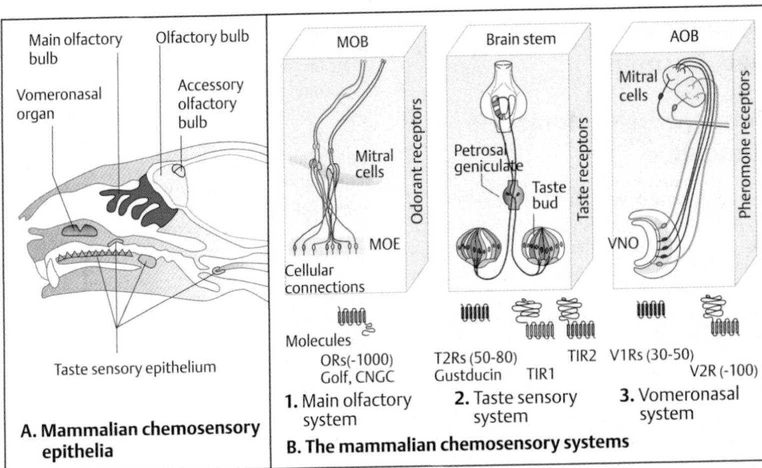


A. Mammalian chemosensory epithelia

B. The mammalian chemosensory systems

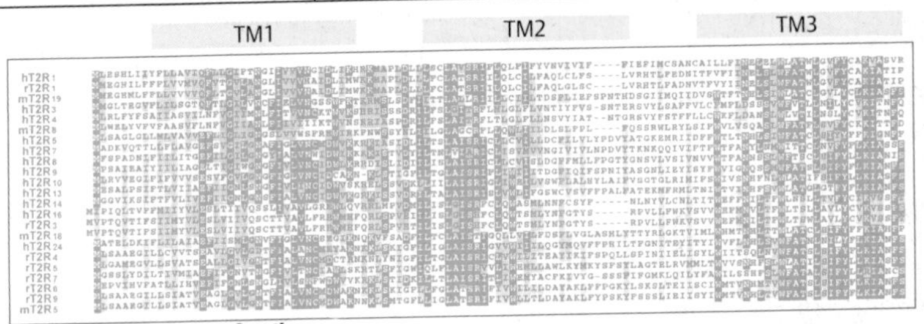


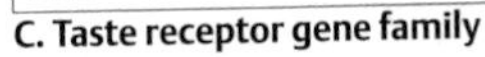
C. Taste receptor gene family

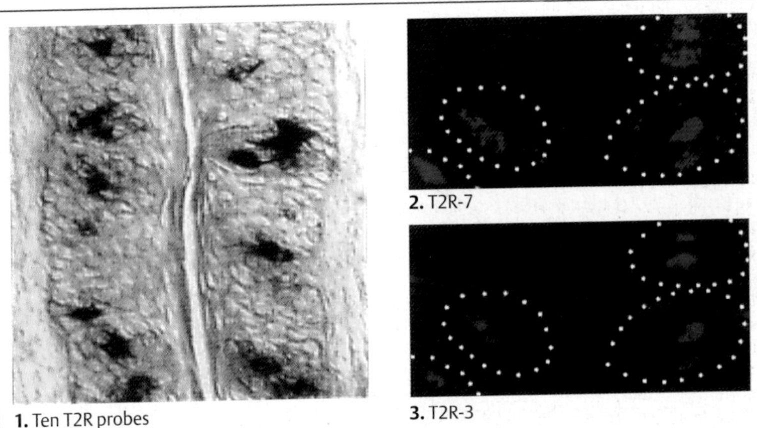

1. Ten T2R probes

2. T2R-7

3. T2R-3

D. Expression of many taste receptor genes in the same cell

# Genes in Embryonic Development

## Developmental Mutants in Drosophila

The embryonic development of an organism is determined by genes that may be active only during specific phases. Analysis of developmental mutants of embryos of the fruit fly *Drosophila melanogaster* has provided insight into the genetic regulation of developmental processes. Early developmental phases in embryos of very different organisms are regulated by similar genes.

### A. The segmental organization of the fruit fly (*Drosophila melanogaster*)

The development of a fruit fly from the fertilized egg cell to the segmented body of the adult organism takes nine days. The larvae hatch after one day and pass through defined stages of embryonic development. The embryo forms a cocoon at five days and, after metamorphosis, emerges as a 2 mm-long adult fly. The head of the adult has three segments (C1 – 3), the thorax three (T1 – 3), and the abdomen eight segments (A1 – 8). A fruit fly has altogether 14 parasegments (P1 – 14), each corresponding to the last half of one and the first half of the next segment. The segmental organization is discernible in the larva.

### B. Embryonic lethal mutations

Embryonic lethal mutations can be identified by appropriate crosses. One-fourth of the progeny of heterozygous flies (A/a) for an embryonic mutant gene (a) are homozygotes (aa) for the mutant allele (1). If a mutation involves maternal genes only (maternal effect), progeny of female homozygotes (bb) are lethally affected (2). Maternal effect genes code for early gene products that determine the polarity of the embryo; see below (C).

### C. Examples of developmental mutants

Many developmental mutations are known in *Drosophila melanogaster.* They can be classified into different hierarchical gene classes. The normal larva (wild type) consists of three head segments, three thorax segments, eight abdominal segments, and the tail end (1). A mutation for anterior maternal effect, *bicoid,* leads to the development of a larva without head or thorax (2). A mutation of a gene called *nanos* affects the posterior end of the early larva. *Gap* genes establish the basic pattern of segmental organization. Mutations of the *gap* genes lead to omissions (gaps) in the segmental construction of the larva. In the *Krüppel* mutant (3), all thoracic and the abdominal segements 1 – 5 are missing; in the *Knirps* mutant (4), abdominal segments 1 – 6 are absent. The genes for *pair-rule* determine the orientation and developmental fate of the 14 parasegments. Some mutations affect every second segment. With *even-skipped* (5), all even-numbered parasegments are missing. Mutation of the gene *fushi tarazu* leads to fewer than normal segments being formed (*fushi tarazu is* Japanese for too few segments). *Segment polarity* genes determine the polarity of each segment (7). There are more than ten *segment polarity* genes. *Homeotic selector* genes (8) determine the ultimate fate of each segment. With the mutant *antennapedia (Ant),* the antenna normally attached immediately under the eye is replaced by a leg (homeotic leg). (Figures adapted from Watson et al., 1992).

### References


Gehring, W.J., et al.: Homeodomain-DNA recognition. Cell **78**:211 – 223, 1994.

Kenyon, C.: If birds can fly, why can't we? Homeotic genes and evolution. Cell **78**:175 – 180, 1994.

Lawrence, P.A.: The Making of a Fly. The Genetics of Animal Design. Blackwell Scientific, Oxford, 1992.

Nüsslein-Volhard, C., Frohnhöfer, H.G., Lehmann, R.: Determination of anterior-posterior polarity in Drosophila. Science **238**:1675 – 1681, 1987.

Watson, J.D., et al.: Recombinant DNA. 2nd ed. W.H. Freeman, Scientific American Books, 1992.

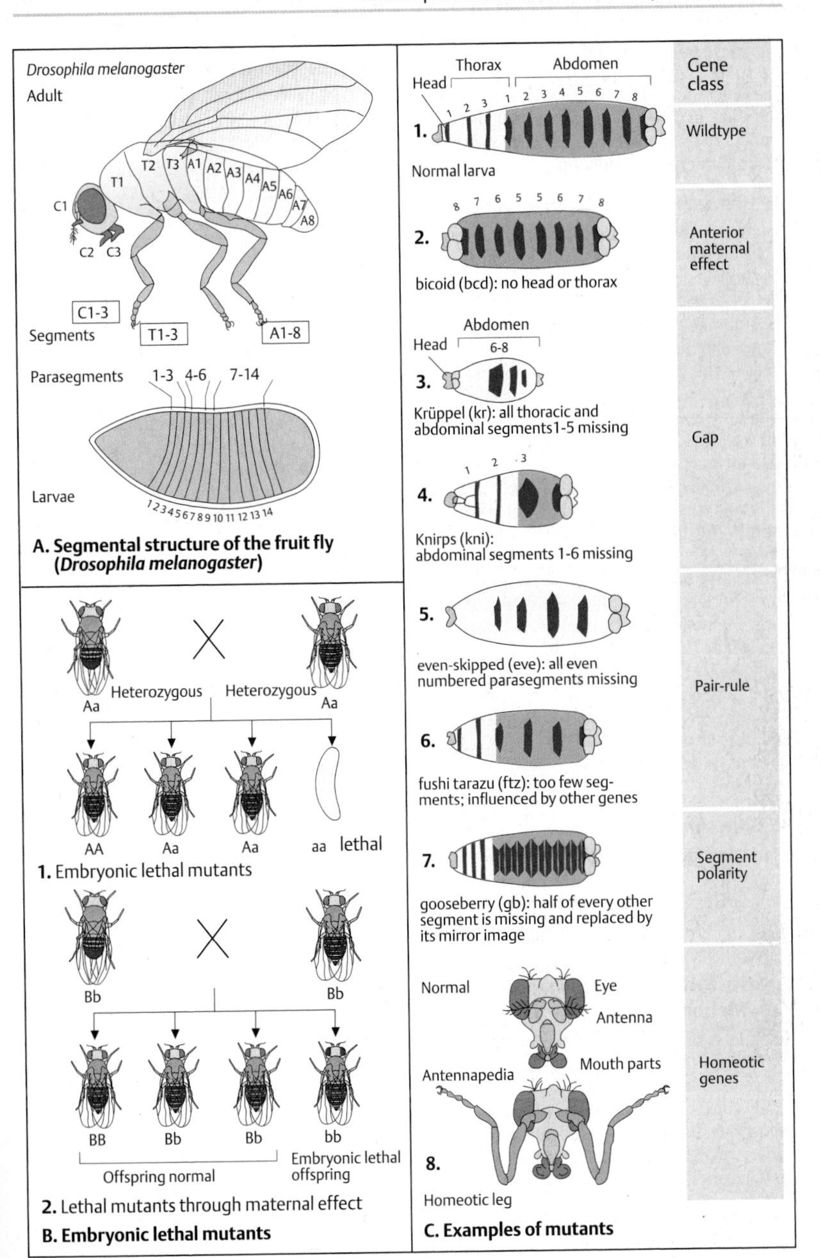


**A. Segmental structure of the fruit fly (*Drosophila melanogaster*)**

**B. Embryonic lethal mutants**

**C. Examples of mutants**

# Homeobox Genes

The genes for embryonic development in *Drosophila* are organized in a functional hierarchy. Similar genes (in some regions even identical) occur in mice and in humans.

## A. Hierarchy of developmental genes

Four independent systems control the embryonic development of *Drosophila melanogaster.* A gradient of maternally derived protein coded for by the *bicoid* gene *(bcd)* determines development in the anterior region (head); likewise for *nanos* in the posterior region. They activate the *gap* genes. The three most important *gap* genes, *hunchback, Knirps,* and *Krüppel,* code for transcription factors of the zinc finger type (p. 218). *Hunchback* is expressed from the anterior part to about the middle of the embryo; *Krüppel* in the region of thoracic segments 4–6 and the first six abdominal segments; *Knirps* further posteriorly (parasegments 10–14). The *gap* genes induce the *pair rule* genes. *Segment polarity* genes determine the correct orientation of the individual segments. After segmentation is complete, *selector* genes determine the further development and the ultimate fate of the segments. They consist of three large gene complexes, the *antennapedia* complex (*ANT*-C), the *bithorax* complex (*BX*-C), and the *ultrabithorax* complex (UBX). Mutations of these gene complexes lead to unusual structures.

## B. Bithorax mutation

A mutant for the *bithorax* complex (*BX*-C) causes the development of an additional thoracic segment with completely developed wings. (Photograph from Lawrence, 1992, after E. B. Lewis.)

## C. Structure of the *antennapedia* gene with homeobox

*Antennapedia* is a gene of the *antennapedia*-complex (*ANT*-C), which is expressed in parasegments 5 and 6. It contains a segment of highly conserved DNA sequences in exon 8, the homeobox. Its sequence is identical in a wide variety of organisms, from *Drosophila* to mammals. It codes for about 60 amino acids (homeodomain). The homeodomain contains four domains of a helical protein with DNA-binding properties and functions as a transcription factor (Gehring et al., 1990).

## D. Homeotic genes in *Drosophila* and *Hox* genes in mouse

Homeobox genes also occur in vertebrates. A series of genes with the same anterior to posterior orientation and corresponding to the ANT and BX complexes (*ANT*-C and *BX*-C) are found in the embryonic brain of the mouse (and in man). The temporal expression corresponds to this orientation.

## E. Homeobox genes (*HOX* genes)

In humans and in mice there are four groups (clusters) of genes that correspond to the homeo genes of *Drosophila* (*HOX* 1–4 in humans, *Hox* 1–4 in the mouse). Since mutations of these genes in mice lead to characteristic disorders, it is assumed that they are of clinical significance in humans also. The posterior end corresponds to the 5′ end and the anterior end to the 3′ end of the coding DNA.

## References


Amores, A., et al.: Zebrafish hox clusters and vertebrate genome evolution. Science **282**:1711–1714, 1998.

Gehring, W.J., et al.: The structure of the homeodomain and its functional implications. Trends Genet. **6**:323–329, 1990.

Krumlauf, R.: Hox genes in vertebrate development. Cell **78**:191–201, 1994.

Lawrence, P.A.: The Making of a Fly. The Genetics of Animal Design. Blackwell Scientific, Oxford, 1992.

Marx, J.: Homeobox genes go evolutionary. Science **255**:399–401, 1992.

Reddihough, G.: Homing in on the homeobox. Nature **357**:643–644, 1992.

Scott, M.P.: A rational nomenclature for vertebrate homeobox (HOX) genes. Nucleic Acids Res. **21**:1687–1688, 1993.

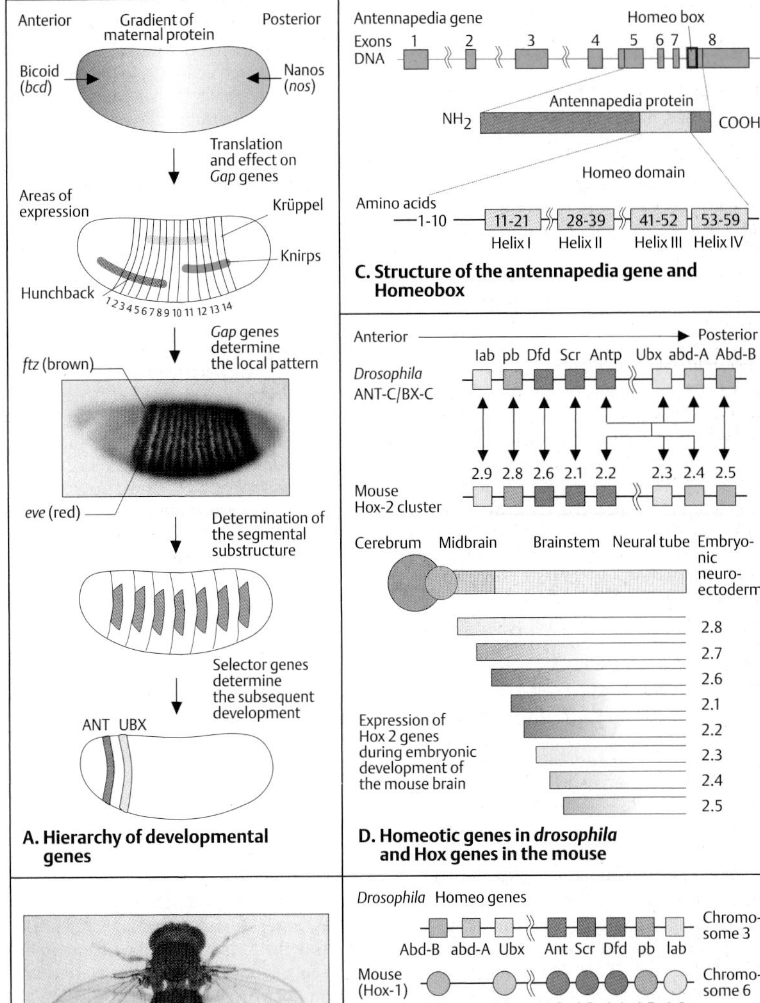


**A. Hierarchy of developmental genes**

**B. Bithorax mutation**

**C. Structure of the antennapedia gene and Homeobox**

**D. Homeotic genes in *drosophila* and Hox genes in the mouse**

**E. Homeobox genes (HOX)**

## Genetics in a Lucent Vertebrate Embryo: Zebrafish

The zebrafish *(Danio rerio)* is the first vertebrate to be studied by systematic mutation search. The mutations found involve early development, the dorsoventral pattern, formation of the notochord and somites, brain and neural development, the eyes, the ears, the internal organs, blood pigmentation, and other areas. Two of the more than a thousand well-characterized mutants are presented below. Detailed information is available in the zebrafish issue of *Development* (123: 1–481, December 1996).

### A. Zebrafish embryos

The zebrafish embryo is optically clear and can easily be studied by dissecting microscopy at maximally 80× magnification. At 29 hours after fertilization (pharyngula period), the main parts of the brain, the forebrain, midbrain, and hindbrain, are clearly visible along with the neural tube, several somites, and the floor plate. At 48 hours (hatching period) pigmentation begins, and the fins, eyes, brain, heart, and other structures become visible. At five days (swimming larva), the outline of a fish begins to show.

### B. Induced mutagenesis

Adult fish (1) are used in a crossing scheme involving the exposure of male fish to 3 mM ethylnitrosourea (ENU) in an aqueous solution for three one-hour periods within one week. Mutagenized males are crossed with wild-type females (2). The first generation of this cross (F1) is heterozygous for one mutagenized genome. The next generation (F2) is raised from sibling matings, resulting in the presence of a mutation **m** in 50% of the F2 fish. Random matings involving two heterozygous parents results in 25% homozygous mutant offspring, which are analyzed. A total of 4264 mutants were identified this way by Haffter et al., 1996.

### C. Skeletal phenotype of the *fused somites (fss)* mutation

Of the 1163 mutants characterized by their phenotypic effects, two examples are given here and in D. The *fss* mutants show irregular somite boundaries 72 hours after fertilization. At the seven-somite stage the only abnormality is the absence of somite boundaries. At later stages malformations of the tail vertebral column are visible (2). In wild-type fish the vertebral centrum has one neural arch dorsally and one hemal arch ventrally. In the *fss* mutant the arches are irregular in shape and, since they grow ectopically, in their relation to each other (2).

### D. The midbrain mutation *no isthmus*

This is one example of the more than 60 mutant genes shown to affect the central nervous system and spinal cord (Haffter et al., 1996; Brand et al., 1996). The *no isthmus (noi)* mutation affects the boundary between midbrain and hindbrain. Normal embryos at the 24–48 h pharyngula stage show a conspicuous constriction at this boundary, whereas mutant *noi* embryos do not. In addition, the cerebellum derived from the posterior part is absent. Brand et al. (1996) studied the expression of two genes in this region, *engrailed (eng)* and wingless *(wnt1)*. Normal embryos *(wt)* at 28 h show strong expression of *eng* between midbrain and hindbrain, while mutant *noi* embryos do not. Eight-somite-stage embryos double stained for eng and krx20 RNA, a marker for rhombomeres 3 and 5 in this region, show *eng* expression and *krox20* expression at the midbrain–hindbrain boundary, but no expression of *eng* in the *noi* mutant. Similarly, *wn1* expression posterior to the tectum (tec, the border of the two brain regions, in the 20-somite-stage embryo is eliminated in the mutant. (Figures adapted from Haffter et al., van Eeden et al., and Brand et al., 1996)

### References


Dodd, A. et al.: Zebrafish: bridging the gap between development and disease. Hum. Mol. Genet. **9**:2443–2449, 2000.

Brand M., et al.: Mutations in zebrafish genes affecting the formation of the boundary between midbrain and hindbrain. Development **123**:179–190, 1996.

Haffter P., et al.: The identification of genes with unique and essential functions in the development of the zebrafish, *Danio rerio*. Development **123**:1–36, 1996.

van Eeden, F.J.M., et al.: Mutations affecting somite formation and patterning in the zebrafish, *Danio rerio*. Development **123**:153–164, 1996.

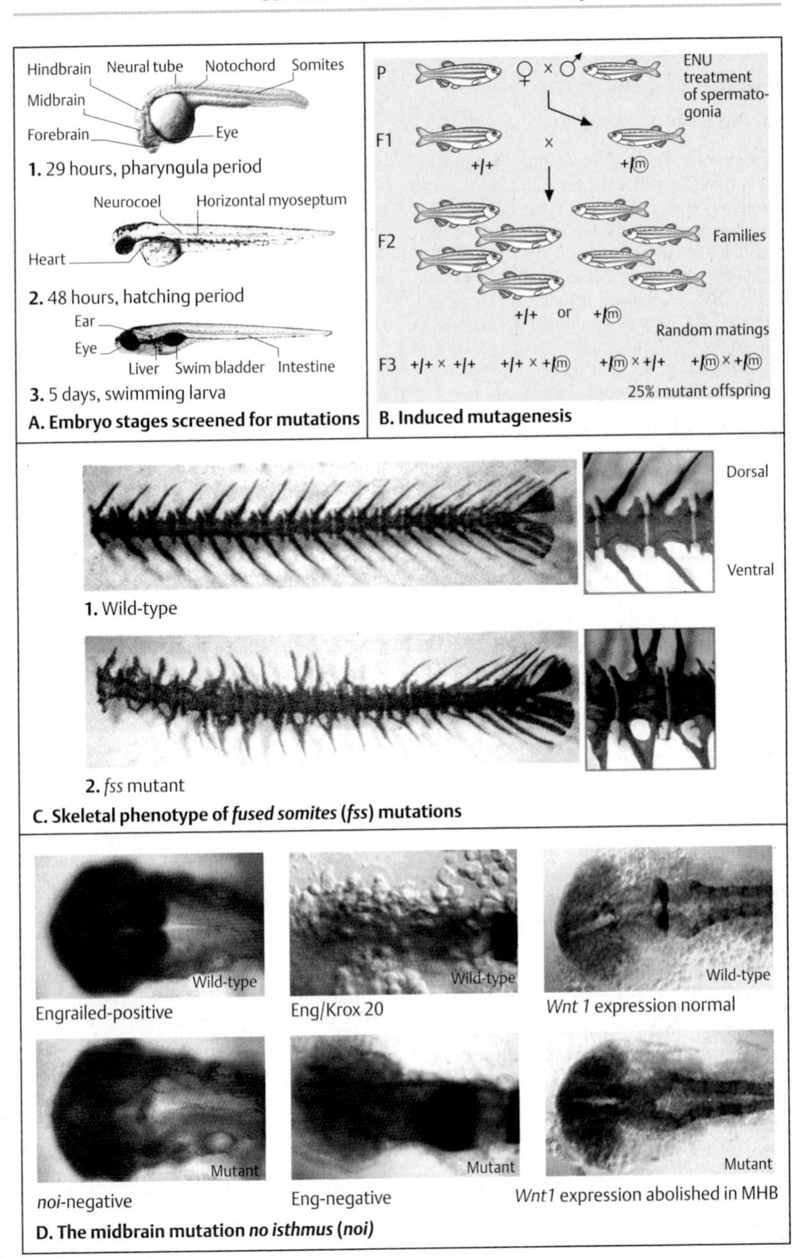


**A. Embryo stages screened for mutations**

**B. Induced mutagenesis**

**C. Skeletal phenotype of *fused somites* (*fss*) mutations**

**D. The midbrain mutation *no isthmus* (*noi*)**

# Developmental Program for Individual Cells in the Nematode C. elegans

Development from a zygote into an adult organism requires that the time and place of cell divisions be organized. Whereas cell origin in complex organisms cannot be determined, it is possible to analyze the origin of each individual cell in the nematode *Caenorhabditis elegans*. In 1965, Sydney Brenner introduced the genetic analysis of *C. elegans* as a model to investigate the interaction of genetic, anatomical, and physiological traits in the development of a relatively simple nervous system.

## A. *Caenorhabditis elegans*

*Caenorhabditis elegans* is a small (about 1 mm long), transparent worm with a life cycle of about 3 days. The basic structure is a bilaterally symmetric elongated body of nerves, muscles, skin, and intestine. It exists as one of two sexes: hermaphrodite or male. Hermaphrodites produce eggs and sperm and can reproduce by self-fertilization. The adult hermaphrodite worm has 959 somatic cell nuclei; the adult male worm has 1031. In addition, there are 1000 to 2000 germ cells.
The complete nucleotide sequence of the *C. elegans* genome was reported in 1998. The 97 Mb DNA contains 19099 predicted genes, 12000 without known function. Coding sequences make up about 27% of the DNA, introns account for 26%. About 32% of the coding sequences are similar to human and about 70% of known human proteins have homologies in *C. elegans*. The largest group of genes are transmembrane receptors (790), in particular chemoreceptors; zinc finger transcription factors (480); and proteins with protein-kinase domains.
(Diagram from Wood, 1988, after Sulston and Horvitz, 1977.)

## B. Embryonic origin of individual cells

The developmental pathway of each individual cell can be traced. The various tissues arise from six founder cells. Of the 959 adult cells, 302 are nerve cells. Except for cells of the intestine and germ line, differentiated tissues stem from different founder cells. Cells with similar function are not necessarily related, while cells with different functions may be of the same origin.
Genetically established rules determine the fate of the two daughter cells at each cell division.

## C. Mutations in the developmental control genes

Many developmental control genes have been identified from analysis of ethyl methanesulfonate-induced point mutations. Some mutants determine an incorrect cell type (e.g., Z instead of B); others divide too early or too late (division mutants) resulting, for example, in C twice, instead of B and C.

## D. Programmed cell death

Programmed cell death (apoptosis) is a normal part of vertebrate and invertebrate development. During embryonic development of *C. elegans*, one in eight somatic cells regularly dies at a defined time and branching point (1). The photographs (2) show the death of a cell (cell p11.aap) over a time span of about 40 minutes (from Wood, 1988, after Sulston and Horvitz, 1977). Mutations that interfere with programmed cell death can lead to severe disturbances. The human *BCL-2* gene corresponds to the *ced-9* gene in *C. elegans*, a gene that prevents apoptosis. The human *BCL-2* gene codes for an inner mitochondrial membrane protein that acts by inhibiting cell loss by apoptosis in pro-B lymphocytes. Disruption of this gene on human chromosome 18 causes follicular lymphoma, a B-cell tumor.

## References


The *C. elegans* Sequencing Consortium: Genome sequence for the nematode *C. elegans*: A platform for investigating biology. Science **282**:2012–2018, 1998.

Culetto, E., Sattelle, D.B.: A role for Caenorhabditis elegans in understanding the function and interactions of human disease genes. Hum. Mol. Genet. 9:869–878, 2000.

W.B. Wood and the Community of C. elegans Researchers: The Nematode Caenorhabditis elegans. Monograph 17, Cold Spring Harbor, New York, 1998.

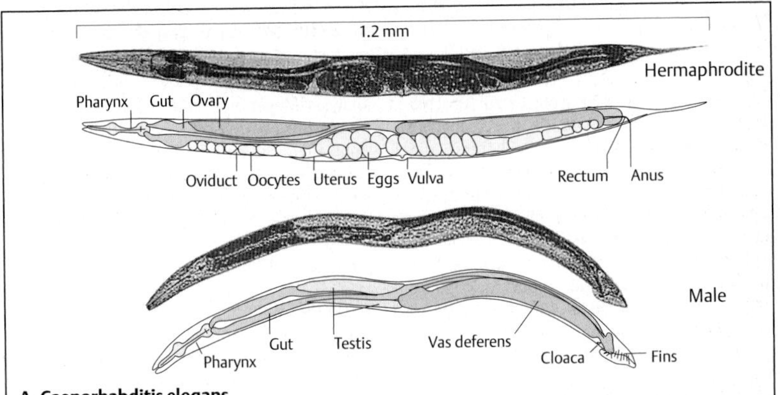


**A. Caenorhabditis elegans**

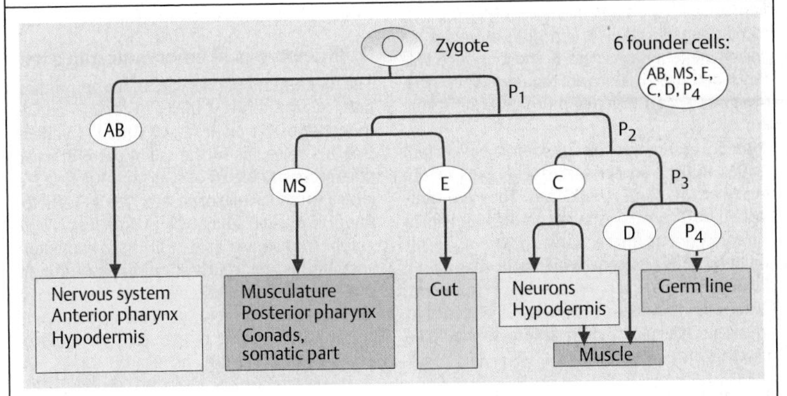


**B. Embryonic origin of individual cells**

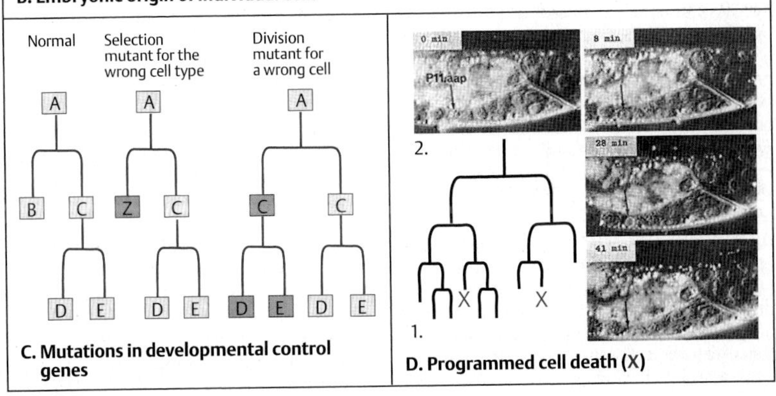


**C. Mutations in developmental control genes**

**D. Programmed cell death (X)**

## Developmental Genes in a Plant Embryo

In plants, as in animals, the basic structural plan is genetically determined. In the plant *Arabidopsis thaliana* (common wall cress), nine genes with numerous alleles determine the organization of the plant embryo along an apical–basal longitudinal axis. They determine a radial pattern and the form (Mayer et al., 1991). This example from plant genetics is included here because *Arabidopsis* is an important object of genetic studies. Its small 130 Mb genome is the first completely sequenced plant (E. Pennisi, Science **290**:32–34, 2000).
The four organs of the flower—sepals, petals, stamens, and carpels—develop under the influence of four classes of genes. Class A genes encode sepals; class A and class B genes together make petals; class B and C induce stamens; and class C genes are required for carpels. The class A and C genes mutually inhibit one another.
Homeotic genes specify the identity of certain tissues in both insects and the organs of the flower of the plant *Arabidopsis*. They are transcribed in a precise spatial pattern and form complex genetic interactions. The overall result is a determined state in which cells maintain a distinct fate irrespective of the environment. Mutations then lead to expression of genes at wrong sites, e.g. legs where antennae should be or petals instead of stamens.

### A. Normal development and structure

The basic structural plan can be understood as an axial and a radial pattern superimposed on each other. An octant stage, a globular stage, and a so-called heart stage can be differentiated before the seedling is formed. The regions A, C, and B of the octant stage correspond to the regions A, C, and B of the heart stage. Region A forms the cotyledon and the meristem; C forms, the hypocotyl region; and B forms the root. The seedling consists of a set of identifiable structures including vessels (v), external epidermis (e), short meristem (s), cotyledons (c), and hypocotyl (h), ground tissue (g), and root primordium (r) at the bottom (labeled v instead of r). In the heart stage, the essential organization of the plant is predetermined.

### B. Deletions in the apical–basal pattern

The mutations can be induced by 0.3% ethyl methanesulfonate. Using complementation analysis, Mayer et al. (1991) determined mutations in three areas of the plant. These mutations affect the apical–basal pattern, the radial pattern, and the shape. Apical–basal deletions involve one of several genes, each leading to a characteristic phenotype: apical deletion *(Gurke)*, central deletion *(Fackel)*, basal deletion *(monopteros)*, and terminal deletion *(Gnom)*.

### C. Wild-type

The normal structure of embryonic *Arabidopsis* results from two basic processes: formation of patterns (apical–basal and radial orientation) and morphogenesis through different cell forms and regional differences in cell division.

### D. Phenotypes of embryonic mutants

The four mutant phenotypes in the apical–basal pattern are Gurke (9 alleles), Fackel (5 alleles), monopteros (11 alleles), and Gnom (15 alleles) (see B). Deletions in the radial pattern lead to phenotypes Keule (9 alleles) and Knolle (2 alleles). Mutants of shape are Fass (12 alleles), Knopf (6 alleles), and Mickey (8 alleles). (Photographs from Mayer et al., 1991.) The *monopteros* gene *(ml)* is apparently very important for apical–basal development. However, it also has an indirect effect on the spatial arrangement of the apical structures. It is not necessary for root development (Berleth and Jürgens, 1993).

### References

Berleth, T., Jürgens, G.: The role of the monopteros gene in organising the basal body region of the Arabidopsis embryo. Development **118**:575–587, 1993.
Jürgens, G.: Memorizing the floral ABC. Nature **386**:17, 1997.
Lin, X., et al.: Sequence and analysis of chromosome 2 of the plant *Arabidopsis thaliana*. Nature **402**:761–768, 1999.
Mayer, U., et al.: Mutations affecting body organization in the *Arabidopsis* embryo. Nature **353**:402–407, 1991.
Pelaz, S., et al.: B and C floral organ identity functions require *SEPALLATA* MADS-box genes. Nature **405**:200–203, 2000.
Sommerville, C., Sommerville, S.: Plant functional genomics. Science **285**:380–383, 1999.

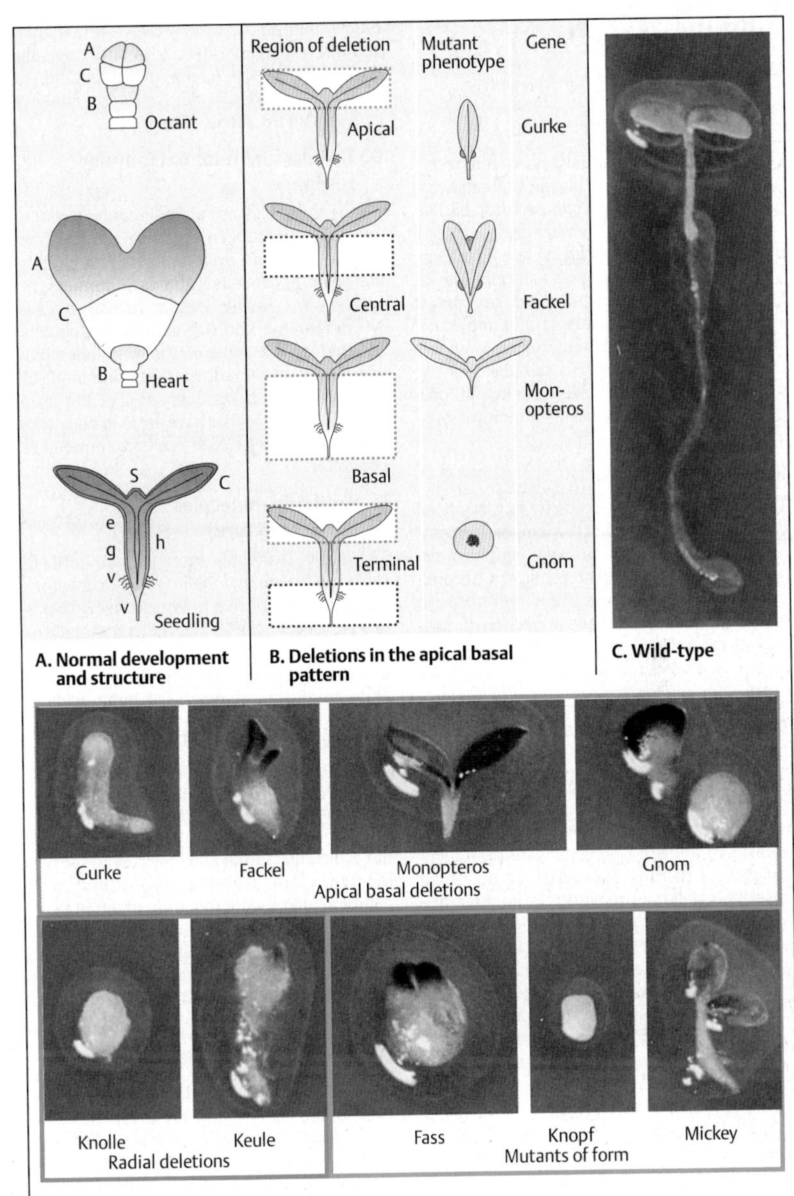


**A. Normal development and structure**

**B. Deletions in the apical basal pattern**

**C. Wild-type**

**D. Phenotypes of embryonic mutations**

# Immune System

## Components of the Immune System

The function of the immune system is to recognize and to eliminate invading foreign organisms. The immune response must be fast, specific, and aimed exclusively against foreign molecules or cells harboring an infective agent, but never against its own normal cells. The effector molecules must be prepared for contact with a variety of organisms and molecular structures. With these requirements, an extremely efficient, genetically complex immune system has evolved. Its development and function are regulated by genes of extraordinary diversity.

### A. Lymphatic organs

The immune system consists of peripheral blood lymphocytes and of the lymphatic organs (lymphoid tissue). The primary lymphoid tissues are the thymus and bone marrow. Secondary lymphoid tissues are the lymph nodes in various regions of the body, especially the nasopharynx, axillae, groins, and intestines. The spleen is considered a secondary lymphatic organ.

### B. Lymphocytes and the immune response

Lymphocytes carry the immune response. This was shown by the following experiment. The normal immune response of a mouse to an administered antigen (foreign organism, foreign molecular structure) was destroyed by a high dose of X-irradiation, to which the immune system is especially sensitive. The immune response was then restored by lymphocytes of a genetically identical (from an inbred strain) unirradiated mouse, but not by any other cells. Other cells are ineffective since only lymphocytes produce an immune response.

### C. T cells and B cells

Lymphocytes exist as one of two functionally and morphologically different types, T lymphocytes and B lymphocytes. T lymphocytes undergo differentiation in the thymus during embryonic and fetal development, thus the designation T lymphocyte or T cell. B lymphocytes differentiate in the bone marrow in mammals and in the bursa of Fabricius in birds (thus the designation B cells). A series of further differentiating steps take place in the lymph nodes (T cells) and in the spleen (B cells).

### D. Cellular and humoral immune response

During the first phase of the immune response induced by an antigen (e.g., a bacterium, virus, fungus, or foreign protein), there is rapid proliferation of B cells (humoral immune response). The B cells mature to plasma cells, which form free antibodies (immunoglobulins) directed at the antigen. The antibodies bind specifically to the antigens. The humoral immune response is rapid, but it does not reach foreign organisms that have invaded body cells. They are the target of the cellular immune response.

### E. Antibody molecules (basic structure)

The basic structural motif of an antibody molecule (immunoglobulin) is a Y-shaped protein composed of two heavy chains (H chains) and two light chains (L chains). They are held together at defined sites by disulfide bonds. L chains and H chains contain regions with variable and constant sequences of amino acids.

### F. Antigen–antibody binding

A foreign molecule, the antigen, is recognized and firmly bound to a specific region of the antibody molecule, the antigen-binding site. Here, the amino acid sequence of the antibody molecule differs from one molecule to the next in three hypervariable regions. The result is the ability to bind a wide spectrum of different antigenic molecules. The three-dimensional structure of this region is known precisely, and important details of the binding process are understood. (Figures adapted from Alberts et al., 1994).

### References

Abbas, A.K., Lichtman, A. H., Pober, J.S.: Cellular and Molecular Immunology, 3rd ed. W.B. Saunders Company, Philadelphia, 1997.

Alberts. B., Bray, D., Lewis, J., Raff, M., Roberts, K., Watson, J.D.: Molecular Biology of the Cell. 3 rd ed. Garland Publ., New York, 1994.

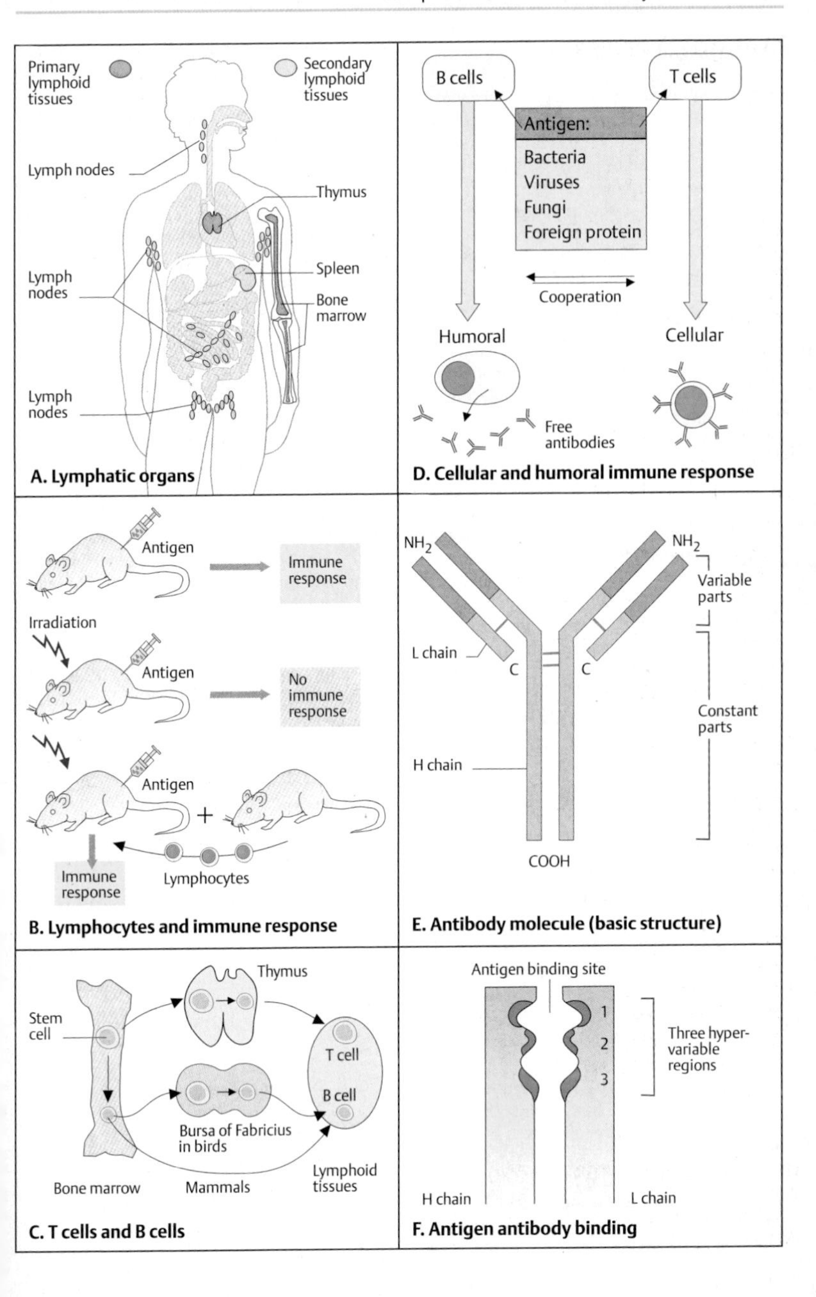
Primary lymphoid tissues
Secondary lymphoid tissues
Lymph nodes
Thymus
Lymph nodes
Spleen
Bone marrow
Lymph nodes
A. Lymphatic organs
B cells
T cells
Antigen:
Bacteria
Viruses
Fungi
Foreign protein
Cooperation
Humoral
Cellular
Free antibodies
D. Cellular and humoral immune response
Antigen
Immune response
Irradiation
Antigen
No immune response
Antigen
Immune response
Lymphocytes
B. Lymphocytes and immune response
$NH_2$
$NH_2$
Variable parts
L chain
C
C
Constant parts
H chain
COOH
E. Antibody molecule (basic structure)
Thymus
Stem cell
T cell
B cell
Bursa of Fabricius in birds
Lymphoid tissues
Bone marrow
Mammals
C. T cells and B cells
Antigen binding site
1
2
3
Three hyper-variable regions
H chain
L chain
F. Antigen antibody binding

## Immunoglobulin Molecules

Immunoglobulins are the effector molecules of the immune system. They occur either as membrane-bound cell surface receptors or as free antibodies. The enormous diversity of their variable regions enables them to bind many very different antigens. Although they differ in details of their function, they share a relatively simple basic pattern, which is derived from a common ancestral molecule (p. 312).

### A. Immunoglobulin G (IgG)

Immunoglobulin G is the prototype of antibody molecules produced by derivatives of B lymphocytes, the plasma cells. The molecule has two H chains and two L chains, held together by disulfide bonds. Each H chain has three constant regions ($C_H1$, $C_H2$, and $C_H3$) and one variable region ($V_H$). Each H chain has a total of 446 amino acids, of which the first 109 belong to the variable region at the N-terminal end. Each L chain has one variable ($V_L$) and one constant ($C_L$) domain and consists of 214 amino acids. In the L chains also, the first 109 amino acids form the variable region. The variable domains of the H chains and the L chains form the antigen-binding sites. The three hypervariable regions within the V region of each chain are also called complementarity determining regions (CDR) because the actual physical contact of molecules based on their complementary structure occurs in these regions. Each domain consists of about 110 amino acid residues. A jointlike area (hinge) between constant region 1 ($C_H1$) and constant region 2 ($C_H2$) of the heavy chain allows some flexibility of the molecule. The H chains are bound to each other and the H to the L chains by disulfide bridges (–S–S–). Furthermore, there are disulfide bridges within the constant and variable domains. The L chains are of one of two types, κ or λ. In addition to immunoglobulin G, there are other types of immunoglobulins, which differ from each other in the constant part of the H chain: IgA ($C_\alpha$), IgD ($C_\delta$), and IgE ($C_\varepsilon$). A very large immunoglobulin, IgM, is made up of five IgG subunits. The different types of H chains are referred to as isotypes.

### B. T-cell receptor (TCR)

Antigen receptors on the surfaces of T cells are heterodimers of covalently bound polypeptide chains, one α and one β chain. The basic structure of a cell surface antigen receptor is similar to that of the secreted immunoglobulins except that the cell surface receptors contain just one constant region. The β chain is the slightly larger chain. The constant regions of the α and β chains (C) each consist of 140–180 amino acids. The variable regions (V) consist of 102–109 amino acids and contain three hypervariable regions, like the immunoglobulin molecules. In addition to those for α and β, genes for a T-cell-receptor γ and δ chain exist. The loci for the TCRα, γ, and δ chains in man lie together on chromosome 14; the locus for TCRβ lies on chromosome 13.

### C. The different domains of an immunoglobulin molecule are encoded by different genes

Each immunoglobulin and receptor molecule is encoded by different DNA sequences, which belong to a large series of genes for the L chains (types κ and λ) and the H chain. Genes for the H chain are located on chromosome 14q32 in humans and chromosome 14 in mice. The genes for the κ light chain are located on the short arm of chromosome 2 (2p12) in humans and chromosome 6 in mice, and for the λ light chain on the long arm of chromosome 22 (22q11) in humans and on chromosome 16 in mice. The genes are rearranged in the developing B and T cells in a pattern that is different and specific in each cell, as seen in the next plate.

#### References

Delves, P.J., Roitt, I.M.: The immune system. Two parts. New Eng. J. Med. **343**:37–49 and 108–117, 2000.

Strominger, J.L.: Developmental biology of T cell receptors. Science **244**:943–950, 1989.

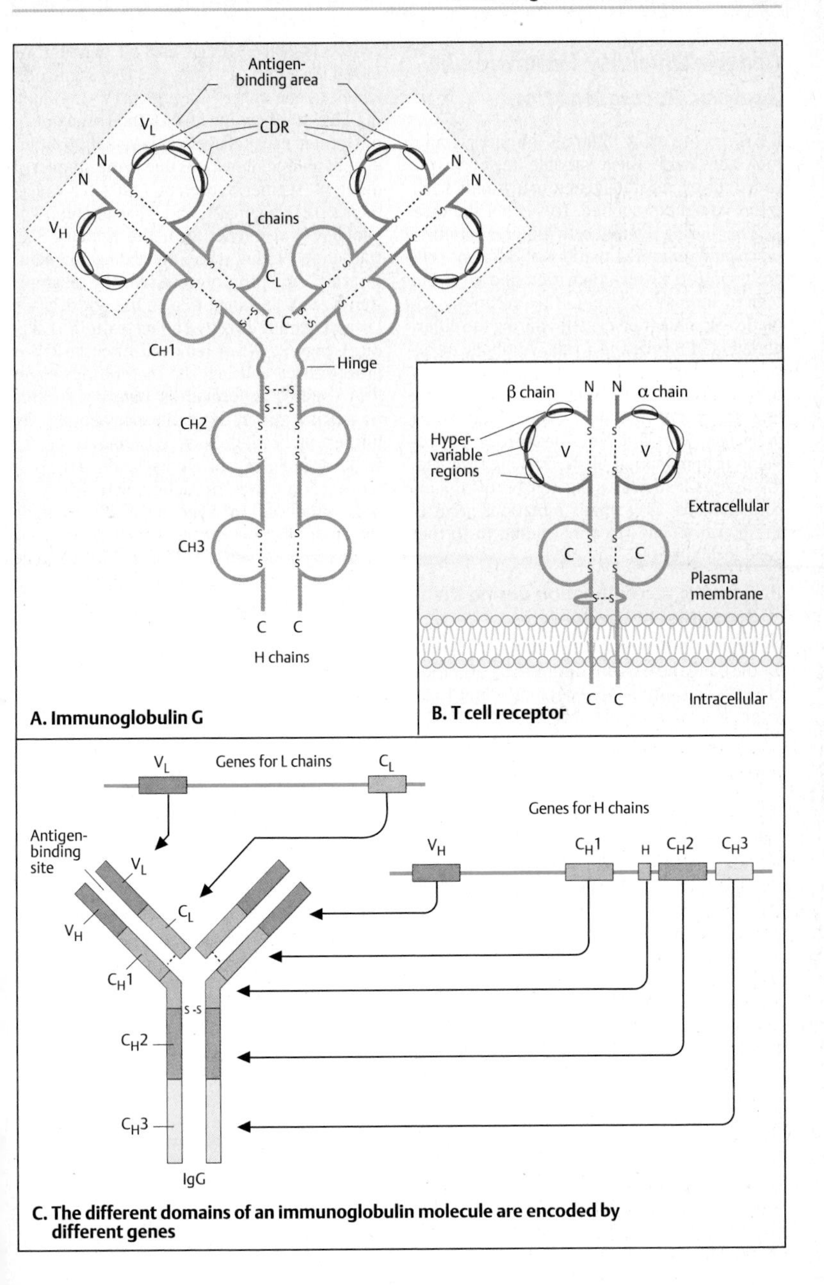


**A. Immunoglobulin G**

**B. T cell receptor**

**C. The different domains of an immunoglobulin molecule are encoded by different genes**

## Genetic Diversity Generated by Somatic Recombination

If each of the many different immunoglobulin molecules with their variable regions were coded for by separate genes, many millions of genes would be required. This is not the case. Rather, during lymphocyte differentiation, a practically unlimited number of different cells are produced by recombination of a large but limited number of genes. This occurs by somatic recombination of genes during the differentiation of B cells and T cells. Antibody diversity arises by the following genetic mechanisms: (1) multiple DNA sequences of the germ-line genes for H and L chains code for Ig molecules with different specificities; (2) somatic recombination of the various DNA segments greatly increases the number of possible combinations; (3) somatic mutations occur in the hypervariable regions, leading to further genetic differences.

### A. Somatic recombination during the formation of lymphocytes

Somatic recombination occurs within the genes for the L and the H chains during the maturation of B cells and within the genes for the four T-cell receptor chains in the maturing T cells. By means of this process, different coding DNA sequences (exons) of the various domains of the particular Ig molecule are in unique combinations in each cell. This provides each molecule with an antigen-binding specificity that differs from that of all other cells. Here, an example of the genetic processes during formation of an immunoglobulin H chain is shown. The exons (V1 to Vn) of the variable region (V) lie at the 5′ end of the IgH locus. They are separated from each other by different lengths of noncoding DNA. A small exon (60–90 bp) that codes for a signal to initiate translation (leader or signal peptide L) lies more than 90 base pairs (bp) in the 5′ direction of the V-region exons. Signal peptides guide the growing polypeptide into the lumen of the endoplasmic reticulum before they are cleaved off. The D genes of the constant region (C) lie at different distances in the 3′ direction from the V genes. Each C segment consists of different exons, corresponding to the domains of the complete C region and different isotypes ($C_\mu$, $C_\delta$), etc.

First, a D gene segment and a J segment are joined (D–J joining). Next, this D–J segment is joined to one of the V segments (V–D–J joining). The combination of V, D, and J segments is different in each cell, with at least 25000 different possible combinations due to the number of different segments present (100–125 V segments, 12 D segments, and 4 J segments). The joined VDJ segments form the primary RNA transcript. At this stage, noncoding segments (introns) are still present. In the example shown, a D2 is joined to a J1, but not to J2–4. J segments not directly connected to a D segment (here, J2–4) are removed. After the RNA is processed by splicing, the mature messenger RNA (mRNA) is formed as template for the translation of an H chain polypeptide. By further processing, such as removal of the leader segment (L) and glycosylation of the protein at certain sites, the definitive H chain is finally produced (of type μ in the example shown). Unlike the H chains, the L chains have no diversity (D) genes, so that a J and a V gene are directly joined by somatic recombination during DNA rearrangement in the lymphocytes. Thus, rearrangement of the genes for the H chain and two types of L chains in the lymphocyte DNA leads to a new combination of genes in each cell.

### References

Abbas, A.K., Lichtman, A.H., Pober, J.S.: Cellular and Molecular Immunology. 3rd ed. W.B. Saunders, Philadelphia, 1997.

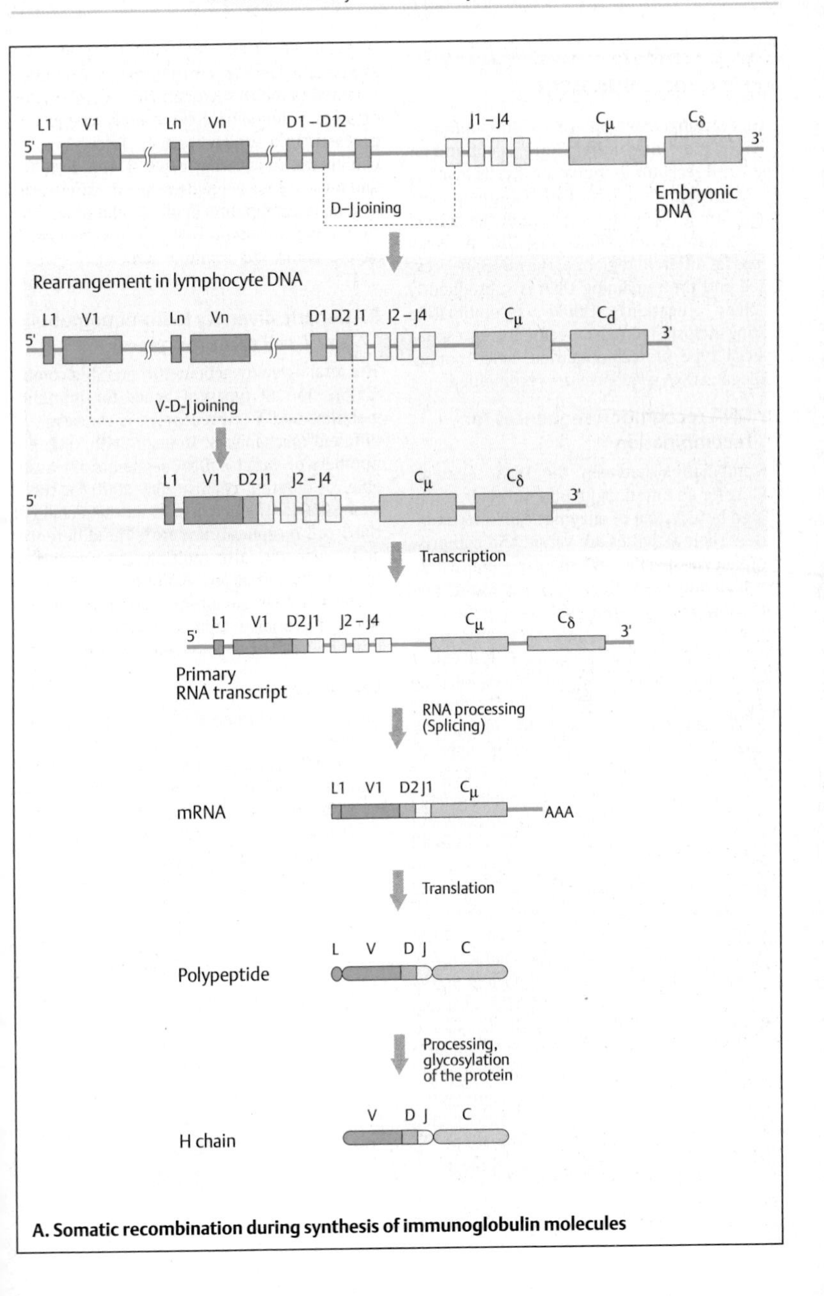


**A. Somatic recombination during synthesis of immunoglobulin molecules**

## Mechanisms in Immunoglobulin Gene Rearrangement

The rearrangement of genes for immunoglobulin molecules in immature B cells and for the T cell receptor in immature T cells involves an excision mechanism. This requires precise recognition to ensure formation of the correct coding information. Noncoding DNA between genes for different regions of the molecule is excised and the remaining DNA is subsequently rejoined (ligation). Unlike recombination during meiosis, IgG rearrangements represent a special type of recombination, since nonhomologous DNA sequences are recombined.

### A. DNA recognition sequences for recombination

Recombination between the DNA segments coding for an immunoglobulin molecule is mediated by a system of enzymes called recombinases. Their activities are varied and comprise lymphocyte-specific and more general activities including exonuclease and ligase functions. The enzymes are controlled by specific DNA recognition sequences. These are located in adjacent noncoding DNA segments at the 3′ end of each V exon (exons for the variable region) and at the 5′ end of each J segment. The D segments are flanked on both sides by recognition sequences. Recognition sequences are noncoding but highly conserved DNA segments of seven base pairs (heptamer) or nine base pairs (nonamer). They are separated by precisely defined intervals, produced by spacers of 23 or 12 base pairs (bp). Upstream (5′ direction) and downstream (3′ direction) from a D segment, the spacers are 12 base pairs long. The sequences of the nucleotide base pairs of the spacers are not conserved. The characteristic rearrangement between neighboring signal sequences for Ig molecules and TCR receptors requires spacers of different lengths, i.e., 12 and 23 base pairs (so-called 12/23 rule). When an H chain is formed, nonhomologous pairing of the heptamer of a D segment and of a J segment occurs. These D and J segments are then joined (D–J joining) by means of recombination: the spacer of 12 or 23 base pairs and all of the intervening DNA forms a loop. This is excised, and the D and J segments are joined. By pairing and recombination of the recognition sequences at the 5′ end of a DJ segment and the recognition sequence at the 3′ end of a V gene, a V segment is joined to the DJ segment. The recombination of genes coding for T-cell receptors (see p. 308) proceeds in a similar manner. Two genes, recombination-activating genes 1 and 2 *(RAG-1* and *RAG-2)* have been identified as stimulating Ig gene recombination in pre-B cells and immature T cells. Mutations in these genes cause severe combined immune deficiency (see p. 314).

### B. Genetic diversity in immunoglobulin and T-cell receptor genes

The total diversity, a bout $10^{18}$ possible combinations for all types of genes for immunoglobulins and T-cell receptors, is the result of different mechanisms. To begin with, different numbers of variable DNA segments are available for different chains (250–1000 for the H chain, 250 for the L chains, 75 for the $\alpha$ chain of the T-cell receptor TCR$\alpha$, etc.). The different D and J segments also multiply the number of possible combinations. Finally, DNA sequence changes (somatic mutations) occur regularly in the hypervariable regions, further increasing the total number of possible combinations.

### References

Abbas, A.K., Lichtman, A.H., Pober, J.S.: Cellular and Molecular Immunology. 3rd ed. W.B. Saunders, Philadelphia, 1997.

Agrawal, A., Schaz, D.G.: RAG1 and RAG2 form a stable postcleavage synaptic complex with DNA containing signal end in V(D)J recombination. Cell **89**:43–53, 1997.

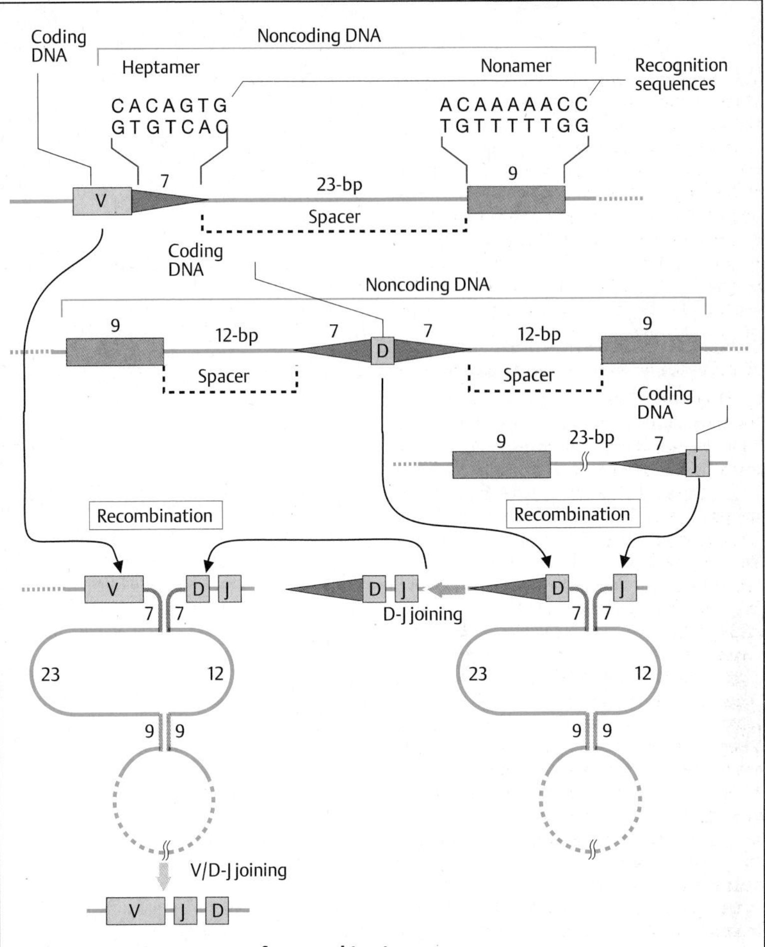


**A. DNA recognition sequences for recombination**

| Mechanism | Immunoglobulin H chain | Immunoglobulin L chains | TCRαβ α | TCRαβ β | TCRγδ γ | TCRγδ δ |
|---|---|---|---|---|---|---|
| Variable domain | 250 – 1000 | 250 | 75 | 25 | 7 | 10 |
| Number of D segments | 12 | 0 | 0 | 2 | 0 | 2 |
| Number of J segments | 4 | 4 | 50 | 12 | 2 | 2 |
| Variable segment combinations | 65 000 – 250 000 | | 1825 | | 70 | |
| Total diversity | $10^{11}$ | | $10^{16}$ | | $10^{18}$ | |

**B. Genetic diversity in immunoglobulin and T-cell receptor genes**

## Genes of the MHC Region

The MHC (*m*ajor *h*istocompatability *c*omplex) region is a region of highly polymorphic genes (about 10–50 alleles per locus). It spans about 3500 kb on the short arm of chromosome 6 in humans and on chromosome 17 in mice. Collectively, these genes are called immune response (Ir) genes. They are expressed on the surface of various cells. Their products can be demonstrated serologically or by cellular reactions in a mixed lymphocyte test. MHC genes control the immune response to antigen proteins by specific binding to T cells.

### A. Basic structure of the MHC gene complex in humans and in mice

The gene loci of the MHC region are grouped into three classes (I–III). Class I in humans includes HLA-A, HLA-B, and HLA-C; in mice, D, L, and K of the H2 system. Many other loci also belong to this class, such as HLA-E to -J. Class II includes HLA-DP, -DQ, and -DR in humans and I-A, and I-E (the letter I, not the Roman numeral) in mice. Alleles are designated according to a numerical system, e.g., HLA-A2, -B5, -DR4, etc. The gene products of the alleles of the HLA system (*h*uman *l*eukocyte *a*ntigens; also said by some to refer to *L*os *A*ngeles, where some of the first basic discoveries were made) can be demonstrated by the toxicity of serum of defined specificities to other leukocytes (serological cytotoxicity). Cytolysis occurs unless the specificities of the serum and of the cells being tested are identical. The gene products of the alleles of the HLA-D system are distinguished by the mixed lymphocyte test, based on lymphocyte proliferation as a reaction to foreign T cells. Strictly speaking, the class III genes do not belong to the MHC loci. They contain genes for different complement proteins and a few other genes.

### B. Genomic organization of the MHC loci

The class II loci are located closest to the centromere. Each consists of a series of genomic subunits. The HLA-D region is about 900 kb long and contains a far greater number of loci than shown here. In addition, some have been renamed. For example HLA-DP has two $\alpha$ and two $\beta$ genes, now called A1, A2 and B1, B2, respectively. The class III loci contain genes for complement factors C2 and C4, steroid 21-hydroxylase, and cytokines (tumor necrosis factor, TNFA; and lymphotoxin, LT, new designation TNFB). Several other class III genes are located here.

### C. Class I and class II MHC molecules

Corresponding to the general organization of the individual class I and class II genes, the class I and class II molecules are basically similar. Both consist of two different polypeptide chains. In class I molecules, an MHC-coded $\alpha$ chain is associated with a non-MHC-coded $\beta$ chain ($\beta_2$-microglobulin). The extracellular portion of the $\alpha$ chain consists of three domains, $\alpha 3$, $\alpha 2$, and $\alpha 1$, each with about 90 amino acids. $\alpha 1$ and $\alpha 2$ form the highly polymorphic peptide-binding region; $\alpha 3$ and $\beta_2$-microglobulin structurally correspond to an immunoglobulin-like region. Elucidation of the crystalline structure of the class I MHC molecules showed that $\alpha 1$ and $\alpha 2$ interact to form a type of platform of eight-stranded $\beta$-folded proteins. The cleft formed between $\alpha 1$ and $\alpha 2$ (25 Å × 10 Å × 11 Å) can bind a protein fragment consisting of 10–20 amino acids. Class II MHC molecules consist of two polypeptide chains, $\alpha$ and $\beta$, each with two domains, i.e., $\alpha 1$, $\alpha 2$ and $\beta 1$, $\beta 2$, each with about 90 amino acids and a transmembrane region of about 25 amino acids. As with the class I molecules, the peptide-binding regions ($\alpha 1$ and $\beta 1$) are highly polymorphic. Unlike the $\beta 1$ domain, $\alpha 1$ does not contain a disulfide bridge.

### References

Klein, J. & Sato, A.: Advances in immunology. The HLA system. New. Engl. J. Med. **343**:702–709 (part I) and 782–786, 2000.

Trowsdale, J.: Genomic structure and function in the MHC. Trend Genet. **9**:117–122, 1993.

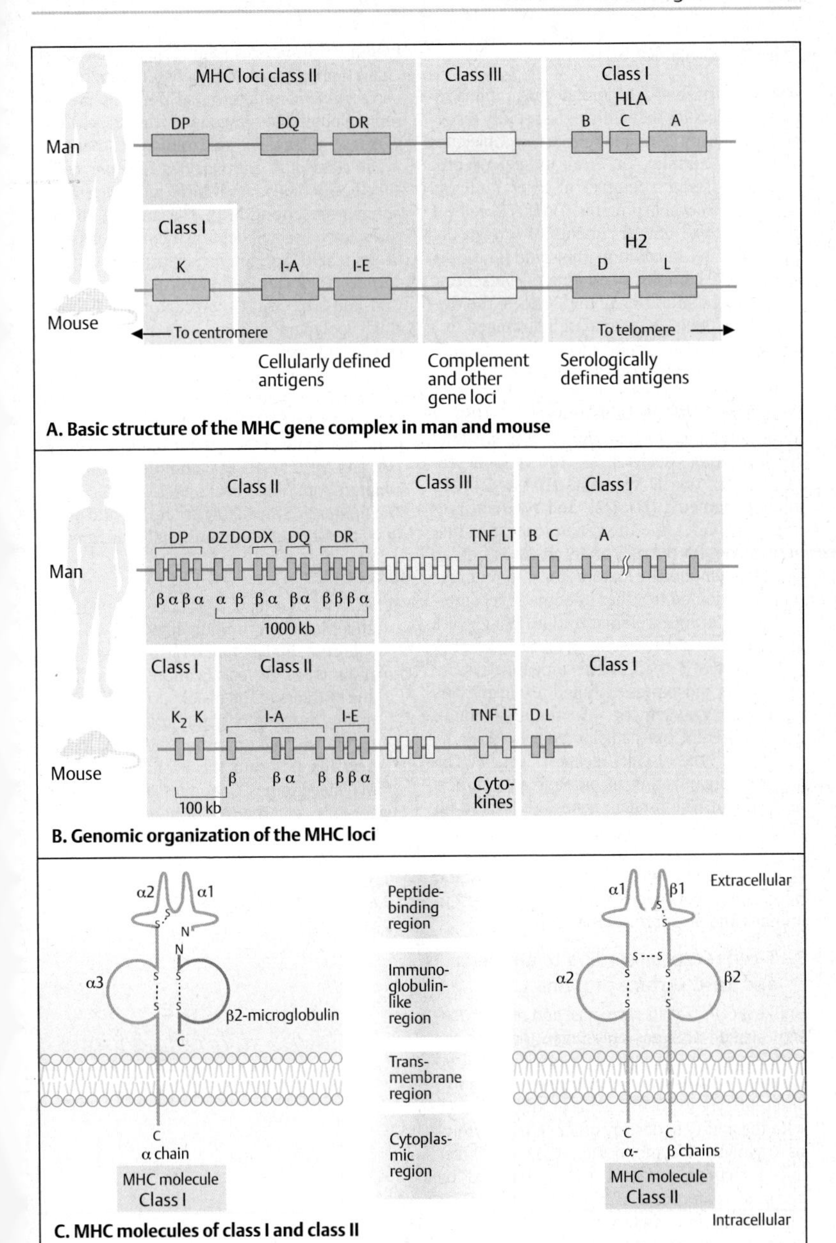


A. Basic structure of the MHC gene complex in man and mouse

B. Genomic organization of the MHC loci

C. MHC molecules of class I and class II

## T-Cell Receptors

The cell surface of a T lymphocyte contains receptor molecules that with high specifity recognize foreign antigens and cell surface molecules of the MHC complex. The T-cell antigen receptor (TCR) consists of a complex of several integral plasma membrane proteins. Unlike B cells, T cells recognize only fragments of foreign antigen proteins. In addition, they bind physically to the MHC complex of an antigen-presenting cell. During maturation of the T cells in the thymus, T-cell gene segments are rearranged in a defined order by somatic recombination, similarly to the formation of immunoglobulins.

### A. T-cell receptor genes (TCR) of man

In the germ line, the genes for the β chain of the T-cell receptor consist of 75–100 variable segments ($V_\beta$), two D segments ($D_\beta 1$, $D_\beta 2$), two joining segments ($J_\beta 1$, $J_\beta 2$), and two constant segments ($C_\beta 1$, $C_\beta 2$). The genomic organization of the genes for the α, δ, and γ chains is similar. During T-lymphocyte maturation, different segments are joined together by somatic recombination as during B-cell maturation. In a given T cell, only one of the two α-chain loci and only one of the two β-chain loci become functionally rearranged and expressed (allelic exclusion). As in the rearrangement of immunoglobulin genes, different mechanisms help produce diversity of the T-cell receptor genes. The genomic organization in humans and mice is very similar. The β-chain genes are located on chromosome 7 in humans and on chromosome 6 in mice. The α- and δ-chain gene loci are located on chromosome 14 in both humans and mice. The γ-chain genes lie on chromosome 7 in humans and on chromosome 13 in mice.

### B. T-cell receptor binding to antigens and MHC surface proteins

Unlike B cells, T cells recognize and react to foreign protein antigens only when the antigens are attached to the surface of other cells. Two different classes of T lymphocytes recognize different types of MHC gene products. T cells with the ability to destroy other cells by cytolysis (cytolytic T lymphocytes, CTLs, or "killer cells") recognize class I MHC molecules by means of a coreceptor, CD8 (formerly T8). CD8 is a membrane-bound glycoprotein of two dimers (α and β), whose extracellular domains are held together by two disulfide bridges (either α-chain homodimers or α–β heterodimers). The extracellular aminoterminal domains are immunoglobulinlike. In general, the structure of CD8 resembles a box with rounded corners. The other class of T lymphocytes ("helper cells") specifically bind class II MHC molecules of the antigen-presenting cells. The antigen consists of a relatively small peptide fragment of foreign protein. With antigen presentation, a series of surface molecules lend physical help, e.g., CD3, CD4, and CD8. CD4 is a coreceptor, a rod-shaped single polypeptide with an extracellular part consisting of four immunoglobulinlike domains. The gp120 protein of the HIV virus reacts with the second domain of CD4. A number of accessory membrane proteins of the CD3 system participate in the specific binding between TCR and MHC (TCR–CD3 complex). (Drawings adapted from Abbas et al., 1997.)

The figures in B are highly schematic and do not show the real three-dimensional structures.

### References


Abbas, A.K., Lichtman, A.H., Pober, J.S.: Cellular and Molecular Immunology. 3rd ed. W.B. Saunders, Philadelphia, 1997.

Amadou, C., et al.: Localization of new genes and markers to the distal part of the human major histocompatibility complex (MHC) region and comparison with the mouse: new insights into the evolution of mammalian genomes. Genomics **26**:9 –20, 1995.

Fugger, L., et al.: The role of human major histocompatibility complex (HLA) genes in disease. pp. 555–585., In: C.R. Scriver, et al., eds., The Metabolic and Molecular Bases of Inherited Disease. 7th ed. McGraw-Hill, New York, 1995

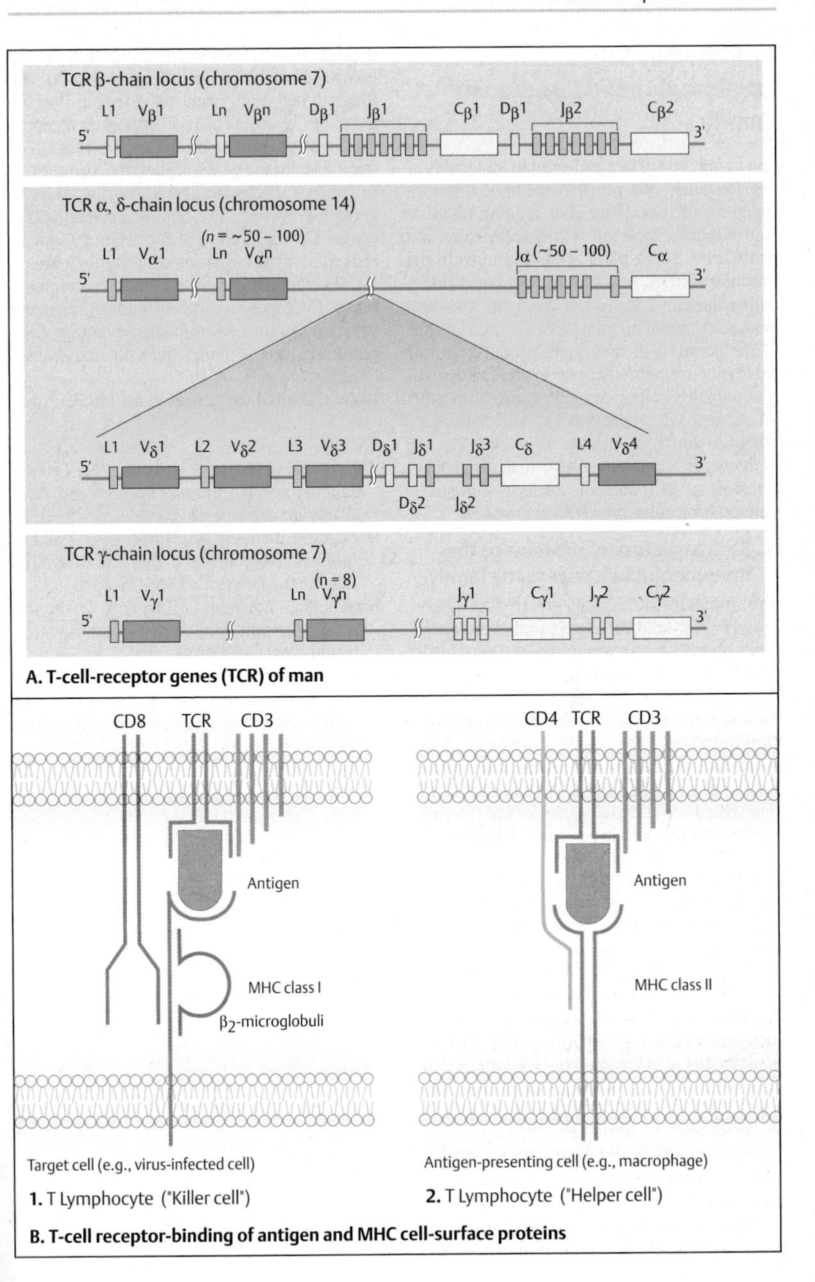


**A. T-cell-receptor genes (TCR) of man**

**1.** T Lymphocyte ("Killer cell")

**2.** T Lymphocyte ("Helper cell")

**B. T-cell receptor-binding of antigen and MHC cell-surface proteins**

## Evolution of the Immunoglobulin Supergene Family

The many cell surface and soluble molecules of the immune system that mediate different functions such as recognition, binding, or adhesion of specific molecules show many structural similarities. Some parts are found outside the immune system. As a group, they constitute a gene superfamily, derived from an ancestral gene common to all members. The homologies of the domains of their gene products and of their gene sequences can be explained by evolutionary origin from a common ancestral gene. The Ig gene family members code for immunoglobulin domains, usually of about 70–110 amino acids homologous with an Ig variable (V) or constant (C) domain. Each Ig domain is derived from conserved DNA sequences.

### A. Basic structure of proteins of the immunoglobulin supergene family

The immunoglobulin molecules of the T-cell receptors (TCR) and the class I and class II MHC molecules (1) are basically similar. They consist of variable Ig-like domains (V), constant Ig-like domains (C), or primordial Ig-like domains (H). Although their genes are located on different chromosomes, the gene products form functional complexes with each other. Others, such as the V, D, and J gene segments of all antigen receptors and their genes for the C domain lie close together in gene clusters. In addition, genes of the MHC loci and for the two CD8 chains lie together. The basic structures of accessory molecules (2) such as CD2, CD3, CD4, CD8, and thymosine 1 (Th-1) are relatively simple. Other members of the Ig superfamily (3) are the Fc receptor II (FcRII); polyimmunoglobulin receptor (pIgR), which transports antibodies through the membranes of epithelial cells; NCAM (neural cell adhesion molecules); and PDGFR (platelet-derived growth factor receptor) (3). (After Hunkapillar and Hood, 1989.)

### B. Evolution of genes of the immunoglobulin supergene family

Distinct evolutionary relationships can be recognized in the homology of genes for Ig-like molecules. A precursor gene for a variant (V) and a constant (C) region must have arisen by duplication and subsequent diversification of a gene for a primordial cell surface receptor. Such a primordial gene could have looked like the gene for thymosine (Thy-1) or poly-Ig receptor. No somatic recombination occurs in these gene families or in the genes of the MHC complex.
In contrast, the rearrangement of lymphocyte germ-line genes by somatic recombination during the maturation of B cells and T cells is the basis for the formation of immunoglobulins, T-cell receptors, and CD8. Somatic recombination of the genes for antigen-binding molecules was an enormous evolutionary advantage. Consequently, this is found even in early vertebrates.
(Figure adapted from Hood et al., 1985).

### References


Abbas, A.K., Lichtman, A.H., Pober, J.S.: Cellular and Molecular Immunology. 3rd ed. W.B. Saunders, Philadelphia, 1997.

Hood, L., Kronenberg, M., Hunkapillar, T.: T-cell antigen receptors and the immunoglobin supergene family. Cell **40**:225, 1985.

Hunkapillar, T., Hood, L.: Diversity of the immunoglobulin gene superfamily. Adv. Immunol. **44**:1–63, 1989.

Shiina, T., et al.: Molecular dynamics of MHC genesis unraveled by sequence analysis of the 1,796,938-bp HLA class I region. Proc. Nat. Acad. Sci. **96**:13282–13287, 1999.

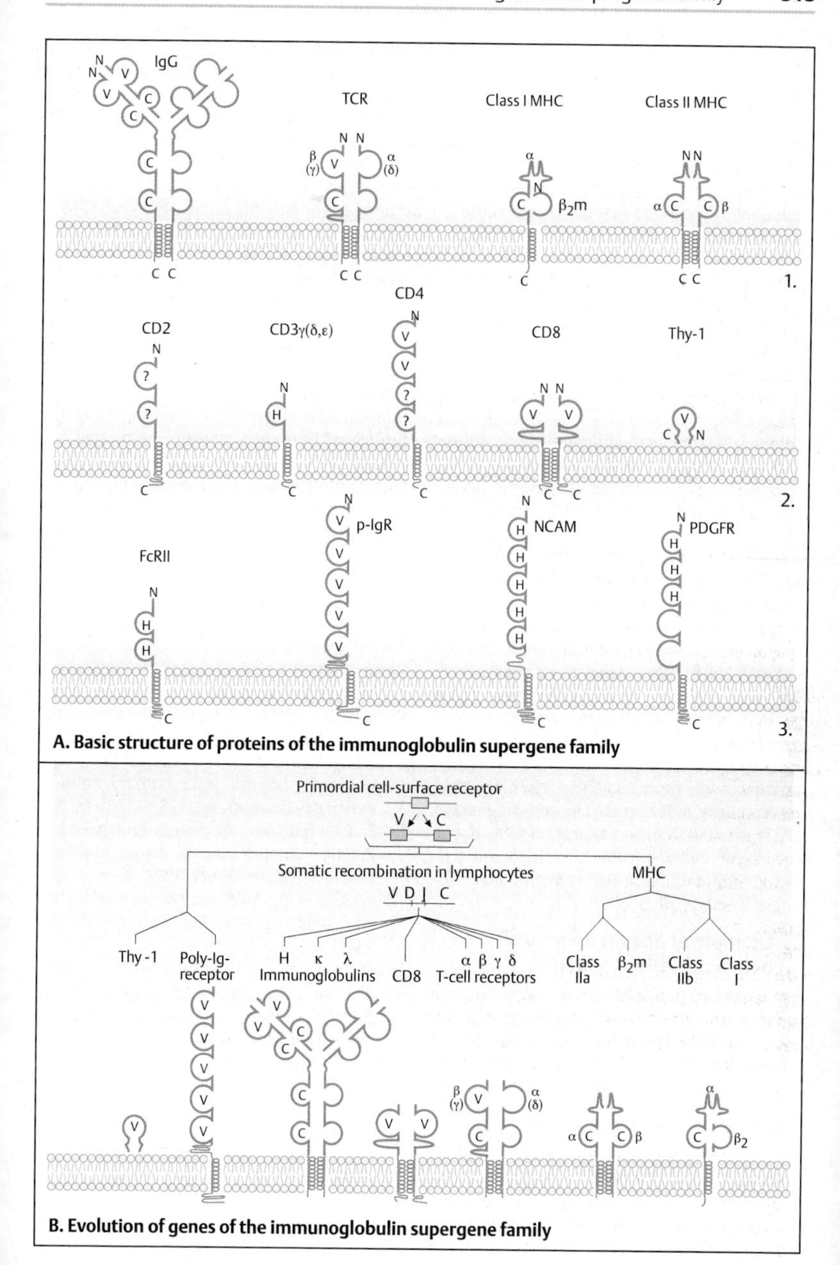


A. Basic structure of proteins of the immunoglobulin supergene family

B. Evolution of genes of the immunoglobulin supergene family

# Hereditary and Acquired Immune Deficiencies

Numerous types of impairment of the immune system exist, congenital (hereditary) and acquired. All result in abnormal susceptibility to infections, often associated with lymphoreticular malignancies, and in some cases autoimmune disease.

## A. Examples of hereditary immune deficiency diseases

Severe combined immune deficiency (SCID) is a heterogeneous group of genetic disorders due to various defects in both B cell and T differentiation.
X-linked agammaglobulinemia type Bruton (McKusick 300300) was the first hereditary immune deficiency described, in 1952 by Ogden Bruton. The first developmental step of B cell differentiation from pre-B to mature B cell is blocked by deficiency of Bruton tyrosine kinase due to mutations in the *BTK* gene on the X chromosome (Xq22). Other forms involve later steps of differentiation (variable immune deficiency) or isolated Ig isotype (subclass) deficiencies. Several T cell immune deficiency diseases exist. The most important is the DiGeorge syndrome (McKusick 188400), characterized by a broad spectrum of highly variable manifestations. The underlying defect involves the third and fourth brachial arch derivatives during embryonic development. A deletion in chromosome region 22q11 is found in most patients, usually as a *de novo* event. Other disorders involve T cell activation and function of one or both major subsets of T cells, CD4 or CD8.

## B. Example of acquired immune deficiency: HIV-1 infection

The global epidemic of the acquired immune deficiency syndrome (AIDS) due to the infection with human immunodeficiency virus type 1 (HIV-1) poses a major public health problem of unprecedented dimensions in modern times. Although it was first noted in 1981 in restricted populations, it now occurs in all populations at all ages throughout the world (United Nations Programme on HIV/AIDS at *http://www.UNAIDS.org/hivaidsinfo/documents-html).* HIV-1 selectively infects T cells of the CD4 type. The first step of infection is specific binding of the extracellular domain of the viral transmembrane glycoprotein gp120 to the CD4 receptor (gp41 is the transmembrane protein). Two chemokine receptors, CCR5 and CXR4, function as major coreceptors (not shown in the figure). Following virus genome uptake, a phase of viral replication by reverse transcription into double-stranded DNA occurs. Provirus DNA is integrated into the cellular DNA by the virus-encoded integrase enzyme. As the infected cell divides, the viral DNA (provirus) is also replicated. The HIV-1 provirus can remain in the cell without being transribed for a long period (latent infection). Another phase of the HIV-1 life cycle involves transcription of viral DNA (provirus) into viral RNA. The full-length RNA transcripts are spliced and transported to the cytoplasm. Here the viral RNA is translated into viral proteins by cellular enzymes; the viral proteins and RNA are then processed and assembled into new virus particles (virus production). These leave the cell as free infectious virions. Usually the cell dies by lysis, but not always. In this case low-level chronic virus production may persist in infected individuals.

## References


Barré-Sinoussi, F.: HIV as the cause of AIDS. Lancet **348**:31–35, 1996.

Buckley, R.H.: Primary immunodeficiency diseases due to defects in lymphocytes. New Eng. J. Med. **343** :1313–1324, 2000.

Cooper, M.D., Lawton, A.R.: Primary immune deficiency diseases, pp. 1783–1791. In: A.S. Fauci, et al., eds., Harrison's Principles and Practice of Internal Medicine. 14th ed. McGraw-Hill, New York, 1998.

Hahn, B.H., et al.: AIDS as a zoonosis: scientific and public health implications. Science **287**:607–614, 2000.

Hong, R.: Inherited immune deficiency, pp. 283–291, In: J.L. Jameson, ed., Principles of Molecular Medicine. Humana Press, Totowa, New Jersey, 1998.

Schwartländer, B., Garnett, G., Anderson, R.: AIDS in a new millenium. Science **289**:64–67, 2000.

Schwartz, K., et al.: RAG mutations in human B cell-negative SCID. Science **274**:97–99, 1996.

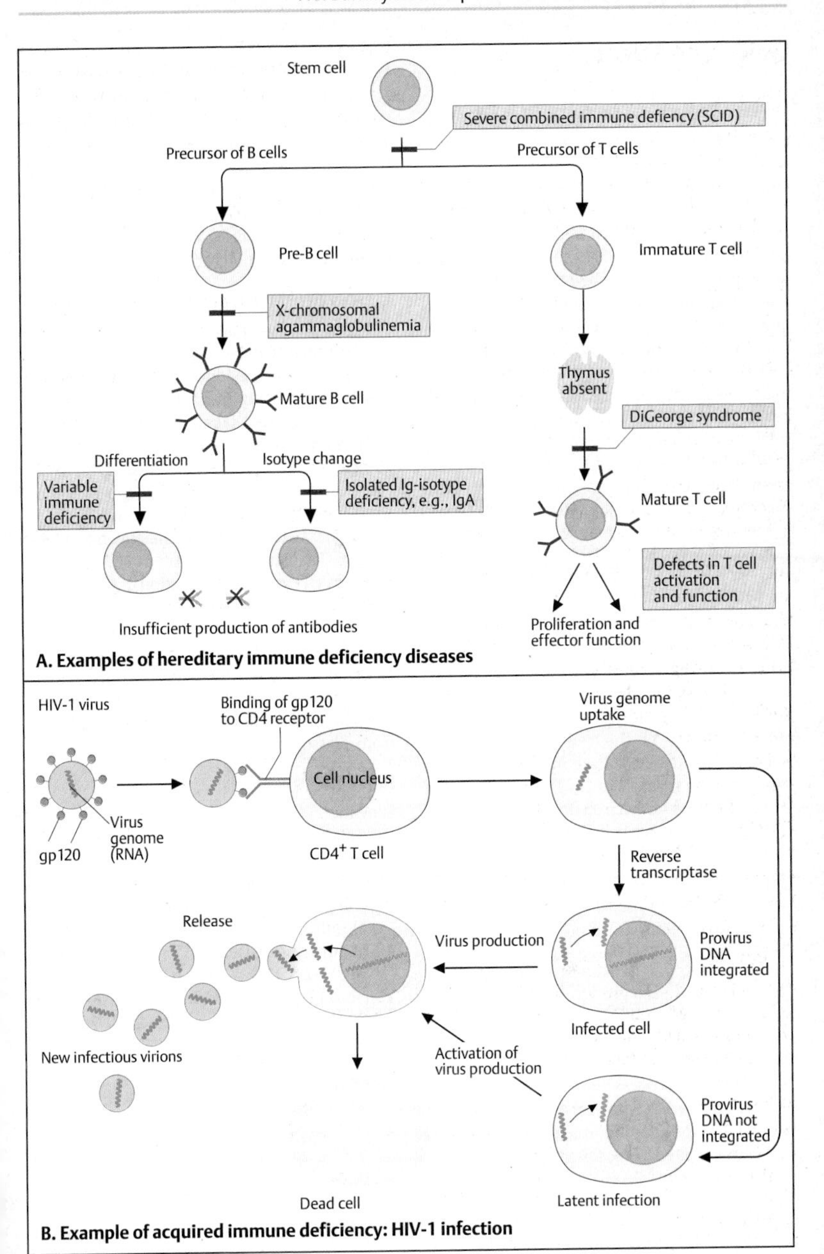


**A. Examples of hereditary immune deficiency diseases**

**B. Example of acquired immune deficiency: HIV-1 infection**

# Origin of Tumors

## *Influence of Growth Factors on Cell Division*

Development, differentiation, and the maintenance of vital functions require exact regulation of the time and location of cell divisions. Rapidly multiplying cells in embryonic tissues must be controlled, just like those in stationary phases in adult tissues. Rapid response to injury or to foreign antigens requires controlled cell division. Multicellular organisms have an extensive repertoire of genetically regulated mechanisms at their disposal for controlling cell division and tissue proliferation. As a group, they are referred to as growth factors. Every growth factor has a specific cell surface receptor. Binding to the receptor initiates (or in some cases blocks) cell division. Most growth factors regulate only certain types of cells and tissues.

### A. Control of cell division by growth factors

Basically, cell division (mitosis) can be controlled by stimulation or inhibition. In the absence of stimulation or with active inhibition, no mitosis occurs. Growth factors have an effect not only on specific types of cells, but also on defined phases of the cell cycle. The most frequently controlled phase of the cell cycle is the transition from $G_0$ to $G_1$. The growth factor group includes growth factors for epidermal cells (EGF), for nerve cells (NGF), for connective tissue or mesenchymal cells (fibroblasts, FGF), and for thrombus-forming cells in the inner lining (endothelium) of blood vessels (PDGF). Their stimulating effect may be opposed by an antagonistic effect (e.g., TGFβ, transforming growth factor, or TNF, tumor necrosis factor). The function of each growth factor is mediated by a specific receptor.

### B. Activation of a growth factor receptor

A growth factor receptor becomes activated by specific extracellular binding to the growth factor. The activated receptor in turn activates a substrate protein.

### C. PDGF-receptor kinases have an effect on numerous substrates

A receptor such as the PDGF (*p*latlet-*d*erived growth *f*actor) receptor can have an effect on numerous substrates. Substrates of the PDGF receptors include the Ras protein (see D), the Src protein (the name is derived from the tumor, a sarcoma, in which it was first found), phospholipase C (a signal transmitter), and others.

### D. Ras proteins as signal transmitters

The Ras proteins play a central role as signal transmitters. They belong to the group of G proteins (guanosine-residue-binding proteins with signal-transmitting functions, see p. 266). The binding of growth factor, e.g., PDGF, activates the Ras protein by stimulating the conversion of associated GDP (guanosine diphosphate) to GTP (guanosine triphosphate) and triggering a short time-limited signal that initiates cell division. The signal is terminated by inactivation of Ras by a GTPase-activating protein (GAP), which converts GTP into GDP. Mutation of the Ras protein or of GAP can remove the time limit of the cell-stimulating signals and result in an active condition with uncontrolled cell division. This can lead to a tumor with uncontrolled growth (malignancy). Several mutations have been defined in the pertinent genes. (Diagrams adapted from Watson et al., 1992.)

### References

Lengauer, C., Kinzler, K.W., Vogelstein, B.: Genetic instabilities in human cancers. Nature **396**:643–649, 1998.

Park, M.: Oncogenes: Genetic abnormalities of cell growth, pp. 589–611. In: C.R. Scriver, et al., eds., The Metabolic and Molecular Bases of Inherited Disease. 7th ed. McGraw-Hill, New York, 1995.

Watson, J.D., et al.: Recombinant DNA. 2nd ed. Scientific American Books, W.H. Freeman, New York, 1992.

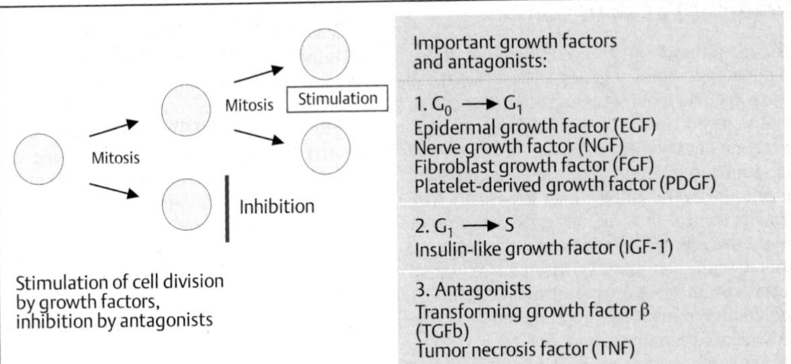

Mitosis
Mitosis
Stimulation
Inhibition
Stimulation of cell division by growth factors, inhibition by antagonists
Important growth factors and antagonists:
1. $G_0 \longrightarrow G_1$
Epidermal growth factor (EGF)
Nerve growth factor (NGF)
Fibroblast growth factor (FGF)
Platelet-derived growth factor (PDGF)
2. $G_1 \longrightarrow S$
Insulin-like growth factor (IGF-1)
3. Antagonists
Transforming growth factor β (TGFb)
Tumor necrosis factor (TNF)


**A. Control of cell division by growth factors**

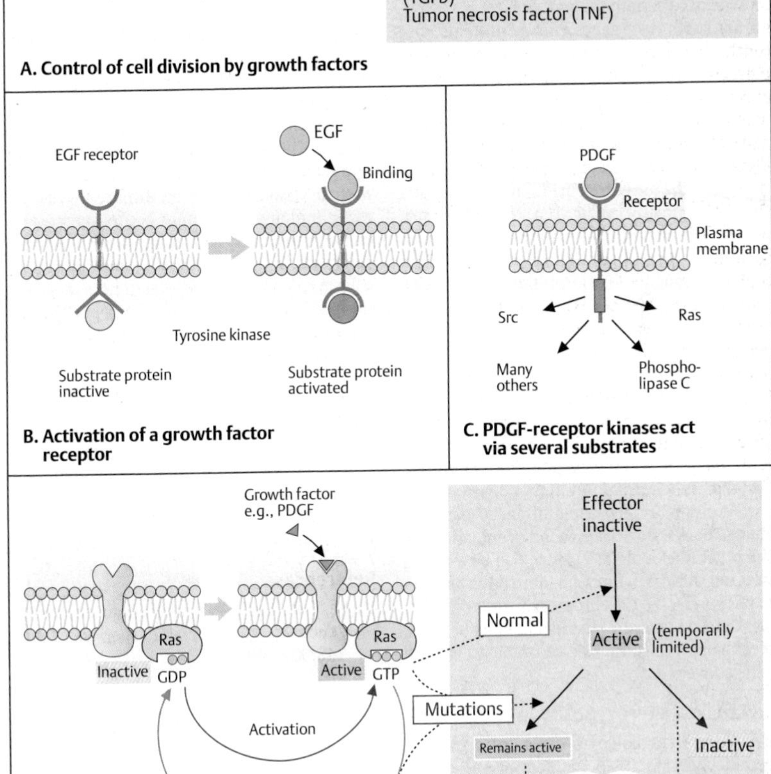


**B. Activation of a growth factor receptor**

**C. PDGF-receptor kinases act via several substrates**

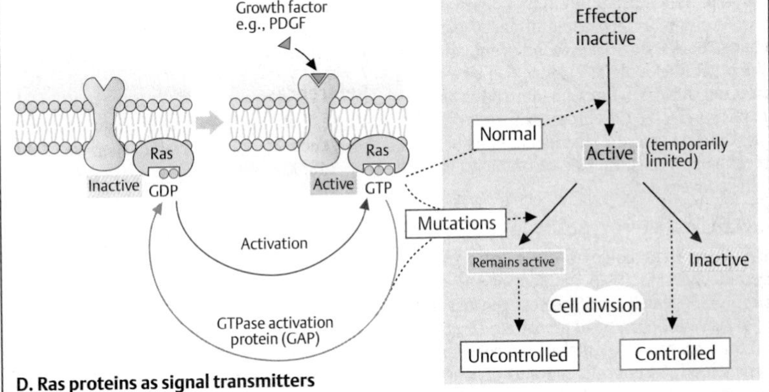


**D. Ras proteins as signal transmitters**

## Tumor Suppressor Genes

Malignant tumors arise as a result of mutations in three basic types of genes, DNA repair genes (see p. 80 DNA repair), tumor suppressor genes, and proto-oncogenes (see next plate). A single mutation does not cause cancer. Rather several mutations in different genes must accumulate in one or several cells, which eventually lose growth control in favor of aggressive growth properties. Most mutations are somatic, i.e., limited to the neoplastic cells. A relatively small subset of mutations are present in the germline (hereditary forms of cancer) and predispose the individual to certain types of cancer.

Tumor suppressor genes encode proteins with function in growth regulation or differentiation pathways. Their name is derived from the observation that one functional allele will suppress tumor development even in the presence of a mutation in the other allele (or its loss). Thus, two mutational events are required to release the growth-controlling function of a tumor suppressor gene (see retinoblastoma, p. 330). The two mutational events in a tumor suppressor gene often become manifest in loss of heterozygosity (LOH) in tumor cells (see B). Tumor suppressor genes can be compared to the brake of a car, cellular oncogenes to the accelerator (see next plate).

### A. Tumor suppressor gene

In contrast to the cellular oncogenes, for which a change in one allele will alter normal function, both alleles of a tumor suppressor gene must lose their function before a tumor develops. The first event is usually a mutation by base exchange or deletion. The second event, affecting the other allele (allele 2), may also be a mutation, but the loss of function more often appears to be from loss of the chromosome after a faulty cell division (mitotic nondisjunction) or other mechanisms (e.g., mitotic recombination with gene conversion).

### B. Loss of heterozygosity in tumor cells

Usually, in about half of the individuals who are heterozygous for DNA markers at the tumor suppressor gene locus of interest, the loss of one allele (event 2) can be demonstrated by Southern blot analysis. In contrast to normal somatic cells (blood), tumor cells contain only one allele (loss of heterozygosity, LOH). The remaining allele carries the mutation. Thus, the mutant allele can be identified by demonstration of LOH. LOH is useful in diagnosis as an indication of the existence of a tumor suppressor gene.

### C. Somatic and germinal mutation

The first mutation in a suppressor gene can either be present in the zygote (germinal mutation, i.e., germ cell mutation due to transmission from an affected parent or due to new mutation) or occur in a single cell of the corresponding tissue (somatic mutation). Loss of function of one allele (corresponding to event 1 in A) predisposes the cell to tumor development.

With a germinal mutation, all cells are predisposed. The tumor arises after loss of function of the second allele. When somatic mutation occurs in a single cell, loss of function of both alleles rarely affects the same cell. But with a germ cells mutation, loss of function of the second allele is frequent, since all cells carry the first mutation, i.e., are predisposed. With somatic mutation, the tumor occurs sporadically (is not hereditary) and arises unifocally from a single cell. In the hereditary form resulting from a germ cell mutation, several tumors may arise from different cells (multifocal tumor). The predisposition for the tumor in the hereditary form shows autosomal dominant inheritance.

### D. Examples of tumor suppressor genes

Numerous types of tumors arise due to loss of function in both alleles of a tumor suppressor gene. Loss of heterozygosity (LOH) can be demonstrated in the tumor cells in about half of these patients.

### References

Skuse, G.R., Ludlow, J.W.: Tumour suppressor genes in disease and therapy. Lancet **345**:902–906, 1995.

Stanbridge, E.J.: Human tumor suppressor genes. Ann Rev Genet. **24**:615–657, 1990.

Weinberg, R.A.: Tumor suppressor genes. Science **254**:1138–1146, 1991.

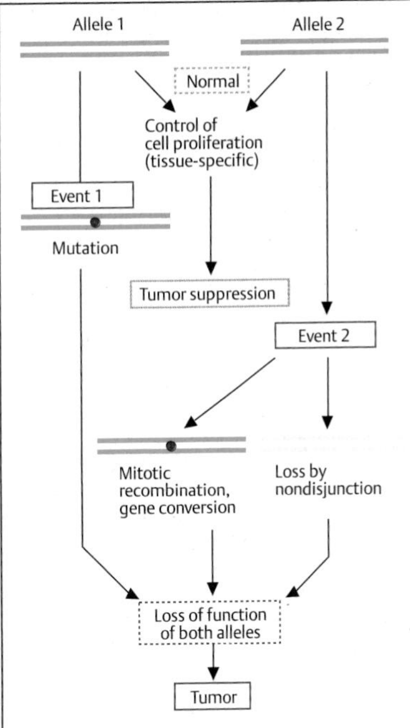


**A. Tumor suppressor gene**

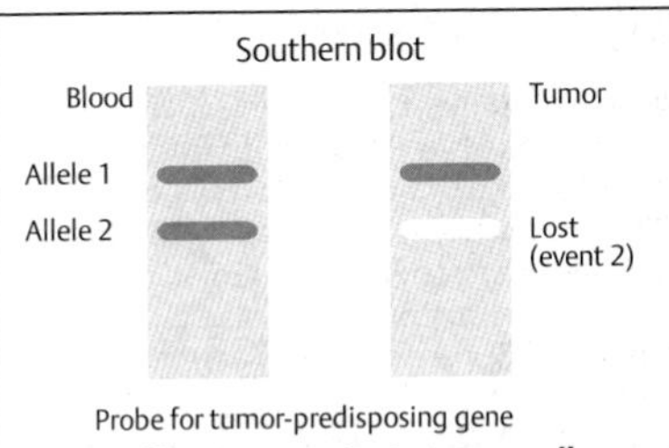


**B. Loss of heterozygosity in tumor cells**

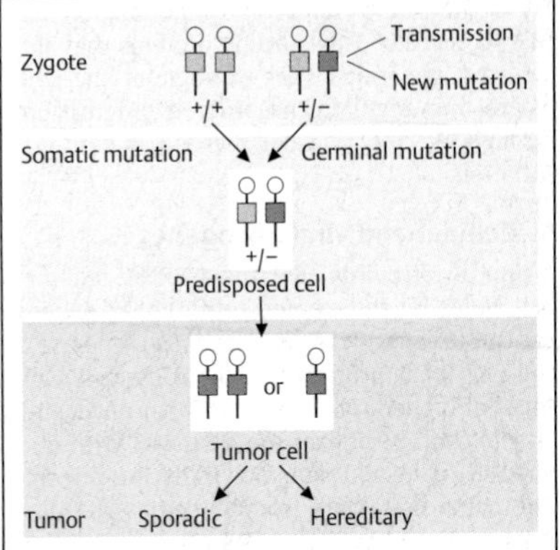


**C. Somatic and germinal mutation**

| Chromosomal localization | Type of tumor |
|---|---|
| 1p | Melanoma; multiple endocrine neoplasia type 2; neuroblastoma, others |
| 1q | Breast cancer (one form) |
| 3p | Kidney cell carcinoma; von Hippel Lindau disease |
| 5q | Familial polyposis coli, colon carcinoma |
| 9p | Familial melanoma |
| 9q | Bladder carcinoma |
| 10q | Astrocytoma, MEN type 2 |
| 11p | Wilms tumor and others |
| 13q | Retinoblastoma; osteosarcoma |
| 17p | Breast cancer, colon carcinoma, and others |
| 17q | Neurofibromatosis type 1 |
| 18q | Colon carcinoma |
| 22q | Neurofibromatosis type 2; meningioma |

**D. Examples of tumor-suppressor genes (loss of heterozygosity in tumor cells)**

## Cellular Oncogenes

Oncogenes (tumor-causing genes) were originally identified in RNA tumor viruses (retroviruses) as genes (v-*onc*) that could transform cells into an altered state of control of cell proliferation, often resulting in a tumor, mainly in chicken, mice, and rats. More than 20 different viral oncogenes are known to have a counterpart in normal cells (c-*onc*), called proto-oncogenes or cellular oncogenes. These cellular genes are highly conserved in evolution because they have important functions in all eukaryotic cells. They encode proteins that are required at defined sites throughout the cell where they regulate the ordered progression through the cell cycle, cell division, and differentiation.

### A. Cellular and viral oncogenes

A typical retrovirus contains an RNA genome that codes for three genes or groups of genes: *gag* (group-specific antigen), *pol* (polymerase), and *env* (coat protein, *env*elope). As with all genes of higher organisms, a cellular oncogene (c-*onc*) consists of exons and introns with defined structure and sequence, as in the gene *src* (the name is derived from sarcoma, a tumor that is induced by a change in this gene). The virus may contain parts of the cellular oncogene (c-*scr*). This is designated viral oncogene (v-*src*) (Rous sarcoma virus). In chickens, it induces a malignant tumor (a sarcoma), first observed by Peyton Rous in 1911. Since many cellular oncogenes are also known in an altered, viral form, it is assumed that the viruses have integrated parts of the respective cellular oncogenes into their own genomes.
Virus-induced tumors are known especially in chickens, rodents, and cats. In man, they do not play a general role in the induction of tumors. Important exceptions are papilloma virus-induced carcinoma of the cervix and carcinoma of the liver secondary to hepatitis virus infection.

### B. Mechanisms of oncogene activation

A cellular oncogene controls cell division. It controls the time and location of the orderly proliferation of cells and tissues (normal growth). Genetic changes can lead to disorders of the regulation of cell divison, increased proliferation of cells, and formation of a tumor. This can be traced back to relatively few mechanisms. A point mutation in a critical region of the gene can lead to disturbances in the regulation of cell division. Examples are mutations in codon 12 or 63 or the H-*ras* gene.
An inactive cellular oncogene may become activated when it is moved by chromosomal translocation to the vicinity of an active gene. In Burkitt lymphoma, an inactive gene is moved to the region of an active gene for the H or L chain of an immunoglobulin. In other cases, the breakpoint of the chromosome translocation may lie within a cellular oncogene and thereby affect its expression. An example is the Philadelphia translocation (see p. 332). Multiplication (amplification) of a gene is a futher mechanism that can lead to altered (usually increased) expression.

### C. Examples of cellular oncogenes and their proteins

The table shows examples of the about 60 known cellular oncogenes, their basic functions, a few tumors induced in man by mutation of the cellular oncogene (c-*onc*), and tumors induced in vertebrates by the homologous viral onogene (v-*onc*). (Data from Lodish et al., 2000; Cannon-Albright et al., 1992.)

### References

Lodish, H., et al.: Molecular Cell Biology. 4th ed. 2000.
Cannon-Albright, L.A., et al.: Assignment of a locusfor familial melanoma, MLM, to chromosome 9p13-p22. Science **258**:1148, 1992.
Levine, A.J., Broach, J.R., eds.: Oncogenes and cell proliferation. Current Opin Genet and Development **5**:1–150, 1995.
Park, M.: Oncogenes, p. 205–228, In: Vogelstein, B., Kinzler, K.W., eds., The Genetic Basis of Human Cancer. McGraw-Hill, New York, 1998.
Park, M.: Oncogenes, p. 589–611. In: Scriver, C.R. et al., eds.: The Metabolic and Molecular Bases of Inherited Disease. 7th ed. McGraw-Hill, New York, 1995.

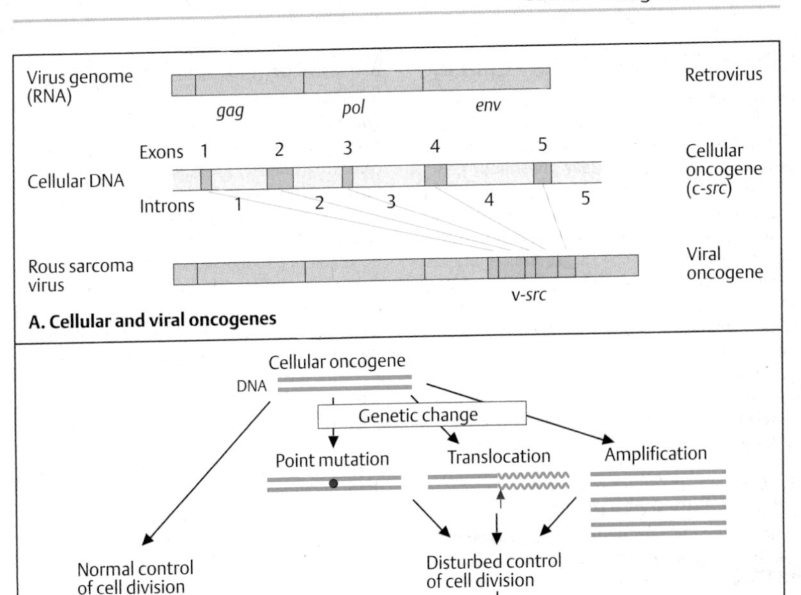


**A. Cellular and viral oncogenes**



**B. Mechanisms of oncogene activation**

| Oncogene | Function | Origin of tumor by c-onc | Origin of tumor by v-onc |
|---|---|---|---|
| *sis* | One form of PDGF | | Simian sarcoma |
| *abl* | Tyrosine protein kinase | Chronic myelogenous leukemia | Abelson mouse leukemia |
| *fes* | Tyrosine protein kinase | | Cat sarcoma |
| *erbB* | Epidermal growth factor | Mammary carcinoma<br>Ovarian carcinoma | Chicken erythroblastosis |
| *src* | Membrane-bound protein kinase | | Rous chicken sarcoma |
| Ha-*ras*<br>Ki-*ras*<br>N-*ras* | Membrane-bound G proteins with GTPase activity | Different carcinomas, neuroblastoma | Harvey mouse sarcoma<br>Kirsten mouse sarcoma |
| *fos* | Transcription factor AP1 | | Osteosarcoma in chicken |
| *myb* | Nuclear protein | Leukemias | Myeloblastosis in chicken |
| *myc* | Nuclear protein | Leukemias | Myelocytosis in chicken |
| N-*myc* | Nuclear protein | Neuroblastoma | |

**C. Examples of cellular oncogenes and their proteins**

## The p53 Protein, a Guardian of the Genome

The p53 protein (named after its molecular weight of 53 kD), a nuclear phosphoprotein, is indispensable for genomic integrity and cell cycle control. It binds to specific DNA sequences and regulates the expression of different regulatory genes involved in growth. It interacts with other proteins in response to DNA damage and mediates apoptosis (cell death) of the cell when the damage is beyond repair. Its basic function is to control entry of the cell into the S phase (see cell cycle control, p. 112). Somatic mutations in the *p53* gene occur in about half of all tumors. Germline mutations lead to a familial form of multiple different cancers (Li–Fraumeni syndrome, see **B**).

### A. The human p53 protein

The active form of the human p53 protein is a tetramer of four identical subunits. Each subunit has 393 amino acids and five highly conserved regions, I–V. Region I is part of the transcription-activation domain; regions II–V belong to sequence-specific DNA-binding domains. The carboxyl end beyond amino acid 300 consists of a nonspecific DNA interaction domain and the tetramerization domain. Proteins encoded by DNA tumor viruses bind to p53 and inhibit its activity. Mutations in the *p53* gene on human chromosome 17 at p13 (spanning 20 kb of DNA and yielding a 2.8 kb mRNA transcript from 11 exons) have the greatest effect when they occur in the conserved regions II–V in codons 129–146 (exon 4), 171–179 (exon 5), 234–260 (exon 7), and 270–287 (exon 8). Particularly vulnerable are the conserved amino acids arginine (R) in positions 175, 248, 249, 273, and 282 and glycine (G) in position 245. Mutations occur mainly as missense, resulting from base-pair substitutions, but some are insertions and deletions and exert a dominant negative effect. Knockout mice develop normally, but develop tumors at a high rate. Activated benzopyrene induces mutations at codons 175, 248, and 275 in cultured bronchial epithelial cells.

### B. Germline mutations of *p53*

In 1969, Li and Fraumeni identified families in whom other members were affected with diverse types of tumors, mainly soft-tissue sarcomas, early-onset breast cancer, brain cancers, cancer of the bone (osteosarcoma) and bone marrow (leukemias), and carcinoma of the lung, pancreas, and adrenal cortex. Similar observations had been reported as "cancer family syndrome" by Lynch. This autosomal dominant cancer syndrome is called the Li–Fraumeni syndrome (McKusick 114480). In the pedigree shown in panel 1, four individuals (II-2, II-3, III-1, III-2) are affected by different types of tumors. A mutation in codon 248 of the *p53* gene (CGG arginine, to TGG, tryptophan) is present in these patients. The mutation is also present in individuals I-1 and III-5. This places these individuals at increased risk for one of the types of cancer mentioned above and shown in panel 2. In contrast, absence of the mutation in individuals III-3 and III-4 indicates that they do not have an increased risk of cancer (Data of D. Malkin). A subset of patients with Li–Fraumeni syndrome does not show p53 mutations.

### C. Model of function of the *p53* gene

Normally, the *p53* gene is inactive (1). *p53* plays an important role in regulating growth in damaged cells (2). DNA damage in cells leads to increased expression of *p53* and interruption of the cell cycle in $G_1$. If DNA repair is successful, the cell can continue its cycle. If repair is not successful, the cell dies (cell death, apoptosis). Damaged cells with p53 protein that is mutant are not arrested in $G_1$. (After Lane, D.P.: Nature **358**:15–15, 1992.)

### References

Bell, D.W., et al.: Heterozygous germline *hCHK2* mutations in Li-Fraumeni syndrome. Science **286**:2828–2831, 1999.

Hanahan, D., Weinberg, R.A.: The hallmarks of cancer. Cell **100**:57–70, 2000.

Lodish, H., et al.: Molecular Cell Biology (with an animated CD-ROM). 4th ed. W.H. Freeman & Co., New York, 2000.

Malkin, D.: The Li-Fraumeni syndrome, pp. 353–407. In: Vogelstein, B., Kinzler, K.W., eds., The Genetic Basis of Human Cancer. McGraw-Hill, New York, 1998.

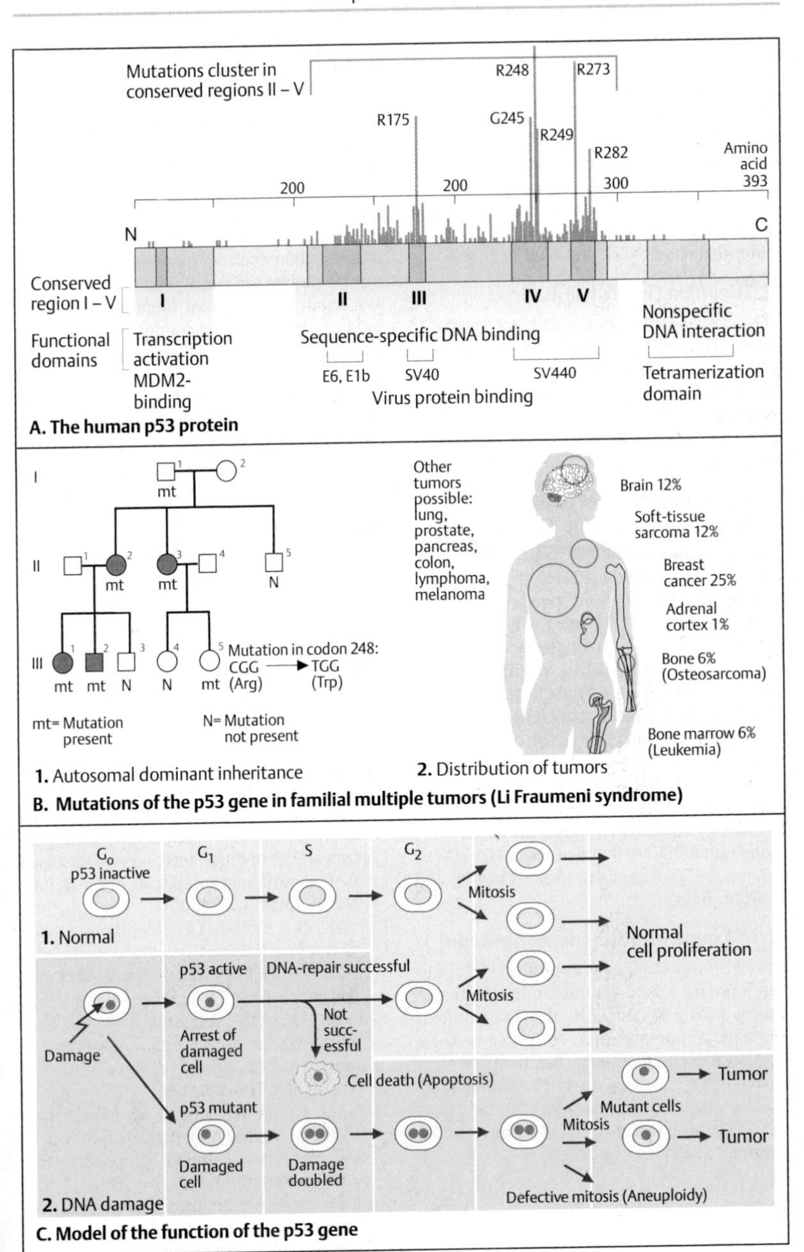


**A. The human p53 protein**

**B. Mutations of the p53 gene in familial multiple tumors (Li Fraumeni syndrome)**

**C. Model of the function of the p53 gene**

## Neurofibromatosis 1 and 2

The neurofibromatoses are clinically and genetically different autosomal dominant hereditary diseases that predispose to benign and malignant tumors of the nervous system. Numerous different forms are known. The most important are neurofibromatosis 1 (NF1, von Recklinghausen disease, MIM 162200) and neurofibromatosis 2 (NF2, MIM 101000).

### A. The main signs of NF1

NF1 is very variable. Lisch nodules of the iris (1) in more than 90% of patients, café-au-lait spots (2) (more than five spots of more than 2 cm diameter are considered diagnostic) in more than 95%, and multiple neurofibromas (3) in more than 90% of patients are the most important signs.

### B. Neurofibromatosis gene *NF1* on human chromosome 17 at q11.2

The localization of the *NF1* gene revealed the gene on a 600 kb *Nru*I restriction fragment. A CpG island (CpG-1) and two translocation breakpoints at t(17;22) and t(1;17) served as important anchor points for gene identification. The *NF1* gene has 79 exons, which span about 335 kb of genomic DNA. Three unrelated genes, *OMGP, EVI2B,* and *EVI2A,* are embedded within the *NF1* gene in intron 35 on the opposite DNA strand. Mutation analysis of the *NF1* gene shows deletions, insertions, base substitutions, and splice mutations leading to truncated and presumably nonfunctional gene products. Currently mutations are found in about 60–70% of patients. (Figure adapted from Claudio and Rouleau, 1998).

### C. *NF1* gene product (neurofibromin)

The *NF1* gene encodes a gene product with 2810 amino acids, called neurofibromin. Between amino acids 840 and 1200, this large protein contains a domain that corresponds to a GTPase-activating protein. The homology includes a gene product in yeast *(S. cerevisiae),* IRA1 (inhibitor of *ras* mutants). Mutations at the *NF1* locus interrupt a signal pathway to the *ras* genes. (After Xu et al., 1990).

### D. Neurofibromatosis gene *NF2* on human chromosome 22 at q12.1

The *NF2* gene was identified in 1993 by Rouleau et al. and Trofatter et al. within a cosmid contig contained in YAC clones (yeast artificial chromosomes). Two deletions observed in unrelated patients aided in finding the almost 100 kb gene with 17 exons. Mutations can be detected in more than 50% of patients (large deletions including the entire gene or several exons and small deletions are frequent). The gene product, called schwannomin, is related to a family of cytoskeleton-membrane proteins (erythrocyte protein 4.1, see p. 374, and the ERM family ezrin, radixin, and moesin) and a family of protein tyrosine phosphatases. The basic function of these proteins is to maintain cellular integrity. (Figure adapted from Claudio and Rouleau, 1998).

### References

Carey, J.C., Viskochil, D.H.: Neurofibromatosis Type 1: a model condition for the study of the molecular basis of variable expressivity in human disorders. Am. J. Med. Genet. (Semin. Med. Genet.) **89**:7–13, 1999.

Claudio, J.O., Rouleau, G.A.: Neurofibromatosis type 1 and type 2, pp. 963–970. In: Principles of Molecular Medicine, J.L. Jameson, ed. Humana Press, Totowa, NJ, 1998.

Huson, S.M.: What level of care for the neurofibromatoses? Lancet **353**:1114–1116, 1999.

Messiaen, L.M., et al.: Exhaustive mutation analysis of the *NF1* gene allows identification of 95% of mutations and reveals a high frequency of unusual splicing defects. Hum. Mutat. **15**:541–555, 2000.

Riccardi, V.M., Eichner, J.E.: Neurofibromatosis. Phenotype, Natural History and Pathogenesis. 2nd ed., Johns Hopkins University Press, Baltimore, 1992.

Rouleau, G.A., et al.: Alteration in a new gene encoding a putative membrane-organizing protein causes neurofibromatosis type 2. Nature **363**:515–521, 1993.

Trofatter, J.A., et al.: A novel Moesin-, Ezrin-, Radixin-like gene is a candidate for the neurofibromatosis 2 tumor suppressor. Cell **72**:791–800, 1993.

Xu, G., et al.: The neurofibromatosis type 1 gene encodes a protein related to GAP. Cell **62**:599–608, 1990.

Neurofibromatosis 1 (NF1)
(von Recklinghausen disease)

Autosomal dominant
Frequency 1 in 3000
Gene locus on 17q11.2
Café-au-lait spots
Lisch nodules in the iris
Multiple neurofibromas
Skeletal anomalies
Predisposition to tumors of the nervous system
50% new mutations

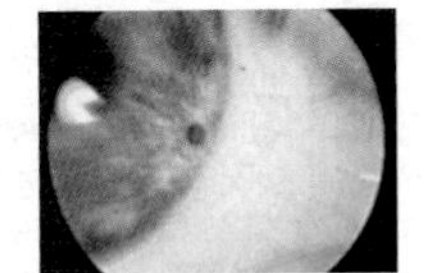

**1.** Lisch nodule

**2.** Café-au-lait spot

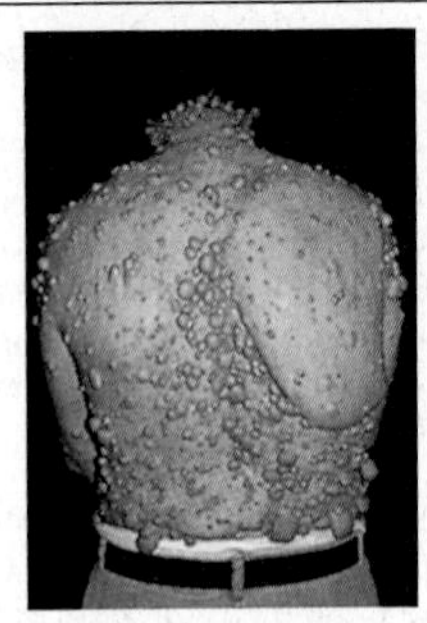

**3.** Neurofibromas

**A. Main manifestations of neurofibromatosis 1**

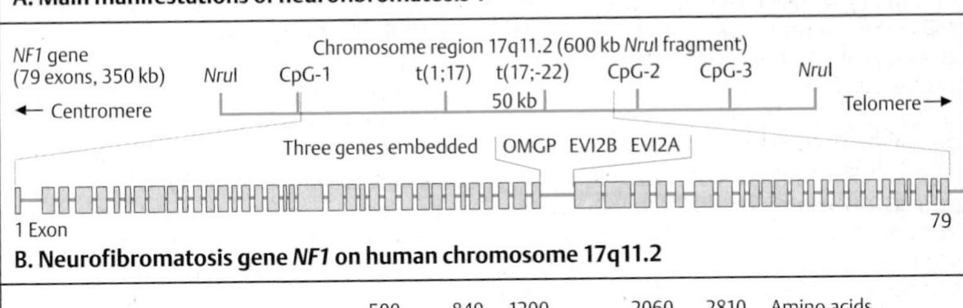


**B. Neurofibromatosis gene *NF1* on human chromosome 17q11.2**

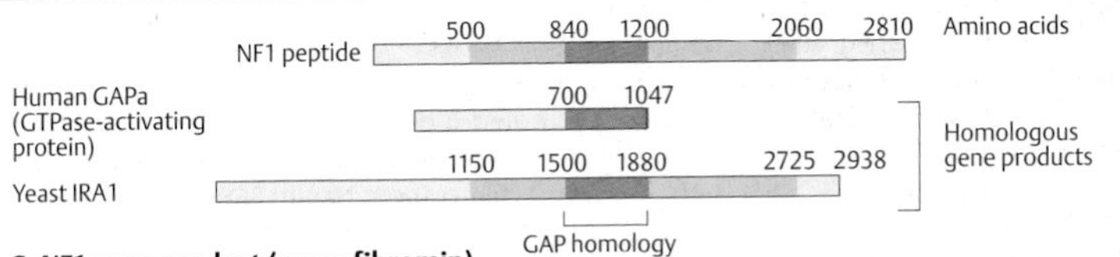


**C. *NF1* gene product (neurofibromin)**

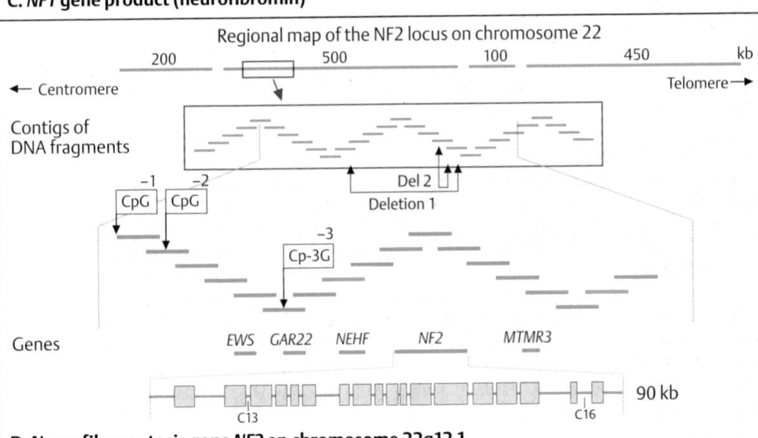


**D. Neurofibromatosis gene *NF2* on chromosome 22q12.1**

# APC Gene in Familial Polyposis Coli

Cancer of the colon and rectum is the second leading cause of death from cancer. About 5% of the population can be expected to develop colorectal cancer. Most colorectal tumors arise from a series of somatic mutations in several genes.

## A. Polyposis coli and colon carcinoma

Familial polyposis (FAP) is an autosomal dominant hereditary disease. In late childhood and early adulthood, up to 1000 and more polyps develop in the mucous membrane of the large intestine (colon) (1). Each polyp can develop into a carcinoma (2). In about 85% of affected persons, small hypertrophic areas that do not affect vision are present in the retina (3).

Hereditary non-polyposis colorectal cancer (HNPCC) affects about one in 200–1000 individuals (3% of all colorectal cancers). It results from germline mutation in one of the DNA mismatch repair genes *hMSH1, hMLH2, hPMS1*, and *hPMS2* or related genes. Microsatellite instability is an important feature of HNPCC. (Photos 1 and 2 from U. Pfeifer, Institut für Pathologie der Universität Bonn; photo 3 from W. Friedl et al., 1991.)

## B. Mutations at different gene loci in polyposis coli and carcinoma of the colon

At least six gene loci are involved in the development of carcinoma of the colon associated with polyposis coli. Somatic mutations may occur in two recessive oncogenes (Ras genes *KRAS1* and *KRAS2*) and in four dominant tumor suppressor genes. Most forms of carcinoma of the colon are not associated with polyposis coli.

## C. The *APC* gene and distributions of mutations

The *APC* gene (adenoma polyposis coli) consists of 8538 bp in 15 exons encoding a 2843-amino-acid protein (not 8535 bp and 2844 amino acids as shown in C). Exon 15 is very large, 6579 base pairs. Over 95% of mutations result in a nonfunctional truncated protein due to nonsense mutations (40%), deletions (41%), insertions (12%), and splice site mutations (7%). The *APC* gene is also involved in sporadic colorectal cancer.

## D. Indirect DNA diagnosis in FAP

Linked DNA markers (RFLPs) near the APC locus (1) can be used for indirect DNA diagnosis. The alleles of three flanking marker pairs (K,k and E,e on the centromere side and A,a on the distal side) form the haplotypes, e.g., e–K–a and E–k–a in individual I-1 in the pedigree (2). The mutation-carrying haplotype must be e–K–a. Since individual III-2 has inherited this haplotype, he is at risk for the disease, whereas individual III-1 is not.

## E. Several mutations in the production of colon carcinoma

Tumor formation goes through several stages. It starts with a somatic or germinal mutation in the *APC* gene. After loss of the other allele (LOH), an adenoma develops with less-differentiated cells and polyp formation. Mutations in other genes lead to malignant transformation and eventually to tumor development. (Diagram after Fearon and Vogelstein, 1990.)

### References


Bronner, C.E., et al.: Mutation in the DNA mismatch repair gene homologue *hMLH1* is associated with hereditary nonpolyposis colon cancer. Nature **368**:258–261, 1994.

de la Chapelle, A., Peltomäki, P.: The genetics of hereditary common cancers. Curr. Opin. Genet. Develop. **8**:298–303, 1998.

Fearon, E.R., Vogelstein, B.: A genetic model for colorectal tumorigenesis. Cell **61**:759–767, 1990.

Fearon, E.R., Cho, K.R.: The molecular biology of cancer, pp. 405–438, In: D.L. Rimoin, J.M. Connor, R.E. Pyeritz, eds., Emery and Rimoin's Principles and Practice of Medical Genetics. 3rd ed. Churchill-Livingstone, Edinburgh, 1996.

Groden, J., et al.: Identification and characterization of the familial adenomatous polyposis coli gene. Cell **66**:589–600, 1991.

Kinzler, K.W., Vogelstein, B.: Colorectal tumors, pp. 565–587. In: B. Vogelstein, K.W. Kinzler, eds., The Genetic Basis of Human Cancer. McGraw-Hill, New York, 1998.

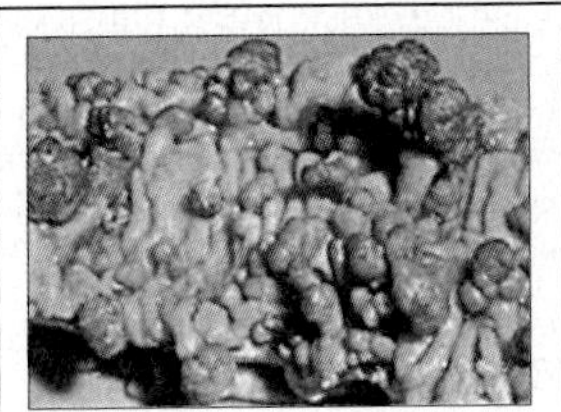

1.

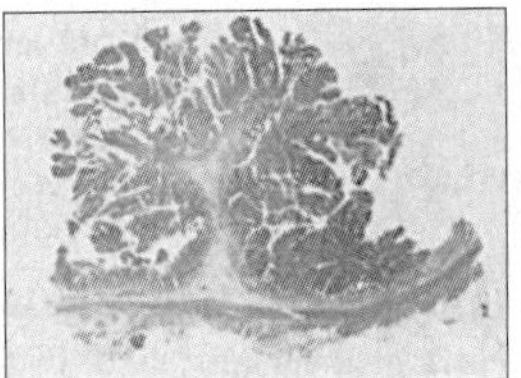

2.

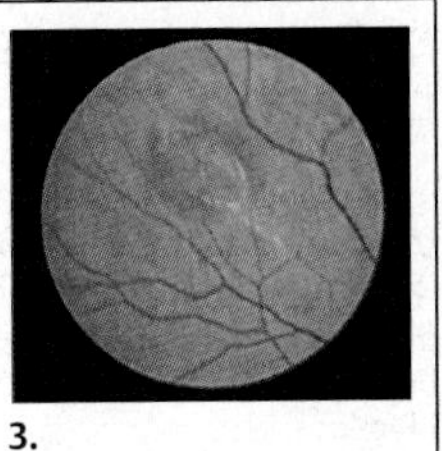

3.

**A. Polyposis coli and colon carcinoma**

| McKusick no. | Disease/gene involved | Gene locus |
|---|---|---|
| 190110 | Colorectal adenoma/carcinoma (*KRAS1*) | 6p12-11 |
| 190070 | Colorectal adenoma/carcinoma (*KRAS2*) | 12p12 |
| 114500 | Isolated colon carcinoma (somatic mutation in the *APC* gene) | 5q21 |
| 175100 | Familial adenomatous polyposis coli (germinal mutation in the *APC* gene) | 5q21 |
| 175100 | *APC* with other tumors: Gardner syndrome (allelic) | 5q21 |
| 159350 | DNA sequences mutated in colon carcinoma (*MCC*) | 5q21-22 |
| 191170 | p53 gene (identical with McKusick *120460) | 17p12-13 |
| 120470 | Deletion in colorectal carcinoma (*DCC*) | 18q21.3 |

**B. Mutations at different gene loci in polyposis coli and colon carcinoma**

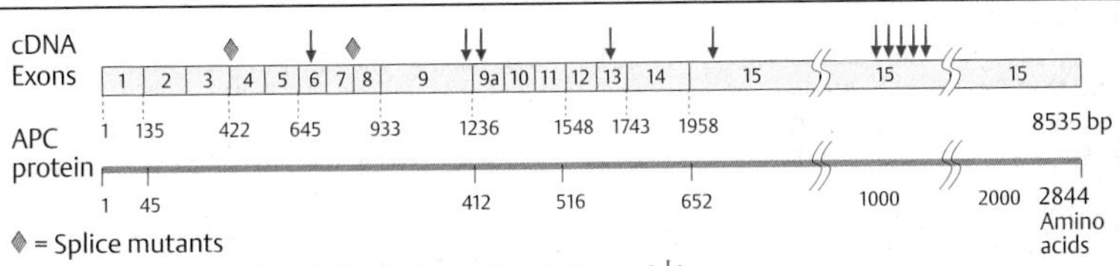


**C. *APC* gene (scheme) and distribution of mutations (↓)**

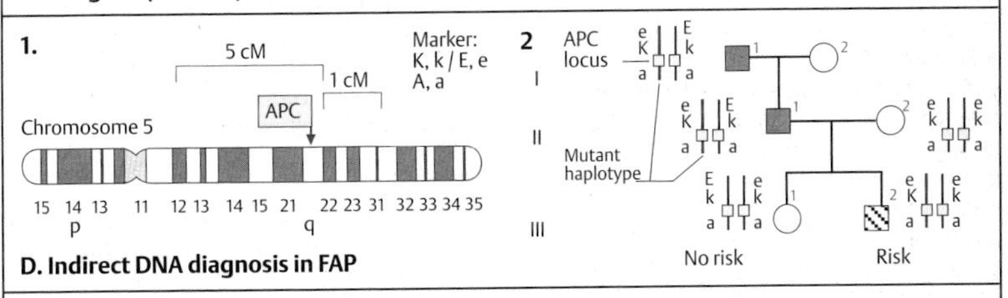


**D. Indirect DNA diagnosis in FAP**

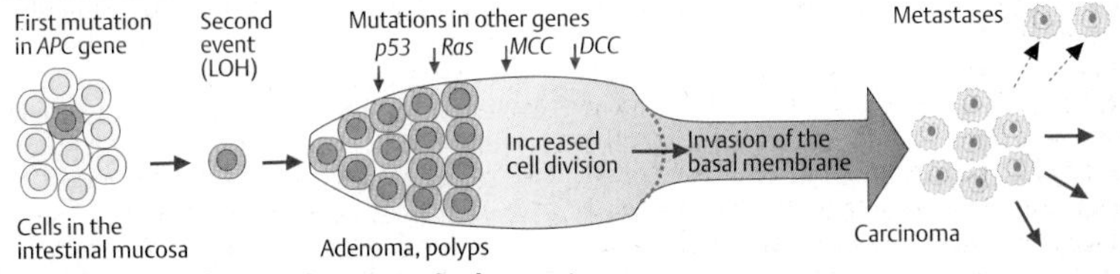


**E. Several mutations in the origin of colon carcinoma**

# Breast Cancer Susceptibility Genes

In 1994 and 1995 two genes were identified that confer susceptibility to breast and ovarian cancer when mutated, the breast cancer genes *BRCA1* and *BRCA2*. Both genes encode multifunctional proteins with important cellular functions in genomic stability, homologous recombination, and double-stranded and transcription-coupled DNA repair (see p. 80). The BRCA1 and BRCA2 proteins interact and play a role in cell cycle control (see p. 112) and in development. An autosomal dominant susceptibility mutant allele in one of these genes is considered the main cause of the cancer in about 5–10% of patients. Mutations in other genes are involved in some cases. The direct causative role of *BRCA1* and *BRCA2* mutations is difficult to assess in individual patients. Different mutations as well as polymorphic variants occur throughout the genes.

## A. The breast cancer susceptibility gene *BRCA1*

The *BRCA1* gene on chromosome 17 at q21.1 consists of 24 exons spanning 80 kb of genomic DNA that encode a 7.8 kb mRNA transcript. The protein has 1863 amino acids. Exon 11 is quite large (3.4 kb). About 55% of all mutations occur in exon 11. Although some mutations occur relatively frequently in other exons, they tend to be evenly distributed throughout the gene (only some mutations are shown). The deletion of an adenine (A) and a guanine (G) in nucleotide position 185 (185delAG) and the insertion of a cytosine in position 5382 (5382insC) account for about 10% of mutations each. These mutations are particularly frequent in the Ashkenazi Jewish population.

The protein has five main functional domains. The RING finger region near the N-terminus at amino acids 1–112 defines a zinc-binding domain of conserved cysteine and histidine residues that mediate protein–protein or protein-DNA interactions. This region is also the site of heterodimerization of BRCA1 and BARD1 (BRCA1-associated RING domain 1). Other functional domains define the central part of the BRCA1 protein. These are two nuclear localization signals (NLS) and two protein-binding domains, one for p53 protein, retinoblastoma (RB) protein, and RAD50 and RAD51. RAD50 and 51 are proteins involved in recombination during mitosis and meiosis, and in recombinational repair of double-stranded DNA breaks. The C-terminus contains a region involved in transcriptional activation and DNA repair.

## B. The breast cancer susceptibility gene *BRCA2*

The *BRCA2* gene on 13q12 comprises 27 exons spanning 80 kb of genomic DNA that encode a 10.4 kb mRNA transcript. Its protein has 3418 amino acids. Exon 11 is large (11.5 kb), as in *BRCA1*. Mutations occur throughout the gene (only some are shown). A deletion of thymine at nucleotide position 6174 (6174delT) is relatively (1%) frequent in the Ashkenazi Jewish population.

The BRCA2 protein has a transcriptional activation domain near the N-terminus and a nuclear location signal (NLS) near the C-terminus. A large central domain consists of eight copies of a 30–80-amino-acid repeat, which are conserved in all mammalian BRCA2 proteins (BRC repeats). Four of these interact with the RAD51 protein.

The *BRCA1* and *BRCA2* genes are expressed ubiquitously with the highest levels of expression in thymus and testis. The spatial and temporal expression patterns of *Brca1* and *Brca2* in the mouse fetal and adult tissues are essentially identical, with highest expression of both in rapidly dividing tissues during differentiation, especially in mammary epithelium. In the mammary gland both genes are expressed during puberty and pregnancy, and their expression is reduced during lacation. (Figures redrawn from Welcsh et al., 2000.)

### References

Miki, Y., et al.: A strong candidate for the breast and ovarian cancer susceptibility gene *BRCA1*. Science **266**:66–71, 1994.

Welcsh, P.L., Schubert, E.L., King, M.C.: Inherited breast cancer: an emerging picture. Clin. Genet. **54**:447–458, 1998.

Welcsh, P.L., Owens, K.N., King, M.C.: Insights into the functions of BRCA1 and BRCA2. Trends Genet: **16**:69–74, 2000.

Wooster, R., et al.: Identification of the breast cancer susceptibility gene *BRCA2*. Nature **378**:789–792, 1995.

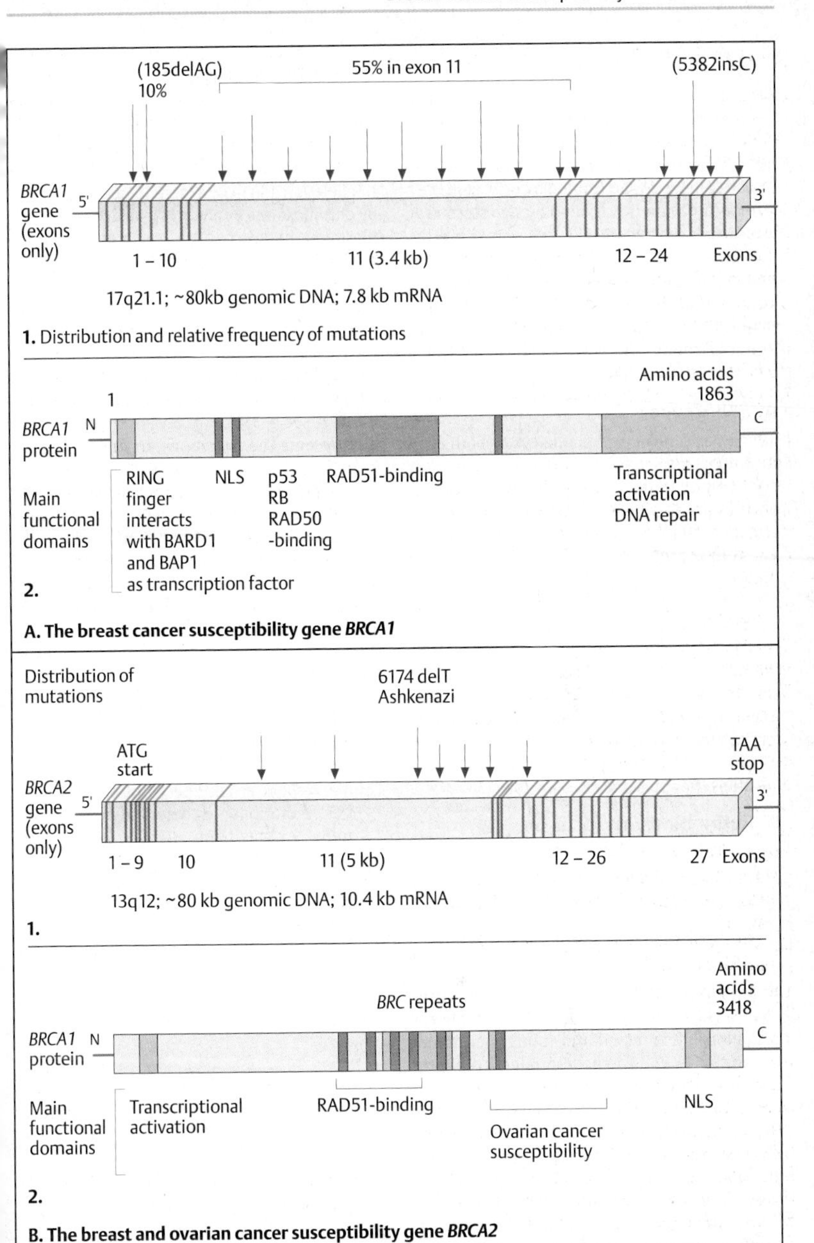


**A. The breast cancer susceptibility gene *BRCA1***

**B. The breast and ovarian cancer susceptibility gene *BRCA2***

# Retinoblastoma

Retinoblastoma (McKusick 180200) is the most frequent tumor of the eye in infancy and early childhood. It occurs in 1 of 15000–18000 live births. This tumor results from loss of function of both alleles of the retinoblastoma gene *RB1*. Tumor initiation is preceded by two steps as A. Knudson predicted in 1971 in his "two-hit" hypothesis (tumor suppressor gene, p. 318). The first predisposing mutation in one allele may occur either in a retinoblast, an undifferentiated retinal cell in the developing embryo, or in the germline. The other allele is inactivated by a second mutation.

## A. Phenotype

Retinoblastoma occurs in one eye or both eyes. An important early sign is the so-called "cat's eye," a white shimmer out of the affected eye (1) or the development of strabismus. One or several tumors originate from the retina (2). The tumor progresses rapidly (3). The relative proportions of the genetic types of retinoblastoma are about 60% somatic mutations (nonhereditary form) and 40% germline mutations, transmitted as an autosomal dominant trait (hereditary form, in about 10–15%, due to transmission from a parent; the remainder due to a new mutation). New mutations usually affect a paternal allele (about 10 : 1). In about 10% of carriers of a germline mutation no tumor develops (nonpenetrance).

## B. Retinoblastoma locus on chromosome 13

The *RB1* locus at 13q14.2 was first defined with cytogenetically visible interstitial deletions.

## C. Retinoblastoma gene *BR-1* and the pRB protein

The *RB1* gene is organized into 27 exons spanning 183 kb of genomic DNA (1). The *RB1* gene is ubiquitously expressed and transcribed into a 4.7 kb mRNA (2). The gene product (pRB protein) has 928 amino acids (3). It is a 100 kD phosphoprotein with important functions in the regulation of the cell cycle. It is activated by phosphorylation (P) during cell cycle progression from $G_0$ to $G_1$ (p. 112) at about 12 distinct serine and threonine residues. Three functional domains, A, B, and C, and a nuclear localization signal (NLS) can be distinguished.

## D. Diagnostic principle

Molecular diagnosis of retinoblastoma greatly contributes to its early recognition and to the correct assessment of individual risks within families. In about 3–5% of patients an interstitial deletion 13q14 or a larger deletion is visible by chromosomal analysis (1). In familial retinoblastoma indirect DNA diagnosis can be achieved by segregation analysis using DNA markers at the *RB1* locus (2). In the example shown, the affected girl (II-1) has inherited haplotype **a** from the unaffected father and haplotype **c** from the unaffected mother. In tumor cells, obtained after one eye had to be removed, haplotype **a** only is present (loss of heterozygosity, LOH, see p. 318). This reveals that haplotype **a** represents the mutation-carrying *RB1* allele. In the family shown (3), I-2 and II-2 are affected (3). Sequence analysis reveals a C-to-T transversion in codon 575 in the two affected individuals (CAA glutamine to TAA stop codon). The mutational spectrum in hereditary retinoblastoma involves deletions (~26%), insertions (~9%), and point mutations (~65%), including splice-site mutations.

(Illustrations courtesy of W. Höpping (A) and D. Lohmann (C and D).)

## References

Lohmann, D.R.: RB1 gene mutations in retinoblastoma. Hum. Mutat. **14**:283–288, 1999.

Lohmann, D.R., et al.: Spectrum of RB1 germline mutations in hereditary retinoblastoma. Am. J. Hum. Genet. **58**:940–949, 1996.

Newsham, I.F., Hadjistilianou, T., Cavenee, W.K.: Retinoblastoma. pp. 363–392. In: B. Vogelstein, K.W. Kinzler, eds. The Genetic Basis of Human Cancer. McGraw-Hill, New York, 1998.

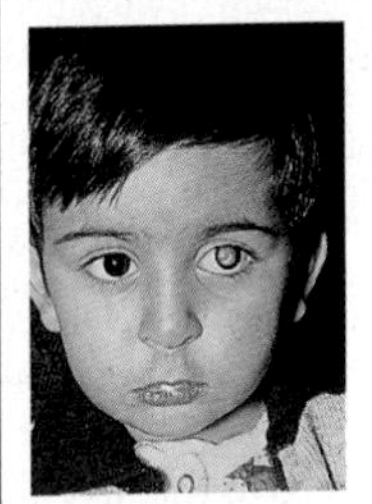
1. So-called cat's eye

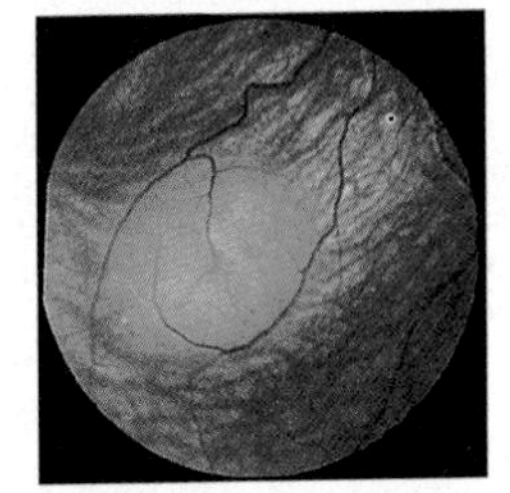
2. Tumor in the retina

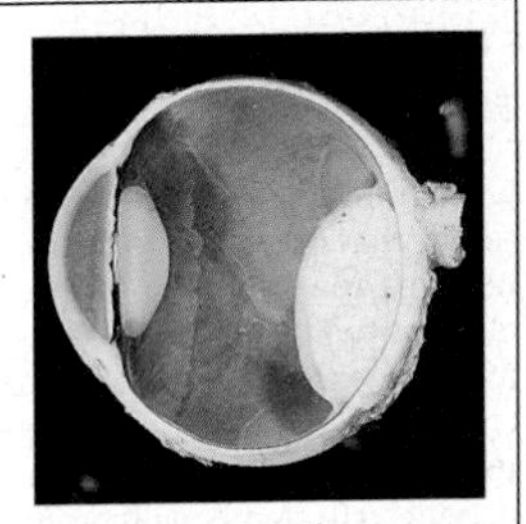
3. Large tumor in the eye

**A. Phenotype**

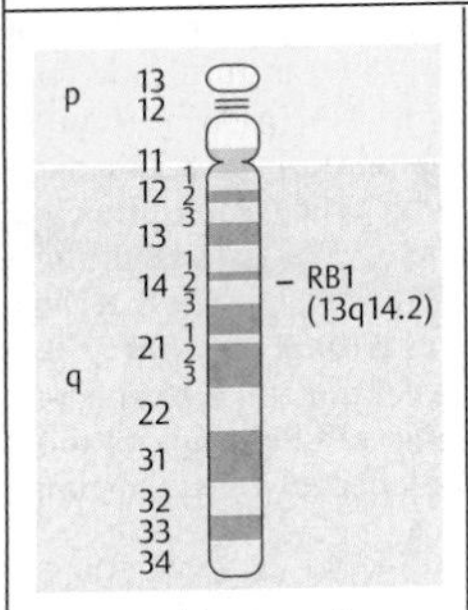


**B. Retinoblastoma locus on chromosome 13**

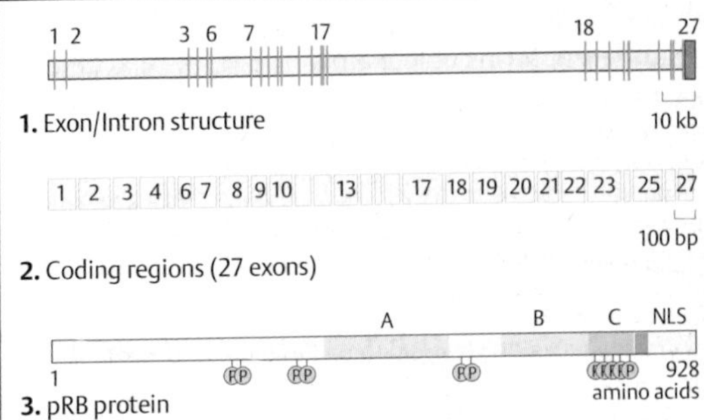


**C. Retinoblastoma gene *RB1* and the pRB protein**

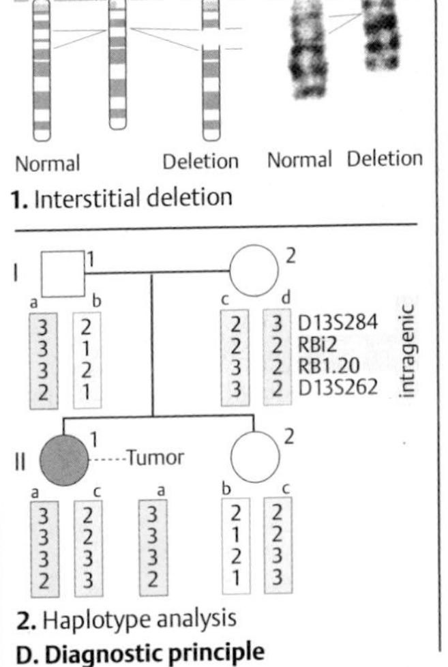


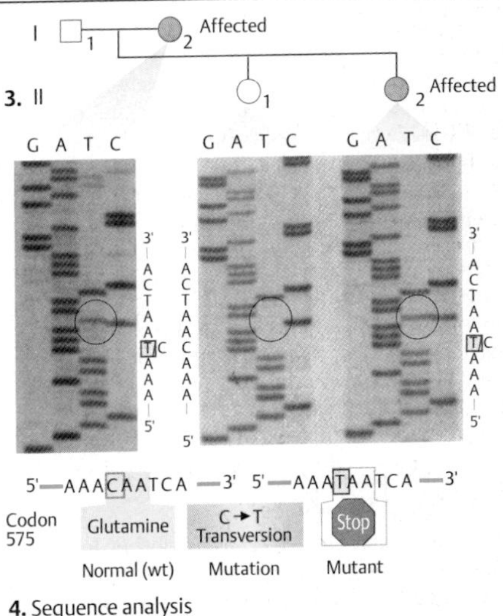


**D. Diagnostic principle**

## Fusion Gene as Cause of Tumors: CML

Chronic myeloid leukemia (CML) is a malignant tumor that originates from a single cell of the bone marrow in adulthood. The number of myelocytes (white blood cells from the bone marrow) is greatly increased. The disease follows a chronic course. Acute crises develop intermittently and terminally. In about 90% of the patients, affected bone marrow cells contain a chromosome 22 with a shortened long arm (22q–, Philadelphia chromosome).

### A. The Philadelphia chromosome ($Ph^1$) in different forms of leukemia

A Philadelphia chromosome is present in the bone marrow cells of most patients with the chronic form of the disease (CML). If it is not present, the illness progresses more rapidly than usual and has a poorer prognosis. In addition, the Philadelphia chromosome may be found in some acute leukemias (acute lymphocytic leukemia, ALL; acute myelocytic leukemia, AML) in adults and in children. Here, $Ph^1$ indicates a poor prognosis, whereas its absence is favorable.

### B. The $Ph^1$ translocation [t(9;22)(q34;q11)]

The Philadelphia chromosome arises by reciprocal translocation between a chromosome 22 and a chromosome 9. The breakpoints are in 9q34 and 22q11. A good half of the long arm of a chromosome 22 is translocated to the long arm of a chromosome 9. A very small segment of the distal long arm of a chromosome 9 (9q34), not visible under the light microscope, is translocated to a chromosome 22. The Philadelphia chromosome (22q–) consists of the short arm and the proximal one-third of the long arm of a chromosome 22 and the small distal segment from the long arm of a chromosome 9. For demonstration of the Philadelphia translocation by in-situ hybridization, see p. 192).

### C. The $Ph^1$ translocation leads to the fusion of two genes

The breakpoints of the $Ph^1$ translocation are located in the *BCR* gene of chromosome 22 and in the *ABL* gene of chromosome 9. The translocation leads to the fusion of these genes. The exact locations of the breakpoints differ from patient to patient, but in the *BCR* gene they are limited to a small region of just 6 kb (thus the designation *BCR*, or *b*reakpoint *c*luster *r*egion). In CML, the breakpoints lie in exons 10–12 of the *BCR* gene; in acute $Ph^1$-positive leukemias (e.g., ALL) they lie further in the 5′ direction in exon 1 or 2. The breakpoint region in the *ABL* gene extends over 180 kb between exons 1a and 1b, which are separated by an intron.

### D. The gene fusion leads to changes in transcription and gene products

The *ABL* gene codes for mRNA transcripts of 7 kb (exon 1b, 2–11) and 6 kb (exon 1a, 2–11) by differential splicing; these in turn code for a protein of about 145 000 Da ($p145^{abl}$). From the fusion of the two genes in CML, an 8.5 kb mRNA transcript results, which codes for a fusion protein of 210 000 Da ($p210^{bcr/abl}$). In the acute form of leukemia (ALL), a transcript results that codes for a fusion protein of 190 000 Da ($p190^{bcr/abl}$). In contrast to the normal protein, it has high tyrosinase activity. This results in uncontrolled cell division in the affected cells and tumor growth.

### References


Bartram, C.R. et al.: Translocation of c-abl oncogene correlates with the presence of a Philadelphia chromosome in chronic myelocytic leukaemia. Nature **306**:277–280, 1983.

Cline, M.J.: The molecular basis of leukemia. New Eng. J. Med. **330**:328–336, 1994.

Faderl, S., et al.: The biology of chronic myeloid leukemia. New Eng. J. Med. **341**:164–172, 1999.

Hentze, B.M., Kulozik, A.E., Bartram, C.R.: Einführung in die medizinische Molekularbiologie. Grundlagen, Klinik, Perspektiven. Springer, Berlin, 1990.

Kurzrok, R., Gutterman, J.U., Talpaz, M.: The molecular genetics of Philadelphia-positive leukemias. New Eng. J. Med. **319**:990, 1988.

Sawyers, C.L.: Chronic myeloid leukemia. New Eng. J. Med. **340**:1330–1340, 1999.

| Chronic myelogenous leukemia (CML) | Other $Ph^1$-positive acute leukemias |
|---|---|
| Myelocytes in peripheral blood increased | Lymphocytes or myelocytes increased |
| Early chronic course | Acute course |
| Adults affected | 20% adults with ALL $Ph^1$-positive |
| Origin from one myeloid cell | 2% adults with AML $Ph^1$-positive |
| In ~ 90% a Philadelphia chromosome ($Ph^1$) | 5% children with ALL $Ph^1$-positive |
| (Translocation 22q to 9q) | Philadelphia translocation as in CML |
| Poor prognosis when $Ph^1$-negative | Poor prognosis when $Ph^1$-positive |

**A. Philadelphia translocation ($Ph^1$) in different forms of leukemia**

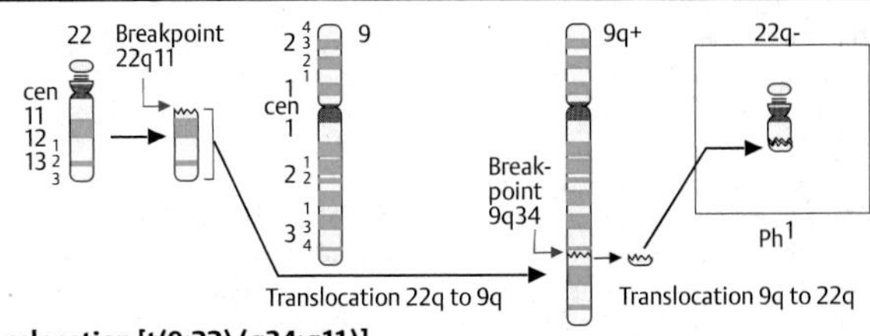


**B. $Ph^1$ translocation [t(9;22) (q34;q11)]**

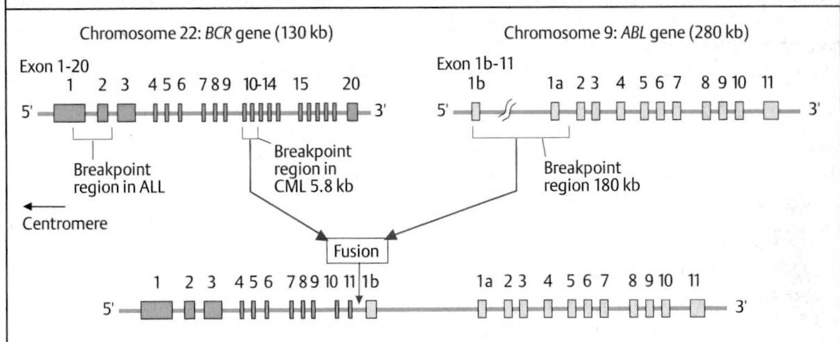


**C. $Ph^1$ translocation causes fusion of two genes**

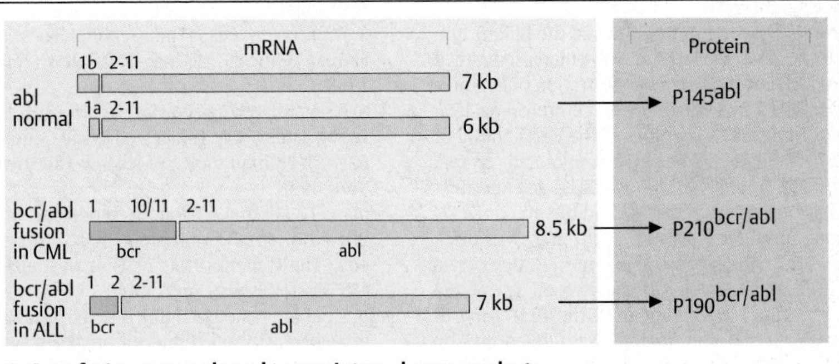


**D. Gene fusion causes altered transcripts and gene products**

## Genomic Instability Syndromes

Genomic instability, visible by light microscopy as breaks and rearrangements in different chromosomes in a variable proportion of metaphase cells, is a hallmark of a group of characteristic hereditary diseases. The underlying genetic defect predisposes patients to different types of cancer. Three important examples are presented here.

### A. Bloom syndrome (BS)

In Bloom syndrome (McKusick 210900) (1), prenatal and postnatal growth deficiency is pronounced (birth weight 2000 g, birth length ~40 cm, adult height around 150 cm). The phenotype (2) includes a narrow face. Usually, but not always, a sunlight-induced erythema develops on the cheeks, eyelids, mouth, ears, and back of the hands (a and b). The photograph on the right (c) shows a boy with Bloom syndrome and acute leukemia. Metaphase cells show about a tenfold increase in the rate of sister chromatid exchanges (SCE), ~60 instead of about 6 per metaphase in normal cells (3). (Sister chromatid exchanges are explained in the glossary, p. 423). Metaphases of patients contain increased breaks in one or both chromatids and exchanges between homologous chromosomes in about 1–2% of cells. In Bloom syndrome patients, different types of malignancies occur in a distribution comparable to that of the general population, but at a much earlier age (mean age 24.7, range 2–48 years). Some patients have multiple primary tumors, which underlines the striking susceptibility to cancer in Bloom syndrome. Chemotherapy is very poorly tolerated.

Homozygosity for mutations in the Bloom syndrome gene (*BLM*) results in an increased rate of somatic mutations, a manifestation of genomic instability. The *BLM* gene on chromosome 15 at q16.1 encodes a member of the RecQ family of DNA helicases. The 1417-amino-acid protein shows homology to the yeast *SGS1* gene product (slow growth suppressor) and the human *WRN* gene product (Werner syndrome, McKusick 277700). Allozygous nonsense mutations (two different mutations in the two alleles of the same gene) are frequent in the *BLM* gene. A characteristic homozygous 6 bp deletion/7 bp insertion at nucleotide 2281 occurs in Ashkenazi Jewish individuals as a result of a founder effect.

### B. Fanconi anemia (FA)

Fanconi anemia (hereditary pancytopenia) (McKusick 227650) is a malformation syndrome (1) with variable clinical expression. Growth deficiency (2), hypoplastic or absent thumbs (3), and short or absent radii are characteristic physical signs.

FA cells are hypersensitive to DNA-crosslinking agents, such as diepoxybutane (DEB). Several complementation groups can be distinguished. Three FA genes have been identified, at chromosome 16q24.3 (FAA), 9q22.3 (FAC), and 3p22–26 (FAD). FAA is the most prevalent group in 60–65% of patients.

### C. Ataxia telangiectasia

Ataxia telangiectasia (McKusick 208900) is a pleiotropic, variable disease due to mutations in the *ATM* gene on chromosome 11q23. Preferential reciprocal translocations between chromosomes 7 and 14 with breakpoints at 7p14, 7q14, 14q11, and 14q32 occur in a small proportion of metaphases. The *ATM* gene has 66 exons spanning more than 150 kb of genomic DNA. From its 13-kb transcript (and smaller alternatively spliced products), a 3056-amino-acid protein kinase ATM (350 kDA) is translated. ATM is activated in response to double-strand DNA breaks. It is part of a network of proteins that regulate cellular responses to DNA damage (p. 80). Clinically different disorders are related at the cellular level (Nijmegen breakage syndrome, McKusick 251260, and others).

### References

Auerbach, A.D., Buchwald, M., Joenje, H.: Fanconi anemia, pp. 317–332. In: B. Vogelstein, K.W. Kinzler, eds., The Genetic Basis of Human Cancer. McGraw-Hill, New York, 1998.

Gatti, R.: Ataxia telangiectasia, pp. 275–300. In: B. Vogelstein, K.W. Kinzler, eds., The Genetic Basis of Human Cancer. McGraw-Hill, New York, 1998.

German, J., Ellis, N.A.: Bloom syndrome, pp. 301–315. In: B. Vogelstein, K.W. Kinzler, eds., The Genetic Basis of Human Cancer. McGraw-Hill, New York, 1998.

Zhao, S., et al.: Functional link between ataxia-telangiectasia and Nijmegen breakage syndrome gene products. Nature **405**:473–477, 2000.

Bloom syndrome

Extreme intrauterine and postnatal growth retardation

Chromosomal instability

Predisposition to leukemias, lymphomas, and other tumors

Immune defects

Sunlight-induced erythema of the face

Hypo- and hyper-pigmented skin areas

Autosomal recessive

Gene locus on chromosome 15q26.1

1. Main features

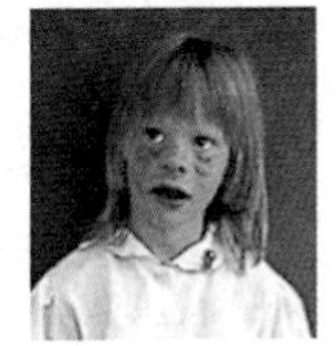

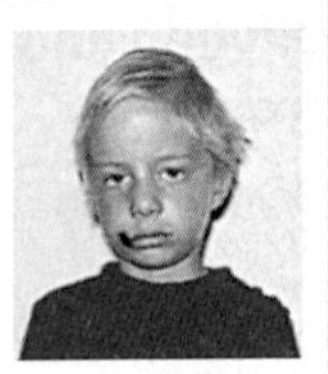

a b c

2. Phenotype

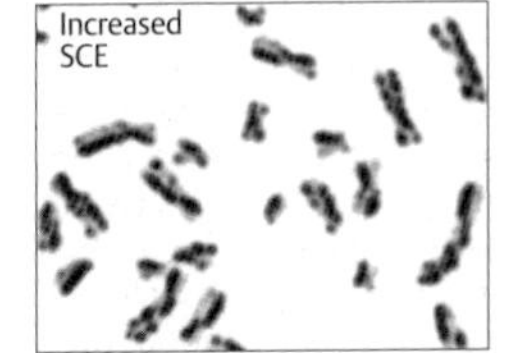


Bloom syndrome

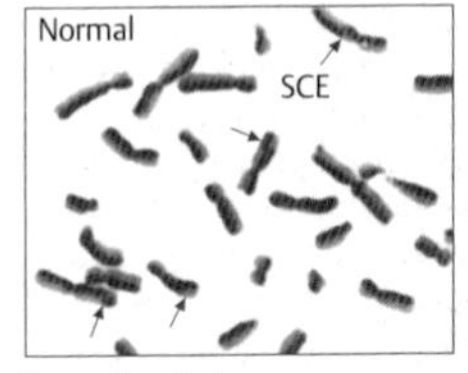


Normal control

3. Increased rate of sister chromatid exchanges (SCE)

**A. Bloom syndrome (BS)**

Fanconi anemia

Growth retardation
Skeletal defects (e.g., radius and thumb)

Bone marrow failure

Skeletal and kidney malformation

Localized pigment changes

Autosomal recessive

Several gene loci

1. Main features

2. Phenotype

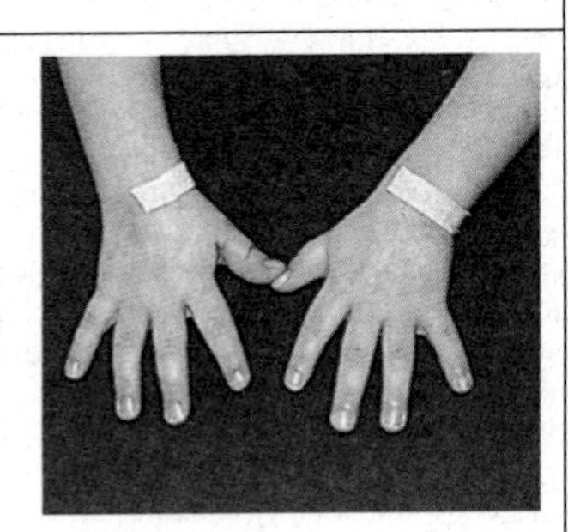

3. Thumb hypoplasia

**B. Fanconi anemia (FA)**

Ataxia telangiectasia

Cerebellar ataxia

Immune defects

Telangiectases of the conjunctivae

Predisposition to tumors (lymphoma, leukemia)

Extreme radiation sensitivity

Autosomal recessive

1. Main features

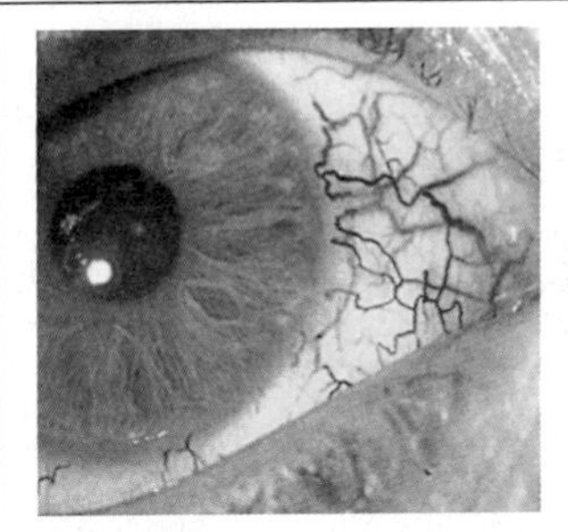

2. Telangiectases

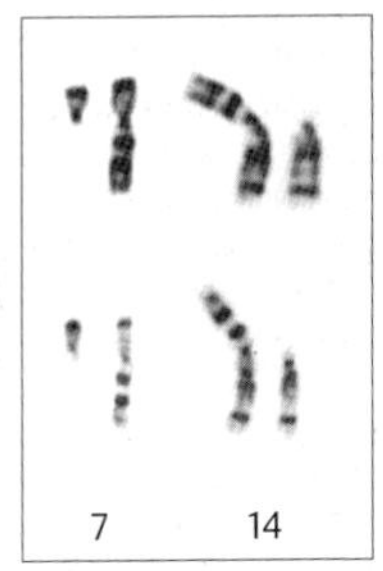


3. Translocation 7q;14p

**C. Ataxia telangiectasia (AT)**

# Oxygen and Electron Transport

## Hemoglobin

Hemoglobin and myoglobin are the oxygen-transporting proteins in vertebrates. Hemoglobin is found in red blood cells, myoglobin in muscle. Hemoglobin arose from myoglobin during the course of evolution. Hemoglobin has four oxygen-binding sites, myoglobin has one. Their genes and their three-dimensional protein structures are completely known in atomic detail. Different types of hemoglobin that are optimally adapted to prenatal and postnatal life have evolved in mammals from an ancestral gene.

### A. Types of hemoglobin

Hemoglobin consists of four subunits (globin chains), two pairs of identical polypeptides, each polypeptide being attached to a heme group with an oxygen-binding site. The different kinds of hemoglobins, which are characteristic of different stages of development (embryonic, fetal, and postnatal), differ in the composition of their chains: the hemoglobin of adults (HbA) contains two α and two β chains ($\alpha_2\beta_2$). A small proportion of adult hemoglobin has two α and two δ chains ($HbA_2$: $\alpha_2\delta_2$). Hemoglobin formed during the fetal period (HbF) contains two α and two γ chains ($\alpha_2\gamma_2$). In the embryonic stage, ζ chains are joined to ε or γ chains (Hb Gower 1: $\zeta_2\varepsilon_2$, and Hb Portland: $\zeta_2\gamma_2$); two α and two ε chains form Hb Gower 2 ($\alpha_2\varepsilon_2$).

### B. Hemoglobins in thalassemia

The thalassemias are a group of genetically determined disorders of hemoglobin synthesis. Thalassemia occurs due to the absence or reduced synthesis of a globin chain, which results in unstable hemoglobin. It affects either the β chain (β-thalassemia) or the α chain (α-thalassemia). Since the α chain is a component of both fetal and adult hemoglobins, the α-thalassemias (α-thal) are especially severe. Hemoglobins with four identical globin chains are completely unstable and incompatible with life (HbH with four β chains, Hb Bart's with four γ chains). Theoretically, a homotetramer of four δ chains could be formed but, because of the normally slow rate of synthesis of δ chains, these are clinically insignificant. The α-thalassemias occur in very different grades of severity because two neighboring genes code for the α chain (see hemoglobin genes, p. 338).

### C. Evolution of hemoglobin

Since hemoglobin has four polypeptide chains, it is a much more efficient oxygen carrier than the single-chained myoglobin molecule. Furthermore, the existence of different globin chains confers a selective advantage in evolution because their slight functional differences result in optimal adaptation to the differences in oxygen concentration before and after birth. The genes for individual hemoglobin chains arose from myoglobin by a series of gene duplications during evolution. The evolutionary age of the individual Hb chains can be estimated from their differences in relation to the rate of mutation. When mammals began to evolve about 100 million years ago, the genes for α and β chains were present, whereas β-like chains (ε, γ, δ) evolved later.

### D. Globin formation in ontogeny

Different types of globin chains are formed at different developmental stages: embryonic hemoglobin during the early embryonic period (to about the 12th week), fetal hemoglobin from about the 12th week until birth, and adult hemoglobin thereafter. They differ in oxygen-binding affinity. Thus, oxygen delivery is optimized for different phases of development. The site of synthesis also differs. During the fetal phase, globin chains are synthesized mainly in the liver, whereas after birth they are synthesized in red blood cell precursors in the bone marrow.

(Figures based on Lehmann & Huntsman, 1974).

### References

Lehmann, H., Huntsman, R.G.: Man's Hemoglobins. North-Holland, Amsterdam, 1974.

Thein S. L., Rochette, J.: Disorders of hemoglobin structure and synthesis, pp. 179–190, In: J.L. Jameson, ed., Principles of Molecular Medicine. Humana Press, Totowa, New Jersey, 1998.

Weatherhall, D.J., Clegg, J.B.: Genetic disorders of hemoglobin. Semin. Hematol. **36**:2–37, 1999.

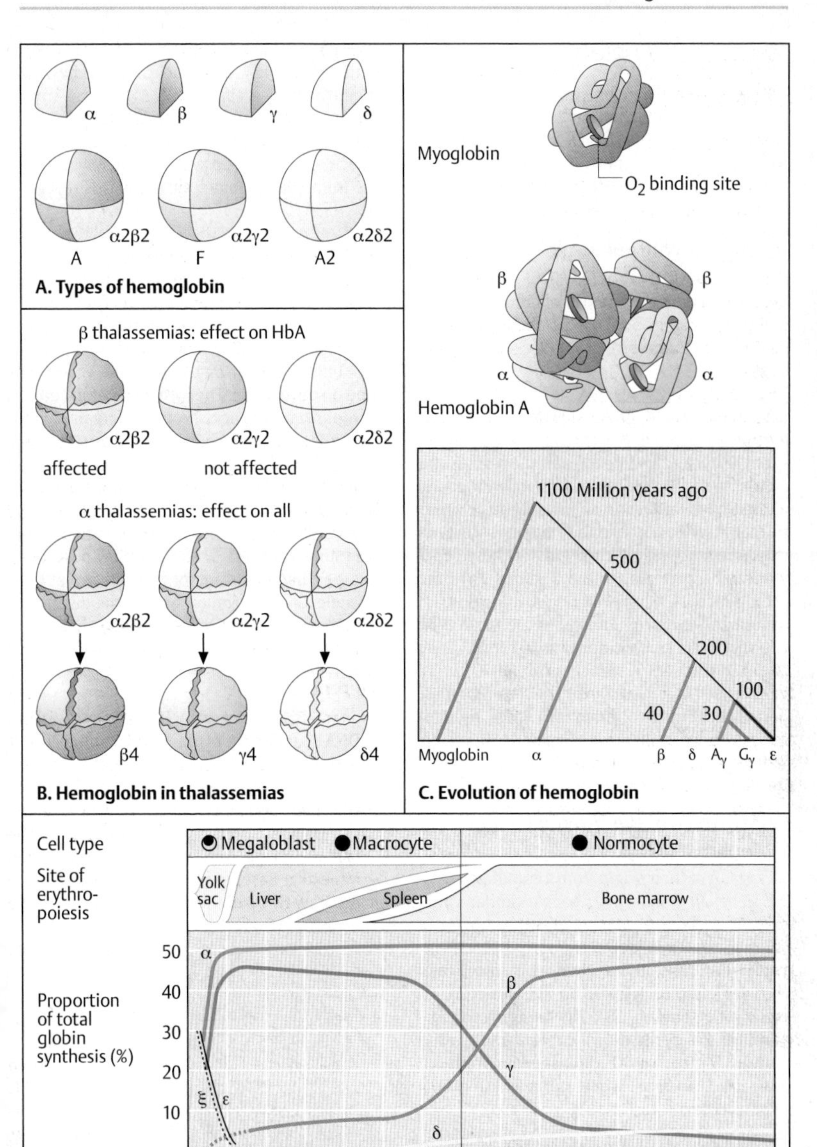


A. Types of hemoglobin

B. Hemoglobin in thalassemias

C. Evolution of hemoglobin

D. Globin synthesis during ontogeny

## Hemoglobin Genes

Each of the globin polypeptide chains is coded for by a gene. In humans and in other mammals, the β-like genes (β, γ, δ) are located together in a cluster on one chromosome, while the α genes are located on another. They are arranged in the order of their activation during ontogeny.

### A. The β-globin and α-globin genes

The β-globin-like genes (ε, γG, γA, δ, β) of man are located on the short arm of chromosome 11 in region 1, band 5.5 (11p15.5). They span about 60000 base pairs (bp), or 60 kb (kilobases), of DNA.

There are two γ genes, γA and γG, which differ only in codon 136. Codon 136 of γA is alanine, and of γG, glycine. A pseudogene ($\psi\beta_1$) is located between the Aγ gene and the δ gene. It is similar to the β gene, but has been permanently altered by deletion and an internal stop codon, so that it cannot code for a functional polypeptide. A region that jointly regulates these genes is located upstream (in the 5' direction) from the β genes (LCR, long-range control region).

In humans, two α-globin genes are located on the short arm of chromosome 16 (16p13.11 to 16p13.33) on a DNA segment of about 30 kb. A ζ gene, which is active only during the embryonic period, lies in the 5' direction. Three pseudogenes: ψζ, $\psi\alpha_2$, and $\psi\alpha_1$ are located in between. A further gene, θ, with unknown function, has been identified in this region.

### B. Structure of the α-globin and β-globin genes

As a result of their origin from a common ancestral gene, all globin genes have a similar structure. Their coding sequences are arranged in three exons. Each globin transcription unit includes nontranslated sequences at the 5' and the 3' ends (see section on the structure of eukaryotic genes, p. 50). The lengths of the β-globin and α-globin exons are similar (e.g., exon 1 of the β gene has 30 codons; exon 1 of the α gene has 31 codons), whereas the lengths of the introns differ.

### C. Tertiary structure of the β-globin chain

The three-dimensional structures of myoglobin and of the hemoglobin α and β chains are very similar, although their amino acid sequences correspond in only 24 of 141 positions. The β chain, with 146 amino acids, is somewhat longer than the α chain, with 141 amino acids. The structural similarity is functionally significant: The oxygen-binding region lies inside the molecule, where it is protected, and oxygen uptake from the aqueous surroundings is reversible.

### D. Domains of the β chain

Three functional and structural domains can be distinguished in all globin chains. They correspond to the three exons of the gene. Two domains, consisting of amino acids 1–30 and 105–146 (coded for by exons 1 and 3), are located on the outside. They are mainly formed of hydrophilic amino acids. A third domain, lying inside the molecule (coded for by exon 2), contains the oxygen-binding site and consists mainly of nonpolar hydrophobic amino acids.

The amino acid sequences of the hemoglobins of more than 60 investigated species are identical in nine positions. These invariant positions are especially important for the function of the molecule. Changes (mutations) in the invariant positions affect function so severely that they are not tolerated.

### References

Antonarakis, S. E., Kazazian, H.H. Jr., Orkin, S. H.: DNA polymorphism and molecular pathology of the human globin gene clusters. Hum. Genet. **69**:1–14, 1985.

Weatherhall, D.J., et al.: The hemoglobinopathies, pp: 3417–3484. In: C.R. Scriver, et al., eds., The Metabolic and Molecular Bases of Inherited Disease. 7th ed. McGraw Hill, New York, 1995.

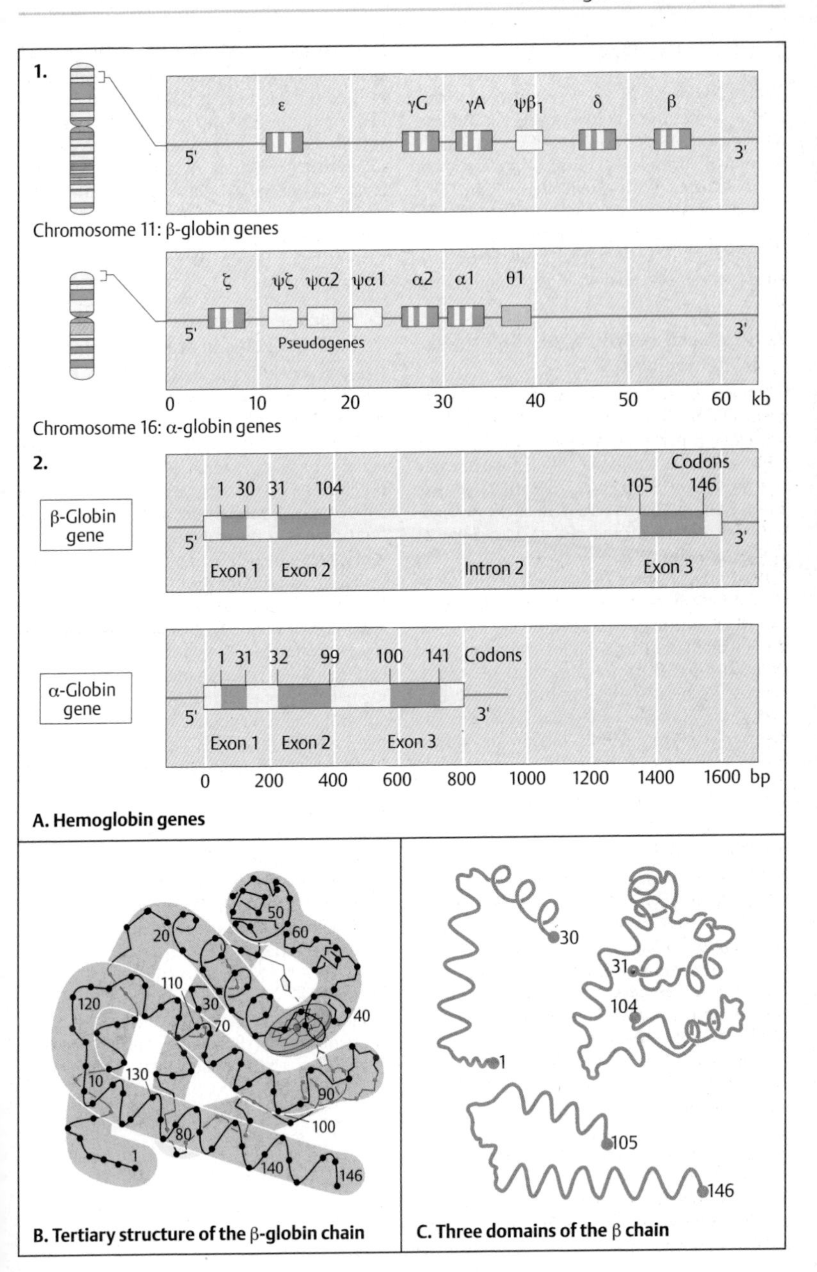


A. Hemoglobin genes

B. Tertiary structure of the β-globin chain

C. Three domains of the β chain

## Sickle Cell Anemia

Sickle cell anemia is a severe progressive disease resulting from homozygosity for a mutation in the β-globin gene. It is especially frequent in Africa and in the black population of North America. With a frequency of 1 in 500, it is an important cause of morbidity and mortality in these regions. The disease is transmitted by autosomal recessive inheritance. Heterozygous carriers can readily be identified (see p. 348).

### A. Sickle cells: erythrocytes deformed by hemoglobin S

In a normal blood smear under the light microscope (1), erythrocytes (red blood cells) appear as regular round disks of about 7 μm diameter. Since a normal red blood cell is nonnucleated and biconcave, the center appears paler than the periphery. The erythrocytes of affected persons are deformed and resemble sickles (2). However, even the non-sickle-shaped red cells display unusual sizes and shapes. In the course of the disease, acute crises called sickle crises (3) occur, during which sickle-like cells are greatly increased and completely dominate the blood picture. Heterozygotes show occasional sickle cells but do not suffer from sickle crises, and at the most have only very mild signs and symptoms. (Figure from Lehmann and Huntsman, 1974.)

### B. Result of a mutation: sickle cell anemia

All manifestations of sickle cell anemia are due to the substitution of a single nucleotide base in the β-globin gene. The sickle cell mutation is the transversion of the second nucleotide base of codon 6, adenine (A), to thymine (T). This changes the codon GAG, for glutamic acid, to GTG, the codon for valine. During the 1950s, Vernon M. Ingram determined the amino acid sequence of hemoglobin and found that the only difference between sickle cell hemoglobin (HbS) and normal adult hemoglobin (HbA) was this exchange in the β chain. This has far-reaching pathophysiological consequences and explains all manifestations of the disease. Sickle cell hemoglobin (HbS) is less soluble than normal hemoglobin and does not allow normal erythrocyte distortion. It crystallizes in the deoxy state and forms small rods. Thus, the erythrocytes become firm and deform into sickle cells. Unlike normal erythrocytes, sickle cells are unable to pass through small arteries and capillaries. These become clogged and cause local oxygen deficiency in the tissues, followed by infection. As a rule, learning disability due to frequent illness occurs. Defective erythrocytes are destroyed (hemolysis). Chronic anemia and its numerous sequelae such as heart failure, liver damage, and infection are the result.

### C. Selective advantage for HbS heterozygotes in areas of malaria

Heterozygotes for the sickle cell mutation are relatively resistant to malarial infection. Erythrocytes of heterozygotes for the sickle cell mutation are a less favorable environment for the malaria parasite than those of normal homozygotes. Thus, heterozygotes develop malaria in a much milder form or not at all. However, this protection is at the expense of the affected homozygotes (HbS/HbS): although they do not contract malaria, they suffer from the severe hemoglobin disorder. The protection against malaria conferred by sickle cell heterozygosity is an advantage in regions where malaria is common. With reduced morbidity and mortality, heterozygotes have a higher probability of survival and of being able to reproduce (selective advantage). This explains the high frequency of the sickle cell gene observed there (see p. 168). The sickle cell mutation has arisen independently in at least four or five different malaria-infested regions and has subsequently spread out in the respective populations. Sickle cell anemia is the best example in humans of a selective advantage in heterozygotes for a mutant allele that leads to severe illness in the homozygous state.

### References

Ashley-Koch, A., Yang, Q., Olney, R.S.: Sickle hemoglobin (HbS) allele and sickle cell disease: a HuGE review. Am. J. Epidemiol. **15**:839–845, 2000.

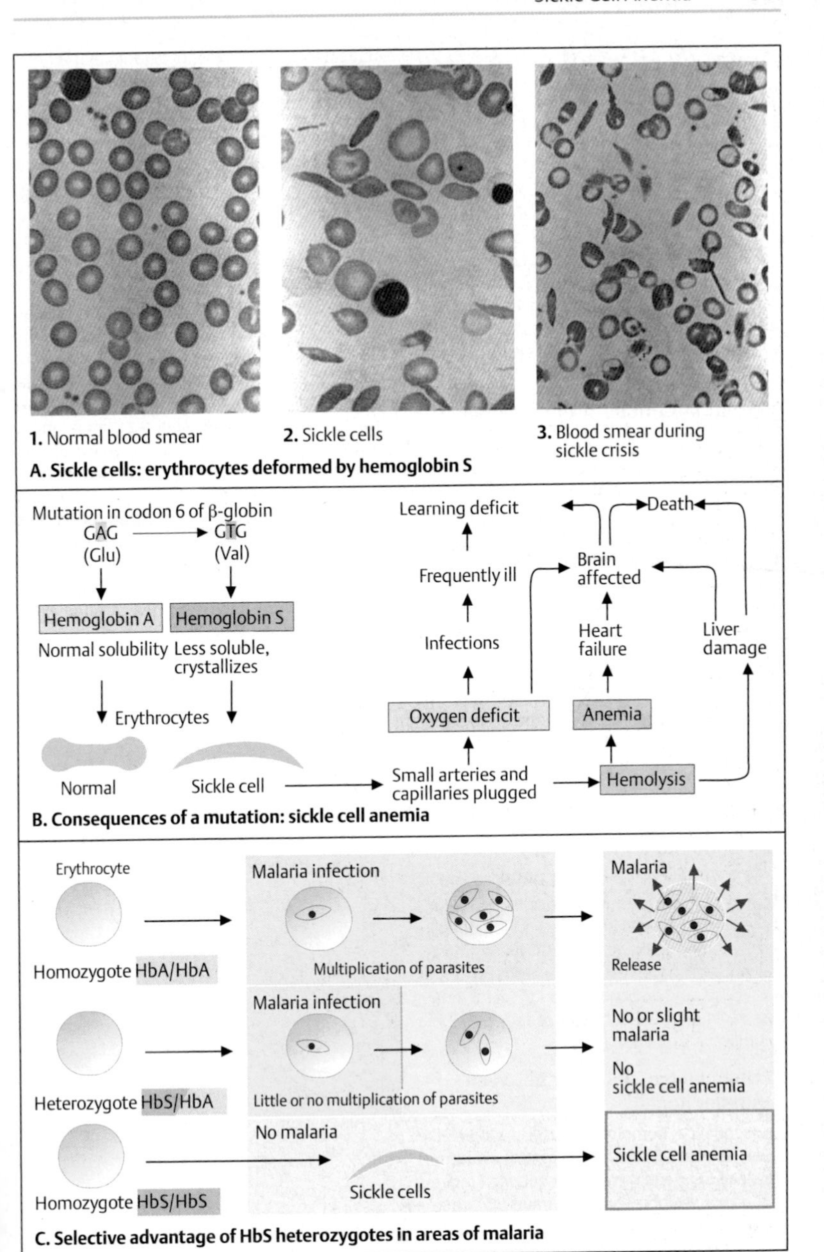


**A. Sickle cells: erythrocytes deformed by hemoglobin S**

**B. Consequences of a mutation: sickle cell anemia**

**C. Selective advantage of HbS heterozygotes in areas of malaria**

## Mutations in Globin Genes

All types of mutations have been demonstrated in the globin genes. The most frequent are point mutations in a single codon. The functional consequences vary, depending on the electrical charge and size of the substituted amino acid and its position in the polypeptide. If one of the hydrophilic amino acids at the surface is replaced by a hydrophobic amino acid (e.g., valine for glutamic acid in the sickle cell mutation), profound physicochemical changes will result. Mutations may decrease the elasticity of the molecule, alter its oxygen affinity, or cause instability.

### A. Point mutations of the β-globin gene

Over 300 point mutations in the β-globin gene and over 100 in one of the α-globin genes have been documented. Two clinically important mutations affect codon 6: the sickle cell mutation, 6 Glu → Val (sickle cell hemoglobin, HbS, resulting in the incorporation of valine instead of glutamic acid) and 6 Glu → Lys (hemoglobin C, HbC, incorporating lysine instead of glutamic acid in codon 6). Compound heterozygotes with the HbS mutation on one chromosome and the HbC on the other (HbSC) are not rare. The marked methemoglobin formation in Hb Zürich and Hb Saskatoon results from substitutions for histidine (His) in codon 63, which alter the oxygen-binding region of the hemoglobin molecule.

### B. Deletion due to unequal crossing-over within a gene

Marked sequence homology of certain regions of the globin genes may lead to nonhomologous pairing and unequal crossing-over during meiosis, e.g., in the regions of codons 90–94 of one DNA strand and codons 95–98 of the other. This explains the deletion of codons 91–95 in hemoglobin Gun Hill.

### C. Unequal crossing-over between similar genes

The sequence homology of the β-globin-like genes (explained by their common evolution) may lead to unequal crossover between regions of the two γ-globin genes (γA and γG), the δ-globin gene, or the β-globin gene. The best known example is partial deletion of the δ and of the β loci (δ–β fusion) in hemoglobin Lepore. The corresponding duplication of the δ/β sequence results in hemoglobin anti-Lepore.

### D. Unstable hemoglobin due to chain elongation

If one of the globin chains is too long, it will destabilize the tetrameric hemoglobin molecule. Hemoglobin Cranston ($Hb^{Cr}$) (1) arises from the insertion of two nucleotide bases (adenine and guanine) into positions 1 and 2 of codon 145 (tyrosine) of the β chain, which leads to a shift of the reading frame. This changes the normal stop codon UAA into AGU, the RNA codon for threonine (Thr). As a result, the normally nontranslated sequences that follow the stop codon are now translated, and a polypeptide is formed that is 11 amino acids too long, extending to position 157. With hemoglobin Constant Spring (2), the α chain is lengthened by mutation of the stop codon UAA to CAA, which codes for glutamine (Gln). The sequences that normally follow the stop codon now become translated, and a peptide that is 31 amino acids too long is formed. A number of other chain-elongation mutations due to similar mechanisms, such as with hemoglobin Ikaria (an α chain with 172 amino acid residues), have been described.

### References

Weatherhall, D.J., et al.: The hemoglobinopathies, pp. 3417–3484, In: C.R. Scriver, et al., eds., The Metabolic and Molecular Bases of Inherited Disease. 7th ed. McGraw-Hill, New York, 1995.

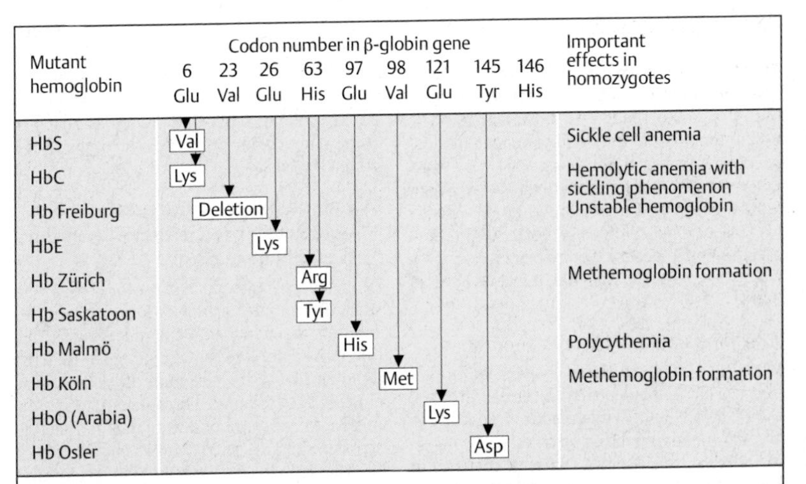


**A. Examples of point mutations in the β-globin gene (10 of 310)**

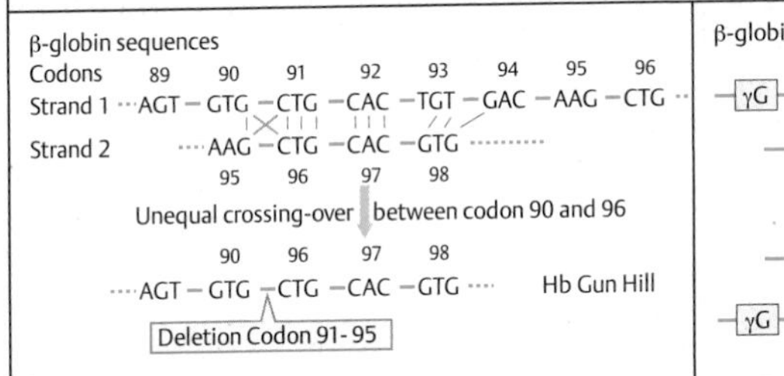


**B. Deletion by unequal crossing-over within a gene**

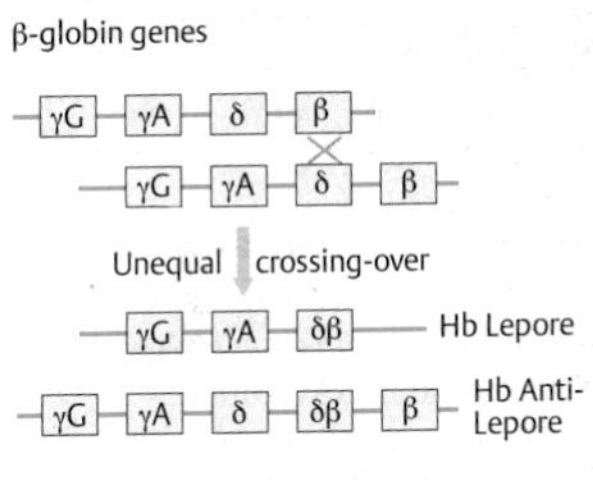


**C. Unequal crossing-over between similar genes**

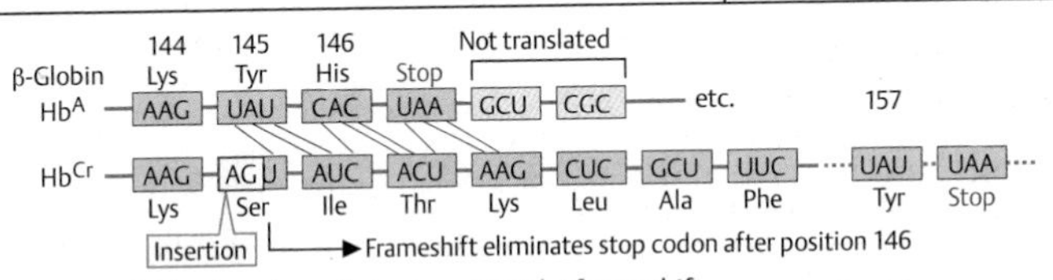


**1.** Hemoglobin Cranston: chain elongation by frameshift

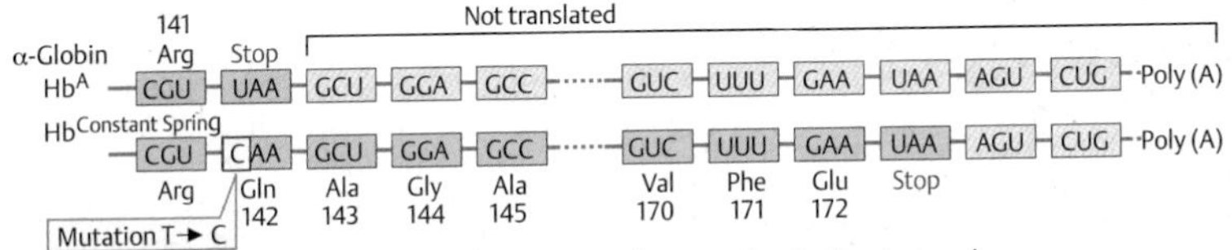


**2.** Hemoglobin Constant Spring: chain elongation by mutation in the stop codon

**D. Unstable hemoglobin due to chain elongation**

## The Thalassemias

The thalassemias are a heterogeneous group of diseases caused by decreased or absent formation of a globin chain. The thalassemias occur predominantly in the Mediterranean region, parts of Africa, and Southeast Asia (the word thalassemia is derived from *thalassa*, the Greek word for sea). In these regions, they are a significant cause of morbidity and mortality, but they are frequent because heterozygotes are protected from severe malarial infection (see p. 340).

### A. Thalassemia, a chronic anemia

Depending on which globin chain is not formed in sufficient amounts, α-thalassemia, β-thalassemia, or δβ-thalassemia results. This leads to chronic anemia, which causes the various manifestations of thalassemia. Oxygen deficiency in the peripheral tissues leads to increased extramedullary (outside the bone marrow) blood formation. A tendency toward infection, undernourishment, and other signs characterize the severe clinical picture. (Photographs from Weatherall and Clegg, 1981).

### B. β-Thalassemia and α-thalassemia

The thalassemias have a wide spectrum of different genotypes and phenotypes (disease manifestations and course). In the β-thalassemias (1), complete absence ($\beta^0$) is distinguished from decreased formation ($\beta^+$) of the β chain. With the α-thalassemias (2), one, two, three, or all four loci for α-globin may be affected. Altogether, there are 12 principal genotypes. In individuals with two mutations at the α-loci (α-thalassemia), the two can lie either on the same chromosome (thal-1) or on different chromosomes (thal-2). Thal-1 occurs mainly in Southeast Asia; thal-2, mainly in Africa. Each α gene is located within a 4 kb region of homology, interrupted by small, nonhomologous regions. The most frequent mechanism for the origin of a chromosome with only one α-globin gene is nonhomologous crossing-over between two α-globin gene loci after mispairing of the homologous chromosomes during meiosis.

### C. β-Thalassemia due to different mutations

Many mutations in the β-globin gene region can lead to β-thalassemia. The mutations may also occur in noncoding sequences (5′ to exon 1 and within introns).

### D. Haplotypes resulting from polymorphic restriction sites in the β-globin gene cluster

From the presence or absence of recognition sites of a number of restriction enzymes (restriction fragment length polymorphism, RFLP), different haplotypes can be distinguished in the β-globin-like gene region (β-globin gene cluster). Each haplotype is characterized by the presence or absence of several polymorphic restriction sites. By establishing the haplotypes of the affected and unaffected individuals within a family, the mutation-carrying haplotype can be identified (indirect genotype analysis). Different mutations have occurred on the background of different haplotypes. Frequently, a particular mutation is linked to a distinct haplotype (linkage disequilibrium). This reflects the time elapsed since the mutation first occurred in the population, where it has been maintained by selection. (Data in C and D from Antonarakis et al., 1985.)

α-Thalassemia may be associated with mental retardation. Two different syndromes can be distinguished: One occurs in patients with a large (1–2 Mb) deletion on the tip of chromosome 16 including the α-globin gene cluster (ATR-16 syndrome). The other is an X-linked disorder with a remarkably uniform phenotype and a mild form of HbH disease without α-globin deletion. A *trans*-acting regulatory factor appears to be encoded on the X chromosome.

### References

Antonarakis, S. E., Kazazian, H.H. Jr., Orkin, S. H.: DNA polymorphism and molecular pathology of the human globin gene clusters: Hum. Genet. **69**:1–14, 1985.

Olivieri, N.F.: The thalassemias. New Eng. J. Med. **341**:99–109, 1999.

Weatherall, D.J., Clegg, J.B.: The Thalassemia Syndromes. 3rd ed. Oxford, 1981.

Weatherall, D.J., Provan, A.B.: Red cells I: inherited anaemias. Lancet **355**:1169–1175, 2000.

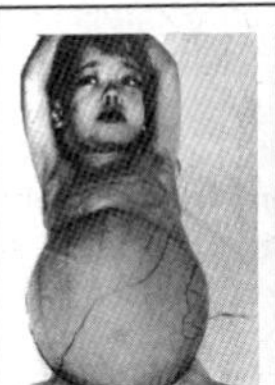

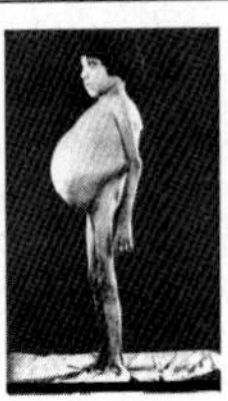

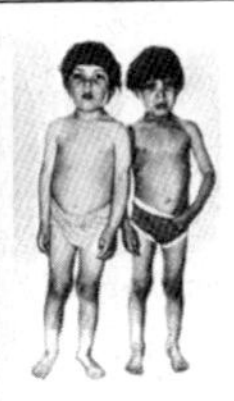

Different forms of thalassemia:

- α: Decreased synthesis of α-globin
- β: Decreased synthesis of β-globin
- δβ: Decreased synthesis of δ-and β-globin

→ Unstable hemoglobin → Chronic anemia

**A. Thalassemia, a chronic anemia**

1. β-Thalassemias

| Genotype | | Phenotype |
|---|---|---|
| + / – | β° heterozygote | Thalassemia minor (asymptomatic) |
| (+) / (+) | β⁺ heterozygote | |
| (+) / (+) | β⁺ homozygote | Thalassemia intermedia (not transfusion dependent) |
| + / – | β° heterozygote | |
| – / – | β° homozygote (β° Thalassemia) | Thalassemia major (transfusion dependent) |
| or | | |
| (+) / – | β⁺/β° homozygote (β⁺ Thalassemia) | |

2. α-Thalassemias

| Genotype | Phenotype |
|---|---|
| α α / α α | Normal |
| α – / α α | "Silent carrier" (normal) |
| – – / α α (thal-1) | Thalassemia |
| α – / α – (thal-2) | |
| α – / – – | HbH Disease (HbH = $\beta_4$) |
| – – / – – | Hydrops fetalis |

**B. β-Thalassemia and α-thalassemia**

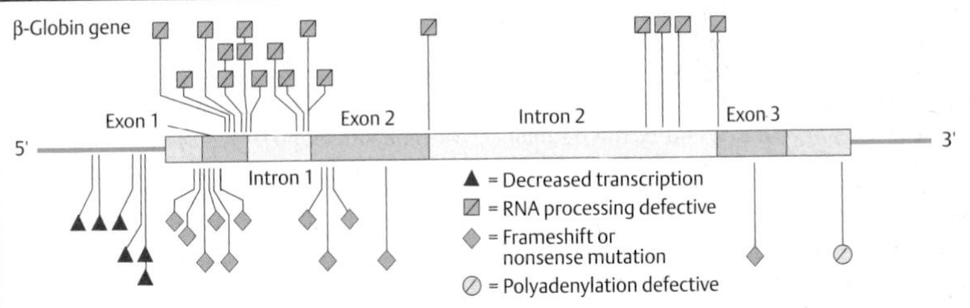


**C. β-Thalassemia due to different mutations**

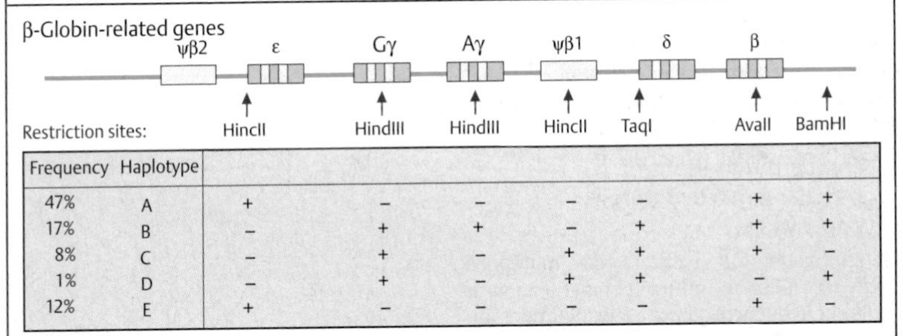


| Frequency | Haplotype | HincII | HindIII | HindIII | HincII | TaqI | AvaII | BamHI |
|---|---|---|---|---|---|---|---|---|
| 47% | A | + | – | – | – | – | + | + |
| 17% | B | – | + | + | – | + | + | + |
| 8% | C | – | + | – | + | + | + | – |
| 1% | D | – | + | – | + | + | – | + |
| 12% | E | + | – | – | – | – | + | – |

**D. Haplotypes resulting from polymorphic restriction sites in the β-globin gene cluster**

## *Hereditary Persistence of Fetal Hemoglobin (HPFH)*

Hereditary persistence of fetal hemoglobin (HPFH) refers to a genetically heterogeneous group of diseases in which the temporal expression of the β-globin genes during development has been altered. Individuals with HPFH produce increased amounts of fetal hemoglobin (HbF). Under some conditions, HbF may be the only β-globin-like gene product formed. Clinically, HPFH is relatively benign, although HbF is not optimally adapted to postnatal conditions. Analysis of HPFH has yielded insight into the control of globin gene transcription and the effect of mutations in noncoding sequences.

### A. Large deletions in the β-globin gene cluster

A numer of very large deletions in the β-globin gene cluster region are known, especially in the 3′ direction. The deletions show different distributions in different ethnic populations, reflecting that they originated at different points in time. δβ-Thalassemia and failure of β-globin production have been the result in some cases.

### B. Mutations in noncoding sequences of the promoter region

Mutations in the noncoding sequences of the promoter region at the 5′ end of the β-globin cluster (on the 5′ side of the γ-globin genes) can also lead to hereditary persistence of fetal hemoglobin. Although the highly conserved sequences CACCC, CCAAT, or ATAAA are not affected, the number of observed mutations substantiates the significance of the remaining noncoding sequences (long-range transcription control). They are probably required for the changes in transcription control of the different gene loci that occur during embryonic and fetal development. (Figure after Gelehrter and Collins, 1990.)

### C. Frequent mutations of β-thalassemia in different populations

Heterozygotes for β-thalassemia mutations occur in different ethnic populations with different frequencies. Since a few mutations are quite frequent in certain populations, preventive diagnostic programs to determine the risk of disease are possible. (Data after Antonarakis et al. 1985.)

According to estimates of the WHO (Bull World Health Org. 1983) about 275 million persons are heterozygotes for hemoglobin diseases worldwide. Substantial numbers are due to the β-thalassemias in Asia (over 60 million), $\alpha^0$-thalassemia in Asia (30 million), HbE/β-thalassemia in Asia (84 million), and sickle cell heterozygosity in Africa (50 million), India, the Caribbean, and the USA (about 50 million). At least 200000 severely affected homozygotes are born annually, about 50%, due to sickle cell anemia and 50% to thalassemia (Weatherall, 1991).

### References


Antonarakis, S. E., Kazazian, H.H. Jr., Orkin, S. H.: DNA polymorphism and molecular pathology of the human globin gene clusters. Hum. Genet. **69**:1–14, 1985.

Gelehrter, T.D., Collins, F.: Principles of Medical Genetics. Williams & Wilkins, Baltimore, 1990.

Orkin, S. H., Kazazian, H.H.: The mutation and polymorphism of the human ß-globin gene and its surrounding DNA. Ann. Rev. Genet. **8**:131–171, 1984.

Stamatoyannopoulos, G., et al., eds.: The Molecular Basis of Blood Diseases. W.B. Saunders, Philadelphia, 1987.

Weatherall, D.J., et al.: The hemoglobinopathies, pp. 3417–3484, In: C.R. Scriver, et al., eds., The Metabolic and Molecular Bases of Inherited Disease. 7th ed. McGraw-Hill, New York, 1995.

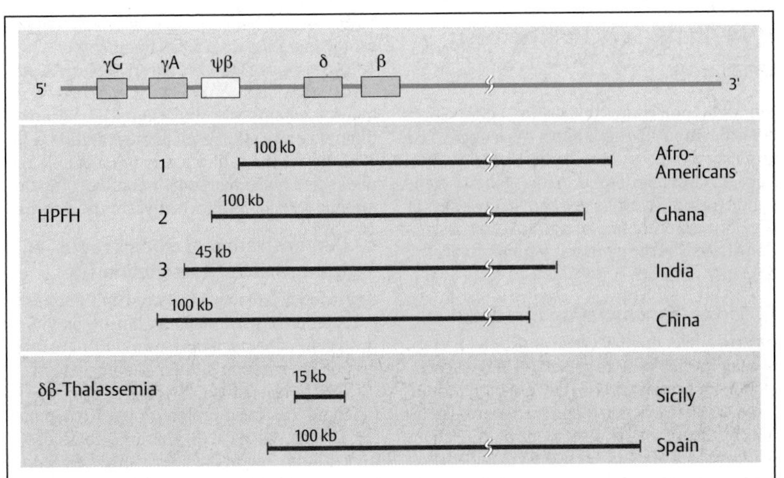


**A. Large deletions in the β-globin cluster**

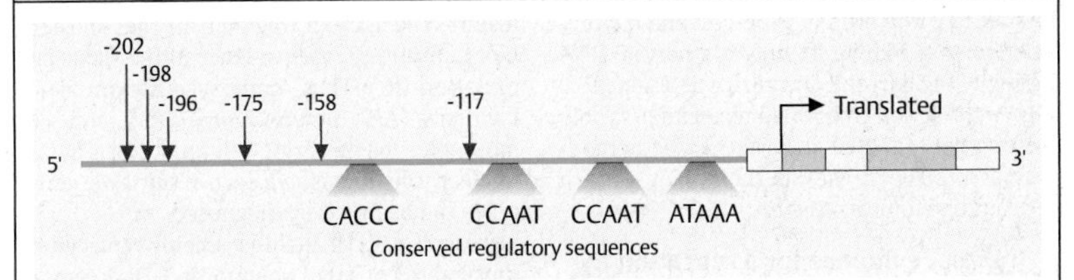


**B. Mutations in noncoding sequences in the promoter of g-globin genes cause hereditary persistence of fetal hemoglobin (HPFH)**

| β-thal mutation | Frequency | Ethnic group | Type |
|---|---|---|---|
| Intron 1 (110 G → A) | 35% | Mediterranian | $\beta^+$ |
| Codon 39 (C → T) | 27% | Mediterranian | $\beta^\circ$ |
| TATA-Box (-29 A → G) | 39% | Afro-Americans | $\beta^+$ |
| Poly A (T → C) | 26% | Afro-Americans | $\beta^+$ |
| Intron 1 (5 G → C) | 36% | India | $\beta^+$ |
| Partial deletion (619 nt) | 36% | India | $\beta^\circ$ |
| Codon 71-72 frameshift | 49% | China | $\beta^\circ$ |
| Intron 2 (654 C → T) | 38% | China | $\beta^\circ$ |

**C. Frequent mutations in β-thalassemia in different populations**

# DNA Analysis in Hemoglobin Disorders

Numerous procedures that do not require direct determination of the altered nucleotide base sequence can be used to demonstrate a mutation. An available probe of the gene or gene region being investigated and knowledge of the normal Southern blot pattern after restriction analysis of the gene are prerequisites (cf. p. 62).

## A. Direct demonstration of a deletion

A partial deletion may cause an altered pattern of a Southern blot. Two genes (α2 and α1) of α-globin are presented (1). They are both located on a 14.5 kb restriction fragment. If partial deletion results in loss of a segment of, e.g., about 4.5 kb that is part of both the α2 and the α1 gene, a fragment of 10.0 instead of 14.5 kb will result in this area (2). Three genotypes are possible (3): two normal genes, α2 and α1, represented by a 14.5 kb fragment; a normal DNA segment (14.5 kb) and one with a deletion (10.0 kb); or a deletion in both gene segments (only one fragment, of 10.0 kb). This can be demonstrated directly in the Southern blot pattern with a probe for the α1 gene (4).

## B. Indirect evidence for a mutation by RFLP analysis

Indirectly, a mutation can be demonstrated if there is an individual difference (polymorphism) in the base sequences of the mutant and the normal gene segments (restriction fragment length polymorphism, RFLP, see p. 64). For instance, if two of the same DNA segments differ in a polymorphism for the recognition sequence of a restriction enzyme, then DNA fragments of different sizes (here, either 7 kb and 6 kb, or 13 kb) result after cleavage with the enzyme (1). If the mutation has occurred within the 13 kb fragment, then this fragment indicates presence of the mutation. In the given DNA segment, there are three possibilities (genotypes): two fragments of 7 kb without mutation; one fragment of 7 kb (normal) and one fragment of 13 kb (which carries the mutation); and two mutation-carrying fragments of 13 kb (3). The Southern blot (4) shows whether the person being examined is homozygous normal (has no 13 kb fragment), is heterozygous (a 7 kb and a 13 kb fragment), or is homozygous for the mutation (two 13 kb fragments).
The prerequisite for this indirect analysis is previous knowledge of which of the DNA fragments contains the mutation. The observed difference is not the result of the mutation, as in A. If the Southern blot pattern of affected and unaffected individuals does not differ, then this method will not be informative for the disorder.

## C. Demonstration of a point mutation from an altered restriction site

A restriction site may be altered by a mutation. For example, a sickle cell mutation in codon 6 of the β gene of hemoglobin (see p. 340) (1) causes loss of a restriction site for the enzyme *Mst*II (CCTN*A*GG instead of CCTN*T*GG) because the A (adenine) has been replaced by a T (thymine) (2). The normal allele ($\beta^A$) in this area produces a 1.15 kb fragment after *Mst*II digestion, whereas the mutation eliminates the restriction site in the middle so that a 1.35 kb fragment results. The 1.35 kb fragment in the Southern blot indicates (3) the presence of the sickle cell mutation ($\beta^S$). Thus, homozygous normal individuals (AA), heterozygotes (AS), and homozygotes for the sickle cell mutation (SS) can be clearly distinguished; each of the three genotypes can be precisely diagnosed.
Increasingly, RFLP analysis is being replaced by the analysis of small polymorphic DNA repeats (microsatellites, see p. 72).

### References

Housman, D.: Human DNA polymorphism. N. Eng. J. Med. **332**:318–320, 1995.

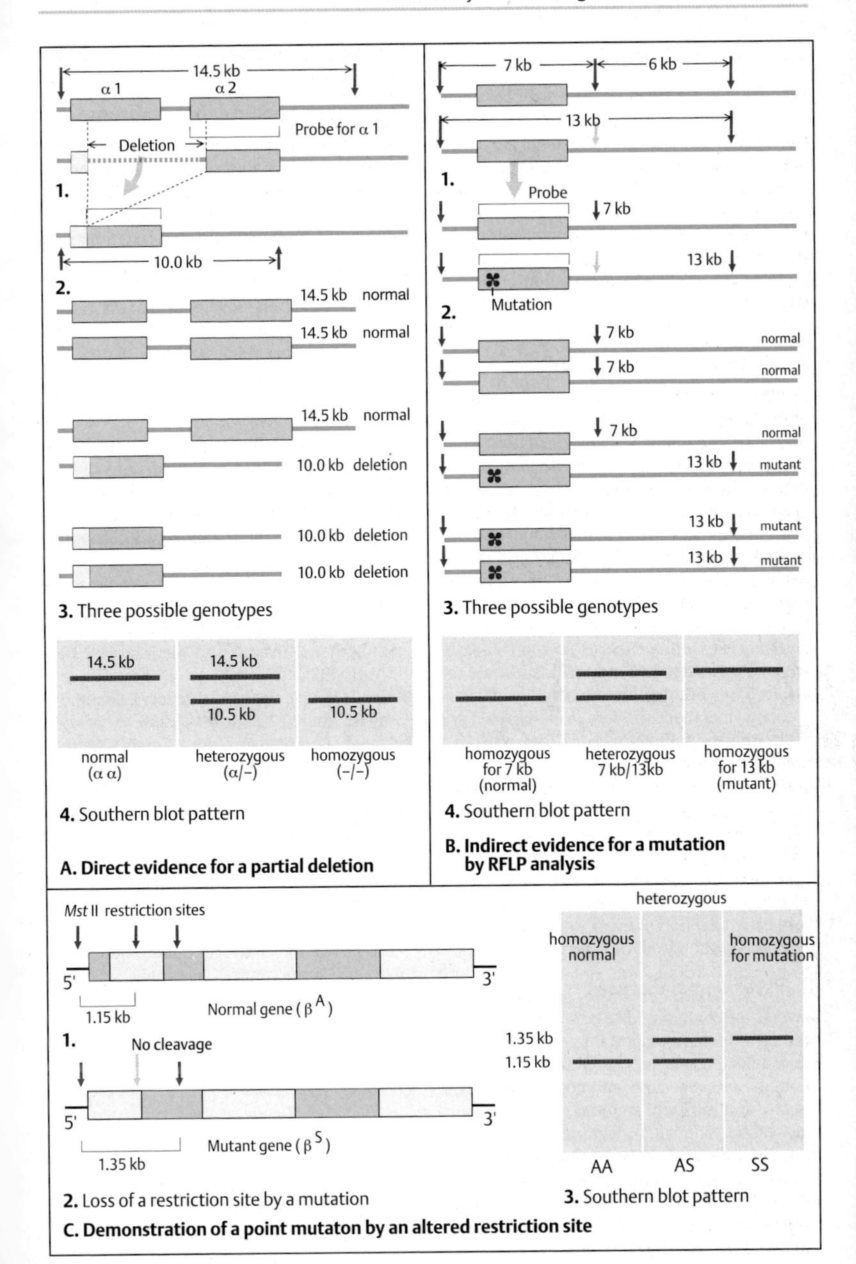


**A. Direct evidence for a partial deletion**

**B. Indirect evidence for a mutation by RFLP analysis**

**C. Demonstration of a point mutaton by an altered restriction site**

# Peroxisomal Diseases

Peroxisomes are small round organelles about 0.5 – 1.0 μm diameter (somewhat smaller than mitochondria). They are found mainly in the cytoplasm of kidney and liver cells. They are the site of some important metabolic functions. The name is derived from hydrogen peroxide, which is formed as an intermediary product of oxidative metabolism in the peroxisomes. A number of defects in peroxisome formation or peroxisome enzymes lead to severe diseases in humans (peroxisomal diseases).

## A. Biochemical reactions in peroxisomes

The electron micrograph (1) shows peroxisomes in a section of rat liver. The dark striated structures within the organelles consist of urates (peroxisomes contain an enzyme that oxidizes uric acid). Peroxisomes have both catabolic (substances are degraded) and anabolic (substances are synthesized) functions (2). Two biochemical reactions are especially important: a peroxisomal respiratory chain and the β-oxidation of very long-chain fatty acids. In the peroxisomal respiratory chain (3), certain oxidases and catalases act together. Specific substrates of the oxidases are organic metabolites of intermediary metabolism. Very long-chain fatty acids are broken down by β-oxidation (4) in a cycle with four enzymatic reactions. Energy production in peroxisomes is relatively inefficient compared with that of mitochondria. While free energy in mitochondria is mainly preserved in the form of ATP (adenosine triphosphate), in peroxisomes it is mostly converted into heat. Peroxisomes are probably a very early adaption of living organisms to oxygen. (Photograph from de Duve, 1986.)

## B. Peroxisomal diseases

Several peroxisomal diseases are known in man; the six most important are listed. All are autosomal recessive hereditary disorders. Patients with neonatal adrenoleukodystrophy do not form sufficient amounts of plasmalogens and cannot adquately degrade phytanic acid and pipecolic acid. When cultured fibroblasts from patients with genetically different types of peroxisomal diseases are fused, the hybrid cells form normal peroxisomes (cells with different defects can correct each other). More than ten complementation groups are known (Raymond et al., 1999).

## C. Cerebro-hepato-renal syndrome type Zellweger

Patients with this autosomal recessive hereditary disease have a characteristic facial appearance (1 – 4), extreme muscle weakness (5), and a number of accompanying manifestations such as calcified stippling of the joints on radiographs (6), renal cysts (7, 8), and clouding of the lens and cornea. The severe form of the disease (type Zellweger) usually leads to death before the age of one year. (Photographs 1 – 5 from Passarge and McAdams, 1967.)

## References


de Duve, C.: Die Zelle. Expedition in die Grundtruktur des Lebens. Spektrum der Wissenschaft, Heidelberg, 1986.

Folz, S.J., Trobe, J.D.: The peroxisome and the eye. Survey of Ophthalmology **35**:353 – 368, 1991.

Lazarow, P.B., Moser, H.W.: Disorders of peroxisome biogenesis, pp. 2287 – 2324, In: C.R. Scriver, et al., eds., The Metabolic Bases of Inherited Disease. 7th ed. McGraw-Hill, New York, 1995.

Moser, H.W.: Peroxisomal disorders. Semin. Pediatr. Neurol. **3**:298 – 304, 1996.

Passarge, E., McAdams, A.J.: Cerebro-hepato-renal syndrome. A newly recognized hereditary disorder of multiple congenital defects, including sudanophilic leukodystrophy, cirrhosis of the liver, and polycystic kidneys. J. Pediat. **71**:691 – 702, 1967.

Raymond, G.V.: Peroxisomal disorders. Curr. Opin. Pediatr. **11**:572 – 576, 1999.

Shimozawa, N., et al.: A human gene responsible for Zellweger syndrome that affects peroxisome assembly. Science **255**:1132 – 1255, 1992.

Warren, D.S., et al.: Phenotype-genotype relationships in *PEX10*-deficient peroxisome biogenesis disorder patients. Hum. Mutat. **15**:509 – 521, 2000.

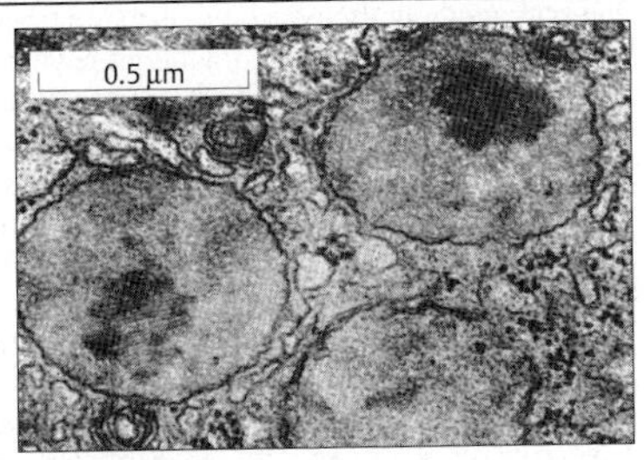


1. Peroxisomes in a rat liver cell

a) Catabolic
- $H_2O_2$-involving cellular respiration
- β-Oxidation of long-chain fatty acids,
- Prostaglandins, cholesterol side chains and others
- Purines, urates
- Pipecolic acid, dicarboxy acids
- Ethanol, methanol

b) Anabolic
- Phospholipids (Plasmalogen)
- Cholesterol, bile acids
- Gluconeogenesis
- Glyoxalate transamination

2. Function of peroxisomes

$O_2$ —Oxidases→ $H_2O_2$ —Catalase→ $2\,H_2O$ and heat (→ $O_2$)

R-$H_2$ → R; R'-$H_2$ → R'

R: D- and L-Amino acids, hydroxy acids, purines, urates, oxalate polyamines, fatty acid derivatives

R': Ethanol, Methanol, Nitrites, Quinones, Formates

3. Peroxisomal respiratory chain

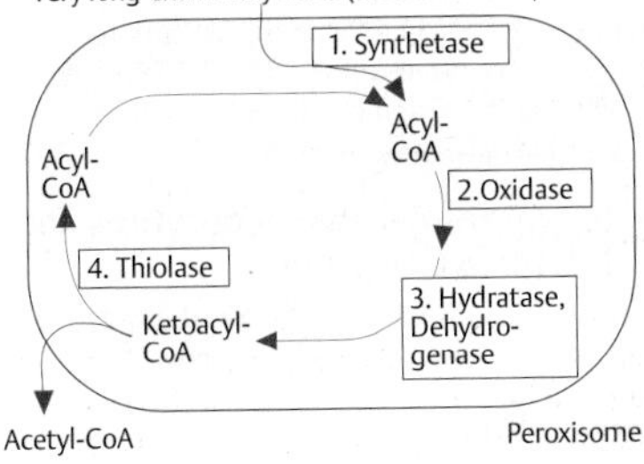


4. β-Oxidation

**A. Biochemical reactions in peroxisomes**

214100 Cerebro-hepato-renal syndrome Zellweger
202370 Neonatal adrenoleukodystrophy
266510 Infantile Refsum disease
239400 Hyperpipecolic acidemia
215100 Rhizomelic chondrodysplasia punctata
259900 Primary hyperoxaluria type I
and others

**B. Examples of peroxisomal diseases**

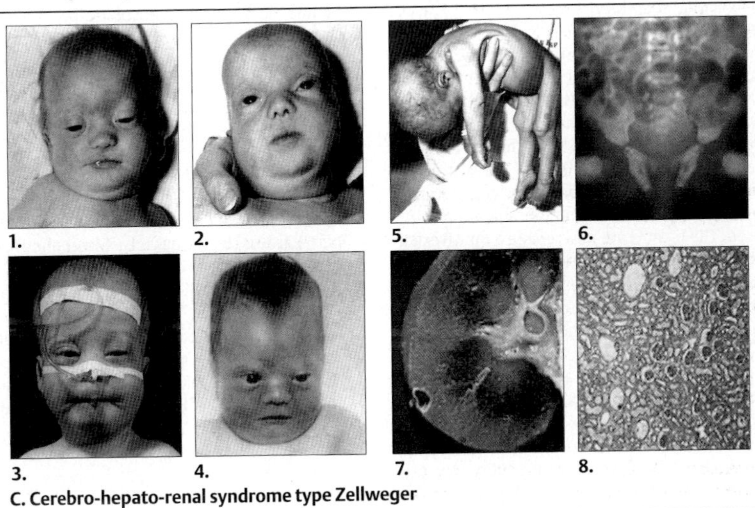


**C. Cerebro-hepato-renal syndrome type Zellweger**

# Lysosomes and LDL Receptor

## Lysosomes and Endocytosis

Lysosomes are membrane-enclosed cytoplasmic organelles with a diameter of 0.05 – 0.5 μm. They contain a wide variety of active hydrolytic enzymes (hydrolases) such as glycosidases, sulfatases, phosphatases, lipases, phospholipases, proteases, and nucleases (lysosomal enzymes) in an acid milieu (pH about 5). Lysosomal enzymes enter a lysosome by means of a recognition signal (mannose 6-phosphate) and a corresponding receptor.

### A. Receptor-mediated endocytosis and lysosome formation

Extracellular molecules to be degraded are taken into the cell by endocytosis. First, the molecules are bound to specific cell surface receptors (receptor-mediated endocytosis). The loaded receptors are concentrated in an invagination of the plasma membrane (coated pit). This separates from the plasma membrane and forms a membrane-enclosed cytoplasmic compartment (coated vesicle). Hormones, growth factors, energy-delivering proteins, and numerous viruses and toxins also enter cells by receptor-mediated endocytosis (see p. 360). The cytoplasmic lining of the vesicle consists of a network of a trimeric protein, clathrin. The clathrin coat is quickly lost within the cell, and an endosome forms, which fuses with membrane vesicles from the Golgi apparatus to form larger endosomal compartments. Here, the receptors are separated from the ligands and are returned to the cell surface in membrane vesicles (receptor recycling). Parts of the membrane are also reused. The ligands are now within a multivesicular body (endolysosomes). Hydrolases (lysosomal enzymes) are transported from the Golgi apparatus to an endolysosome in clathrin-enclosed vesicles after they become equipped with a recognition signal (mannose-6-phosphate receptor), required for uptake into the endolysosome and for normal functioning of the lysosome.

There are different classes of endolysosomes, defined according to relative acidity, receptor content, biochemical composition, morphological appearance, and other characteristics. The acid milieu in the lysosomes is maintained by a hydrogen pump in the membrane, which hydrolyzes ATP and uses the energy produced to move $H^+$ ions into the lysosome. Some of the mannose-6-phosphate receptors are transported back to the Golgi apparatus.

### B. Mannose-6-phosphate receptors

There are two types of mannose-6-phosphate receptor molecules, which differ in their binding properties and their cation dependence. They consist of either 2 or 16 extracellular domains with different numbers of amino acids. The cDNA of Ci-MPR (cation-independent mannose-6-phosphate receptor) is identical with insulinlike growth factor II (IGF-2). Thus, Ci-MPR is a multifunctional binding protein.

### C. Biosynthesis of the recognition signal

Two enzymes are essential for the formation of mannose-6-phosphate recognition signals: a phosphate transferase and a phosphoglycosidase. The phosphate is delivered by uridine-diphosphate-*N*-acetylglucosamine (UDP-GlcNAc) (uridine-5′-diphosphate-*N*-acetylglucosamine-glycoprotein-*N*-acetylglucosaminyl-phosphotransferase). A second enzyme, (*N*-acetylglucosamine-1-phosphodiester-*N*-acetylglucosaminidase) cleaves off the *N*-acetylglucosamine, leaving the phosphate residue at position 6 of the mannose.

(Figures after Sabatini and Adesnik, 1995, and C. de Duve, 1986; the relative sizes of the individual structures are not to scale.)

### References

de Duve, C.: A Guided Tour of the Living Cell. Vol. I and II. Scientific American Books, Inc., New York, 1984.

Gilbert-Barness, E., Barness, L.: Metabolic Diseases. Foundations of Clinical Management, Genetics, and Pathology. Vol. I + II, Eaton Publishing, Natick, MA, 2000.

Sabatini, D.D., Adesnik, M. B.: The biogenesis of membranes and organelles, pp. 459 – 553. In: C.R. Scriver, et al., eds. The Metabolic and Molecular Bases of Inherited Disease. 7th ed. McGraw-Hill, New York, 1995.

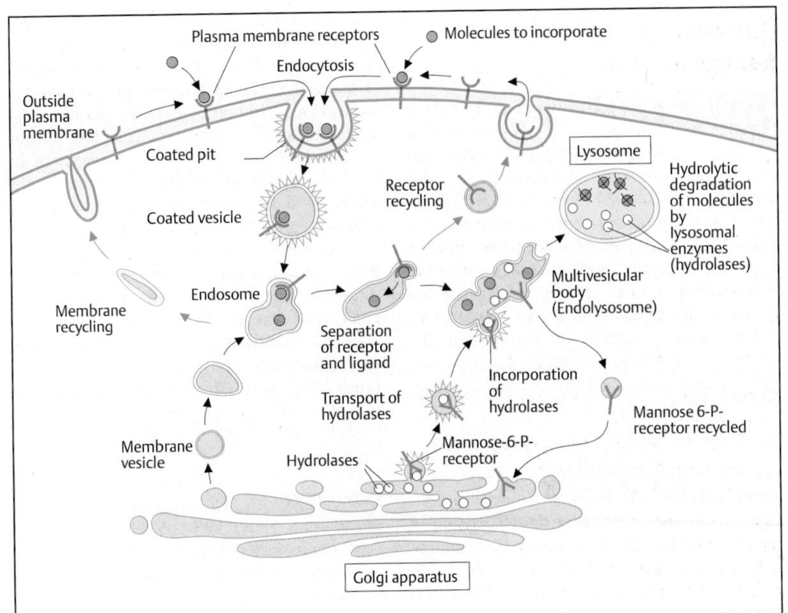


A. Receptor-mediated endocytosis and lysosome formation

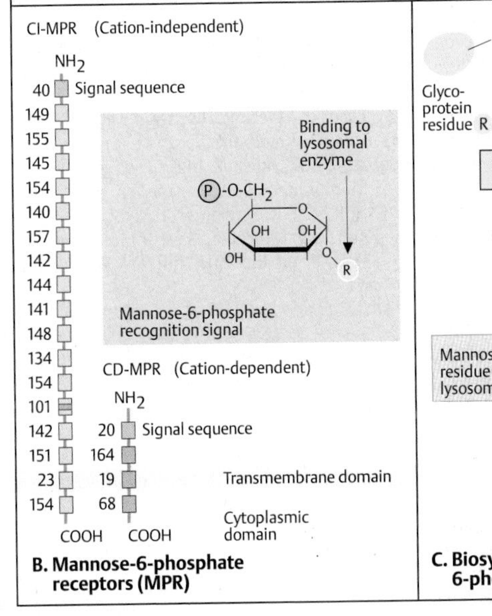


B. Mannose-6-phosphate receptors (MPR)

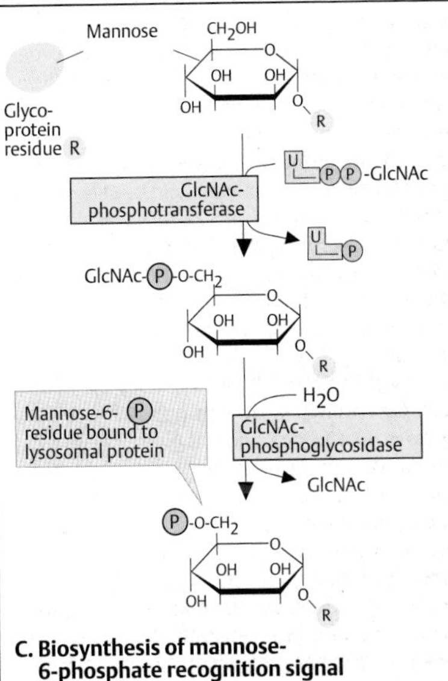


C. Biosynthesis of mannose-6-phosphate recognition signal

# Diseases Due to Lysosomal Enzyme Defects

Mutations in genes for enzymes that degrade complex macromolecules in lysosomes (lysosomal enzymes) lead to disease. Clinical signs and biochemical and cellular manifestations depend on the altered enzyme's specificity in lysosomal metabolism. With such an enzyme defect, macromolecules that normally would be degraded are stored (storage disease). This occurs at different rates, so that each disease has its own characteristic course. Twelve groups of diseases due to genetically determined disorders of specific lysosomal function are known, each with about three to ten individually defined diseases.

## A. Defective uptake of enzymes into lysosomes: I-cell disease (mucolipidosis type II)

Due to a mutation of the gene on chromosome 4 for the phosphotransferase needed to form the mannose-6-phosphate recognition signal (see p. 352), hydrolases cannot be taken up into lysosomes. Unlike normal cultured fibroblasts (1), those of patients (2) contain numerous densely packed cytoplasmic *i*nclusion bodies (thus, the name I-cell disease). The vesicular inclusions consist of hydrolases that cannot enter the lysosomes because the mannose-6-phosphate recognition signal is absent. Numerous enzymes are missing from the lysosomes, while their concentration in other parts of the cells and in body fluids is increased. Patients (3) show a severe progressive clinical picture, with the first signs usually apparent in the first half-year of life.

## B. Degradation of heparan sulfate by eight lysosomal enzymes

Heparan sulfate is an example of a macromolecule that is degraded stepwise by different lysosomal enzymes. Lysosomal enzymes are bond-specific, not substrate-specific. Thus, they also degrade other glycosaminoglycans, such as dermatan sulfate, keratan sulfate, and chondroitin sulfate (mucopolysaccharides). Specific enzyme defects cause the mucopolysaccharide storage diseases (see next page).
The first step in mucopolysaccharide degradation is the removal of sulfate from the terminal iduronate group by an iduronate sulfatase. A defect in the gene that codes for this enzyme leads to mucopolysaccharide storage disease type II (Hunter). The gene is located on the X chromosome, so that Hunter disease is transmitted by X-chromosomal inheritance. All other mucopolysaccharidoses are autosomal recessive. In the next step (2), the terminal iduronate is split off by an α-L-iduronidase. A mutation of the gene coding for this enzyme in the homozygous state leads to mucopolysaccharidosis (MPS) type I (Hurler/Scheie). In the next three steps a mutation (in the homozygous state) of a gene coding for one of the enzymes causes mucopolysaccharidosis type III (Sanfilippo). The four genetically and enzymatically different types (III-A to III-D) cannot be distinguished clinically. MPS type VII (Sly), due to a defect of β-glucuronidase, has a further characteristic clinical picture.

### References

Kornfeld, S., Sly, W.S.: I-cell disease and Pseudo-Hurler polydystrophy: Disorders of lysosomal enzyme phosphorylation and localization, pp. 2495–2508. In: C.R. Scriver, et al., eds., The Metabolic and Molecular Bases of Inherited Disease. 7th ed. McGraw-Hill, New York, 1995.

Neufeld, E.F., Muenzer, J.: The mucopolysaccharidoses, pp. 2465–2494. In: C.R. Scriver, et al., eds., The Metabolic and Molecular Bases of Inherited Disease. 7th ed. McGraw-Hill, New York, 1995.

Sabatini, D.D., Adesnik, M.B.: The biogenesis of membranes and organelles, pp. 459–553. In: C.R. Scriver, et al., eds., The Metabolic and Molecular Bases of Inherited Disease. 7th ed. McGraw-Hill, New York, 1995.

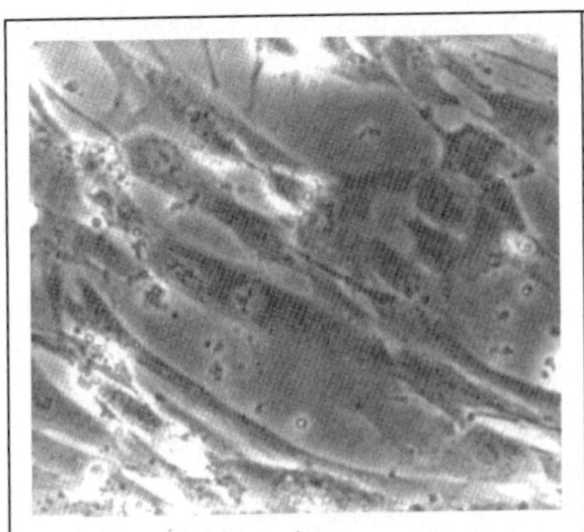

**1.** Normal fibroblast culture

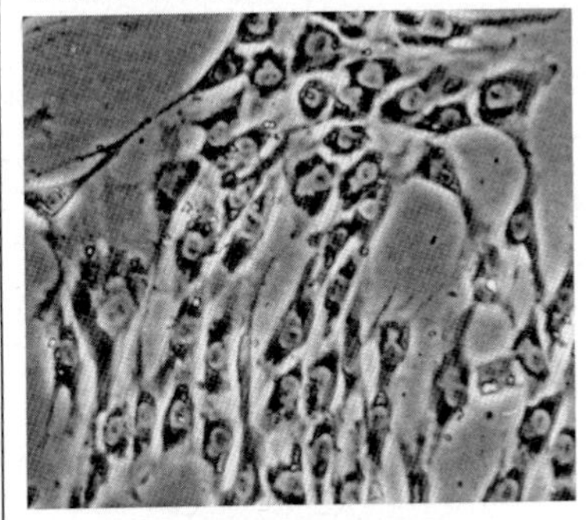

**2.** Fibroblast culture in I-cell disease

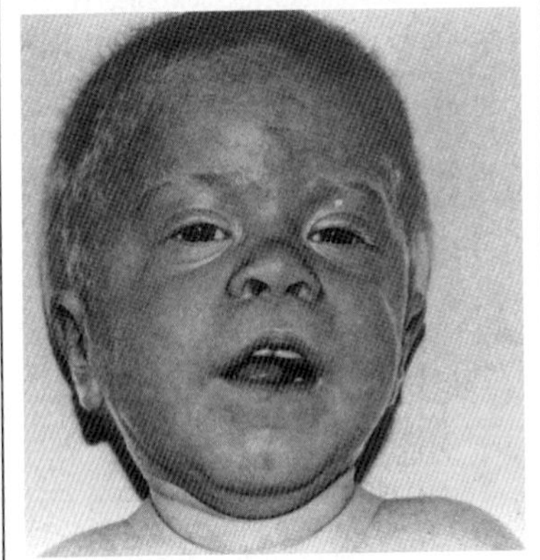

**3.** Patient with I-cell disease

**A. Defective uptake of enzymes in lysosomes: I-cell disease**

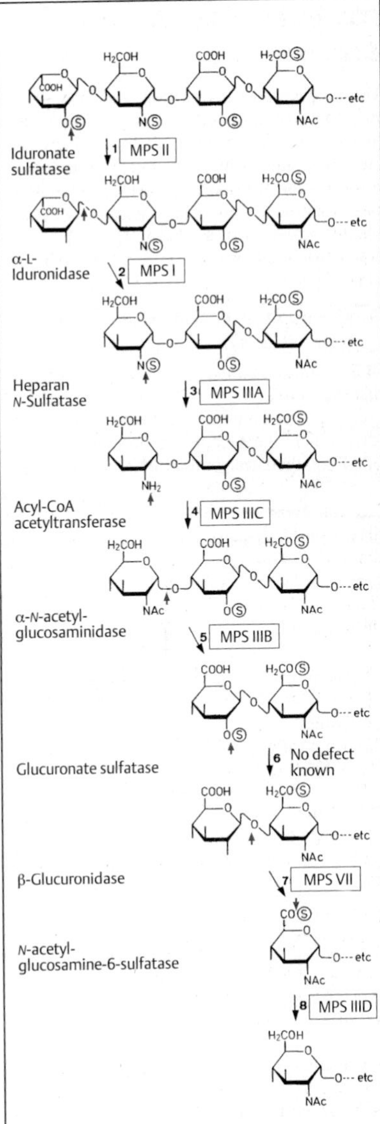


**B. Degradation of heparan sulfate by eight lysosomal enzymes**

## Mucopolysaccharide Storage Diseases

The mucopolysaccharide storage diseases (the mucopolysaccharidoses) are a clinically and genetically heterogeneous group of lysosomal storage diseases caused by defects in different enzymes for mucopolysaccharide degradation (glycosaminoglycans). Except for mucopolysaccharide storage disease type II (Hunter), all are transmitted by autosomal recessive inheritance.

### A. Mucopolysaccharide storage disease type I (Hurler)

At first almost inapparent, the early signs of the disease occur at about 1 – 2 years of age, with increasing coarsening of the facial features, retarded mental development, limited joint mobility, enlarged liver, umbilical hernia, and other signs. Radiographs show coarsening of skeletal structures (dysostosis multiplex). The photographs show the same patient at different ages (own data).

### B. Mucopolysaccharide storage disease type II (Hunter)

This type of mucopolysaccharidosis is transmitted by X-chromosomal inheritance. Four cousins from one pedigree are shown. Clinically, the disease is similar to, but less rapidly progressive than, MPS type I. (Photos from Passarge et al., 1974). Molecular diagnosis is possible in most cases.

### References

McKusick, V.A.: Mendelian Inheritance in Man. 12th ed. 1998.

Passarge, E., et al.: Krankheiten infolge genetischer Defekte im lysosomalen Mucopolysaccarid-Abbau. Dtsch Med Wschr. 99 : 144 – 158, 1974.

Classification of the mucopolysaccharide storage diseases (MPS)

| MPS Type | Enzyme Defect | Important Manifestations |
|---|---|---|
| IH (Hurler) | α-L-Iduronidase | Dysostosis multiplex, severe developmental disorder, corneal clouding |
| IS (Scheie) | α-L-Iduronidase | Stiff joints, corneal clouding, normal mental development |
| II (Hunter) (X-chromosomal) | Iduronate sulfatase | Dysostosis multiplex, no corneal clouding, developmental retardation |
| III (Sanfilippo) | | |
| A | Heparan *N*-sulfatase | Severe psychomotor retardation beginning about age 6 – 8 years, relatively mild somatic signs. |
| B | α-*N*-Acetylglucosaminidase | |
| C | Acetyl-CoA: α-glucosaminide *N*-acetyltransferase | |
| D | *N*-acetylglucosamine-6-sulfate sulfatase | |
| IV (Morquio) | | |
| A | Galactose-6-sulfatase | Corneal clouding, severe skeletal changes, short stature, |
| B | β-Galactosidase | odontoid process hypoplasia, normal mental development |
| VI (Maroteaux – Lamy) | *N*-acetylgalactosamine-4-sulfatase (aryl-sulfatase B) | Dysostosis multiplex, corneal clouding, normal mental development |
| VII (Sly) | β-Glucuronidase | Dysostosis multiplex, corneal clouding |

(After McKusick, 1998)

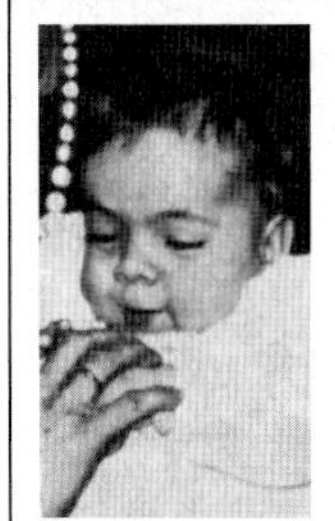

8 weeks

7 months

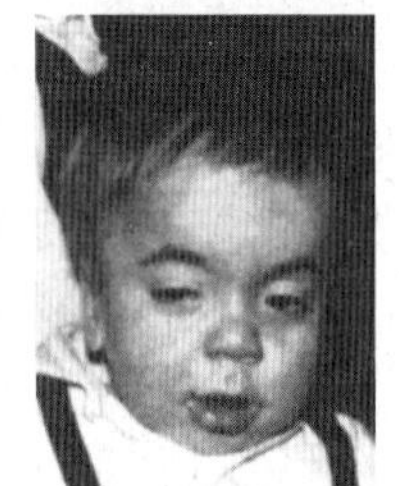

2 1/4 years

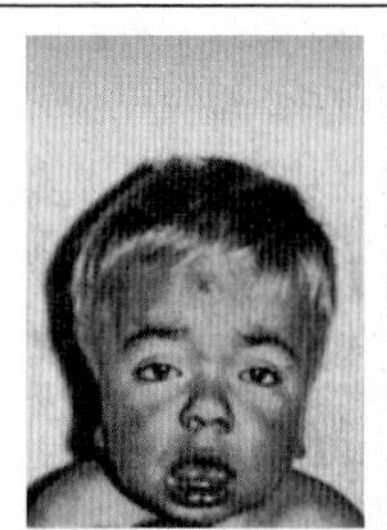

3 3/4 years

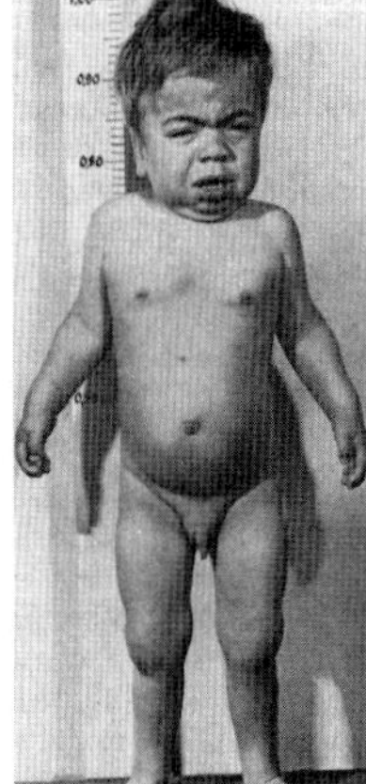

5 years

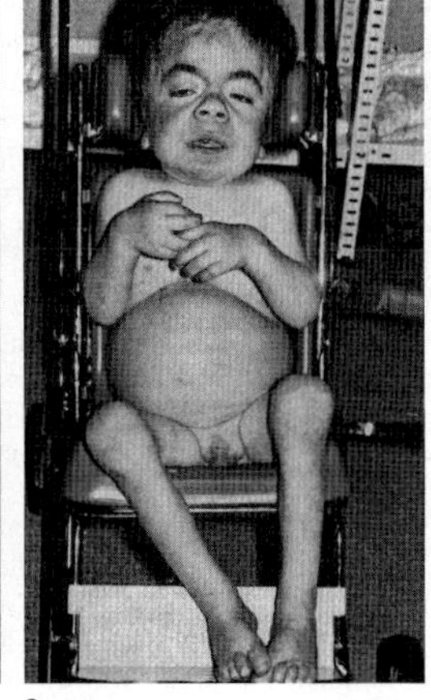

8 years

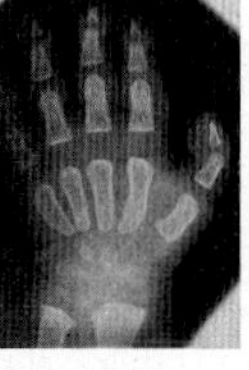

Dysostosis multiplex

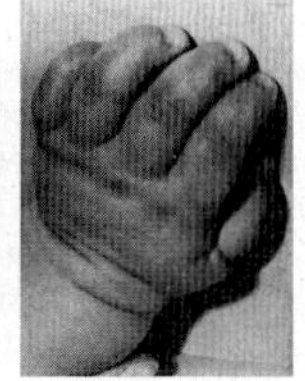

Joint contractures

**A. Mucopolysaccharide storage disease type I (Hurler)**

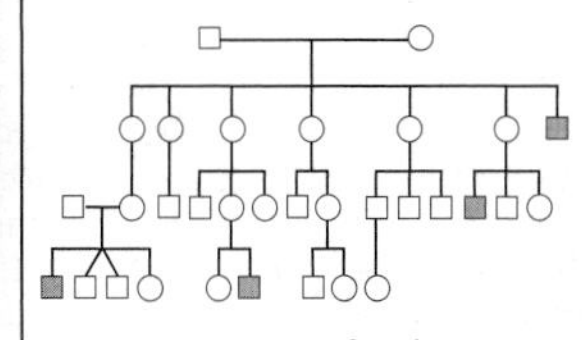

□ = male ○ = female

■ = Hunter syndrome

X-Chromosomal inheritance

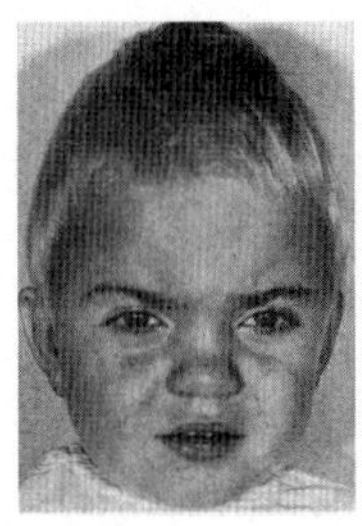

4 1/2 years

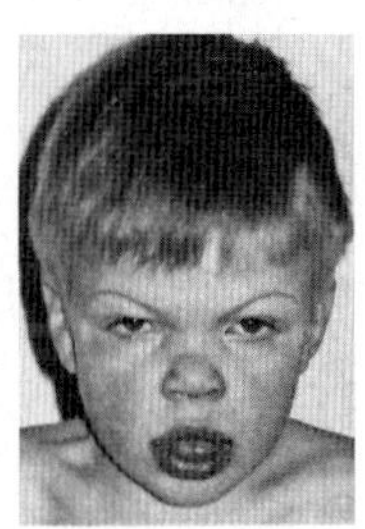

10 years

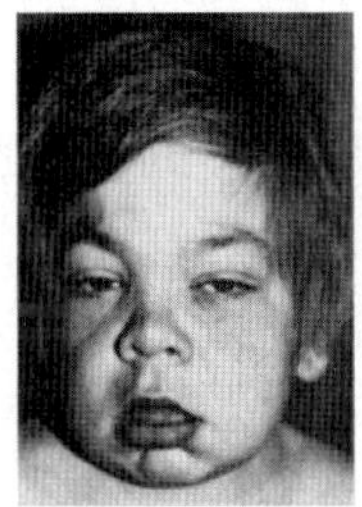

13 years

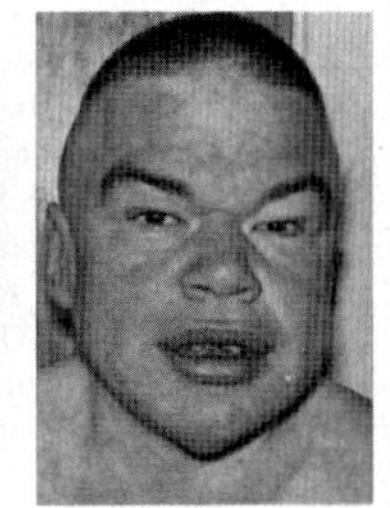

21 years

**B. Mucopolysaccharide storage disease type II (Hunter)**

# Familial Hypercholesterolemia

Familial hypercholesterolemia (FH) is a hereditary disorder of intracellular lipid metabolism. Several different genetic forms exist, each characterized by the step of the metabolic pathway involved and the type of mutation. Mixed forms, caused by multigenic and environmental factors, and monogenic forms can be distinguished.

## A. The disease phenotype

Familial hypercholesterolemia (1) (McKusick 144400) occurs in about 1 in 500 persons in the heterozygous state. In the rare homozygous state (1 in $10^6$) it is a devastating disease usually leading to death during the first or second decade of life. The heterozygous form is characterized by early signs of atherosclerosis (2). The number of functional LDL receptors per cell is decreased by about 50% (3). Deposits of cholesterol esters in the tendons, especially the achilles tendon, and the skin (xanthomas) are common in heterozygotes (4). A characteristic sign is lipid deposits in the eye in front of the iris (arcus senilis, 5).

## B. The LDL receptor

The LDL receptor is a cell surface receptor for low-density lipoprotein (LDL), which contains apoB-100, the protein that carries most of the cholesterol ester in human plasma. This receptor mediates endocytosis of the extracellular ligand. The LDL receptor is a membrane-bound protein of 839 amino acids with five domains: three extracellular domains, one transmembrane domain, and one intracellular, with the carboxyl end. The extracellular domains consist of one domain with seven cysteine-rich units of 40 amino acids each, the ligand-binding region; a domain with epidermal growth factor (EGF) precursor homology; and a small serine- und threonine-rich domain linked to oligosaccharides. The transmembrane domain contains 22 hydrophobic amino acids. The fifth domain with the intracellular COOH terminus consists of 50 amino acids. It controls the interaction of the receptor with the coated pit during endocytosis. The corresponding gene consists of 18 exons that span 45 kb genomic DNA on human chromosome 19p13.1 – 13.3. In addition to the main locus on 19p, two additional loci for autosomal dominant hypercholesterolemia exist (Varret et al., 1999; Haddad et al., 1999). (Figure adapted from H. Schuster, Berlin).

## C. LDL receptor-mediated endocytosis

The LDL receptor mediates the endocytosis of LDL. The receptors loaded with LDL accumulate in a coated pit (a), which separates from the plasma membrane and forms an endocytotic vesicle (b). This transports LDL molecules to a lysosome. (Photograph from Anderson et al., 1977).

## D. Homology with other proteins

The mammalian LDL receptor is one of a five-member family including the LDL receptor itself, the VLDL receptor (very low density lipoprotein), the ApoE receptor 2 (ApoER2), the LDL receptor-related protein (LRP), and megalin. LRP and megalin are multifunctional and bind diverse ligands such as lipoproteins, protease and their inhibitors, peptide hormones, and carrier proteins of vitamins (Krieger and Herz, 1994). The proximal halves of the extracellular domains of the LDL receptor family are structurally related to the epidermal growth factor family (EGF). These are related to protease of the blood coagulation system, factors IX and X, protein C, and complement C9.

## References


Anderson, R.G.W., Brown, M.S., Goldstein, J.L.: Cell. **10**:351, 1977.

Brown, M.S., Goldstein, J.L.: A receptor-mediated pathway for cholesterol homeostasis. Science **232**:34 – 47, 1986.

Goldstein, J.L., Brown, M.S.: Familial hypercholesterolemia, pp. 1981 – 2030. In: C.R. Scriver, et al., eds., The Metabolic and Molecular Bases of Inherited Disease. 7th ed. McGraw-Hill, New York, 1995.

Haddad, L., et al.: Evidence for a third genetic locus causing familial hypercholestolemia. A non-LDLR, non-APOB kindred. J. Lipid. Res. **40**:1113 – 1122, 1999.

Norman, D., et al.: Characterization of a novel cellular defect in patients with phenotypic homozygous familial hypercholesterolemia. J. Clin. Invest. **104**:619 – 628, 1999.

Varret, M., et al.: A third major locus for autosomal dominant hypercholesterolemia maps to 1p34.1-p32. Am. J. Hum. Genet. **64**:1378 – 1387, 1999.

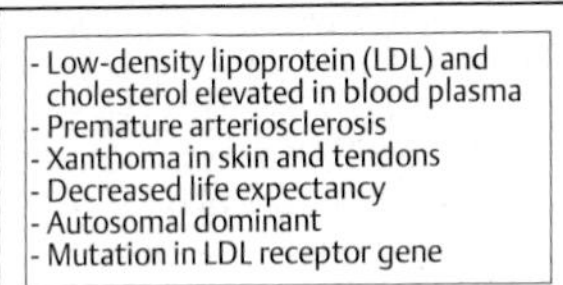

- Low-density lipoprotein (LDL) and cholesterol elevated in blood plasma
- Premature arteriosclerosis
- Xanthoma in skin and tendons
- Decreased life expectancy
- Autosomal dominant
- Mutation in LDL receptor gene


**1.** General features

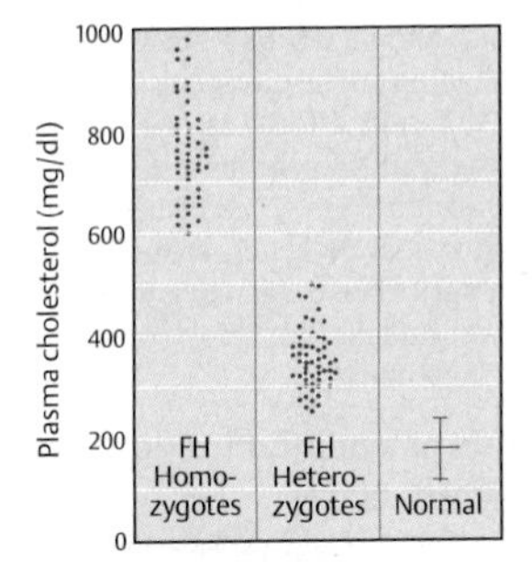


**2.** Hypercholesterolemia

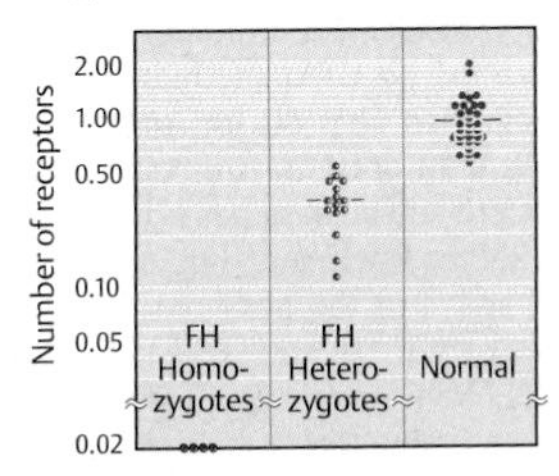


**3.** LDL receptors decreased

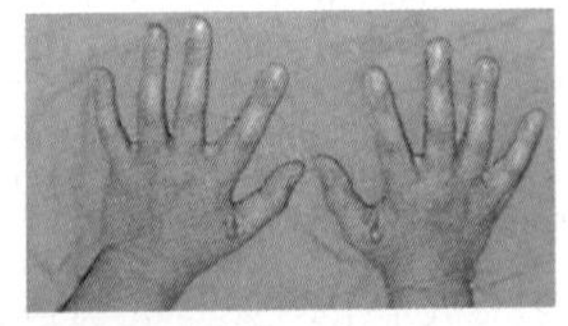

**4.** Xanthoma formation

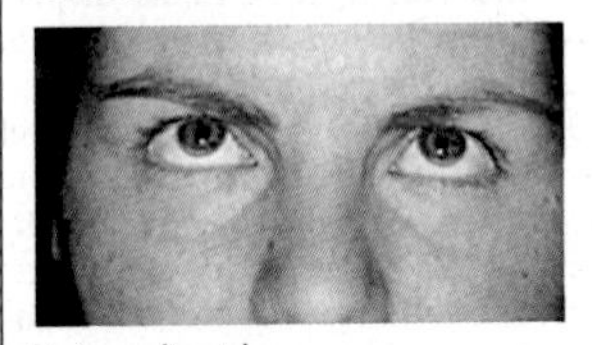

**5.** Arcus lipoides

**A. Familial hypercholesterolemia**

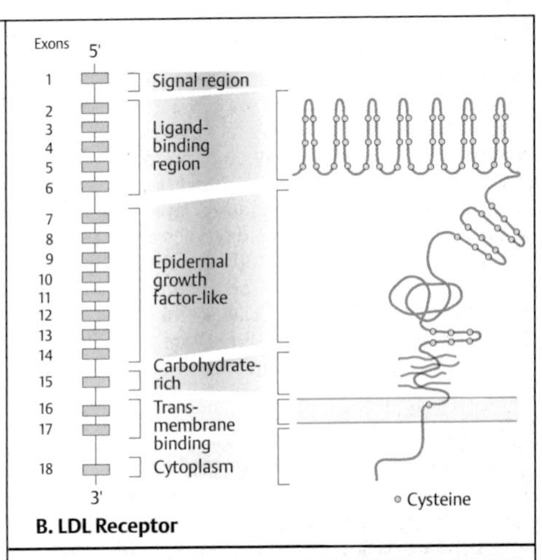


**B. LDL Receptor**

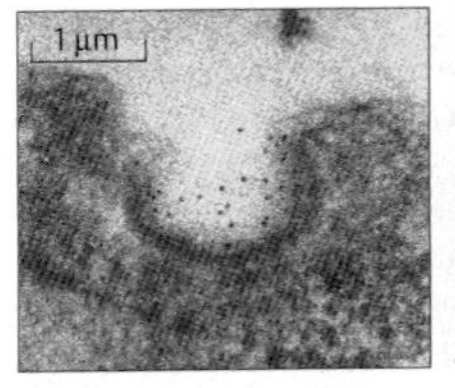


a Coated pit

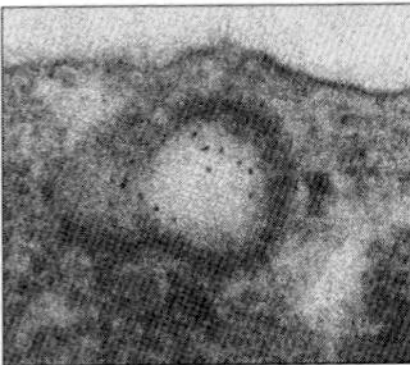

b Endocytotic vesicle

Electron micrographs of fibroblasts in culture that have taken up LDL molecules (black dots, made visible by binding to ferritin).

**C. Receptor-mediated endocytosis of LDL**

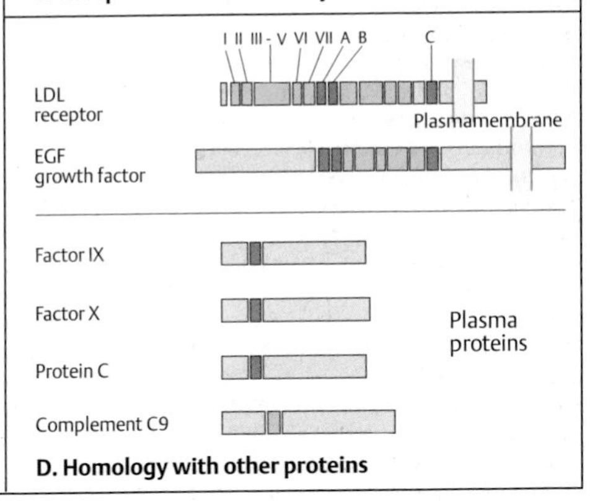


**D. Homology with other proteins**

## Mutations in the LDL Receptor

Low-density lipoprotein (LDL) is the main carrier of cholesterol in the blood. An LDL particle has a diameter of 22 nm and a molecular mass of about 3000 kDa. Its hydrophobic core contains about 1500 esterified cholesterol molecules surrounded by an outer layer of phospholipids and unesterified cholesterols containing a single apoB-100 lipoprotein molecule. LDL delivers cholesterol to peripheral tissues and regulates *de novo* cholesterol synthesis there. Mutations in the LDL receptor gene or in the ligand apoB-100 lipoprotein result in hypercholesterolemia.

### A. Intracellular LDL receptor metabolism and classes of mutation

Five principal classes of LDL receptor mutations can be distinguished: (1) receptor null mutations ($R^{\circ}$) due to lack of receptor protein synthesis in the endoplasmic reticulum (ER); (2) defective intracellular transport to the Golgi apparatus; (3) defective extracellular ligand binding; (4) defective endocytosis ($R^{+}$ mutations); and (5) failure to release the LDL molecules inside the endosome (recycling-defective mutations). The receptor–LDL complex enters the cell by endocytosis. In the endosome the LDL including of apoB-100 is separated from the receptor. In the lysosome the LDL is broken down into amino acids and cholesterol. Free cholesterol activates the enzyme acetyl-CoA cholesterol transferase (ACAT), which catalyzes the esterification. The LDL receptor is recycled to the cell surface in a recycling vesicle. The key enzyme for endogenous cholesterol synthesis is 3-hydroxy-3-methylglutaryl-CoA reductase (HMG-CoA reductase). This enzyme is downregulated by exogenous LDL uptake. LDL receptor mutations interrupt this control feedback mechanism and result in increased endogenous cholesterol synthesis. HMG-CoA reductase also downregulates LDL receptor protein synthesis to prevent overloading with cholesterol.

### B. Mutational spectrum in the LDL receptor gene

About 350 mutations have been recorded in the LDL receptor gene (Varret et al., 1998). Of these, 63% are missense mutations. Mutations occur in all parts of the gene, but there is a relative excess of mutations in exons 4 and 9. Exons 13 and 15 are involved less often than expected. A high proportion of mutations (74%) located in the ligand-binding domain (exons 2–6) involve amino acids conserved in evolution. (Varret et al., 1998; data also accessible at *http://www.umd.necker.fr*). In addition to point mutations, several deletions of various sizes and locations, and insertions have been described. Depending on the intragenic location of a mutation, different effects can be observed, including absent mRNA synthesis, defective intracellular transport due to abolished binding (1) or receptor recycling (2), reduced membrane anchorage (4), and defective internalization (5). Alu repeats may be involved as a cause of intragenic deletions.

### C. Diagnosis of a point mutation in the LDL receptor gene

Direct sequencing demonstrates a mutation in exon 9. First, exon 9 is amplified by PCR (P1 and P2 = primers 1 and 2). The mutation in codon 408, GTG (valine) to GTA (methionine), produces a recognition site (N) for *Nla*III (GATC) that is not normally present. This results in two fragments of 126 and 96 base pairs (bp) instead of the usual 222 bp fragment. Thus, affected individuals (1 and 3 in the pedigree) have two smaller fragments of 126 and 96 kb (2) in addition to the 222 kb fragment. Sequence analysis of the patient (individual 1 in the pedigree) demonstrates the mutation by the presence of an additional adenine (A) next to the normal guanine (3). With this knowledge about the mutation, the latter can be indirectly demonstrated within a family by the additional recognition site for a restriction enzyme. (Data from H. Schuster, Berlin).

### References


Goldstein, J.L., Brown, M.S.: Familial hypercholesterolemia, pp. 1981–2030, In: C.R. Scriver, et al., eds., The Metabolic and Molecular Bases of Inherited Disease. 7th ed. McGraw-Hill, New York, 1995.

Varret, M., et al.: LDLR database (second edition): new additions to the database and the software, and results of the first molecular analysis. Nucleic Acids Res. **26**:248–252, 1998.

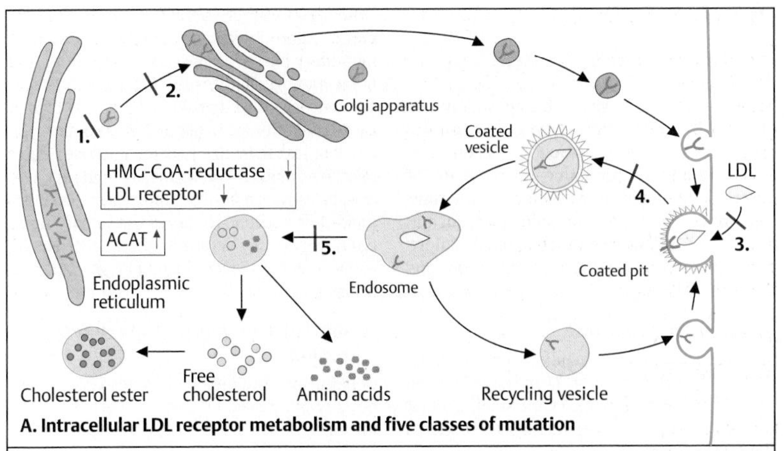


A. Intracellular LDL receptor metabolism and five classes of mutation

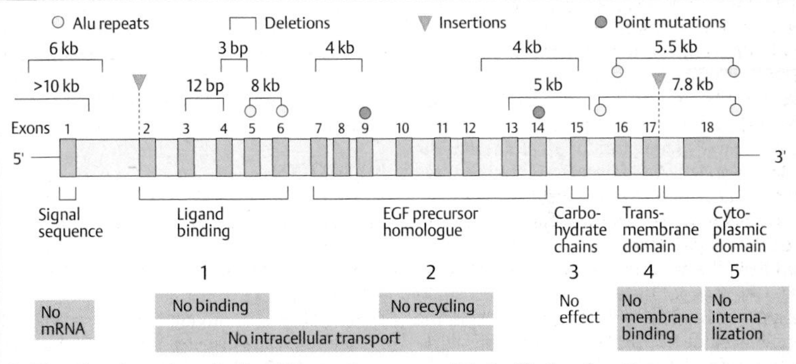


B. Mutational spectrum in the LDL receptor gene and their effect on function

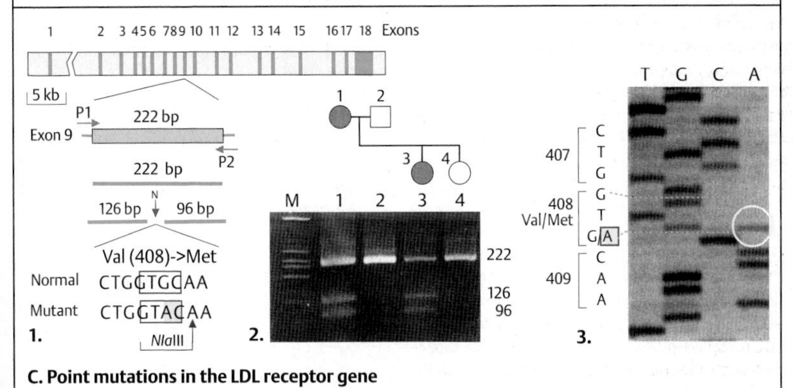


C. Point mutations in the LDL receptor gene

# Homeostasis

## Insulin and Diabetes Mellitus

Diabetes mellitus is one of the most common diseases of the Western world, occurring in about 1–2% of the population. The blood sugar is abnormally elevated due to a variety of causes, including genetic factors. With time, this leads to numerous complications, such as myocardial infarction, stroke, renal failure, vascular damage leading to amputation, and blindness.

### A. Insulin production

In humans, the gene for insulin is located on the short arm of chromosome 11 in region 1, band 5.5. With 1430 base pairs, it is a small gene. It consists of a signal sequence (L, leader) and two exons. The gene is expressed exclusively in β-cells of the islands of Langerhans of the pancreas. A β-cell-specific enhancer is located at the 5′ end of the gene, and a variable number of tandem repeats (VNTR) are located further upstream. The primary transcript is spliced to produce the mRNA template for preproinsulin. The signal peptide (24 amino acids) is removed, and the B chain and A chain are joined by two disulfide bridges. Proper binding and three-dimensional structuring require the presence of a connecting peptide (C peptide). The complete insulin molecule consists of an A chain of 21 amino acids and a B chain of 30 amino acids. The signal peptide of the insulin molecule is required for secretion.

### B. Insulin receptor

Insulin initiates its physiological effect by binding to a receptor (insulin receptor). When bound to insulin, the insulin receptor functions as an enzyme and phosphorylates tyrosine in the target proteins. This is the intracellular signal for the metabolic processes induced by insulin.

### C. Diabetes mellitus (simplified model)

Diabetes mellitus is classified into two basic types: type I (insulin-dependent diabetes mellitus, IDDM) and type II (non-insulin-dependent, NIDDM). The majority of diabetes type I cases are caused by external factors, such as certain viral infections, on a background of genetic susceptibility. Diabetes type II is mainly due to genetic factors, but also in part to overnourishment. Apart from an autosomal dominant hereditary form with onset in young adults, it is not a monogenic disorder. Monozygotic twins are concordant for type II in about 40–50% of cases and for type I in about 25%, as opposed to a risk of less than 10% for type I in first-degree relatives (about 2–7% according to family relationship and age at onset of disease). Diabetes mellitus is a secondary manifestation of a number of genetically determined diseases, e.g., insulin receptor defect (insulin resistance syndrome).

### D. Influence of genes of the HLA-D region

Genetic susceptibility to diabetes type I is especially influenced by certain alleles of class I MHC genes (see p. 308). The presence of alleles DR3 and DR4, especially in DR3/DR4 heterozygotes, is associated with susceptibility to diabetes type I. DR2 confers relative resistance to diabetes. Genes conferring susceptibility to diabetes have been located at several sites of the genome.

### References

Bennett, S.T., et al.: Susceptibility to human type I diabetes at *IDDM2* is determined by tandem repeat variation at the insulin gene minisatellite locus. Nature Genet. **9**:284–292, 1995.

Davies, J.L., et al.: A genome-wide search for human type I diabetes susceptibility genes. Nature **371**:130–136, 1994.

Lowe, W.L. Jr.: Diabetes mellitus, pp. 433–442, In: J.L. Jameson, ed., Principles of Molecular Medicine. Humana Press, Totowa, New Jersey, 1998.

Schwartz, M.W., et al.: Leptin- and insulin-receptor signalling. Nature **404**:663, 2000.

Taylor, S.I.: Diabetes mellitus, pp. 843–896, In: C.R. Scriver, et al., eds., The Metabolic and Molecular Bases of Inherited Diseases. 7th ed. McGraw-Hill, New York, 1995.

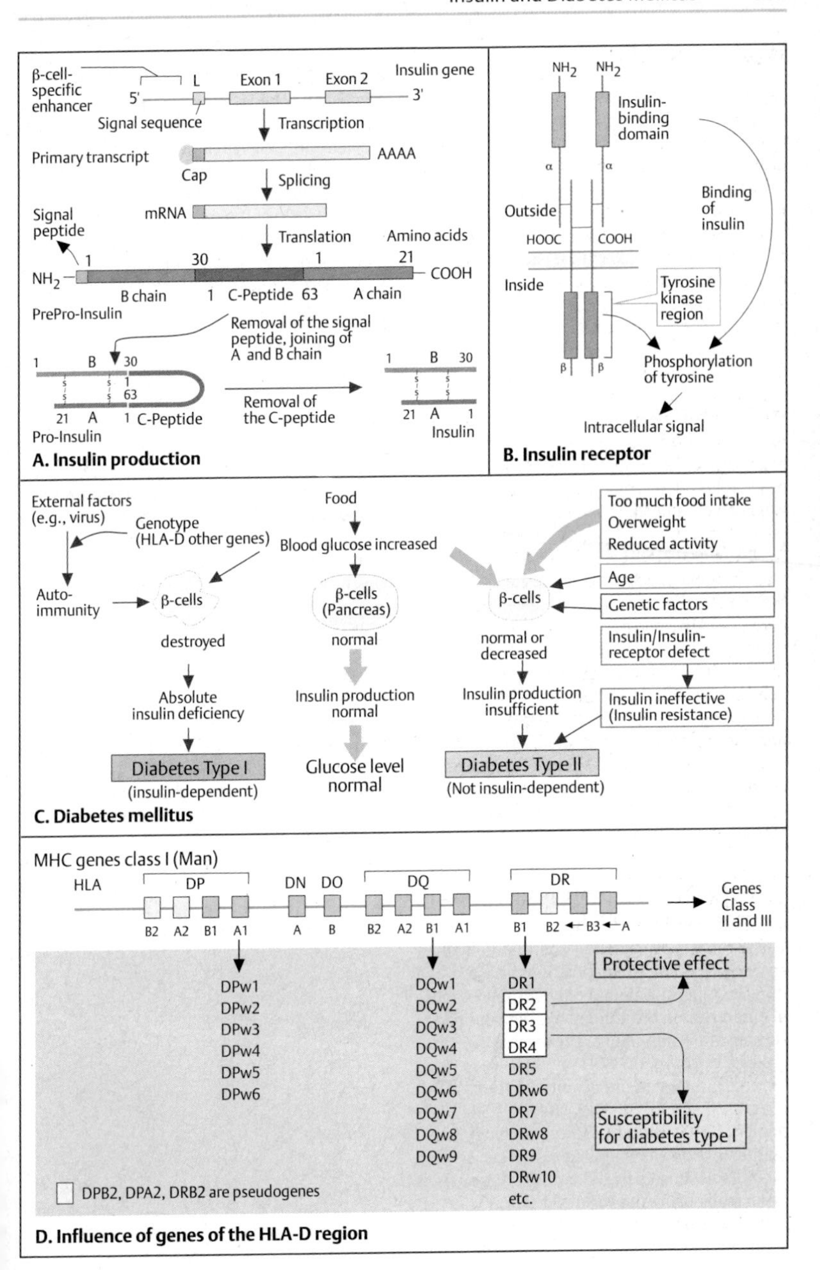
β-cell-specific enhancer
L
Exon 1
Exon 2
Insulin gene
5'
3'
Signal sequence
Transcription
Primary transcript
AAAA
Cap
Splicing
mRNA
Signal peptide
Translation
Amino acids
1
30
1
21
NH2
COOH
B chain
1 C-Peptide 63
A chain
PrePro-Insulin
Removal of the signal peptide, joining of A and B chain
Removal of the C-peptide
Pro-Insulin
Insulin
A. Insulin production
NH2
Insulin-binding domain
Binding of insulin
Outside
HOOC
COOH
Inside
Tyrosine kinase region
Phosphorylation of tyrosine
Intracellular signal
B. Insulin receptor
External factors (e.g., virus)
Genotype (HLA-D other genes)
Auto-immunity
β-cells
destroyed
Absolute insulin deficiency
Diabetes Type I
(insulin-dependent)
Food
Blood glucose increased
β-cells (Pancreas)
normal
Insulin production normal
Glucose level normal
Too much food intake
Overweight
Reduced activity
Age
Genetic factors
Insulin/Insulin-receptor defect
β-cells
normal or decreased
Insulin production insufficient
Insulin ineffective (Insulin resistance)
Diabetes Type II
(Not insulin-dependent)
C. Diabetes mellitus
MHC genes class I (Man)
HLA
DP
DN
DO
DQ
DR
B2 A2 B1 A1
A B
B2 A2 B1 A1
B1 B2 B3 A
Genes Class II and III
DPw1
DPw2
DPw3
DPw4
DPw5
DPw6
DQw1
DQw2
DQw3
DQw4
DQw5
DQw6
DQw7
DQw8
DQw9
DR1
DR2
DR3
DR4
DR5
DRw6
DR7
DRw8
DR9
DRw10
etc.
Protective effect
Susceptibility for diabetes type I
DPB2, DPA2, DRB2 are pseudogenes
D. Influence of genes of the HLA-D region

# *Protease Inhibitor $\alpha_1$-Antitrypsin*

$\alpha_1$-Antitrypsin ($\alpha_1$-AT) is an essential protease inhibitor in blood plasma. It binds to a wide range of proteases, such as elastase, trypsin, chemotrypsin, thrombin, and bacterial proteases. Its most important physiological effect is the inhibition of leukocyte elastase, a protease that breaks down the elastin of the pulmonary alveolar walls. Deficiency of $\alpha_1$-antitrypsin leads to increasing destruction of the pulmonary alveoli, obstructive emphysema of the lungs, and in newborns, a form of hepatitis.

## A. $\alpha_1$-Antitrypsin

$\alpha_1$-Antitrypsin (1) in humans is a glycoprotein composed of 394 amino acids and 12% carbohydrate. It is coded for by a 10.2 kb gene with five exons on chromosome 14 (14q32.1).

## B. $\alpha_1$-Antitrypsin deficiency

The uninhibited action of leukocyte elastase on the elastin of the pulmonary alveoli leads to chronic obstructive pulmonary emphysema (1). (Radiograph from N. Konietzko, Essen.) The most frequent deficiency allele is Pi(Z). The plasma concentration of $\alpha_1$-AT with genotype PiZZ (homozygote) is usually about 12–15% of normal (with the normal allele M). MZ heterozygotes have 64%, and MS heterozygotes 86% of MM homozygote activity. $\alpha_1$-Antitrypsin deficiency in the lung can be corrected by intravenous administration of $\alpha_1$-antitrypsin. (After R. H. Ingram.)

## C. $\alpha_1$-Antitrypsin: protein, gene, and important mutations

The $\alpha_1$-antitrypsin protein has three oligosaccharide side chains at positions 46, 83, and 247. The protein is highly polymorphic because of differences in the amino acid sequence and in carbohydrate side chains. The reactive site is located at position 358/359 (methionine/serine). Clinically, the most important mutations affect codons 213 (PiZ), 256 (PiP), 264 (PiS), 342 (PiZ), and 357 (Pi[Pittsburgh]). The gene contains variant restriction enzyme sites, which can be used for reliable diagnosis. Today the diagnosis is often made using the PCR reaction.

## D. Synthesis of $\alpha_1$-antitrypsin

The $\alpha_1$-AT gene is expressed in liver cells (hepatocytes). The gene product is channeled through the Golgi apparatus and released from the cell (secreted). The Z mutation leads to aggregation of the enzyme in the liver cells, with too little of it being secreted. The S mutation leads to premature degradation. About 2–4% of the population in Central and Northern Europe are MZ heterozygotes.

## E. Reactive center of protease inhibitors

$\alpha_1$-Antitrypsin is one member of a family of protease inhibitors that show marked homology, especially at their reactive centers. Oxidizing substances have an inhibitory effect and inactivate the molecule. Smokers have a much more rapid course of $\alpha_1$-AT deficiency disease (onset of dyspnea at 35 years of age instead of 45–50).
(Figures in C–E adapted from Cox, 1995, and Owen et al., 1983).

## References

Cox, D.W.: $\alpha_1$-Antitrypsin deficiency, pp. 4125–4158. In: C.R. Scriver, et al, eds., The Metabolic and Molecular Bases of Inherited Disease. 7th ed. McGraw-Hill, New York, 1995.

Lancet editorial: $\alpha_1$-Antitrypsin Z and the liver. Lancet **340**:402–403, 1992.

Lomas, D.A., et al.: The mechanism of Z $\alpha_1$-antitrypsin accumulation in the liver. Nature **357**:605–607, 1992.

Owen, M.C., et al.: Mutation of antitrypsin to antithrombin: $\alpha_1$-antitrypsin Pittsburgh (358 -> Arg), a fatal bleeding disorder. N. Engl. J. Med. **309**:694–698, 1983.

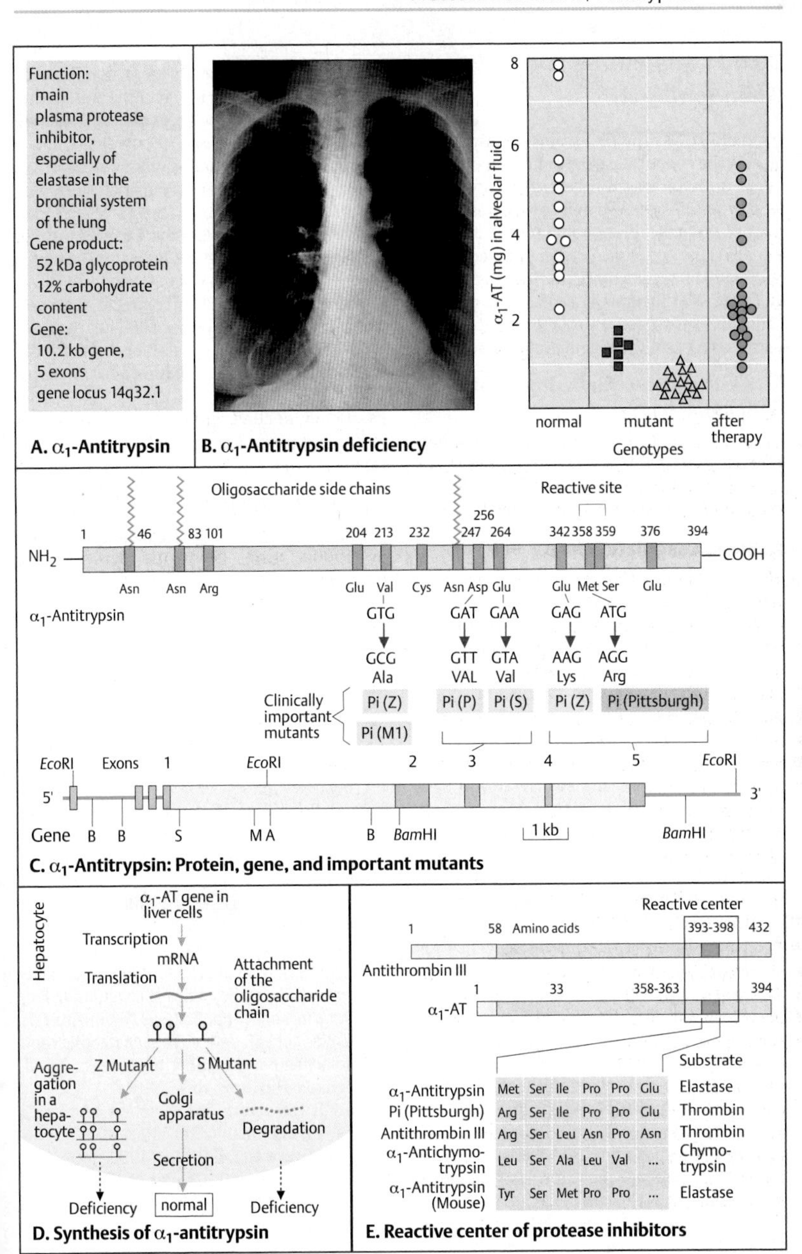


**A. $\alpha_1$-Antitrypsin**

**B. $\alpha_1$-Antitrypsin deficiency**

**C. $\alpha_1$-Antitrypsin: Protein, gene, and important mutants**

**D. Synthesis of $\alpha_1$-antitrypsin**

**E. Reactive center of protease inhibitors**

# *Blood Coagulation Factor VIII (Hemophilia A)*

Hemophilia was the first major disease recognized to be genetically determined. The Talmud refers to its increased occurrence in males in certain families, corresponding with X-chromosomal inheritance. Hemophilia A results from the deficiency of blood coagulation factor VIII; hemophilia B results from a deficiency of factor IX. Factor VIII functions as a cofactor in the activation of factor X to factor Xa during the intermediate phase of the coagulation cascade.

## A. X-chromosomal inheritance of hemophilia A

A classic example of the X-chromosomal inheritance of hemophilia A was seen in some royal families in Europe in the nineteenth century and the first half of the twentieth century.

## B. Blood coagulation factor VIII

When activated by thrombin, factor VIII protein consists of five subunits (A1, A2, A3, C1, C2) held together by calcium ions (1). The inactive factor VIII protein (2) contains three domains (A, B, C). Domain A occurs in three homologous copies (A1, A2, A3), domain C in two (C1, C2), and domain B in one copy. In humans, the gene for factor VIII (3) maps to the distal long arm of the X chromosome in region 2, band 8 (Xq28). It consists of 26 exons and spans 186000 base pairs (186 kb), corresponding to about 0.1% of the whole X chromosome. Noteworthy in this gene are the large exon 14 (3106 base pairs), which codes for the B domain, and a large intron of 32000 base pairs between exons 22 and 23. Most point mutations occur in DNA sequences involving TCGA, the recognition sequence for the restriction enzyme *TaqI*. It contains the dinucleotide CG, which is easily mutated. Since the cytosine of this dinucleotide is frequently methylated and deamination of methyl cytosine leads to C-to-T transition, mutations in CG dinucleotide regions are frequent. Mutation of TCGA to TTGA creates a stop codon (TGA), resulting in a truncated factor VIII protein. Even a stop codon at position 2307 leads to severe hemophilia, although only the last 26 amino acids are missing (Gitschier et al., 1985).
Polymorphic restriction sites (RFLPs, restriction fragment length polymorphisms) can be utilized for molecular genetic diagnosis of hemophilia A (4). When present, a variant recognition sequence (B*) for the restriction enzyme *BclI* in the region of exons 17 and 18 produces a fragment of 879 base pairs and a fragment of 286 base pairs; when it is absent, a single fragment of 1165 base pairs results. This can be used in RFLP diagnosis (5): The index patient (II-1) with hemophilia A carries the 879 bp fragment. This fragment indicates the mutation. His sister (II-2) has an affected son (III-2) who also carries the 879 bp fragment, inherited from his mother. A brother (III-1) carries the 1165-bp fragment and thus is not at risk for the disease, because this is not linked to the mutation.
In addition to point mutations, factor VIII gene rearrangements involving the long intron 22 are frequent.

## C. Severity and factor VIII activity

Hemophilia occurs with a frequency of about 1 in 10000 male newborns. Severity and frequency of bleeding are dependent on the degree of residual factor VIII activity.

## References

Dahlbäck, B.: Blood coagulation. Lancet **355**:1627–1632, 2000.
Daly, M., et al.: Coagulation disorders, pp. 209–218, In: J.L. Jameson, ed., Principles of Molecular Medicine, Humana Press, Totowa, New Jersey, 1998.
Fuentes-Prior, P., et al.: Structural basis for the anticoagulant activity of the thrombin-thrombomodulin complex. Nature **404**:518–525, 2000.
Gitschier, J. et al.: Detection and sequence of mutations in the factor VIII gene of haemophiliacs. Nature **315**:427–430, 1985.
Kazazian, H.H. Jr., et al.: Hemophilia A and parahemophilia: Deficiencies of coagulation factors VIII and V, pp. 3241–3267. In: In: C.R. Scriver, et al., eds., The Metabolic and Molecular Bases of Inherited Disease. 7th ed. McGraw-Hill, New York, 1995.
Pratt, K.P.: Structure and function of the C2 domain of human factor VIII at 1.5 Å resolution. Nature **402**:439–441, 1999.
Saxena, R, Mohanty, S., Choudhry, V.P.: Prenatal diagnosis of haemophilia. Indian. J. Pediatr. **65**:645–649, 1998.

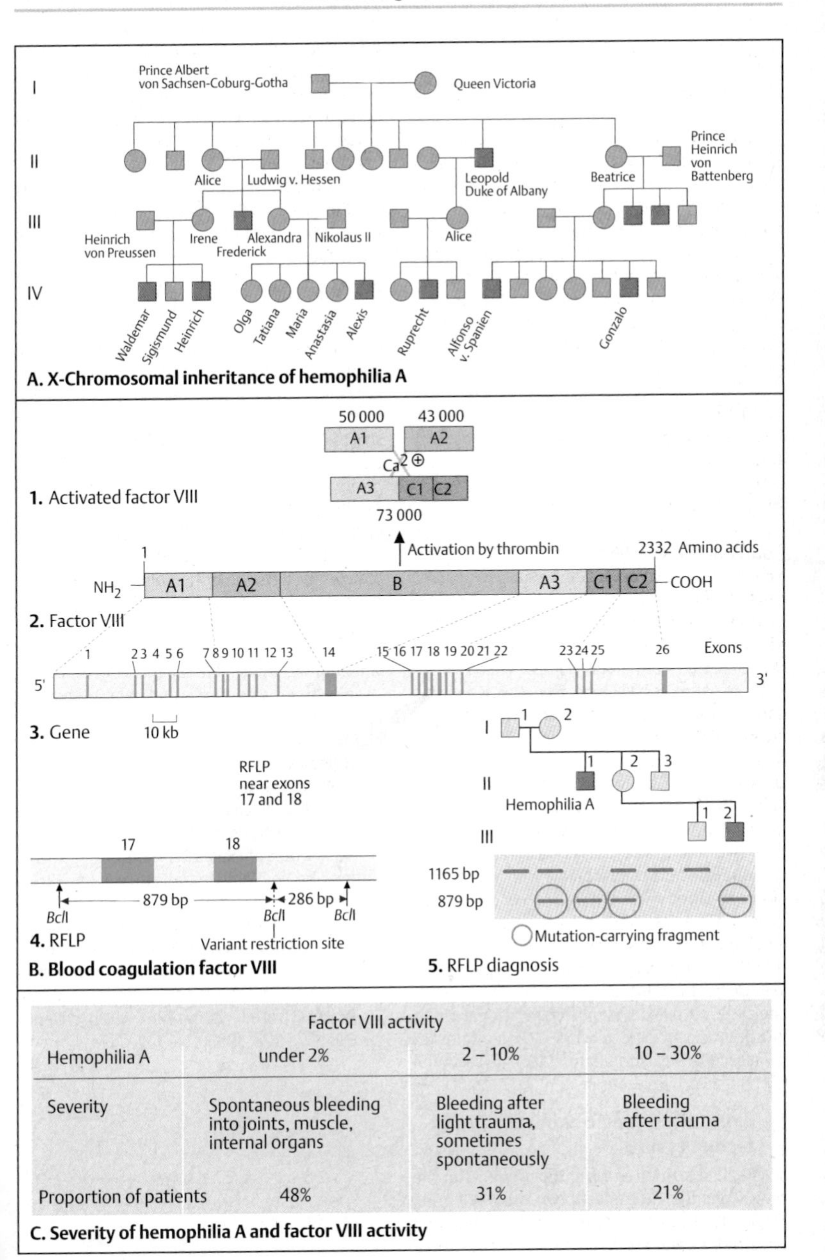


A. X-Chromosomal inheritance of hemophilia A

B. Blood coagulation factor VIII

| | Factor VIII activity | | |
|---|---|---|---|
| Hemophilia A | under 2% | 2 – 10% | 10 – 30% |
| Severity | Spontaneous bleeding into joints, muscle, internal organs | Bleeding after light trauma, sometimes spontaneously | Bleeding after trauma |
| Proportion of patients | 48% | 31% | 21% |

C. Severity of hemophilia A and factor VIII activity

# Von Willebrand Factors

Von Willebrand factor (vWF) is a complex multimeric protein found in plasma, platelets, and subendothelial connective tissue. It has two basic biological functions: it binds to specific receptors on the surface of platelets and subendothelial connective tissue, and it forms bridges between platelets and damaged regions of a vessel. Furthermore, it binds to clotting factor VIII and stabilizes it. Deficiency of vWF leads to decreased or absent platelet adhesion and to secondary deficiency of factor VIII (von Willebrand disease or von Willebrand–Jürgens syndrome). Hereditary deficiency of vWF is the most common bleeding disorder in man, with a frequency of about 1 : 250 for all forms, including the mild ones, and about 1 : 8000 for severe forms.

## A. Von Willebrand cDNA and prepropeptide

Von Willebrand factor is formed in endothelial cells, in megakaryocytes, and possibly in some other tissues and is coded for by a large (178 kb) gene with 52 exons of various sizes on chromosome 12 (12p12-pter). Several polymorphic restriction sites (red arrows) exist. The cDNA of vWF is about 8.7 kb long. The corresponding mRNA codes for a primary peptide (prepro-vWF) of 2813 amino acids, including a signal peptide of 22 amino acids, a segment of 741 amino acids (vW antigen II), and a subunit of four different domains (A–D), which together make up more than 90% of the sequence. The three A domains (A1–A3) in the middle contain binding sites for collagen, heparin, and thrombocytes. Three small B domains are on the carboxy side of the D4 domain, before the two C domains. vWF contains 8.3% cysteine (234 of 2813 amino acids), concentrated at the amino and carboxy ends, whereas the three A domains are cysteine-poor. After posttranslational modification, the mature plasma vWF contains 12 oligosaccharide side chains (19% of the total weight is carbohydrate).

## B. Biosynthesis of the von Willebrand factors (vWF)

vWF is first formed as a prepropeptide. After the signal peptide is removed, two pro-vWF units attach to each other at their carboxy ends by means of numerous disulfide bridges to form a dimer. The dimers represent the repetitive units, or protomers, of mature vWF. The pro-vWF dimers are transported to the Golgi apparatus, where the pro-vWF (vW antigen II or vWagII) is removed. Mature vWF and vWagII are stored in Weibel–Palade bodies in epithelial cells. The mature subunits and vWagII contain binding sites for factor VIII, heparin, collagen, ristocetin + platelets, and thrombin-activated platelets.

## C. Classification of von Willebrand diseases

Von Willebrand disease is a heterogeneous group of disorders divided into several subtypes. In types I and III, the defect is quantitative; in type II, qualitative. Dominant and recessive phenotypes with vWF deficiency often cannot be readily distinguished because heterozygosity may not be manifest and can only be determined by laboratory tests. Type I with subtypes A and B is the most frequent group (70% of all patients). vWF deficiency may simulate platelet dysfunction or hemophilia.
(Figures adapted from Sadler, 1995).

### References


Lillicrap, D.: Molecular diagnosis of inherited bleeding disorders and thrombophilia. Semin. Hematol. **36**:340–351, 1999.

Manusco, D.J., et al.: Structure of the gene for human von Willebrand factor. J. Biol. Chem. **264:**19514–19527, 1989.

Mohlke, K.L., Nochols, W.C., Ginsburg, D.: The molecular basis of von Willebrand disease. Int. J. Clin. Lab. Res. **29**:1–7, 1999.

Sadler, J.E.: Von Willebrand disease, pp. 3269–3287, In: C.R. Scriver, et al., eds., The Metabolic and Molecular Bases of Inherited Disease. 7 th ed. McGraw-Hill, New York, 1995.

Wise, R.J., et al.: Autosomal recessive transmission of hemophilia A due to a von Willebrand factor mutation. Hum. Genet. 91 : 367–372, 1993.

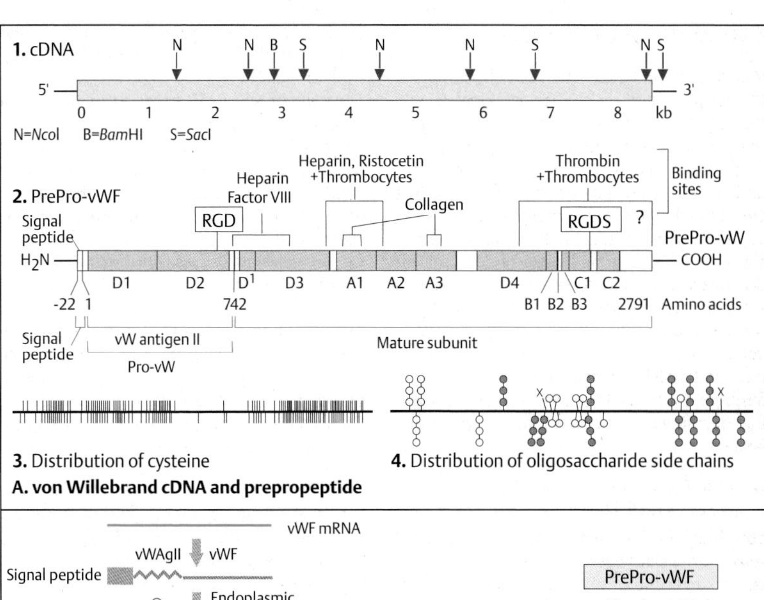


**A. von Willebrand cDNA and prepropeptide**

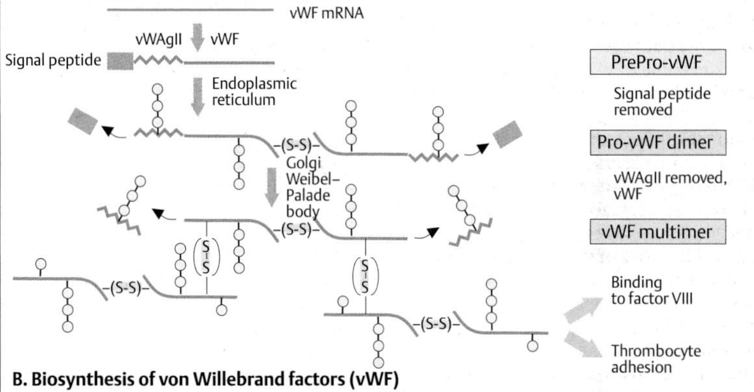


**B. Biosynthesis of von Willebrand factors (vWF)**

| von Willebrand disease | Genetics | vWF antigen | Factor VIII | Multimer structure |
|---|---|---|---|---|
| Type I | AD | decreased | decreased | normal |
| IIA | AD | decreased or normal | decreased or normal | large and intermediary absent |
| IIB | AD | decreased or normal | decreased or normal | large absent in plasma, normal thrombocytes |
| IIC | AR | decreased or normal | decreased or normal | large absent in plasma and in thrombocytes |
| IID | AD | normal | normal | large absent |
| IIE | AD | decreased | normal | large absent |
| III | AR | absent | greatly decreased | absent |

**C. Classification of von Willebrand diseases**

## Cytochrome P450 Genes

Complex chemical substances, such as drugs or plant toxins, are degraded by an oxidation system (monooxygenases) in the endoplasmic reticulum of liver cells. These enzymes (collectively referred to as cytochrome P450) absorb light maximally at 450 nm after binding to CO. Cytochrome P450 is the last enzyme in the essential electron-transporting chain in microsomes of the liver and mitochondria of the adrenal cortex. A large system of evolutionarily related genes code for the different P450 proteins in mammals.

### A. Cytochrome P450 system

The cytochrome P450 system (1) consists of oxidizing enzymes (mixed monooxygenases). They represent the first phase of detoxification: a substrate (RH) is oxidized to ROH utilizing atmospheric oxygen ($O_2$), with water ($H_2O$) being formed as a byproduct. A reductase delivers hydrogen ions ($H^+$) either from NADPH or NADH. A characteristic feature of P450 enzymes (2) is that a single chemical substrate can frequently be degraded by several P450 enzymes and that a single P450 protein can oxidize a number of structurally different chemical substances. The capacity to metabolize and detoxify a wide range of chemical substances is considerable. However, the enzyme activities of phase I and phase II must be well coordinated, since toxic intermediates with undesirable side effects occasionally arise in the initial stages of phase II.

### B. Debrisoquin metabolism

Debrisoquin is an isoquinoline-carboxamidine. It was used to treat high blood pressure until it was found to cause severe side effects in 5–10% of the population. These persons have reduced activity of a degrading enzyme, debrisoquin-4-hydroxylase. A number of other medications, including β-adrenergic blockers, antiarrhythmics, and antidepressives, are also degraded by this enzyme and may also cause untoward reactions in persons with low activity. Individuals with a slow rate of degradation show an increased ratio of debrisoquin/4-hydrodebrisoquin (1). The enzyme is coded for by the *450-db1* gene, a member of the cytochrome P450-IID family (CYP2D). Mutations may cause aberrant splicing and produce a variant pre-mRNA containing an additional intron (Gonzalez et al., 1988).

### C. CYP gene superfamily (cytochrome P450 genes)

The cytochrome P450 genes in mammals are designated CYP genes. They make up a superfamily of genes that resemble each other in exon/intron structure and that code for similar gene products. An evolutionary pedigree has been derived based on comparisons of their cDNA sequences. According to this pedigree, the CYP gene family arose during the last 1500–2000 million years. It is assumed that the CYP-2 family in particular developed in response to toxic substances in plants that had to be detoxified by animal organisms. At least 30 gene duplications and gene conversions have led to an unusually diverse repertoire of CYP genes.
(Figures adapted from Gonzales et al, 1988, and Gonzales & Nebert, 1990).

### References

Ayesh, R., et al.: Metabolic oxidation phenotypes as markers for susceptibility to lung cancer. Nature **312**:169–170, 1984.

Gonzalez, F.J., et al.: Characterization of the common genetic defect in humans deficient in debrisoquine metabolism. Nature **331**:442–446, 1988.

Gonzalez, F.J., Nebert D.W.: Evolution of the P450 gene superfamily: animal-plant “warfare”, molecular drive, and human genetic differences in drug oxidation. Trends Genet. **6**:182–186, 1990.

Nebert, D.W., Gonzalez, F.J.: P450 genes. Structure, evolution, and regulation. Ann. Rev. Biochem. **56**:955–994, 1987.

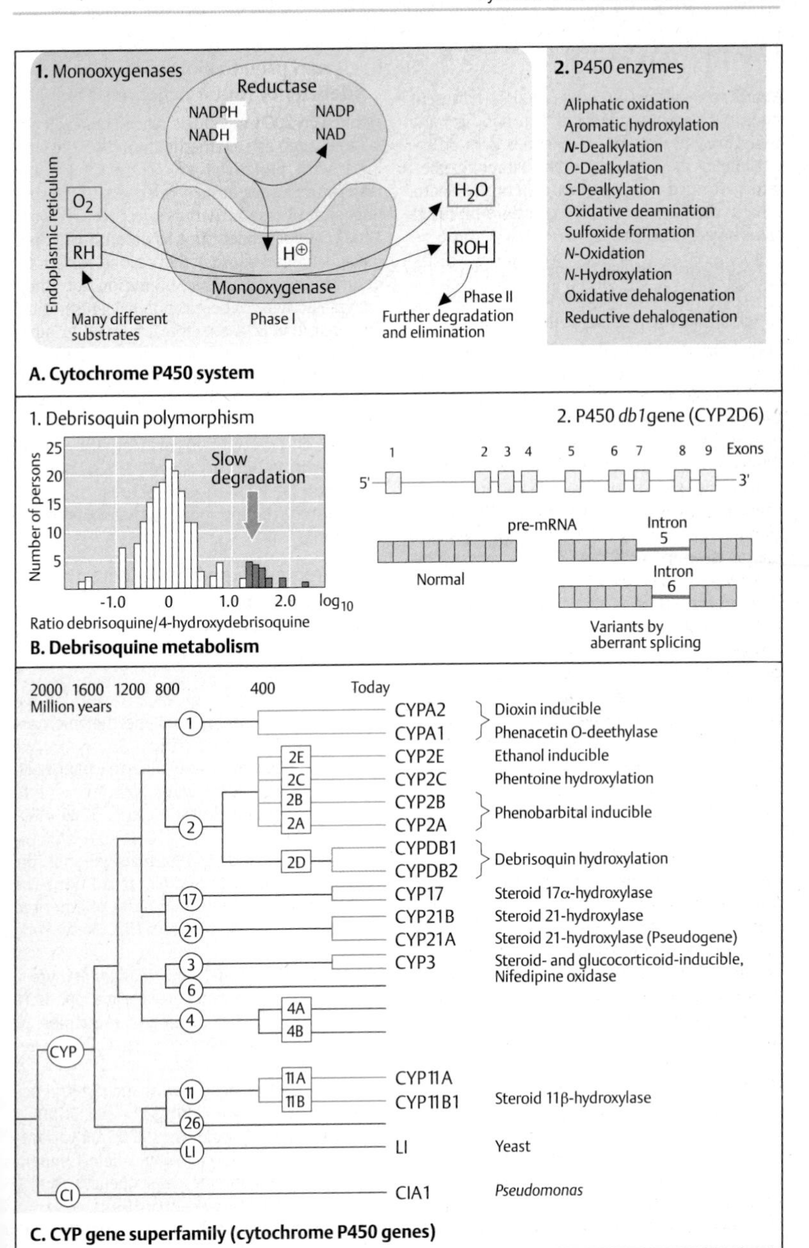


**A. Cytochrome P450 system**

**B. Debrisoquine metabolism**

**C. CYP gene superfamily (cytochrome P450 genes)**

# Pharmacogenetics

Many medications are degraded at different rates in different individuals. This has a genetic basis. Enzymes coded for by genes with different alleles may have different catabolic rates, which in turn can result in genetically determined differences in the reaction to drugs (pharmacogenetics).

## A. Malignant hyperthermia due to abnormal regulation of a calcium channel in muscle cells

Malignant hyperthermia is a severe, life-threatening complication of anesthesia that may occur in persons with extreme hypersensitivity to halothane and similar agents used in general anesthesia. Normally, a nerve impulse depolarizes the plasma membrane of a nerve ending at the nerve–muscle endplate (1) (motor endplate), and the volt-gated calcium channel in the plasma membrane of the nerve ending is temporarily opened. The massive influx of calcium into the cell (the extracellular $Ca^{2+}$ concentration is about 1000 times higher than the intracellular) triggers the release of acetylcholine. Binding of the latter to the acetylcholine receptor of the muscle cell temporarily opens the receptor-controlled cation ($Na^{+}$) channels. This opens calcium channels located in the sarcoplasmic reticulum of the muscle cell. The resulting rapid increase in $Ca^{2+}$ concentration in the cytosol causes the myofibrils in the muscle cell to contract. The calcium channels in the sarcoplasmic reticulum are regulated by a receptor (ryanodin receptor) (2). Ryanodin (an alkaloid) binds to the calcium channel. The ryanodin receptor is a protein with four transmembrane domains. Mutations in the ryanodin receptor lead to greatly increased sensitivity to halothane and other anesthetic agents (3), which cause muscle spasm, drastic elevation of temperature (hyperthermia), acidosis, and cardiac arrest (4). Malignant hyperthermia is inherited as an autosomal dominant trait (5). One gene in man lies on chromosome 19 at 19q13.1 (MacLennan and Phillips, 1992). Additional loci are on 7q, 17q, and 3q13.1 (Subrak et al., 1995). The mutant haplotype of a given family can be determined by segregation analysis. A ryanodin receptor mutation has been demonstrated in porcine malignant hyperthermia.

## B. Serum pseudocholinesterase deficiency (butyrylcholinesterase)

About 1 in 200 individuals reacts to muscle relaxants, such as suxamethonium (succinylcholine), with prolonged muscle relaxation and respiratory arrest. In such persons, serum pseudocholinesterase activity is decreased. Persons at risk cannot be identified by determining their pseudocholinesterase activity alone (1), but by determining dibucaine inhibition of their enzyme activity. Whereas homozygous normal persons show 80% enzyme activity after dibucaine administration, persons at risk show only 20%. Individuals with intermediate values of 60% are regarded as heterozygotes (2). A number of different alleles can lead to different degrees of reduced enzyme activity. (Figure after Harris, 1975.) This enzyme is now referred to as butyrylcholinesterase because it hydrolyzes butyrylcholine more readily than acetylcholine.

## C. Examples of genetically determined adverse reactions to drugs

See table opposite. (Note: Hemoglobin H involves deletions of the α loci, not a loci.)

### References

Denborough, M.: Malignant hyperthermia. Lancet **352**:1131–1136, 1998.

Evans, W.E., Relling, M.V.: Pharmacogenetics: Translating functional genomics into rational therapeutics. Science **286**:487–491, 1999.

Kalow, W., Grant, D.M.: Pharmacogenetics, pp. 293–326, In: C.R. Scriver, et al., eds., The Metabolic and Molecular Bases of Inherited Disease. 7th ed. McGraw-Hill, New York, 1995.

McLennan, D.H., Britt, B.A.: Malignant hyperthermia and central core disease, pp. 949–954. In: J.L. Jameson, ed., Principles of Molecular Medicine, Humana Press, Totowa, New Jersey, 1998.

Roses, A.D.: Pharmacogenetics and the practice of medicine. Nature **405**:857–865, 2000.

Sachse, C., et al.: Cytochrome P450 2 D6 variants in a Caucasian population: allele frequencies and phenotypic consequences. Am. J. Hum. Genet. 60:284–295, 1997.

Subrak, R., et al.: Am J Hum Genet. 56:684–691, 1995.

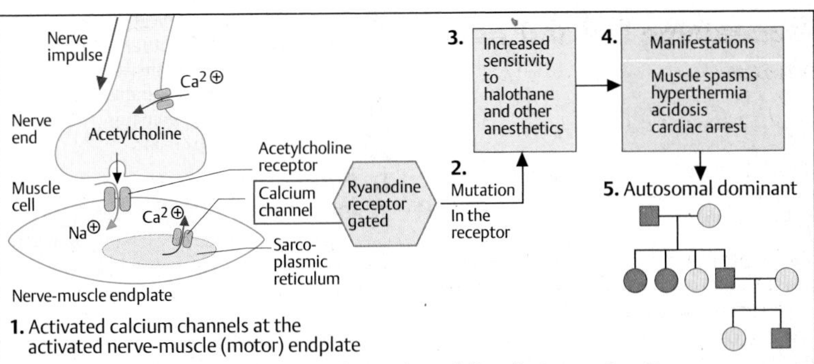


**A. Malignant hyperthermia due to a calcium channel disorder in muscle cells**

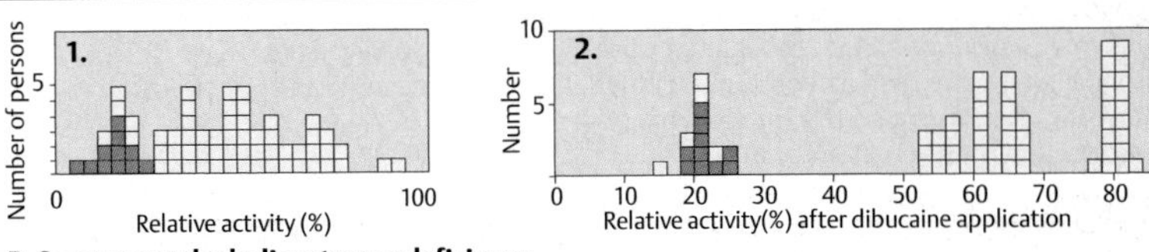


**B. Serum pseudocholinesterase deficiency**

| Defect | Relevant chemical substance | Clinical consequence | Frequency | Pathogenesis | Genetics |
|---|---|---|---|---|---|
| Coumarin resistance | Coumarin (warfarin) | Ineffective anticoagulation therapy | Rarer than 1:80 000 | Increased vitamin K affinity due to enzyme or receptor defect | Autosomal dominant |
| Increased sensitivity to isoniazid | Isoniazid, Sulfamethazine, Phenelzine, Hydralazine, etc. | Polyneuritis, lupus-like reaction | In about 50% | Decreased activity of liver Isoniazid acetylase | Autosomal recessive |
| Isoniazid ineffective | Isoniazid, Sulfamethazine, Phenelzine, Hydralazine | Reduced antituberculous effect | | Increased INH excretion | Autosomal dominant |
| Glucose-6-phosphate dehydrogenase (G6PD) deficiency | Sulfonamides, Antimalarial drugs, Nitrofurantoin, Vicia faba | Hemolysis | Rare in Europeans, frequent in Africa | G6PD deficiency in erythrocytes | X-chromosomal (many mutant forms) |
| Hemoglobin Zürich | Sulfonamides | Hemolysis | Rare | Unstable hemoglobin due to point mutation in β-globin (arginine instead of histidine in position 63) | Autosomal dominant |
| Hemoglobin H | Sulfonamides | Hemolysis | Rare | Unstable hemoglobin of 4 β chains due to deletion of the a loci | Autosomal dominant |
| Glaucoma in adults (some forms) | Corticoids | Glaucoma | Frequent | Unknown | Possibly autosomal dominant |

**C. Examples of genetically determined reactions to pharmaceuticals**

# Maintaining Cell and Tissue Shape

## Cytoskeletal Proteins in Erythrocytes

Eukaryotic cells contain supportive proteins that confer stability while allowing flexibility (cytoskeletal proteins). The membrane skeleton is a network of structural proteins underlying the plasma membrane and partly associated with it. Erythrocytes must meet extreme requirements: about a half million times during a 4-month lifespan, they traverse small capillaries with diameters less than that of the erythrocytes themselves. Membrane flexibility is also essential for muscle cell function. Thus, it is not surprising that the cytoskeletal proteins of erythrocytes and muscle cells are similar.

### A. Erythrocytes

A normal erythrocyte is maintained in a characteristic biconcave discoid form by the cytoskeletal proteins. Genetic defects in different cytoskeletal proteins lead to characteristic erythrocyte deformations: as ellipses (elliptocytes), as spheres (spherocytes), or as cells with a mouthlike area (stomatocytes) or thornlike projections (acanthocytes). The various forms are the result of defects of different proteins. (Scanning electron micrographs from Davies and Lux, 1989.)

### B. Skeletal proteins in erythrocytes

SDS polyacrylamide gel electrophoresis differentiates numerous membrane-associated erythrocyte proteins. Each band of the gel is numbered, and the individual proteins are assigned to them. The main proteins include α- and β-spectrin, ankyrin, an anion-channel protein (band-3 protein), proteins 4.1 and 4.2, actin, and others. The chromosomal localization of their genes and associated diseases due to mutations are known for man and mouse.

### C. α- and β-Spectrin

The main component of cytoskeletal proteins is spectrin, a long protein composed of a 260 kDa α chain and a 225 kDa β chain. The chains consist of 20 (α chain) and 18 (β chain) subunits, each with 106 amino acids. Each subunit is composed of three α-helical protein strands running counter to one another. Subunit 10 and subunit 20 of the α chain consist of five, instead of three, parallel chains. The individual subunits are assigned to different domains (I–V in the α chain and I–IV in the β chain).

### D. Proteins of the erythrocyte membrane

The rod-shaped spectrin proteins, which run parallel to the erythrocyte plasma membrane, are attached to the anion channels by ankyrin and to the glycophorin molecules by protein 4.1. The anion channels in erythrocytes are important for $CO_2$ transport. Glycophorins (A, B, C) are transmembrane proteins with several carbohydrate units. Actin is the main protein for muscle contraction and cell flexibility.
(Figures after Luna and Hitt, 1992).

### References


Becker, P.S., Lux, S.E.: Hereditary spherocytosis and hereditary elliptocytosis, pp. 3513–3560. In: C.R. Scriver, et al., eds., The Metabolic and Molecular Bases of Inherited Disease. 7th ed. McGraw-Hill, New York, 1995.

Davies, K.A., Lux, S.E.: Hereditary disorders of the red cell membrane skeleton. Trends Genet. **5**:222–227, 1989.

Delaunay, J.: Disorders of the red cell membrane, pp.191–196, In: J.L. Jameson, ed., Principles of Molecular Medicine. Humana Press, Totowa, New Jersey, 1998.

Luna, E.J., Hit, A.L.: Cytoskeleton plasma membrane interactions. Science 258:955–964, 1992.

Tse, W.T., Lux, S.E.: Red blood cell membrane disorders. Br. J. Haematol. **104**:2–13, 1999.

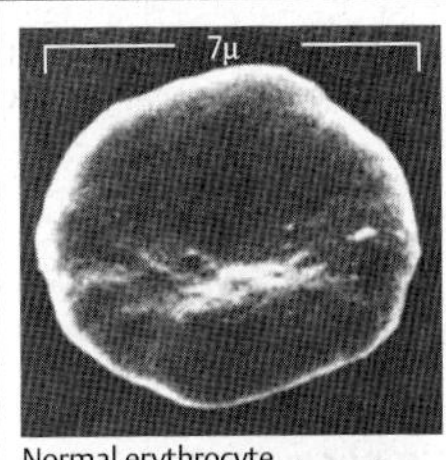

Normal erythrocyte

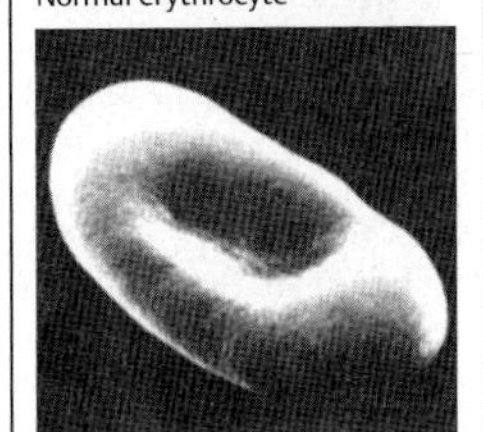
Elliptocyte

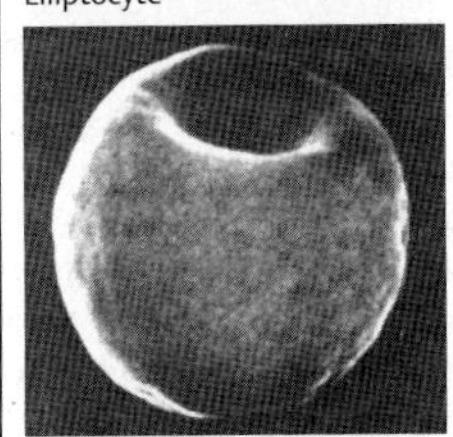
Spherocyte

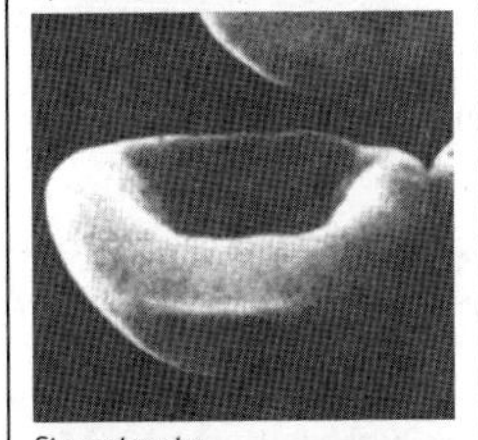
Stomatocyte

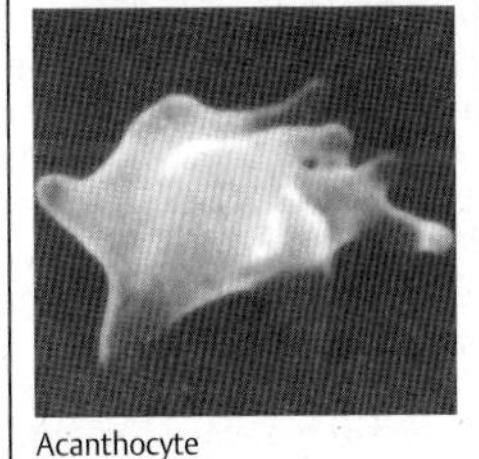
Acanthocyte

**A. Erythrocytes**

| Band | SDS gel | Protein | Chromosomal localization | Disease |
|---|---|---|---|---|
| 1 | | α-Spectrin | 1q22 – 25 | Elliptocytosis-1 |
| 2 | | β-Spectrin | 14q23 – 24 | Spherocytosis-2 |
| | | Ankyrin | 8p11 – 21 | Spherocytosis-1 |
| 3 | | Anion channel | 17 | Acanthocytosis |
| 4.1 | | Protein 4.1 | 1q22 – 25 | Elliptocytosis-2 |
| 4.2 | | Protein 4.2 | | |
| 5 | | Actin | 7pter – q22 | |
| 6 | | Glycerol-aldehyde-3-P dehydrogenase | 12p13 | |
| 7 | | Tropomyosin (non-muscle) | 1q31 – 41 | Stomatocytosis |

**B. Erythrocyte skeleton proteins**

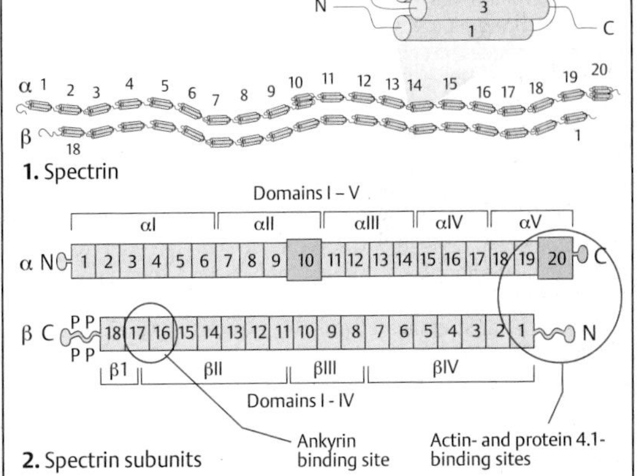


**C. α- and β-spectrin**

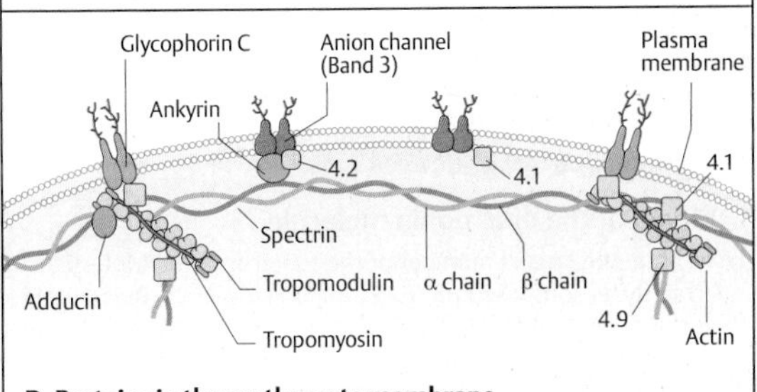


**D. Proteins in the erythrocyte membrane**

## Hereditary Muscle Diseases

Spontaneous degeneration of muscle fibers and death of muscle cells (muscular dystrophy) is a common cause of muscle disease in infants, children, and adults. Muscular dystrophies are genetically heterogeneous and clinically variable. About 50 different forms are listed in McKusicks's catalogue Mendelian Inheritance of Man Online (OMIM at *http://www3.ncbi.nih.gov/Omim/*).

### A. The dystrophin–glycan complex

A complex system of interconnected noncovalently bound proteins in the sarcolemma (plasma membrane) of muscle cells lends the cell stability under the extreme exertion of contraction and relaxation. They connect the extracellular matrix and the intracellular myofibrils, elongated protein molecules aligned in parallel chains (myofilaments). The largest of the interconnected proteins, α-dystroglycan (156 kDa), is located outside the cell. It is connected to the extracellular matrix by a heterotrimeric protein, laminin-2. β-Dystroglycan (43 kDa) is embedded in the sarcolemma and connected to a series of other cytoskeletal proteins, which are divided into the sarcoglycan and syntrophin subcomplexes. Several members of the sarcoglycan complex are related to specific types of muscular dystrophies due to mutations in the corresponding genes.

Dystrophin, a large, elongated protein, provides a bridge between the intracellular cytoskeleton involved in the contractile myofilaments and the extracellular matrix. Two dystrophin molecules connect neighboring dystrophin–glycan complexes. The N-terminal end of dystrophin is connected to the thin myofilament F-actin (filamentous actin). The C-terminal end of dystrophin is connected to β-dystroglycan and the syntrophins. (This highly schematic illustration is adapted from Pasternak, 1998; Duggan et al., 1997; and Nova-Castra Immunohistochemistry Laboratories, Newcastle upon Tyne, UK.).

### B. Model of the dystrophin molecule

Dystrophin, the largest member of the spectrin superfamily, is composed of 3685 amino acids (molecular mass 427 kDa) which form four functional domains: (1) the N-terminal actin-binding domain of 336 amino acids; (2) 24 long repeating units, each consisting of 88- to 126-amino-acid triple-helix segments as in spectrin; (3) a 135-amino-acid cysteine-rich domain, which binds to the sarcolemma proteins; and (4) the C-terminal domain of 320 amino acids with binding sites to syntrophin and dystrobevin. The triple helix segments form the central rod domain, which is 100–125 nm long. (Diagram adapted from Koenig et al., 1988).

### C. The dystrophin gene

The human *dystrophin* gene *(DMD)* is located on the short arm of the X chromosome in region 2, band 1.1 (Xp21.1). *Dystrophin* is by far the largest known gene in man, spanning 2.4 million base pairs (2.4 Mb or 2400 kb) in 79 exons (2). The large DMD transcript has 14 kb. The dystrophin gene contains at least seven intragenic promoters. The primary transcript is alternatively spliced into a variety of different mRNAs that encode smaller proteins expressed in other tissues than muscle cells, especially in the central nervous system.

### D. Distributions of deletions in the dystrophin gene

The frequent deletions in the *DMD* gene (60% of patients) are unevenly distributed. Most frequently involved are exons 43–55 and exons 1–15, roughly corresponding to the F-actin binding site and the dystroglycan-binding site. Duplications (in 6% of patients) and point mutations also occur. (Figure kindly provided by Professor C. R. Müller-Reible, Würzburg).

### References


Ahu, A.W., Kunkel, L.M.: The structural and functional diversity of dystrophin. Nature Genet. **3**:283–291, 1993.

Duggan, D.J., et al.: Mutations in the sarcoglycan genes in patients with myopathy. New Eng. J. Med. **336**:618–624, 1997.

Koenig, M., Monaco, A.P., Kunkel, L.M.: The complete sequence of dystrophin predicts a rod-shaped cytoskeletal protein. Cell **53**:219–228, 1988.

Pasternak, J.J.: An Introduction to Human Molecular Genetics. Fitzgerald Science Press, Bethesda, Maryland, 1998.

Worton, R.: Muscular dystrophies: diseases of the dystrophin-glycoprotein complex. Science **270**:755–756, 1995.

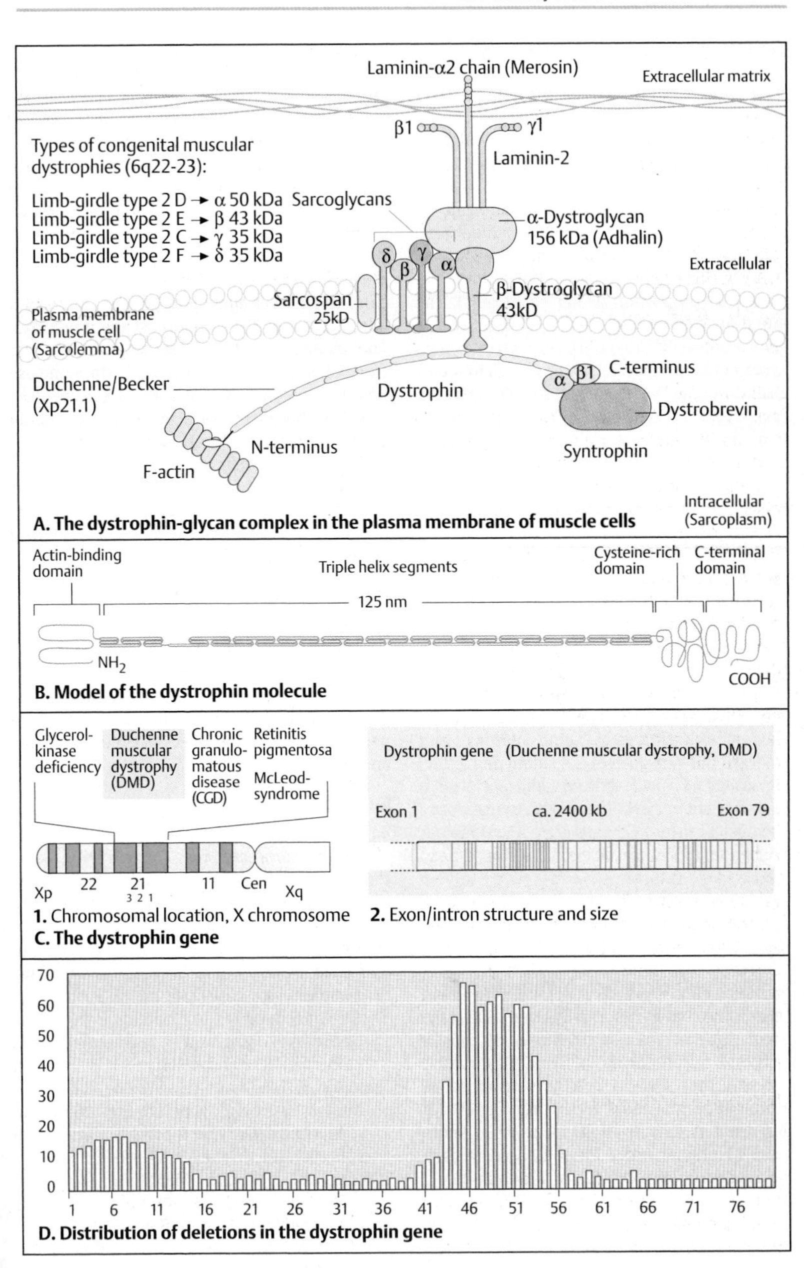


**A. The dystrophin-glycan complex in the plasma membrane of muscle cells**

**B. Model of the dystrophin molecule**

**1.** Chromosomal location, X chromosome **2.** Exon/intron structure and size

**C. The dystrophin gene**

**D. Distribution of deletions in the dystrophin gene**

# Duchenne Muscular Dystrophy

Duchenne muscular dystrophy (DMD, McKusick 310200) is the most common of the more than 10 clinically and genetically distinct muscular dystrophies. It is caused by mutations in the *DMD* gene. It occurs in 1 of 3500 live born males either by a new mutation or by transmission of the mutation from a heterozygous mother or a mother with germline mosaicism (i.e., the mother carries a *DMD* mutation in a variable proportion of her germ cells). The mutation rate is high, probably because the gene is unusually large and has a high rate (10%) of recombination within it. The French neurologist Guillaume Duchenne (1806 – 1875) was the first to report this disease, in 1861.

A clinically milder variant, Becker muscular dystrophy (BMD), is an allelic disorder due to in-frame mutations in the same gene that allow residual function of the dystrophin protein.

## A. Clinical signs

DMD is the most distinctive progressive proximal muscular dystrophy. The age of onset is usually less than 3 years and signs are evident at 4 – 5 years; the patient requires a wheelchair by 12 years and usually succumbs to the disorder by age 20 years. Progressive muscular weakness of the hips, thighs, and back cause difficulties in walking and using steps. Lumbar lordosis and enlarged but weak calves (pseudohypertrophy) are visible (1). The affected child performs a characteristic series of maneuvers to rise from a kneeling position (Gower's sign, 2). (Drawings Duchenne, 1861, and Gowers, 1879, from Emery, 1987).

In the Becker type, the age of onset is later and the disease progresses much more slowly than in DMD.

## B. Dystrophin analysis in muscle cells

Dystrophin, which normally lies along the plasma membrane (sarcolemma) of muscle cells (1), is absent in patients (2). Female heterozygotes show a patchy distribution of groups of normal and defective muscle cells (3) as a result of X inactivation (see p. 228). (Photographs kindly provided by Dr. R. Gold, Department of Neurology, University of Würzburg).

## C. Investigation of a family with DMD

Deletions occur in certain regions in 55% of cases and duplications in 5%. However, point mutations (in 40%) are not always detectable. In this situation, indirect DNA analysis can be performed. The panel shows a simplified example of a two-allele system (marker DXS7). (Data kindly provided by Dr. C. R. Müller-Reible, Institute of Human Genetics, University of Würzburg). Since the affected patients III-1 and III-2 carry the marker allele 1 at the DMD locus, allele 1 indicates the presence of the mutation. The unaffected male II-4 confirms that allele 2 does not represent the mutation. The females I-2, II-1, and II-2 are obligate heterozygotes (2 – 1). In this example the males III-3 and III-4 are not affected. This can be explained by recombination in their mother, II-5. In current practice one uses a set of several linked markers flanking the disease locus to avoid an erroneous diagnosis due to recombination. Female heterozygotes show mild clinical signs in 2 – 3%. About 23% of mothers of isolated patients are noncarriers.

## D. Other forms of muscular dystrophy

Several other forms of genetically determined muscular dystrophy are known in man. Course, diagnosis, and molecular genetic analysis depend on the basic disorder. Selected examples are listed.

## References

Emery, A.E.H.: Duchenne Muscular Dystrophy. Oxford University Press, Oxford, 1987.

Hoffman, E.P.: Muscular dystrophies, pp. 859 – 868. In: J.G. Jameson, ed., Principles of Molecular Medicine. Humana Press, Totowa, N.J., 1998.

Pasternak, J.J.: Molecular genetics of muscular disorders, pp. 229 – 256, In: J.J. Pasternak, An Introduction to Human Molecular Genetics. Fitzgerald Science Press, Bethesda, Maryland, 1998.

Tennyson, C.N., Klamut, H.J., Worton, R.G.: The human dystrophin gene requires 16 hours to be transcribed and is cotranscriptionally spliced. Nature Genet. **9**:184 – 190, 1995.

Worton, R.G., Brooke, M.H.: The X-linked muscular dystrophies, pp. 4195 – 4226. In: C.R. Scriver, et al., eds., The Metabolic and Molecular Bases of Inherited Disease. 7th ed., McGraw-Hill, New York, 1995.

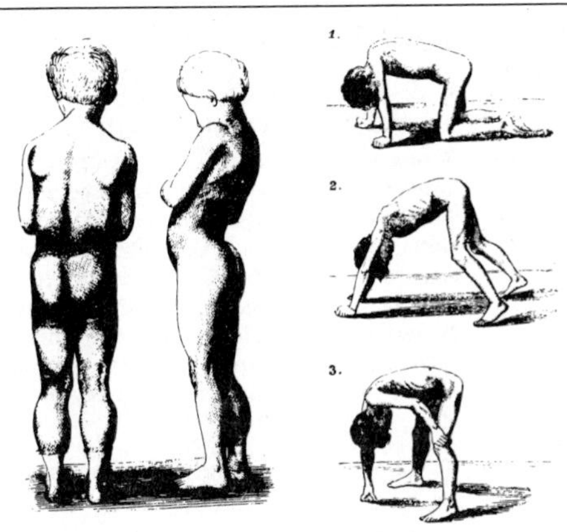


1. Calf hypertrophy and lordosis

2. Difficulty in rising (Gower's sign)

**A. Clinical signs of Duchenne muscular dystrophy**

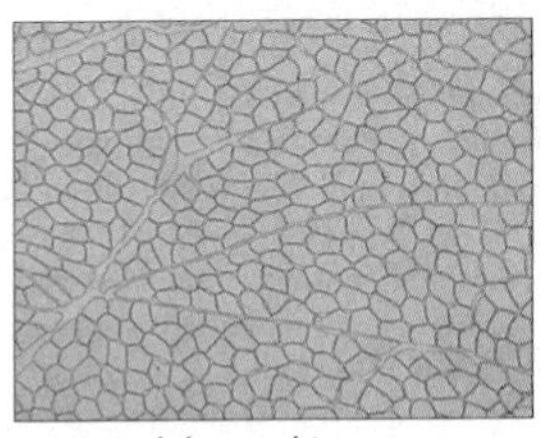

**1.** Normal dystrophin

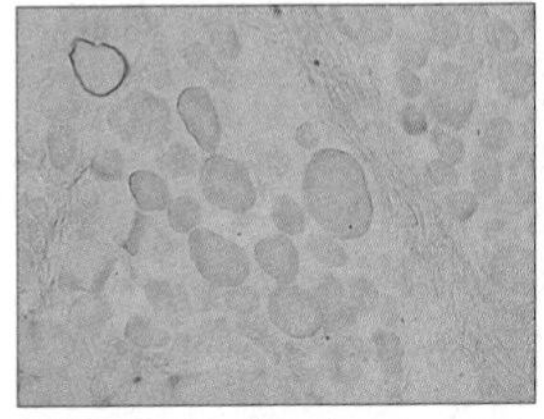

**2.** Dystrophin absent

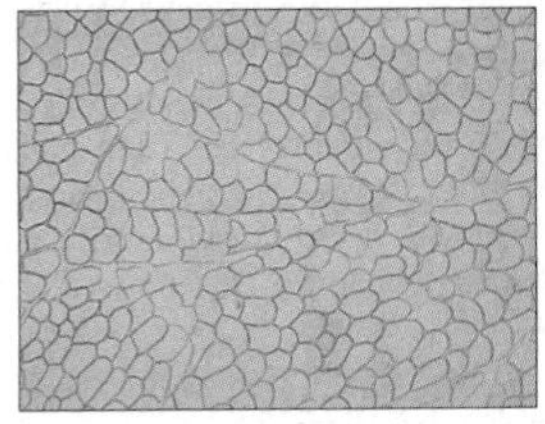

**3.** Areas lacking dystrophin in heterozygotes

**B. Dystrophin analysis in muscle cells**

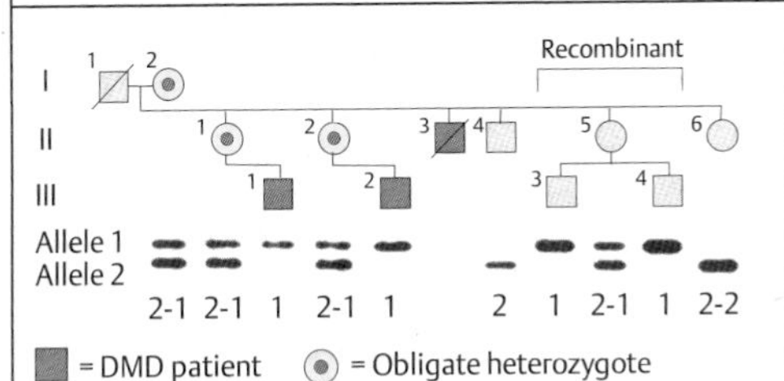


■ = DMD patient ⊙ = Obligate heterozygote

**C. Investigation of a family with DMD by DNA markers**

| Disease | Chromosomal location | McKusick No. |
|---|---|---|
| X-chromosomal: | | |
| Muscular dystrophy Duchenne | Xp21.2 | 310200 |
| Muscular dystrophy Becker (allelic with DMD) | Xp21.2 | 310200 |
| Muscular dystrophy Emery–Dreifuss | Xq28 | 310300 |
| Autosomal dominant: | | |
| Myotonic dystrophy | 19q13 | 160900 |
| Facioscapulo-humeral dystrophy | 4q35 – qter | 158900 |
| Oculo-pharyngeal muscular dystrophy | Unknown | 164300 |
| Autosomal recessive: | | |
| Duchenne-like muscular dystrophy | 13q12 – 13 | 253700 |
| Congenital muscular dystrophy-type Fukuyama | 9q31 – 33 | 253800 |
| Limb-girdle muscular dystrophy (several types) | 15q15 – q22, other loci | 253600 |

**D. Important forms of hereditary muscular dystrophy in man**

# Collagen Molecules

Collagen, the most abundant protein in mammals, constitutes about one-quarter of the total body protein. It occurs in skin, bones, tendons, cartilage, blood vessels, teeth, basement membranes, the corneas and vitreous bodies, and supporting tissues of the internal organs. Collagen forms interlinked, insoluble threads (fibrils) of unusual strength. A fiber of 1 mm diameter can hold a weight of almost 10 kg. More than ten distinct human diseases are caused by mutations in one of the genes encoding collagen. Collagen genes form a multigene family with more than 28 members, their genes being located on 12 different chromosomes.

## A. Collagen structure

The amino acid sequence of collagen is simple and periodic (1). Every third amino acid is glycine (Gly). Other amino acids alternate between the glycines. The general structural motif is $(Gly-X-Y)_n$. X is either proline or hydroxyproline; Y is either lysine or hydroxylysine (2). Three chains of collagen form a triple helix (3). In collagen type I, the helix is composed of two identical α1 chains and an α2 chain. It is first formed as a precursor molecule, procollagen (4).
Procollagen peptidases remove peptides at the N-terminal and C-terminal ends to form tropocollagen (5). Tropocollagen molecules are connected by the numerous hydroxylated proline and lysine residues to form collagen fibrils (6). Each fibril consists of staggered, parallel rows of end-to-end-tropocollagen molecules, separated by gaps (7). Collagen fibrils are visible as transverse stripes under the electron microscope (8). (Photograph from Stryer, 1995.)

## B. Prototype of a gene (*COL2 A1*) for procollagen type II (α1[II])

A procollagen molecule is encoded by a gene consisting of 52 exons. The translated part of exon 1 (85 base pairs) codes for a signal peptide necessary for secretion. Exon sizes differ, with one exon coding for 5, 6, 11, 12, or 18 periodic Gly–X–Y units. The genes for procollagen types I, II, and III differ in that some exons are fused, but otherwise they are similar, especially for the three main fibrillar collagen types (I, II, III).

## C. Gene structure and procollagen type α1(I)

The 52 exons of the *COL1A1* gene correspond to the different domains (A to G) of procollagen α1(I). The *COL1A2* gene for procollagen α2(I) is about twice as large (~40 kb) as the *COL1A1* gene because the introns between the exons are on average twice as long as in *COL1A1*.

## References

Byers, P.H.: Disorders of collagen synthesis and structure, pp. 4029–4077. In: C.R. Scriver, et al., eds., The Metabolic and Molecular Bases of Inherited Disease. 7th ed. McGraw-Hill, New York, 1995.

De Paepe, A.: Heritable collagen disorders: from phenotype to genotype. Verh. K. Acad. Geneeskd. Belg. **65**:463–482, 1998.

Chu, M.-L., Prockop, D.J.: Collagen gene structure, pp. 149–165. In: P.M. Broyce, B. Steinmann, eds., Connective Tissue and Its Heritable Disorders. Wiley-Liss, New York, 1993.

Examples of different types of collagen (four of twenty-three)

| Collagen Type | Gene | Localization | Chain | Molecule | Tissue distribution |
|---|---|---|---|---|---|
| I | *COLIA1*<br>*COLIA2* | 17q21–22<br>7q21–22 | α1(I) | $[\alpha1(I)]_2\ \alpha2(I)$ | Skin, tendons, bone, arteries |
| II | *COL2 A1* | 12q13–14 | α1(II) | $[\alpha1(II)]_3$ | Cartilage, vitreous body |
| III | *COL3 A1* | 2q31–32 | α1(III) | $[\alpha1(III)]_3$ | Skin, arteries, uterus |
| IV | *COL4 A1*<br>*COL4 A2*<br>(6 α chains) | 13q33–34 | α1(IV)<br>α2(IV) | $[\alpha1(IV)]_2\ \alpha2(IV)$<br>$[\alpha1(IV)]_3$<br>$[\alpha2(IV)]_3$ | Basement membranes |

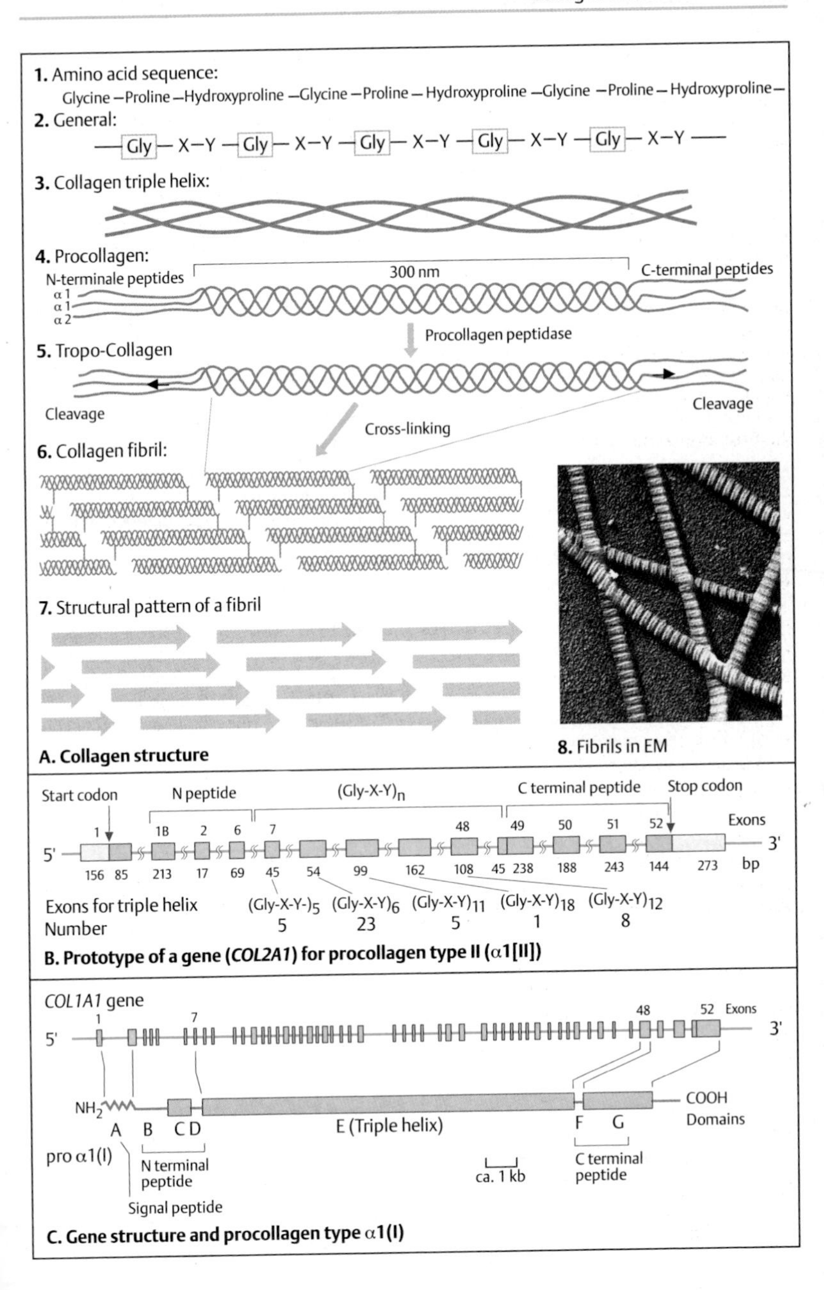


**A. Collagen structure**

**B. Prototype of a gene (*COL2A1*) for procollagen type II (α1[II])**

**C. Gene structure and procollagen type α1(I)**

## Osteogenesis Imperfecta

Osteogenesis imperfecta (OI) ("brittle bone disease") is a heterogeneous group of clinically and genetically different types of diseases with a total frequency of at least 1 in 10 000 individuals. Spontaneously occurring bone fractures, bone deformity, small stature, defective dentition (dentinogenesis imperfecta), hearing impairment due to faulty formation of the auditory ossicles, and blue sclerae (the fully developed conjunctiva of the eye is thinner than normal, so that refracted light is shifted toward blue) occur in the various forms to different extents and with different grades of severity, depending on the type of mutation. OI has been demonstrated in a seventh-century Egyptian mummy (cited from Byers, 1993).

### A. Molecular mechanisms in osteogenesis imperfecta

Some types of mutation may lead to reduced production of proα1(I) (1 and 2), e.g., deletion of a *COL1A1* allele, a transcription or splicing defect, or faulty formation of collagen fibrils. The relative excess of proα2(I) molecules becomes degraded. Thus, less procollagen than normal is formed, but it is not defective. Numerous other types of mutations can lead to defective procollagen (3). Mutations in the proα1(I) gene are more severe than mutations in the proα2(I) gene because a greater amount of defective collagen is formed.
(Figure after Wenstrup et al., 1990).

### B. Mutations and phenotype

The location of a mutation in the gene influences the phenotype. Generally, mutations in the 3′ region are more serious than mutations in the 5′ region (position effect). Mutations of the proα1(I) chain are more severe than those in the proα2(I) chain (chain effect). The substitution of a larger amino acid for glycine, which is indispensable for the formation of the triple helix, leads to severe disorders (size effect). Different types of mutations may occur, such as deletions, mutations in the promoter or enhancer, and splicing mutations. The codons (AAG, AAA) for the amino acid lysine, which occurs frequently in collagen, are readily transformed into a stop codon by substitution of the first adenine by a thymine (TAG or TAA), so that a short, unstable procollagen is formed. Splicing mutations may lead to the loss of exons (exon skipping). (Figure from Byers, 1990).

### C. Different forms of osteogenesis imperfecta

Osteogenesis imperfecta may be classified according to severity into four basic phenotypes (Sillence classification). Although the classification does not correspond to the types of mutation, it has in general proved clinically useful. OI types I and IV are less severe than type II (lethal in infancy) and type III. Three radiographs show a relatively mild (but for the patient nevertheless very disabling) deformity of the tibia and fibula in OI type IV (1); severe deformities in the tibia and fibula in OI type III (2); and the distinctly thickened and shortened long bones in the lethal OI type II (3). Mutations in OI are autosomal dominant, the severe forms being due to *de novo* mutations. Germline mosaicism has been shown to account for rare instances of affected siblings being born to unaffected parents.

### References

Byers, P.H.: Brittle bones—fragile molecules: disorders of collagen gene structure and expression. Trends Genet. **6**:293–300, 1990.

Byers, P.H.: Osteogenesis imperfecta, pp. 137–350, In: P.M. Broyce, B. Steinmann, eds., Connective Tissue and Its Heritable Disorders. Wiley-Liss, New York, 1993.

Byers, P.H.: Disorders of collagen synthesis and structure, pp. 4029–4077. In: C.R. Scriver, et al., eds., The Metabolic and Molecular Bases of Inherited Disease. 7th ed. McGraw-Hill, New York, 1995.

Chu, M.-L., Prockop, D.J.: Collagen gene structure, pp. 149–165. In: Broyce, P.M., Steinmann, B., eds., Connective Tissue and Its Heritable Disorders. Wiley-Liss, New York, 1993.

Kocher, M.S., Shapiro, F.: Osteogenesis imperfecta. J. Am. Acad. Orthop. Surg. **6**:225–236, 1998.

Wenstrup, R.J., et al.: Distinct biochemical phenotypes predict clinical severity in nonlethal variants of osteogenesis imperfecta. Am. J. Hum. Genet. **46**:975–982, 1990.

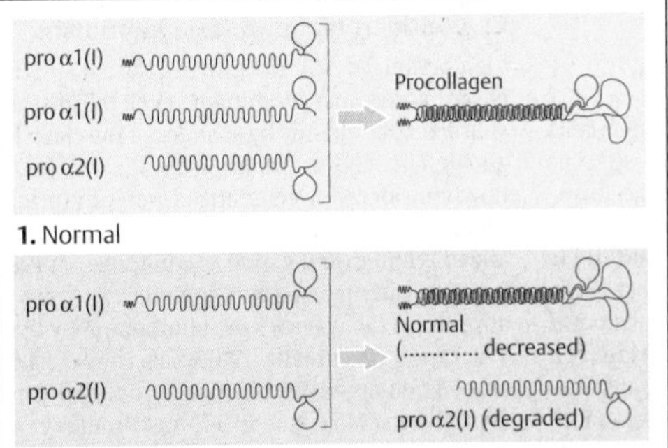


**1.** Normal

**2.** Decreased synthesis of procollagen α1(I)

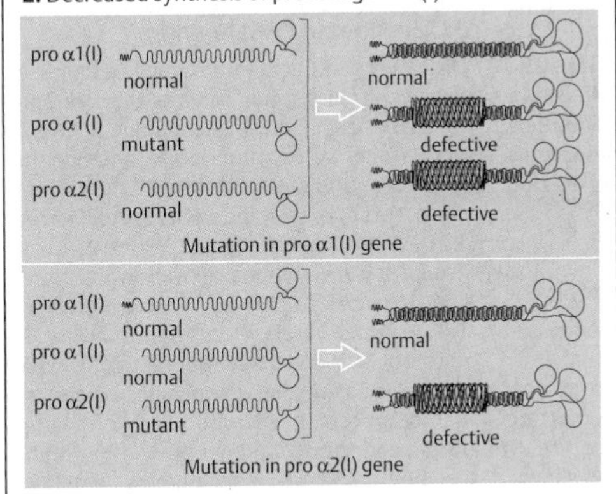


**3.** Defective procollagen due to a mutation

**A. Molecular mechanisms in osteogenesis**

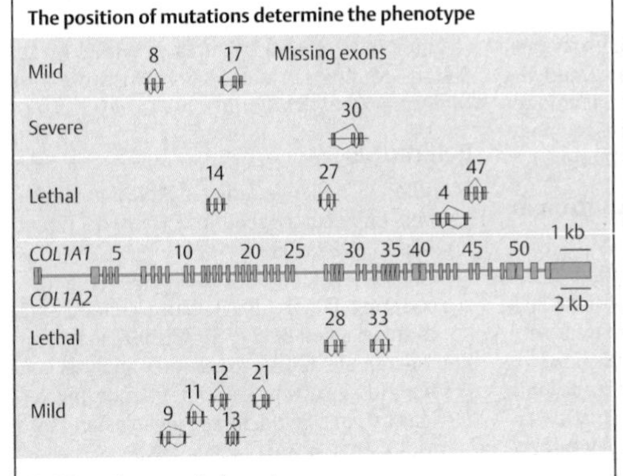


**B. Mutations and phenotype**

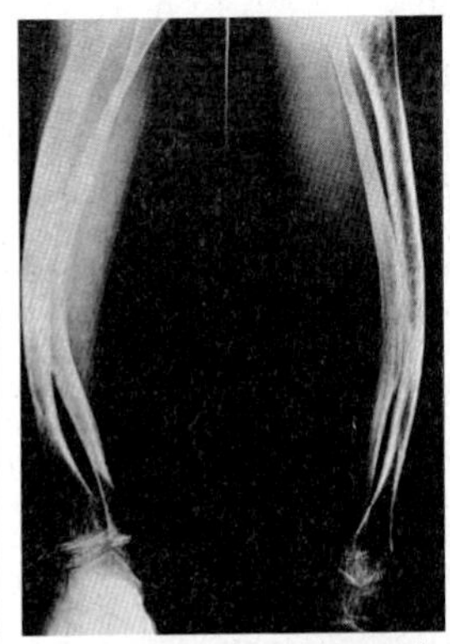

**1.** Bone deformation (OI type IV)

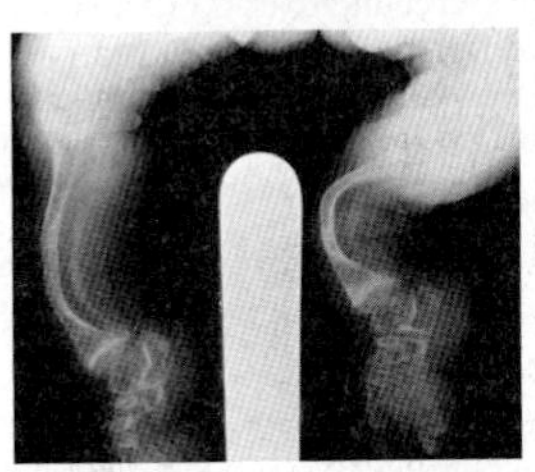

**2.** Severe deformation (OI type III)

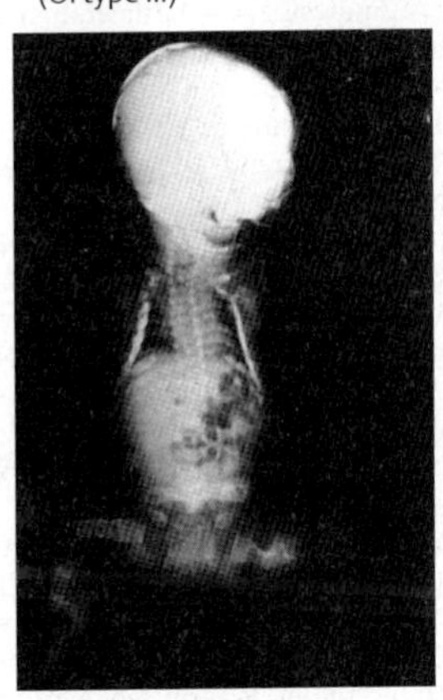

**3.** Lethal form (OI type II)

**C. Different forms of osteogenesis imperfecta**

# Molecular Basis of Bone Development

The skeleton develops from mesodermal cells committed to differentiate into three specialized cell types: chondrocytes (cartilage-forming cells), obsteoblasts (bone-forming cells), osteoclasts (bone-degrading cells), and their precursor cells. Osteoblasts produce most of the proteins for the extracellular bone matrix and control its mineralization. The osteoblast cell lineage involves osteoblast-specific transcription factors (OSFs). One such transcription factor was identified in 1997 as a major regulator of osteoblast differentiation, the core-binding factor Cbfal.

The mouse Cbfal transcription factor (the human counterpart is referred to as CBFA1) is a member of the *runt*-domain family. The *runt*-domain is a DNA-binding domain homologous to that produced by the *Drosophila* pair-rule gene *runt*.

## A. Effects of homozygous *Cbfa1* mutations on the mouse skeleton

Targeted disruption of the *Cbfa1* gene in the homozygous state (–/–) results in a severe phenotype. Homozygous mutant mice are small and die from respiratory failure at birth. In contrast to normal mice (+/+, 1), the mutant mice (–/–, 2) completely lack bone development, as shown by lack of alizarin red staining. Just before birth, normal +/+ mouse embryos (3) show well-developed bones (stained red) in the upper extremities, including the clavicles (arrow) and the tuberositas humeri (circle). Heterozygous mice (+/–, 4) show severe hypoplasia and reduced ossification of the long bones. Homozygous mutant mice (–/–, 5) totally lack any bone formation, as shown by lack of red staining.

## B. Mice heterozygous for a mutation in the *Cbfa1* gene

Mice heterozygous (+/– ) for a mutation in the *Cbfa1* gene on mouse chromosome 17 show lack of ossification of the skull (2) compared with normal mice (1). Normal calcified bone is stained red by alizarin red, here at embryonic day 17.5, three and a half days before birth. Cartilage is stained blue by alcian blue. Heterozygous mice lack a clavicle (4, arrows) in contrast to normal mice (3).

## C. Cleidocranial dysplasia in humans

Cleidocranial dysplasia (CCD, McKusick 118980) is an autosomal dominant skeletal disease characterized mainly by absence of the clavicles and deficient bone formation of the skull. CCD is now considered a generalized bone dysplasia. Radiological findings are consistent with generalized underossification (Mundlos, 1999). Patients can oppose their shoulders (1) due to absence of the clavicles (2; photograph by Dr. J. Warkany, Cincinnati). The calvarium (skull case) is enlarged with a poorly ossified midfrontal area (3). (Figure in 3 from Mundlos et al., 1997).

## D. The human *CBFA1* gene

The *CBFA1* gene, at chromosome location 6p21, encodes a transcription factor of the core-binding factor (CBF) family. The human *CBFA1* gene has two alternative transcription initiation sites with two promoters (P1 and P2) and seven exons. Part of exon 1 and exons 2 and 3 encode the DNA-binding *runt* domain; exons 4, 5, 6, and 7 code for transcriptional activation and repression domains. The nuclear localization signal (NLS) is located at the 5′ end of exon 3. Exon 6 is alternatively spliced and unique to *CBFA1*. The role of the *CBFA1* gene also includes a major regulatory function in chondrocyte differentiation during endochrondral bone formation (Mundlos, 1999). As such, it functions as a "master gene" in bone development.

A wide range of mutations have been found: All mutations result in loss of function, i.e., haploinsufficiency causes the CCD phenotype. (Figures in A, B, and D kindly provided by Dr. Stefan Mundlos, Max-Planck Institute for Molecular Genetics, Berlin).

### References

Mundlos, S.: Cleidocranial dysplasia: clinical and molecular genetics. J. med. Genet. **36**:177–182, 1999.

Mundlos S., et al.: Mutations involving the transcription factor CBFA1 cause cleidocranial dysplasia. Cell **89**:773–779, 1997.

Zou, G., et al.: *CBFA1* mutations analysis and functional correlation with phenotypic variability in cleidocranial dysplasia. Hum. Mol. Genet. **8**:2311–2316, 1999.

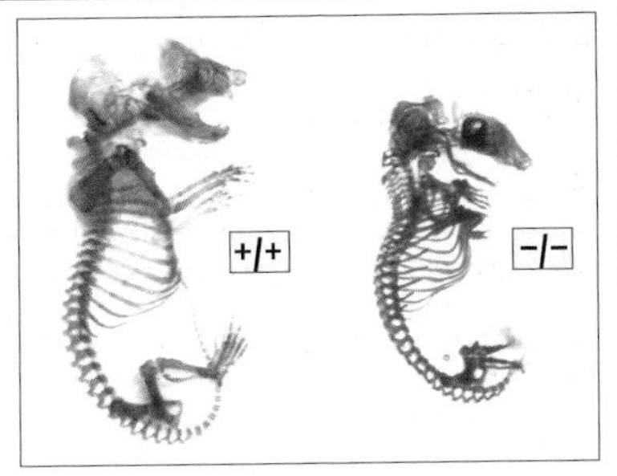


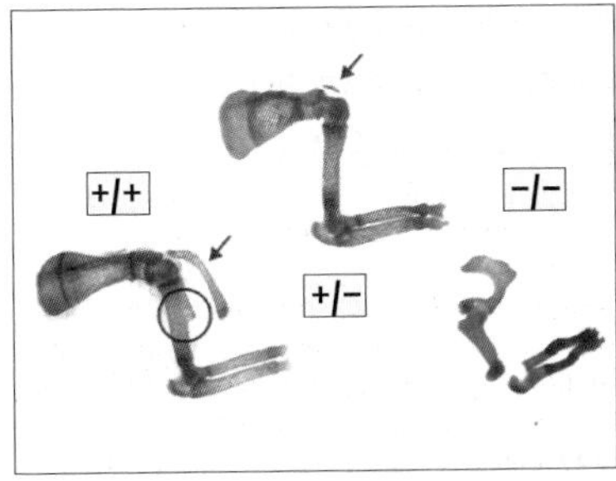


**1.** Normal skeleton **2.** Homozygous mutant, lack of bones **3.** Normal humerus **4.** Hypoplasia **5.** Lack of bone development

**A. Effect of homozygous *Cbfa1* mutations on the mouse skeleton**

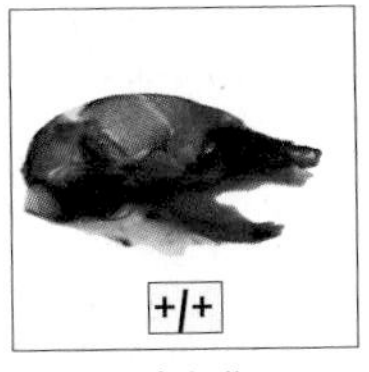


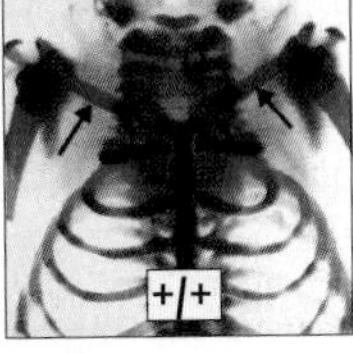


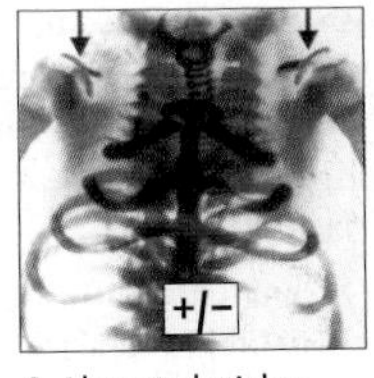


**1.** Normal skull **2.** Lack of ossification **3.** Normal thorax **4.** Absent clavicles

**B. Mice heterozygous for a mutation in the *Cbfa1* gene**

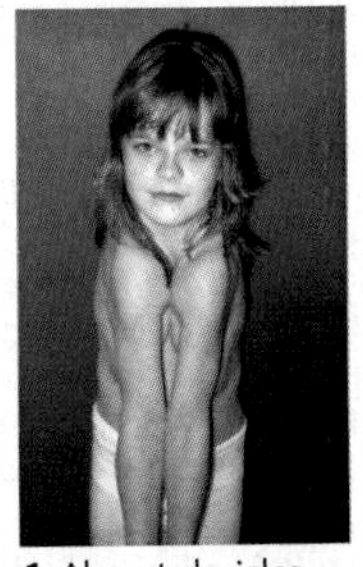

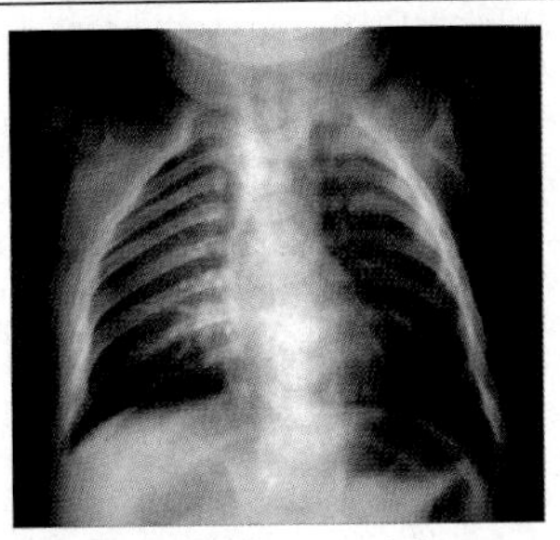

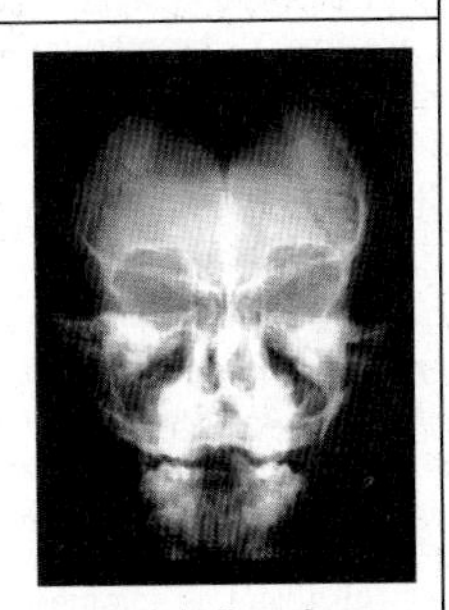

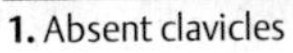

**1.** Absent clavicles **2.** X-ray: narrow thorax and absent clavicles **3.** Lack of skull ossification

**C. Cleidocranial dysplasia in human**



**D. The human *CBFA1* gene**

# Mammalian Sex Determination and Differentiation

## Sex Determination

In the 1940's, the French embryologist Alfred Jost observed that when the undifferentiated gonads were removed from a male rabbit fetus before male development had begun, it developed as a female. In 1959, chromosomal analysis of two disorders in man, Turner syndrome and Klinefelter syndrome, yielded the first evidence that genetic factors on the Y chromosomes of mammals are important in determining male sex. A specific gene on the mammalian Y chromosome (*SRY*, sex-related Y) induces male sex development during embryogenesis (sex determination).

### A. Determination of male phenotype by the Y chromosome

Individuals with Turner syndrome have only one X chromosome (no Y chromosome) and a female phenotype, although incompletely developed and usually accompanied by malformations. Individuals with Klinefelter syndrome have two X chromosomes, a Y chromosome, and a male phenotype, although also incompletely developed (p. 402).

### B. Sex-determining region SRY on the Y chromosome

The relevant region in man lies in the distal short arm of the Y chromosome at Yp11.32. The short arm and the proximal half of the long arm of the Y chromosome have been divided into seven intervals (2). The most distal region of the short arm is the pseudoautosomal region 1 (PAR 1). This region is homologous to the distal segment of the short arm of the X chromosome. Homologous pairing occurs here with crossing-over during meiosis. The physical map of the pseudoautosomal region and the proximal half of interval 1 (1A1 – 1B) span somewhat more than 2500 kb in man (3). Intervals 2 – 7 contain no genes for male sex determination. The crucial portion of the Y chromosome for male sex determination in man is about 35 kb (in the mouse about 14 kb) of a region designated Sry (sex-related Y chromosomal region) in the interval 1A1 proximal to the pseudoautosomal region (1). (After U. Wolf et al., 1992).

### C. Male development of an XX mouse transgenic for the *Sry* gene

Clinical observations and experimental evidence indicate that the presence of *SRY* induces male development, irrespective of the presence of the remainder of the Y chromosome. A chromosomally female transgenic mouse (XX) shows normal male development after the 14 kb DNA fragment carrying the *Sry* region of a mouse Y chromosome is implanted into its blastocyst. (Figure from Koopman et al., 1991).

### D. *Sry* expression during embryonic gonadal development of the mouse

During embryonic development of an XY mouse, *Sry* is expressed only between days 10.5 and 12.5. The subsequent events leading to male development are initiated during this short time of expression. (Figure from Koopman and Gubbay, 1991).

### References


Cameron, F., Sinclair A.H.: Mutations in SRY and SOX9: testis-determining genes. Hum. Mutat. **9**:388 – 395, 1997.

Goodfellow, P.N., Camerino, G.: *DAX-1*, an "anti-testis" gene. Cell Mol. Life Sci. **55**:857 – 863, 1999.

Hargraeve, T.B.: Understanding the Y chromosome. Lancet **354**:1746 – 1747, 1999.

Koopman, P., Gubbay J.: The biology of Sry. Seminars Develop. Biol. **2**:259 – 264, 1991.

Koopman, P., et al.: Male development of chromosomally female mice transgenic for *Sry*. Nature **351**:117 – 121, 1991.

McElreavey, K., Fellous, M.: Sex determination and the Y chromosome. Am. J. Med. Genet. (Semin. Med. Genet.) **89**:176 – 185, 1999.

Roberts, L.M., Shen, J., Ingraham, H.A.: New solutions to an ancient riddle: Defining the differences between Adam and Eve. Am. J. Hum. Genet. **65**:933 – 942, 1999.

Swain, A., Lovell-Badge, R.: Mammalian sex determination: a molecular drama. Genes Dev. **13**:755 – 767, 1999.

Wolf, U., Schempp, W., Scherer, G.: Molecular biology of the human Y chromosome. Rev. Physiol. Biochem. Pharmacol. **121**:148 – 213, 1992.

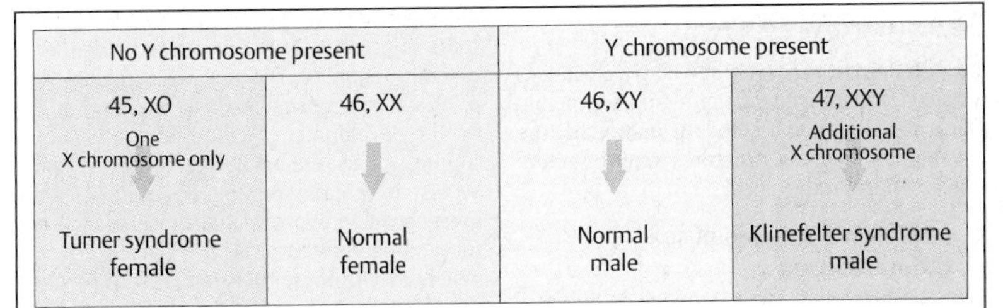

No Y chromosome present
Y chromosome present
45, XO
One X chromosome only
Turner syndrome female
46, XX
Normal female
46, XY
Normal male
47, XXY
Additional X chromosome
Klinefelter syndrome male


**A. Determination of the male phenotype by the Y chromosome**

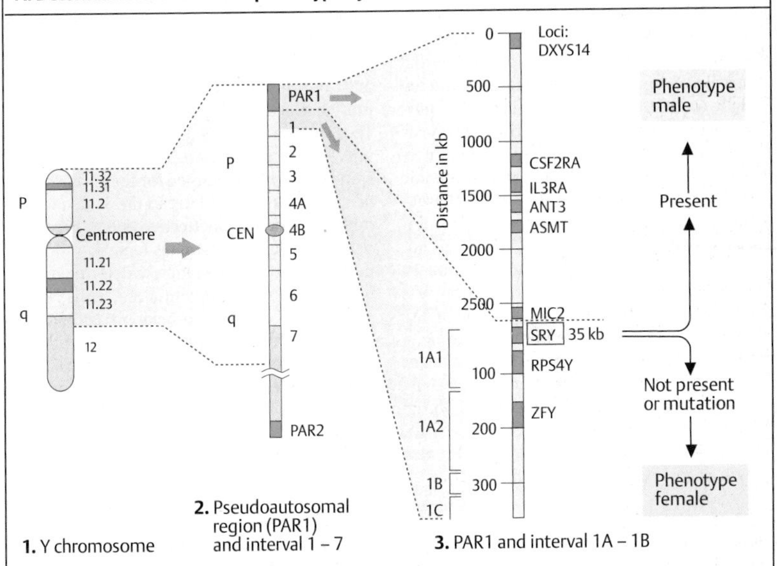


**1.** Y chromosome **2.** Pseudoautosomal region (PAR1) and interval 1 – 7 **3.** PAR1 and interval 1A – 1B

**B. Sex-determining region SRY on the Y chromosome**

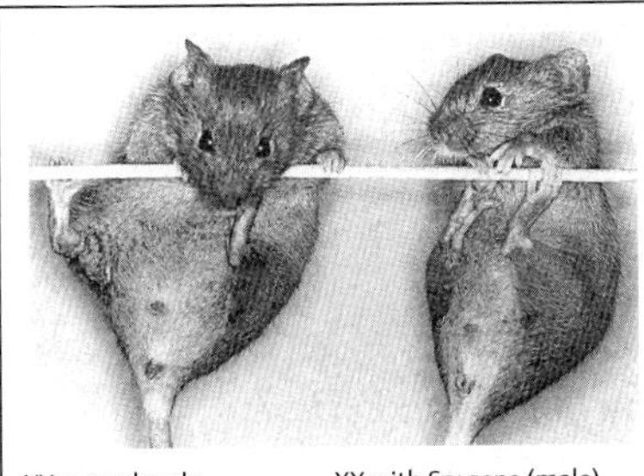

XY normal male XX with *Sry* gene (male)

**C. Male development of an XX mouse transgenic for the *Sry* gene**

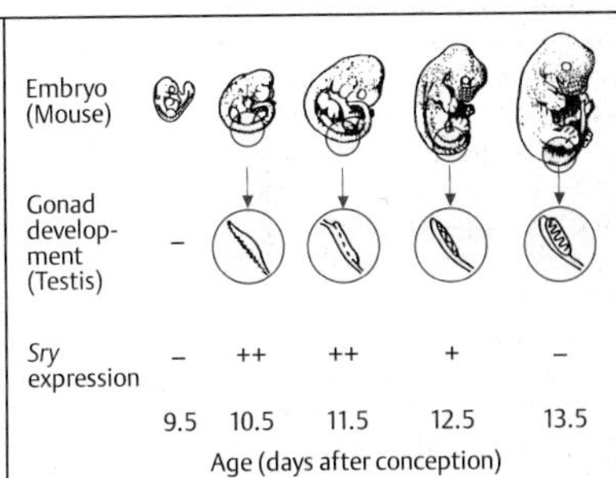


**D. *Sry* expression during embryonal gonad development**

## Sex Differentiation

Sex differentiation (development of a given sex) consists of many genetically regulated, hierarchical developmental steps. In mammals, the development of male structures requires induction by appropriate genes.

### A. Indifferent anlagen of sex differentiation

The gonads (1), the efferent ducts (mesonephric and paramesonephric) (2), and the external genitalia (3) all develop from an indifferent stage. At about the end of the sixth week of pregnancy in humans, after the primordial germ cells of the embryo have migrated into the initially undifferentiated gonads, an inner portion (medulla) and an outer portion (cortex) of the gonads can be distinguished. When a normal Y chromosome is present, early embryonic testes develop at about the 10th week of pregnancy under the influence of a testis-determining factor (TDF). If a normal Y or TDF (SRY) is not present, ovaries develop. The wolffian ducts, the precursors of the male efferent ducts (vas deferens, seminal vesicles, and prostate), develop under the influence of testosterone, a male steroid hormone formed in the fetal testis. At the same time, the müllerian ducts—precursors of the fallopian tubes, the uterus, and the upper vagina—are suppressed by a hormone, the Müllerian Inhibition Factor (MIF; also known as anti-müllerian hormone, AMH).
When testosterone is absent or ineffective, the wolffian ducts degenerate. The müllerian ducts develop under the influence of estradiol, a hormone produced by the fetal ovaries. The external genitalia (3) in humans do not develop until relatively late, starting in the 15th to 16th week. Full development of male external genitalia depends on a derivative of male-inducing testosterone, 5-dihydrotestosterone, a metabolite of testosterone produced by the enzymatic action of 5α-reductase.

### B. Sequence of events in sex differentiation

Sex differentiation proceeds in a cascadelike manner, with a series of temporally regulated successive steps at different levels of differentiation. After the primordial germ cells migrate into the undifferentiated gonads, early embryonic testes develop under the influence of testis-determining factor (TDF) if a Y chromosome is present. TDF is identical with the Y-specific sequences of the SRY region (see p. 386). During normal male differentiation, the further development of the müllerian ducts is suppressed by the *m*üllerian *i*nhibitor *f*actor. Testosterone can exert its effect only in the presence of an appropriate intracellular receptor (androgen receptor TFM, see p. 390).
When a Y chromosome is not present or when the SRY region is missing or altered by mutation, testes are not formed. In this case the wolffian ducts cease to develop. In the absence of testes, ovaries develop from the undifferentiated gonads; the wolffian ducts degenerate; and the müllerian ducts differentiate into uterine tubes, uterus, and the upper vagina.
Testosterone also has an effect on the central nervous system ("brain imprinting"). It is assumed that this is required for the psychosexual orientation apparent later in life. When testosterone is absent or ineffective due to a receptor defect, gender orientation is female.
In the majority of genetically determined disorders of sexual differentiation, gonadal and genital sex do not correspond (pseudohermaphroditism). In true hermaphroditism, where the gonads consist of both testicular and ovarian tissues, male and female structures exist side by side.

### References

Arango, N.A., Lovell-Badge, R., Behringer, R.R.: Targeted mutagenesis of the endogenous mouse *Mis* gene promoter: in vivo definition of genetic pathways of vertebrate sexual development. Cell **99**:409–419, 1999.

Ferguson-Smith, M.A., Goodfellow, P.N.: SRY and primary sex-reversal syndromes, pp. 739–749, In: C.R. Scriver, et al., eds., The Metabolic and Molecular Bases of Inherited Disease. 7th ed. McGraw-Hill, New York, 1995.

Wilson, J.D., Griffin, J.E.: Disorders of sexual differentiation, pp. 2119–2131. In: A.S. Fauci, et al., eds., Harrison's Principles and Practice of Internal Medicine. 14th ed.. McGraw-Hill, New York, 1998.

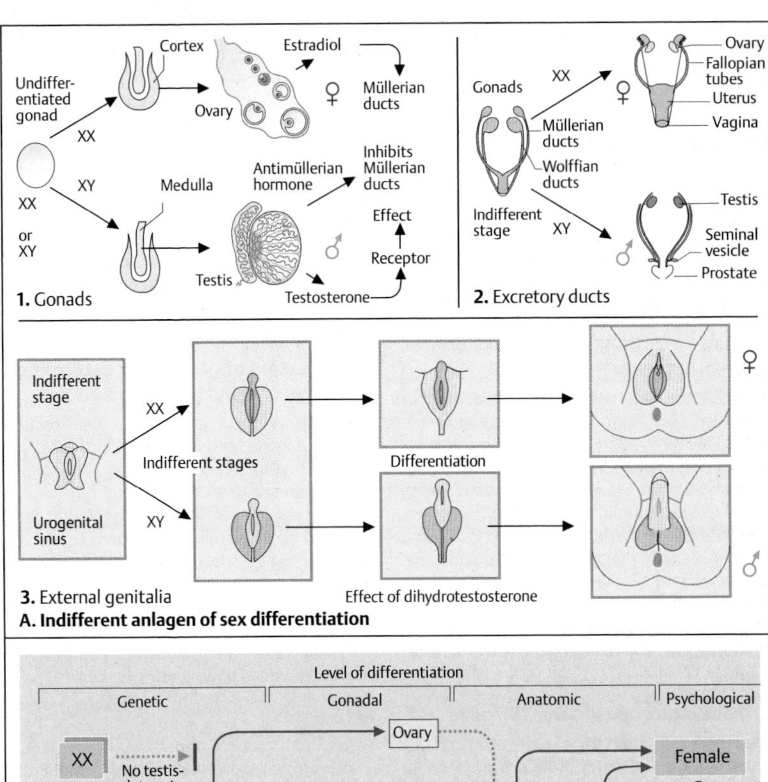


**A. Indifferent anlagen of sex differentiation**

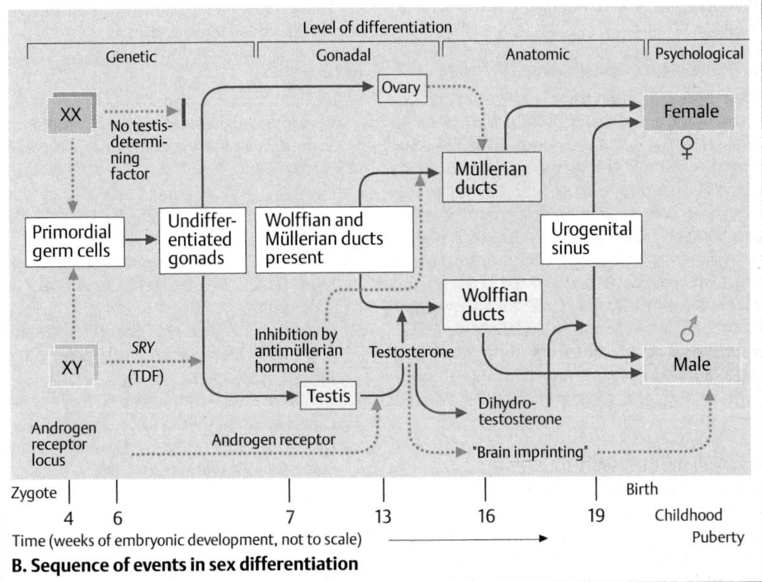


**B. Sequence of events in sex differentiation**

# Disorders of Sexual Development

Normal sexual development is the result of numerous genes. Mutation or chromosomal rearrangements of any of these genes cause partial or total failure of sex differentiation. The classification of genetically determined disorders of sexual development takes the different developmental processes into account. Pinpointing the basic defect is a prerequisite for diagnosis and treatment.

## A. Male-determining region *SRY* on the Y chromosome

Normally, the male-determining Y-specific DNA sequences (*SRY*) remain on the Y chromosome during the homologous pairing and crossing-over during meiosis. However, since the male-determining region *SRY* is located very close to the pseudoautosomal region (PAR), crossing-over in the PAR border region may result in a transfer of the *SRY* region to the X chromosome. This results in a male individual with an XX karyotype (XX male). Conversely, if the *SRY* region is missing from a Y chromosome, a female phenotype with XY chromosomes (XY female) results.

## B. Point mutations in the *SRY* gene

The human *SRY* gene has a single exon and encodes a 204-amino-acid protein from a 1.1 kb transcript. The middle section of the SRY protein consists of 79 highly conserved amino acids with DNA-bending and DNA-binding capability, the HMG box (high mobility group protein). Complete or partial gonadal dysgenesis results from point mutations and deletions in the *SRY* gene, in particular the HMG box. (Figure adapted from Wolf et al., 1992; for an update of mutations see McElreavey and Fellous, 1999). Sex reversal also results from mutations in the *SOX9* gene on chromosome 17 at q24 in campomelic dysplasia. (Foster et al., 1994; Wagner et al., 1994).

## C. Androgen receptor

The fetal testis produces testosterone, the hormone that induces male sexual differentiation. Testosterone is taken up by cells of the target tissues (wolffian ducts and urogenital sinus) (1). In the urogenital sinus, testosterone is converted into dihydrotestosterone (DHT) by the enzyme 5$\alpha$-reductase. Both testosterone and dihydrotestosterone bind to an intracellular receptor (androgen receptor). The activated hormone–receptor complex (TR* or DR*) acts as a transcription factor for genes that regulate the differentiation of the wolffian ducts and the urogenital sinus. Thus, normal male fetal development is dependent on normal biosynthesis of testosterone and normal receptors. Androgen receptor mutations lead to disorders of sexual development (2) with X-chromosomal inherited complete or incomplete androgen resistance (testicular feminization, TFM).

## Classification of genetically determined disorders of sexual development

1. Defects of sex determination due to mutation or structural aberration of the *SRY* region on the Y chromosome (e.g., XY gonadal dysgenesis, XX males, and others)
2. Defects of androgen biosynthesis (e.g., adrenogenital syndrome due to 21-hydroxylase deficiency, see p. 392)
3. Defects of androgen receptors (testicular feminization)
4. Defects of the müllerian inhibition substance (so-called hernia uteri syndrome)
5. XO/XY gonadal dysgenesis

## References

Foster, J.W., et al.: Campomelic dysplasia and autosomal sex reversal caused by mutations in an SRY-related gene. Nature **372**: 525–530, 1994.

Goodfellow, P.N., Camerino, G.: DAX-1, an "anti-testis" gene. Cell Mol. Life Sci. **55**:857–863, 1999.

Gottlieb, B., et al.: Androgen insensitivity. Am. J. Med. Genet. (Semin. Med. Genet.) **89**:210–217, 1999.

McElreavey, K., Fellous, M.: Sex determination and the Y chromosome. Am. J. Med. Genet. **89**:176–185, 1999.

Wagner, T., et al.: Autosomal sex reversal and campomelic dysplasia are caused by mutations in and around the SRY-related SOX9. Cell **79**:1111–1120, 1994.

Wilson, J.D., Griffin, J.E.: Disorders of sexual differentiation, pp. 2119–2131. In: A.S. Fauci, et al., eds., Harrison's Principles and Practice of Internal Medicine. 14th ed., McGraw-Hill, New York, 1998.

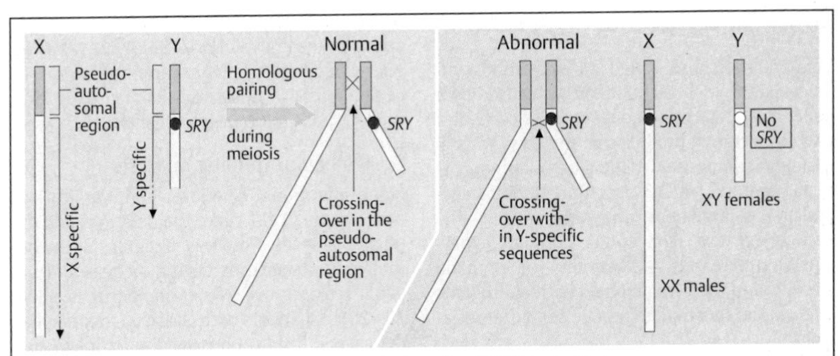


**1.** SRY remains on the Y chromosome **2.** Transfer of SRY to the X chromosome

**A. Male-determining region *SRY* on the Y chromosome**

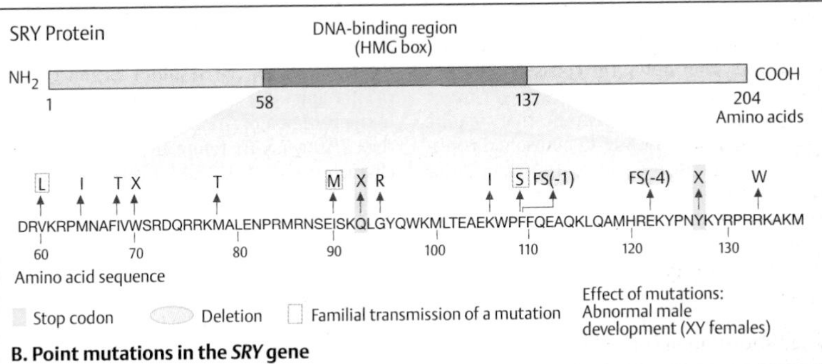


**B. Point mutations in the *SRY* gene**

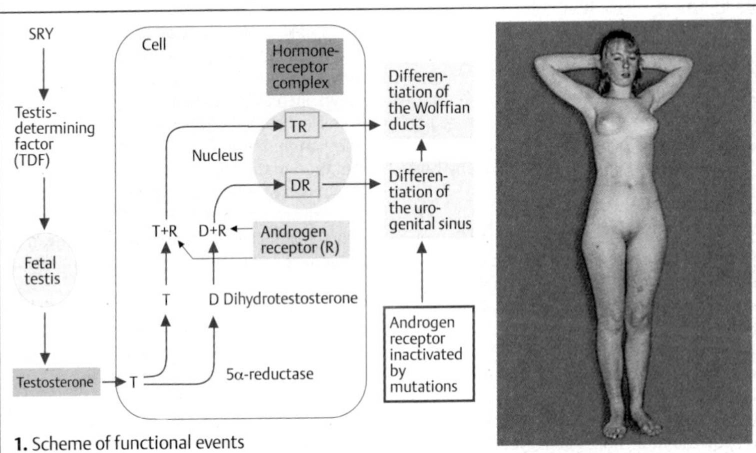


**1.** Scheme of functional events **2.** Phenotype

**C. Androgen receptor and TFM**

## Congenital Adrenal Hyperplasia

This disorder, also called adrenogenital syndrome (AGS, McKusick 201910), is caused by a genetically determined deficiency of cortisol, a steroid hormone produced in the fetal adrenal cortex. A compensatory increase in adrenocortical hormone (ACTH) excretion leads to secondary enlargement (hyperplasia) of the adrenal cortex (congenital adrenal hyperplasia), increased production of prenatal steroids and their metabolites with androgenic effects, and incomplete female sex differentiation.

### A. Clinical phenotype and genetics

Girls are born with ambiguous or virilized genitalia (1). The adrenal cortex is enlarged (2). Increased production of androgenic metabolites causes masculinization. The cortisol deficiency (3) leads to life-threatening crises due to loss of sodium chloride (salt-wasting) that require prompt treatment. AGS is an autosomal recessive heritable disorder (4). Untreated girls develop a male physical appearance (5). In boys, the early signs are limited to salt-wasting. Initially, skeletal maturation is accelerated and the children are tall for their age; however, they stop growing prematurely and eventually are too short. Besides the classic form of the disorder with a frequency of 1:5000, there are other forms with less pronounced masculinization due to different mutations.

### B. Biochemical defect

The enzymatic conversion of progesterone to deoxycortisol (DOC) by hydroxylation at position 21 (steroid 21-hydroxylase) is decreased. As a result, the concentration of 17-hydroxyprogesterone is increased.

### C. Gene locus and gene structure

21-Hydroxylase is encoded by the *CYP21* gene (formerly called *CYP21B* and *21-OHB*), a member of the cytochrome P450 oxidase gene family. This gene is located within the class III genes of the major histocompatibility complex on the short arm of human chromosome 6. It is part of a tandem paired arrangement of three other genes: active *C4A* and *C4B* genes and a 96–98% homologous inactive *CYP21P* gene, a pseudogene due to intragenic deletions resulting in stop codons (formerly called *CYP21A* or *21-OHA*). These genes originated from a duplication event in evolution. The *CYP21* (*21-OHB*) gene consists of 10 exons spanning almost 6 kb of genomic DNA (the actual distance of 30 kb to the *C4 A* and *C4 B* genes is not shown to scale).

### D. Molecular genetic analysis

Point mutations, deletions, and duplications occur in the *CYP21* gene. The deletions and duplications result from misalignment of the homologous chromatids during meiosis and unequal crossing-over. Deletions occur in about 20–25% of patients with classic 21-hydroxylase deficiency. Duplications have no clinical consequences. Deletions and duplications can be easily detected by Southern blot analysis. The most frequent type of deletion is loss of a 30 kb region including the 3' part of the *CYP21P* pseudogene, the entire *C4B* gene, and the 5' part of the *CYP21* gene. The resulting fusion gene of *CYP21P* and *CYP21* carries a *Taq*I restriction site in the 5' region of *CYP21P* that is not present in the *CYP21* gene. Therefore, the fusion gene has a characteristic 3.2 kb *Taq*I fragment. This distinguishes the rearrangement from the normal *CYP21* gene, which has a characteristic 3.7 kb fragment. In the example shown, the *CYP21* gene (*21-OHB*) is represented by a 3.7 kb DNA fragment, the pseudogene *CYP21P* (*21-OHA*) by a 3.2 kb fragment after *Taq*I digestion (1). Thus, the normal pattern is a 3.7 kb and a 3.2 kb fragment (2). Homozygous deletion of either of the genes may be apparent by lack of either of the two fragments (3, 4). A heterozygous deletion shows reduced intensity (5) and a duplication shows increased intensity (6). (Figure adapted from New et al., 1989).

Another common mechanism for the origin of mutation in the *CYP21* gene is gene conversion. This involves nonreciprocal exchange between the closely linked *CYP21* and *CYP21P* genes, which results in transfer of mutations from the pseudogene *CYP21P* to *CYP21*.

### References

New, M.I., et al.: The adrenal hyperplasias, pp. 1881–1917. In: C.R. Scriver, et al., eds., The Metabolic Basis of Inherited Disease. 6th ed. McGraw-Hill, New York, 1989.

Wilson, R.C., New, M.I.: Congenital adrenal hyperplasia, pp. 481–493. In: J.L. Jameson, ed., Principles of Molecular Medicine. Humana Press, Totowa, New Jersey, 1998.

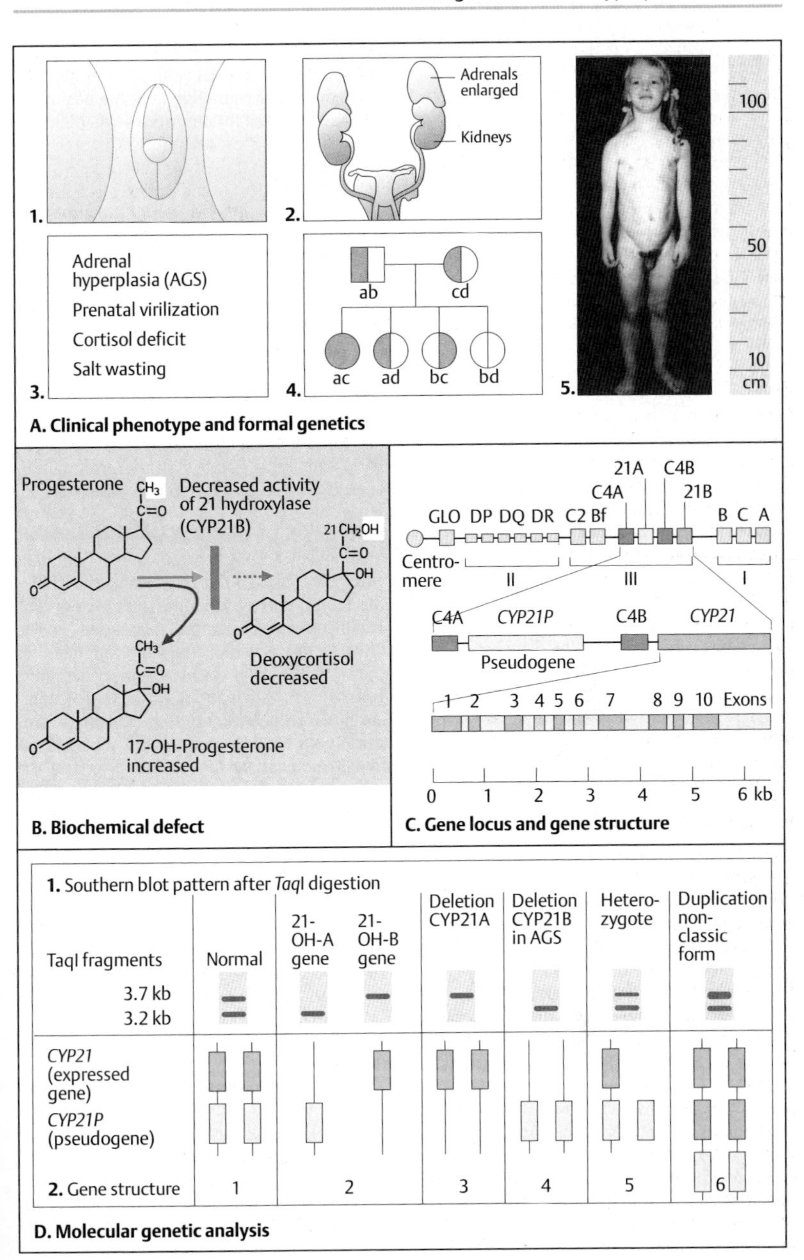


A. Clinical phenotype and formal genetics

B. Biochemical defect

C. Gene locus and gene structure

D. Molecular genetic analysis

# Atypical Inheritance Pattern

## *Unstable Number of Trinucleotide Repeats*

Heritable changes in the number of repeated groups of three nucleotides each (trinucleotide or triplet repeat) represent a new class of mutations in man for which there is no parallel in other organisms. They either occur within the gene and are translated or occur outside the gene in an untranslated region, and they are unstable during transmission through the germline. Unaffected persons may carry a premutation, which may be converted to a full mutation when passed through the germline to the next generation. Therefore, the effects of the mutation differ in severity in affected members within the same family. Occasionally, there is regression and a generation is skipped.

### A. Genetic diseases with increased numbers of trinucleotides

Some important genetically determined diseases are based on a greater than normal number of trinucleotides: Huntington disease, fragile X syndrome, myotonic dystrophy, spinobulbar muscular atrophy type Kennedy, and spinocerebellar ataxia type 1. A total of 14 defined trinucleotide repeat diseases have been described (see Cummings and Zoghbi, 2000).

### B. Huntington disease

Huntington disease is a progressive disease of the brain. Within 5–10 years, it leads to complete loss of motor control and intellectual abilities (1). It usually begins around age 40–50 with uncoordinated movements (chorea, St. Vitus' dance), excitation, hallucinations, and psychological changes. The disease is transmitted by autosomal dominant inheritance and shows complete penetrance. It presents an affected family with two difficult problems: (i) due to its late onset, carriers of the mutation have usually completed their family planning before the disease is manifest, and (ii) children of affected persons first learn as young adults that they are at a 50% risk of developing the disease later in life. Thus, the introduction of a direct predictive DNA diagnostic procedure is very important. However, before such a genetic test is carried out, it must be established through genetic counseling that the persons at risk have decided for themselves whether they want to have the test performed. The gene is located on the distal short arm of chromosome 4 (2). It spans 210 kb and codes for a protein (called huntingtin) of important function. The 5′ end of the gene contains numerous copies of a trinucleotide sequence consisting of cytosine, adenine, and guanine (CAG), a codon for the amino acid glutamine. Normally the gene has 10–34 CAG repeats; in patients there are 42–100. The diagnostic test (3) demonstrates that affected individuals (here, individuals 1, 2, and 4) have enlarged DNA fragments due to expanded CAG repeats. (Findings of the Institut für Humangenetik of the Universität Göttingen with kind permission by Prof. W. Engel; Zühlke et al., Hum. Mol. Genet. **2**:1467–1469, 1993).

### C. Myotonic dystrophy (MDY1)

Myotonic dystrophy is an autosomal dominant hereditary disease that predominantly affects the central nervous and muscular systems (1). The myotonia causes a masklike facies (2). The disease is very variable and in many families shows increasing severity in consecutive generations (anticipation). An increased number of CTG repeats, more than 50 copies compared with 5–35 in normal individuals (3), is found immediately beyond the 3′ end of the gene in affected persons. This is demonstrated in a Southern blot as an enlarged DNA fragment (4). (Schematic representation of a Southern blot at the gene locus D19S95, probe pBBO.7 after DNA cleavage with *Eco*RI. After Harley et al., Lancet **339**:1125–1128, 1992).

### References

Cummings, C.J., Zoghbi, H.Y.: Fourteen and counting: unraveling trinucleotide repeat diseases. Hum. Mol. Genet. **9**:909–916, 2000.

Harley, H.G. et al.: Unstable DNA sequence in myotonic dystrophy. Lancet **339**:1125–1130, 1992.

Zoghbi, H.Y.: Spinocerebellar ataxia and other disorders of trinucleotide repeats, p. 913–920, In: Jameson, J.L., ed., Principles of Molecular Medicine. Humana Press, Totowa, New Jersey, 1998.

| Disease (Examples) | Gene | Frequency | Trinucleotide | Normal Number | Mutant Allele | Chromosome |
|---|---|---|---|---|---|---|
| Huntington disease | *HD* | 1:10 000 | $(CAG)_n$ | 0–26 | 36–121 | 4p16.3 |
| Fragile X syndrome | *FMR1* | 1:5 000 | $(CGG)_n$ | 6–50 | 52–500 | Xq27.3 |
| Myotonic dystrophy | *DMPK* | 1:8 000 | $(CTG)_n$ | 5–37 | 50–500 | 19q13.2 |
| Spinal-bulbar muscular atrophy (Kennedy) | *SBMA* | <1:50 000 | $(CAG)_n$ | 11–31 | 36–65 | Xq11-12 |

**A. Genetic diseases due to increased number of trinucleotides**

Severe progressive disease of the central nervous system
Loss of motor and intellectual control
Onset age 25–60
Autosomal dominant
CAG repeat size increased
Predictive diagnosis possible, but problematic

**1.** Main manifestations

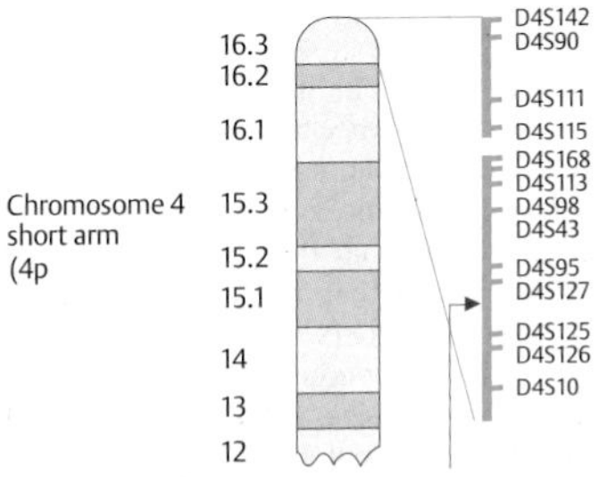


**2.** Localization of the gene

Affected individuals 1, 2, and 4 have expanded CAG repeats

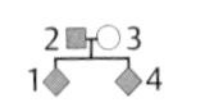


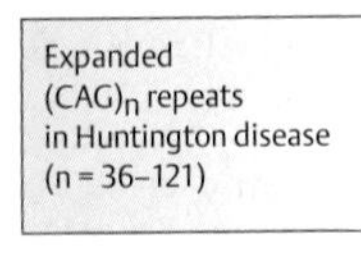

Expanded
$(CAG)_n$ repeats
in Huntington disease
(n = 36–121)


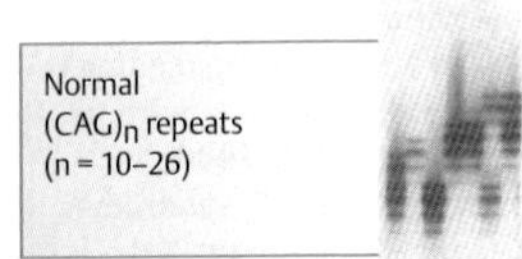


**3.** Diagnostic test

Control

**B. Huntington disease**

Muscle weakness,
Myotonia, mask-like face,
Cataract, alopecia,
Variable expression,
Autosomal dominant,
CTG repeat increased

1. Main manifestations

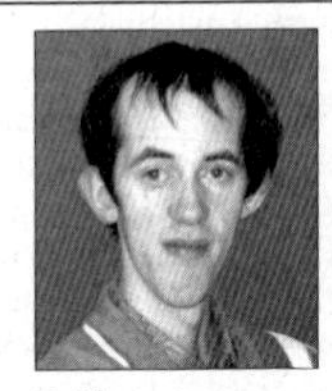

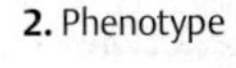

**2.** Phenotype

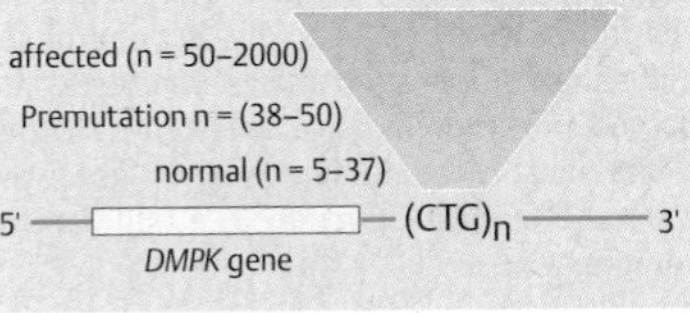


**3** Expanded CTG repeat in Myotonic Dystrophy

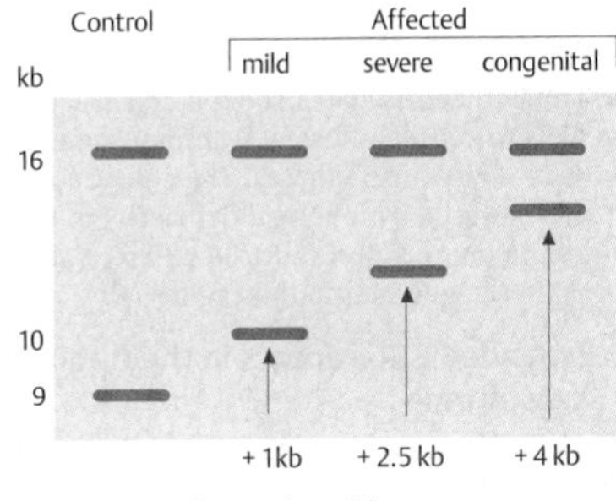


Diagram of a Southern blot at gene locus D19S95 (probe pBB0.7)

**4.** Correlation with degree of severity

**C. Myotonic Dystrophy**

# Fragile X Syndrome

The fragile X syndrome (McKusick 309550; other designations: fraX syndrome, X-chromosomal mental retardation with fragile site on the X chromosome, Martin–Bell syndrome) is the most frequent form of hereditary mental retardation in males, with a frequency of about 1:2000–4000 individuals. The responsible mutation usually consists of an increased number of unstable trinucleotide repeats. Unlike in classic X-chromosomal inheritance, there are males without manifestations, and a large proportion of female carriers show partial manifestations.

The unstable expansion of a trinucleotide repeat (CGG) is located in the 5'-untranslated region of the *FMR1* gene. Recent findings indicate that an increase beyond 200 repeats impedes the migration of the 40S ribosomal subunit. This causes translational suppression.

## A. Phenotype

The phenotype is very variable. The mental retardation varies; there is no distinct neurological dysfunction. In adult males, the testes are enlarged (macroorchidism). Affected individuals can usually be integrated well into the family and learn to function in a familiar environment.

## B. Fragile site Xq27.3

The gene locus (FRAXA) for the gene (*FMR1*) is located on the distal long arm of the X chromosome in region 2, band 7.3 (Xq27.3). In this region the great majority of patients and some of the female heterozygotes show a constriction (fragile site) in the affected X chromosome in about 2–25% of metaphases. The constriction must be induced by folic acid deficiency in the culture medium, and it must be differentiated from other fragile sites in this region.

## C. Expanded CGG repeats in the fragile X syndrome

The heritable unstable sequences explain two unusual characteristics of the fraX syndrome; (i) the transition from a premutation (about 60–200 CGG repeats) without clinical manifestation into a full mutation (more than 200 CGG repeats) during transmission through the germline, and (ii) differences in the FRAXA locus within a given family.

Fragile X syndrome is heritable as an X-linked dominant trait. The risk of transmission and clinical manifestation varies according to the type of mutation (premutation or full mutation), the gender of the patient and of the parent carrying an expanded trinucleotide repeat, and the relationship within the family. Males with the full mutation are mentally retarded and do not reproduce. Heterozygous females for the full mutation have a risk of variable mental retardation of 50%. They transmit the full mutation to 50% of their offspring. A premutation present in a male ("normal male transmitter") is transmitted to all daughters and none of the sons. Female carriers of the premutation or full mutation have a 50% risk of transmitting the mutant allele. The actual risk of manifest fragile X syndrome depends on the number of CGG repeats and varies between 10% (60–69 repeats) and 50% (more than 100 repeats) for sons (Gene Clinics at *http: www.geneclinics.org*).

The number of CGG repeats is variable within a family. In the pedigree shown (1), individuals II-3 and III-1 have more than 200 repeats and have fragile X syndrome. Individuals I-2, I-3, II-1, and III-3 are carriers of the premutation with 79–82 repeats. The normal number of CGG repeats is 6 to about 50, the premutation is defined by about 55 to 100, and the full disease-causing mutation by more than 200 repeats (2).

The different numbers of CGG repeats can be demonstrated in Southern blots as DNA fragments of different sizes (3). The normal gene is represented by a small DNA fragment (S). A premutation leads to slightly enlarged fragments. The full mutation is characterized by large fragments (L). With this procedure, a reliable diagnosis of the genotype is possible. (Photograph of a Southern blot: *Hind*III digestion and hybridization with probe Ox1.1; P. Steinbach, Ulm).

### References

Eichler, E.E., et al.: Length of uninterrupted CGG repeats determines instability in the *FMR1* gene. Nature Genet. **8**:88–94, 1994.

Jin, P., Warren, S.T.: Understanding the molecular basis of fragile X syndrome. Hum. Mol. Genet. **9**:901–908, 2000.

Nelson, D.L.: Fragile X syndrome, pp. 1063–1067. In: J.L. Jameson, ed., Principles of Molecular Medicine. Humana Press, Totowa, NJ, 1998.

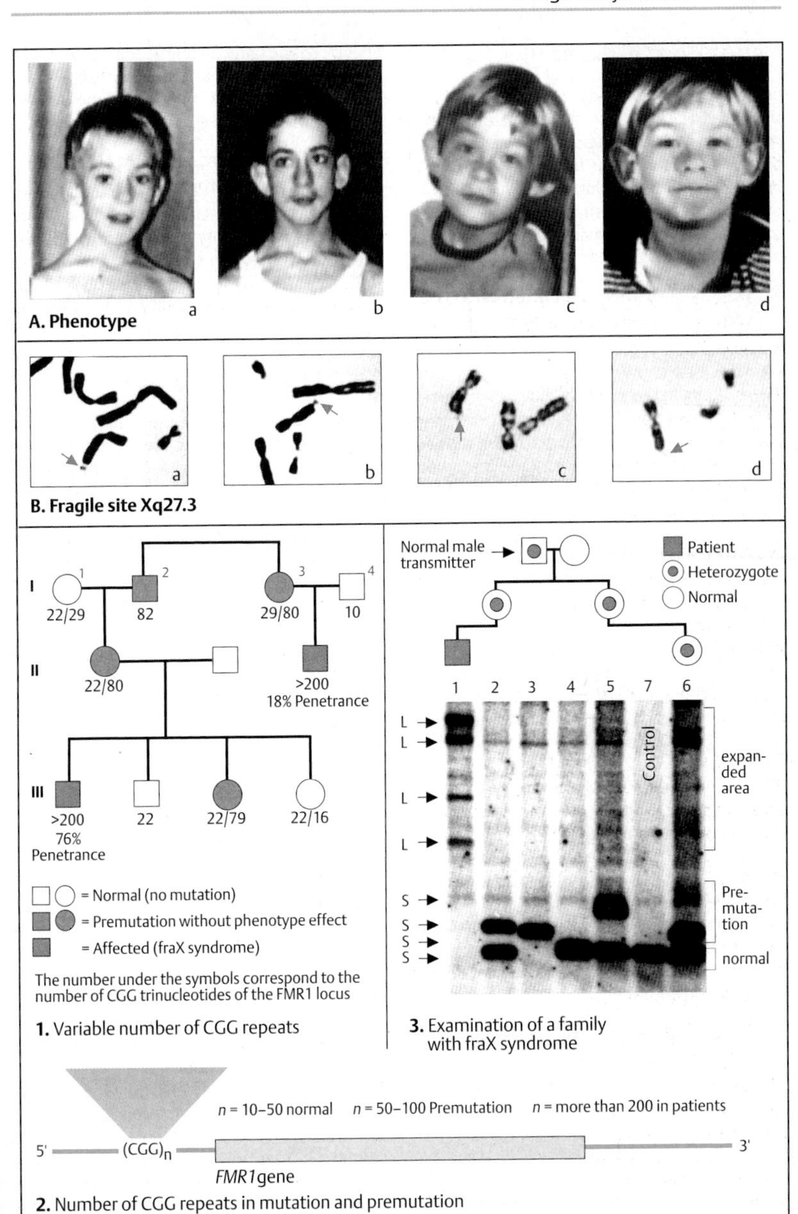

a
b
c
d
A. Phenotype
a
b
c
d
B. Fragile site Xq27.3
I
II
III
1
2
3
4
22/29
82
29/80
10
22/80
>200
18% Penetrance
>200
76%
Penetrance
22
22/79
22/16
= Normal (no mutation)
= Premutation without phenotype effect
= Affected (fraX syndrome)
The number under the symbols correspond to the number of CGG trinucleotides of the FMR1 locus
1. Variable number of CGG repeats
Normal male transmitter
Patient
Heterozygote
Normal
1 2 3 4 5 7 6
L
L
L
L
S
S
S
S
Control
expan-ded area
Pre-muta-tion
normal
3. Examination of a family with fraX syndrome
n = 10–50 normal
n = 50–100 Premutation
n = more than 200 in patients
5'
(CGG)n
3'
FMR1gene
2. Number of CGG repeats in mutation and premutation

**C. Expanded CGG repeat in fragile X syndrome**

## Imprinting Diseases

Prader–Willi syndrome (PWS) and Angelman syndrome (AS) are two distinct neurogenetic developmental disorders that result from different genetic lesions in a 3–4 Mb contiguous region of human chromosome 15 (15q11–13). This region is imprinted, i.e., genes on the maternal or the paternal allele only are expressed.

### A. Two syndromes associated with the same chromosomal region

Prader–Willi syndrome is characterized by neonatal muscular weakness and feeding difficulties, followed in early childhood by reduced or lack of satiation control leading to massive obesity in many patients. Several other, variable features occur, such as mental retardation, characteristic facial features, short stature, hypopigmentation, behavioral problems, and other findings. In Angelman syndrome the developmental retardation is usually very severe. Nearly total lack of speech development, an abnormal electroencephalogram with tendency to seizures, and hyperactivity are almost always present.

### B. Parental origin of the deletion

PWS results when the deleted chromosome involves the chromosome 15 of paternal origin (loss of one paternal allele 2 in the diagram of a Southern blot on the left). AS results when the deletion involves the chromosome 15 of maternal origin (loss of one maternal allele 1 in the scheme on the right).

### C. Uniparental disomy

Uniparental disomy (UPD) is the presence of two chromosomes or genes from the same parent. The diagram of a Southern blot shows two different types of UPD in PWS: isodisomy and heterodisomy. In isodisomy the two parental alleles are identical (two maternal alleles 1 shown in the diagram). In heterodisomy the two alleles are of the same parental origin, but differ (one allele 1 and one allele 2, both of maternal origin, as shown).

### D. Parent-of-origin deletion and uniparental disomy

A deletion and uniparental disomy have the same functional result, i.e., loss of the genetic activity of one parental allele. The frequency of a deletion is about the same for PWS and AS (70% each), whereas the frequency of UPD differs considerably: 29% in PWS, 1% in AS.

### E. Chromosomal region 15q11–13 and imprinting center

Five genes known to date are transcribed from the paternal allele only, not from the maternal allele where they are constitutively repressed (blue squares). From the more distal gene *UBE3A* (a ubiquitin-protein ligase E3) only the maternal allele is transcribed. About 25% of cases of Angelman syndrome are caused by deletion or mutation of this gene. The breakpoints of the common large deletions occur predominantly in three breakpoint cluster regions. The imprinting center (IC) was originally defined by small deletions outside of the known imprinted genes. About 1% of patients with PWS and 4% with AS have imprinting center defects. The imprinted region on 15q11–13 shows a difference in methylation pattern between the maternal and the paternal allele. This is the basis of a diagnostic test.

(Data in E. kindly provided by Dr. Karin Buiting).

Other imprinted chromosomal regions are also associated with human diseases, e.g., Beckwith–Wiedemann syndrome and some patients with Russell–Silver syndrome, among others.

### References

Buiting, K., et al.: Inherited microdeletions in the Angelman and Prader-Willi syndromes define an imprinting centre on human chromosome 15. Nature Genet. **9**:395–400, 1995.

Gillessen-Kaesbach, G., et al.: DNA methylation based testing of 450 patients suspected of having Prader-Willi syndrome. J. med. Genet. **32**:88–92, 1995.

Horsthemke, B., Dittrich, B., Buiting, K.: Imprinting mutations on human chromosome 15. Human Mutat. **10**:329–337, 1997.

Nicholls, R.D.: Prader-Willi and Angelman Syndromes, pp. 1053–1062. In: J.L. Jameson, ed., Principles of Molecular Medicine. Humana Press, Totowa, New Jersey, 1998.

Nicholls, R.D., Saitoh, S., Horsthemke, B.: Imprinting in Prader-Willi and Angelman syndromes. Trends Genet. **14**:194–200, 1998.

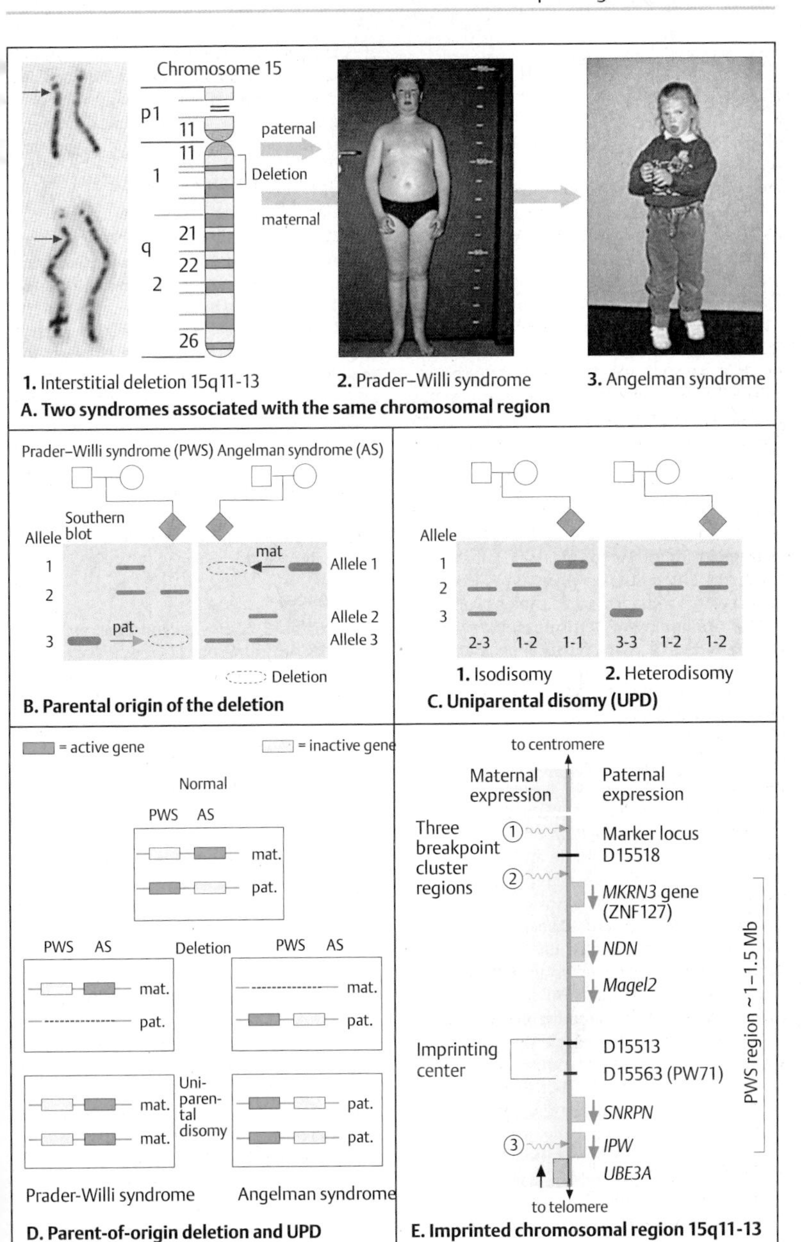


**1.** Interstitial deletion 15q11-13 **2.** Prader–Willi syndrome **3.** Angelman syndrome

**A. Two syndromes associated with the same chromosomal region**

**B. Parental origin of the deletion**

**C. Uniparental disomy (UPD)**

**D. Parent-of-origin deletion and UPD**

**E. Imprinted chromosomal region 15q11-13**

# Karyotype – Phenotype Correlation

## Autosomal Trisomies

A trisomy (the presence of three homologous chromosomes instead of the usual two) arises prezygotically during meiosis due to faulty distribution (nondisjunction) of a chromosome pair. It may also arise after fertilization (postzygotic) during somatic cell division (mitosis); in this case, trisomy is present in a certain proportion of cells (chromosomal mosaicism). Trisomy leads to a phenotype characteristic for the particular chromosome, although in humans most trisomies are lethal in early embryonic development.

### A. Trisomy in jimsonweed (*Datura stramonium*)

In 1922, Blakeslee observed that triploid and tetraploid jimsonweed plants (*Datura stramonium*) differ little in phenotype. However, when plants contained three copies of only one of the 12 chromosomes (trisomy), and two each of the others, a characteristic appearance resulted for each of the trisomies (from Blakeslee, 1922).

### B. Trisomies in the mouse

During the 1970s, A. Gropp and co-workers investigated the effect of trisomies on the development of the mouse. Trisomic mice, resulting from the segregation of translocations, had a developmental profile and certain morphological changes characteristic for each trisomy (1). Embryos with a chromosome missing (monosomies) died very early in gestation. (Figure from A. Gropp, 1982). A mouse embryo with trisomy 12 shows an open skull cap and other malformations on the 14th day of development (2), unlike other embryos of the same age (H. Winking, Lübeck, 1991; Boué et al., 1985). Only trisomy 19 is compatible with survival until birth (day 21), but the brain is too small (3). These animals die shortly after birth.

### C. Autosomal trisomies in man

Of the 22 autosomes in man, only three occur regularly as trisomies in live-born infants: trisomy 21, trisomy 18, and trisomy 13. They differ in phenotype and course of disease. Other trisomies are not observed in live-born infants because they are lethal in early embryonic life, and not compatible with life at birth (see p. 402). Trisomy 21 causes the clinical picture of Down syndrome (formerly called mongolism).

### D. Nondisjunction as a cause of trisomy

Especially in trisomy 21, the frequency of nondisjunction depends on the age of the mother at the time of conception (1). The age of the father has very little or no influence.

Nondisjunction may occur during the first or the second maturation division (meiosis I or meiosis II, p. 116) (2). The difference can be established by appropriate chromosomal markers. If nondisjunction occurs in meiosis I, the three chromosomes will be different (1 + 1 + 1), whereas if nondisjunction occurs during meiosis II, two of the three chromosomes will be identical (2 + 1). In humans, about 70% of nondisjunctions occur in meiosis I, and 30% in meiosis II.

### References


Antonarakis, S. E.: Down syndrome, pp. 1069–1078, In: Jameson, J.L., ed., Principles of Molecular Medicine. Humana Press, Totowa, New Jersey, 1998.

Blakeslee, A.F.: Variation in Datura due to changes in chromosome number. Am. Naturalist **56**:16–31, 1922.

Boué, A., Gropp, A., Boué, J.: Cytogenetics of pregnancy wastage. Adv. Hum. Genet. **14**:1–57, 1985.

Epstein, C.J.: Down syndrome (trisomy 21), pp. 749–794. In: C.R. Scriver, et al., eds., The Metabolic and Molecular Bases of Inherited Disease. 7th ed. McGraw-Hill, New York, 1995.

Gropp, A.: Value of an animal model for trisomy, Virchows Arch. Pathol. Anat. **395**:117–131, 1982.

Therman, E., Susman, M.: Human Chromosomes. Structure, Behavior, Effects. 3rd ed. Springer Verlag, Heidelberg, 1993.

Traut, W.: Chromosomen. Klassische und Molekulare Cytogenetik. Springer Verlag, Heidelberg, 1991.

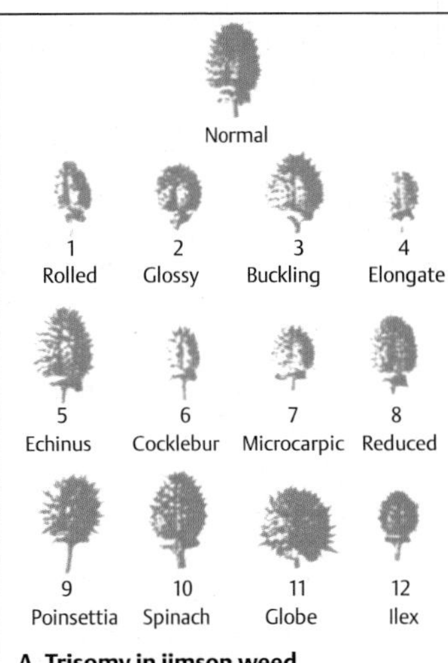

**A. Trisomy in jimson weed (*Datura stramonium*)**

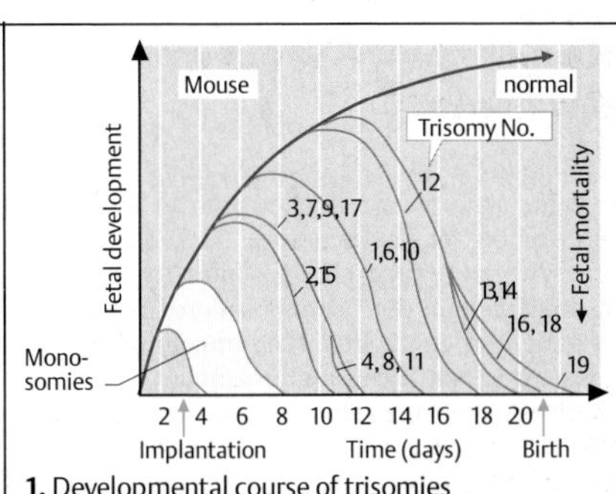


**1.** Developmental course of trisomies

Control Trisomy 12

**2.** Mouse embryo

Trisomy 19 Control

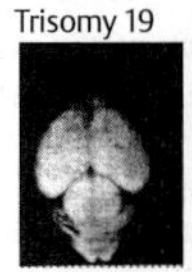
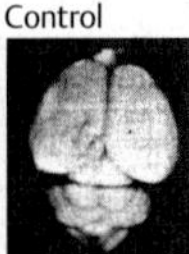

**3.** Brain

**B. Trisomies in the mouse**

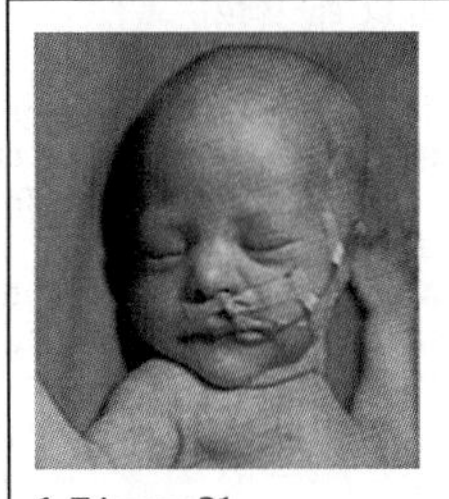

**1.** Trisomy 21

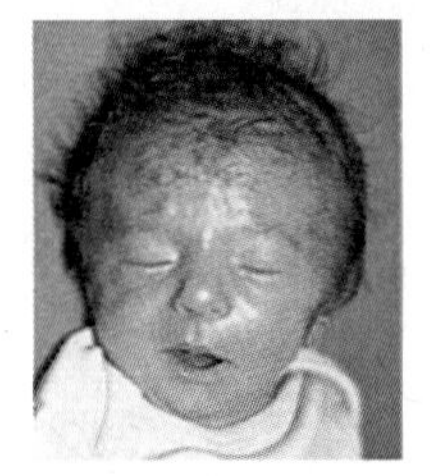

**2.** Trisomy 18

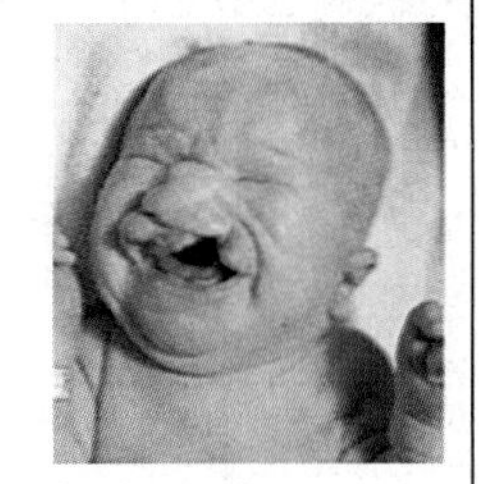

**3.** Trisomy 13

**C. Trisomies in man**

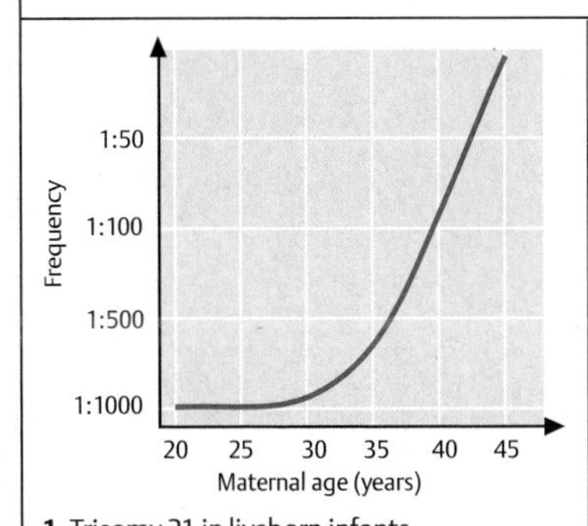


**1.** Trisomy 21 in liveborn infants

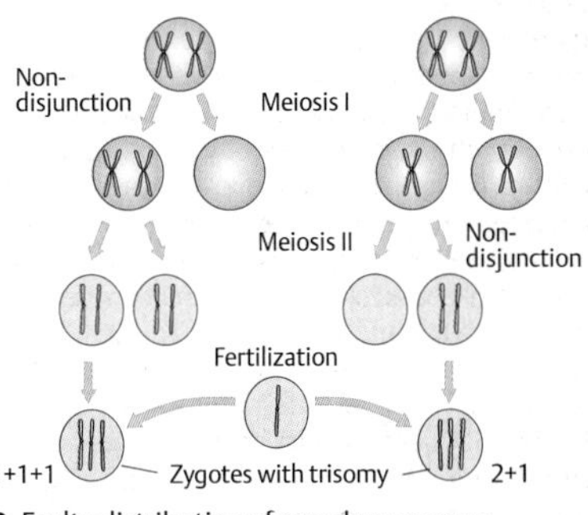


**2.** Faulty distribution of one chromosome

**D. Nondisjunction as cause of trisomy**

## Other Numerical Chromosomal Deviations

In addition to the autosomal trisomies, there are other conditions associated with an abnormal number of chromosomes. They involve either the entire set of chromosomes (triploidy or tetraploidy) or the X chromosome or Y chromosome. Deviations from the normal number of X or Y chromosomes comprise about half of all chromosomal aberrations in man (total frequency about 1:400).

### A. Triploidy

Triploidy is one of the most frequent chromosomal aberrations in man (1). Possible causes include a diploid spermatocyte, a diploid oocyte, or fertilization of an egg cell by two spermatozoa (dispermy, p. 196). Triploidy usually leads to spontaneous miscarriage within the first four months of pregnancy. The fetus shows numerous severe malformations (2), such as cardiac defects, cleft lip and palate, skeletal defects, and others. The additional chromosome set may be of either maternal or paternal origin, with different clinical consequences.

### B. Monosomy X (Turner syndrome)

Monosomy X (karyotype 45,XO) is a frequent chromosomal aberration, representing about 5% of those in humans at conception. However, of 40 zygotes with monosomy X, only one will develop to birth. The phenotypic spectrum is very wide. During the fetal stage, (1) lymphedema of the head and neck result in cystic hygroma, large multilocular thin-walled lymphatic cysts. Congenital cardiovascular defects, especially involving the aorta, and kidney malformations are frequent. An important component of the disease is the absence of ovaries, which develop only as connective tissue (streak gonads). Small stature is always a feature (average adult height about 150 cm). In newborns, webbing of the neck (pterygium colli) may be present as a residual of the lymphedema (clinical picture of Ullrich–Turner syndrome). On the other hand, the manifestations may be mild (2). Very frequently, pure monosomy is not present, but rather chromosomal mosaicism with normal cells (45,XO/46,XX) or a structurally altered X chromosome (deletion of the short arm, isochromosome of the long or short arm, ring chromosome, or other).

### C. Additional X or Y chromosomes

An additional X chromosome in males (47,XXY) leads to the clinical picture of Klinefelter syndrome after puberty when untreated (1). This includes tall stature, absent or decreased development of male secondary sex characteristics, and infertility due to absent spermatogenesis. With an additional Y chromosome (47,XYY) no unusual phenotype results (2). Girls with three X chromosomes (47,XXX) are also physically unremarkable (3). However, learning disorders and delayed speech development have been observed in some of these children.

### D. Wide spectrum of chromosomal aberrations in human fetuses

The relative proportions of the various trisomies observed in fetuses after spontaneous abortion differ. The most frequent is trisomy 16, which accounts for about 5% of all autosomal trisomies. Autosomal monosomies lead to death of the embryo within the first days or weeks. (Data after Lauritsen, 1982).

### References

DeGrouchy, J., Turleau, C.: Clinical Atlas of Human Chromosomes. 2nd ed. John Wiley & Sons, New York, 1984.

Lauritsen, J.G.: The cytogenetics of spontaneous abortion. Res. Reproduct. **14**:3–4, 1982.

Schinzel, A.: Catalogue of Unbalanced Chromosome Aberrations in Man. W. de Gruyter, Berlin, 1984.

**Triploidy**

- Most frequent chromosomal aberration (15%) in fetuses following spontaneous abortion
- Severe growth retardation, early lethality
- Occasional liveborn infant with severe malformation
- Dispermia a frequent cause

1.

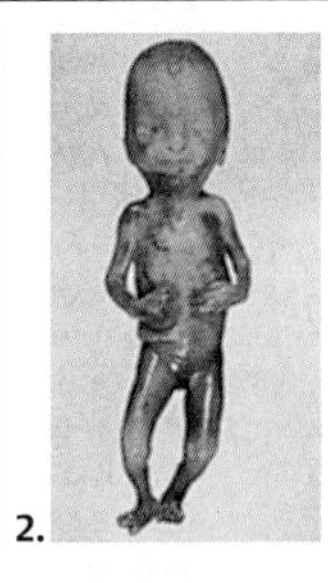

2.

3.

**A. Triploidy**

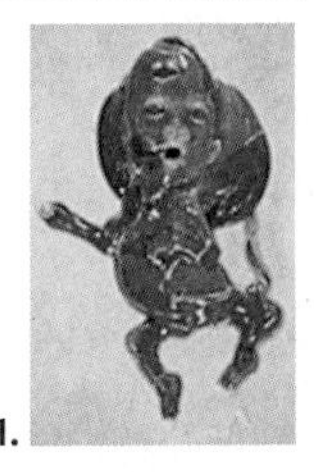

1.

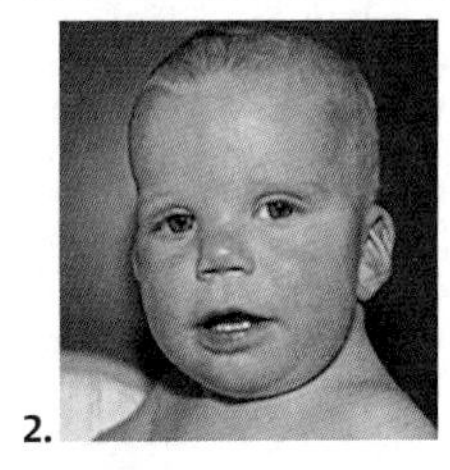

2.

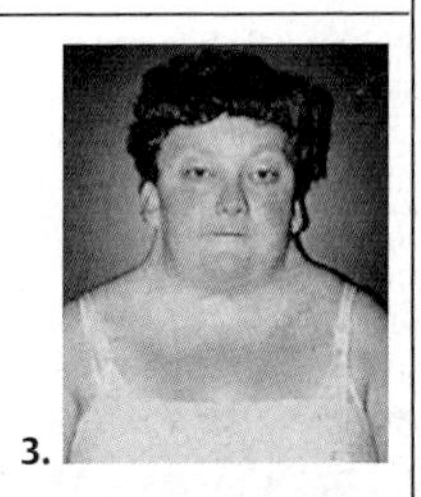

3.

**B. Monosomy X (Turner syndrome; 45, XO)**

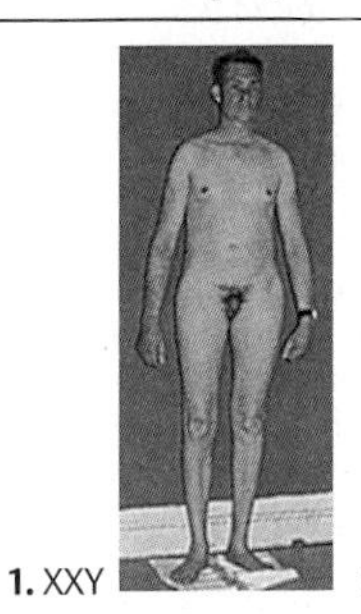

**1.** XXY

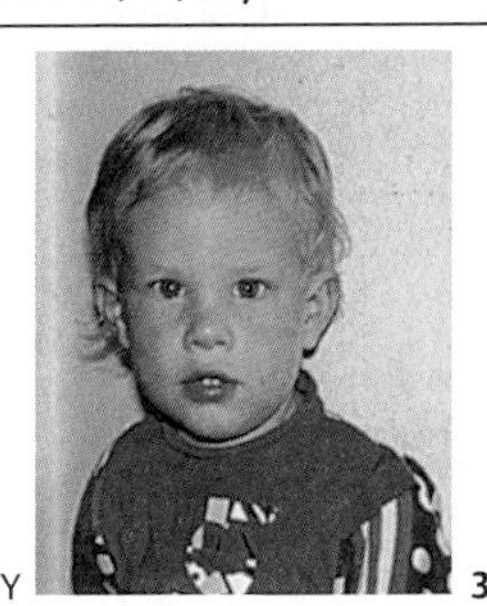

**2.** XYY

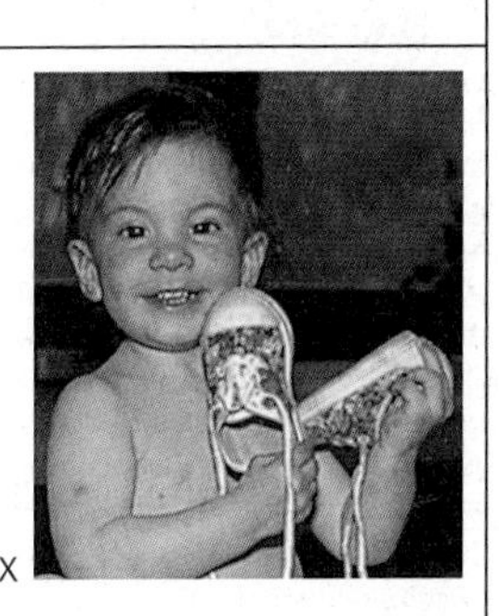

**3.** XXX

**C. Additional X or Y chromosome**

Proportion of autosomal trisomies in 669 trisomic fetuses following spontaneous abortion

| Trisomic chromosomes | 1 | 2 | 3 | 4 | 5 | 6 | 7 | 8 | 9 | 10 | 11 | 12 | 13 | 14 | 15 | 16 | 17 | 18 | 19 | 20 | 21 | 22 |
|---|---|---|---|---|---|---|---|---|---|---|---|---|---|---|---|---|---|---|---|---|---|---|
| Proportion (%) | 0.0 | 4.9 | 0.6 | 2.5 | 0.2 | 0.5 | 4.0 | 3.9 | 2.7 | 2.0 | 0.3 | 1.0 | 4.6 | 4.6 | 7.7 | 32.3 | 0.6 | 5.1 | 0.2 | 2.7 | 9.4 | 10.2 |

**D. Wide spectrum of chromosomal aberrations in human fetuses**

## Deletions and Duplications

Deletions and duplications are important structural aberrations of chromosomes. Deletion, which causes hemizygosity and functional haploinsufficiency for the loci involved, may occur *de novo* or be the result of the meiotic segregation of a parental balanced reciprocal translocation (see p. 198). Duplication of a chromosomal segment leads to partial trisomy, resulting in functional imbalance of the genes contained in the involved region.

### A. Deletion 5p–: Cri-du-chat syndrome

In 1963, Lejeune and his co-workers in Paris described children with a partial deletion of the short arm of a chromosome 5 (5p–) and retarded mental and physical development. About 15% of the parents show a translocation of chromosome 5. In these cases, the risk of recurrence of the disorder is increased. Affected infants have prolonged, high-pitched crying resembling that of a kitten (cri-du-chat, cat cry).

### B. Deletion 4p–: Wolf–Hirschhorn syndrome

Described in 1964 independently by U. Wolf in Freiburg and K. Hirschhorn in New York and their co-workers, this is a characteristic phenotype resulting from a partial deletion of chromosomal material of the short arm of a chromosome 4. Variable but considerable mental and statomotoric retardation is associated with characteristic facial features (1, 2) and with midline defects (cleft palate, hypospadias), coloboma of the iris, congenital heart defects, and other malformations. In some patients the deletion can only be detected by FISH. The simplified map of 4p16 (3) shows the critical chromosomal region (WHSCR, Wolf–Hirschhorn critical region). (Figure adapted from Wright et al., 1999).

### C. Microdeletion syndromes

Of the more than 20 different microdeletion syndromes (for review see Spinner and Emanuel, 1996; Budarf and Emanuel, 1997) three are presented here. The Williams–Beuren syndrome (1, McKusick 194050, 130160) usually presents with characteristic facial features ("elfinlike"), infantile hypercalcemia, supravalvular aortic stenosis, growth retardation, and impaired mental development. The underlying deletion involves the long arm of chromosome 7 at q11.23. The gene for elastin (*ELN*) is lost most frequently. Deletion of 22q11 leads to a group of clinically different but overlapping disorders (DiGeorge syndrome, McKusick 188400), characterized by absence or hypoplasia of the thymus and the parathyroid glands and malformations of the aortic arch; velocardiofacial syndrome, McKusick 192430; conotruncal cardiac defects, McKusick 217095; and others (2). The Rubinstein–Taybi syndrome (McKusick 180849) is characterized by typical facial features (3), broad thumbs and toes, and their associated radiological changes, mental retardation, and other features. A deletion of 16p13.3 is detectable in about 12% of patients. Point mutations in the CREB-binding gene (*CBP* gene) cause this disorder.

### D. Phenotype of duplication 5q at different ages

A unique duplication illustrates the similar facial phenotypes at different ages: in a fetus at 22 weeks gestation (1), in a 5-month-old infant (2), and in an 8-year-old child. The affected individuals are siblings in one family. The partial duplication 5q33-qter resulted from a paternal reciprocal translation (Passarge et al., 1982). A number of other duplications are associated with characteristic phenotypes.

### References


Budarf, M.L., Emanuel, B.S.: Progress in the autosomal segmental aneusomy syndromes (SASs): single or multi-locus disorders? Hum. Mol. Genet. **6**:1657–1665, 1997.

Meng, X., et al.: Complete physical map of the common deletion region in Williams syndrome and identification and characterization of three novel genes. Hum. Genet. **103**: 590–599, 1998.

Passarge, E., et al.: Fetal manifestation of a chromosomal disorder: partial duplication of the long arm of chromosome 5 (5q33-qter). Teratology **25**:221–225, 1982.

Petrij, F., et al.: Diagnostic analysis of the Rubinstein-Taybi syndrome: five cosmids should be used for microdeletion detection and low number of protein truncating mutations. J. Med. Genet. **37**:168–176, 2000.

Wright, T.J., et al.: Comparative analysis of a novel gene from the Wolf-Hirschhorn/Pitt-Rogers-Danks syndrome critical region. Genomics **59**:203–212, 1999.

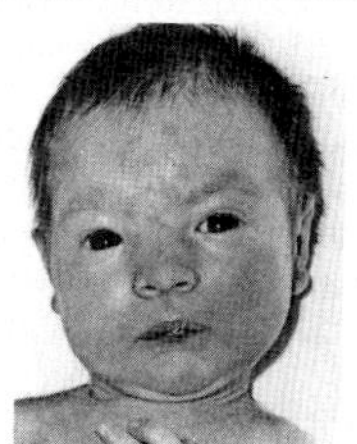

7 days

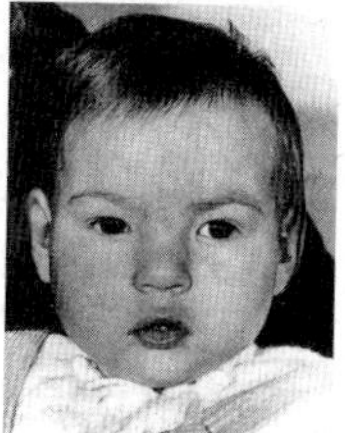

9 months

3 years

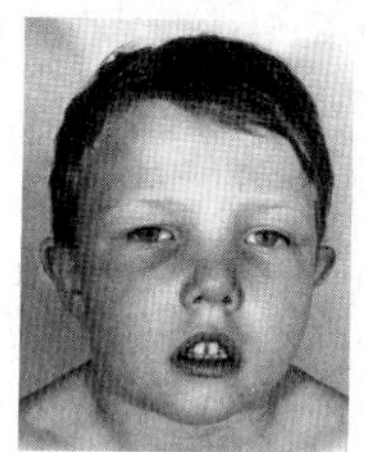

6 years

**A. Deletion 5p–: Cri-du-chat syndrome**

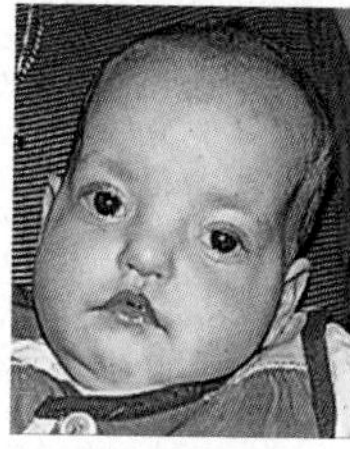

**1.** Age: 1 1/4 years

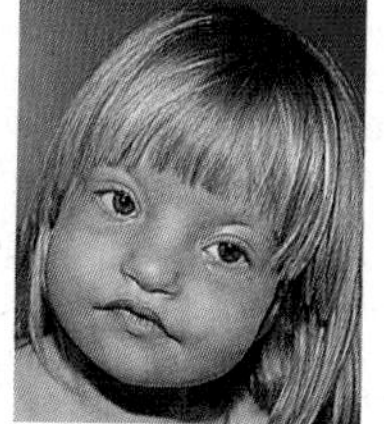

**2.** Age: 4 years

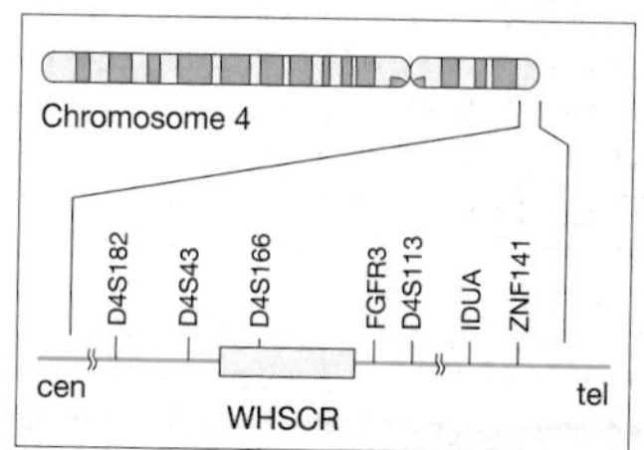


**3.** Scheme of physical map of 4p16

**B. Deletion 4p–: Wolf–Hirschhorn syndrome**

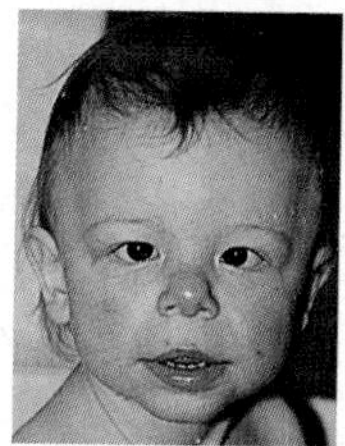

**1.** Williams–Beuren

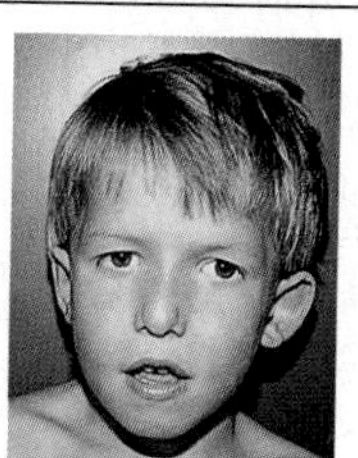

**2.** Del22q11

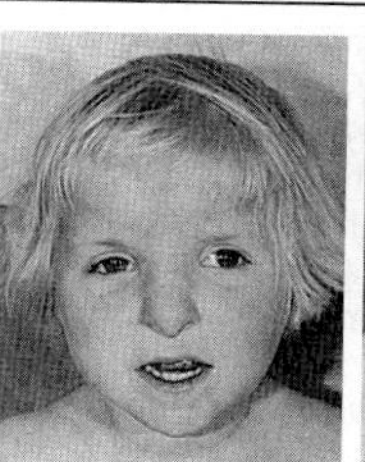

**3.** Rubinstein–Taybi syndrome

**C. Other microdeletion syndromes (examples)**

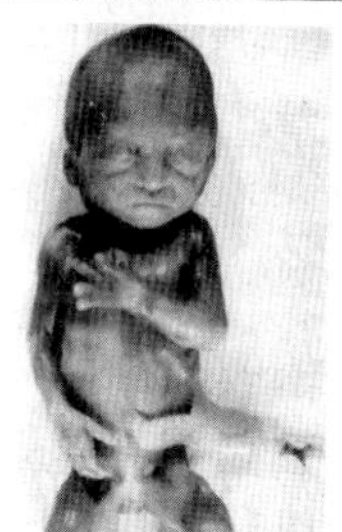

**1.** Fetus: 22nd week

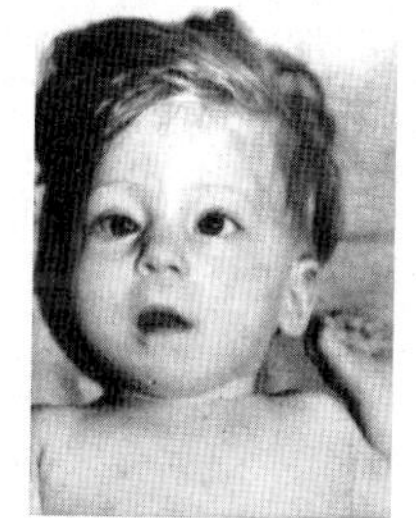

**2.** 5 months

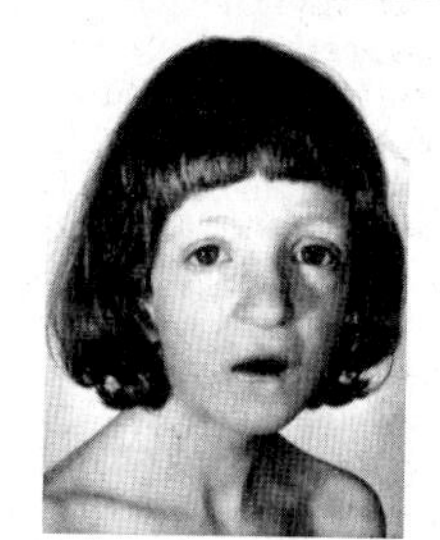

**3.** 8 years

**D. Phenotype of duplication 5q at different ages**

# A Brief Guide to Genetic Diagnosis

## Principles

The diagnosis of a genetic disease requires a systematic approach that takes clinical and genetic considerations into account. Whereas clinical medicine tends to classify diseases according to organ system, age of onset, gender, or primary method of detection (radiology, imaging techniques), medical genetics, like pathology, is oriented towards the basic cause or lesion, in this case the gene or genes affected by a relevant genetic change. Genetic diagnosis is based on an interdisciplinary analysis of all clinical and laboratory data from a genetic perspective.

### A. Genetic diagnosis, a multistep procedure

The phenotype, which is the clinical manifestations including individual and family history in medical terms, is the starting point. The first decision the medical geneticist must make is whether a pattern of manifestations can be recognized. Tools that can assist in this decision include training and personal experience; appropriate textbooks and other literature; and online search systems such as OMIM, MEDLINE, PubMed, POSSUM, London Dysmorphology Data Base for congenital malformations, and cytogenetic databases.

If a disease pattern can be recognized, the next decision concerns the category of disease. Although difficult to establish in practice, the disease category is important for the next steps to be taken. For this purpose, the McKusick catalog of human genes and diseases, Mendelian Inheritance in Man (MIM) and its online version (OMIM) are indispensable. The possibility of genetic heterogeneity must be considered at this stage. The term genetic heterogeneity refers to a phenotype (disease) that has different causes. A particular phenotype may be caused by mutations at different loci (*locus heterogeneity*) or by different mutant alleles at the same locus (*allele heterogeneity*). All genetic diagnostic procedures should be preceded by genetic counseling, which properly includes obtaining the (informed) consent of the persons involved.

### B. Genotype analysis by PCR typing of a polymorphic restriction site

Genotype analysis by PCR typing of a polymorphic restriction site is preferred to the more laborious Southern blot hybridization (see p. 62). (Figure adapted from Strachan and Read, 1999).

### C. Protein truncation test (PTT)

This is a test for frameshift, splice, or nonsense mutations that leads to a truncated protein due to an early stop codon created downstream of the mutation. The truncated protein is detected in an assay based on an in-vitro translation system. The translation will be interrupted at a premature stop codon resulting from the mutation. The size of the newly translated protein is determined by gel electrophoresis. PTT detects the approximate location of the mutation as reflected by the size of the mutant protein. PTT is useful in studying genes with frequent nonsense mutations, such as the *APC*, *BRCA1*, and *BRCA2* genes. However, it cannot be applied for genes with frequent missense mutations. The figure shown here is highly schematic. (Adapted from Strachan and Read, 1999; and Beaudet, 1998).

### References


Aase, J.M.: Diagnostic Dysmorphology. Plenum Medical Book Company, New York, 1990.

Beaudet, A.L.: Genetics and disease, pp. 365–395. In: Fauci, A.S., et al., eds., Harrison's Principles of Internal Medicine. 14th ed. McGraw-Hill, New York, 1998.

Jones, K.L.: Smith's Recognizable Patterns of Human Malformation. 5th ed. W.B. Saunders, Philadelphia, 1997.

McKusick, V.A.: Mendelian Inheritance in Man. A Catalog of Human Genes and Genetic Disorders. 12th ed. Johns Hopkins University Press, Baltimore, 1998.

Strachan, T., Read, A.P.: Human Molecular Genetics. 2nd ed. Bios Scientific Publishers, Oxford, 1999.

van der Luijit, R., et al.: Rapid detection of translation terminating mutations at the adenomatous polyposis (APC) gene by direct protein truncation test. Genomics 20:1–4, 1994.

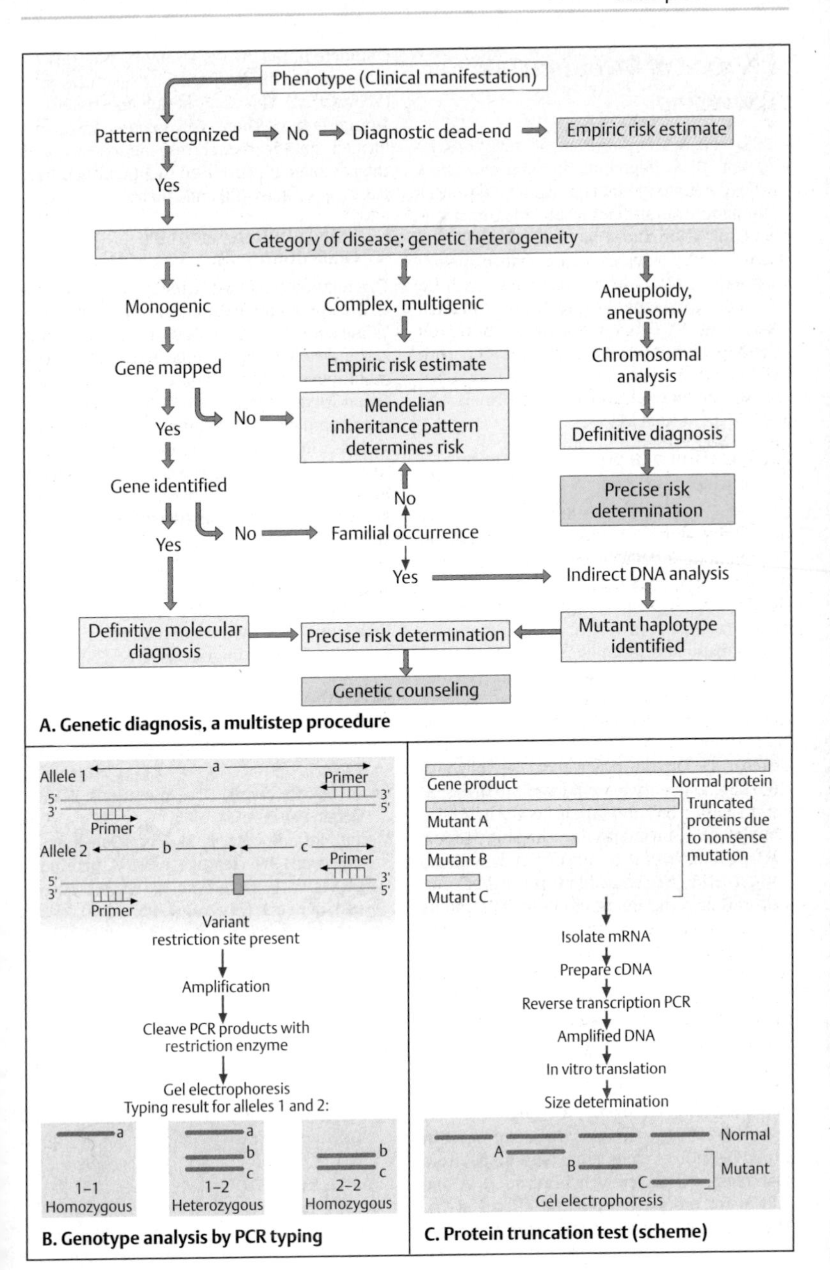
Phenotype (Clinical manifestation)
Pattern recognized
No
Diagnostic dead-end
Empiric risk estimate
Yes
Category of disease; genetic heterogeneity
Monogenic
Complex, multigenic
Aneuploidy, aneusomy
Gene mapped
Empiric risk estimate
Chromosomal analysis
Yes
No
Mendelian inheritance pattern determines risk
Definitive diagnosis
Gene identified
No
Precise risk determination
Yes
No
Familial occurrence
Yes
Indirect DNA analysis
Definitive molecular diagnosis
Precise risk determination
Mutant haplotype identified
Genetic counseling
Allele 1
a
Primer
5'
3'
Primer
Allele 2
b
c
Primer
5'
3'
Primer
Variant restriction site present
Amplification
Cleave PCR products with restriction enzyme
Gel electrophoresis
Typing result for alleles 1 and 2:
1–1 Homozygous
1–2 Heterozygous
2–2 Homozygous
Gene product
Normal protein
Mutant A
Mutant B
Mutant C
Truncated proteins due to nonsense mutation
Isolate mRNA
Prepare cDNA
Reverse transcription PCR
Amplified DNA
In vitro translation
Size determination
A
B
C
Normal
Mutant
Gel electrophoresis

**A. Genetic diagnosis, a multistep procedure**

**B. Genotype analysis by PCR typing**

**C. Protein truncation test (scheme)**

# Detection of Mutations without Sequencing

In addition to the detection of mutations by different DNA fragments in Southern blots (p. 62), there are methods based on differences in the hybridization of mutated and normal segments of DNA. Incomplete hybridization is determined by using short segments of single-stranded DNA (oligonucleotides) with a sequence complementary to the investigated region (see A). Other methods are based on demonstrating incomplete hybridization with mRNA (see B) or on the fact that a hybridized segment of normal and mutant DNA is less stable than normal DNA.

## A. Detection of a point mutation by oligonucleotides

Short segments of DNA (oligonucleotides) are used to determine whether there is a mutation in a segment of DNA (1, normal DNA; 2, mutation from G to A). An oligonucleotide is a synthetically produced DNA segment about 20 nucleotides long; its sequence is complementary to a corresponding segment of the investigated gene. It hybridizes completely with its complementary segment (3). If a mutation, here from G to A (1), is located in this region, hybridization will not be perfect at this site (mismatch) (4). On the other hand, an oligonucleotide that is complementary to the DNA segment with the mutation will hybridize completely (allele-specific oligonucleotide, ASO) (5). This hybridizes incompletely with the normal DNA (6). By parallel use of both nucleotides, mutant and nonmutant DNA can be differentiated. The test results (7) show the hybridization of mutated DNA and of control DNA with the allele-specific oligonucleotides (ASO 1 for the control, ASO 2 for the mutation). Hybridization is indicated by a signal (dot-blot analysis).

## B. Demonstration of a point mutation by ribonuclease A cleavage

The basis for this method is that a normal DNA strand hybridizes completely with mRNA from that region. Completely hybridized DNA and mRNA are protected from the effects of the RNA-splitting enzyme ribonuclease A (ribonuclease protection assay). Hybridization is incomplete in the area of a mutation. In this region, mRNA will be cleaved by ribonuclease A (RNAase A). This can be demonstrated by Southern blot. There will be two fragments formed that together correspond to the size of the completely hybridized fragment (600 base pairs (bp), versus 400 and 200 bp).

## C. Denaturing gradient gel electrophoresis

This method exploits differences in the stability of DNA segments with and without mutation. While double-stranded DNA of a control person is completely complementary (homoduplex), a mutation leads to a mismatch at the site of mutation (heteroduplex). This DNA is less stable than completely complementary DNA strands (it has a lower melting point). If normal DNA (control) and DNA with the mutation are placed in a gel with an increasing concentration gradient of formamide (denaturing gradient gel), the mutant and normal DNA can subsequently be differentiated in a Southern blot. The normal DNA remains stable to higher concentrations of formamide and migrates farther than mutant DNA, which dissociates earlier and therefore does not migrate as far.

## References

Caskey, C.T.: Disease diagnosis by recombinant DNA methods. Science **236**:1223–1229, 1987.

Dean, M.: Resolving DNA mutations. Nature Genet. **9**:103–104, 1995.

Mashal, R.D., Koontz, J., Sklar, J.: Detection of mutations by cleavage of DNA heteroduplexes with bacteriophage resolvases. Nature Genet. **9**:177–183, 1995.

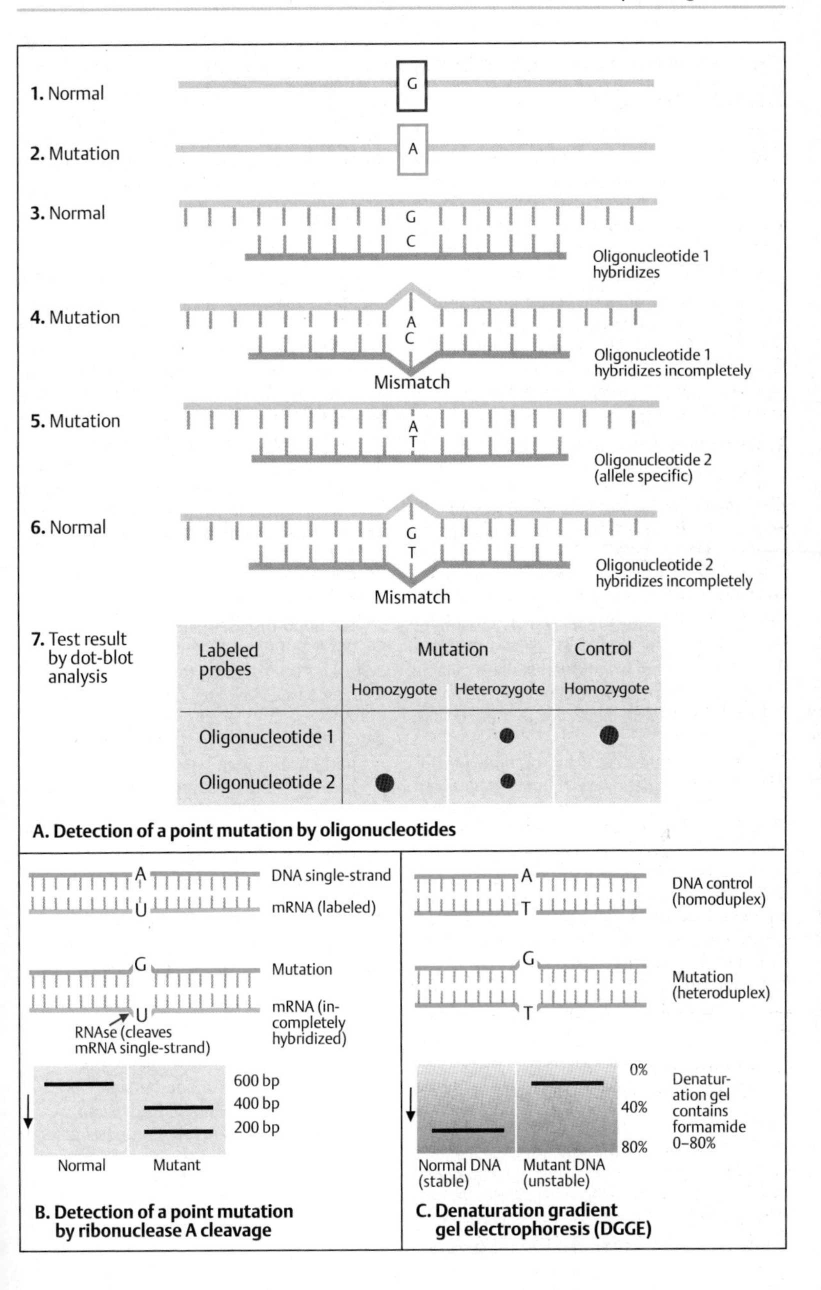


**A. Detection of a point mutation by oligonucleotides**

**B. Detection of a point mutation by ribonuclease A cleavage**

**C. Denaturation gradient gel electrophoresis (DGGE)**

# Chromosomal Location of Monogenic Diseases

## *Chromosomal Location of Human Genetic Diseases*

Nowhere is the growth of knowledge about disease-causing mutations in human genes more apparent than in the mapping of their locations to specific sites on individual chromosomes. This progress is documented in twelve published editions of Mendelian Inheritance in Man. A Catalog of Human Genes and Disorders (MIM) by Victor A. McKusick, M.D., at the Johns Hopkins University School of Medicine. Initiated in the early 1960s, its first edition in 1966 contained a total of 1487 entries without a single autosomal gene mapped. This was achieved in 1968 at the time of the second edition (1545 entries). The subsequent editions reveal an entry-doubling time of about 15 years (3368 in the 6th edition (1983), 5710 entries in the 10th edition (1992), 8587 entries in the 12th edition (1998) and 10848 on 21 September 1999 (Hamosh et al., 2000). Since 1987 the McKusick catalog is internationally available online from the National Library of Medicine (Online Mendelian Inheritance in Man, OMIM, see references).

Regularly updated, OMIM has become a major source of information on human genes and genetic diseases. Each entry has a unique 6-digit identifying number and is assigned to one of five catalogs according to genetic category: (1) autosomal dominant, (2) autosomal recessive, (3) X-chromosomal, (4) Y-chromosomal, and (5) mitochondrial. Autosomal entries initiated since 1994 begin with the digit 6. The McKusick catalog has provided a systematic basis for the genetics of man comparable to the first periodic table of chemical elements by Dimitrij I. Mendelyev in 1869 or to the "Chronologisch-thematisches Verzeichnis sämtlicher Tonwerke Wolfgang Amade Mozarts" by Ludwig Alois Ferdinand Köchel in 1862.

A special feature of the McKusick catalog is a map of disease-related gene loci assigned to specific chromosomal sites, called The Morbid Anatomy of the Human Genome. Since this first appeared in 1971 (3rd edition) on a single page, the complete information can no longer be presented in a readable printed version. The map of disease loci presented here on the next five pages, therefore, represents selected entries. For complete information, the reader is referred to the network of data available through OMIM. However, the maps shown on the following pages do provide an overview. More than 1000 disease genes have been mapped as of early 2000.

The McKusick catalog also reflects an important difference between customary clinical medicine and medical genetics. Whereas medicine classifies diseases according to their main manifestations, organ systems, age, gender, and other criteria related to the phenotype, medical genetics focuses on the genotype. The gene locus involved, the type of mutation, and genetic heterogeneity provide the basis for disease classification. This expands the concept of disease beyond the clinical manifestation and age of onset (see Childs, 1999).

### References


Antonarakis, S. E., McKusick, V.A.: OMIM passes the 1,000-disease-gene mark. Nature Genet. **25**:11, 2000.

Boguski, M. S.: Hunting for genes in computer data bases. New Eng. J. Med. **333**:645 – 647, 1995.

Boyadiv, S. A., Jabs, E.W.: Online Mendelian Inheritance in Man (OMIM) as a knowledge base for human developmental disorders. Clin. Genet. **7**:253 – 266, 2000.

Childs, B.: Genetic Medicine. Johns Hopkins Univ. Press, Baltimore, 1999.

Hamosh, A., et al.: Online Mendelian Inheritance in Man (OMIM). Hum. Mutat. **15**:57 – 61, 2000.

Lawrence, S., Giles, C.L.: Assessibility of information in the Web. Nature **400**:107 – 109, 1999.

McKusick, V.A.: Mendelian Inheritance in Man. Catalog of Human Genes and Genetic Disorders. 12th ed. Johns Hopkins Univ. Press, Baltimore, 1998. Online Version OMIM at: (http://www.ncbi.nlm.nih.gov/Omiml).

Pelz, J., Arendt, V., Kunze, J.: Computer assisted diagnosis of malformation syndromes: an evaluation of three databases (LDDB, POSSUM, and SYNDROC). Am. J. Med. Genet. **63**:257 – 267, 1996.

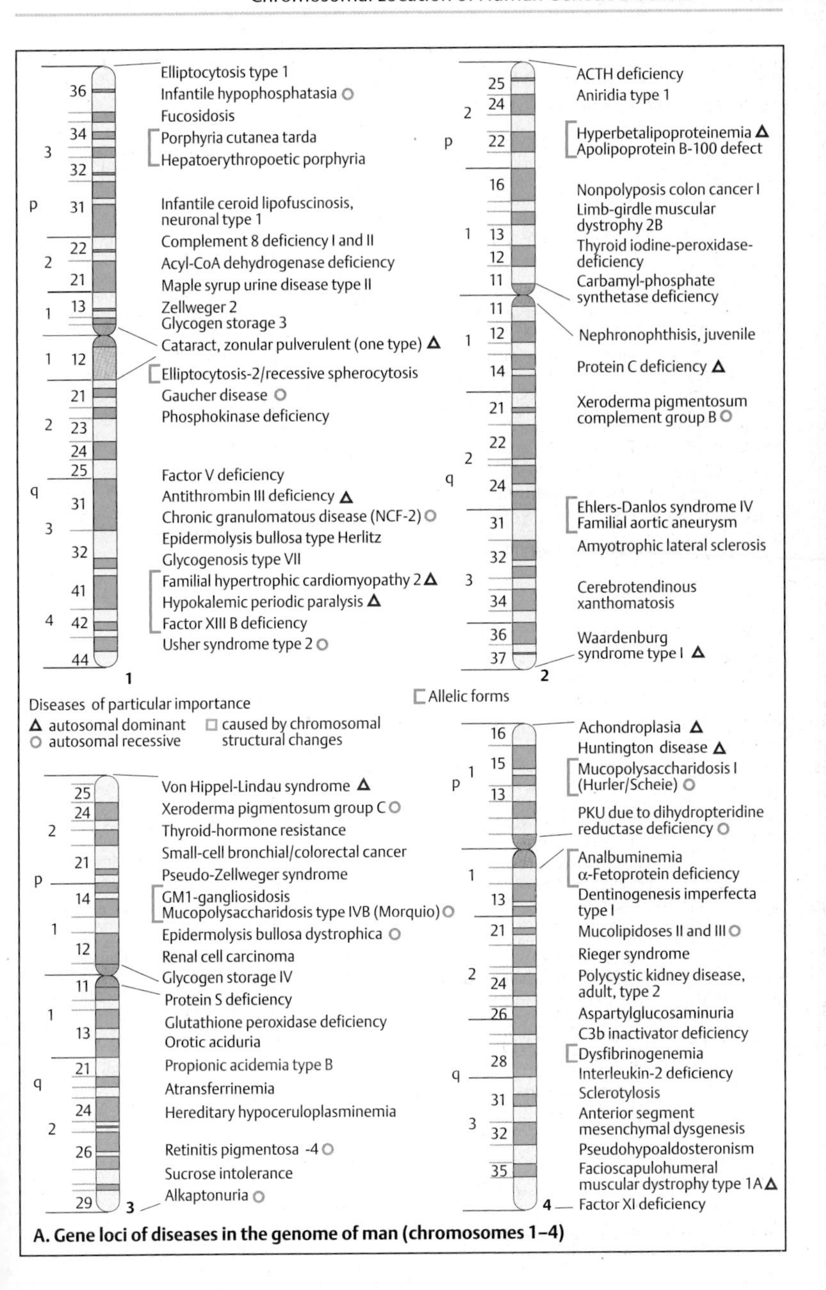


**A. Gene loci of diseases in the genome of man (chromosomes 1–4)**

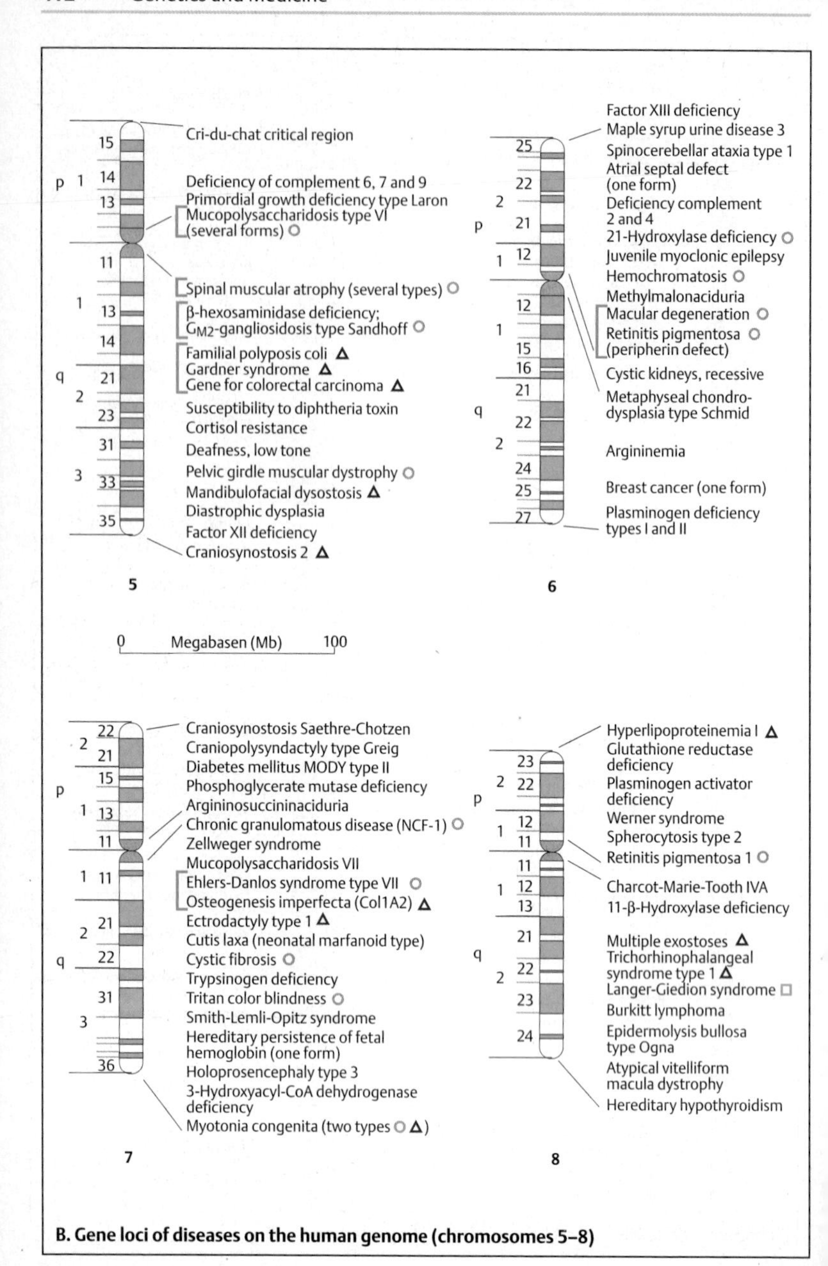

Cri-du-chat critical region
Deficiency of complement 6, 7 and 9
Primordial growth deficiency type Laron
Mucopolysaccharidosis type VI (several forms)
Spinal muscular atrophy (several types)
β-hexosaminidase deficiency; GM2-gangliosidosis type Sandhoff
Familial polyposis coli
Gardner syndrome
Gene for colorectal carcinoma
Susceptibility to diphtheria toxin
Cortisol resistance
Deafness, low tone
Pelvic girdle muscular dystrophy
Mandibulofacial dysostosis
Diastrophic dysplasia
Factor XII deficiency
Craniosynostosis 2
5
Factor XIII deficiency
Maple syrup urine disease 3
Spinocerebellar ataxia type 1
Atrial septal defect (one form)
Deficiency complement 2 and 4
21-Hydroxylase deficiency
Juvenile myoclonic epilepsy
Hemochromatosis
Methylmalonaciduria
Macular degeneration
Retinitis pigmentosa (peripherin defect)
Cystic kidneys, recessive
Metaphyseal chondrodysplasia type Schmid
Argininemia
Breast cancer (one form)
Plasminogen deficiency types I and II
6
0 Megabasen (Mb) 100
Craniosynostosis Saethre-Chotzen
Craniopolysyndactyly type Greig
Diabetes mellitus MODY type II
Phosphoglycerate mutase deficiency
Argininosuccininaciduria
Chronic granulomatous disease (NCF-1)
Zellweger syndrome
Mucopolysaccharidosis VII
Ehlers-Danlos syndrome type VII
Osteogenesis imperfecta (Col1A2)
Ectrodactyly type 1
Cutis laxa (neonatal marfanoid type)
Cystic fibrosis
Trypsinogen deficiency
Tritan color blindness
Smith-Lemli-Opitz syndrome
Hereditary persistence of fetal hemoglobin (one form)
Holoprosencephaly type 3
3-Hydroxyacyl-CoA dehydrogenase deficiency
Myotonia congenita (two types)
7
Hyperlipoproteinemia I
Glutathione reductase deficiency
Plasminogen activator deficiency
Werner syndrome
Spherocytosis type 2
Retinitis pigmentosa 1
Charcot-Marie-Tooth IVA
11-β-Hydroxylase deficiency
Multiple exostoses
Trichorhinophalangeal syndrome type 1
Langer-Giedion syndrome
Burkitt lymphoma
Epidermolysis bullosa type Ogna
Atypical vitelliform macula dystrophy
Hereditary hypothyroidism
8

**B. Gene loci of diseases on the human genome (chromosomes 5–8)**

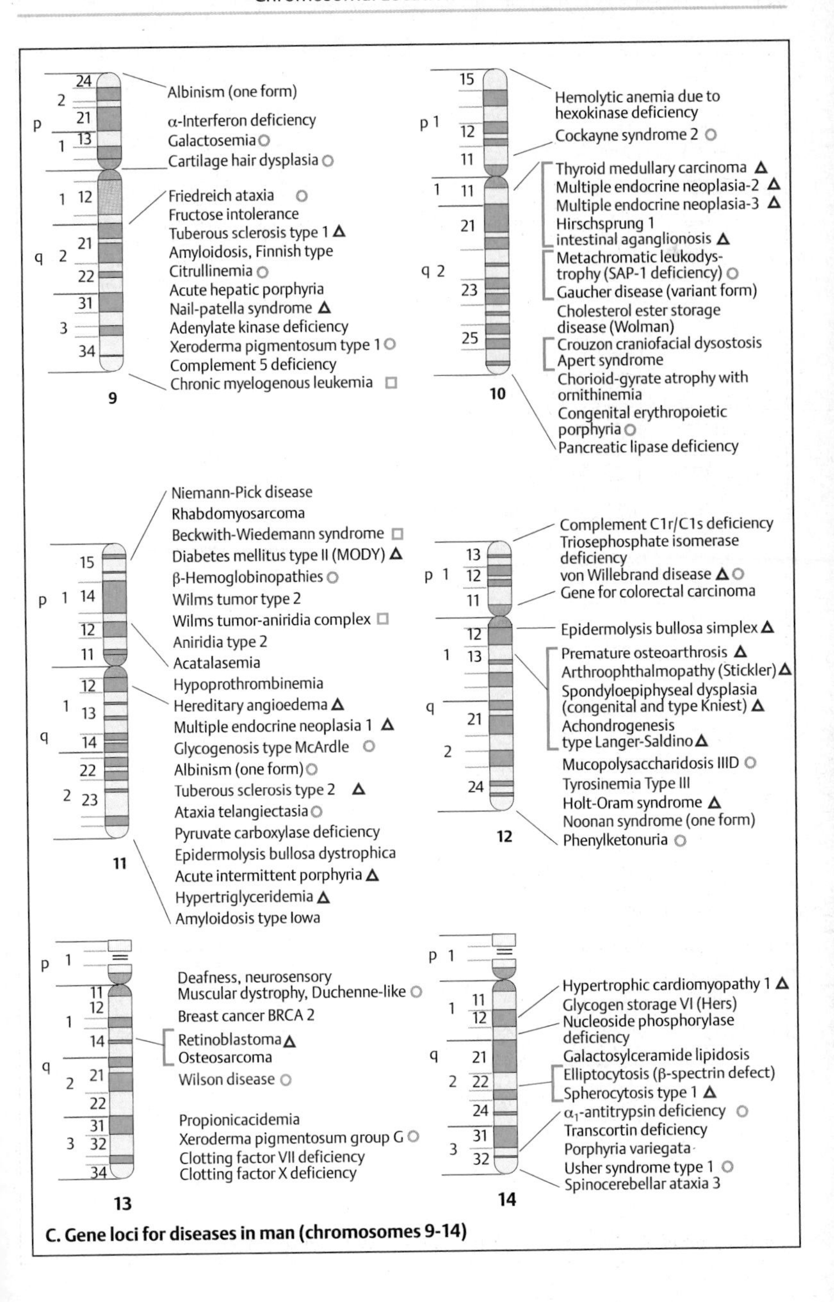

9
p
q
24
2
21
13
1
12
1
21
2
22
31
3
34
Albinism (one form)
α-Interferon deficiency
Galactosemia
Cartilage hair dysplasia
Friedreich ataxia
Fructose intolerance
Tuberous sclerosis type 1
Amyloidosis, Finnish type
Citrullinemia
Acute hepatic porphyria
Nail-patella syndrome
Adenylate kinase deficiency
Xeroderma pigmentosum type 1
Complement 5 deficiency
Chronic myelogenous leukemia
10
p 1
15
12
11
1
11
21
q 2
23
25
Hemolytic anemia due to hexokinase deficiency
Cockayne syndrome 2
Thyroid medullary carcinoma
Multiple endocrine neoplasia-2
Multiple endocrine neoplasia-3
Hirschsprung 1 intestinal aganglionosis
Metachromatic leukodys-trophy (SAP-1 deficiency)
Gaucher disease (variant form)
Cholesterol ester storage disease (Wolman)
Crouzon craniofacial dysostosis
Apert syndrome
Chorioid-gyrate atrophy with ornithinemia
Congenital erythropoietic porphyria
Pancreatic lipase deficiency
11
p 1
15
14
12
11
q
1
12
13
14
22
2
23
Niemann-Pick disease
Rhabdomyosarcoma
Beckwith-Wiedemann syndrome
Diabetes mellitus type II (MODY)
β-Hemoglobinopathies
Wilms tumor type 2
Wilms tumor-aniridia complex
Aniridia type 2
Acatalasemia
Hypoprothrombinemia
Hereditary angioedema
Multiple endocrine neoplasia 1
Glycogenosis type McArdle
Albinism (one form)
Tuberous sclerosis type 2
Ataxia telangiectasia
Pyruvate carboxylase deficiency
Epidermolysis bullosa dystrophica
Acute intermittent porphyria
Hypertriglyceridemia
Amyloidosis type Iowa
12
p 1
13
12
11
q
1
12
13
21
2
24
Complement C1r/C1s deficiency
Triosephosphate isomerase deficiency
von Willebrand disease
Gene for colorectal carcinoma
Epidermolysis bullosa simplex
Premature osteoarthrosis
Arthroophthalmopathy (Stickler)
Spondyloepiphyseal dysplasia (congenital and type Kniest)
Achondrogenesis type Langer-Saldino
Mucopolysaccharidosis IIID
Tyrosinemia Type III
Holt-Oram syndrome
Noonan syndrome (one form)
Phenylketonuria
13
p 1
q
11
12
1
14
2
21
22
31
3
32
34
Deafness, neurosensory
Muscular dystrophy, Duchenne-like
Breast cancer BRCA 2
Retinoblastoma
Osteosarcoma
Wilson disease
Propionicacidemia
Xeroderma pigmentosum group G
Clotting factor VII deficiency
Clotting factor X deficiency
14
p 1
q
1
11
12
21
2
22
24
31
3
32
Hypertrophic cardiomyopathy 1
Glycogen storage VI (Hers)
Nucleoside phosphorylase deficiency
Galactosylceramide lipidosis
Elliptocytosis (β-spectrin defect)
Spherocytosis type 1
$\alpha_1$-antitrypsin deficiency
Transcortin deficiency
Porphyria variegata
Usher syndrome type 1
Spinocerebellar ataxia 3

**C. Gene loci for diseases in man (chromosomes 9-14)**

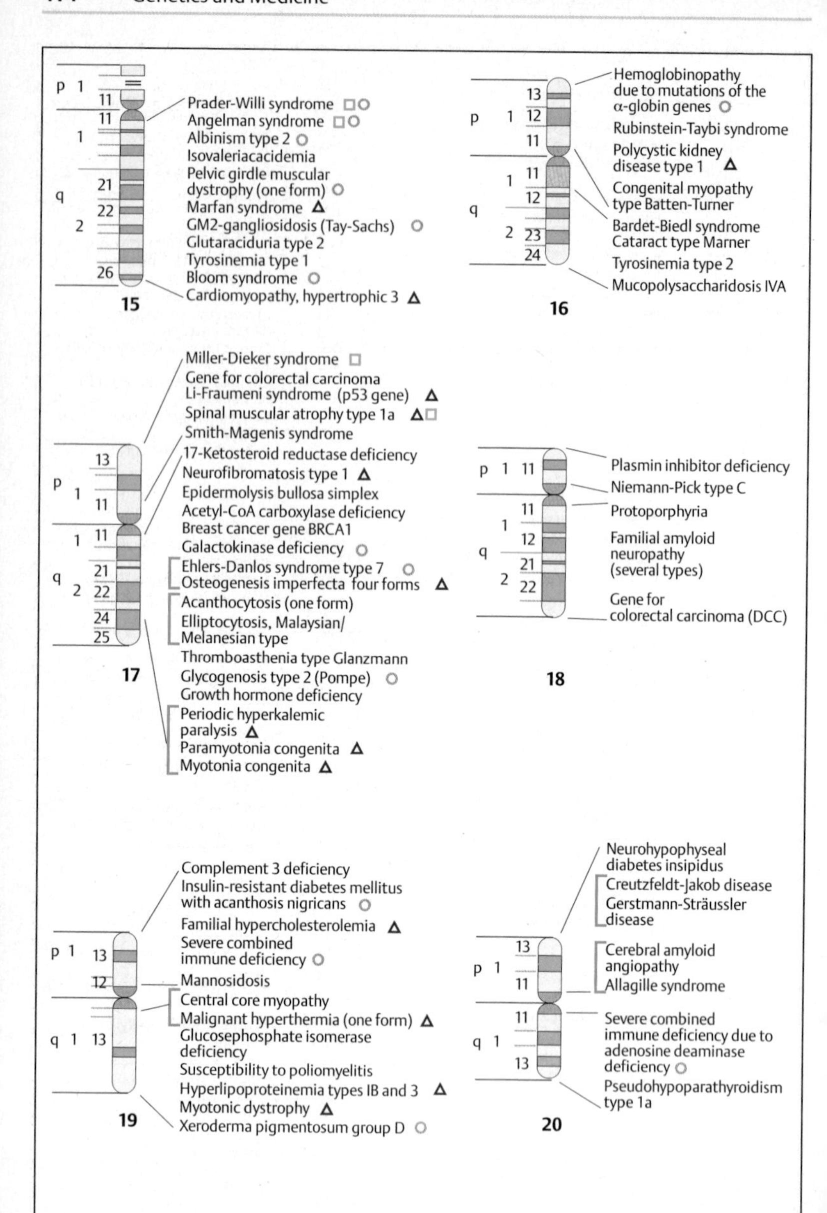

15
p 1 11
q 1 11 21 22 2 26
Prader-Willi syndrome
Angelman syndrome
Albinism type 2
Isovaleriacacidemia
Pelvic girdle muscular dystrophy (one form)
Marfan syndrome
GM2-gangliosidosis (Tay-Sachs)
Glutaraciduria type 2
Tyrosinemia type 1
Bloom syndrome
Cardiomyopathy, hypertrophic 3
16
p 1 13 12 11
q 1 11 12 2 23 24
Hemoglobinopathy due to mutations of the α-globin genes
Rubinstein-Taybi syndrome
Polycystic kidney disease type 1
Congenital myopathy type Batten-Turner
Bardet-Biedl syndrome
Cataract type Marner
Tyrosinemia type 2
Mucopolysaccharidosis IVA
17
p 1 13 11
q 1 11 2 21 22 24 25
Miller-Dieker syndrome
Gene for colorectal carcinoma
Li-Fraumeni syndrome (p53 gene)
Spinal muscular atrophy type 1a
Smith-Magenis syndrome
17-Ketosteroid reductase deficiency
Neurofibromatosis type 1
Epidermolysis bullosa simplex
Acetyl-CoA carboxylase deficiency
Breast cancer gene BRCA1
Galactokinase deficiency
Ehlers-Danlos syndrome type 7
Osteogenesis imperfecta four forms
Acanthocytosis (one form)
Elliptocytosis, Malaysian/Melanesian type
Thromboasthenia type Glanzmann
Glycogenosis type 2 (Pompe)
Growth hormone deficiency
Periodic hyperkalemic paralysis
Paramyotonia congenita
Myotonia congenita
18
p 1 11
q 1 11 12 2 21 22
Plasmin inhibitor deficiency
Niemann-Pick type C
Protoporphyria
Familial amyloid neuropathy (several types)
Gene for colorectal carcinoma (DCC)
19
p 1 13 12
q 1 13
Complement 3 deficiency
Insulin-resistant diabetes mellitus with acanthosis nigricans
Familial hypercholesterolemia
Severe combined immune deficiency
Mannosidosis
Central core myopathy
Malignant hyperthermia (one form)
Glucosephosphate isomerase deficiency
Susceptibility to poliomyelitis
Hyperlipoproteinemia types IB and 3
Myotonic dystrophy
Xeroderma pigmentosum group D
20
p 1 13 11
q 1 11 13
Neurohypophyseal diabetes insipidus
Creutzfeldt-Jakob disease
Gerstmann-Sträussler disease
Cerebral amyloid angiopathy
Allagille syndrome
Severe combined immune deficiency due to adenosine deaminase deficiency
Pseudohypoparathyroidism type 1a

**D. Gene loci of diseases on the human genome (chromosomes 15-20)**

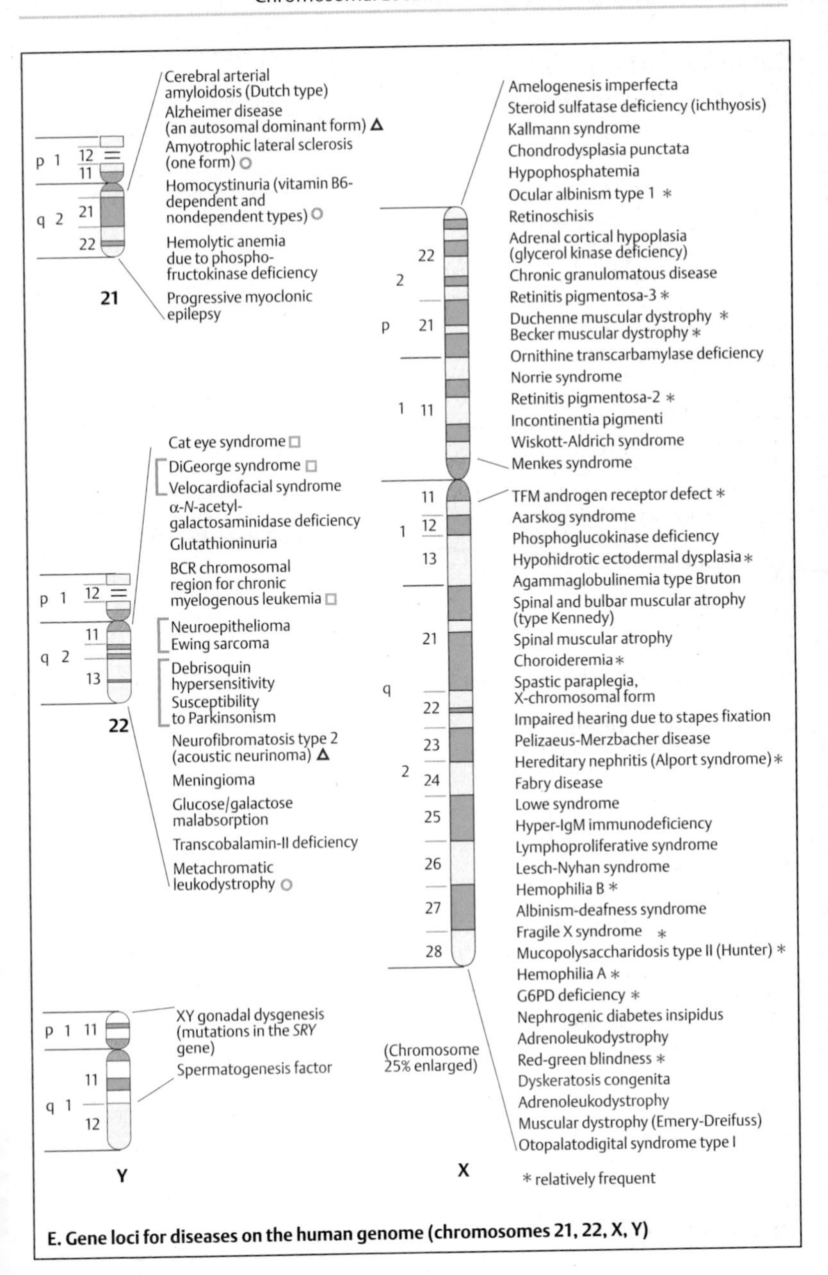


**E. Gene loci for diseases on the human genome (chromosomes 21, 22, X, Y)**

# Chromosomal Location of Human Genetic Diseases

(Alphabetic list to the maps on pp. 410–415)

Aarskog syndrome (X ch.)
Acanthocytosis (one form) (ch. 17)
Acatalasemia (ch. 11)
Acetyl-CoA carboxylase deficiency (ch. 17)
Achondroplasia (ch. 4)
ACTH deficiency (ch. 2)
Acute hepatic porphyria (ch. 9)
Acute intermittent porphyria (ch. 11)
Acyl-CoA dehydrogenase deficiency (ch. 1)
Adenylate kinase deficiency (ch. 9)
Adrenal cortical hypoplasia with glycerol kinase deficiency (X ch.)
Adrenoleukodystrophy (X ch.)
α-Interferon deficiency (ch. 9)
Agammaglobulinemia (X ch.)
Alagille syndrome (ch. 20)
Albinism (one form) (ch. 9)
Albinism (one form) (ch. 11)
Albinism type 2 (ch. 15)
Albinism–deafness syndrome (X ch.)
Alkaptonuria (ch. 3)
Alpha-N-acetylgalactosaminidase deficiency (ch. 22)
Alpha-1-antitrypsin deficiency (ch. 14)
Alpha-fetoprotein deficiency (ch. 4)
Alzheimer disease (one form) (ch. 21)
Amelogenesis imperfecta (X ch.)
Amyloidosis, Finnish type (ch. 9)
Amyloidosis type Iowa (ch. 11)
Amyotrophic lateral sclerosis (one form) (ch. 21)
Amyotrophic lateral sclerosis, juvenile (ch. 2)
α-N-acetylgalactosaminidase deficiency (ch. 22)
Analbuminemia (ch. 4)
Androgen receptor defect (X ch.)
Angelman syndrome (ch. 15)
Aniridia type 1 (ch. 2)
Aniridia type 2 (ch. 11)
Anterior segmental mesenchymal dysgenesis (ch. 4)
Antithrombin III deficiency (ch. 1)
Apert syndrome (ch. 10)
Apolipoprotein B-100 defect (ch. 2)
Argininemia (ch. 6)
Argininosuccinicaciduria (ch. 7)
Arthroophthalmopathy (Stickler syndrome) (ch. 12)
Aspartylglucosaminuria (ch. 4)
Ataxia telangiectasia (one form) (ch. 11)
Atransferrinemia (ch. 3)
Atrial septal defect (one form) (ch. 6)
Atypical vitelliform macular dystrophy (ch. 8)
Bardet–Biedl syndrome (ch. 16)
BCR chromosomal region for chronic myelogenous leukemia (ch. 22)
Becker muscular dystrophy (X ch.)
Beckwith–Wiedemann syndrome (ch. 11)
Beta-hemoglobinopathies (ch. 11)
Beta-hexosaminidase deficiency; $GM^2$–gangliosidosis type Sandhoff (ch. 5)
11-Beta-hydroxylase deficiency (ch. 8)
Bloom syndrome (ch. 15)
Breast cancer gene BRCA1 (ch. 17)
Breast cancer gene BRCA 2 (ch. 13)
Breast cancer (one form) (ch. 6)
Burkitt lymphoma (ch. 8)
Carbamylphosphate synthetase–I deficiency (ch. 4)
Cardiomyopathy, familial hypertrophic type 3 (ch. 15)
Cartilage–hair dysplasia (ch. 9)
Cat eye syndrome (ch. 22)
C3b-inactivator deficiency (ch. 4)
Central core myopathy (ch. 19)
Cerebral amyloid angiopathy (ch. 20)
Cerebral arterial amyloidosis (Dutch type) (ch. 21)
Cerebrotendinosis xanthomatosis (ch. 2)
Charcot–Marie–Tooth neuropathy type1b (ch. 1)
Charcot–Marie–Tooth neuropathy type II (ch. 1)
Charcot–Marie–Tooth neuropathy type IVa (ch. 8)
Cholesteryl ester storage disease (Wolman) (ch. 10)
Chondrodysplasia punctata (X ch.)
Choroid gyrate atrophy with ornithinemia (ch. 10)
Choroideremia (X ch.)
Chronic granulomatous disease (NCF-1) (ch. 7)
Chronic granulomatous disease (NCF-2 deficiency) (ch. 1)
Chronic granulomatous disease (X ch.)
Chronic myelogenous leukemia (ch. 9)
Citrullinemia (ch. 9)
Clotting factor VII deficiency (ch. 13)
Clotting factor X deficiency (ch. 13)

Cockayne syndrome 2 (ch. 10)
Colon cancer, familial nonpolyposis type 1 (ch. 2)
Colorectal adenocarcinoma (ch. 12)
Colorectal carcinoma (ch. 5 and 18)
Colorectal carcinoma/Li – Fraumeni syndrome (ch. 17)
Complement 2 and 4 deficiency (ch. 6)
Complement 3 deficiency (ch. 19)
Complement 5 deficiency (ch. 9)
Complement 6, 7, and 9 deficiency (ch. 5)
Complement 8 deficiency 1 and 2 (ch. 1)
Complement C1r/C1s deficiency (ch. 12)
Congenital erythropoietic porphyria (ch. 10)
Congenital myopathy type Batten – Turner (ch. 16)
Cortisol resistance (ch. 5)
Craniopolysyndactyly type Greig (ch. 7)
Craniosynostosis Saethre – Chotzen (ch. 7)
Craniosynostosis type 2 (ch. 5)
Creutzfeldt – Jakob disease (ch. 20)
Cri du chat critical region (ch. 5)
Crigler – Najjar syndrome (ch. 1)
Crouzon craniofacial dysostosis (ch. 10)
Cutis laxia (neonatal marfanoid type) (ch. 7)
Cystic fibrosis (ch. 7)
Deafness, low-tone (ch. 5)
Deafness, neurosensory (ch. 13)
Debrisoquin hypersensitivity (ch. 22)
Dentinogenesis imperfecta type I (ch. 4)
Diabetes mellitus type MODY (ch. 11)
Diabetes mellitus, MODY type II (ch. 7)
Diastrophic dysplasia (ch. 5)
DiGeorge syndrome (ch. 22)
Duchenne muscular dystrophy (X ch)
Dysfibrinogenemia (ch. 4)
Dyskeratosis congenita (X ch.)
Ectrodactyly type 1 (ch. 7)
Ehlers – Danlos syndrome type 4 (ch. 2)
Ehlers – Danlos syndrome type 7 (ch. 7)
Ehlers – Danlos syndrome type 7A1 (ch. 17)
Elliptocytosis (β-spectrin defect) (ch. 14)
Elliptocytosis, Malaysian/Melanesian type (ch. 17)
Elliptocytosis-2/recessive spherocytosis (ch. 1)
Epidermolysis bullosa dystrophica (ch. 3)
Epidermolysis bullosa simplex (ch. 12)
Epidermolysis bullosa simplex (ch. 17)
Epidermolysis bullosa type Herlitz (ch.1)
Epidermolysis bullosa type Ogna (ch. 8)
Ewing sarcoma (ch. 22)
Fabry disease (X ch.)
Facioscapulohumeral muscular dystrophy (ch. 4)
Factor V deficiency (ch. 1)
Factor XI deficiency (ch. 4)
Factor XII deficiency (ch. 5)
Factor XIIIa deficiency (ch. 6)
Factor XIII B deficiency (ch. 1)
Familial amyloid neuropathy (several types) (ch. 18)
Familial aortic aneurysm (ch. 2)
Familial hypercholesterolemia (ch. 19)
Familial hypertrophic cardiomyopathy (ch. 1)
Familial polyposis coli (ch. 5)
Fragile X syndrome (X ch.)
Friedreich's ataxia (ch. 9)
Fructose intolerance (ch. 9)
Fucosidosis (ch. 1)
G6PD deficiency (X ch.)
Galactokinase deficiency (ch. 17)
Galactose epimerase deficiency (ch. 1)
Galactosemia (ch. 9)
Galactosylceramide lipidosis (ch. 14)
Gardner syndrome (ch. 5)
Gaucher disease (ch. 1)
Gaucher disease (variant form) (ch. 10)
Gerstmann – Sträussler disease (ch. 20)
Glucose/galactose malabsorption (ch. 22)
Glucosephosphate isomerase deficiency (ch. 19)
Glutaricaciduria type 2 (ch. 15)
Glutathione peroxidase deficiency (ch. 3)
Glutathione reductase deficiency (ch. 8)
Glutathionuria (ch. 22)
Glycogenosis type VII (ch. 1)
Glycogenosis type McArdle (ch. 11)
Glycogenosis type 2 (Pompe) (ch. 17)
Glycogen storage type 3 (ch. 1)
Glycogen storage type 4 (ch. 3)
Glycogen storage VI (Hers) (ch. 14)
GM1-gangliosidosis (ch. 3)
GM2-gangliosidosis (Tay – Sachs) (ch. 15)
GM2-gangliosidosis type Sandhoff (ch. 5)
Growth hormone deficiency (ch. 17)
Hemochromatosis (ch. 6)
Hemoglobinopathies due to mutations of the α-globin genes (ch. 16)
Hemolytic anemia due to hexokinase deficiency (ch. 10)
Hemolytic anemia due to phosphofructokinase deficiency (ch. 21)
Hemophilia A (X ch.)
Hemophilia B (X ch.)
Hepatoerythropoietic porphyria (ch. 1)
Hereditary angioedema (ch. 11)
Hereditary congenital hypothyroidism (ch. 8)

Hereditary hypoceruloplasminemia (ch. 3)
Hereditary nephritis (Alport syndrome) (X ch.)
Hereditary persistence of fetal hemoglobin (one form) (ch. 7)
Hirschsprung disease (chs. 10 and 13)
Holoprosencephaly type 3 (ch. 7)
Holt–Oram syndrome (ch. 12)
Homocystinuria (B6-responsive and B6-nonresponsive forms) (ch. 21)
Huntington disease (ch. 4)
3-Hydroxyacyl-CoA dehydrogenase deficiency (ch. 7)
21-Hydroxylase deficiency (ch. 6)
Hyperbetalipoproteinemia (ch. 2)
Hyper-IgM immune deficiency (X ch.)
Hyperlipoproteinemia type 1 (ch. 8)
Hyperlipoproteinemia type 1b (ch. 19)
Hyperlipoproteinemia type 3 (ch. 19)
Hypertriglyceridemia (ch. 11)
Hypertrophic cardiomyopathy (ch. 14)
Hypochondroplasia (ch. 4)
Hypohidrotic ectodermal dysplasia (X ch.)
Hypophosphatemia (X ch.)
Hypoprothrombinemia (ch. 11)
Immune deficiency, severe combined (ch. 19)
Immunodeficiency due to ADA deficiency (ch. 20)
Impaired hearing (lower frequencies) (ch. 5)
Impaired hearing due to stapes fixation (X ch.)
Infantile ceroid lipofuscinosis, neuronal type (ch. 1)
Infantile hypophosphatasia (ch. 1)
Insulin-resistant diabetes mellitus with acanthosis nigricans (ch. 19)
Interleukin 2 deficiency (ch. 4)
Intestinal aganglionosis (Hirschsprung) (ch. 10 and 13)
Isovalericacidemia (ch. 15)
Juvenile myoclonic epilepsy (ch. 6)
Kallmann syndrome (X ch.)
17-Ketosteroid reductase deficiency (ch. 17)
Lamellar cataract (one type) (ch. 1)
Langer–Giedion syndrome (ch. 8)
Lesch–Nyhan syndrome (X ch.)
Li–Fraumeni syndrome (ch. 17)
Limb–girdle muscular dystrophy 2b (ch. 2)
Lowe syndrome (X ch. )
Lymphoproliferative syndrome (X ch.)
Macular degeneration (ch. 6)
Malignant hyperthermia (ch. 19, others)
Mandibulofacial dysostosis (Franceschetti–Klein syndrome) (ch. 5)
Mannosidosis (ch. 19)
Maple syrup urine disease type 2 (ch. 1)
Maple syrup urine disease type 3 (ch. 6)
Marfan syndrome (ch. 15)
Meningioma (ch. 22)
Menkes syndrome (X ch.)
Metachromatic leukodystrophy (ch. 22)
Metachromatic leukodystrophy (SAP–1 deficiency) (ch. 10)
Metaphyseal chondrodysplasia type Schmid (ch. 6)
Methylmalonicaciduria (ch. 6)
Miller–Dieker syndrome (ch. 17)
Morquio syndrome B (ch. 3)
Mucolipidosis types II and III (ch. 4)
Mucopolysaccharidosis type I (Hurler/Scheie) (ch. 4)
Mucopolysaccharidosis type II (X ch.)
Mucopolysaccharidosis type IVa (ch. 16)
Mucopolysaccharidosis type IVb (ch. 3)
Mucopolysaccharidosis type VI (Maroteaux–Lamy) (ch. 5)
Mucopolysaccharidosis type VII (ch. 7)
Multiple endocrine neoplasia type 1 (ch. 11)
Multiple endocrine neoplasia type 2 (ch. 10)
Multiple endocrine neoplasia type 3 (ch. 10)
Multiple exostoses (ch. 8)
Muscular dystrophy, Duchenne-like (ch. 13)
Muscular dystrophy type Becker (X ch.)
Muscular dystrophy type Duchenne (X ch.)
Muscular dystrophy type Emery–Dreifuss (X ch.)
Myotonia congenita (ch. 17)
Myotonia congenita (two types) (ch. 7)
Myotonic dystrophy (ch. 19)
Myotubular myopathy (X ch.)
Nail–patella syndrome (ch. 9)
Nephrogenic diabetes insipidus (X ch.)
Nephronophthisis, juvenile (ch. 2)
Neuroepithelioma (ch. 22)
Neurofibromatosis type 1 (ch. 17)
Neurofibromatosis type 2 (acusticus neurinoma) (ch. 22)
Neurohypophyseal diabetes insipidus (ch. 20)
Niemann–Pick disease (ch. 11)
Niemann–Pick type C (ch. 18)
Noonan syndrome (one locus) (ch. 12)
Norrie syndrome (X ch.)
Nucleoside phosphorylase deficiency (ch. 14)
Ocular albinism (X ch.)
Ornithine transcarbamylase deficiency (X ch.)
Oroticacidemia (ch. 3)
Osteogenesis imperfecta (ch. 17)

Osteogenesis imperfecta (COL1A2) (ch. 7)
Osteosarcoma (ch. 13)
Otopalatodigital syndrome type 1 (X ch.)
Pancreatic lipase deficiency (ch. 10)
Paramyotonia congenita (ch. 17)
Pelizaeus – Merzbacher disease (X ch.)
Pelvic girdle muscular dystrophy (ch. 5 and ch. 15)
Periodic hyperkalemic paralysis (ch. 17)
Phenylketonuria (PKU) (ch. 12)
Phosphoglucokinase deficiency (X ch.)
Phosphoglycerate mutase deficiency (ch. 7)
Phosphokinase deficiency (ch. 1)
PKU due to dihydropteridine reductase deficiency (ch. 4)
Plasmin inhibitor deficiency (ch. 18)
Plasminogen activator deficiency (ch. 8)
Plasminogen deficiency, types I and II (ch. 6)
Polycystic kidney disease (ch. 16 and 4)
Porphyria cutanea tarda (ch. 1)
Porphyria variegata (ch. 14)
Prader – Willi syndrome (ch. 15)
Primordial growth deficiency type Laron (ch. 5)
Progressive myoclonic epilepsy (ch. 21)
Propionicacidemia type A (ch. 13)
Propionicacidemia type B (ch. 3)
Protein C deficiency (ch. 2)
Protein S deficiency (ch. 3)
Protoporphyria (ch. 18)
Pseudoaldosteronism (ch. 4)
Pseudohypoparathyroidism type 1a (ch. 20)
Pseudo-Zellweger syndrome (ch. 3)
Red – green blindness (X ch.)
Renal cell carcinoma (ch. 3)
Retinitis pigmentosa (perpherin defect) (ch. 6)
Retinitis pigmentosa type 1 (ch. 8)
Retinitis pigmentosa type 2 (X ch.)
Retinitis pigmentosa type 3 (X ch.)
Retinitis pigmentosa type 4 (ch. 3)
Retinoblastoma (ch. 13)
Retinoschisis (X ch.)
Rhabdomyosarcoma (ch. 11)
Rieger syndrome (ch. 4)
Rubinstein – Taybi syndrome (ch. 16)
Sclerotylosis (ch. 4)
Sex reversal (XY females due to mutation in the SRY gene) (Y ch.)
Small cell bronchial carcinoma/colorectal cancer (ch. 3)
Smith – Lemli – Opitz syndrome (ch. 7)
Smith – Magenis syndrome (ch. 17)
Spastic paraplegia (X chromosomal form) (X ch.)
Spermatogenesis factor (Y ch.)
Spherocytosis type 1 (ch. 14)
Spherocytosis type 2 (ch. 8)
Spinal muscular atrophy (X ch.)
Spinal muscular atrophy Ia (ch. 17)
Spinal muscular atrophy IVa (ch. 8)
Spinal muscular atrophy Werdnig – Hoffmann and other types (ch. 5)
Spinocerebellar ataxia type 1 (ch. 6)
Spinocerebellar ataxia type 3 (ch. 14)
Spondyloepiphyseal dysplasia (congenital type) (ch. 12)
Spondyloepiphyseal dysplasia (type Kniest) (ch. 12)
Steroid sulfatase deficiency (ichthyosis) (X ch.)
Sucrose intolerance (ch. 3)
Susceptibility to diphtheria toxin (ch. 5)
Susceptibility to Parkinsonism (ch. 22)
Susceptibility to poliomyelitis (ch. 19)
T cell leukemia/lymphoma (ch. 14)
TFM androgen receptor defect (X ch.)
Thrombasthenia type Glanzmann (ch. 17)
Thyroid hormone resistance (ch. 3)
Thyroid iodine peroxidase deficiency (ch. 2)
Thyroid medullary carcinoma (ch. 10)
Transcobalamin II deficiency (ch. 22)
Transcortin deficiency (ch. 14)
Trichorhinophalangeal syndrome type 1 (ch. 8)
Triosephosphate isomerase deficiency (ch. 12)
Tritan color blindness (ch. 7)
Trypsinogen deficiency (ch. 7)
Tuberous sclerosis type 1 (ch. 9)
Tuberous sclerosis type 2 (ch. 11)
Tuberous sclerosis type 3 (ch. 12)
Tyrosinemia type 1 (ch. 14)
Tyrosinemia type 2 (ch. 16)
Usher syndrome type 1 (ch. 14)
Usher syndrome type 2 (ch. 1)
Velocardiofacial syndrome (ch. 22)
Vitelline macular dystrophy (ch. 6)
Von Hippel – Lindau syndrome (ch. 3)
Von Willebrand disease (ch. 12)
Waardenburg syndrome type 1 (ch. 2)
Werner syndrome (ch. 8)
Williams – Beuren syndrome (ch. 4)
Wilms tumor – aniridia complex (ch. 11)
Wilms tumor type 2 (ch. 11)
Wilson's disease (ch. 13)
Xeroderma pigmentosum complementation group B (ch. 2)
Xeroderma pigmentosum group C (ch. 3)
Xeroderma pigmentosum group D (ch. 19)
Xeroderma pigmentosum group G (ch. 13)

Xeroderma pigmentosum type 1 (ch. 9)
XY gonadal dysgenesis (Y ch.)
Zellweger syndrome (ch. 7)
Zellweger syndrome type 2 (ch. 1)
Zonular cataract (ch. 1)

**Caveat:** There are numerous similar disorders caused by mutations of genes at other loci, sometimes with other modes of inheritance.
This list and the corresponding maps are not complete, but give examples. For a complete map and list, see McKusick's catalog, 12th and OMIM edition, 1998.

# General References


Alberts, B., Bray, D., Lewis, J., Raff, M., Roberts, K., Watson, J.D.: Molecular Biology of the Cell. 3rd ed. Garland Publishing Co., New York, 1994.

Alberts, B., Bray, D., Johnson, A., Lewis, J., Raff, M., Roberts, K., Walter, P.: Essential Cell Biology. An Introduction to the Molecular Biology of the Cell. Garland Publishing Co., New York, 1998.

Ayala, F.J., Kiger, J.A.: Modern Genetics. 2nd ed. Benjamin Cummings Publishing Co., Menlo Park, CA, 1984.

Beaudet, A.L.: Genetics and disease. p. 365–394. In: Harrison's Principles of Internal Medicine. 14th ed., A.S. Fauci et al., eds. McGraw-Hill, New York, 1998.

Brown, T.A.: Genomes. Bios Scientific Publishers., Oxford, 1999.

Burns, G.W., Bottinger, P.J.: The Science of Genetics. 6th ed. Macmillan Publishing Co., New York and London, 1989.

Campbell, N.A.: Biology, 2nd ed. Benjamin Cummings Publishing Co., Menlo Park, California, 1990.

Cavalli-Sforza, L.L., Menozzi, P., Piazza, A.: The History and Geography of Human Genes. Princeton University Press, Princeton, New Jersey, 1994.

Childs, B.: Genetic Medicine. A Logic of Disease. Johns Hopkins University Press, Baltimore, 1999.

Connor, J.M., Ferguson-Smith, M.A.: Essential Medical Genetics, 4th ed., Blackwell Scientific Publishers, Oxford, 1993.

Dobzhansky, T.: Genetics and the Evolutionary Process. Columbia University Press, New York and London, 1970.

Emery A.E.H. Methodology in Medical Genetics. 2nd ed. Edinburgh, Churchill Livingstone, 1986.

Falconer, D.S.: Introduction to Quantitative Genetics, 2nd ed. Longman, London, 1981.

Fauci, A.S., Braunwald, E., Isselbacher, K.J., Wilson, J.D., Martin, J.B., Kasper, D.L. Hauser, S.L., Longo, D.L., eds.: Harrison's Principles of Internal Medicine, 14th ed., McGraw-Hill, New York, 1998.

Gehring, W.: Master Control Genes in Development and Evolution. The Homeobox Story. Yale University Press, New Haven, 1999.

Gelehrter, T.D., Collins, F.S., Ginsburg, D.: Principles of Medical Genetics, 2nd ed., Williams & Wilkins, Baltimore, 1998.

Gilbert-Barness, E., Barness, L.: Metabolic Diseases. Foundations of Clinical Management, Genetics and Pathology. Eaton Publishing, Natick, MA 01760 USA, 2000.

Glover, D., Hames, B.D., eds: DNA Cloning: A Practical Approach. Core Techniques. Vol. 1, IRL Press, Oxford, 1995.

Griffith, A.J.F., Suzuki, D.T., Miller, J.H., Lewontin, R.C., Gelbart, W.M.: An Introduction to Genetic Analysis. 7th ed. W.H. Freeman & Co., New York, 2000.

Harper, P.S.: Practical Genetic Counselling. 5th ed., Butterworth-Heinemann, Oxford, 1998.

Jameson, J.L., ed: Principles of Molecular Medicine. Humana Press, Totowa, New Jersey, 1998.

Jorde, L.B., Carey, J.C., White, R.L.: Medical Genetics. 2nd ed. C.V. Mosby, St. Louis, 2000.

King, R., Rotter, J., Motulsky, A.G., eds: The Genetic Basis of Common Disorders. Oxford University Press, Oxford, 1992.

Knippers, R.: Molekulare Genetik, 7. Aufl. Georg Thieme Verlag, Stuttgart–New York, 1997.

Koolman, J., Röhm, K.-H.: Color Atlas of Biochemistry. Thieme, Stuttgart – New York, 1996.

Korf, B.R.: Human Genetics: A Problem-Based Approach, 2nd ed., Blackwell Science, Malden, MA, 2000.

Kulozik, A.E., Hentze, M.W., Hagemeier, C., Bartram, C.R.: Molekulare Medizin. de Gruyter, Berlin–New York, 2000.

Levitan, M.: Textbook of Human Genetics, 3rd ed., Oxford University Press, Oxford, 1988.

Lewin, B.: Genes VII. Oxford University Press, Oxford, 2000.

Lewontin, R.C.: The Genetic Basis of Evolutionary Change. Columbia University Press, New York, London, 1974.

Lodish, H., Burk, A., Zipursky, S.L., Matsudaira, P., Baltimore, D., Darnell, J.: Molecular Cell Biology (with an animated CD-ROM). 4th ed. W.H. Freeman & Co., New York, 2000.

McKusick, V.A.: Mendelian Inheritance in Man. A Catalog of Human Genes and Genetic Disorders. 12th ed., Johns Hopkins University Press, Baltimore, 1998. (Also see online version OMIM).

Muller, R.F., Young, I.D.: Emery's Elements of Medical Genetics. 10th ed., Churchill Livingstone, Edinburgh, 1998.

Murphy, E.A., Chase, G.E.: Principles of Genetic Counseling. Year Book Medical Publishers, Chicago, 1975.

Ohno, S.: Evolution by Gene Duplication. Springer Verlag, Heidelberg, 1970.

Ott, J.: Analysis of Human Genetic Linkage. Johns Hopkins University Press, Baltimore, 1991.

Passarge, E.: Genetische Grundlagen der Inneren Medizin, p. 21–54. In: Die Innere Medizin, 10. Auflage, W. Gerok, Ch. Huber, Th. Meinertz, H. Zeidler, eds., Schattauer Verlag Stuttgart, 2000.

Passarge, E.: Genetik, p. 5–56. In: Klinische Pathophysiologie, 8. Auflage, W. Siegenthaler, ed., Thieme Verlag Stuttgart, 2001.

Pasternak, J.J.: An Introduction to Human Molecular Genetics. Fitzgerald Science Press, Bethesda, Maryland, 1998.

Purves, W.K. Orians, G.H., Heller, H.C.: Life. The Science of Biology. 4th ed. Sinauer Associates, Inc. and W.H. Freeman & Co., Salt Lake City, 1995.

Rimoin, D.L., Connor, J.M., Pyeritz, R.E., eds: Emery and Rimoin's Principles and Practice of Medical Genetics, 3rd ed., Churchill-Livingstone, Edinburgh, 1996.

Rothwell, N.V.: Understanding Genetics. A Molecular Approach. Wiley-Liss, New York, 1993.

Schmidtke, J.: Vererbung und Vererbtes - Ein humangenetischer Ratgeber. Rowohlt Taschenbuch Verlag, Reinbek bei Hamburg, 1997.

Scriver, C.R., Beaudet, A.L., Sly, W.S., Valle, D., eds: The Metabolic and Molecular Bases of Inherited Disease. 8th ed., McGraw-Hill, New York, 2001.

Segal, N.: Entwined Lives: Twins and what they tell us about human behavior. Dutton Books, New York, 1999.

Snustad, D.P., Simmons, M.J., Jenkins, J.B.: Principles of Genetics. John Wiley & Sons, Inc., New York, 1997.

Stent, G., Calendar, R.: Molecular Genetics. An Introductory Narrative. 2nd ed. W.H. Freeman & Co., San Francisco, 1978.

Strachan, T., Read, A.P.: Human Molecular Genetics. 2nd ed. Bios Scientific Publishers, Oxford, 1999.

Stryer, L.: Biochemistry. 4th ed. W.H. Freeman, New York, 1995.

Sutton, M.E.: An Introduction to Human Genetics, 4th ed., Harcourt Brace Jovanovich Publishers, San Diego, 1988.

Terwilliger, J., Ott, J.: Handbook for Human Genetic Linkage. Johns Hopkins University Press, Baltimore, 1994.

Thompson, M.W., McInnes, R.R., Willard, H.F.: Genetics in Medicine, 5th ed., W.B. Saunders, Philadelphia, 1991.

Tomasello, M.: The Cultural Origins of Human Cognition. Harvard University Press, Cambridge, MA, 1999.

Vogel, F., Motulsky, A.G.: Human Genetics. Problems and Approaches. 3rd ed., Springer Verlag Heidelberg–New York, 1997.

Watson, J.D., Hopkins, N.H., Roberts, J.W., Steitz, J.A., Weiner, A.M., eds: Molecular Biology of the Gene. 4th ed. The Benjamin Cummings Publishing. Co., Menlo Park, California, 1987.


## Selected Websites for Access to Genetic Information

Online Mendelian Inheritance in Man, OMIM (TM). McKusick-Nathans Institute for Genetic Medicine. Johns Hopkins University (Baltimore, Maryland) and the National Center for Biotechnology Information, National Library of Medicine (Bethesda, Maryland), 2000, at World Wide Web URL: (http://www.ncbi.nlm.nih.gov/Omim/).

GeneClinics, a clinical information resource relating genetic testing to the diagnosis, management, and genetic counseling of individuals and families with specific inherited disorders: (http://www.geneclinics.com).

Information on Individual Human Chromosomes and Disease Loci: Chromosome Launchpad: (http://www.ornl.gov/hgmis/launchpad).

National Center of Biotechnology Information Genes and Disease Map: (http://www.ncbi.nlm.nih.gov/disease/).

Medline: (http://www.ncbi.nlm.nim.nih.gov/PubMed/).

MITOMAP: A human mitochondrial genome database: (http://www.gen.emory.edu/mitomap.html), Center for Molecular Medicine, Emory University, Atlanta, GA, USA, 2000.

# Glossary

## Definitions of Genetic Terms

**Acentric**—refers to a chromosome or chromatid without a centromere.

**Acrocentric** (White, 1945)—refers to a chromosome with a centromere that lies very close to one of the ends, dividing the chromosome into a long and a very short arm.

**Active site**—the region of a protein that is mainly responsible for its functional activity.

**Alkyl group**—covalently joined carbon and hydrogen atoms as in a methyl or ethyl group.

**Allele** (Johannsen, 1909) or **allelomorph** (Bateson and Saunders, 1902)—one of several alternative forms of a gene at a given gene locus.

**Alu sequences**—a family of related DNA sequences, each about 300 base pairs long and containing the restriction site for the Alu restriction enzyme; about 500 000 copies of Alu sequences are dispersed throughout the human genome.

**Amber codon**—the stop codon UAG.

**Ames test**—a mutagenicity test carried out with a mixture of rat liver and mutant bacteria.

**Aminoacyl tRNA**—a transfer RNA carrying an amino acid.

**Amplification**—production of additional copies of DNA sequences.

**Anaphase** (Strasburger, 1884)—a stage of mitosis and of meiosis I and II. Characterized by the movement of homologous chromosomes (or sister chromatids) toward opposite poles of the cell division spindle.

**Aneuploid** (Täckholm, 1922)—having either fewer or more than the normal number of chromosomes.

**Aneusomy**—deviation from the normal presence of homologous chromosomal segments. Aneusomy by recombination refers to the duplication/deficiency resulting from crossing-over within an inversion (inverted region).

**Annealing**—to cause complementary single strands of nucleic acid to pair and form double-stranded molecules (DNA with DNA, RNA with RNA, or DNA with RNA).

**Antibody**—a protein (immunoglobin) that recognizes and binds to an antigen as part of the immune response.

**Antigen**—a substance with a molecular surface structure that triggers an immune response, i.e., the production of antibodies, and/or that reacts with (its) specific antibodies (antigen-antibody reaction).

**Antisense RNA**—an RNA strand that is complementary to mRNA. Thus it cannot be used as a template for normal translation. The term antisense is generally used to refer to a sequence of DNA or RNA that is complementary to mRNA.

**Apoptosis**—programmed cell death, characterized by a series of regulated cellular events resulting in cell death, to eliminate a damaged cell or a normal cell no longer needed during development.

**Attenuator**—terminator sequence regulating the termination of transcription, involved in controlling the expression of some operons in bacteria.

**Australopithecus**—the genus of fossil Hominidae from Eurasia. Walked erect; brain size between that of modern man and other modern primates; large and massive jaw. Lived about 4–5 million years ago.

**Autonomously replicating sequence (ARS)**—a DNA sequence that permits replication.

**Autoradiography** (Lacassagne and Lattes, 1924)—photographic detection of a radioactive substance incorporated by metabolism into cells or tissue. The distribution of the radioactively labeled substance can be demonstrated, e.g., in tissue, cells, or metaphase chromosomes by placing a photographic film or photographic emulsion in close contact with the preparation.

**Autosome** (Montgomery, 1906)—any chromosome except a sex chromosome (the latter usually designated X or Y). Autosomal refers to genes and chromosomal segments that are located on autosomes.

**Auxotrophic** (Ryan and Lederberg, 1946)—refers to cells or cell lines that cannot grow on minimal medium unless a certain nutritive substance is added (cf. prototrophic).

**Backcross**—cross of a heterozygous animal with one of its homozygous parents. In a double backcross, two heterozygous gene loci are involved.

**Bacteriophage**—a virus that infects bacteria. Usually abbreviated as phage.

**Banding pattern** (Painter, 1939)—staining pattern of a chromosome consisting of alternating light and dark transverse bands. Each chromosomal segment of homologous chromosomes shows the same specific banding pattern, characterized by the distribution and size of the bands, which can be used to identify that segment. The term was introduced by Painter in 1939 for the linear pattern of strongly and weakly staining bands in polytene chromosomes of certain diptera (mosquitoes, flies). Each band is defined relative to its neighboring bands. The sections between bands are interbands.

**Barr body**—see X chromatin.

**Base pair (bp)**—in DNA, two bases—one a purine, the other a pyrimidine—lying opposite each other and joined by hydrogen bonds. Normal base pairs are A with T and C with G in a DNA double helix. Other pairs can be formed in RNA.

**B cells**—B lymphocytes.

**Bimodal distribution**—refers to a frequency distribution curve with two peaks. If the frequency distribution curve of a population trait is bimodal, it is frequently evidence of two different phenotypes distinguished on a quantitative basis.

**Bivalent** (Haecker, 1892)—pairing configuration of two homologous chromosomes during the first meiotic division. As a rule the number of bivalents corresponds to half the normal number of chromosomes in diploid somatic cells. Bivalents are the cytogenetic prerequisite for crossing-over of nonsister chromatids. During meiosis, a trisomic cell forms a trivalent of the trisomic chromosomes.

**Breakpoint**—site of a break in a chromosomal alteration, e.g., translocation, inversion, or deletion.

**CAT (or CAAT) box**—a regulatory DNA sequence in the 5' region of eukaryotic genes; transcription factors bind to this sequence.

**cDNA**—complementary DNA synthesized by the enzyme reverse transcriptase from RNA as the template.

**Cell cycle**—life cycle of an individual cell. In dividing cells, the following four phases can be distinguished: G1 (interphase), S (DNA synthesis), G2, and mitosis. Cells that do not divide are said to be in the Go phase.

**Cell hybrid**—in culture, a somatic cell generated by fusion of two cells of different species. It contains the complete or incomplete chromosome complements of the parental cells. Cell hybrids are an important tool in gene mapping.

**Centimorgan**—a unit of length on a linkage map [100 centimorgans (cM) = 1 Morgan]. The distance between two gene loci in centimorgans corresponds to their recombination frequency expressed as percentage, i.e., one cM corresponds to one percent recombination frequency. Named after Thomas H. Morgan (1866–1945), who initiated the classic genetic experiments on drosophila in 1910.

**Centromere** (Waldeyer, 1903)—chromosomal region to which the spindle fibers attach during mitosis or meiosis. It appears as a constriction at metaphase.

**Chaperone**—a protein needed to assemble or fold another protein correctly.

**Chiasma** (Janssens, 1909)—cytologically recognizable region of crossing-over in a bivalent. In some organisms the chiasmata move toward the end of the chromosomes (terminalization of the chiasmata) during late diplotene and diakinesis (see meiosis). The average number of chiasmata in autosomal bivalents is about 52 in human males, about 25–30 in females. The number of chiasmata in man was first determined in 1956 in a paper that confirmed the normal number of chromosomes in man (C. E. Ford and J. L. Hamerton, Nature 178 : 1020, 1956).

**Chimera** (Winkler, 1907)—an individual or tissue that consists of cells of different genotypes of prezygotic origin.

**Chromatid** (McClung, 1900)—longitudinal subunit of a chromosome resulting from chromosome replication; two chromatids are held together by the centromere and are visible during early prophase and metaphase of mitosis and between diplotene and the second metaphase of meiosis. Sister chromatids arise from the same chromosome; nonsister chromatids are the chromatids of homologous chromosomes.

After division of the centromere in anaphase, the sister chromatids are referred to as daughter chromosomes. A *chromatid break* or a chromosomal aberration of the chromatid type affects only one of the two sister chromatids. It arises after the DNA replication cycle in the S phase (see cell cycle). A break that occurs before the S phase affects both chromatids and is called an isolocus aberration (*isochromatid break*).

**Chromatin** (Flemming, 1882)—the stained material that can be observed in interphase nuclei. It is composed of DNA, basic chromosomal proteins (histones), nonhistone chromosomal proteins, and small amounts of RNA.

**Chromomere** (Wilson, 1896)—each of the linearly arranged thickenings of the chromosome visible in meiotic and under some conditions also in mitotic prophase. Chromomeres are arranged in chromosome-specific patterns.

**Chromosome** (Waldeyer, 1888)—the gene-carrying structures, which are composed of chromatin and are visible during nuclear division as threadlike or rodlike bodies. *Polytene chromosomes* (Koltzhoff, 1934; Bauer, 1935) are a special form of chromosomes in the salivary glands of some diptera larvae (mosquitoes, flies).

**Chromosome walking**—sequential isolation of overlapping DNA sequences in order to find a gene on the chromosome studied.

**Cis-acting**—refers to a regulatory DNA sequence located on the same chromosome (cis) as opposed to trans-acting on a different chromosome.

**Cis/trans** (Haldane, 1941)—in analogy to chemical isomerism, refers to the position of genes of double heterozygotes (heterozygotes at two neighboring gene loci) on homologous chromosomes. When two certain alleles, e.g., mutants, lie next to each other on the same chromosome, they are in the *cis* position. If they lie opposite each other on different homologous chromosomes, they are in the *trans* position. The cis/trans test (Lewis, 1951; Benzer, 1957) uses genetic methods (*genetic complementation*) to determine whether two mutant genes are in the cis or in the trans position. With reference to genetic linkage, the expressions cis and trans are analogous to the terms coupling and repulsion (q.v.).

**Cistron** (Benzer 1957)—a functional unit of gene effect as represented by the cis/trans test. If the phenotype is mutant with alleles in the cis position and the alleles do not complement each other (genetic complementation), they are considered alleles of the same cistron. If they complement each other they are considered to be nonallelic. This definition of Benzer was later expanded (Fincham, 1959): accordingly, a cistron now refers to a segment of DNA that codes for a unit of gene product. Within a cistron, mutations in the trans position do not complement each other. Functionally, the term cistron can be equated with the term gene.

**Clade**—a group of organisms evolved from a common ancestor.

**Clone** (Webber, 1903)—a population of molecules, cells or organisms that have originated from a single cell or a single ancestor and are identical to it and to each other.

**Cloning efficiency**—a measure of the clonability of individual mammalian cells in culture.

**Cloning vector**—a plasmid, phage, or yeast artificial chromosome (YAC) used to carry a foreign DNA fragment for the purpose of cloning (producing multiple copies of the fragment).

**Coding strand of DNA**—the strand of DNA that bears the same sequence as the RNA strand (mRNA) that is used as a template for translation (sense RNA). The other strand of DNA, which directs synthesis of the mRNA, is the template strand (see antisense RNA).

**Codominant**—see dominant.

**Codon** (Crick, 1963)—a sequence of three nucleotides (a triplet) in DNA or RNA that codes for a certain amino acid or for the terminalization signal of an amino acid sequence.

**Complementation, genetic** (Fincham, 1966)—complementary effect of (restoration of normal function by) double mutants at different gene loci. An example, genetic complementation for xeroderma pigmentosum, is illustrated on page 82.

**Concatemer**—an association of DNA molecules with complementary ends linked head to tail and repeated in tandem. Formed during replication of some viral and phage genomes.

**Concordance**—the occurrence of a trait or a disease in both members of a pair of twins (mono- or dizygotic).

**Conjugation**—the transfer of DNA from one bacterium to another.

**Consanguinity**—blood relationship. Two or more individuals are referred to as consanguineous (related by blood) if they have one or more ancestors in common. A quantitative expression of consanguinity is the coefficient of inbreeding (q.v.).

**Consensus sequence**—a corresponding or identical DNA sequence in different genes or organisms.

**Contig**—a series of overlapping DNA fragments (contiguous sequences).

**Cosmid**—a plasmid carrying the cos site (q.v.) of λ phage in addition to sequences required for division. Serves as a cloning vector for DNA fragments up to 40 kb.

**Cos site**—a restriction site required of a small strand of DNA to be cleaved and packaged into the λ phage head.

**Coupling** (Bateson, Saunders, Punnett, 1905)—cis configuration (q.v.) of double heterozygotes.

**Covalent bond**—a stable chemical bond holding molecules together by sharing one or more pairs of electrons (as opposed to a noncovalent hydrogen bond).

**Crossing-over** (Morgan and Cattell, 1912)—the exchange of genetic information between two homologous chromosomes by chiasma formation (q.v.) in the diplotene stage of meiosis I; leads to genetic recombination of neighboring (linked) gene loci.

Crossing-over may also occur in somatic cells during mitosis (Stern, 1936). Unequal crossing-over (Sturtevant, 1925) results from mispairing of the homologous DNA segments at the recombination site. It results in structurally altered DNA segments or chromosomes, with a duplication in one and a deletion in the other.

**Cyclic AMP (cAMP)**—cyclic adenine monophosphate, a second messenger produced in response to stimulation of G-protein-coupled receptors.

**Cyclin**—a protein involved in cell cycle regulation.

**Cytokine**—a small secreted molecule that can bind to cell surface receptors on certain cells to trigger their proliferation or differentiation.

**Cytoplasmic inheritance**—transmission of genetic information located in mitochondria. Since sperm cells do not contain mitochondria, the transmitted information is of maternal origin.

**Cytoskeleton**—network of stabilizing protein in the cytoplasm and cell membrane.

**Dalton**—a unit of atomic mass, equal to 1/12 the mass of the of the $^{12}$C nuclide ($1.657 \times 10^{-24}$gm).

**Deficiency** (Bridges, 1917)—loss of a chromosomal segment resulting from faulty crossing-over, e.g., by unequal crossing-over or by crossing-over within an inversion (q.v.) or within a ring chromosome (q.v.). It arises at the same time as a complementary duplication (q.v.). This is referred to as duplication/deficiency.

**Deletion** (Painter & Muller, 1929)—loss of part of or a whole chromosome or loss of DNA nucleotide bases.

**Denaturation of DNA**—separation of double-stranded nucleic acid molecules into single strands. Rejoining of the complementary single strands is referred to as renaturation.

**Diakinesis** (Haecker, 1897)—a stage during late prophase I of meiosis.

**Dicentric** (Darlington, 1937)—refers to a structurally altered chromosome with two centromeres.

**Dictyotene**—a stage of fetal oocyte development during which meiotic prophase is interrupted. In human females, oocytes attain the stage of dictyotene about 4 weeks before birth; further development of the oocytes is arrested until ovulation, at which time meiosis is continued.

**Diploid** (Strasburger, 1905)—cells or organisms that have two homologous sets of chromosomes, one from the father (paternal) and one from the mother (maternal).

**Diplotene**—a stage of prophase I of meiosis.

**Discordance**—the occurrence of a given trait or disease in only one member of a pair of twins.

**Disomy, uniparental (UPD)**—presence of two chromosomes of a pair from only one of the parents. One distinguishes UPD due to isodisomy, in which the chromosomes are identical, and heterodisomy, in which they are homologous.

**Dispermy**—the penetration of a single ovum by two spermatozoa.

**Dizygotic**—twins derived from two different zygotes (fraternal twins), as opposed to monozygotic (identical) twins, derived from the same zygote.

**D loop**—a region in mitochondrial DNA where one strand is paired with RNA, which displaces the other strand.

**DNA (deoxyribonucleic acid)**—the molecule containing the primary genetic information in the form of a linear sequence of nucleotides in groups of threes (triplets) (see codon).

*Satellite DNA* (sDNA) (Sueoka, 1961; Britten & Kohne, 1968)—contains tandem repeats of nucleotide sequences of different lengths. sDNA can be separated from the main DNA by density gradient centrifugation in cesium chloride, after which it appears as one or several bands (satellites) separated from that of the main body of DNA. In eukaryotes, light (AT-rich) and heavy (GC-rich) satellite DNA can be distinguished.

Microsatellites are small (2–10) tandem repeats of DNA nucleotides; minisatellites are tandem repeats of about 20–100 base pairs; classical satellite DNA consists of large repeats of 100–6500 bp (see p. 353).

**DNA library**—a collection of cloned DNA molecules comprising the entire genome (genomic library) or of cDNA fragments obtained from mRNA produced by a particular cell type (cDNA library).

**DNA polymerase**—a DNA-synthesizing enzyme. To begin synthesis, it requires a primer of RNA or a complementary strand of DNA.

**DNase (deoxyribonuclease)**—an enzyme that digests DNA.

**Domain**—a distinctive region of the tertiary structure of a protein or a particular region of a chromosome.

**Dominant** (Mendel, 1865)—refers to a genetic trait that can be observed in the heterozygous state. The terms "dominant" and "recessive" refer to the effects of the alleles at a given gene locus. The effects observed depend in part on the accuracy of observation. When the effects of each of two different alleles at a (heterozygous) locus can be observed, the alleles are said to be codominant. At the DNA level, allelic genes at two homologous loci are codominant.

**Downstream**—the 3' direction of a gene.

**Drift, genetic** (Wright, 1921)—random changes in gene frequency of a population. Especially relevant in small populations, where random differences in the reproductive frequency of a certain allele can change the frequency of the allele. Under some conditions an allele may disappear completely from a population (loss) or be present in all individuals of a population (fixation).

**Duplication** (Bridges, 1919)—addition of a chromosomal segment resulting from faulty crossing-over (see deficiency). It may also refer to additional DNA nucleotide base pairs. Duplication of genes (gene duplication) played an important role in the evolution of eukaryotes.

**Electrophoresis** (Tiselius, 1937)—separation of molecules by utilizing their different speeds of migration in an electrical field. As support medium, substances in gel form such as starch, agarose, acrylamide, etc. are used. Further molecular differences can be detected by modifications such as two-dimensional electrophoresis (electric field rotated 90° for the second migration) or cessation of migration at the isoelectric point (isoelectric focusing).

**Elongation**—addition of amino acids to a polypeptide chain.

**Elongation factor**—one of the proteins that associate with ribosomes while amino acids are added; EF in prokaryotes and eEF in eukaryotes.

**Endocytosis**—specific uptake of extracellular material at the cell surface. The material is surrounded by an invagination of the cell membrane, which pinches off to form a membrane-bound vesicle containing the material.

**Endonuclease**—a heterogeneous group of enzymes that cleave bonds between nucleotides of single- or double-stranded DNA or of RNA.

**Endoplasmic reticulum**—a complex system of membranes in the cytoplasm.

**Endoreduplication** (Levan & Hauschka, 1953)—chromosome replication during interphase without actual mitosis. Endoreduplicated chromosomes in metaphase consist of four chromatids lying next to each other, held together by two neighboring centromeres.

**Enhancer**—a cis-acting regulatory DNA segment that contains binding sites for transcription factors. An enhancer is located at various distances from the promoter. It causes an increase in the rate of transcription (about tenfold).

**Enzyme** (E. Büchner, 1897)—a protein that catalyzes a biochemical reaction. Enzymes consist of a protein part (apoenzyme), responsible for the specificity, and a nonprotein part (coenzyme), needed for activity. Enzymes bind to their substrates, which become metabolically altered or combined with other substances during the train of the reaction. Most of the enzymatically catalyzed chemical reactions can be classified into one of six groups:

(1) hydrolysis (cleavage with the addition of $H_2O$), by *hydrolases*;

(2) transfer of a molecular group from a donor to a receptor molecule, by *transferases*;

(3) oxidation and reduction, by *oxidases* and *reductases* (transfer of one or more electrons or hydrogen atoms from a molecule to be oxidized to another molecule that is to be reduced);

(4) isomerization, by *isomerases* (rearranging the position of an atom or functional group within a molecule);

(5) joining of two substrate molecules to form a new molecule, by *ligases* (*synthetases*);

(6) nonhydrolytic cleavage with formation of a double bond on one or both of the two molecules formed, by *lyases*.

**Epigenetic influence**—a factor that changes the phenotype without altering the genotype.

**Episome** (Jacob & Wollman, 1958)—a plasmid (q.v.) that can exist either independently in the cytoplasm or as an integrated part of the genome of its bacterial host.

**Epistasis** (Bateson, 1907)—interaction of genes at the same gene locus (allelic) or at different gene loci (nonallelic) to alter phenotypic expression.

**Epitope**—the part of an antigen molecule that binds to an antibody.

**EST** (expressed sequence tag)—a sequenced site from an expressed gene that "tags" a stretch of unsequenced cDNA next to it; used to map genes (see STS).

**Euchromatin** (Heitz, 1928)—chromosome or chromosomal segment that stains less intensely than heterochromatin (q.v.). Euchromatin corresponds to the genetically active part of chromatin that is not fully condensed in the interphase nucleus.

**Eukaryote** (Chatton, 1925)—plants and animals with cells that have a chromosome-containing nucleus, which divides during mitosis and meiosis, in contrast to prokaryotes.

**Euploid** (Täckholm, 1922)—refers to cells, tissues, or individuals with the complete normal chromosomal complement characteristic of that species (cf. aneuploid, heteroploid, polyploid).

**Excision repair**—repair of bulk lesions in DNA in which a stretch of nucleotides (about 14 in prokaryotes and about 30 in eukaryotes) is excised from the affected strand and replaced by the normal sequence (resynthesis).

**Exocytosis**—specific process by which nondiffusable particles are transported through the cell membrane to be discharged into the cellular environment.

**Exon** (Gilbert, 1978)—a segment of DNA that is represented in the mature mRNA of eukaryotes (cf. intron).

**Exonuclease**—an enzyme that cleaves nucleotide chains at their terminal bonds only, either the 5' or 3' end (cf. endonuclease).

**Expression**—the observable effects of an active gene.

**Expression vector**—a cloning vector containing DNA sequences that can be transcribed and translated.

**Expressivity** (Vogt, 1926)—refers to the degree of phenotypic expression of a gene or genotype. Absence of expressivity is also called nonpenetrance.

**Fibroblast**—type of connective tissue cell. Can be propagated in culture flasks containing suitable medium (*fibroblast cultures*).

**Fingerprint**—a characteristic pattern of fragments of DNA or proteins.

**Fitness, biological**—refers to the probability (between 0.0 and 1.0) that a gene will be passed on to the next generation. For a given genotype and a given environment, the biological (or reproductive) fitness is determined by survival rate and fertility.

**Founder effect**—presence of a particular allele in a population due to a mutation in a single ancestor.

**Gamete** (Strasburger, 1877)—a haploid germ cell, either a spermatozoon (male) or an ovum (female). In mammals, males are heterogametic (XY) and females homogametic (XX). In birds, females are heterogametic (ZW) and males homogametic (ZZ).

**Gene** (Johannsen, 1909)—a hereditary factor that constitutes a single unit of hereditary material. It corresponds to a segment of DNA that codes for the synthesis of a single polypeptide chain (cf. cistron).

**Gene amplification** (Brown & David, 1968)—selective production of multiple copies of a given gene without proportional increases of other genes.

**Gene bank**—a collection of cloned DNA fragments that together represent the genome they are derived from (gene library).

**Gene cluster** (Demerec & Hartman, 1959)—a group of two or more neighboring genes of similar function, e.g., the HLA system or the immunoglobulin genes.

**Gene conversion** (Winkler, 1930; Lindgren, 1953)—interaction of alleles that leads to unequal genetic exchange during meiosis. Gene conversion refers to a process of nonreciprocal transfer of genetic information. One gene serves as a sequence donor, remaining unaffected, while the other gene receives sequences and undergoes variation.

**Gene dosage**—refers to the quantitative degree of expression of a gene. Also used to refer to the number of copies of a gene in the genome.

**Gene family**—a set of evolutionarily related genes by virtue of identity or great similarity of some of their coding sequences.

**Gene flow** (Berdsell, 1950)—transfer of an allele from one population to another.

**Gene frequency**—the frequency of a given allele at a given locus in a population (allele frequency).

**Gene locus** (Morgan, Sturtevant, Muller, Bridges, 1915)—the position of a gene on a chromosome.

**Gene map**—the position of gene loci on chromosomes. A physical map refers to the absolute position of gene loci, their distance from each other being expressed by the number of base pairs between them. A genetic map expresses the distance of genetically linked loci by their frequency of recombination.

**Gene product**—the polypeptide or ribosomal RNA coded for by a gene (see protein).

**Genetic code**—the information contained in the triplets of DNA nucleotide bases used to incorporate a particular amino acid into a gene product.

**Genetic marker**—a polymorphic genetic property that can be used to distinguish the parental origin of alleles.

**Genetics** (Bateson, 1906)—the science of heredity and the hereditary basis of organisms; derived from Gk. *genesis* (origin).

**Genome** (Winkler, 1920)—all of the genetic material of a cell or of an individual.

**Genome scan**—a search with marker loci on all chromosomes for linkage with an unmapped locus.

**Genomics**—the scientific field dealing with the structure and function of the entire genome (see part II, Genomics).

**Genotype** (Johannsen, 1909)—all or a particular part of the genetic constitution of an individual or a cell (cf. phenotype).

**Germ cell**—a cell able to differentiate into gametes by meiosis (as opposed to somatic cells).

**Germinal**—refers to germ cells, as opposed to somatic cells.

**G6PD**—glucose-6-phosphate dehydrogenase.

**Gyrase**—a topoisomerase that unwinds DNA.

**Haploid** (Strasburger, 1905)—refers to cells or individuals with a single chromosome complement; gametes are haploid.

**Haplotype** (Ceppellini et al., 1967)—a combination of alleles at two or more closely linked gene loci on the same chromosome, e.g., in the HLA system (q.v.).

**Helicase**—an enzyme that unwinds and separates the two strands of the DNA double helix by breaking the hydrogen bonds during transcription or repair.

**Helix-loop-helix**—a structural motif in DNA-binding proteins, such as some transcription factors.

**Hemizygous**—refers to genes and gene loci that are present in only one copy in an individual, e.g., on the single X chromosome in male cells (XY) or because the homologous locus has been lost.

**Heritability** (Lush, 1950; Falconer, 1960)—the ratio of additive genetic variance to the total phenotypic variance. Phenotypic variance is the result of the interaction of genetic and nongenetic factors in a population.

**Heterochromatin** (Heitz, 1928)—a chromosome or chromosomal segment that remains darkly stained in interphase, early prophase, and late telophase because it remains condensed, as all chromosomal material is in metaphase. This contrasts with euchromatin, which becomes invisible during interphase. Heterochromatin corresponds to chromosomes or chromosome segments showing little or no genetic activity. *Constitutive* and *facultative* heterochromatin can be distinguished. An example of facultative heterochromatin is the heterochromatic X chromosome resulting from inactivation of one X chromosome in somatic cells of female mammals. An example of constitutive heterochromatin is the centric heterochromatin at centromeres that can be demonstrated as C bands.

**Heterodisomy**—presence of two homologous chromosomes from one parent only (cf. isodisomy and UPD).

**Heteroduplex**—refers to a region of a double-stranded DNA molecule with noncomplementary strands that originated from different duplex DNA molecules.

**Heterogametic** (Wilson, 1910)—producing two different types of gametes (q.v.), e.g., X and Y in (male) mammals or ZW in female birds.

**Heterogeneity, genetic** (Harris, 1953; Fraser, 1956)—an apparently uniform phenotype being caused by two or more different genotypes.

**Heterokaryon** (Ephrussi and Weiss, 1965; Harris and Watkins, 1965; Okada and Murayama, 1965)—a cell having two or more nuclei with different genomes.

**Heteroploid** (Winkler, 1916)—refers to cells or individuals with an abnormal number of chromosomes.

**Heterosis** (Shull, 1911)—increased reproductive fitness of heterozygous genotypes compared with the parental homozygous genotypes, in plants and animals.

**Heterozygous** (Bateson and Saunders, 1902)—having two different alleles at a given gene locus (cf. homozygous).

**Hfr cell**—a bacterium that possesses DNA sequences that lead to a high frequency of DNA transfer at conjugation.

**HGPRT**—hypoxanthine-guanine-phosphoribosyl transferase. An enzyme in purine metabolism that is inactive in Lesch-Nyhan syndrome.

**Histocompatibility**—tissue compatibility. Determined by the major histocompatibility complex MHC (see HLA).

**Histone** (Kossel, 1884)—chromosome-associated protein of the nucleosome. Histones H1-H4 form a nucleosome (q.v.).

**HLA** (Dausset, Terasaki, 1954)–human leukocyte antigen system A. HLA is also said by some to refer to Los Angeles, where Terasaki made essential discoveries.

**Hogness box**–a nucleotide sequence that is part of the promoter in eukaryotic genes.

**Homeobox**–a highly conserved DNA segment in homeotic genes.

**Homeotic gene**–one of the developmental genes in drosophila that can lead to the replacement of one body part by another by mutation.

**Homologous**–refers to a chromosome or gene locus of similar maternal or paternal origin.

**Homozygosity mapping**–mapping genes by identifying chromosomal regions that are homozygous as a result of identity by descent from a common ancestor in consanguineous matings (see identity by descent, IBD).

**Homozygous** (Bateson and Saunders, 1902)–having identical alleles at a given gene locus.

**Hox genes**–clusters of mammalian genes containing homeobox sequences. They are important in embryonic development.

**Hybridization**–cross between two genotypically different plants or animals belonging to the same species. The term is often used in more narrow definitions: fusion of two single complementary DNA strands (DNA/DNA hybridization), fusion of complementary DNA and RNA strands (DNA/RNA hybridization), or the in vitro fusion of cultured cells of different species (cell hybridization).

**Hydrogen bond**–a noncovalent weak chemical bond between an electronegative atom (usually oxygen or nitrogen) and a hydrogen atom; important in stabilizing the three-dimensional structure of proteins or base pairing in nucleic acids.

**Identity by descent (IBD)**–refers to homozygous alleles at one gene locus that are identical because they are inherited from a common ancestor (see consanguinity).

**Immunoglobulin** – an antigen-binding molecule.

**Imprinting, genomic**–different expression of an allele or chromosomal segment depending on the parental origin.

**Inbreeding coefficient** (Wright, 1929)–measure of the probability that two alleles at a gene locus of an individual are identical by descent, i.e., that they are copies of a single allele of an ancestor common to both parents (IBD, identity-by-descent). Also, the proportion of loci at which the individual is homozygous.

**Incidence**–the rate of occurrence of a disease in a population. In contrast, prevalence is the percentage of a population that is affected with a particular disease at a particular time.

**Inducer**–a molecule that induces the expression of a gene.

**Initiation factor**–a protein that associates with the small subunit of a ribosome when protein synthesis begins (IF in prokaryotes, eIF in eukaryotes).

**Insertion**–insertion of chromosomal material of nonhomologous origin into a chromosome without reciprocal translocation (q.v.).

**Insertion sequence** (IS)–a small bacterial transposon carrying genes for its own transposition (q.v.).

**In silico**–a process taking place within a computer; for example, analysis of biological data.

**Intercalating agent**–a chemical compound that can occupy a space between two adjacent base pairs in DNA.

**Interphase**–the period of the cell cycle between two cell divisions (see mitosis).

**Intron** (Gilbert, 1978)–a segment of noncoding DNA within a gene (cf. exon). It is transcribed, but removed from the primary RNA transcript before translation.

**Inversion** (Sturtevant, 1926)–structural alteration of a chromosome through a break at two sites with reversal of direction of the intermediate segment and reattachment. A *pericentric inversion* includes the centromere in the inverted segment. A *paracentric inversion* does not involve the centromere. An inversion per se does not cause clinical signs, but it represents a potential genetic risk because crossing-over may occur in the region of the inversion and lead to aneusomy in offspring (aneusomy by recombination). Chromosomal inversions played an important role in evolution.

**Inverted repeat**—two identical, oppositely oriented copies of the same DNA sequence. They are an important feature of retroviruses.

**In vitro**—a biological process taking place outside a living organism or in an artificial environment in the laboratory.

**In vivo**—within a living organism.

**Isochromatid break**—see chromatid.

**Isochromosome** (Darlington, 1940)—a chromosome composed of two identical arms connected by the centromere, e.g., two long or two short arms of an X chromosome. Implies duplication of the doubled arm and deficiency of the absent arm. An isochromosome may have one or two centromeres.

**Isodisomy**—presence of two identical chromosomes from one of the parents (cf. heterodisomy).

**Isolate, genetic** (Waklund, 1928)—a physically or socially isolated population that has not interbred with individuals outside of that population (no panmixis).

**Isotype**—closely related chains of immunoglobulins.

**Isozyme or isoenzyme** (Markert and Möller, 1959; Vesell 1959)—one of multiple distinguishable forms of enzymes of similar function in the same organism. Isoenzymes are a biochemical expression of genetic polymorphism.

**Karyotype** (Levitsky, 1924)—the chromosome complement of a cell, an individual, or a species.

**Knockout** (of a gene)—intentional inactivation of a gene in an experimental organism in order to obtain information about its function (same as targeted gene disruption).

**Lagging strand of DNA**—the new strand of DNA replicated from the 3' to 5' strand. It is synthesized in short fragments in the 5' to 3' direction (Okazaki fragments), which are subsequently joined together.

**Lampbrush chromosome** (Rückert, 1892)—a special type of chromosome found in the primary oocytes of many vertebrates and invertebrates during the diplotene stage of meiotic division and in drosophila spermatocytes. The chromosomes show numerous lateral loops of DNA that are accompanied by RNA and protein synthesis.

**Lariat**—an intermediate form of RNA during splicing when a circular structure with a tail is formed by a 5' to 2' bond.

**Leader sequence**—a short N-terminal sequence of a protein that is required for directing the protein to its target.

**Leaky mutant**—a mutation causing only partial loss of function.

**Leptotene**—a stage of meiosis (q.v.).

**Lethal equivalent** (Morton, Crow, and Muller, 1956)—a gene or combination of genes that in the homozygous state is lethal to 100% of individuals. This may refer to a gene that is lethal in the homozygous state, to two different genes that each have 50% lethality, to three different genes each with 33% lethality, etc. It is assumed that each individual carries about 5–6 lethal equivalents.

**Lethal factor** (Bauer, 1908; Hadorn, 1959)—an abnormality of the genome that leads to death in utero, e.g., numerous chromosomal anomalies.

**Library**—see DNA library.

**Ligand**—a molecule that can bind to a receptor and thereby induce a signal in the cell, e.g., a hormone.

**LINE (long interspersed nuclear element)**—long interspersed repetitive DNA sequences (see p. 352).

**Linkage disequilibrium** (Kimura, 1956)—nonrandom association of alleles at closely linked gene loci that deviates from their individual frequencies as predicted by the Hardy-Weinberg equilibrium. Linkage disequilibrium is usually due to the founder effect (q.v.).

**Linkage, genetic** (Morgan, 1910)—localization of gene loci on the same chromosome close enough to cause deviation from independent segregation.

**Linkage group**—gene loci on the same chromosome which are so close together that they usually are inherited together without recombination.

**Linker DNA**—a synthetic DNA double strand that carries the recognition site for a restriction enzyme and that can bind two DNA fragments. Also, the stretch of DNA between two nucleosomes.

**Locus**—see gene locus.

**Lymphocyte**—cell of the immune system, of one of two general types: B lymphocytes from the bone marrow and thymus-derived T lymphocytes.

**Lysosome** (deDuve et al., 1955)—small cytoplasmic organelle containing hydrolytic enzymes.

**Map distance**—distance between gene loci, expressed in either physical terms (number of base pairs, e.g. kb 1000 bp or Mb million bp) or genetic terms (recombination frequency, expressed as cM, centimorgan. One cM corresponds to 1%).

**Mapping**—various methods to determine the position of a gene on a chromosome (physical map) or its relative distance to other gene loci and their order (genetic map).

**Marker, genetic**—an allele used to recognize a particular genotype.

**Megabase (Mb)**—1 million base pairs.

**Meiosis** (Farmer and Moore, 1905)—the special division of a germ cell nucleus that leads to reduction of the chromosome complement from the diploid to the haploid state. Prophase of the first meiotic division is especially important and consists of the following stages: leptotene, zygotene, pachytene, diplotene, diakinesis.

**Mendelian inheritance** (Castle, 1906)—inheritance according to the laws of Mendel as opposed to extrachromosomal inheritance, under the control of cytoplasmic hereditary factors (mitochondrial DNA).

**Metabolic cooperation** (Subak-Sharpe et al., 1969)—correction of a phenotype in cells in culture by contact with normal cells or cell products. An example of metabolic correction is the cross correction of cultured cells of different mucopolysaccharide storage diseases or correction of HGPRT-deficient cells by normal cells.

**Metacentric**—refers to chromosomes that are divided by the centromere into two arms of approximately the same length.

**Metaphase** (Strasburger, 1884)—stage of mitosis in which the contracted chromosomes are readily visible.

**MHC (major histocompatibility complex)** (Thorsby, 1974)—the principal histocompatibility system, consisting of class I and class III antigen genes of the HLA system including the class II genes.

**Missense mutation**—a mutation that alters a codon to one for a different amino acid (see nonsense mutation).

**Mitosis** (Flemming, 1882)—nuclear division during the division of somatic cells, consisting of prophase, metaphase, anaphase, and telophase.

**Mitosis index** (Minot, 1908)—the proportion of cells present that are undergoing mitosis.

**Mixoploidy** (Nemec, 1910; Hamerton, 1971)—a tissue or individual having cells with different numbers of chromosomes (chromosomal mosaic).

**MLC**—*m*ixed *l*ymphocyte *c*ulture (Bach and Hirschhorn, Bach and Lowenstein, 1964). MLC is a test for differences in HLA-D phenotypes.

**Modal number** (White, 1945)—the number of chromosomes of an individual or a cell.

**Monoclonal antibody**—an antibody representing a single antigen specificity, produced from a single progenitor cell.

**Monolayer** (Abercrombie and Heaysman, 1957)—the single-layered sheet of cultured diploid cells on the bottom of a culture flask.

**Monosomy** (Blakeslee, 1921)—absence of a chromosome in an otherwise diploid chromosomal complement.

**Monozygotic**—pertaining to uniovular twins (cf. dizygotic).

**Morphogen**—a protein present in embryonic tissues in a concentration gradient that induces a developmental process.

**Mosaic**—tissue or individuals made up of genetically different cells, as a rule of the same zygotic origin (cf. chimera).

**mRNA** (Brenner, Jacob, and Meselson, 1961; Jacob and Monrod, 1961)—messenger RNA.

**mtDNA**—mitochondrial DNA.

**Multigene family**—a group of genes related by their common evolution.

**Mutagen**—a chemical or physical agent that can induce a mutation.

**Mutation** (de Vries, 1901)—permanent alteration of the genetic material. Different types include point mutations from exchange, loss, or insertion of base pairs within a gene and chromosomal mutation with alteration of the chromosome structure. A *missense mutation* is an alteration resulting in a gene product containing a substitution for a wrong amino acid. A *nonsense mutation* is an alteration that produces a stop codon in the midst of a genetic message so that a totally inadequate gene product is formed.

**Mutation rate**—the frequency of a mutation per locus per individual per generation.

**Noncovalent bond**—a (weak) chemical bond between an electonegative atom (usually oxygen or nitrogen) and a hydrogen atom (see hydrogen bond), but not involving sharing of electrons.

**Nondisjunction** (Bridges, 1912)—faulty distribution of homologous chromosomes at meiosis. In mitotic nondisjunction, the distribution error occurs during mitosis.

**Nonsense codon**—a codon that does not have a normal tRNA molecule. Any of the three triplets that terminate translation (UAG, UAA, UGA).

**Nonsense mutation**—a mutation that results in lack of any genetic information, e.g., a stop codon (see missense mutation).

**Northern blot**—transfer of RNA molecules to a membrane by a procedure similar to that of a Southern blot (q.v.).

**Nucleoside**—compound of a purine or a pyrimidine base with a sugar (ribose or deoxyribose) (cf. nucleotide).

**Nucleosome** (Navashin, 1912; Kornberg, 1974)—a subunit of chromatin consisting of DNA wound around histone proteins in a defined spatial configuration.

**Nucleotide**—Single monomeric building block of a polynucleotide chain that makes up nucleic acid. A nucleotide is a phosphate ester consisting of a purine or a pyrimidine base, a sugar (ribose or deoxyribose as a pentose), and a phosphate group (see p. 28).

**Ochre codon**—the stop codon UAA.

**Okazaki fragment**—a short nucleotide sequence that is synthesized on the lagging strand of DNA during replication (q.v.) (see p. 42).

**Oncogene** (Heubner and Todaro, 1969)—a DNA sequence of viral origin that can lead to malignant transformation of a eukaryotic cell after being integrated into the cell genome (see proto-oncogene).

**Open reading frame (ORF)**—a DNA sequence of variable length that does not contain stop codons and therefore can be translated (see p. 48).

**Operator** (Jacob and Monod, 1959)—the recognition site of an operon at which the negative control of genetic transcription takes place by binding to a repressor.

**Operon** (Jacob et al., 1960)—in prokaryotes, a group of functionally and structurally related genes that are regulated together.

**Origin of replication (ORI)**—site of start of DNA replication.

**Ortholog**—a homologous DNA sequence or a gene that has evolved from a common ancestor between species, e. g., the α- and β-globin genes (see paralog).

**Pachytene** (de Winiwarter, 1900)—stage of prophase meiosis I.

**Palindrome** (Wilson and Thomas, 1974)—adjacent inverted repetitive linear base sequences of DNA that can help to form a hairpinlike structure by base-pairing of complementary sequences.

**Panmixis** (Weismann, 1895)—pairing system with random partner selection, as opposed to assortative mating.

**Paralog**—a DNA sequence or a gene that has evolved from a common ancestor within a species, e. g., the two α loci in humans (see ortholog).

**Parasexual** (Pontocorvo, 1954)—refers to genetic recombination by nonsexual means, e.g., by hybridization of cultured cells (see hybridization).

**PCR (polymerase chain reaction)** (Mullis 1985)—technique for in vitro propagation (amplification) of a given DNA sequence. It is a repetitive thermal cyclic process consisting of denaturation of genomic DNA of the sequence of interest, annealing the DNA to appropriate oligonucleotide primers, and replication of the DNA segment complementary to the primer.

**Penetrance** (Vogt, 1926)—the frequency or probability of expression of an allele (cf. expressivity).

**Peptide**—a compound of two or more amino acids joined by peptide bonds.

**Phage**—see bacteriophage.

**Phenocopy** (Goldschmidt, 1935)—a nonhereditary phenotype that resembles a genetically determined phenotype.

**Phenotype** (Johannsen, 1909)—the observable effect of one or more genes on an individual or a cell.

**Pheromone**—a signaling molecule that can alter the behavior or gene expression of other individuals of the same species.

**Phosphodiester bond**—the chemical bond linking adjacent nucleotides of DNA or RNA.

**Phytohemagglutinin (PHA)**—a protein substance obtained from kidney beans (*Phaseolus vulgaris*). It is used to separate red from white blood cells. Nowell (1960) discovered its ability to induce blastic transformation (see Transformation) and cell division in lymphocytes. It is the basis of phytohemagglutinin-stimulated lymphocyte cultures for chromosomal analysis.

**Plasmid** (Lederberg, 1952)—autonomously replicating circular DNA structures found in bacteria. Although they are usually separate from the actual genome, they may become integrated into the host chromosome.

**Plastids**—any of several types of organelles found in plant cells, e.g., chloroplasts.

**Pleiotropy** (Plate, 1910)—expression of a gene with multiple, seemingly unrelated phenotypic features.

**Point mutation**—alteration of the genetic code within a single codon. The possible types are the exchange of a base: a pyrimidine for another pyrimidine (or a purine for another purine) as a *transition* (Frese, 1959), i.e., thymine for cytosine (or adenine for guanine); or the exchange of a pyrimidine by a purine or visa versa: *transversion*, i.e., thymine by adenine or visa versa (Frese, 1959). Besides the two types of exchange, a point mutation may be due to the insertion of a nucleotide base or the deletion of one or several base pairs.

**Polar body** (Robin, 1862)—an involutional cell arising during oogenesis that does not develop further as an oocyte.

**Polyadenylation**—the addition of multiple adenine residues at the 3' end of eukaryotic mRNA after transcription.

**Polycistronic messenger**—mRNA including coding regions from more than one gene (in prokaryotes).

**Polygenic** (Plate, 1913; Mather, 1941)—refers to traits that are based on several or numerous genes whose effects cannot be individually determined. The term multigenic is sometimes used instead.

**Polymerases**—enzymes that catalyze the combining of nucleotides to form RNA or DNA (genetic transcription and DNA replication).

**Polymorphism, genetic** (Ford, 1940)—existence of more than one normal allele at a gene locus, with the rarest allele exceeding a frequency of 1%. A polymorphism may exist at several levels, i.e., variants in DNA sequence, amino acid sequence, chromosomal structure, or phenotypic traits (pp. 164 ff).

**Polypeptide**—see peptide.

**Polyploid** (Strasburger, 1910)—refers to cells, tissues, or individuals having more than two copies of the haploid genome, e.g., three (triploid) or four (tetraploid). In man, triploidy and tetraploidy in a conceptus are usually lethal and as a rule lead to spontaneous abortion.

**Polytene** (Koltzoff, 1934; Bauer, 1935)—refers to a special type of chromosome resulting from repeated endoreduplication of a single chromosome. Giant chromosomes arise in this manner (cf. chromosome).

**Population** (Johannsen, 1903)—individuals of a species that interbreed and constitute a common gene pool (cf. race).

**Positive negative effect**—a mutation exerting a negative functional effect in one allele.

**Premature chromosomal condensation** (Johnson and Rao, 1970)—induction of chromosomal condensation in an interphase nucleus after fusion with a cell in mitosis. Condensed S-phase chromosomes appear pulverized (so-called chromosomal pulverization).

**Prevalence**—see incidence.

**Pribnow box**—part of a promoter (TATAAT sequence 10 bp upstream of the gene) in prokaryotes.

**Primary transcript**—the original RNA transcribed from a eukaryotic gene before processing (splicing, addition of the cap, and polyadenylation).

**Primer**—a DNA or RNA oligonucleotide that after hybridization to an inversely complementary DNA has a 3'-OH end to which nucleotides can be added for synthesis of a new chain by DNA polymerase.

**Prion**—proteinaceous infectious particles that cause degenerative disorders of the central nervous system.

**Proband**—see Propositus.

**Probe**—a defined DNA or RNA fragment used to identify complementary sequences by specific hybridization.

**Prokaryote**—one-celled organism without a cell nucleus or intracellular organelles.

**Promoter**—a defined DNA region at the 5' end of a gene that binds to transcription factors and RNA polymerase during the initiation of transcription. The −10 sequence is the consensus sequence TATAATG about 10 bp upstream of a prokaryotic gene (Pribnow box).

**Prophage**—a viral (phage) genome integrated into the bacterial (host) genome.

**Propositus, proband**—the individual in a pedigree that has brought a family to attention for genetic studies.

**Protein**—one or more polypeptides with a specific amino acid sequence and a specific three-dimensional structure. Proteins are biomolecules representing the key structural elements of living cells and participating in nearly all cellular and biochemical reactions (see gene product).

**Proteome**—the complete set of all protein-encoding genes or all the proteins produced by them.

**Proto-oncogene (cellular oncogene)**—a eukaryotic gene. It may be present in truncated form in a retrovirus, where it may behave as an oncogene.

**Prototrophic**—refers to cells or cell lines that do not require a special nutrient added to the culture medium (cf. auxotrophic).

**Provirus**—duplex DNA derived from an RNA retrovirus and incorporated into a eukaryotic genome.

**Pseudogene**—DNA sequences that closely resemble a gene but are without function due to an integral stop codon, deletion, or other structural change. A processed pseudogene consists of DNA sequences that resemble the mRNA copy of the parent gene, i.e., it does not contain introns.

**Pseudohermaphroditism**—a condition in which an individual has the gonads of one sex and phenotypic features of the opposite sex.

**Quadriradial figure**—the configuration assumed when homologous segments of chromosomes involved in a reciprocal translocation pair at meiosis. Rarely, such a figure may occur during mitosis.

**Race**—a population (q.v.) that differs from another population in the frequency of some of its gene alleles (L. C. Dunn: Heredity and Evolution in Human Populations, Harvard University Press, Cambridge, Mass., 1967). Accordingly, the concept of race is flexible and relative, defined in relation to the evolutionary process. The term race can be used to classify groups, whereas the classification of individuals is often uncertain and of biologically dubious value.

**Reading frame**—sequence of DNA nucleotides that can be read in triplets to code for a peptide (cf. open reading frame).

**Receptor**—a transmembrane or intracellular protein involved in transmission of a cell signal.

**Recessive** (Mendel, 1865)—refers to the genetic effect of an allele (q.v.) at a gene locus that is manifest as phenotype in the homozygous state only (q.v.).

**Reciprocal translocation**—mutual exchange of chromosome parts.

**Recombinant DNA**—A DNA molecule consisting of parts of different origin.

**Recombination** (Bridges and Morgan, 1923)—the formation of new combinations of genes as a result of crossing-over between homologous chromosomes during meiosis.

**Recombination frequency**—frequency of recombination between two or more gene loci. Expressed as the theta ($\theta$) value. A $\theta$ of 0.01 (1% recombination frequency) corresponds to 1 centimorgan (cM).

**Regulatory gene**—a gene coding for a protein that regulates other genes.

**Renaturation of DNA**—combining of complementary single strands of DNA to form double-stranded DNA (cf. denaturation).

**Repair** (Muller, 1954)—correction of structural and functional DNA damage.

**Replication**—identical duplication of DNA.

**Replication fork**—the unwound region of the DNA double helix in which replication takes place.

**Replicon** (Huberman and Riggs, 1968)—an individual unit of discontinuous DNA replication in eukaryotic DNA.

**Reporter gene**—a gene used to analyze another gene, especially the regulatory region of the latter.

**Repulsion** (Bateson, Saunders, and Punnett, 1905)—term to indicate that the mutant alleles of neighboring heterozygous gene loci lie on opposite chromosomes, i.e., in trans configuration (see cis/trans).

**Resistance factor**—a plasmid gene causing antibiotic resistance.

**Restriction enzyme, or restriction endonuclease** (Meselson and Yuan, 1968)—endonuclease that cleaves DNA at a specific base sequence (restriction site or recognition sequence).

**Restriction map**—a segment of DNA characterized by a particular pattern of restriction sites.

**Restriction site**—a particular sequence of nucleotide bases in DNA that allow a particular restriction enzyme to cleave the DNA molecule at, or close to, that site (recognition site).

**Retrotransposon**—a mobile DNA sequence that can insert itself at a different position by using reverse transcriptase (see transposon).

**Retrovirus**—a virus with a genome consisting of RNA that multiplies in a eukaryotic cell by conversion into duplex DNA.

**Reverse transcriptase**—an enzyme complex that occurs in RNA viruses and that can synthesize DNA from an RNA template.

**RFLP**—restriction fragment length polymorphism. The production of DNA fragments of different lengths by a given restriction enzyme, due to inherited differences in a restriction site.

**Rho factor**—a protein involved in termination of transcription in E. coli.

**Ribosome** (Roberts, 1958; Dintzis et al., 1958)—complex molecular structure in pro- and eukaryotic cells consisting of specific proteins and ribosomal RNA in different subunits. The translation of genetic information occurs in ribosomes.

**Ring chromosome**—a circular chromosome. In prokaryotes the normal chromosome is ring-shaped. In mammals it represents a structural anomaly and implies that chromosomal material has been lost.

**RNA (ribonucleic acid)**—a polynucleotide with a structure similar to that of DNA except that the sugar is ribose instead of deoxyribose.

**Satellite** (Navashin, 1912)—small mass of chromosomal material attached to the short arm of an acrocentric chromosome (q.v.) by a constricted appendage or stalk. It is involved in the organization of the nucleolus. The stalk region can be stained by specific silver stain (NOR stain, nucleolus-organizing region). The size of the satellite, the length of its stalk, and the intensity of its fluorescence after staining with acridine are polymorphic cytogenetic markers.

**Satellite DNA (sDNA)** (Sueoka, 1961; Kit, 1961; Britten and Kohne, 1968)—DNA that is either heavier (GC-rich) or lighter (AT-rich) than the main DNA (see DNA). Not to be confused with the satellite regions of acrocentric chromosomes.

**SCE (sister chromatid exchange)** (Taylor, 1958)—an exchange between the two chromatids of a metaphase chromosome. After two replication cycles in a cell culture in the presence of a halogenated base analog (e.g., 5-bromodeoxyuridine), both DNA strands of one chromatid will be substituted with the halogenated base analog, whereas only one DNA strand of the other chromatid will be substituted. As a result, the two chromatids differ in staining intensity, and it is possible to determine where crossing-over of the two chromatids has occurred (see p. 334).

**Segregation** (Bateson and Saunders, 1902)—the separation of alleles at a gene locus at meiosis and their distribution to different gametes. Segregation accounts for the 1 : 1 distribution of allelic genes to different chromosomes.

**Selection** (Darwin, 1858)—preferential reproduction or survival of different genotypes under different environmental conditions.

**Selection coefficient**—quantitative expression (from 0 to 1) of the disadvantage that a genotype has (compared with a standard genotype) in transmitting genes to the next generation. The selection coefficient (s) is the numerical amount by which biological fitness (1-s) is decreased; i.e., a selection coefficient of 1 indicates a complete lack of biological fitness.

**Selective medium**—a medium that supports growth of cells in culture containing a particular gene.

**Semiconservative** (Delbrück and Stent, 1957)—the normal type of DNA replication. One DNA strand is completely retained; the other is synthesized completely anew.

**Signal sequence**—the N-terminal amino acid sequence of a secreted protein; required for transport of the protein to the right destination in the cell.

**Silencer sequence**—a eukaryotic DNA sequence that blocks the access of gene activity proteins required for transcription, by forming condensed heterochromatin in that particular area.

**SINE (short interspersed nuclear element)**—short repetitive DNA sequences (cf. LINE).

**snRNPs (small nuclear ribonucleoprotein particles)**—complexes of small nuclear RNA molecules and proteins.

**Somatic**—refers to cells and tissues of the body, as opposed to germinal (referring to germ cells).

**Somatic cell**—any cell of an organism that does not undergo meiosis and does not form gametes (as opposed to germ cells).

**Somatic cell hybridization**—formation of cell hybrids in culture.

**Southern blot** (Southern, 1975)—method of transferring DNA fragments from an agarose gel to a membrane after the fragments have been separated according to size by electrophoresis.

**Speciation** (Simpson, 1944)—formation of species during evolution. One of the first steps toward speciation is the establishment of a reproductive barrier against genetic exchange. A frequent mechanism is chromosomal inversion.

**Species** (Ray, 1670)—a natural population in which there is interbreeding of the individuals, which share a common gene pool.

**S phase** (Howard and Pelc, 1953)—phase of DNA synthesis (DNA replication) between the G1 and the G2 phase of the eukaryotic cell cycle.

**Spliceosome**—an aggregation of different molecules that can splice RNA.

**Splicing**—a step in processing a primary RNA transcript in which introns are excised and exons are joined.

**Splice junction**—the sequences at the exon/intron boundaries.

**Stem cell**—a cell able to renew itself by division, retaining the potential for differentiation within a particular developmental pathway. One distinguishes omnipotent and pluripotent stem cells.

**STS (sequence tagged site)**—a short segment of DNA of known sequence.

**Submetacentric**—refers to a chromosome consisting of a short and a long arm because of the position of its centromere.

**Synapsis** (Moore, 1895)—the pairing of homologous chromosomes during meiotic prophase.

**Synaptonemal complex** (Moses, 1958)—parallel structures associated with chiasmata formation during meiosis, visible under the electron microscope.

**Syndrome**—within human genetics, a group of clinical and pathological characteristics that are etiologically related, regardless of whether the details of their relationship have yet to be identified.

**Synteny** (Renwick, 1971)—refers to gene loci that are located on the same chromosome, whether or not they are linked.

**Tandem duplication**—short identical DNA segments adjacent to each other.

**TATA box**—a conserved, noncoding DNA sequence about 25 bp in the 5' region of most eukaryotic genes. It consists mainly of sequences of the TATAAAA motif. Also known as Hogness box (cf. Pribnow box, of prokaryotes).

**T cells**—T lymphocytes.

**Telocentric** (Darlington, 1939)—refers to chromosomes or chromatids with a terminally located centromere, without a short arm or satellite. They do not occur in man.

**Telomerase**—a ribonucleoprotein enzyme that adds nucleotide bases at the telomere.

**Telomere** (Muller, 1940)—the terminal areas of both ends of a chromosome containing specific consensus sequences (see p. 180).

**Template**—the molecule that determines the nucleotide sequence for the formation of another, similar (complementary) molecule (see DNA and RNA).

**Teratogen** (Ballantyne, 1894)—chemical or physical agent that leads to disturbances of embryological development and malformations.

**Termination codon**—one of the three triplets signaling the end of translation (UAG, UAA, UGA).

**Terminator**—a DNA sequence that signals the end of transcription.

**Tetraploid** (Nemec, 1910)—having a double diploid chromosome complement, i.e., four of each kind of chromosome are present (4n instead of 2n).

**Topoisomerase**—a class of enzymes that can control the three-dimensional structure of DNA by cutting one DNA strand, rotating it about the other, and resealing it (class I) or cutting and resealing both ends (class II). Used to unwind the DNA double helix at transcription.

**Transcript**—an RNA copy of a segment of the DNA of an active gene.

**Transcription**—the synthesis of messenger RNA (mRNA), the first step in relaying the information contained in DNA.

**Transcription unit**—all of the DNA sequences required to code for a given gene product (operationally corresponding to a gene). Includes promoter and coding and noncoding sequences.

**Transduction** (Zinder and Lederberg, 1952)—transfer of genes from one cell to another (usually bacteria) by special viruses, the bacteriophages.

**Transfection**—introduction of pure DNA into a living cell (cf. transformation).

**Transformation**—this term has several different meanings in biology. In genetics, three main types of transformation are distinguished: 1) *malignant transformation*, the transition of a normal cell to a malignant state with loss of control of proliferation; 2) *genetic transformation* (Griffith, 1928; Avery et al. 1944), a change of genetic attributes of a cell by transfer of genetic information; and 3) *blastic transformation*, the reaction of lymphocytes to mitogenic substances (e.g., phytohemagglutinin or specific antigens) leading to cell division.

**Transgenic**—refers to an animal or a plant into which a cloned gene has been introduced and stably incorporated. It reveals information about the biological function of the (trans)gene.

**Transition**—Replacement of a purine with another purine or a pyrimidine with another pyrimidine (see mutation).

**Translation**—the second step in the relay of genetic information. Here the sequence of triplets in mRNA are translated into a corresponding sequence of amino acids to form a polypeptide as the gene product.

**Translocation**—transfer of all or part of a chromosome to another chromosome. A translocation is usually *reciprocal*, leading to an exchange of nonhomologous chromosomal segments. A translocation between two acrocentric chromosomes that lose their short arms and fuse at their centromeres is called a *fusion* type translocation (Robertsonian translocation).

**Transposon**—a DNA sequence with the ability to move and become inserted at a new location of the genome.

**Transversion**—replacement of a purine with a pyrimidine or vice versa (see mutation).

**Triplet**—a sequence of three nucleotides comprising a codon of a nucleic acid and representing the code for an amino acid (triplet code, see codon).

**Trisomy** (Blakeslee, 1922)—an extra chromosome, in addition to a homologous pair of chromosomes.

**UPD**—uniparental disomy (see disomy).

**Upstream**—5' direction of a gene.

**Variation**—the differences among related individuals, e. g., parents and offspring, or among individuals in a population.

**Variegation**—different phenotypes within a tissue.

**Vector**—a molecule that can incorporate and transfer DNA.

**Virion**—a complete extracellular viral unit or particle.

**Virus**—DNA or RNA of defined size and sequence, enclosed in a protein coat encoded by its genes and able to replicate in a susceptible host cell only (see Part I).

**Western blot**—technique to identify protein antigens, in principle similar to the Southern blot method (q.v.).

**Wild-type**—refers to the genotype or phenotype of an organism found in nature or under standard laboratory conditions, roughly meaning "normal."

**X chromatin** (formerly called Barr body or sex chromatin) (Barr and Bartram, 1949)—darkly staining condensation in the interphase cell nucleus representing an inactivated X chromosome

**Xenogenic**—refers to transplantation between individuals of different species.

**X-inactivation** (Lyon, 1961)—inactivation of one of the two X chromosomes in somatic cells of female mammals during the early embryonic period by formation of X chromatin.

**X-linked**—refers to genes on the X chromosome.

**YAC (yeast artificial chromosome)**—a yeast chromosome into which foreign DNA has been inserted for replication in dividing yeast cells. YACs can incorporate relatively large DNA fragments, up to about 1000 kb.

**Y chromatin** (F body; Pearson, Bobrow, Vosa, 1970)—the brightly fluorescing long arm of the Y chromosome visible in the interphase nucleus.

**Z DNA**—alternate conformation of DNA. Unlike normal B DNA, (Watson-Crick model), the helix is left-handed and angled (zigzag, thus Z DNA).

**Zinc finger**—A finger-shaped region found in many DNA-binding regulatory proteins, the "finger" being held together by a strategically placed zinc atom.

**Zoo blot**—a Southern blot containing conserved DNA sequences from related genes of different species. It is taken as evidence that the sequences are coding sequences from a gene (see p. 250).

**Zygote** (Bateson, 1902)—the new diploid cell formed by the fusion of the two haploid gametes, an ovum and a spermatozoon, at fertilization. The cell from which the embryo develops.

**Zygotene** (de Winiwarter, 1900)—a stage of prophase of meiosis I.

## References for the Glossary

Bodmer, W.F., Cavalli-Sforza, I.L.: Genetics and the Evolution of Man. W. H. Freeman & Co., San Francisco, 1976.

Brown, T.: Genetics. A Molecular Approach, 2nd ed. Chapman & Hall, London, 1992.

Brown, T.A.: Genomes. BIOS Scientific Publishers, Oxford, 1999.

Dorland's Illustrated Medical Dictionary, 28th ed. W.B. Saunders Co., Philadelphia, London, Toronto, Montreal, Sydney, Tokyo, 1994.

Muller, R.F., Young, I.D.: Emery's Elements of Medical Genetics, 10th ed. Churchill Livingstone, Edinburgh, 1998.

Griffiths, A.J.F., et al.: An Introduction to Genetic Analysis, 7th ed. W.H. Freeman, New York, 2000.

Lewin, B.: Genes VII. Oxford University Press, Oxford, 2000.

Lodish, H., et al: Molecular Cell Biology, 4th ed. W.H. Freeman, New York, 2000.

Passarge, E.: Definition genetischer Begriffe (Glossar), pp. 311–323. In: Elemente der Klinischen Genetik. G. Fischer, Stuttgart, 1979.

Rieger, R., Michaelis, A., Green, M.M.: Glossary of Genetics and Cytogenetics, 5th ed. Springer Verlag, Berlin, Heidelberg, New York, 1979.

Rothwell, N.Y.: Understanding Genetics. A Molecular Approach. Wiley-Liss, New York, 1993.

Tanaka, Y., Macer, D.: Sense, nonsense and antisense. Trends Genet. **10**:417, 1994.

Watson, J.D.: Molecular Biology of the Gene, 3rd ed. W.A. Benjamin, Menlo Park, California, 1976.

Whitehouse, H.L.K.: Towards the Understanding of the Mechanisms of Heredity, 3rd ed. Edward Arnold, London, 1973.

### Website:

Glossary of Genetic Terms, National Institute of Human Genome Research (http://www.nhgri.nih.gov/DIR/VIP/Glossary/).

# Index

Page numbers in **bold** signify main entries.
Human diseases are shown in **bold**.

## D

## G

## H

# I

## T

## U



## V